MUTATION, CANCER, AND MALFORMATION

ENVIRONMENTAL SCIENCE RESEARCH

Recent Volumes in this Series

MUTATION, CANCER, AND MALFORMATION

Edited by
Ernest H. Y. Chu

University of Michigan Medical School
Ann Arbor, Michigan

and
Walderico M. Generoso

Oak Ridge National Laboratory
Oak Ridge, Tennessee

PLENUM PRESS • NEW YORK AND LONDON

Library of Congress Cataloging in Publication Data

International Workshop on Principles of Environmenal Mutagenesis, Carcinogenesis,
  and Teratogenesis (1983: Shanghai, China)
  Mutation, cancer, and malformation.

  (Environmental science research; v. 31)
  "Proceedings of an International Workshop on Principles of Environmental
Mutagenesis, Carcinogenesis, and Teratogenesis, held May 25–June 1, 1983, in
Shanghai, People's Republic of Chinca"–T.p. verso.
  Includes bibliographies and index.
  1. Genetic toxicology–Congresses. 2. Mutagenesis–Congresses. 3. Carcinogenesis
–Congresses. 4. Teratogenesis–Congresses. 5. Environmentally induced diseases–
Congresses. I. Chu, Ernest H. Y. II. Generoso, W.M. III. Title. IV. Series. [DNLM: 1.
Carcinogens, Environmental–congresses. 2. Environmental Exposure–congresses. 3.
Mutagens–congresses. 4. Mutation–congresses. 5. Teratogens–congresses. W1
EN986F v.31 / QH 465.A1 I61m 1983]
RA1224.3.I58   1983   616.99′2   84-17868
ISBN-13: 978-1-4612-9463-4   e-ISBN-13: 978-1-4613-2399-0
DOI: 10.1007/978-1-4613-2399-0

Proceedings of an international workshop on Principles of Environmental
Mutagenesis, Carcinogenesis, and Teratogenesis, held May 25–June 1, 1983,
in Shanghai, People's Republic of China

©1984 Plenum Press, New York
Softcover reprint of the hardcover 1st edition 1984

A Division of Plenum Publishing Corporation
233 Spring Street, New York, N.Y. 10013

FOREWORD

During the early 1930s, when I was a graduate student and later
a post-doctoral researcher at the National Research Council for the
University of Wisconsin at Madison, we had the opportunity to get
acquainted with many graduate students from China who were sent to
the University for training in modern basic sciences as well as
social sciences. The University of Wisconsin continues to graduate
a large number of Chinese students.

Economic conditions in the 1930s were very precarious for the
United States and other parts of the world. Many of us students
grew closer together because we were living on similarly tight
budgets. As a matter of fact, we subleased a part of our apartment
in Madison to some Chinese graduate students. This was a very nice
opportunity for us to learn about the scientific and cultural background of our Chinese friends. Many of them came from the interior
of China and had had very little opportunity to become acquainted
with people from a western culture. Living with these students was
a very pleasant and educational experience which gave us a good picture of the cultural life and educational system of China at that
time--an intimate picture that one normally would not see without
travelling in that country. Many Chinese students were also anxious
to make friends, to get an idea of how westerners lived, to learn
our cultural arts, our family structure, and our general philosophy
of life. It was especially instructive for us, because the Chinese
students came from a considerably longer historical tradition than
we, as most of us had migrated to the United States only a few years
before entering the University.

Many of the friendships established during those years lasted
for many thereafter, until the political upheaval in China made it
impossible to stay in contact; it is very difficult for us now to
trace many of the colleagues with whom we were so closely associated.

Since the early 1930s, a number of political upheavals, World
War II, and other changes have not only influenced the economic
development of China and of our country but have also had a profound

influence on the development and direction of scientific issues.

My interest in molecular structure and physical chemistry has shifted to basic biology and genetics, since so many interesting developments have taken place in the field of biology in the last 20 to 30 years, and even more intensely in the last 10 to 15 years. It was for this reason that I was most pleased to have an opportunity, at the meeting arranged by the Carnegie Institution of Washington, to become acquainted again with Dr. C. C. Tan, with whose work I was very familiar. Through conversations and letters, we set up a cooperative effort. He was especially interested in bringing to China more up-to-date discussions on the recognition of environmental mutagens, carcinogens, and teratogens. These are areas that we have developed through workshops in the United States and in many countries abroad, such as Latin America, India, Egypt, the Philippines, and others.

The Council for Research Planning in Biological Sciences, Inc. was incorporated in 1981, and one of its major objectives is to help our colleagues in developing countries to become proficient with the techniques of the quickly developing field of environmental mutagenesis. Since there is great overlap between environmental mutagenesis, carcinogenesis, and teratogenesis, it was a natural development to consider all three together. Dr. Tan received the suggestion with great enthusiasm, and promised to initiate activity on the proposal as soon as he returned to China. We were also most fortunate to have been working in close cooperation with Dr. Ernest Chu, whose fatherland is China, and who has been lecturing and teaching as a guest professor at a number of universities in China. Also very closely associated with us has been Dr. Walderico Generoso, from the Biology Division of the Oak Ridge National Laboratory, who has worked with us on a number of the training programs and workshops that we developed in environmental mutagenesis. The leadership in developing this cooperation with Dr. Tan was taken by Drs. Chu and Generoso and led to these proceedings which I, unfortunately, was not able to attend.

Judging from all the reports I have received from scientists who attended the workshop, it turned out most successfully with the additional international participation of European and Japanese geneticists, who are also interested in this field. Much of the workshop's success was due to the very generous support of the National Cancer Institute, the World Health Organization, the Exxon Corporation, the Genetics Society of China, and the China Association for Science and Technology.

As a result of this meeting, the Chinese Environmental Mutagen Society has been founded, a group which we hope will continue to provide leadership in developing the fields of environmental mutagenesis, carcinogenesis, and teratogenesis.

This volume was prepared under the editorship of Drs. Chu and Generoso who should be commended for the preparation of this very comprehensive work, since it covers many areas that will be beneficial to developing countries. It will also be a valuable text for individuals who are just starting their training in the areas of environmental mutagenesis, carcinogenesis, and teratogenesis.

These are very quickly developing fields, and it is not yet possible to determine which aspects will become most important in the future. Exposure to chemical mutagens and carcinogens is unavoidable in our environment, in our food, and in the air we breathe. Many of us believe that these important fields require much further investigation. We are therefore most pleased to have the cooperation of our Chinese colleagues in this effort.

Alexander Hollaender

Council for Research Planning in
  Biological Sciences, Inc.
1717 Massachusetts Avenue, N.W.
  Suite 600
Washington, D.C. 200036-2077 U.S.A.

PREFACE

Protection of the environment in the interest of human health
and well-being is a global concern. National efforts and
international cooperation are essential to bring about a better
world to live in. In the People's Republic of China, where one-
fourth of the world's population live, the health impact of the
rapid technological and industrial modernization that are currently
underway can be great, not only at the national level but also
worldwide. Fortunately, there has been increasing emphasis on
research and training in the areas of carcinogenesis, mutagenesis
and teratogenesis, and in the various aspects of toxicology and
epidemiology in the People's Republic of China. However, although
some glimpses on the organization of research programs and
scientific accomplishments in China have been reported by recent
visitors, the information is sporadic and not widely available. In
the meantime, Chinese scientists have been eager to establish
contacts with colleagues in other countries; they have been keenly
interested in learning about recent advances in science and
technology and have expressed willingness to share their own
experiences and discoveries.

In response to a Chinese initiative under the leadership of
Professor Chia-Chen Tan of Fudan University, an International
Workshop on the Principles of Environmental Mutagenesis,
Carcinogenesis and Teratogenesis was organized, with the dual
purpose of introducing to the Chinese colleagues the latest
information in these areas in an organized manner, while affording
the opportunity for the visitors, and thereby the world scientific
community, to learn about the progress that is being made in the
host country. It was hoped that the Workshop would provide a forum
for information exchange on both the scientific basis of toxicology
and the methodologies and experiences in dealing with the human
risk problems in different countries.

The Workshop was held in Shanghai from May 25 to June 1, 1983.
It was cosponsored by the Genetics Society of China, the Shanghai
Association for Science and Technology, the International
Association of Environmental Mutagen Societies and the

International Programme on Chemical Safety (World Health
Organiation/International Labor Organization/United Nations
Environmental Programme). This Workshop is also Number 5 of the
IPCS Joint Symposia Series. Financial assistance came from the
United States Public Health Service (grant CA 34165), the
International Programme on Chemical Safety, the China Association
for Science Technology and the EXXON Corporation, to support the
travel of participants and the cost of the preparation of these
proceedings. In addition, governmental, industrial and private
sources of support from various countries including Japan, the
United Kingdom, Italy, West Germany, Switzerland and the U.S.
enabled the participation of several speakers in the Workshop.

In addition to some 45 formal presentations, a poster session
was held consisting of Chinese contributed papers. Thirty-seven
full length papers and 41 selected abstracts have been included in
this volume. The Workshop was attended by 35 scientists from 11
countries and more than 200 scientists representing over 130
research and educational institutions in China. One important
outcome of the Workshop was the founding of the Chinese Society of
Environmental Mutagens. Thus the Workshop more than accomplished
the expected goal of information exchange at the international
level.

The Organizing Committee consisted of Alexander Hollaender
(Chairman), E. H. Y. Chu, W. M. Generoso, C. Ramel, C. C. Tan and
Y. Tazima. It is our pleasure to acknowledge the invaluable advice
from numerous individuals in many countries. We are particularly
grateful to Wang Xinnan (the Shanghai Association for Science and
Technology), Xue Shouzhen (Shanghai First Medical College), Xue
Jinlun (Fudan University) and their colleagues and coworkers for
their unfailing efforts leading to a smooth and successful
conclusion of the Workshop. In addition to all scientific
contributors, we thank Mrs. Mhairi Gehlhar for copy-editing,
Ms. Lisa Campeau and Mrs. Mary Kellogg for typing, and the
publisher for bringing out the present volume in a relatively short
time.

E. H. Y. Chu

W. M. Generoso

CONTENTS

REPRODUCTIVE TOXICOLOGY

ENVIRONMENTAL TOXICOLOGY, MUTAGENESIS, AND CARCINOGENESIS

ENVIRONMENTAL MUTAGENESIS AND DISEASE IN HUMAN POPULATIONS[1]

Arno G. Motulsky

Departments of Medicine and Genetics, and Center for
Inherited Diseases, University of Washington
Seattle, Washington 98195

SUMMARY

Environmental chemicals can affect the genetic material and
cause a variety of different mutations. Mutations in somatic
tissues can lead to cancer while germinal mutations can cause
various genetic diseases. The impact of germinal mutations on
health will depend upon their frequency; their nature (point
mutation vs. chromosomal change, dominant vs. recessive); and upon
the mechanisms maintaining a given mutation in the population.
Mutations causing early prenatal lethality have fewer public health
effects than genetic diseases associated with prolonged medical and
social problems. Differences between and within species in
metabolism of environmental chemicals and in DNA repair make
mutational estimates in humans imprecise. Results on mutation
frequency in somatic cells cannot be readily transferred to
conclusions regarding germinal mutations until appropriate
comparisons have been made. Studies on atom bomb survivors suggest
an increased mutational frequency but such results failed to reach
conventional statistical significance. Current estimates of the
role of induced germinal mutation in human populations have wide
confidence limits. An accurate assessment of the potential hazards
of environmental human mutagenesis requires better fundamental
understanding of human genetics and continued attention to studies
on humans and their tissues and fluids. Crash programs on
environmental mutagenesis at the expense of other biomedical
research appear unwarranted.

---

[1]Supported by grant GM 15253 from the U.S. National Institutes of
Health.

1

INTRODUCTION

Modern environments differ from those of earlier times in human exposure to many different manmade contaminants. Pesticides, fungicides, food additives, synthetic drugs, and atmospheric and water pollutants did not exist until the advent of the industrial age some five human generations ago. 70,000 of such substances exist and more than 25,000 are in common use in the United States. More compounds are being synthesized every year and human exposure to them is increasing. Many chemicals have been shown to be mutagenic in lower species and concern regarding the implications of chemicals on human health is therefore warranted. A detailed consideration of environmental mutagenesis in man has been provided recently in comprehensive publications (7,14,18).

CANCER AND MUTATIONS (4)

Mutations of the genetic material of somatic cells can lead to malignant neoplasms. Such carcinogenic effects are much delayed between the initial mutation and the actual diagnosis of a clinical cancer. The latency period ranges from about 5 years for certain chronic leukemias to 15-20 years for solid cancers. It will, therefore, take many years before carcinogenic effects are apparent. A variety of environmental chemicals have already been clearly implicated in cancer development [e.g., asbestos: lung and pleural cancer; benzene: marrow cancers; vinyl chloride: certain liver malignancies and others (4)]. A major environmental hazard concerns tobacco. The evidence relating cigarette smoking to cancer of the lung is overwhelming. Mortality continues to increase in the United States and pulmonary cancer is now the most frequent lethal malignancy in males (20). Mortality from lung cancer in women, whose widespread smoking began at a later date, has also risen continuously over the past 15 years. No other common cancers have shown such trends in either sex. Age-specific mortality has remained at similar rates for the past 30 years for common cancers such as carcinoma of the breast in women and of the colon and rectum in men (20).[2] Considering the long latent period between exposure and development of clinical cancers, these data suggest that these cancers are not related to novel environmental carcinogens introduced between 1930 and 1960. It is conceivable that new chemical substances introduced since the 1960s could raise the frequency of these cancers and other mutants in the next few years. Continued careful monitoring of the mortality of different cancers is therefore essential.

---

[2]Carcinoma of the colon and rectum in women has slightly decreased in frequency.

Another noteworthy epidemiologic trend bearing on environmental carcinogenesis is the continued decline in cancer of the stomach in developed countries (20). This cancer was the most common lethal malignancy 50 years ago and now has become much less frequent. While the exact cause of this decline has not been elucidated, it is likely that better refrigeration and possibly the addition of food preservatives has reduced the frequency of food contaminated with carcinogens acting on the gastric mucosa. Thus, the choice of a pre-industrial life style with emphasis on organic and natural foods may not necessarily be "healthier" than eating a modern diet!

The facts of increasing rates of lung cancer due to tobacco, decreasing rates of stomach cancer, and fairly steady rates for most other cancers need to be recalled in the formulation of public health policies for cancer control. The overwhelming role of tobacco in current lung cancer mortality must be squarely faced. A very large number of lives could be saved by intensive public health education to discourage smoking. The data for the other common cancers suggest that a new mass endemic of these malignancies is unlikely. Crash programs to detect ubiquitous novel carcinogens are therefore not required. At the same time, careful monitoring of various groups of individuals who may have high exposure to one or another potential carcinogenic chemical needs to be encouraged. Collaborative studies need to be established to collect a sufficiently large number of cases that allow a meaningful assessment of the data. Many such studies are unreliable because of the vagaries of small numbers. An increase in a given cancer in a small occupational grouping would not be detected by broad registries of the general population.

GERMINAL MUTATIONS IN HUMANS

Germ cell mutations, i.e., those affecting eggs and sperm, are more insidious in humans. A cancer due to a somatic cell mutation will kill a single individual but the original mutation will be extinguished with the death of the patient. In contrast, germinal mutations may remain hidden for many generations. Thus, recessive mutations require the homozygous state for clinical expression and many induced mutations are likely to be recessive in nature. The phenotypic effects of mutations affecting genes in polygenic systems are difficult to predict and no definite statements regarding such mutations can be made until better understanding of the various genes underlying such phenotypes becomes available. Only dominant mutations and certain chromosomal defects will have clinical effects in the first generation. The persistence of mutations depends upon the biologic fitness of the phenotype. If carriers of a given mutation have no offspring because of lethality or infertility, the mutation will only persist for one generation.

Since the reproductive performance of the carriers of recessive
mutations in the heterozygote state will be similar to that of
normals, such mutations will persist for many generations.

Considerable problems in recognition of environmentally-
induced mutations are posed by the high probability that new
mutations will present with phenotypes similar to those that exist
in the populations because of spontaneous mutations. It will,
therefore, be difficult to recognize slight to moderate increases
in human diseases caused by mutations induced by environmental
agents. Background frequencies of existing tumors and genetic
diseases are such that marked increases of a given phenotype need
to occur before a causal relationship to an environmental mutagen
is apparent. Careful epidemiologic monitoring of the existing
genetic disease load is therefore necessary to detect moderate
increases. Etiologic heterogeneity, i.e., different genes or
nongenetic factors causing the same disease phenotype, is a further
complication. Even if an increase of a given disease entity is
found by epidemiologic monitoring, the underlying cause for the
increase will not necessarily be obvious and will require detection
of the offending agent. The search for induced mutations in human
populations is thus very difficult.

It should be recognized that current quantitative assessments
of human mutational risks are based on inferences with wide
confidence limits (1,7,14). In fact, there is as yet no direct
evidence for any induced germinal mutation having caused genetic
disease in humans. Not even the extensive studies on atomic bomb
survivors in Japan have furnished unambiguous evidence for a
clearcut increase in radiation-induced mutations. There were no
statistically significant effects between irradiated populations
and controls in untoward pregnancy outcomes, survival through
childhood, X chromosomal aneuploidy, and electrophoretically
detectable novel biochemical variants (16). However, the
differences between the irradiated and control groups were all in
the expected direction of increased mutations as is consistent with
data from lower organisms.

GENETIC VARIABILITY IN MUTAGEN METABOLISM (12,17)

Considerable species differences exist in the metabolism of
many xenobiotics. Such differences may affect quantitative rates
of metabolism as well as qualitative routes of metabolism. Thus,
different quantities of a given mutagen may be found following
administration of a standard dose in animals of different species.
Furthermore, a given mutagenic chemical may be detected in one but
not in another species because of qualitative differences in
metabolism between species.

Metabolic variation occurs frequently within one species. Human twin studies have shown a high heritability for the metabolism of most xenobiotics suggesting strong genetic determinants in biochemical makeup involved in the metabolism of chemicals. In some instances, specific enzyme differences under monogenic control have been detected. Some of these enzyme deficiencies affect the action of specific drugs and cause unexpected drug reactions. The study of such reactions evolved into the field of pharmacogenetics (8). The realization that not only therapeutic drugs but the metabolism and response to any kind of foreign chemical might be under genetic control led to the concept of ecogenetics, i.e., the role of genetic factors in response to various environmental agents (9). The demonstration in recent years that about 5-8% of populations of European origin are homozygotes for a recessive mutation affecting a specific subcomponent of the P450 system that causes marked differences in oxidation of many different drugs is of great interest (6). This system is currently being studied to assess its significance for differences in susceptibilities to cancer of the lung from smoking (Idle, personal communication). Earlier studies with other components of the P450 system in humans failed to demonstrate monogenic effects although undefined genetic factors have usually been demonstrated (17).

The implications of pharmacogenetic and ecogenetic data are far-reaching for human mutagenesis. Inert substances may become mutagenic following metabolic activation and in vitro testing may give a false sense of security. Conversely, mutagenic substances may become "detoxified" so that they no longer are mutagenic. If a monogenic trait with low frequency mediates a critical metabolic step, a small proportion of the population may be at higher risk. Racial and ethnic differences in gene frequencies are common and some populations may be at higher danger of mutational injury than others (12). The results from one population cannot necessarily be transferred to another. Alleles with different functional capacity (isoalleles) may also exist at certain genetic loci. Those individuals within the normal range who, because of isoallelic variation, carry out a given biochemical reaction less efficiently may be at higher risk. If multiple genes are involved in a given metabolic reaction, those individuals at the extremes of the distribution curve may be in significantly greater jeopardy (10). Additional sources of variation in the human population relate to potential differences in DNA repair. Some persons may be heterozygotes for DNA repair enzyme deficiencies that in the homozygote state cause certain diseases such as Bloom's syndrome and others. Isoallelic variation of DNA repair (as described above) within the normal range may also exist. Differences in oncogene genotypes may also contribute to cancer susceptibility and are currently under intense study (23).

The exact significance of mutations induced by chemicals in bacteria, mammalian cell cultures, and lower organisms for a definitive and quantitative assessment of mutational damage in humans is not clear. It is sobering to realize that definitely proven harmful teratogenic chemicals in humans have not usually been found by predictive laboratory tests nor by epidemiologic studies but were discovered by alert, careful clinical observers who discerned previously unrecognized clinical patterns and related them to a specific teratogenic exposure (19). Examples include rubella embryopathy, fetal alcohol syndrome, and fetal injury caused by dilantin and dicumarol. The lessons for carcinogenesis and mutagenesis are clear.

GENETIC DISEASES IN MAN

Chromosomal Malformations

Most autosomal chromosomal anomalies such as the trisomies and triploids are lethal prenatally. A significant increase of such aberrations does not cause health impairment except for a possible increase in the rate of spontaneous abortions. In some cases, early embryonal death may only be perceived as a delayed menstrual period. Such effects would be difficult to detect. Prenatal lethality could also be caused by genomic imbalances secondary to structural chromosomal errors that are potentially inducible by chemical agents. However, some unbalanced chromosomal errors cause birth defects in the postnatal period. The most viable cytogenetic autosomal chromosome error is Down's syndrome (trisomy 21) with frequent survival into adulthood but associated with severe mental retardation. Patients with common X chromosomal errors such as XXY (Klinefelter's syndrome), XO (Turner's syndrome), XXX, and XYY are less severely affected clinically and survive to adulthood. XXY and XO individuals are sterile in the vast majority of cases. XYY males are usually fertile while XXX females appear to be subfertile, although exact fertility data are lacking. Significant psychosocial effects are found in many patients affected with these aneuploidies affecting the X chromosome.

The role of environmental agents in the induction of human chromosomal malformations remains unclear. Whether nondisjunction, e.g., trisomy 21, can be induced by radiation remains equivocal (5). While chromosomal breakage caused by radiation can occur, there is yet no direct evidence for the role of chemicals in such defects.

Monogenic Chemicals

Mendelian or monogenic diseases may manifest as autosomal dominants, autosomal recessives, or X-linked traits. The mutational mechanisms of many Mendelian diseases---particularly autosomal dominant diseases---are largely unknown. From inferences in the hemoglobin system, a variety of mutations (including single nucleotide substitutions, deletions, frameshifts, nonsense mutations, transcription control errors, and splicing errors) can occur (13). Many recessive diseases are caused by various enzyme deficiencies. Different mutations may be found in different families and many affected patients are compound heterozygotes for two different mutations affecting the same gene rather than true homozygotes. While much progress has been made in understanding mutations at the DNA level and at the visible chromosomal level, minor chromosomal changes that are not visible with current techniques are also likely. Many advances have already been made by finding "microcytogenetic" alterations in a variety of tumors and genetic diseases. Such defects that are larger than those demonstrable by molecular methods and smaller than those detectable by cytologic technique are likely to be uncovered with continued advances in methodology.

Since Mendelian dominant mutations manifest in the first generation, it has been suggested that certain rare dominant traits be used as "sentinel phenotypes" to monitor mutational frequency. While such a scheme has theoretical merit, there are logistical difficulties. Most clinical phenotypes of this type require expert diagnosis by specialists. Since the collaboration of many medical facilities and different experts is required for assessment of a sufficiently large number of cases of a rare disease, this approach has not yet been successfully applied.

Some recessive diseases are common because some heterozygote traits (such as Hb S and the various thalassemias) had a biologic advantage in the past and reached high frequencies because of selection. Other genetic traits may be frequent because of genetic drift or "founder" effects. An increase of mutations affecting these traits will have little effect on the total frequency of such conditions. Furthermore, outbreeding in modern societies causes a reduction of the frequency of all recessive diseases, making it difficult to detect mutationally-caused increases.

Multifactorial Diseases

Genes play a significant role in many different common chronic diseases such as diabetes, hypertension, atherosclerosis, allergies, schizophrenia, affective disorders, as well as in some birth defects. The specific genetic mechanisms in these diseases have not been worked out (11). If many different genes are the

underlying biologic substrate of such conditions, the effect of
mutations will be minimal. If, however, one or several major genes
contribute to the underlying genetic variability, mutational
effects on such genes would play a larger role in shaping a given
disease frequency. Since the total impact of genetics on public
health mediated by such common diseases and birth defects is great,
investigations on the underlying genetics are important. It should
also be remembered that the finding of a single gene effect in
these disorders does not necessarily mean that such a gene has a
major importance in disease etiology. Such identifiable single
genes may be constituents of the total genetic background and must
be distinguished from major gene effects, i.e., genes that
contribute preponderantly to the underlying disease phenotype.

WHAT APPROACHES ARE PROMISING FOR ASSESSMENT OF THE ROLE OF
MUTATIONS IN HUMANS? (2)

Many of the test systems used in lower organisms are covered
elsewhere in this volume and will not be discussed here.

In Vitro Studies in Humans (2,7,14)

A number of promising cytogenetic tests in humans require more
study. These include structural abnormalities in chromosomes
obtained from lymphocyte cultures and the significance of sister
chromatid exchanges. The significance of Y bodies in sperm, i.e.,
two Y chromosomes originating from nondisjunction, merits further
exploration. Biochemical test systems are being extensively
explored. Many genetic loci can be sampled from a single blood
specimen since many different enzymes and proteins can be
visualized on various supporting media, such as starch gels, to
look for variant gene products (15). For instance, a search for 50
different proteins in 10,000 (diploid) persons allows assessment of
1 million genetic loci. However, these approaches have not yet
shown definite mutations induced by radiation (7) or chemicals.
Two-dimensional gel electrophoresis of proteins from blood and body
fluids allows the visualization of many more protein "spots" but
the differentiation between primary gene products and secondary
modifications of proteins is difficult and often uncertain.

Somatic cell mutations have been searched for in red cells by
looking for the very rare erythrocyte with a new mutant hemoglobin
(21). The technique is difficult and not yet standardized. Even
if perfect, the results of this method or any other technique
measuring somatic mutation rates would need to be compared against
germinal mutation rates to assess their significance as an assay
for germinal mutations. DNA studies of sperm and other cells are
awaited with interest, including work on DNA repair and unscheduled
DNA synthesis.

In Vivo Studies (2)

A variety of studies have been carried out, but none has been successful in detecting critical differences between the exposed and nonexposed populations. Abortion frequencies are difficult to measure. Sex ratio changes to look for lethal X-linked mutants are difficult to assess because of other nongenetic factors affecting the sex ratio. Pregnancy outcomes are not easy to standardize and their interpretation is not always clear. Few "sentinel" phenotypes of dominant mutations exist to carry out appropriate studies. Birth defects of various sorts have many different causes and their genetic significance is not always clear.

CONCLUSIONS

Somatic mutation as the cause of cancer and germinal mutations as the cause of genetic diseases are clearly established as a biological phenomenon. Chemicals have produced cancer in humans and germinal mutations in other species. Because of the difficulties of studying the frequencies of genetic diseases in human populations, definite proof for the induction of genetic disease by chemicals does not exist. Nevertheless, data from other species on mutagenesis by environmental agents suggest that concern is warranted regarding somatic and germinal mutations potentially inducible by chemicals. Appropriate research in this area, therefore, requires support. In my opinion, however, the problem does not require a large-scale diversion of scarce research resources in the relevant sciences to work in environmental mutagenesis—particularly in countries with few trained research workers and limited research facilities.

Germ cell mutations are more difficult to detect than somatic mutations manifesting as malignancies. Knowledge of mutational frequency and the role of environmental mutagens in the etiology of human genetic diseases and birth defects is, therefore, imperfect. Many fundamental data regarding mutational mechanisms and mutation frequencies are lacking and assessments of the role of induced mutations in human populations are usually based on inferences. Better understanding and meaningful predictions require improved knowledge of human and medical genetics at all levels.

REFERENCES

1. BEIR (1980) The Effects on Populations of Exposures to Low Levels of Ionizing Radiation, National Academy of Sciences, Washington, D.C.
2. Bloom, A.D., Ed. (1981) Guidelines for Studies of Human Populations Exposed to Mutagenic and Reproductive Hazards,

March of Dimes Birth Defects Foundation, White Plains, New York.

3. Carter, C.O. (1982) Contribution of gene mutations to genetic disease in humans. Progr. Mutation Res. 3:1-38.

4. Doll, R., and R. Peto (1981) The Causes of Cancer. Quantitative Estimates of Avoidable Risks of Cancer in the United States Today. Oxford University Press, New York.

5. Hook, E.B., and I.H. Porter (1977) Human population cytogenetics. Comments on racial differences on frequency of chromosome abnormalities, putative clustering of Down's syndrome, and radiation studies, In Population Cytogenetics: Studies in Humans, E.B. Hook, and I.H. Porter, Eds., Academic Press, New York, pp. 353-365.

6. Idle, J.R., and R.L. Smith (1979) Polymorphisms of oxidation at carbon centers of drugs and their clinical significance. Drug Metab. Rev. 9:301-317.

7. International Commission for Protection Against Environmental Mutagens and Carcinogens (1983) Estimation of genetic risks and increased incidence of genetic disease due to environmental mutagens. Mutation Res. 225:255-291.

8. Motulsky, A.G. (1964) Pharmacogenetics. Progr. Med. Genet. 3:49-74.

9. Motulsky, A.G. (1977) Ecogenetics: Genetic variation in susceptibility to environmental agents, In Human Genetics, Proceedings 5th International Congress of Human Genetics, Mexico City. Excerpta Medica, Amsterdam, pp. 375-385.

10. Motulsky, A.G. (1978) Multifactorial inheritance and heritability in pharmacogenetics. Human Genet. Suppl. 1:7-11.

11. Motulsky, A.G. (1982) Genetic approaches to common diseases, In Human Genetics, Part B: Medical Aspects, B. Bonne-Tamir, Ed., Alan R. Liss, New York, pp. 89-95.

12. Motulsky, A.G. (1982) Interspecies and human genetic variation, problems of risk assessment in chemical mutagenesis and carcinogenesis. Progr. Mutation Res. 3:75-83.

13. Motulsky, A.G., and J.C. Murray. Current concepts in hemoglobin genetics, In Distribution and Evolution of Hemoglobin and Globin Loci, J. Bowman, Ed., Elsevier, New York.

14. National Research Council Committee on Chemical Environmental Mutagens (1983) Identifying and Estimating the Genetic Impact of Chemical Mutagens. National Academy Press, Washington, D.C.

15. Neel, J.V. (1981) In quest of better ways to study human mutation rates, In Population and Biological Aspects of Human Mutation, E.B. Hook, and I.H. Porter, Eds., Academic Press, New York, pp. 361-375.

16. Neel, J.V., W.J. Schull, and M. Otake (1982) Current status of genetic follow-up studies in Hiroshima and Nagasaki. Progr. Mutation Res. 3:39-51.

17. Pelkonen, O., E.A. Sotaniemi, and N.T. Karki (1982) Human
    metabolic variability in xenobiotic biotransformation:
    Implications for genotoxicity. Progr. Clin. Biol. Res.
    109:61-73.
18. Sankaranarayanan, K. (1982) Genetic Effects of Ionizing
    Radiation in Multicellular Eukaryotes and the Assessment of
    Genetic Radiation Hazards in Man, Elsevier Biomedical Press,
    Amsterdam.
19. Shepard, T.H. (1982) Detection of human teratogenic agents. J.
    Pediat. 101:810-815.
20. Silverberg, E. (1983) Cancer statistics, 1983. Ca-A Cancer
    J. Clin. 33:9-25.
21. Stamatoyannopoulos, G., and P.E. Nute (1981) Screening of human
    erythrocytes for products of somatic mutation: An approach and
    a critique, In Population and Biological Aspects of Human
    Mutation, E.B. Hook, and I.H. Porter, Eds., Academic Press,
    New York, pp. 265-273.
22. Vogel, F., and A.G. Motulsky (1979) Human Genetics. Principles
    and Practices, Springer-Verlag, Berlin.
23. Weinberg, R.A. (1983) A molecular basis of cancer. Sci. Am.
    249:126-144.
24. Yunis, J.J. (1983) The chromosomal basis of human neoplasia.
    Science 221:227-240.

CLINICAL GENETICS OF HUMAN CANCER

John J. Mulvihill

Clinical Genetics Section, Clinical Epidemiology
Branch, National Cancer Institute, NIH
Bethesda, Maryland 20205

SUMMARY

In a sense, the clinical genetics of cancer, especially of childhood cancer, is the intersection of human mutagenesis, carcinogenesis, and teratogenesis. Examples include: two or more mutations seem necessary for the initiation of many cancers (Knudson hypothesis); some of the same chemical and physical agents that cause cancer also cause birth defects, sometimes in the same patient (e.g., diethylstilbestrol and hydantoin); and treatment for cancer can break chromosomes, and cause cancer and birth defects. Clinical observations continue to initiate studies of disease etiology; they also validate findings from epidemiologic and laboratory research and provide opportunities to prevent birth defects, cancer and, possibly, other genetic diseases. Rare patients or families with various combinations of cytogenetic defects, single mutant genes, birth defects, and cancer (singly or in combinations) can improve understanding of another (e.g., carcinogenesis). Current efforts to bring appropriate human specimens to laboratory scientists should lead to identification of gene markers (e.g., restriction fragments polymorphisms), assignments of cancer gene loci including oncogenes, and detection of intermediate metabolites (adducts of DNA with chemicals) important in mutagenesis, carcinogenesis, and teratogenesis.

"Know syphilis in all its manifestations and relations, and all other things will be added unto you." Osler, 1897 (37)

ATAXIA-TELANGIECTASIA: A PROTOTYPE

In 1963, Robert W. Miller noted five pairs of siblings who
had died of childhood leukemia years before. He suggested that
clinicians recall one family to look for genetic traits recognized
to predispose to leukemia such as Down syndrome, Bloom syndrome,
and Fanconi anemia. In one particular family, he proposed the
diagnosis of ataxia-telangiectasia because the skimpy records
showed that the leukemic siblings, as well as a surviving brother,
had some neurologic problem. Three thousand miles away in Oregon,
Hecht and others diagnosed ataxia-telangiectasia in the brother,
confirming Dr. Miller's astute "bedside" diagnosis. (Citations of
original reports, as well as updates, are available [8]).

Like Osler's view of syphilis and general medicine, current
knowledge about ataxia-telangiectasia touches on many facets of
human cancer biology and cancer genetics. For example, in that
boy, who eventually died of chronic lung infection, Hecht and
coworkers documented the clonal evolution of a marker chromosome,
subsequently found to be chromosome 14 in other patients. The
percentage of the marker clone in peripheral lymphocytes increased
with time, as if it were a cell line that would become frankly
malignant. Unfortunately, the boy died of intercurrent lung
problems, but patients with the same translocation who progressed
to leukemia have since been described. Hecht further confirmed
that ataxia-telangiectasia was one of the chromosomal breakage
syndromes. Recently, translocations, like those of B-cell
lymphomas known to complicate ataxia-telangiectasia, have been
shown to involve the approximation of gene loci for immunoglobulin
heavy chains on 14q32 and the human myc gene on 8q24 to 8qter
(11).

An additional insight came from Birmingham, England, where a
patient with ataxia-telangiectasia developed lymphoma. After the
usual doses of radiotherapy, acute radiation toxicity occurred.
Two similarly radiosensitive patients with ataxia-telangiectasia
had been reported, but what was different in Birmingham was the
presence of radiobiologist David G. Harnden, who could assay
radiosensitivity in vitro. He found the patient's cells were
unduly sensitive to gamma-radiation. Other researchers
subsequently reported a defect in the endonuclease involved in
repairing DNA damage after exposure to X-rays.

Finally, epidemiologists have added to the understanding of
ataxia-telangiectasia by quantifying the frequency of cancer in
affected (homozygous) patients, who have two doses of the mutant
gene, and by suggesting an excess of cancer in heterozygotes, that
is, clinically normal individuals with one mutant gene.
Immunologists documented abnormalities of humoral and cell-
mediated immunity in patients with ataxia-telangiectasia and

uniform elevations of at least one carcinoembryonic antigen, alpha-fetoprotein.

In summary, a few patients studied intensively by alert clinicians and collaborating laboratory scientists have provided great insight into the interrelationships of gamma-radiation, cancer, immunodeficiency, and single gene defects of DNA repair.

CANCER CYTOGENETICS

The principle that most cancers arise from chromosomal imbalance, proposed by Boveri in 1914 (ref. 7a), has remained a reliable unifying concept strengthened by each technical advance in cytogenetics. Since encyclopedic reviews (7,27,38) and a journal, Cancer Genetics and Cytogenetics, are available, just three points deserve emphasis.

Emerging Generalizations

Several principles emerge in summarizing the cytogenetic abnormalities associated with human cancers (Table 1).

1. To date, leukemia and lymphoma have dominated the known associations of cytogenetic defects with cancer, despite their rarity in the population. This phenomenon is likely an artifact of the feasibility and ease of repeated blood and bone marrow sampling. As samples of solid tumors become easy to obtain repeatedly and to assay, they are increasingly found to have non-random chromosomal abnormalities, as Boveri hypothesized. Hence, improved techniques for karyotyping uncultured solid tumor cells are urgently needed.

2. Among the leukemias and lymphomas, the chief type of abnormality is an acquired translocation which does not seem to result in a loss of genetic material and which is seen only in the malignant tissue. Hence, the best known marker, the Philadelphia chromosome, is a translocation of material from chromosome 22, often to chromosome 9, and is seen only in bone marrow cells. This cytogenetic defect is not associated with congenital anomalies or a constitutionally abnormal karyotype. The primary etiologic significance of such cytogenetic abnormalities may be debated, but their role in the selective proliferative advantage of that progenitor cell and·its clone seems clear. Hence, an environmental insult to a cell, e.g., radiation, viruses, or chemicals, may well cause a translocation or monosomy that confers a selective advantage on it and its progeny.

3. In contrast, the cytogenetic abnormalities of solid tumors, especially embryonal neoplasms, a) represent a net loss of

genetic material, b) are constitutional (present in every body
cell and presumably in the zygote or early embryo), and c) are
associated with birth defects, often of the same organ system that
eventually becomes malignant. Hence, the Miller syndrome of
aniridia, growth and mental retardation, urogenital anomalies, and
Wilms tumor is seen in infants with a constitutional loss of part
of the short arm of chromosome 11 (11p13-) (24).

The neoplastic associations with aneuploidies of the Y
chromosome remain enigmatic. Individuals with the XXY Klinefelter
syndrome have a risk of breast cancer that approaches that of
normal XX women, and an even greater relative risk for rare
extragonadal germ cell tumors, especially mediastinal teratomas
(40). It seems that the Y-bearing germ cells have a special
predisposition to neoplasia, in particular when found in ectopic
sites. Further evidence for this hypothesis comes from the known
risk of malignancy in cryptorchid testes, even after orchiopexy
and even in the normally descended contralateral testis. Finally,
the various syndromes of gonadal dysgenesis, with or without the
Turner phenotype, carry a high risk of gonadoblastoma in the
dysgenetic tissue, but only when it has a Y chromosome.
Histologically identical tissue without a Y chromosome does not
seem predisposed to cancer.

Mapping Human Genes Associated with Neoplasia

The human DNA is sufficient for 50,000 to 100,000 genes.
Identified mostly through their disease-producing mutants, some
3368 loci have been catalogued (25). About 9% of these disease
traits have neoplasia as a feature or rare complication (31).

Sophisticated laboratory techniques of molecular and cellular
hybridization and karyotyping have surpassed inefficient family
studies of genetic linkage, resulting in some evidence of the
chromosomal locations of about 500 human genes (24). Among these
are some of the 200 genes associated with neoplasia,
retinoblastoma being a cardinal example (Figure 1). Assignment of
the retinoblastoma gene to the long arm of chromosome 13 was first
made from finding a constitutional deletion in a score of patients
with retinoblastoma and multiple birth defects, and then, in some
laboratories, by directly seeing similar defects in extended
peripheral blood chromosomes and in presumed retinoblastoma cells
of children with normal lymphocytic karyotypes (4,30). (Other
laboratories do not see 13q- in tumor cells or prophase
chromosomes [13,14].) A separate approach, using classic family
studies, linked the retinoblastoma trait closely to the gene for
the enzyme esterase D (41). This demonstration is especially
important because it supports the hope that looking for minute
chromosomal deletions--even though they represent hundreds of
genes--can lead the way to isolating individual "cancer" genes.

Such localization by karyotyping, even though lacking in
resolution, may point to areas of the genome that should be
studied by restriction endonucleases to discover actual base-pair
alterations; without such guideposts, DNA mapping may be too
cumbersome to apply.

Fitting in Human Oncogenes

Despite progress in mapping genes on human chromosomes, long
regions have few assigned genes, e.g., 3p, 5q, 8q, and 13q, all
areas associated with neoplasia (Figure 1). Until late 1982, 8q
was conspicuous, for it had no assigned gene loci despite being
involved in various translocations associated with leukemia,
lymphoma, mixed salivary gland tumor, and renal cell carcinoma.
But now sequences of DNA on 8q have been found to be complementary
to nucleotide sequences of two genes known to be among the few
genes of well studied retroviruses. Specifically, by DNA
hybridization techniques, v-myc (the onc gene of the avian
myelocytomatosis-29 virus) and v-mos (from the Moloney and Gazdar
murine sarcoma viruses) (10) have homologous sequences on the long
arm of human chromosome 8 (24). For convenience, these homologous
stretches of human DNA are also called oncogenes, c-myc and c-mos
(c, for cellular), although their normal function is unknown. Of
15 v-onc genes, at least nine homologous c-onc genes have been
assigned to human chromosomes (Table 1, Figure 1 in bold face
type).

Since viral oncogenes are so critical in experimental animal
carcinogenesis, it is plausible to infer that the homologous human
sequences are likewise important in the origins of human cancers.
In the best reported example, the 8;14 translocation in B-cell
lymphoma cell lines (11), c-myc is in the area of the break point
on chromosome 8 and, owing to the resultant translocation, becomes
near to an immunoglobulin gene on chromosome 14. The functional
consequence could be that c-myc acts as a promoter to allow
transcription to proceed through genes not normally active at that
time in the organism's life.

Many details are wanting, and the present interpretation may
prove simplistic; but the dramatic confluence of knowledge gained
from sustained animal research of viral oncology, from improved
cytogenetic techniques, and from mapping human genes has produced
a surge of hope that a huge advance in understanding human cancer
is at hand.

One recent interpretation of these facts and theories
emphasizes the crucial role of chromosomal breakage and
rearrangement, whether by translocation, inversion, insertion, or
deletion. It has been hypothesized that when one chromosomal
segment, which may contain a c-onc gene, is no longer adjacent to

LYMPHOPROLIFERATIVE DISORDERS

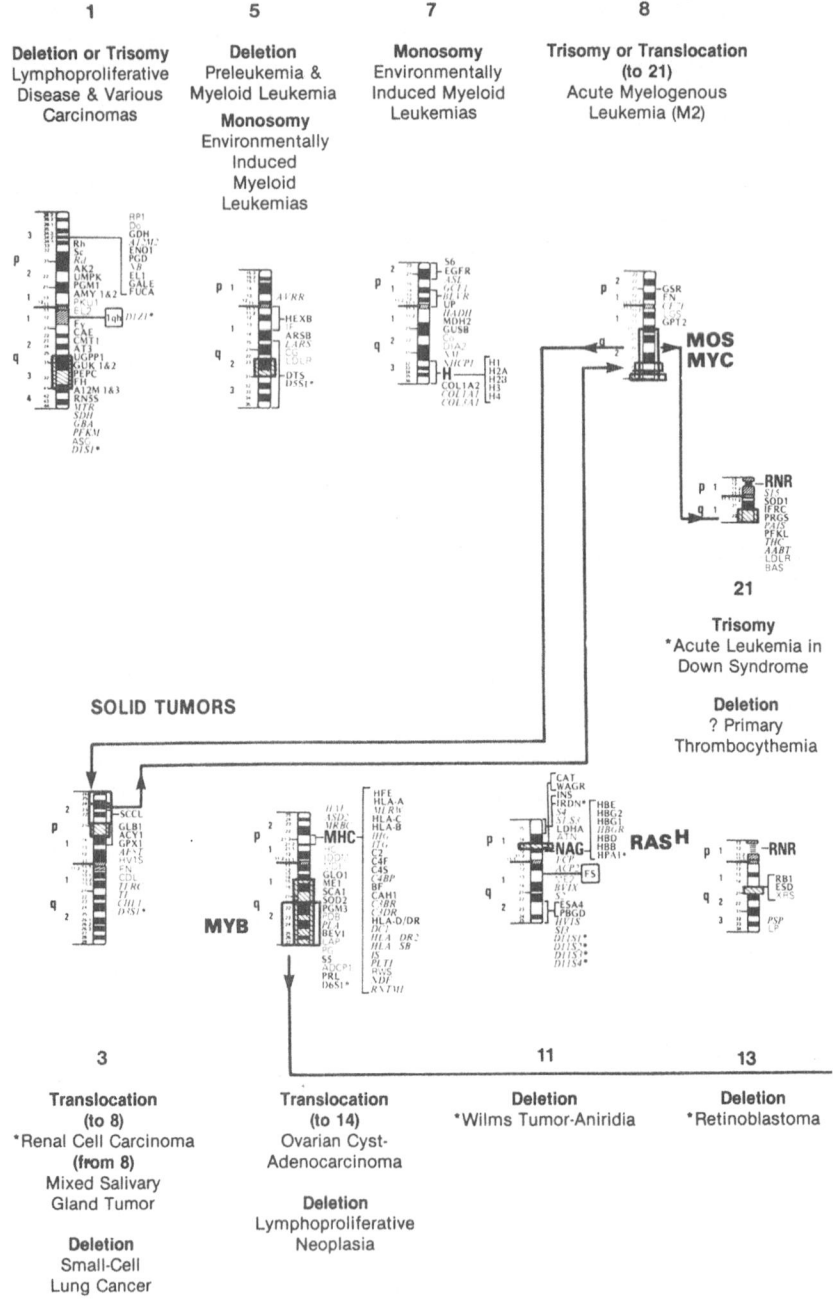

1

Deletion or Trisomy
Lymphoproliferative
Disease & Various
Carcinomas

5

Deletion
Preleukemia &
Myeloid Leukemia

Monosomy
Environmentally
Induced
Myeloid
Leukemias

7

Monosomy
Environmentally
Induced Myeloid
Leukemias

8

Trisomy or Translocation
(to 21)
Acute Myelogenous
Leukemia (M2)

21

Trisomy
*Acute Leukemia in
Down Syndrome

Deletion
? Primary
Thrombocythemia

SOLID TUMORS

3

Translocation
(to 8)
*Renal Cell Carcinoma
(from 8)
Mixed Salivary
Gland Tumor

Deletion
Small-Cell
Lung Cancer

Translocation
(to 14)
Ovarian Cyst-
Adenocarcinoma

Deletion
Lymphoproliferative
Neoplasia

11

Deletion
*Wilms Tumor-Aniridia

13

Deletion
*Retinoblastoma

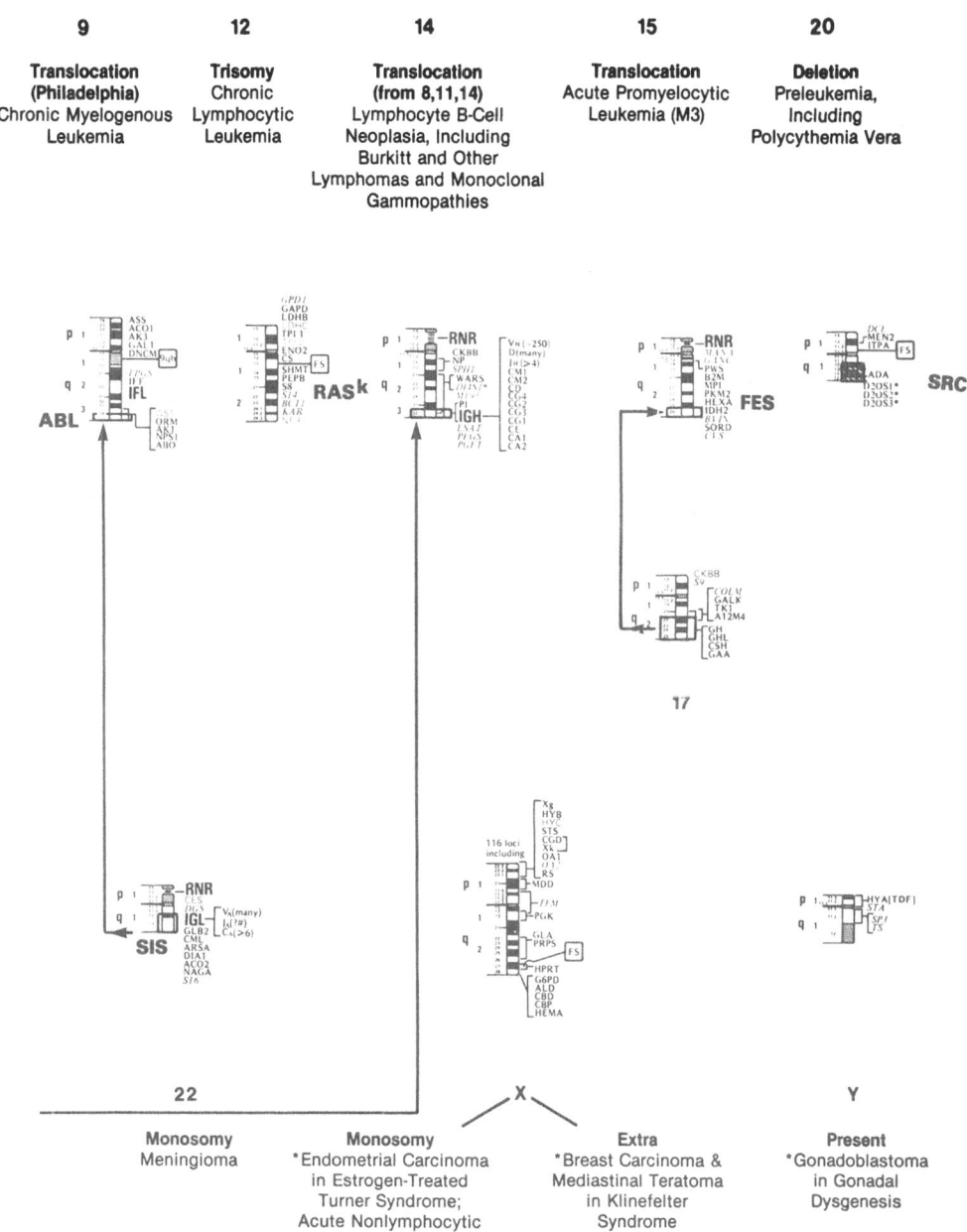

(Figure 1 Legend on following page)

Figure 1. This diagram, slightly modified from a prior version
     (35), superimposes the cytogenetic abnormalities associated
     with human neoplasia (7,27,38) on the map of single gene
     graits. V. A. McKusick supplied the basic gene map as of
     February 1982; an update has since appeared (24); but the
     assigned oncogenes have been added in bold type.
     Constitutional (inborn) chromosomal defects have asterisks;
     the remaining abnormalities are confied to the neoplastic
     tissue or closely related cell lines. The type of
     abnormality is in bold face type below the chromosome number:
     deletions are shaded, regions involved with translocation are
     indicated by open boses connected with arrows. Selection was
     usually confined to associations reported by at least two
     laboratories.

a second segment that is supposed to suppress the onc gene except
during embryogenesis, then the onc gene can initiate neoplastic
cell proliferation (39).

SINGLE GENES PREDISPOSING TO CANCER

Retinoblastoma

     Of the assigned genes for human neoplasia, retinoblastoma
continues to serve as the epitome of a human cancer gene in
several ways.

     1. The patterns of age at occurrence of retinoblastomas
provide evidence for various models of the pathogenesis of
hereditary tumors in general. The median age at diagnosis for
bilateral, familial retinoblastoma patients was one year; for
unilateral, sporadic retinoblastoma, three years. Knudson (15)
melded these and other observations into a compelling mutational
model, which resurrected prior multiple-hit theories of
carcinogenesis. He theorized that all retinoblastomas (and, by
extension, other cancers) arise from at least two mutations. The
second mutation always occurs after conception, that is,
postzygotically, as a result of an environmental or unknown agent.
The first mutation may be postzygotic, as in sporadic cases, or
prezygotic, as in hereditary cases.

Table 1. Chromosomes and Cancer--1983(7,27,38)

Chromosome	Oncogene	Cell Type
A. Leukemia - Lymphoma		
*Trisomy 21 (Down syndrome)		Acute leukemia
*14q- (Ataxia-telangiectasia)		Leukemia and lymphoma
Philadelphia (9;22 translocation)	abl, sis	Chronic myelogenous leukemia
8: translocation, aneuploidy	mos, myc	Acute myelogenous leukemia
15;17 translocation	fes	Acute progranulocytic leukemia
14q+ (translocations)		Burkitt and other lymphomas
12 trisomy		Chronic lymphocytic leukemia
5 or 7 monosomy		Environmental leukemias
B. Solid Tumors		
*13q14-		Retinoblastoma
*11p13-	ras^H	Wilms tumor
*3;8 translocation	mos, myc	Renal cell carcinoma
*XXY (Klinefelter syndrome)		Breast cancer, germ cell tumor
*Y (gonadal dysgenesis)		Gonadoblastoma
3p-		Small cell lung carcinoma
22-	sis	Meningioma
6;14 translocation		Ovarian adenocarcinoma

*Constitutional (inborn)

In short, features of sporadic cases are late age of onset, unilaterality, and negative family history. Hereditary cases, on the other hand, tend to have an early age of onset, multiple or bilateral sites, positive family history, and a predisposition for other tumors. These four characteristics have become hallmarks of a cancer of genetic origin.

2. The usual therapy for bilateral retinoblastoma is enucleation of the more involved eye and radiotherapy to the other eye. Survivors of such therapy have an excess of certain malignancies (Figure 2). Osteosarcomas occur far from irradiated areas, e.g., around the knee, the usual site of osteosarcoma of childhood, at a rate 50 times that of the general population; but the excess is seen only in patients with hereditary bilateral retinoblastoma. Such patients also have a high risk of

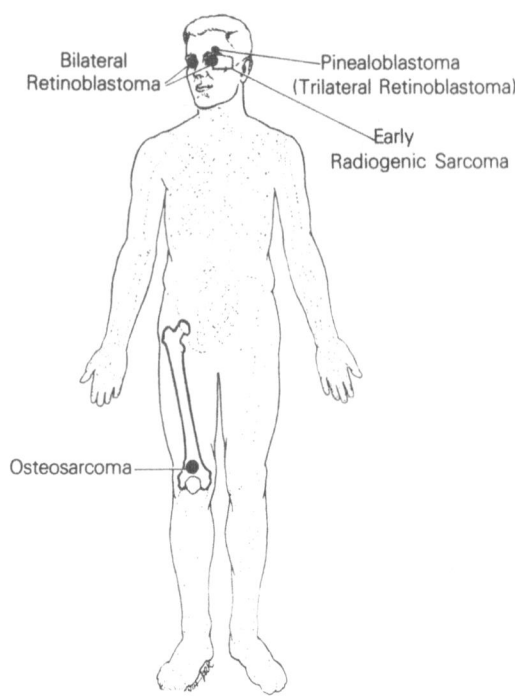

Figure 2. The pleiotropic effect of the
retinoblastoma genes.

pinealoblastoma, an embryonal tumor of the pineal gland, the so-
called "trilateral retinoblastoma" (3).

    3. Another pleiotropic effect of the retinoblastoma gene is
bony and soft-tissue sarcomas occurring in the field of radiation
(again in survivors of hereditary retinoblastoma). The incidence
of radiogenic sarcomas is not as unusual as their shorter latent
period, in comparison with tumors after radiation for non-
hereditary childhood cancers (42).

    4. This short latency period suggests that a further
manifestation of the retinoblastoma gene could be unusual
sensitivity to ionizing radiation. Indeed, one laboratory
reported significantly decreased colony survival in fibroblasts
from patients with hereditary but not sporadic retinoblastoma
(36).

    Other recent promising advances in studies of retinoblastoma
include description of reliable techniques for in vitro
propagation of most fresh human tumor cell lines (6) and for an in
vivo rat model using a cell line inoculum (16).

Some laboratories found neither 13q- in retinoblastoma cells (13,19) nor significant in vitro radiosensitivity (12,29); but the promise persists that, since a mutation exists in every cultured cell, a reliable in vitro phenotype can be found. As before, studies of mechanisms of carcinogenesis in rare cancers may provide insights into the origins of common adult cancers.

Polyposis and Cancer of the Colon

Despite wide publicity about diet as a cause of one common cancer, colorectal carcinoma, the most potent cause of this tumor is the single mutant gene for multiple polyposis of the colon or adenomatosis of the colorectum. By age 40 years, half of the individuals with the mutant gene have colon cancer; if they live long enough, all will develop the tumor (45).

Although the presence of many polyps is a sufficient clinical marker of the mutant gene, the disorder can be studied in vitro to understand the pathogenesis of colon cancer without polyposis (17). Tetraploidy, that is, four copies of each chromosome, has been found in epithelioid skin cultures and in colonic mucosal cells from patients with polyposis. Both are tissues that can become malignant in the polyposis syndromes, especially in Gardner syndrome; chromosomes in peripheral lymphocytes and fibroblasts (which do not express malignancy in the syndrome, but do, of course, contain the mutant gene) are normal. Stools from polyposis patients, compared with those of normal controls on similar diets, have more cholesterol and bile salts and fewer cholesterol metabolites, a pattern seen in other populations at high risk of colon cancer, perhaps for dietary reasons. Unusual features of polyposis fibroblasts, found by Kopelovich of Memorial Sloan-Kettering Cancer Center and awaiting confirmation, are a) a low serum requirement for growth, b) lack of contact inhibition, c) increased plasminogen activator, d) increased agglutination with concanavalin A, e) easy transformability by Kirsten murine sarcoma virus and by tetranodecanoyl phorbol-13-acetate, and f) disrupted actin organization. The first four traits are known to be in vitro manifestations of malignant transformation. Disorganization of intracellular actin-rich microfilaments may be directly related to the action of the polyposis gene.

Public Health Implications for Cancer Prevention

Prophylactic surgery. Despite the valid research strategy of seeking in vitro phenotypes of the mutant gene for polyposis, the fact remains that multiple polyps are sufficient to identify the gene clinically. The current recommendation is that, when the syndrome is diagnosed, colectomy should be done before age 20 years to prevent colon cancer.

As recounted elsewhere (32), primary cancer prevention by surgical removal of the organ at risk is achieved by a) colectomy in polyposis coli and familial colon cancer, b) thyroidectomy in the multiple mucosal neuroma syndrome, c) mastectomy or d) oophorectomy in familial breast or ovarian cancer, respectively, and e) gonadectomy in gonadal dysgenesis. Similarly, orchiopexy is done in cryptorchidism, in part to prevent malignancy. In view of these practices, it is a misconception to say that a genetic predisposition to cancer is untreatable.

Cancer risk in heterozygotes? A second public health implication of Mendelian traits which predispose to cancer arises from the Hardy-Weinberg principle. It states that the frequency of heterozygotes is much higher than the frequency of homozygotes. For example, if homozygotes with a recessive disorder occur in $1/x$ individuals, homozygotes number $2/\bar{x}$. For ataxia-telangiectasia, the ratio of affected heterozygotes to single gene carriers is $1/40,000$ to $1/100$, or 400 to 1. The implications in a population would be serious if a gene which predisposes to cancer in the double dose also did so in the single dose. To date, only Swift has addressed the problem; his diverse results need independent confirmation (43). For heterozygotes with the ataxia-telangiectasia or xeroderma pigmentosum gene, he finds some risk for some cancers in some age groups; for Fanconi anemia, results are largely negative despite an earlier positive study. The studies lost much power because gene carriers cannot be identified with certainty; Swift had to calculate the probability of their being heterozygous. Perhaps additional studies should await the development of a reliable practical laboratory assay for gene carriers.

Cancer Ecogenetics

Finally, the rare single gene traits which predispose to cancer have an important contribution to make to any hypothesis about the origins of human cancer. Rather than claiming, beyond from knowledge, that 90% of cancers are due to environmental factors, common sense dictates that one should expect people to differ in their response to carcinogens. It is well known that patients differ in their response to drugs; for example, there are fast and slow acetylators of isoniazid owing to genetic differences among individuals in a metabolizing enzyme. Study of genetic differences in response to drugs is called pharmacogenetics. By analogy, "ecogenetics" can be considered the study of genetic variations in response to any environmental agent (9,18,26,28,33, Table 2). Although several disorders in Table 2 have been discussed above, cutaneous albinism serves to crystallize the concept of cancer ecogenetics. If a negro with albinism gets skin cancer in sunny Africa, while her normally

Table 2. Genetic-Environmental Interactions
(Ecogenetics) in Human Malignancy

Environmental agent	Genetic Trait	Tumour or outcome
Ionizing radiation	Ataxia-telangiectasia with lymphoma	Radiation toxicity
	Retinoblastoma	Sarcoma
	Nevoid basal cell carcinoma syndrome	Basal cell carcinoma
Ultraviolet radiation	Xeroderma pigmentosum	Skin cancer, melanoma
	Cutaneous albinism	Skin cancer
	?Dysplastic nevus syndrome	Melanoma
Stilbestrol	XO Turner syndrome	Adenosquamous endometrial carcinoma
Androgen	Fanconi pancytopenia	Hepatoma
?Diet	Polyposes coli	Colonic carcinoma
	Lewis antigen (a-b-) (fucosyltransferase activity)	Alimentary tract carcinomas[26]
Iron	Hemochromatosis	Hepatocellular carcinoma
Tyrosine	Tyrosinemia	Hepatocellular carcinoma
Epstein-Barr virus	X-linked lympho-proliferative syndrome	Burkitt and other lymphomas
	HLA-A2, Bw46	Nasopharyngeal carcinoma
Papillomavirus type 5	Epidermodysplasia verruciformis	Skin cancer
N-substituted aryl compounds	N-acetyltransferase activity	Urinary bladder cancer[9,26,28]

* Updated from a prior version[33]

pigmented brother does not, which is the carcinogen: ultraviolet radiation or the mutant gene?

FAMILIAL CANCER

Towards a Definition

The notion of "cancer family" remains poorly defined, because "cancer" include many diseases and, like the concept "family" itself, cannot be precisely limited. Because one in three or four Americans will develop cancer in a lifetime, every family will have some members with cancer; by chance, some will have many affected relatives. As a working definition, a cancer family can be considered to comprise an excessive aggregation of one or more tumor types within a kindred group related by blood or residence.

An excess could, in theory, be statistically defined--but that is not an easy task. An excess frequency of cancer in a family is suggested by the occurrence of histologically proved malignancy:

in multiple generations,

in a high percentage of siblings,

at an unusual age,

in an atypical sex, or

in association with other cancers, genetic disorders, or birth defects in the same individual.

For example, breast cancer in two elderly sisters is not unusual; if it afflicted two sisters in their thirties, however, it would be remarkable. If two brothers were affected in their twenties the probability of a chance occurrence would be vanishingly small; it would be an excess beyond question.

Although family size obviously influences the number of relatives eligible for cancer, only 1.5 to 6% of individuals have three or more relatives with cancer (2,22).

One known specific pattern of cancer in families has been designated "the cancer family syndrome of Lynch" (21). Features include adenocarcinoma of the colon and endometrium occurring at an earlier age than usual, transmission in an apparently dominant fashion, and multiple primary tumors in more than 20% of affected individuals.

Environmental Familial Cancer

Cancer may run in families for either environmental or genetic reasons or both. Environmental factors rarely seem conspicuous, but the effect of environment on one family was obvious: a 60-year-old shipyard worker, occupationally exposed to asbestos, was diagnosed with asbestosis of the lungs (20). Two years later, his wife died of asbestos-induced mesothelioma. A daughter developed pleural mesothelioma. The women's only known carcinogenic exposure was to the asbestos dust that the worker brought home in his clothes.

Other environmental familial cancers have been attributed to benzene exposure; fathers passed to sons their livelihood as cobblers and, with it, exposure to benzene, which causes leukemia (1).

Familial Ovarian Cancer

In some families, an excess of cancer seems to be confined to a single tumor type. In a follow-up of 16 families with two or more women with ovarian cancer, 28 women had elected prophylactic oophorectomy (44). Nonetheless, three of these developed intraabdominal carcinomatosis one, five and eleven years after prophylactic surgery; pathologically the malignant tissue was indistinguishable from ovarian adenocarcinoma. The authors' preferred interpretation was that the tumors arose from residual peritoneal tissue that also had a predisposition to malignancy, probably genetic in origin.

Radiosensitivity in Cancer Families

Identifying genetic factors as determinants of familial cancer is aided by finding laboratory markers of pathogenetic significance in normal cells from family members with or at high risk of cancer. A few single gene traits, classically xeroderma pigmentosum and ataxia-telangiectasia, show abnormal response to radiation in vitro. For several years, the Clinical Epidemiology Branch of the U.S. National Cancer Institute has partially supported the Health Sciences Division, Atomic Energy of Canada, and supplied its investigators with fibroblast strains. In selecting specimens, families are sought with some clinical features that suggest an unusual response to radiation or radiomimetic agents. As reviewed elsewhere (33), results to date include the following:

1. In the recently described dysplastic nevus syndrome, affected individuals have pigmented nevi that greatly differ from ordinary moles because they have diverse shapes, sizes and colors, tend to arise throughout life, and have a high risk of becoming cutaneous malignant melanoma. The dysplastic nevus seems to be precursor to a sizable fraction of sporadic occurrences of melanoma and to many, if not all, cases of melanoma in patients immunosuppressed for renal transplantation. Also, the dysplastic nevus syndrome is often familial, is transmitted in an apparently autosomal dominant mode, and accounts for most instances of familial melanoma. The development of nevi is thought to be somewhat potentiated by sunlight, since the nevi are more frequent on sun-exposed skin than on protected areas. Fibroblasts on six melanoma patients from five families with the dysplastic nevus syndrome showed ultraviolet (but not ionizing) radiosensitivity, and four were also sensitive to 4-nitroquinolone-1-oxide, an ultraviolet radiomimetic chemical, as measured by colony survival.

2. A woman lost four of six children to acute myelogenous leukemia, and herself had rectal carcinoma 14 years after (and probably as a result of) radiotherapy for uterine cervical

carcinoma (Figure 3a). In correlation with prior results of
abnormal transformation following exposure to simian virus 40,
cells from the woman and from the two available leukemic daughters
were moderately sensitive to ionizing radiation, in the range of
heterozygotes with the ataxia-telangiectasia gene. Her husband,
two remaining health sons, and a sister with breast cancer had
normal colony survival after radiation.

　　　3.   Two members of a family with diverse malignancies gave a
history of exposure to radiation (Figure 3b). A teenage boy had a
vertebral osteosarcoma in the field of radiotherapy given for

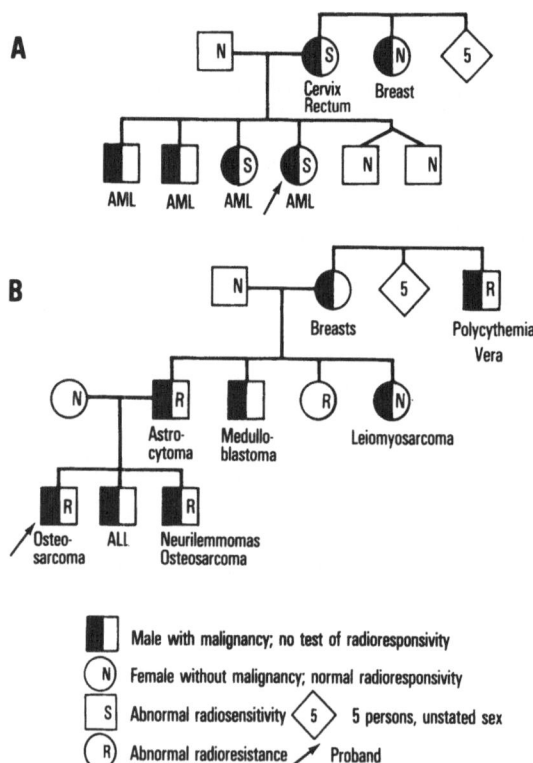

Figure 3.   Cancer families with some radiogenic neoplasms and
　　　unusual response in vitro of skin fibroblasts to ionizing
　　　radiation. A (NCI family 375): Abnormal radiosensitivity in
　　　vitro in a woman with radiogenic rectal cancer and four of
　　　six children with acute myelogenous leukemia (AML). B (NCI
　　　family 165): Abnormal radioresistance in a family with
　　　diverse malignancies (ALL, acute lymphoblastic leukemia).
　　　Reprinted with permission from (33).

bilateral malignant schwannomas 12 years before; his paternal great-uncle, with polycythemia vera, had had some occupational exposure during the manufacture of heavy water. Of eight family members tested, unusual radioresistance was observed in fibroblasts from four blood relatives with cancer and one without, but not in a blood relative with synovial sarcoma or in two normal spouses. This novel finding (greater cell survival after radiation than seen with normal cells) could account for a cancer predisposition: perhaps the abnormal cells survive a radiation exposure well enough to manifest a mutation that causes cancer, whereas normal cells, given the same dose, would die.

MUTAGENESIS AND TERATOGENESIS IN CANCER PATIENTS

     Are cancer and its treatment bad for genes, embryos, and fetuses? Perhaps cancer patients are the human population now exposed to the largest doses of potential physical and chemical mutagens and teratogens, that is, radiotherapy and drugs. Of course, cancer treatment is usually designed to interact specifically with DNA and to block its function.

     All pregnancies initiated by cancer patients are of research interest for diverse reasons. Pregnancies occurring before cancer serve as a control experience and as a measure of cancer predisposition. Pregnancy during a woman's cancer therapy measures teratogenicity and, if treatment antedated conception, mutagenicity as well. Pregnancies after cancer treatment measure viability of the reproductive tract and potential germinal mutagenicity.

     Experience to date is limited, but three recent reports from a cooperative clinical trial group (23) and single large cancer centers (5,14), included 155 pregnancies during or after chemotherapy in 109 women with cancer and in the wives of seven men with cancer (Table 3). Abortion was elected in 16% of the pregnancies, an expected high frequency. Of the remaining gestations, 74% ended with normal liveborns, 8% with spontaneous abortion or stillborns, 8% in premature or small-for-dates babies, and 10% with some "abnormality". Whereas the Stanford group saw no birth defects (14), the two studies I collaborated in (5,23) as considered abnormalities such very minor anomalies as hair patch, pelvic asymmetry, ear tags, sacral and pilonidal dimples, febrile seizure, and nevus. Most clinicians would deem these seven children normal, especially since they had isolated defects. Three major malformations--hydrocephaly, cleft lip and palate, and polydactyly with abruptio placentae and stillbirth--occurred in offspring of women undergoing chemotherapy during pregnancy and

Table 3. Outcomes of Pregnancies During or
After Cancer Chemotherapy

Tumor type	Total Cases	Preg-nan-cies	Elective abor-tions	Normal	Fetal Death	"Abnormal"
Hodgkin disease[44]	37	58	9	27	7	9 (+6 LBW)*
Other cancers[34]	29	31	3	21	2	2 (+1 LBW)
Several cell types[5]	30	42	10	24	2	4
(males only	7	12	2	6	1	3)
Hodgkin disease[14]	20	28	5	21	0	0 (+3 LBW)
TOTAL	116	160	25	93	11	15 (+10 LBW)
	[% of 133 pregnancies not aborted:			70%	8%	11% (+7%)]

*LBW = low birth weight

more probably represent teratogenicity than mutagenicity. Only
two anomalies in the 130 pregnancies could possibly represent
mutational disease. Hydrocephaly occurred in a baby conceived one
year after the mother stopped chemotherapy for Hodgkin disease.
The other was congenital hip dysplasia in a baby born six years
after completion by the mother of chemotherapy for Ewing sarcoma.
But neither disease is among the sentinel phenotypes or
cytogenetic syndromes that are considered useful in monitoring
populations for mutagens.

CONCLUSION

Further investigations of human carcinogenesis,
teratogenesis, and mutagenesis are best rooted in the concept of
ecogenetics, and are most likely to be advanced by bringing

etiologic observations of clinicians to the attention of imaginative laboratory scientists and epidemiologists.

REFERENCES

1. Aksoy, M., S. Erdem, G. Erdogan, and G. Dincol (1976) Combination of genetic factors and chronic exposure to benzene in the aetiology of leukaemia. Hum. Hered. 26:149-153.
2. Albert, S., and M. Child (1977) Familial cancer in the general population. Cancer 40:1674-1679.
3. Bader, J.L., A.T. Meadows, L.E. Zimmerman, L.B. Rorke, P.A. Voute, L.A.A. Champion, and R.W. Miller (1982) Bilateral retinoblastoma with ectopic intracranial retinoblastomas: Trilateral retinoblastoma. Cancer Genet. Cytogenet. 5:203-213.
4. Balaban, G., F. Gilbert, W. Nichols, A.T. Meadows, and J. Shields (1982) Abnormalities of chromosome #13 in retinoblastomas from individuals with normal constitutional karyotypes. Cancer Genet. Cytogenet. 6:213-221.
5. Blatt, J., J.J. Mulvihill, J.L. Ziegler, R.C. Young, and D.G. Poplack (1980) Pregnancy outcome following cancer chemotherapy. Am. J. Med. 69:828-832.
6. Bogenmann, E., and C. Mark (1983) Routine growth and differentiation of primary retinoblastoma cells in culture. J.N.C.I. 70:95-104.
7. Boiron, M., Ed. (1982) Chromosomes and hematology. Pathol. Biol. 30:745-816.
7a. Boveri, T. (1914) Zur Frage der Entstenung Maligner Tumoren, Gustav Fischer, Jena.
8. Bridges, B.A., and D.G. Harnden, Eds. (1982) Ataxia-telangiectasia: A cellular and molecular link between cancer, neuropathology, and immune deficiency, John Wiley, New York.
9. Cartwright, R.A., H.J. Rogers, D. Barham-Hall, R.W. Glashan, R.A. Ahmad, E. Higgins, and M.A. Kahn (1982) Role of N-acetyltransferase phenotypes in bladder carcinogenesis: A pharmacogenetic epidemiological approach to bladder cancer. Lancet 2:842-846.
10. Coffin, J.M., H.E. Varmus, J.M. Bishop, M. Essex, W.D. Hardy, Jr., G.S. Martin, N.E. Rosenberg, E.M. Scolnick, R.A. Weinberg, and P.K. Vogt (1981) Proposal for naming host cell-derived inserts in retrovirus genomes. J. Virol. 40:953-957.
11. Dalla-Favera, R., S. Martinotti, R.C. Gallo, J. Erikson, and C.M. Croce (1983) Translocation and rearrangements of the c-myc oncogene locus in human undifferentiated B-cell lymphomas. Science 219:963-967.

12. Ejima, Y., M.S. Sasaki, H. Utsumi, A. Kaneko, and H. Tanooka
    (1982) Radiosensitivity of fibroblasts from patients with
    retinoblastoma and chromosome-13 anomalies. Mut. Res.
    103:177-184.
13. Gardner, H.A., B.L. Gallie, L.A. Knight, and R.A. Phillips
    (1982) Multiple karyotypic changes in retinoblastoma tumor
    cells: Presence of normal chromosome No. 13 in most tumors.
    Cancer Genet. Cytogenet. 6:201-211.
14. Horning, S.J., R.T. Hoppe, H.S. Kaplan, and S.A. Rosenberg
    (1981) Female reproductive potential after treatment for
    Hodgkin's disease. N. Engl. J. Med. 304:1377-1382.
15. Knudson, A.G., Jr. (1978) Retinoblastoma: A prototypic
    hereditary neoplasm. Semin. Oncol. 5:57-60.
16. Kobayashi, M., N. Mukai, S.P. Solish, and M.E. Pomeroy (1982)
    A highly predictable animal model of retinoblastoma. Acta
    Neuropathol. 57:203-208.
17. Kopelovich, L. (1983) Prevention of hereditary large bowel
    cancer: Tissue culture assays and logistics. In Prevention
    of Hereditary Large Bowel Cancer. J.R.F. Ingall, and
    A.J. Mastromarino, Eds., Alan R. Liss, Inc., New York,
    pp. 131-145.
18. Koprowski, H., M. Blaszczyk, Z. Steplewski, M. Brockhaus,
    J. Magnani and V. Ginsburg (1982) Lewis blood-type may affect
    the incidence of gastrointestinal cancer. Lancet
    1:1332-1333.
19. Kusnetsova, L.E., E.L. Prigogina, H.E. Pogosianz, and
    B.M. Belkina (1982) Similar chromosomal abnormalities in
    several retinoblastomas. Hum. Genet. 61:201-204.
20. Li, F.P., J. Lockich, J. Lapey, W.B. Neptune, and
    E.W. Wilkins, Jr. (1978) Familial mesothelioma after intense
    asbestos exposure at home. J. Am. Med. Assn. 240:467.
21. Lynch, H.T. (1974) Familial cancer prevalence spanning eight
    years: Family N. Arch. Intern. Med. 134:931-938.
22. Lynch, H.T., F.D. Brodkey, P. Lynch, J. Lynch, K. Maloney,
    L. Rankin, C. Kraft, M. Swartz, T. Westercamp, and
    H.A. Guirgis (1976) Familial risk and cancer control. J. Am.
    Med. Assn. 236:582-584.
23. McKeen, E.A., J.J. Mulvihill, F. Rosner, and M.H. Zarrabi
    (1979) Pregnancy outcome in Hodgkin's disease. Lancet 2:590.
24. McKusick, V.A. (1982) The human gene map 20 October 1982.
    Clin. Genet. 22:359-391.
25. McKusick, V.A. (1983) Mendelian Inheritance in Man: Catalogs
    of Autosomal Dominant, Autosomal Recessive, and X-Linked
    Phenotypes, Sixth Edition, Johns Hopkins University Press,
    Baltimore.
26. Miller, M.E. (1982) Acetylator phenotype in bladder cancer.
    Lancet 2:1348.
27. Mitelman, F., and G. Levan (1981) Clustering of aberrations to
    specific chromosomes in human neoplasms. IV. A survey of
    1,871 cases. Hereditas 95:79-139.

28. Mommsen, S., A. Sell, and N. Barfod (1982) N-acetyltransferase phenotypes of bladder cancer patients in a low-risk population. Lancet 2:1228.

29. Morten, J.E.N., D.G. Harnden, and A.M.R. Taylor (1981) Chromosome damage in G$_0$ X-irradiated lymphocytes from patients with hereditary retinoblastoma. Cancer Res. 41:3635-3638.

30. Motegi, T. (1982) High rate of detection of 13q14 deletion mosaicism among retinoblastoma patients (using more extensive methods). Hum. Genet. 61:95-97.

31. Mulvihill, J.J. (1977) Genetic repertory of human neoplasia, In: Genetics of Human Cancer, J.J. Mulvihill, R.W. Miller, and J.F. Fraumeni, Jr., Eds., Raven Press, New York, pp. 137-143.

32. Mulvihill, J.J. (1981) Cancer control through genetics, In: Genes, Chromosomes, and Neoplasia, F.E. Arrighi, P.N. Rao, and E. Stubblefield, Eds., Raven Press, New York, pp. 501-510.

33. Mulvihill, J.J. (1982) Clinical genetics of human cancer. In: Host Factors in Human Carcinogenesis, Bartsch, H., and B. Armstrong, Eds., International Agency for Research on Cancer, Lyon, pp. 107-117.

34. Mulvihill, J.J. (1982) Towards documenting human germinal mutagens: Epidemiologic aspects of ecogenetics in human mutagenesis, In: Environmental Mutagens and Carcinogens, T. Sugimura, S. Kondo, and H. Takebe, Eds., University of Tokyo, Tokyo, pp. 625-637.

35. Mulvihill, J.J., and S.M. Robinette (1982) Neoplasia of man (Homo sapiens), In Genetic Maps, Volume 2, S.J. O'Brien, Ed., National Cancer Institute, Bethesda, Maryland, pp. 356-359.

36. Nove, J., W.W. Nichols, R.R. Weichselbaum, and J.B. Little (1981) Abnormalities of human chromosome 13 and in vitro radiosensitivity: A study of 19 fibroblast strains. Mut. Res. 84:157-167.

37. Osler, W. (1932) Aequanimitas, Third Edition, Blakiston's Son, Philadelphia, p. 134.

38. Sandberg, A.A. (1980) The Chromosomes in Human Cancer and Leukemia, Elsevier, New York.

39. Sandberg, A.A. (1983) A chromosomal hypothesis of oncogenesis. Cancer Genet. Cytogenet. 8:277-285.

40. Schimke, R.N., C.M. Madigan, B.J. Silver, C.J. Fabian, and R.L. Stephens (1983) Choriocarcinoma, thyrotoxicosis, and the Klinefelter syndrome. Cancer Genet. Cytogenet. 9:1-8.

41. Sparkes, R.S., A.L. Murphree, R.W. Lingua, M.C. Sparkes, L.L. Field, S.J. Funderburk, and W.F. Benedict (1983) Gene for hereditary retinoblastoma assigned to human chromosome 13 by linkage to esterase D. Science 219:971-973.

42. Strong, L.C. (1977) Theories of pathogenesis: Mutation and cancer, In Genetics of Human Cancer, J.J. Mulvihill,

R.W. Miller, and J.F. Fraumeni, Jr., Eds., Raven Press, New York, pp. 401-414.

43. Swift, M. (1982) Disease predisposition of ataxia-telangiectasia heterozygotes. In Ataxia-Telangiectasia: A Cellular and Molecular Link Between Cancer, Neuropathology, and Immune Deficiency, Bridges, B.A., and D.G. Harnden, Eds., John Wiley, New York, pp. 355-361.

44. Tobacman, J.K., M.A. Tucker, R. Kase, M.H. Greene, J. Costa, and J.F. Fraumeni, Jr. (1982) Intra-abdominal carcinomatosis after prophylactic oophorectomy in ovarian-cancer-prone families. Lancet 2:795-797.

45. Veale, A.M.O. (1965) Intestinal Polyposis. In Eugenics Laboratory Memoirs Series 40. Cambridge University Press, London.

CHROMOSOME ABNORMALITIES IN CANCER DEVELOPMENT[1]

Masao S. Sasaki

Radiation Biology Center, Kyoto University
Yoshida-konoecho, sakyo-ku
Kyoto 606, Japan

SUMMARY

Recent advances in the cytogenetics of human cancers have
provided us much important information for the understanding of
the mechanisms of initiation and progression of tumors.
Chromosome abnormalities in human tumors may be classified into
two categories: those specific to particular tumors and those
relatively common to all tumors. The former involve chromosome
rearrangements or deletions at a particular part of a particular
chromosome. This type of abnormality seems to be closely
associated with the initiation of cancer. The latter involves
changes in the particular part of the genome or genome
reorganizations. Those genome reorganizations are assumed to be
related to the progression or propagation of tumors.

Complementary to the information obtained from the chromosome
abnormalities in tumor cells, there are currently pieces of
information on the prezygotic chromosome abnormalities that are
associated with specific tumors. They are particularly
interesting in that such chromosome defects are also related to
the subvisible germinal mutations that are dominantly expressed to
develop cancers in a specific tissue. Retinoblastoma (RB) is a
malignant tumor of the eye in children. It occurs either by
somatic mutation in retinoblasts, or by inherited germinal
mutations. Some of the germinal mutations have been unequivocally
demonstrated to be a deletion or functional inactivation at a

---

[1]Work supported in part by the Princess Takamatsu Cancer Research
Fund.

particular site (q14) of chromosome 13. There is also some
evidence to suggest that the mutant RB genes are associated with
some gene-control elements. These lines of evidence point to the
possibility that many tumors originate from uncontrolled- or
maldifferentiation of the tissue. Chromosome abnormalities in
retinoblastoma tumor cells suggest that, in addition to the
initiation by mutation toward cancer, further genome
reorganizations are necessary for the progression or propagation
of tumors.

INTRODUCTION

Chromosome aberration is per se a major genetic damage, and
its genesis constitutes one of the most important topics in
genetic risk evaluation from human exposure to environmental
mutagens. In addition, evidence has been accumulating indicating
the involvement of chromosome aberrations as a source of
mutational change (16) and cancer (81,83,85), thus providing the
cytogenetic basis of the somatic mutation theory of
carcinogenesis.

Since the first proposal by Boveri (9) in 1914, the role of
chromosome abnormalities in carcinogenesis has long been a subject
of repeated discussion and controversy. From the findings of von
Hansemann (101) on the abnormal mitoses in tumor cells, Boveri (9)
developed the idea that chromosomal imbalance resulting from the
abnormal mitosis might be responsible for the malignant change of
the cells. This somatic mutation theory of carcinogenesis was
later developed into a more generalized form by Bauer (6) in 1928
and received a world wide support. However, little was known
about the origin, types and nature of mutation itself which leads
the normal cell to malignancy. It is now well documented that
many carcinogens are mutagens and that there is a strong
correlation between carcinogenicity and mutagenicity of agents.
However, such parallelism does not a priori mean that mutation is
actually involved in carcinogenesis. Recently, defects in DNA
repair have been extensively studied in human monogenic syndromes
with high cancer propensity. In 1968, an autosomal recessive
disease, xeroderma pigmentosum (XP) was shown by Cleaver (12) to
be defective in the repair of ultraviolet light (UV)-induced DNA
damage. Hypermutability of the XP cells was unequivocally
demonstrated by Maher et al. (49) in UV-irradiated cultured
fibroblasts derived from XP patients. These lines of experimental
evidence again strongly support the somatic mutation theory of
carcinogenesis. However, it should be noted that they do not
necessarily answer the primary intriguing question: what is the
mutation that is responsible for carcinogenicity?

In this respect, chromosome abnormality found in tumor cells has attracted attention as a type of somatic mutation. It has long been a subject of cancer research, whether chromosome abnormalities are primarily responsible for carcinogenesis, or are the result of carcinogenesis. There is now a substantial body of data indicating that chromosome changes in tumor cells are not random. Moreover, a number of human tumors are specifically associated with particular chromosome abnormalities (reviewed in refs. 81,83,85). These findings are strongly indicative that chromosome alterations are directly related to the initiation and progression of cancer, rather than the consequence of carcinogenesis. On the basis of recent cytogenetic findings in tumor cells and genetically determined cancer syndromes in man, the possible origin and nature of mutations toward cancer will be discussed.

CHROMOSOMES IN HUMAN CANCERS

Based on the extensive studies by himself and others on the chromosomes of transplantable animal tumors, Makino (50,51) in 1952 concluded that each tumor was characterized by a specific chromosome constitution, indicating that the abnormal mitosis itself might not necessarily be characteristic of cancer, but rather the specific chromosome constitution, even it was abnormal, was conserved by normal mitosis. He proposed the so-called "stemline theory" for the development of cancer. Such clonal proliferation of cells with an abnormal karyotype had also been demonstrated in human tumors (52,53). Then, a question arose as to whether the chromosome abnormality in tumor cells is the cause or the result of malignancy.

In 1960, Nowell and Hungerford (67) described the first consistent chromosome abnormality in human tumor; they found that a part of the long arm of a chromosome belonging to group G was missing in leukemic cells from patients with chronic myelocytic leukemia (CML). This unusually small chromosome was called the Philadelphia chromosome, or Ph^1 chromosome. This finding was particularly interesting in that a specific chromosomal change was associated with a specific type of tumor, hence strongly implicating the etiological involvement of a chromosomal change in the development of a specific tumor.

Aided by the development of chromosome banding techniques in the early 1970s, rapid progress was made in cancer cytogenetics and many details of chromosome mutations in human tumors have been revealed; some chromosome changes are specific to specific cancers and the others are common to many cancers. The nature of chromosome reorganization found in human cancers poses a new

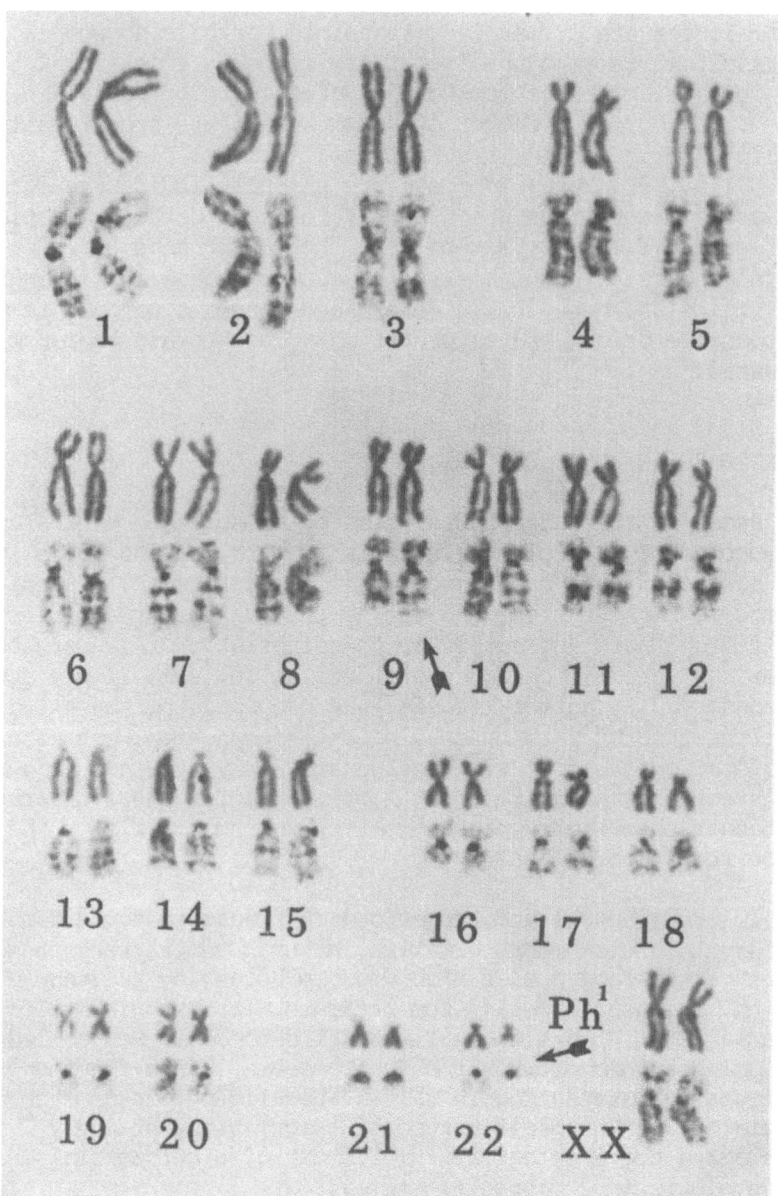

Figure 1. Karyotype of a chronic myelogenous leukemia (CML) cell.
Chromosomes aligned in the upper rows are stained without
banding and those in the lower rows are G-banded. Arrows
indicates translocation between chromosome 9 and 22, and Ph[1]
chromosome.

question as to how mutagens and carcinogens may play a role in such genomic changes toward cancer.

Chromosome Changes Specific to Specific Cancers

The Ph[1] chromosome was previously thought to be a terminal deletion of chromosome 21. However, the question of the origin of the Ph[1] chromosome was solved by the banding technique; the chromosome involved is chromosome 22 rather than 21 (11,69) and the Ph[1] chromosome results from a reciprocal translocation between chromosomes 9 and 22 (77) (Fig. 1). The Ph[1] chromosome is found in leukemic cells from approximately 85% patients with CML, and about 95% of the Ph[1] chromosomes are the result of translocation between chromosomes 9 and 22 (92). The break points are q11 on chromosome 22 and q34 on chromosome 9. The rest of the Ph[1] positive cases are variant translocations resulting from exchanges between chromosome 22 and a chromosome other than 9 (Fig. 2), or a cyclic exchange among chromosomes 9, 22 and a third chromosome. Although low in frequency, the variant forms of Ph[1] translocations are particularly informative about the nature of the chromosomal translocation leading to CML. Among the variant forms, translocations between two chromosomes are equal in frequency to those among three chromosomes, suggesting that the variant translocation between two chromosomes is actually cyclic exchange among three chromosomes (72). In this process, the distal part of a chromosome 22 translocates to a third chromosome, the distal segment of which then goes to a chromosome 9, and finally the distal part of chromosome 9 goes to a chromosome 22. The identification of the cyclic exchange involving three chromosomes depends upon the size of the transposing material coming from the third chromosome. If this is true, the transposition of the distal part of a chromosome 9 to a chromosome 22 is essential for the formation of Ph[1] chromosome and hence the etiology of CML.

The use of chromosome banding techniques has revealed many other chromosome changes which characterize specific tumors. Figure 3 shows the types of chromosome rearrangements found in the literature on the chromosomes involved in acute nonlymphocytic leukemia (ANLL). Among these, 8;21 translocation, t(8;21)(q22;q22), has been identified to be specific for acute myelocytic leukemia belonging to M2 of the FAB classification (78). Another specific chromosome association in ANLL is the 15;17 translocation, t(15;17)(q25;q22), in acute promyelocytic leukemia, which is classified as M3 according to the FAB classification (80).

In lymphomas, chromosome 14 is frequently involved in the formation of chromosome rearrangements, as often identified as the 14q+ anomaly. The donor chromosomes vary among cases (84) of non-

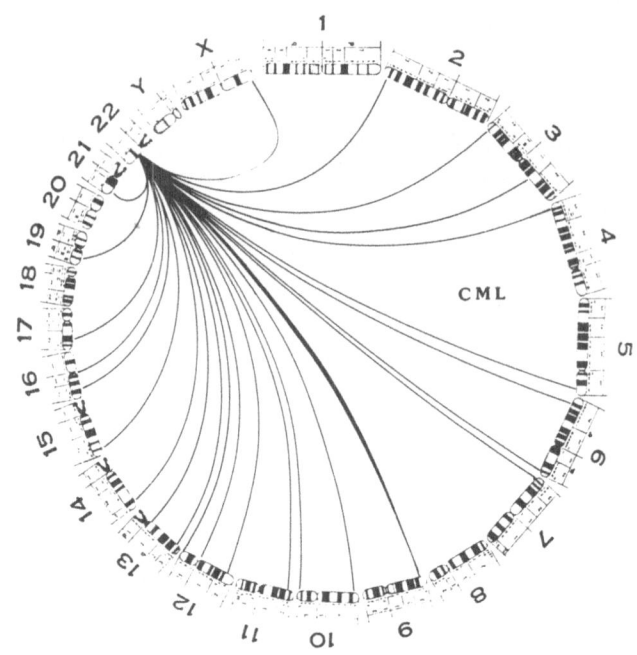

Figure 2. Types of chromosome rearrangements involved in the
   formation of Ph^I chromosome in chronic myelogenous leukemias
   (CML). Lines indicate reciprocal translocations between two
   chromosomes at indicated sites.

Burkitt's lymphomas (Fig. 4), but chromosome 8 is the most common.
In many cases, chromosome 14 at band q32 serves as the recipient
of the translocation. In Burkitt's lymphoma, whether endemic or
nonendemic, the donor chromosome is usually chromosome 8,
resulting in 8;14 translocation. The 8;2 and 8;22 translocations
have also been found as variant forms (7,61,99). Regarding the
origin of these specific translocations, it is interesting to note
that these translocations occur in association with the
immunoglobulin genes. The structural gene for heavy chain
immunoglobulin has been located on chromosome 14 (18) and those of
k and λ light chains have been assigned to the chromosomes 2 and
22, respectively (28,54,59). Moreover, the break point in
chromosome 14 has been specified, i.e., the region between the
heavy chain constant region (C_H) and the variable region (V_H).
The V_H gene distal to the C_H gene is translocated to q34 of
chromosome 8 in Burkitt's lymphoma cells (26,27).

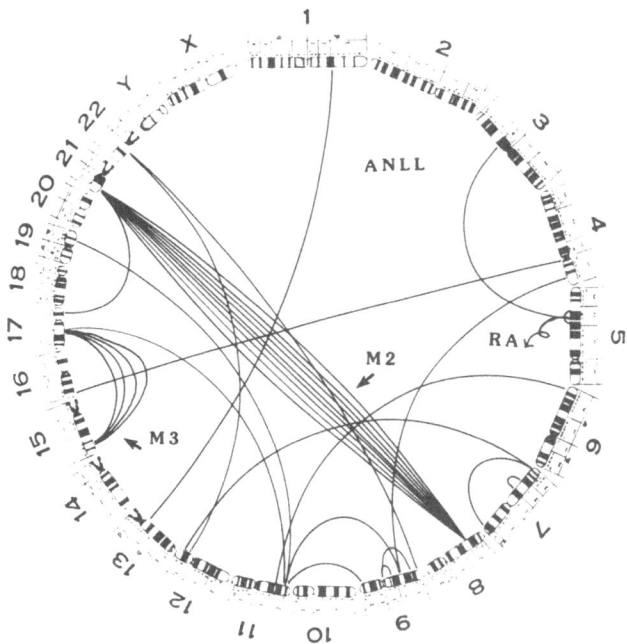

Figure 3. Types of euploid rearrangements of chromosomes found in
acute nonlymphocytic leukemias (ANLL). M2, M3: Leukemias
belonging to class M2 and M3, respectively. RA: Refractory
anemia.

Due to technical difficulties, little information is
available for the T-cell malignancy (30). In patients with the
genetic disease ataxia telangiectasia, which predisposes the
affected persons to the development of cancers particularly of the
lymphoreticular tissues, lymphocytes with abnormal karyotypes are
frequently found in the peripheral blood. These cells have been
considered to be leukemic T cells. In these cells, chromosome 14
is frequently involved in chromosome rearrangements. However, the
break point is 14q12 (Fig. 5), while those in B-cell lymphomas are
at 14q32.

Figure 6 shows the types and location of the typical
chromosomal changes which are specific to the specific type of
human tumors. Such tumor specific chromosome changes are, in many
cases, either reciprocal translocations or interstitial deletions.
The cancer gene map tells us much about the nature of the
chromosome mutations leading to malignancy.

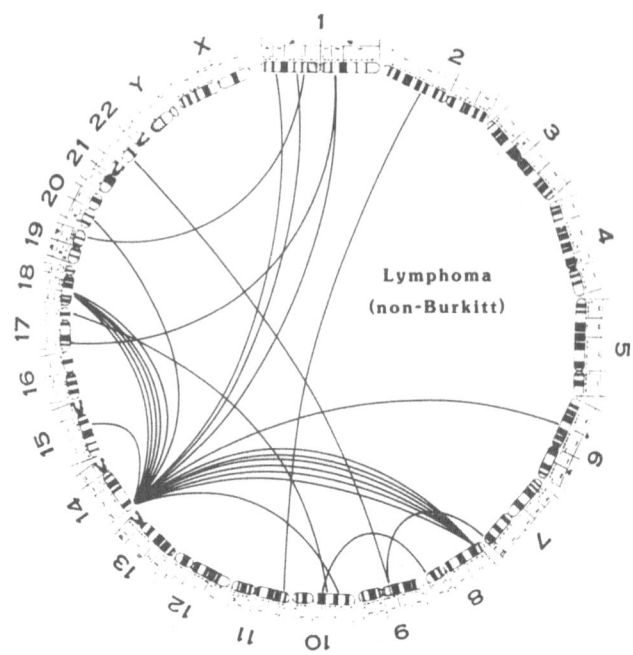

Figure 4. Types of chromosome rearrangements in non-Burkitt
lymphoma cells. chromosome mutations leading to malignancy.

First, tissue specificity tends to be integrated into mutation at
specific loci. Second, deletion and gene transposition associated
with cancer are expressed dominantly, suggesting that they belong
to the regulatory-type mutations. Third, because of tissue- or
tumor-specificity, it is reasonable to assume that specific
chromosome changes directly relate to the initiation toward
cancer, or at least to very early stages of carcinogenesis.

Chromosome Changes Common to Cancers

Chromosome changes found in tumors are generally complicated.
It is not necessarily simple to distinguish between chromosome
changes which are specific to particular tumors and those
relatively common to a variety of tumors. In CML, the Ph^1
chromosome or 9;22 translocation is a specific change which
characterizes the disease. However, in the blastic phase of CML,
a variety of additional chromosome changes can be found, such as
addition of chromosome 8, 19, 1q, 17q, Ph^1, and loss of chromosome
7, Y, and/or additional rearrangements of chromosomes. These
additional changes are also found in other acute leukemias as well

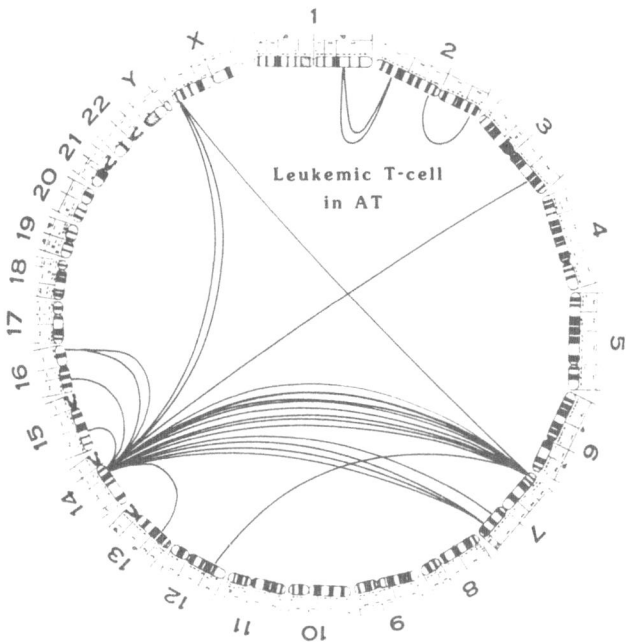

Figure 5. Types of chromosome rearrangements found in leukemic T
    lymphocytes in patients with ataxia telangiectasia (AT).
    Data from 5 Japanese patients.

as in other tumors. These abnormalities are nonrandom, and the
clustering of aberrations must have some biological significance.
Chromosome changes relatively common to human tumors involve
polyploidization, loss or gain of particular chromosomes (e.g.,
1q+, i(6p), i(17q)). However, the lack of cancer specificity
suggests that these chromosome changes are not directly related to
the initiation of cancer, but rather that they are important
changes associated with the progression of cancers. These changes
may reflect the presence of loci in these chromosome regions that
are concerned with the growth control, and the malignant cell
growth may be attained by the loss of genes or by gene
amplification associated with such genome rearrangements.

The special cases of gene amplification are double minutes
(DM) and homogeneously stained regions (HSR). They are the result
of selective multiplication of a particular gene (46), and can be

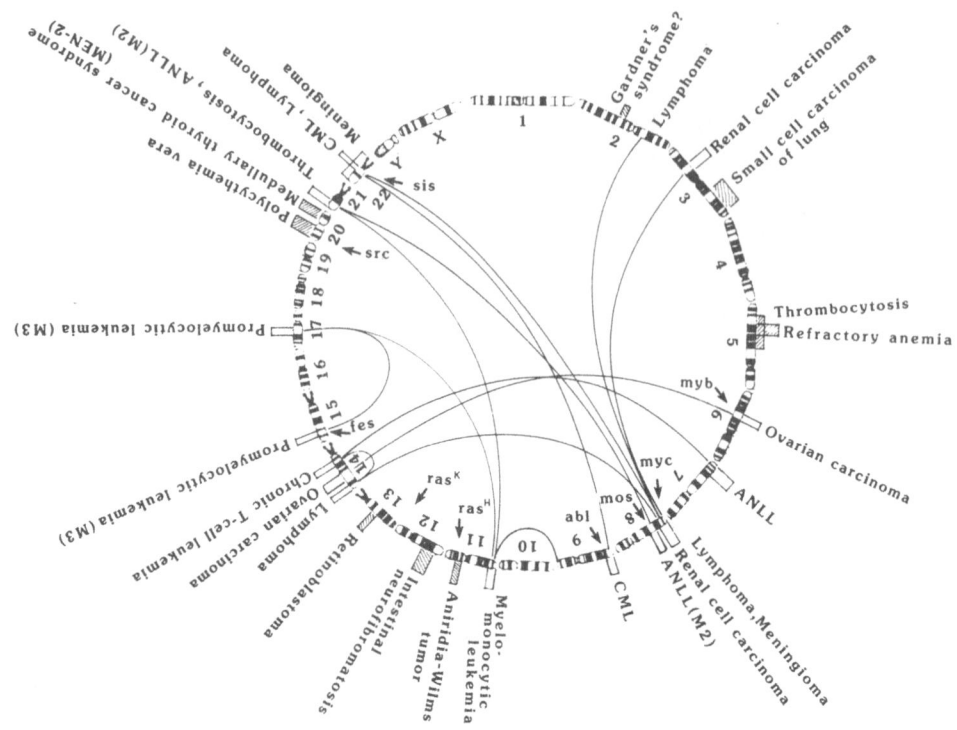

Figure 6. Chromosome abnormalities associated with the specific
        tumors in man. Lines connecting two chromosomes represent
        chromosome rearrangements at indicated sites. The shaded
        areas show interstitial deletions. The chromosomal sites of
        the cellular oncogenes are also presented.

recognized as chromosomes or chromosome region by their
homogeneous staining with Giemsa. DM chromosomes lack
centromeres, segregate unpredictably at cell division and may be
the predecessor to HSR (91,97). DM chromosomes and HSR are
chromosome abnormalities frequently found in malignant human
tumors.

Hypotetraploid constitutions frequently found in a variety of
solid tumors and hyperhaploid constitutions found in leukemia and
other cancers may be the alternate ways to attain gene
amplification, or a deviation from the genetic control of cell
growth and differentiation. In hypotetraploid tumor cells, some
chromosomes are disomic or trisomic, while others remain
tetrasomic, which is the condition analogous to the gain or loss
of certain chromosomes to triploidy. Similarly, in the

hyperhaploid cancer cells, some chromosomes remain disomic. These chromosome conditions found in malignant human cancers are essentially the same as the hyperdiploid conditions, suggesting that the deviation from regulation of cell growth and differentiation is under the control of genome balance. Unlimited growth may be attained by amplifying the particular genes or the chromosomes carrying those genes, as well as by losing its suppressor or inhibitor genes.

These lines of cytogenetic evidence suggest that the chromosome changes relatively common to cancers are those connected with the progression of tumor cells that have been already initiated toward malignancy. To date, the cytogenetic effects of environmental mutagens and carcinogens have been mainly discussed in relation to their potential to the initiation step, but genomic changes associated with the progression of tumor must also play an important role in the development of cancer. Cytogenetic insights along this line should therefore be encouraged.

Chromosome Changes and Cellular Oncogenes

Very significant discoveries in the past few years might provide a new link between chromosome changes and cellular oncogenes in human cancers. In 1981, Klein (42) suggested that the events leading to malignancy in some human and murine B-cell lymphomas might be intimately related to the genetic rearrangements of immunoglobulin genes during the normal differentiation of B-lymphocytes, and that transposition of the active promoter might result in the increased expression or activation of normal cellular genes, probably the cellular oncogenes. In the past few years, rapid expansion of molecular analysis of cellular oncogenes in normal and tumor cells has revealed an intimate relation between the break points involved in chromosome translocations and the chromosomal location of cellular oncogenes in a variety of human tumors (reviewed in ref. 82). To date, nine cellular oncogenes (c-onc) have been assigned to human chromosomes. Figure 6 shows the location of oncogenes on the specific chromosomes and their relevance to the chromosome abnormalities in human cancers.

As predicted by Klein (42), c-myc was located on chromosome 8 in man. Moreover, in Burkitt's lymphoma cells with 8;14 translocation, the c-myc gene was found to be translocated from its normal position on chromosome 8 to a new location distal to the constant region of immunoglobulin heavy chain genes on chromosome 14 (19,64,96). The variable region on chromosome 14 exchanges reciprocally with the c-myc containing region in the distal part of chromosome 8 (19). Analogous translocations have been found in mouse plasmacytomas, in which the distal region of

chromosome 15 is translocated to the end of chromosome 12, or less
frequently to chromosome 6 (68).   In the mouse, the immunoglobulin
heavy chain locus is assigned to chromosome 12 and the light chain
locus is assigned to chromosome 6 (23).   The c-myc sequence is
present on chromosome 15 and exchanges its position with
immunoglobulin genes by chromosomal translocation (17).

Another oncogene, c-mos, has been located on human chromosome
8 at band q22 (64), which is the breakpoint involved in 8;21
translocation of acute myelocytic leukemia belonging to M2.
Recently, De Klein (21) located the oncogene c-abl on chromosome 9
and found that, in association with the Ph^I translocation in CML,
c-abl had been translocated to chromosome 22.   In the mouse, c-abl
is present on chromosome 2 (36), where clustering of break points
of chromosome aberrations has been found in spontaneous and X-ray-
induced myelocytic leukemias (38).   Concordance of the chromosomal
locations of cellular oncogenes and the sites of chromosome
aberrations in human tumors has been shown to be true for other
oncogenes.   The c-ras oncogene has been located on the short arm
of chromosome 11 (22); the deletion of that region is known to
predispose to the aniridia-Wilms' tumor (32).   The assignment of
c-myb to chromosome 6, c-fes to chromosome 15, c-sis to chromosome
22, and c-src to chromosome 20 are also interesting, since they
suggest possible involvement in the tumor-specific chromosome
abnormalities such as 6;14 translocation in ovarian tumors, 15;17
translocations in acute promyelocytic leukemia, 8;22 translocation
in meningioma and 20q- abnormality in myeloproliferative disease,
respectively.

The mechanisms of such site-specific chromosome
rearrangements and the involvement of cellular oncogenes as well
as their link to cancer are not known.   However, the involvement
of immunoglobulin genes in lymphomas and plasmacytomas suggests
that gene rearrangements during cellular differentiation are
intimately related.   It becomes clear that c-myc gene translocated
to the C$_H$-region segment is not necessarily activated to
transcription.   The c-myc gene is translocated in the opposite
transcriptional direction from that of the immunoglobulin genes
(1,19).   Moreover, molecular analysis of the c-myc sequences in
plasmacytomas and Burkitt's lymphomas indicates that the
translocation disrupts the normal c-myc function by separating the
c-myc gene from its normal promoter (1).   Adams et al. (1)
suggested that plasmacytoma- and B-lymphoma carcinogenesis involve
two steps: translocation of c-myc and activation of another
oncogene.   The relationship between the two events is not known.

Several attempts have been made to identify the gene products
of DM chromosomes and HSR in cancer cells (8,47,76).   Recently,
Alitalo et al. (2) demonstrated in human neuroendocrine tumor
cells in culture that c-myc genes are amplified to a large number

of copies in the HSR region and possibly also in DM chromosomes. Enhanced expression by amplification of c-myc sequences may contribute to the malignancy in these tumor cells. The role of enhanced expression or multiplication of copies of cellular oncogenes in the development of cancer is not clear, but such transforming-gene products are characterized by the protein kinase activity to phosphorylate tyrosine moiety (29,48), which resembles that of growth factors (14,98). Such a process may be related to the progression of tumors by enhancing cellular growth. The role of cellular oncogenes in the development of cancer remains unclear, but their association with chromosome abnormalities in tumor cells encourages further studies.

PREZYGOTIC CHROMOSOME MUTATIONS IN RELATION TO CANCER

Prezygotic Chromosome Mutations Associated with the Specific Cancers

There is a type of human tumor whose development is associated with prezygotic or constitutional chromosome abnormalities. In persons who have inherited such particular chromosome lesions, the lesions appear to link to the development of particular types of tumors in particular tissues, and usually at particular stages of development. These chromosome lesions are especially important in relation to the etiologic involvement of chromosome mutations in carcinogenesis, as well as to the nature of the somatic mutation leading to malignancy.

Retinoblastoma (RB) is a childhood malignant eye tumor. Approximately 40% of RB tumors are hereditary: patients inherit the RB genes transmitted over generations, or those newly arisen in parental germ cells. In these patients, all the somatic cells carry RB genes but the tumor develops in their eye(s). The occurrence of tumors in both eyes (bilateral) is frequent in these patients (102). The remaining 60% of RB cases are non-hereditary, arising as a consequence of the somatic mutations occurring in the retinal cells of the child. In these patients, only one eye is affected (unilateral).

By analyzing the age of onset in the hereditary and non-hereditary forms of RB, Knudson (43) proposed the theory that RB arises as the result of two successive mutations. According to this theory, the hereditary RB has already one mutation as a prezygotic mutation and only one additional mutation is necessary for malignant transformation, whereas unilateral non-hereditary cases require two postzygotic mutations to develop tumors. The difference between the two types of RB is manifested as an early onset and more frequent bilateral occurrence in the former, in contrast to a later onset and unilateral involvement in the

latter. However, in family studies of RB, Matsunaga (55,56)
reported that in familial RB the expressivity, as determined by
tumor laterality, could not be adequately explained by a simple
stochastic hit model as predicted by Knudson (43). Rather, the
expressivity of the carrier child was influenced by that in his
carrier parent. According to this "host resistance" model, a
single mutation is enough to initiate malignancy and the
expressivity is under the control of the inherited host resistance
level, which is under multigenic control.

Some RB patients have constitutional chromosome abnormalities
involving chromosome 13, as manifested by a partial deletion of
its long arm (13q-) (reviewed in ref. 57). The association of
specific chromosome abnormality to specific types of embryonal
tumor is particularly interesting. In a series of cytogenetic
surveys of patients with RB, we found five patients with
constitutional chromosome abnormalities. They all involved
chromosome 13: two patients with del(13)(q12;q22), one with
del(13)(q14), one with mosaicism of normal and del(13)(q14), and
one with t(13;X)(q12;q22). Figure 7 summarizes the deleted
regions in chromosome 13 found in our study together with those in
the literature. It is evident that the deletion of a region at
band q14 in chromosome 13 is critical for the development of RB.
In a case with 13;X translocation, the translocated X chromosome
was late replicating in cultured skin fibroblasts (25) and the

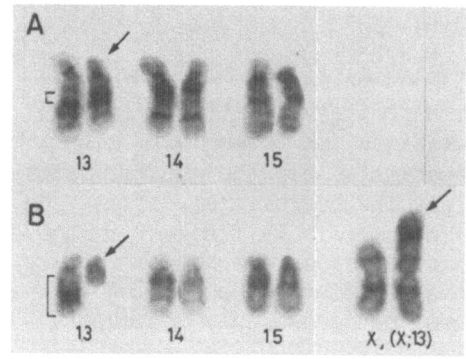
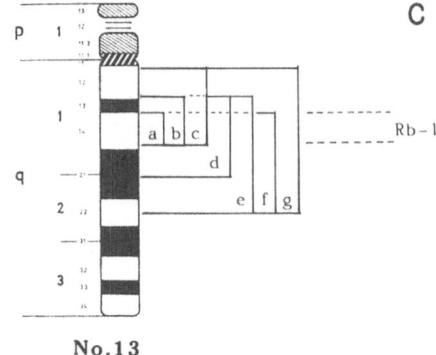

Fig. 7. Left: Abnormalities of chromosome 13 in retinoblastoma
    (RB) patients. A, interstitial deletion of chromosome 13; B,
    translocation between chromosome 13 and X. Arrows indicate
    abnormal chromosomes. Right: Schematic representation of
    chromosome 13 summarizing the regions involved in the
    interstitial deletions found in RB patients. Brackets
    indicate the regions involved: (a): refs. 41, 62, 87, 106.
    (b): refs. 103, 104.   (c): refs. 33, 40.   (d): refs. 20, 41.
    (e): refs. 74, 87.   (f): Refs. 44, 70, 87, 94, 104, 105, 106.
    (g): ref. 73.

gene expression in 13q14 was found to be suppressed, probably by
the spreading of the inactivation of X chromosome as indicated by
the reduced expression of esterase D gene, which is also assigned
to 13q14. A comparable case has been reported by Nichols et
al. (66). If such inactivation had also occurred in retinal
cells, these observations indicate that the functional monosomy or
the turning off of the normal expression of gene(s) at q14 of
chromosome 13 would be enough to lead retinal cells to malignancy
or maldifferentiation (65,85). Such abnormal expression might
also be possible by translocation at band q14. The 13;18
translocation has been reported in RB patients (63). The second
case has been recently found in our laboratory. The RB patient
had a balanced 13;10 reciprocal translocation. Cohen et al. (13)
showed that the t(3;8)(q12;q24) balanced translocation was
associated with the clear cell carcinoma of the kidney in ten
members of a family over three generations.

Recently, the major gene for RB has been assigned to q14 of
chromosome 13 (95). Knudson et al. (44) predicted that the RB
gene might be a deletion of a submicroscopic region at band q14.
However, we have some experimental evidence which suggests that
the RB gene is essentially a different type of mutation from
interstitial deletion, although the mechanism linking to the
initiation of RB might be common to both mutations, possibly
through a disruption or repression of the normal expression of
particular gene(s) at q14. As is the case for some other
hereditary neoplasms with a dominant mode of inheritance (88,90),
generation of non-constitutional chromosome rearrangements is seen
in cultured skin fibroblasts from patients with hereditary RB.
This is not the case for patients with 13q- (86). Furthermore,
cultured skin fibroblasts from patients with hereditary RB are
highly sensitive to transformation by murine sarcoma virus, while
fibroblasts from RB patients with 13q- show normal sensitivity to
the virus (60). It is of particular interest that there seems to
be a correlation between the expressivity and susceptibility to a
virus. The nature of the RB gene, its control mechanism of the
virus susceptibility, and their relevance to the RB development
are yet to be elucidated.

A condition analogous to the 13q- anomaly in RB has been
found in aniridia-Wilms' tumor. The constitutional chromosome
abnormality found in patients with aniridia-Wilms' tumor is an
interstitial deletion of chromosome 11 (11p-). Fig. 8 shows the
sites involved in the deletions, as reported in the literature
including our own studies. The critical involvement of a region
at band p13 of chromosome 11 is highly suggestive.

Although the information is still fragmentary, many other
cases have been suggested for a possible link between particular
prezygotic chromosome lesions and development of particular tumors

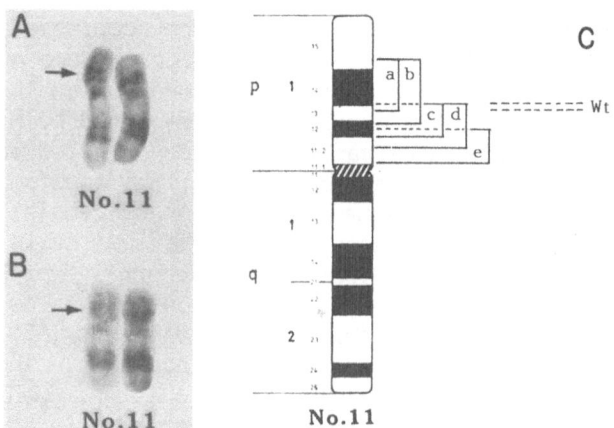

Fig. 8. _Left_: Abnormalities of chromosome 11 in patient with
aniridia-Wilms' tumor. _Right_: Schematic representation of
chromosome 11 summarizing the regions involved in the
interstitial deletions found in patients with aniridia-Wilms'
tumor. Brackets indicate the regions involved: (a): refs.
15, 31, 58, 107. (b): ref. 75. (c): ref. 3. (d): ref. 39.
(e): ref. 31.

(34,100). The presence of the constitutional or prezygotic
chromosome abnormalities in procancer syndromes thus provides
strong evidence for the etiologic involvement of chromosome
structural changes in malignant transformation.

Chromosome Changes in Hereditary Tumors

The spectrum of chromosome abnormalities in hereditary tumors
might provide much information about the role of chromosome
changes in cancer development. In some RB tumor cells, chromosome
aberrations involving chromosome 13 have been reported (4,37).
Similarly, 11p- anomaly was found in tumor cells of Wilms' tumor
(93). These findings are consistent with the idea that the tumor
gene is recessive in its character and deletion at its homologous
site results in the expression of the tumor gene at hemizygous
condition. In other words, a primary predisposition may exist,
but malignant transformation is triggered by subsequent loss of a
part of chromosome 13. However, in RB, chromosome abnormalities
involving chromosome 13 are not as common as expected (5,35).
Gardner et al. (35) studied chromosomes of ten RB tumors and found
that chromosome aberrations involving chromosome 13 were
relatively infrequent. Instead, duplication of the long arm of
chromosome 1 and the long arm of chromosome 17 were most common.
We also studied the chromosomes of ten RB tumors and came to the
same conclusion (24,89). Chromosome abnormalities were detected
in nine out of ten tumors, but chromosome 13 was not specifically
involved. Duplication of the long arm of chromosome 1 was the

most common, and i(18q) and i(17q) were also frequent. These
chromosome abnormalities were also common in other malignant human
tumors (45,79), and they may be categorized into genomic changes
that are associated with the progression of tumors rather than
their initiation.

     As suggested from the presence of the 13q- anomaly as the
prezygotic mutation, a deletion of a region 13q14 in retinal cells
might also lead to RB even the patient does not have genetic
predisposition. Therefore, in such case the 13q- anomaly might
not necessarily be the second mutation in the Knudson's sequential
two-mutation model. The cytogenetic studies in RB tumors indicate
that two mutations, if required, are not necessarily at the
homologous sites, but rather suggest that one mutation, probably
first at region 13q14, will be enough to initiate malignancy.
However, the second mutation, often manifested as genome
reorganization, is also required to provide growth or survival
advantage to the tumor cells.

CONCLUSIONS

     I have reviewed the data on chromosome abnormalities in
relation to the development of human tumors. While the kinds of
mutational events have a wide spectrum, varying from a single base
change to an alteration of the genome as a whole, it is surprising
that mutations at the chromosomal level are recognized in so many
human tumors. Some of these chromosome mutations, frequently
interstitial deletions or reciprocal translocations, are
apparently tumor specific, suggesting etiological involvement.
Evidence has also been accumulated to indicate that the generation
of such tumor-specific and site-specific chromosome mutations is
intimately related to the normal function of the gene(s) in its
particular tissue, such as growth and differentiation. Prezygotic
chromosome lesions leading the specific tissue to a specific type
of tumor elucidate the nature of the mutations linking to the
malignant transformation. Such a class of precancer, prezygotic
chromosome lesions often involve deletions or translocations at
particular sites of chromosomes, thus suggesting that deletion,
repression or disruption of the normal expression of particular
gene(s) may be a critical event in carcinogenesis.

     In addition to tumor specific mutations, there are also
chromosome mutations relatively common to a variety of tumors.
They include genome reorganization such as gene amplification, or
elimination of a particular part of the genome. The high
frequency of their occurrence suggests that these abnormalities
are primarily related to the progression or propagation of the
tumor cells, rather than of etiological significance in
carcinogenesis. Analyses of chromosome abnormalities in malignant

human tumors indicate that both types of chromosome abnormalities, those involved in the initiation and those related to the progression of tumors, are important in the development of tumors.

Our current knowledge on the chromosomal changes in human tumors raises the question of the mechanisms of the involvement of the environmental mutagens and carcinogens in the development of tumors. The site-directed chromosomal mutations in tumor cells indicate that mutations are produced as a consequence of the biological response of the living cells to the action or products of mutagens and carcinogens. Studies of indirect mutagenesis should also be encouraged in research on environmental mutagenesis and carcinogenesis.

REFERENCES

1.   Adams, J.M., S. Gerondakis, E. Webb, L.M. Corcoran, and S. Cory (1983) Cellular myc oncogene is altered by chromosome translocation to an immunoglobulin locus in murine plasmacytomas and is rearranged similarly in human Burkitt lymphomas. Proc. Natl. Acad. Sci. USA 80:1982-1986.
2.   Alitalo, K., M. Schwab, C.C. Lin, H.E. Varmus, and J.M. Bishop (1983) Homogeneously staining chromosomal regions contain amplified copies of an abundantly expressed cellular oncogene (c-myc) in malignant neuroendocrine cells from a human colon carcinoma. Proc. Natl. Acad. Sci. USA 80:1707-1711.
3.   Anderson, S.R., P. Geertinger, H.-W. Larsen, M. Mikkelsen, A. Parving, S. Vestermark, and M. Warburg (1978) Aniridia, cataract and gonadoblastoma in a mentally retarded girl with deletion of chromosome 11. Ophthalmologica 86:171-177.
4.   Balaban-Malenbaum, G., F. Gilbert, W.W. Nichols, R. Hill, J. Shields, and A.T. Meadows (1981) A deleted chromosome 13 in human retinoblastoma cells: Relevance to tumorigenesis. Cancer Genet. Cytogenet. 3:243-250.
5.   Balaban-Malenbaum, G., F. Gilbert, W. Nichols, A.T. Meadows, and J. Shields (1982) Abnormalities of chromosome #13 in retinoblastomas from individuals with normal constitutional karyotypes. Cancer Genet. Cytogenet. 6:213-221.
6.   Bauer, H. (1928) Mutationstheorie der Geschwulst-Enstehung. Ubergang von Korperzellen in Geschwulstzellen durch Gen-Anderung. Springer-Verlag, Berlin.
7.   Berger, R., A. Bernheim, H.-J. Weh, G. Flandrin, M.T. Daniel, J.-C. Brouet, and N. Colbert (1979) A new translocation in Burkitt's tumor cells. Human Genet. 53:111-112.
8.   Biedler, J.L., R.A. Ross, S. Shanske, and B.A. Spengler (1980) Human neuroblastoma cytogenetics: Search for significance of homogeneously staining regions and double

minute chromosomes, In Advances in Neuroblastoma Research,
A.E. Evans, Eds., Raven Press, New York, pp. 81-96.

9.   Boveri, T. (1914) Zur Frage der Entstehung malignen Tumoren.
     Fisher, Jena (The Origin of Malignant Tumors, M. Boveri,
     Trans., William and Wilkins Co., Baltimore, 1929).

10. Cairns, J. (1981) The origin of human cancers. Nature
     289:353-357.

11. Casperson, T., G. Gahrton, J. Lindsten, and L. Zech (1970)
     Identification of Philadelphia chromosome as a number 22 by
     quinacrine mustard fluorescence analysis. Exp. Cell
     Res. 63:238-244.

12. Cleaver, J.E. (1968) Defective repair replication of DNA in
     xeroderma pigmentosum. Nature 218:652-656.

13. Cohen, A.J., F.P. Li, S. Berg, D.S. Marchetto, S. Tsai,
     S.C. Jacobs, and R.S. Brown (1979) Hereditary renal-cell
     carcinoma associated with a chromosomal translocation. New
     Engl. J. Med. 301:592-607.

14. Cohen, S., G. Carpenter and L. King, Jr. (1980) Epidermal
     growth factor-receptor-protein kinase interaction. J. Biol.
     Chem. 255:4834-4842.

15. Coltier, E., M. Rose, and S.A. Moel (1978) Aniridia,
     cataracts, and Wilms' tumor in monozygous twins. Am. J.
     Ophthalmol. 86:129-132.

16. Cox, R., and W.K. Masson (1978) Do radiation-induced
     thioguanine resistant mutants of cultured mammalian cells
     arise by HGPRT gene mutation of X chromosome rearrangement?
     Nature 276:629-630.

17. Crews, S., R. Barth, L. Hood, J. Prehn, and K. Calame (1982)
     Mouse c-myc oncogene is located on chromosome 15 and
     translocated to chromosome 12 in plasmacytoma. Science
     218:1319-1321.

18. Croce, C.M., M. Shander, J. Martinis, L. Cicurel,
     G.G. D'Anconna, T.W. Dolby, and H. Koprowski (1979)
     Chromosomal location of the genes for human immunoglobulin
     heavy chains. Proc. Natl. Acad. Sci. USA 76:3416-3419.

19. Dalla-Favera, R., M. Bregni, J. Erikson, D. Patterson,
     R.C. Gallo, and C.M. Croce (1982) Human c-myc onc gene is
     located on the region of chromosome 8 that is translocated in
     Burkitt lymphoma cells. Proc. Natl. Acad. Sci. USA
     79:7824-7827.

20. De Grouchy, J., C. Turleau, M.O. Cabanis, and J.M. Richardet
     (1980) Retinoblastome et deletion intercalaire du chromosome
     13. Arch. fr. Pediat. 37:531-535.

21. De Klein, A., A.G. van Kessel, G. Grosveld, C.R. Bartram,
     A. Hagemeijer, D. Bootsma, N.K. Spurr, N. Heisterkamp,
     J. Groffen, and J.R. Stephenson (1982) A cellular oncogene is
     translocated to the Philadelphia chromosome in chronic
     myelocytic leukemia. Nature 300:765-767.

22. De Martinville, B.J. Giacalone, C. Shih, R.A. Weinberg, and
     U. Francke (1983) Oncogene from human EJ bladder carcinoma is

located on the short arm of chromosome 11. <u>Science</u>
219:498-501.

23. D'Eustachio, P., D. Pravtcheva, K. Marcu, and F.H. Ruddle
    (1980) Chromosomal location of the structural gene cluster
    encoding murine immunoglobulin heavy chain. <u>J. Exp. Med.</u>
    151:1545-1550.

24. Ejima, Y., M.S. Sasaki, A. Kaneko, and H. Tanooka (1982)
    Cytogenetic study on the development of retinoblastoma.
    <u>Proc. Jap. Cancer Assoc.</u> 41st Ann. Meeting, p. 58.

25. Ejima, Y., M.S. Sasaki, A. Kaneko, H. Tanooka, Y. Hara, and
    Y. Kinoshita (1982) Possible inactivation of part of
    chromosome 13 due to 13qXp translocation associated with
    retinoblastoma. <u>Clin. Genet.</u> 21:357-361.

26. Erikson, J., A. Ar-Bushdi, H.L. Drwinga, P.C. Nowell, and
    C.M. Croce (1983) Transcriptional activation of the
    translocated c-myc oncogene in Burkitt lymphoma. <u>Proc. Natl.
    Acad. Sci. USA</u> 80:820-824.

27. Erikson, J., J. Finan, P.C. Nowell, and C.M. Croce (1982)
    Translocation of immunoglobulin V$_H$ genes in Burkitt lymphoma.
    <u>Proc. Natl. Acad. Sci. USA</u> 79:5611-5615.

28. Erikson, J., J. Martinis, and C.M. Croce (1981) Assignment of
    the genes for human λ immunoglobulin chains to chromosome 22.
    <u>Nature</u> 294:173-175.

29. Erikson, R.L., M.S. Collet, E. Erikson, A.F. Purchio, and
    J.S. Brugge (1980) Protein phosphorylation mediated by
    partially purified avian sarcoma virus transforming-gene
    product. <u>Cold Spring Harbor Symp. Quant. Biol.</u> 44:907-917.

30. Finan, J., R. Daniele, D. Rowlands, Jr., and P.C. Nowell
    (1978) Cytogenetics of chronic T cell leukemia, including two
    patients with a 14q+ translocation. <u>Virchows
    Arch.</u> 29:121-127.

31. Francke, U., D.L. George, M.G. Brown, and V.M. Riccardi
    (1977) Gene dose effect: Intraband mapping of the LDH A locus
    using cells from four individuals with different interstitial
    deletion of 11p. <u>Cytogenet. Cell Genet.</u> 19:197-207.

32. Francke, U., L.B. Holmes, L. Atkins, and V.M. Riccardi (1979)
    Aniridia-Wilms' tumor association: Evidence for specific
    deletion of 11p13. <u>Cytogenet. Cell Genet.</u> 24:185-192.

33. Francke, U., and F. Kung (1976) Sporadic bilateral
    retinoblastoma and 13q- chromosomal deletion. <u>Med. Pediat.
    Oncol.</u> 2:379-385.

34. Gardner, E.J., S.W. Rogers, and S. Woodward (1982) Numerical
    and structural chromosome aberrations in cultured lymphocytes
    and cutaneous fibroblasts of patients with multiple adenomas
    of the colorectum. <u>Cancer</u> 49:1413-1419.

35. Gardner, H.A., B.L. Gallie, L.A. Knight, and R.A. Phillips
    (1982) Multiple karyotypic changes in retinoblastoma tumor
    cells: Presence of normal chromosome No. 13 in most tumors.
    <u>Cancer Genet. Cytogenet.</u> 6:201-211.

36. Goff, S.P., P. D'Eustachio, F.H. Ruddle, and D. Baltimore (1982) Chromosomal assignment of the endogenous proto-oncogene c-abl. Science 218:1317-1319.
37. Hashem, N., and S.H. Khalifa (1975) Retinoblastoma: A model of hereditary fragile chromosomal regions. Human Genet. 25:35-49.
38. Hayata, I., T. Ishihara, K. Hirashima, T. Sado, and J. Yamagiwa (1979) Partial deletion of chromosome 2 in myelocytic leukemia of irradiated C3H/He and RFM mice. J. Natl. Cancer Inst. 63:843-848.
39. Hittner, H.M., V.M. Riccardi, and U. Francke (1979) Aniridia caused by a heritable chromosome 11 deletion. Ophthalmologica 86:1173-1183.
40. Howard, R.O., W.R. Breg, D.M. Albert, and R.L. Lesser (1974) Retinoblastoma and chromosome abnormality. Arch. Ophthalmol. 92:490-493.
41. Johnson, M.P., N. Ramsay, J. Cervenka, and N. Wang (1982) Retinoblastoma and its association with a deletion in chromosome 13: A survey using high resolution chromosome techniques. Cancer Genet. Cytogenet. 6:29-37.
42. Klein, G. (1981) The role of gene dosage and genetic transpositions in carcinogenesis. Nature 294:313-318.
43. Knudson, A.G., Jr. (1971) Mutation and cancer: statistical study of retinoblastoma. Proc. Natl. Acad. Sci. USA 68:820-823.
44. Knudson, A.G., Jr., A.T. Meadows, W.W. Nichols, and R. Hill (1976) Chromosomal deletion and retinoblastoma. New Engl. J. Med. 295:1120-1123.
45. Kovacs, G.Y. (1978) Letter to the editor. Lancet i:555.
46. Let, F.W., R.E. Kellems, J.R. Bertino, and R.T. Schimke (1978) Selective multiplication of dihydrofolate reductase genes in methotrexate-resistant variants of cultured murine cells. J. Biol. Chem. 253:1357-1370.
47. Levan, A., G. Levan, and F. Mitelman (1977) Chromosomes and cancer. Hereditas 86:15-30.
48. Levinson, A.D., H. Opperman, L. Levinstow, H.E. Vermus, and J.M. Bishop (1978) Evidence that the transforming gene of avian sarcoma virus encode a protein kinase associated with a phosphoprotein. Cell 15:561-569.
49. Maher, V.M., D.J. Darney, A.L. Mendrala, B. Konze-Thomas, and J.J. McCormick (1979) DNA excision repair processes in human cells can eliminate the cytotoxic and mutagenic consequences of ultraviolet irradiation. Mutation Res. 43:117-138.
50. Makino, S. (1952) A cytological study of the Yoshida sarcoma, an ascites tumor of rats. Chromosoma 4:649-674.
51. Makino, S. (1956) Further evidence favoring the concept of stem cell in ascites tumors of rats. Ann. N.Y. Acad. Sci. 63:818-830.

52. Makino, S., T. Ishihara, and A. Tonomura (1959) Cytological
     studies of tumors, XXVII. The chromosomes of thirty human
     tumors. Z. Krebsforsch. 63:184–208.
53. Makino, S., M.S. Sasaki, and A. Tonomura (1964) Cytological
     studies of tumors, XL. Chromosome studies in fifty-two human
     tumors. J. Natl. Cancer Inst. 32:741–777.
54. Malcolm, S., P. Barton, C. Murphy, M.A. Ferguson-Smith,
     D.L. Bentley, and T.H. Rabbitts (1982) Localization of human
     immunoglobulin light chain variable region genes to the short
     arm of chromosome 2 by in situ hybridization. Proc. Natl.
     Acad. Sci. USA 79:4957–4961.
55. Matsunaga, E. (1978) Hereditary retinoblastoma: Delayed
     mutation or host resistance? Am. J. Human Genet. 30:406–424.
56. Matsunaga, E. (1979) Hereditary retinoblastoma: Host
     resistance and age at onset. J. Natl. Cancer
     Inst. 63:933–939.
57. Matsunaga, E. (1980) Retinoblastoma: Host resistance and
     13q– chromosomal deletion. Human Genet. 56:53–58.
58. Maurer, H.S., T.W. Pendergrass, W. Borges, and G.R. Horig
     (1979) The role of genetic factors in the etiology of Wilms'
     tumor. Cancer 43:205–208.
59. McBride, O.W., P.A. Hieter, G.F. Hollis, D. Swan, M.C. Otey,
     and P. Leder (1982) Chromosomal location of human kappa and
     lambda immunoglobulin light chain constant region genes. J.
     Exp. Med. 155:1480–1490.
60. Miyaki, M., N. Akamatsu, T. Ono, and M.S. Sasaki (1983)
     Susceptibility of skin fibroblasts from patients with
     retinoblastoma to transformation by murine sarcoma virus.
     Cancer Letters 18:137–142.
61. Miyoshi, I., S. Hiraki, I. Kimura, K. Miyamoto, and J. Sato
     (1979) 2/8 translocation in a Japanese Burkitt's lymphoma.
     Experientia 35:742–743.
62. Motegi, T. (1981) Lymphocyte chromosome survey in 42 patients
     with retinoblastoma: Effort to detect 13q14 deletion
     mosaicism. Hum. Genet. 58:168–173.
63. Motegi, T., M. Komatsu, Y. Nakazawa, M. Ohuchi, and K. Minoda
     (1982) Retinoblastoma in a boy with a de novo mutation of a
     13/18 translocation: The assumption that the retinoblastoma
     locus is at 13q14, particularly at the distal portion of it.
     Human Genet. 60:193–195.
64. Neel, B.G., S.C. Jhanwar, R.S.K. Chaganti, and W.S. Hayward
     (1982) Two human c-onc genes are located on the long arm of
     chromosome 8. Proc. Natl. Acad. Sci. USA 79:7842–7846.
65. Nichols, W.W. (1982) Status of prezygotic chromosome lesions
     in relation to cancer. Cytogenet. Cell Genet. 33:179–184.
66. Nichols, W.W., R.C. Miller, M. Sobel, R.S. Sperkes,
     T. Mohandas, I. Veomett, and J.R. Davis (1980) Further
     observations on a 13qXp translocation associated with
     retinoblastoma. Am. J. Ophthalmol. 89:621–627.

67. Nowell, P.C., and D.A. Hungerford (1960) A minute chromosome in human chronic granulocytic leukemia. Science 132:1497.

68. Ohno, S., M. Babonits, F. Wiener, J. Spira, G. Kline, and M. Potter (1979) Nonrandom chromosome changes involving Ig gene-carrying chromosome 12 and 6 in pristane-induced mouse plasmacytomas. Cell 18:1000-1007.

69. O'Riordan, M.L., J.A. Robinson, K.E. Buckton, and H.J. Evans (1971) Distinguishing between the chromosomes involved in Down's syndrome (trisomy 21) and chronic myeloid leukemia (Ph^1) by fluorescence. Nature 230:167-168.

70. Orye, E., M.J. Delbeke, and B. Vandenabeale (1974) Retinoblastoma and long arm deletion of chromosome 13: Attempts to define the deleted segment. Clin. Genet. 5:457-464.

71. Ozawa, H. (1978) Retinoblastoma and D-chromosome deletion (13q-). Jap. J. Ophthalmol. 22:320-325.

72. Pasquali, F., R. Casalone, D. Francesconi, D. Peretti, M. Fraccaro, C. Bernasconi, and M. Lazzarino (1979) Transposition of 9q34 and 22(q11-qter) region has a specific role in chronic myelocytic leukemia. Human Genet. 52:55-67.

73. Petit, P., and J.P. Fryns (1979) Interstitial deletion of 13q associated with retinoblastoma and congenital malformation. Ann. Genet. 22:106-107.

74. Riccardi, W.M., H.M. Hittner, U. Francke, S. Pippin, G.P. Holmquist, F.L. Kretzer, and R. Ferral (1979) Partial triplication and deletion of 13q: Study of a family presenting with bilateral retinoblastomas. Cell Genet. 15:332-345.

75. Riccardi, V.M., E. Sujanski, A.C. Smith, and U. Francke (1978) Chromosomal imbalance in the aniridia-Wilms' tumor association: 11p interstitial deletion. Pediatrics 61:604-610.

76. Ross, R.A., T.H. Joh, D.J. Reis, B.A. Spengler, and J.L. Biedler (1980) Neurotransmitter-synthesizing enzymes in human neuroblastoma cells: Relationship to morphological diversity, In Advances in Neuroblastoma Research, A.E. Evans, Ed., Raven Press, New York, pp. 151-160.

77. Rowley, J.D. (1973) A new consistent chromosomal abnormality in chronic myelogenous leukemia identified by quinacrine fluorescence and Giemsa staining. Nature 243:290-293.

78. Rowley, J.D. (1973) Identification of a translocation with quinacrine fluorescence in a patient with acute leukemia. Ann. Genet. 16:109-112.

79. Rowley, J.D. (1977) Mapping human chromosomal regions related to neoplasia: Evidence from chromosome 1 and 17. Proc. Natl. Acad. Sci. USA 74:5729-5733.

80. Rowley, J.D. (1977) A possible role for nonrandom chromosomal changes in human hematologic malignancies, In Chromosomes Today, Vol. 6, A. de la Chapelle, and M. Sorsa, Eds.,

Elsevier/North Holland Biomedical Press, Amsterdam, pp. 345-359.

81. Rowley, J.D. (1981) Nonrandom chromosome changes in human leukemia, In Genes, Chromosomes, and Neoplasia, F.A. Arrighi, P.N. Rao, and E. Stubblefield, Eds., Raven Press, New York, pp. 273-296.

82. Rowley, J.D. (1983) Human oncogene locations and chromosome aberrations. Nature 301:290-291.

83. Sandberg, A.A. (1980) The Chromosomes in Human Cancer and Leukemia, Elsevier/North Holland, New York.

84. Sandberg, A.A., and N. Wake (1981) Chromosomal changes in primary and metastatic tumors and in lymphomas: Their nonrandomness and significance, In Genes, Chromosomes, and Neoplasia, F.E. Arrighi, P.N. Rao, and E. Stubblefield, Eds., Raven Press, New York, pp. 297-333.

85. Sasaki, M. (1982) Role of chromosomal mutation in the development of cancer. Cytogenet. Cell Genet. 33:160-168.

86. Sasaki, M.S. (1982) Dominantly expressed procancer mutations and induction of chromosome rearrangements. Prog. Mutation Res. 4:75-84.

87. Sasaki, M.S. (unpublished data).

88. Sasaki, M.S., and Y. Ejima (1981) Procancer class of genes and generation of chromosome mutation. Gann Monograph on Cancer Research 27:85-94.

89. Sasaki, M.S., A. Kaneko, and H. Tanooka (1981) Chromosomal changes in retinoblastoma tumors. Proc. Jap. Cancer Assoc. 40th Ann. Meeting, p. 57.

90. Sasaki, M.S., Y. Tsunematsu, J. Utsunomiya, and J. Utsumi (1980) Site-directed chromosome rearrangements in skin fibroblasts from persons carrying genes for hereditary neoplasms. Cancer Res. 40:4796-4803.

91. Schimke, R.T., Ed. (1982) Gene Amplification, Cold Spring Harbor Laboratory, Cold Spring Harbor, New York.

92. Second International Workshop on Chromosomes in Leukemia -1979- (1980), Cancer Genet. Cytogenet. 2:93-113.

93. Slater, R.M., and J. de Kraker (1982) Chromosome number 11 and Wilms' tumor. Cancer Genet. Cytogenet. 5:237-245.

94. Sperkes, R.S., H. Müller, and I. Klisak (1979) Retinoblastoma with 13q- chromosome deletion associated with maternal paracentric inversion of 13q. Science 203:1027-1029.

95. Sperkes, R.S., A.L. Murphree, R.W. Lingua, M.C. Sperkes, L.L. Field, S.J. Funderburk, and W.F. Bendict (1983) Gene for hereditary retinoblastoma assigned to human chromosome 13 by linkage to esterase D. Science 219:971-973.

96. Taub, R., I. Kirsch, C. Morton, G. Lenoir, D. Swan, S. Tronick, S. Aaronson, and P. Leder (1982) Translocation of the c-myc gene into the immunoglobulin heavy chain locus in human Burkitt lymphoma and murine plasmacytoma cells. Proc. Natl. Acad. Sci. USA 79:7837-7841.

97. Tyler-Smith, C., and C.J. Bostock (1981) Gene amplification in methotrexate-resistant mouse cells, III. Interrelationships between chromosome changes and DNA sequence amplification or loss. J. Mol. Biol. 153:237-256.
98. Ushiro, H., and S. Cohen (1980) Identification of phosphotyrosine as a products of epidermal growth factor-activated protein kinase in A-431 cell membrane. J. Biol. Chem. 30:406-424.
99. Van Den Berghe, H., C. Parlois, S. Gosseye, V. Englebienne, G. Cornu, and G. Sokal (1979) Variant translocation in Burkitt lymphoma. Cancer Genet. Cytogenet. 1:9-14.
100. Van Dyke, D.L., C.E. Jackson, and V.R. Babu (1982) Localization of autosomal, dominant multiple endocrine neoplasia 2 syndrome (MEN2) to 20p12.2. Cytogenet. Cell Genet. 32:324.
101. Von Hansemann, D. (1880) Uber asymetrishe Zellteilung in Epithelkrebsen und deren biologishe Bedeutung. Arch. Pathol. Anat. Physiol. 119:299-326.
102. Vogel, F. (1979) Genetics of retinoblastoma. Human Genet. 52:1-54.
103. Walbaum, R., P. Francois, J.P. Farriaux, and M. Woillez (1978) Un cas de retinoblastome bilateral avec monosomie 13 partielle (q12-q14). Hum. Genet. 44:219-226.
104. Wilson, M.G., A.J. Ebbin, J.W. Towner, and W.H. Spencer (1977) Chromosome abnormalities in patients with retinoblastoma. Clin. Genet. 12:12-8.
105. Wilson, M.G., J.W. Towner, and A. Fujimoto (1973) Retinoblastoma and D-chromosome deletions. Am. J. Hum. Genet. 25:57-61.
106. Yunis, J.J., and N. Ramsay (1978) Retinoblastoma and subband deletion of chromosome 13. Am. J. Dis. Child 132:161-163.
107. Yunis, J.J., and N.K.C. Ramsay (1980) Familial occurrence of the aniridia-Wilms' tumor syndrome with deletion 11p13-14.1. J. Pediat. 96:1027-1030.

MUTATIONS OF CELLULAR ONCOGENES AS A BASIS FOR

NEOPLASTIC CHANGE

Harold E. Varmus

Department of Microbiology and Immunology
University of California
San Francisco, CA   94143

SUMMARY

All RNA tumor viruses (retroviruses) replicate through a DNA
intermediate, called a provirus, and many induce tumors swiftly
through the action of oncogenes (v-onc's) that are the transduced
versions of normal cellular genes (c-onc's). There are at present
at least 18 distinguishable v-onc genes, several of which encode
protein kinases specific for tyrosine residues.

Several kinds of evidence, from many laboratories, now
support the idea that normal cellular genes, most of them first
identified as homologues of retroviral oncogenes, are substrates
for the mutational events that lead to neoplasia caused by various
agents. (i) Many tumorigenic retroviruses do not carry oncogenes;
instead, retroviral proviruses can act as insertional mutagens to
augment expression of cellular genes. In some cases, the
activated genes have been found to be members of the c-onc family.
For example, in B cell lymphomas induced by the avian leukosis
virus, a gene called c-myc, the progenitor of the viral oncogene
(v-myc) of highly oncogenic avian leukemia viruses, is mutated by
proviral insertions that enhance the level of gene expression. In
other cases, e.g., mammary carcinomas induced by the mouse mammary
tumor virus, proviral insertion mutations affect cellular genes
not represented among retroviral oncogenes. Study of these
insertion mutations has revealed the capacity of retroviral DNA to
activate expression of cellular genes by at least two mechanisms:
the provision of a strong viral promoter element for transcription
of "downstream" genes or the indirect action of a viral
"enhancing" element upon cellular sequences competent to serve as
transcriptional initiation sites. (ii) Translocations occur

commonly in tumor cells; in some cases, the break points for the translocations have been located nearby and even with c-onc loci, including c-myc. (iii) Gene amplification, manifest cytogenetically by double minute chromosomes or by homogeneously staining regions, is frequently found in tumor cells; in several tumor lines of human and murine origin the amplified domains have been shown to include c-onc genes that are also expressed at a correspondingly high level. Again, the c-myc locus is frequently involved. (iv) DNA from human tumors and tumor cell lines can often transform cultured mouse cells; the transforming activities have been identified in several cases as mutated versions of c-onc's.

These observations suggest that many of the targets for oncogenic mutations have been identified. The information compiled to date indicates that mutations of certain genes are likely to contribute to oncogenesis in certain cell lineages. In addition, several tumors have been shown to contain at least two mutant oncogenes, reaffirming the dictum that oncogenesis is a multi-step process.

INTRODUCTION

The idea that mutations of cellular genes in somatic or germinal cells are the primary lesions in neoplasia has been accepted for many years by most students of cancer (5,34). However, candidate genetic substrates for key mutational events have been at hand for only a few years, and only in the very recent past have we obtained glimpses of the kinds of mutations that may be involved. Those genes that are thought to be the targets for cancer-inducing mutations are generally referred to as "cellular oncogenes". More properly, they might be called "pre-" or "proto-oncogenes", because virtually all of them exhibit characteristics expected of genes that are normally valuable, rather than detrimental, to the organisms that inherit them: the genes are highly conserved during evolution and expressed in many tissues, often with provocative differences in level of expression among cell lineages (2).

In this essay, I will outline the lines of experimentation that led—and are still leading—to the identification of cellular oncogenes, and I will summarize the kinds of mutations that have been observed to afflict those genes. Particular emphasis will be placed upon insertion mutations caused by the proviral form of retroviral genomes, since that is a subject of special concern in our own laboratory.

ACTIVATION OF ONCOGENES

   Some of the work that will be discussed below suggests that
there are many ways in which benign cellular oncogenes (i.e., pre-
oncogenes) can be converted to active oncogenes. Basically, these
varied mechanisms are believed to affect the efficiency of
expression of an oncogene, the nature of its gene product, or
both. Although epigenetic changes could conceivably produce these
effects (e.g., by altering rates of transcription or modifying
proteins chemically), attention has thus far been directed almost
exclusively towards several kinds of mutations (Table 1). The
most complex of these--the transduction of cellular oncogenes by
RNA tumor viruses (retroviruses)--is of fundamental importance,
because the products, potent viral oncogenes under the control of
efficient viral regulatory signals, have been the most useful
reagents for the discovery of cellular oncogenes (see below). The
transduced oncogenes differ in two ways from their cellular
progenitors: they are expressed at higher levels, and they exhibit
at least minor--and sometimes major--structural changes. The
contributions of these quantitative and qualitative differences to
the oncogenicity of viral oncogenes is an issue of great
importance in contemporary tumor virology (3). A second category
of mechanism also involves retroviruses. Integration of viral DNA
lacking its own oncogenes in the vicinity of a cellular oncogene
can activate the tumorigenic potential of that oncogene, primarily
(it appears) by enhancing expression of the gene. Two other kinds
of genetic rearrangement, in addition to proviral insertion
mutations, have been shown to afflict cellular oncogenes:
translocations that join DNA on one chromosome to sites on another
chromosome, and amplifications that convert unique sequence DNA to
DNA present as many as 50-100 times per cell, with a corresponding
augmentation in expression. Often such amplifications are
signaled to the karyologist by double minute chromosomes (DMs) or
homogeneously staining regions (HSRs). More conventional
mutations have also been found to affect cellular oncogenes; for
example, a single base substitution in an oncogene in a human
bladder carcinoma cell line renders that gene capable of producing
neoplastic changes when introduced into an appropriate recipient
cell (36,42). A final type of mutation is a distinctly artificial
one, performed directly in the laboratory. When the promoter for
a cellular oncogene that has been molecularly cloned in bacteria
is replaced by a much stronger viral promoter, the recombinant
gene is sometimes competent to cause neoplastic transformation of
cultured cells (4).

Table 1. The Mechanisms That Alter The Structure and
          Function of Cellular Oncogenes
          See Text for Explanation

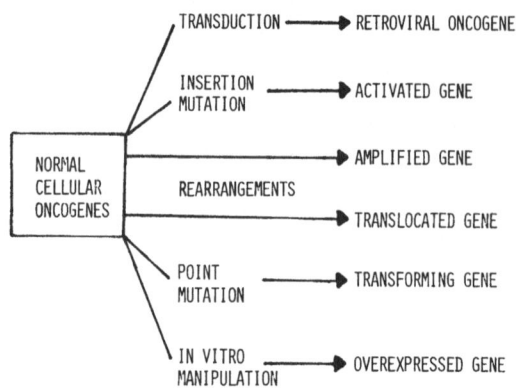

METHODS FOR DISCOVERY OF ONCOGENES

        Some of these mutational changes are reflected in the ways in
which cellular oncogenes have been identified. The first and most
prolific route to the discovery of cellular oncogenes has been to
isolate components of retroviral genomes responsible for the swift
induction of tumors and the neoplastic transformation of cultured
cells. These genetic domains, called viral oncogenes (or v-
onc's), now number at least eighteen distinguishable types among
the many tumorigenic retroviruses; in each case, the viral
oncogene has been shown to be closely-related to—and apparently
derived from—a highly conserved cellular gene, or c-onc (2,3,6;
see Table 2). Some of these genes can now be grouped into
families, based upon homologies between amino acid sequences of
their gene products that were not apparent in nucleic acid
hybridizations tests. Moreover, several of the homologous
proteins exhibit protein kinase activities specific for tyrosine
residues and are associated with the plasma membranes of cells.
Despite this modest degree of uniformity, many oncogenic proteins
do not appear to encode protein kinases, some are found in the
nucleus, and it is likely that several biochemical mechanisms for
neoplastic change will be found exemplified among these genes.

        A second and similarly prolific means to detect cellular
oncogenes depends upon the ability of cultured mouse fibroblasts
(NIH/3T3 cells) to undergo neoplastic change in response to
dominant mutations in the DNA extracted from various types of
tumors, including tumors of non-viral etiology (9,40). In many
cases, the "transforming genes" have proved to be mutated versions

Table 2. Retroviral oncogenes (v-onc's) derived from cellular genes (c-onc's). This list includes all the well-studied retroviral oncogenes, each derived from a normal cellular gene (c-onc); a few more recently isolated v-onc's have been omitted. In some cases (fes and fps, Ha-ras and bas), the same gene has been given different names because the genes were independently identified in viruses isolated from different animal species. The list groups those genes that share the property of encoding protein kinases specific for tyrosine residues and those genes that encode proteins with homologous amino acid sequences (see Bishop (2) for further details).

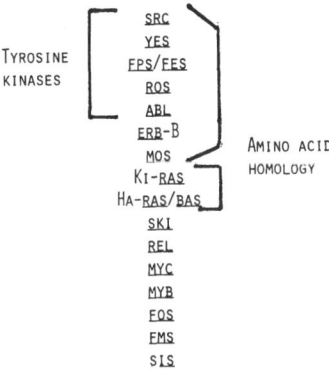

of normal cellular genes first identified as oncogenes by virtue of their homology with retroviral oncogenes (Table 3; refs. 14,30,35,37,41). Others appear to be novel oncogenes, some of which are distantly related to previously identified c-onc's (35,41).

Table 3. Cellular Genes Implicated in Tumorigenesis
By the 3T3 Cell Transformation Assay

Gene	Tumor	Alteration
c-Ha-ras	Bladder CA	Single base substitution
c-Ki-ras	Lung CA, colon CA, others	?
c-N-ras	Neuroblastoma	?
Several unnamed	Various leukemias and solid tumors	?
c-mos	Plasmacytoma	Proviral insertion
B-lym	B cell lymphoma	?

A third method, illustrated below in considerable detail, has been to seek genes whose activity is affected by insertional mutations caused by retroviral proviruses. In these instances, the initiating event in tumorigenesis appears to be an insertion of proviral DNA, with its regulatory apparatus, near a cellular oncogene. Since molecular probes for the mutagen (proviral DNA) are available, putative oncogenes can be identified and isolated by their linkage to the provirus. Again, some but not all of the oncogenes identified in this fashion were previously known as progenitors of retroviral oncogenes (Table 4).

A presently laborious additional approach to the identification of new oncogenes is suggested by the recent discoveries that known oncogenes are present in amplified DNA or at the breakpoints in chromosomal translocations (Table 5; refs. 1,8,13,20,24,38,39). Molecular cloning of amplified DNA (e.g., from DMs) or of chromosomal domains that span translocation breakpoints is likely to be used in the near future to find potential oncogenes.

At the moment, it seems probable that the available experimental approaches can bring to our attention many, but not all, of the host genes involved in various forms of cancer. Other avenues—e.g., precise mapping of the affected loci in certain familial cancer syndromes—may be necessary to allow identification of other members of the general class of cellular oncogenes. There is, on the other hand, the encouraging evidence that the available strategies frequently detect the same genes (Tables 2-5); this suggests that the number of oncogenes is not unworkably large and that the procedures—all unconventional from the viewpoint of a classical geneticist—may in some sense validate each other. To date, there is at least one gene, c-myc, that has been encountered as a homologue of a viral transforming gene, as a cellular gene linked to tumor-inducing proviruses, as an amplified gene, and as a gene at the site of a chromosomal translocation. Similarly, c-Ki-ras has been found as a homologue

Table 4. Cellular Genes Implicated in Oncogenesis By
Integrated Proviruses Without Oncogenes

GENE	VIRUSES	TUMOR
c-MYC	ALV, CSV, MAV	B-CELL LYMPHOMA
c-ERB-B	ALV	ERYTHROLEUKEMIA
INT-1	MMTV (C3H)	MAMMARY CA (C3H)
INT-2	MMTV (R III)	MAMMARY CA (R III)
UNNAMED	MLV	T CELL LYMPHOMA

Table 5. Cellular Genes Implicated in Non-Viral Human
and Murine Tumors by Rearrangements

GENE	TUMOR	ALTERATION
c-MYC, c-MOS	BURKITT'S LYMPHOMA PLASMACYTOMA	TRANSLOCATIONS
c-MYC	MYELOID LEUKEMIA APUDOMA	AMPLIFICATION
c-ABL	MYELOID LEUKEMIA ERYTHROID LEUKEMIA	TRANSLOCATION AMPLIFICATION
c-KI-RAS	ADRENOCORTICAL CA COLONIC CA	AMPLIFICATION
"N-MYC"	NEUROBLASTOMAS	AMPLIFICATION

of a viral oncogene, as an amplified gene, and as a mutated gene
in human tumors capable of transforming NIH/3T3 cells. Specific
examples involving these genes will be described below.

VIRUSES WITHOUT ONCOGENES AND INSERTIONAL MUTAGENESIS

    The pathological spectrum of retroviruses is vast (see 43),
but tumorigenic viruses can be conveniently placed in two
categories: those viruses that carry onc genes transduced from
normal cells, and those that appear to contain only sequences
implicated in viral replication. Two general hypotheses have
received the most attention as explanations for the oncogenic
activity of the latter class of viruses: that some component of
the viral genome may serve as an oncogene, although it is not
derived from cellular genes, and that proviruses may act as
mutational agents, thereby affecting the behavior of host genes.

    Retroviruses could instigate mutations in infected cells in
several ways. As agents that introduce their genomes as DNA
proviruses more or less randomly into host chromosomes, they are
capable of inactivating host genes by interrupting them. This
capacity is exemplified by the insertional mutations produced by
murine leukemia virus (MLV) proviruses that have been integrated
within a Rous sarcoma virus (RSV) provirus responsible for the
transformed phenotype of the host (44). An insertional mutation
is also apparently responsible for the dilute coat-color allele in
certain inbred mouse strains (19). However, mutations that
inactivate genes are likely to be recessive lesions, and the
target gene for an oncogenic mutation of this type would have to
be sex-linked or inactivated on both autosomal chromosomes to
produce the tumor phenotype. A more promising possibility is that
a provirus would activate the expression of host genes,
particularly genes adjacent to the integration site through the

agency of viral regulatory signals encoded in long terminal
repeats (LTRs; see 45, for review). Such lesions would be
dominant in most cases, permitting a single insertion to induce a
tumor.

LYMPHOMAGENESIS BY AVIAN LEUKOSIS VIRUS

Of the tumors induced by viruses without oncogenes, avian B
cell lymphomas have probably been the most informative
mechanistically and serve as the paradigm for insertional
mechanisms in neoplasia. These tumors arise in certain
susceptible lines of chickens 6-12 months after newly-hatched
birds have been inoculated with various strains of avian leukosis
virus (ALV). During the preneoplastic stage, large numbers of
cells in the target organ, the bursa of Fabricius, become
infected, as evidenced by the introduction of ALV DNA into
different sites in the genomes of many cells. Analysis of virus-
specific DNA from tumors reveals a pattern consistent with the
clonal origin of the tumor cells: proviral DNA appears to be in
the same site in most or all of the cells (15,25,26,31). All
tumors that have been examined in this fashion retain at least a
part of one provirus, suggesting that proviral DNA is important
for maintenance of the tumor phenotype; however, several tumors
contain only proviral DNA afflicted by deletions and not expressed
at detectable levels (25,31), implying that products of viral
genes are not necessary to maintain tumorous growth. The
experiments of Neel et al. (25) and Payne et al. (31) further
demonstrated that the ALV proviruses, even the defective ones, are
able to stimulate transcription of flanking host DNA; mapping
experiments suggested that the flanking sequences might be the
same in several tumors. Using reagents available for several onc
sequences, Hayward et al. (18) then showed that the insertion site
in most tumors was within or adjacent to the c-myc locus and that
cells with interrupted loci contained elevated levels of c-myc
RNA. Meanwhile, Cooper and Neiman (9) had found that DNA from
similar tumors was able to transform NIH/3T3 cells; however, the
transforming component appeared to be separate from either ALV DNA
or the interrupted c-myc domains (10), suggesting that a mutation
at another locus was present in the tumor cells. This second
locus has been named Blym, and it appears to encode a polypeptide
of about 7 Kd (16).

The following sequence of events could produce the phenomena
described in the preceding paragraph: (i) The oncogenic process is
initiated in one of the rare ALV-infected B cells that acquires a
provirus near a c-myc locus, provided that the inserted DNA is
suitably placed to enhance expression of c-myc and that the host
cell is at a stage susceptible to the oncogenic consequences of
heightened expression; (ii) development of the tumor phenotype

requires additional genetic events, one of which occurs at the
Blym locus and is reflected in the transforming activity of tumor
DNA; and (iii) outgrowth of tumorigenic clones is favored by
mutations (e.g., deletions) that improve the ability of proviral
DNA to enhance c-myc expression or impair its ability to direct
synthesis of viral antigens to which the host immune system is
responsive. These conjectures raise an additional series of
questions: What is responsible for the enhanced expression of c-
myc? Are similar oncogenic mechanisms operative in other tumors
induced by viruses without oncogenes? If so, what are the
determinants of the oncogenic spectra of various viruses? Can
this approach be used to identify new cellular oncogenes,
including genes important in human tumors and in tumors arising in
the absence of viruses?

HOW DO PROVIRUSES ENHANCE EXPRESSION OF ONCOGENES?

     Most of the B-cell lymphomas examined to date exhibit an
arrangement of ALV DNA and c-myc conducive to the use of a viral
promoter for efficient transcription of the c-myc gene (18). In
these tumors, proviral DNA is positioned on the 5' side of the two
well-defined exons of c-myc, in the same transcriptional
orientation; because the long terminal repeat (LTR) at the 5' end
of proviral DNA is often deleted or inactivated, the 3' LTR is
presumed to provide the initiation site for transcription,
producing abundant species with viral sequences linked to c-myc
sequences (18,32). Since these proviruses are usually, if not
always, partially deleted, it appears that two rare mutational
events—integration next to c-myc and deletion formation—may be
required for maximal expression of c-myc. In at least one
instance, the second mutation appears to involve homologous
recombination between LTRs, because a solitary LTR is all that
remains of proviral DNA in the c-myc locus (D. Westaway,
unpublished). In several of the tumors in our collection,
proviral DNA is found on the 5' side of c-myc, but in the opposite
transcriptional orientation (32). We have also examined one tumor
in which part of an ALV provirus is present on the 3' side of c-
myc in the same transcriptional orientation (32). The residual
LTR appears to provide the polyadenylation site for the abundant
c-myc RNA; however, again, the mechanism for enhanced expression
is unknown. It seems likely, based upon these unexpected
arrangements of viral DNA and c-myc, that LTRs confer upon
adjacent regions of the chromosome the property of being
transcribed efficiently from normally used or novel promoters.
This property could depend upon the topology of the local domain
of DNA, upon the arrangement of chromatin proteins, or upon
concentrations of RNA polymerase along the chromosome. Whatever
the mechanism, similar effects have been observed in cultured
cells microinjected with plasmids containing a viral thymidine

kinase gene when the plasmid contains a region of retroviral DNA
containing the putative enhancer sequence (23).

EVIDENCE FOR INSERTIONAL MUTATION DURING ONCOGENESIS BY MOUSE MAMMARY TUMOR VIRUS (MMTV)

MMTV lacks onc sequences and induces mammary adenocarcinomas
derived from only one or a few of the many infected cells in the
preneoplastic gland (7). Furthermore, the tumors arise after
considerable latency (the animals are usually infected shortly
after birth by virus in milk), and the DNA from tumors has been
reported to transform NIH/3T3 cells (21). To determine the role
of the viral genome in mammary tumorigenesis, we identified a
tumor in C3H mice that contained only one MMTV provirus, in
addition to the proviruses endogenous to the germ line. We cloned
the 3' third of this provirus with about 15 kilobases (kb) of
flanking cellular DNA, prepared probes for unique-sequence DNA
from over 30 kb of the flanking region retrieved from a library of
normal mouse DNA, and examined DNA from additional tumors (most
bearing multiple new proviruses) for evidence of insertions in the
same chromosomal domain (28). In this survey, 18 of 26 tumors in
C3H mice were found to contain MMTV proviruses within a region of
20 kb, a frequency much too high to be explained by chance.

This chromosomal domain, which we call int-1, appears to
contain a novel oncogene since it is not homologous to known
retroviral oncogenes, or to the NIH/3T3 cell transforming gene
described by Lane et al. (21); it is expressed in mammary tumors
with int-1 insertions but not in normal mammary tissue; and it is
highly conserved among vertebrates (29). Activation of int-1
transcription appears to proceed solely by an enhancement
mechanism, rather than provision of a promoter, since all of the
proviral insertions are either upstream from the transcribed
region in the opposite transcriptional orientation or downstream
in the same orientation (29).

WHAT IS THE RELATIONSHIP OF VIRUSES, INSERTION SITES, AND TUMOR TYPE?

Many retroviruses can induce multiple varieties of tumors,
closely related retroviruses often cause tumors in different
organs, and unrelated viruses appear responsible for the same type
of tumor (17,43). Although such phenomena are not yet understood,
it is now possible to ask whether different viruses cause a single
type of tumor by a single mechanism, and whether the same virus
uses a single mechanism to produce tumors in different organs.
For example, two viruses other than ALV have been found to produce
insertional mutations in c-myc during the induction of B-cell

lymphomas (Table 4). Several tumors induced in chickens by the chicken syncytial virus (CSV), a reticuloendotheliosis virus unrelated to ALV, show evidence of proviral insertions near c-myc, deletions affecting cellular and perhaps proviral DNA, and sometimes amplification of the interrupted c-myc locus (27). Lymphomas produced in quail by myeloblastosis-associated virus-1 (MAV-1), a virus differing from ALV principally in the LTR domain, also contain insertions of MAV-specific DNA adjacent to c-myc (D. Westaway and C. Moscovici, unpublished results). Although c-myc RNA was not measured in these two instances, it seems likely that three different viruses--ALV, CSV, and MAV--initiate B-cell lymphomagenesis by an insertional activation of c-myc.

On the other hand, ALV can induce erythroblastosis, as well as B cell lymphomas, and insertion mutations are found in the c-erb-B locus rather than the c-myc locus (T. Fung and H.J. Kung, personal communication). Similarly, we have examined DNA from over a dozen renal tumors, mainly nephroblastomas, induced by MAV strains, and have failed to observe any disruptions of the c-myc locus (D. Westaway and C. Moscovici, unpublished results). Thus, very similar or identical viruses appear to employ different mechanisms to cause tumors in different organs.

MMTV-induced tumors illustrate another variation on this theme. Although about 75% of mammary tumors in C3H mice have insertions in int-1, in mice derived from the RIII strain another site (called int-2), unrelated and unlinked to int-1, is the common target for proviral insertion mutations (33). It remains to be seen whether the tumors that result from insertions at these two loci arise in different target cells or whether the viruses that infect the different mouse strains have different target specificities.

CELLULAR ONCOGENES AND OTHER GENETIC REARRANGEMENTS

The proviral insertion mutations discussed in the preceding sections represent one type of genetic rearrangement that can affect both the efficiency of expression of a cellular oncogene and the composition of its protein product. These possibilities are echoed in other types of chromosomal rearrangements recently found to affect cellular oncogenes (Fig. 1).

The most startling of these discoveries again concern B cell neoplasms and the c-myc locus. In several murine plasmacytomas and human Burkitt's lymphoma cell lines, translocations (12;15 in the mouse cells; 8;2, 8;14, or 8;22 in the human cells) have been shown to join the 5' region of an immunoglobulin gene to the 5' region of a c-myc gene (20). These mutations are structurally

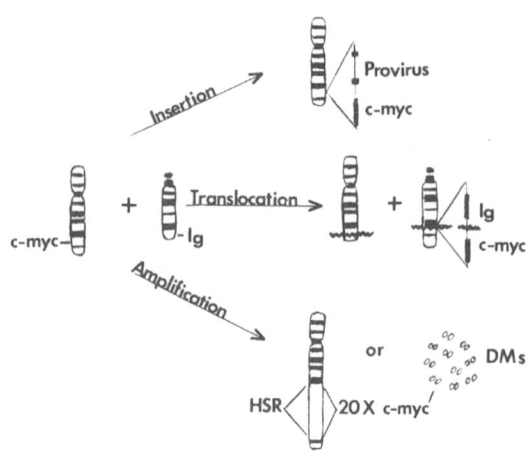

Figure 1. Rearrangements affecting cellular oncogenes in tumors
    are illustrated with the c-myc gene on an arbitrary
    chromosome. The upper diagram shows a proviral insertion
    mutation within the c-myc locus, as occurs in ALV-induced B
    cell lymphomas; the middle diagram shows a chromosomal
    translocation that joins an immunoglobulin gene to c-myc, as
    occurs in murine plasmacytomas and Burkitt's lymphoma; and
    the bottom diagram shows amplification of a unit that
    encompasses the c-myc gene, accompanied by the appearance of
    a homogeneously staining region (HSR) or double minute
    chromosomes (DMs), as occurs in certain human tumors
    described in the text. cellular oncogenes. See text for
    explanation.

similar to ALV proviral insertions positioned on the 5' side of c-
myc in the opposite transcriptional orientation (32). However,
the effects of the translocations upon the expression of c-myc and
the c-myc protein are still under study.

The c-myc gene has also been found amplified in a human
myeloid leukemia cell line called HL60 (8,13) and a human
neuroendocrine tumor line called COLO320 (1). Similarly, the c-
Ki-ras gene is amplified in a murine adrenocortical tumor line
called Y1 (38) and a human colonic carcinoma line called SW480
(24). The amplified genes are overexpressed in concert with the
extent of their amplification.

In both the COLO320 and Y1 lines, sibling cultures show HSR
or DMs, karyological evidence of amplification (see 12). The
amplified oncogenes have been located within the HSR by in situ
hybridization to metaphase chromosomes or among the DMs by
annealing to DNA enriched for DMs (1,39). The HSR in COLO320 is

present on a marker chromosome containing parts of chromosomes 8 and X; hence c-myc has been translocated as well amplified in this line (K. Alitalo and M. Schwab, unpublished).

GENETIC VALIDATION OF AN ONCOGENIC ROLE FOR PUTATIVE ONCOGENES

Classic genetic techniques have been used to show that some retroviral oncogenes are solely responsible for the initiation and maintenance of neoplastic transformation. For example, various mutants of Rous sarcoma virus, including non-conditional and temperature-sensitive mutations of the src gene, demonstrate that a src gene product with protein kinase activity is continuously required and sufficient for cell transformation (22). However, it is more difficult to define the roles of mutated cellular oncogenes in naturally occurring tumors. In most instances, the evidence that the mutated genes are instrumental in oncogenesis is circumstantial: the same genes are often affected (e.g., by insertion mutations or other rearrangements) in several tumors of the same type, and the affected genes have often been identified as potential oncogenes by more than one method (Tables 2-5). Mutations that render genes transforming in NIH/3T3 cells are, of course, detected by an assay that would appear to be a test of their oncogenicity. However, conclusions must be tempered by the realization that the indicator cells are not strictly normal and that the test often scores genes derived from a different cell type and a different animal species.

The situation may be complicated by the presence of multiple mutations in tumor cells (see Table 6 and the above discussions of ALV-induced bursal lymphomas and MMTV-induced mammary tumors). If a number of mutant genes contribute in concert to the genesis of a tumor cell, it may be extremely difficult to determine the function of each mutation in the neoplastic process. In this

Table 6. Evidence for multiple mutations in tumorigenesis.

Tumor	Proviral Insertions	Translocations	Amplifications	Transforming Genes
Avian B cell lymphomas	c-myc			B-lym
Burkitt's lymphoma		c-myc		B-lym
CML (HL60)			c-myc	+
Neuroblastoma			"N-myc"	N-ras
Mouse mammary CA	int-1/int-2			+

situation, we may have to rely upon observations that lack conventional genetic rigor to gain useful insights into carcinogenesis.

ACKNOWLEDGEMENTS

Work in this laboratory is supported by grants Nos. CA 19287, CA 12705 and CA 09043 from the National Institutes of Health and MV 481 from the American Cancer Society to the author and J.M. Bishop.

REFERENCES

1. Alitalo, K., M. Schwab, C.C. Lin, H.E. Varmus, and J.M. Bishop (1983) Homogeneously staining chromosomal regions contain amplified copies of an abundantly expressed cellular oncogene (c-myc) in malignant neuroendocrine cells from a human colon carcinoma. Proc. Natl. Acad. Sci. USA 80:1707-1711.
2. Bishop, J.M. (1983) Cellular oncogenes and retroviruses. Ann. Rev. Biochem. (in press).
3. Bishop, J.M., and H.E. Varmus (1982) Functions and origins of retroviral transforming genes, In Molecular Biology of Tumor Viruses Part III RNA Tumor Viruses, R.A. Weiss, N. Teich, H.E. Varmus, and J. Coffin, Eds., Cold Spring Harbor Press, New York, pp. 999-1103.
4. Blair, D.G., M. Oskarsson, T.G. Wood, W.L. McClements, P.J. Fischinger, and G.G. Vande Woude (1981) Activation of the transforming potential of a normal cell sequence: A molecular model for oncogenesis. Science 212:941-943.
5. Cairns, J. (1980) The origins of human cancers. Nature 289:353-356.
6. Coffin, J.M., H.E. Varmus, J.M. Bishop, M. Essex, W.D. Hardy, G.S. Martin, N.E. Rosenberg, E.M. Scolnick, R. Weinberg, and P.K. Vogt (1981) A proposal for naming host cell-derived inserts in retrovirus genomes. J. Virol. 40:953-962.
7. Cohen, J.C., P. Shank, V.L. Morris, R. Cardiff, and H.E. Varmus (1979) Integration of the DNA of mouse mammary tumor virus in virus-infected normal and neoplastic tissues of the mouse. Cell 16:333-343.
8. Collins, S., and M. Groudine (1982) Amplification of endogenous myc-related DNA sequences in a human myeloid leukemia cell line. Nature 298:679-681.
9. Cooper, G.M., and P.E. Neiman (1980) Transforming genes of neoplasms induced by avian lymphoid leukosis viruses. Nature 287:659-661.

10. Cooper, G.M., and P.E. Neiman (1981) Two distinct candidate transforming genes of lymphoid leukosis virus-induced neoplasms. Nature 292:857–859.

11. Cooper, G.M., S. Okenquist, and L. Silverman (1980) Transforming activity of DNA of chemically transformed and normal cells. Nature 284:418–422.

12. Cowell, J.K. (1982) Double minutes and homogeneously staining regions: Gene amplification in mammalian cells. Ann. Rev. Genet. 16:21–59.

13. Dalla Favera, R., F. Wong-Staal, and R.C. Gallo (1982) Onc gene amplification in promyelocytic leukaemic cell line HL-60 and primary leukaemic cells of the same patient. Nature 299:61–62.

14. Der, C.J., T.G. Krontiris, and G.M. Cooper (1982) Transforming genes of human bladder and lung carcinoma cell lines are homologous to the ras genes of Harvey and Kirsten sarcoma viruses. Proc. Natl. Acad. Sci. USA 79:3637–3641.

15. Fung, Y.K.T., A.M. Fadly, L.B. Crittenden, and H.J. Kung (1981) On the mechanism of retrovirus-induced avian lymphoid leukosis: Deletion and integration of the proviruses. Proc. Natl. Acad. Sci. USA 78:3418–3422.

16. Goubin, G., D.S. Goldman, J. Luce, P.E. Neiman, and G.M. Cooper (1983) Molecular cloning and nucleotide sequence of a transforming gene detected by transfection of chicken B-cell lymphoma DNA. Nature 302:114–119.

17. Gross, L. (1970) Oncogenic Viruses, Pergamon Press, New York.

18. Hayward, W.S., B.G. Neel, and S.M. Astrin (1981) Activation of a cellular onc gene by promoter insertion in ALV-induced lymphoid leukosis. Nature 290:475–480.

19. Jenkins, N.A., N.G. Copeland, B.A. Taylor, and B.K. Lee (1981) Dilute (d) coat colour mutation of DBA/2J mice is associated with the site of integration of an ecotropic MuLV genome. Nature 293:370–373.

20. Klein, G. (1983) Specific chromosomal translocations and the genesis of B cell derived tumors in mice and men. Cell 32:311–315.

21. Lane, M.A., A. Sainten, and G.M. Cooper (1981) Activation of related transforming genes in mouse and human mammary carcinomas. Proc. Natl. Acad. Sci. USA 78:5185–5189.

22. Linial, M., and D. Blair (1982) Genetics of retroviruses, In Molecular Biology of Tumor Viruses Part III RNA Tumor Viruses, R.A. Weiss, N. Teich, H.E. Varmus, and J. Coffin, Eds., Cold Spring Harbor Press, pp. 649–784.

23. Luciw, P.A., J.M. Bishop, H.E. Varmus, and M.R. Capecchi (1983) Location and function of retroviral and SV40 sequences that enhance biochemical transformation after microinjection of DNA. Cell 33:707–716..

24. McCoy, M.S., J.J. Toole, J.M. Cunningham, E.H. Chang, D.R. Lowy, and R.A. Weinberg (1983) Characterization of a human colon/lung carcinoma oncogene. Nature 302:79–82.

25. Neel, B.G., W.S. Hayward, H.L. Robinson, J. Fang, and
    S.M. Astrin (1981) Avian leukosis virus-induced tumors have
    common proviral integration sites and synthesize discrete
    viral RNAs: Oncogenesis by promoter insertion. Cell
    23:323-332.
26. Neiman, P., L.N. Payne, and R.A. Weiss (1980) Viral DNA in
    bursal lymphomas induced by avian leukosis viruses. J.
    Virol. 34:178-184.
27. Noori-Daloii, M.R., R.A. Swift, H.I. Kung, L.B. Crittenden,
    and R.L. Witter (1981) Specific integration of REV proviruses
    in avian bursal lymphomas. Nature 294:574-576.
28. Nusse, R., and H.E. Varmus (1982) Many tumors induced by the
    mouse mammary tumor virus contain a provirus integrated in
    the same region of the host genome. Cell 31:99-109.
29. Nusse, R., A. van Ooyen, D. Cox, Y.W. Fung, and H.E. Varmus
    (1983) Mode of proviral activation of a putative mammary
    oncogene (int-1) on mouse chromosome 15. Nature (in press).
30. Parada, L.F., C.J. Tabin, C. Shih, and R.A. Weinberg (1982)
    Human EJ bladder carcinoma oncogene is homologue of Harvey
    sarcoma virus ras gene. Nature 297:474-478.
31. Payne, G.S., S.A. Courtneidge, L.B. Crittenden, A.M. Fadly,
    J.M. Bishop, and H.E. Varmus (1981) Analysis of avian
    leukosis virus DNA and RNA in bursal tumors: Viral gene
    expression is not required for maintenance of the tumor
    state. Cell 23:311-322.
32. Payne, G.S., J.M. Bishop, and H.E. Varmus (1982) Multiple
    arrangements of viral DNA and an activated host oncogene in
    bursal lymphomas. Nature 295:209-217.
33. Peters, G., S. Brookes, R. Smith, and C. Dickson (1983)
    Tumorigenesis by mouse mammary tumor virus: Evidence for a
    common region for provirus integration in mammary tumors.
    Cell 33:369-377.
34. Ponder, B.-J. (1980) Genetics and cancer. Biochim. Biophys.
    Acta Revs. on Cancer 605:369-414.
35. Pulciani, S., E. Santos, A.V. Lauver, I.K. Long,
    S.A. Aaronson, and M. Barbacid (1982) Oncogenes in solid
    human tumours. Nature 300:539-541.
36. Reddy, E.P., R.K. Reynolds, E. Santos, and M. Barbacid (1982)
    A point mutation is responsible for the acquisition of
    transforming properties by the T24 human bladder carcinoma
    oncogene. Nature 300:149-150.
37. Santos, E., S.R. Tronick, S.A. Aaronson, S. Pulciani, and
    M. Barbacid (1982) T24 human bladder carcinoma oncogene is an
    activated form of the normal human homologue of BALB- and
    Harvey-MSV transforming genes. Nature 298:343-347.
38. Schwab, M., K. Alitalo, H.E. Varmus, J.M. Bishop, and
    D. George (1983) A cellular oncogene (c-Ki-ras) is amplified,
    overexpressed, and located within karyotypic abnormalities in
    mouse adrenocortical tumor cells. Nature 303:497-501.

39. Schwab, M., K. Alitalo, K.-H. Klempnauer, H.E. Varmus,
    J.B. Bishop, F. Gilbert, G. Brodeur, M. Goldstein, and
    J. Trent (1983) Amplified DNA with limited homology to myc
    cellular oncogene is shared by human neuroblastoma cell lines
    and a neuroblastoma tumor. Nature 305:245-248.
40. Shih, C., B.-Z. Shilo, M.P. Goldfarb, A. Dannenberg, and
    R.A. Weinberg (1979) Passage of phenotypes of chemically
    transformed cells via transfection of DNA and chromatin.
    Proc. Natl. Acad. Sci. USA 76:5714-5718.
41. Shimizu, K., M. Goldfarb, Y. Suard, M. Perucho, Y. Li,
    T. Kamata, J. Feramisco, E. Stavnezer, J. Fogh, and
    M.H. Wigler (1983) Three human transforming genes are related
    to the viral ras oncogenes. Proc. Natl. Acad. Sci. USA
    80:2112-2116.
42. Tabin, C.J., S.M. Bradley, C.I. Bargmann, R.A. Weinberg,
    A.G. Papageorge, E.M. Scolnick, R. Dhar, D.R. Lowy, and
    E.H. Chang (1982) Mechanism of activation of a human
    oncogene. Nature 300:143-148.
43. Teich, N., A. Bernstein, T. Mak, J. Wyke, and W. Hardy (1982)
    Retroviral diseases, In Molecular Biology of Tumor Viruses
    Part III RNA Tumor Viruses, R.A. Weiss, N. Teich,
    H.E. Varmus, and J. Coffin, Eds., Cold Spring Harbor Press,
    New York, pp. 785-997.
44. Varmus, H.E., N. Quintrell, and S. Ortiz (1981) Retroviruses
    as mutagens: Insertion and excision of a nontransforming
    provirus alter expression of a resident transforming
    provirus. Cell 25:23-36.
45. Varmus, H.E. (1982) Form and function of retroviral
    proviruses. Science 216:812-820.

HUMAN GENE MAPPING IN THE ANALYSIS OF ONCOGENES

Fa-Ten Kao

Eleanor Roosevelt Institute for Cancer Research and the
Department of Biochemistry, Biophysics and Genetics
University of Colorado Health Science Center
Denver, Colorado 80262

SUMMARY

Recent advances in human gene mapping have assigned more than
420 genes to various human chromosomes. Regional mapping of genes
to specific sites within the chromosome has also been achieved.
In addition, more than 130 random human DNA segments have been
mapped. This accomplishment has mainly resulted from somatic cell
genetics and more recently from recombinant DNA technology and in
situ hybridization. Human/rodent cell hybrids can be constructed
for retaining specific human chromosomes. Useful hybrid clone
panels have been established which contain unique combinations of
human chromosomes for convenient synteny analysis. Recombinant
DNA techniques have provided means for a direct assay of the
structural genes in the human genome by Southern blot analysis.
Using these methods, at least 9 oncogenes have been successfully
assigned to specific human chromosomes. Recent findings that
certain oncogenes are located in the chromosomal regions involved
in consistent translocations found in certain cancers have
prompted extensive research activities attempting to elucidate the
role of oncogenes in eliciting these malignancies. Genetic
predisposition to certain forms of cancer has been shown to
involve specific chromosomal deletions, suggesting a recessive
nature of the genes involved. It appears essential to identify
genes in these deleted regions. Together with gene mapping
methodologies and genetic fine structure analysis, alterations on
the nucleotide level involving structural or regulatory sequences
may be detected which may underlie the molecular etiology of
certain human neoplasia.

INTRODUCTION

The perpetuation of cancer cells through transmission of the unrestrained growth characteristic from parent to progeny cells has led to the suggestion that changes in the chromosomal DNA may be the initial step in the etiology of cancer. Somatic mutations occurring either in the structural gene or in the regulatory DNA sequences may alter normal control mechanisms in cell division. A variety of cancer-causing agents such as viruses, chemicals, radiations, etc. may act on the same target in the cellular genome and result in cancerous growth of the cell. In humans, predisposition to retinoblastoma and Wilms tumor has been attributed to specific chromosomal deletions in chromosome 13 (13q14) (97) and chromosome 11 (11p13) (76), respectively. Consistent chromosomal translocations in Burkitt lymphoma and chronic myelogenous leukemia also suggest the involvement of cancer-causing genes for these malignancies (44,79). Recent advances in molecular genetic methodologies (61) coupled with somatic cell genetic approaches (72,73) have made important progress in the identification and localization of oncogenes in the human genome (12,78). A basic cellular genetic concept unifying various routes of carcinogenesis has emerged (5,6,12,44). Genetic fine structure analysis of the human genome may uncover the subtle molecular changes at the nucleotide level leading to the cancerous growth. Eventually, the prevention, diagnosis and therapy of cancers may be achieved through the development of molecular and genetic approaches and methodologies.

OVERVIEW OF HUMAN GENE MAPPING

The human genome contains at least 50,000 to 100,000 structural genes coding for the amino acid sequence of proteins (59). These genes are distributed among 22 autosomes and one of the sex chromosomes, occupying a genome of approximately 3×10^9 base pairs. About 1,600 genes have been confidently identified (57), of which more than 420 have been mapped to specific human chromosomes (36). In the following are briefly summarized the various methods for mapping human genes.

Pedigree Analysis

This is the classic genetic method for studying linkage relationships among various human traits, and for establishing relative distances among linked genes. A human trait segregating in families in a Mendelian fashion is analyzed. Linked genes can be identified by statistical analysis of the pedigree data derived from informative families. The most useful method has been the lod (log of the odds) score method (29,65). Linkage analysis will not, however, assign genes directly to a particular chromosome.

The first human gene assigned to a specific human chromosome is
the color blindness gene assigned to the X chromosome by Wilson in
1911. However, the progress in gene assignment by this method was
very slow. Between 1911 and 1967, only about 100 genes were
assigned and they were largely assigned to the X chromosome; very
few autosomal gene assignments were made.

Somatic Cell Hybrid Method in Gene Mapping

The breakthrough in human gene mapping occurred when an
entirely different approach was used, namely somatic cell
genetics. In 1955, Puck and Marcus (74) developed single cell
plating and cloning techniques in cultured mammalian cells,
marking the beginning of somatic cell genetics. The mammalian
somatic cells grown in culture can be manipulated as with
bacterial cells (72). Subsequently, various mammalian cell
mutants were induced and isolated for use in genetic experiments
with high resolving power (10,43). Initial cytogenetic advances
had identified the human chromosome number to be 46 (93). More
refined techniques were developed for unequivocal identification
of each individual human chromosome by its characteristic banding
patterns (9). Recently, high resolution chromosome analysis has
increased the number of identified chromosome bands from about 300
to more than 1,000 (98). This method has increased cytogenetic
resolving power in detecting small deletions and in assigning
genes to smaller areas of the chromosome.

The discovery of cell fusion has greatly expanded the genetic
analysis in mammalian cells and particularly in gene mapping.
Through the efforts of several laboratories, including Okada (66),
Barski (4), Ephrussi (89), Littlefield (52), and Harris (32), cell
fusion techniques have now become a very important laboratory
procedure for a variety of cellular and genetic studies (77),
including the formation of hybridoma for monoclonal antibody
production (46).

When two rodent cell lines are fused together, the resulting
hybrids usually retain the combined chromosome number of the two
parental cells, with only small loss of chromosomes after long
term growth (40,52). However, Weiss and Green (94) observed that
in human-mouse cell hybrids, human chromosomes were preferentially
lost. This finding marked the beginning of the rapid progress in
human gene mapping using somatic cell hybrids. The first gene
assigned to an autosome by this method was the thymidine kinase
(TK) gene on human chromosome 17 (62,63). Subsequently,
impressive progress has been made in mapping a large number of
human genes using this new approach (36,59). Further refined and
more efficient procedures are being developed, especially in
combined use with the powerful recombinant DNA techniques
(73,80,88).

Human-rodent cell hybrids usually can be isolated about 2-3 weeks after fusion using the HAT selective system (52,88,94). Various human chromosomes are retained in different cell hybrids in a more or less random fashion for most of the human chromosomes. After the initial chromosome loss, the remaining human chromosomes are fairly stable and the hybrids can be characterized for their human chromosome content.

Two major techniques, cytogenetic and isozyme analysis, have been used for identifying each of the 24 different human chromosomes in the hybrids. The development of the alkaline Giemsa-11 differential staining method (3,7,25) has also facilitated cytogenetic analysis, especially for identifying chromosomal translocations between rodent and human chromosomes (25,45). Isozyme analysis can distinguish the species origin of enzymes synthesized in the hybrids (31). The most commonly used procedure involves the separation of isozymes by electrophoresis followed by simple histochemical staining to reveal the different location of human and rodent isozymes in situ (31). A gene can be assigned to a specific human chromosome if it always segregates together with a particular human chromosome, or with an isozyme marker with known chromosomal location. If two genes are shown to be on the same chromosome, they are syntenic (75).

Panels of cell hybrids are now available which contain unique combinations of human chromosomes (16,19,21,35,39,55,67). Thus, by assaying a particular human gene or its product in these hybrids, chromosomal assignment can be conveniently determined. Such hybrids have been used in mapping cellular oncogenes in the human genome as described below.

Another approach to establishing hybrids for efficient and systematic gene mapping is to construct hybrids containing only a single human chromosome. Various auxotrophic mutants of the CHO-K1 cells have been fused with human cells and grown in the selective medium F12D (43). The surviving hybrids must retain the human chromosome carrying the complementing gene for the auxotrophic deficiency. Specific hybrids have been isolated and shown to retain a complete CHO genome plus a single human chromosome 12 in the gly⁻A-human fusion (41,50), and a single chromosome 21 in the ade⁻C-human fusion (64). It is hoped that eventually a complete set of hybrids will be established each containing a single, different human chromosome. These hybrids will be useful not only for very efficient gene mapping with little ambiguity, but also for fine structure analysis of specific human chromosomes (26,49).

After a gene is assigned to a specific chromosome, the next step is to find its location within the chromosome. Various human cells or cell hybrids with well-defined deletions, duplications or

translocations of particular segments of human chromosomes can be
used for regional mapping purposes. A gene can be regionally
mapped if it can be correlated with such defined chromosome
segments (24). Thus, a series of cell hybrids containing various
deletions of a single human chromosome can be established for
systematic regional mapping of genes assigned to that chromosome.
We have established such hybrids for human chromosome 11 (38).
Cell surface antigen markers on human chromosome 11 have been used
to facilitate selection for subclones with deletions between the
marker gene and the centromere (42). These deletion hybrids have
been used for a quick regional mapping of the genes assigned to
chromosome 11 (27,28,38,60).

Recombinant DNA Technology in Gene Mapping

The advances in recombinant DNA technology have provided
purified gene probes which can be used in molecular hybridization
for detecting homologous DNA sequences in the human genome and in
the cell hybrids containing human chromosomes (61,73,80,88). The
filter hybridization technique using Southern blot (90) has
greatly improved the specificity and sensitivity of such analysis.
The method involves use of restriction enzymes to digest the DNA
from human and rodent parents and cell hybrids, agarose gel
electrophoretic separation of the digested DNA, transfer of DNA
from gel to nitrocellulose filter, hybridization of the filter
with the labeled gene probe, and autoradiographic detection of the
duplex hybridization bands. By selecting appropriate restriction
enzymes, the cleaved DNA fragments characteristic to human and
rodent sequences can be produced and used for identifying the
human gene sequences in the cell hybrids. Thus, gene probes that
cross-hybridize between human and rodent gene sequences can also
be used in these experiments.

The hybridization results revealing the presence or absence
of a particular human gene sequence in the cell hybrids can be
matched with the human chromosome content in the same cell
hybrids. The human chromosome with the highest concordance
frequency with the gene probe tested will define the chromosomal
assignment of that gene. This procedure is essentially the same
as that used for analyzing a specific isozyme marker or a gene
product using cell hybrid method. Similarly, regional mapping of
the gene probe can also be achieved using cell hybrids containing
various deletions of the chromosome, as demonstrated in the
assignment of the β-globin gene complex to the short arm of
chromosome 11, in the region 11q12 (28). It should be pointed out
that this method will map structural gene sequences in the human
genome regardless of whether or not the gene is expressed in the
cell hybrids. This feature is particularly important in mapping
genes encoding differentiated functions which in most cases are
not expressed in the cell hybrids (18).

In Situ Hybridization in Gene Mapping

The in situ hybridization method was first developed by
Pardue and Gall (69) and has recently been extended from detecting
locations of multiple gene copies arranged in tandem to detecting
singly located genes (30). This method offers promise for a
general method for chromosomal and regional assignment of genes
prepared in pure forms. Due to the spread of the grains detected
in autoradiographic preparations, this method cannot determine the
order of closely linked genes with resolution higher than the
current cytogenetic banding techniques. However, it is highly
useful in regional mapping without relying on the availability of
specific chromosomal deletions or translocations. Recent
improvement in using biotin-conjugated, instead of ^{3}H-labeled,
nucleotide probes followed by fluorescent or histochemical
staining (48) will produce sharper bands for a better definition
of the chromosomal sites than the radioactivity technique.

MAPPING OF CELLULAR ONCOGENES IN THE HUMAN GENOME

Retroviruses are RNA viruses, some of which carry oncogenes
which can transform cells in vitro and induce malignant growth in
vivo (5,6,12). Recent application of gene transfer techniques
(96) has led to the identification and isolation of human genomic
sequences homologous to a variety of viral oncogenes
(12,47,86,87). Using gene mapping procedures described above, at
least 9 of these human homologous oncogenes have been assigned to
specific human chromosomes. Table 1 lists the chromosomal
locations of the cellular oncogenes in the human genome. From
these assignments, it is clear that these genes are not clustered
in the human genome, but are scattered over several different
chromosomes.

Recently, the human gene mapping approach has provided
important leads in the molecular etiology of certain cancers.
Burkitt lymphoma involves consistent translocation between
chromosomes 8 and 14 (44). It has been shown that the chromosome
segment translocated from 14 to 8 carries part of the DNA
sequences of the heavy chain variable region of the immunoglobulin
gene complex which was previously mapped to chromosome 14 by
somatic cell hybrid method (14) and was also confirmed using
immunoglobulin gene probes (35). The remaining portion of the
variable region and the constant region of the heavy chain gene
complex are still in the non-translocated chromosome 14 (23).
Even more interesting is the finding that the piece of chromosome
8 translocated to chromosome 14 carries the human gene c-myc
homologous to the avian myelocytomatosis virus

Table 1. Chromosomal assignment of human
cellular genes homologous to viral oncogenes

Human cellular oncogenes	Viral oncogenes	Human chromosome assignment	References
c-myb	Avian myeloblastosis virus	6	16
c-mos	Moloney murine sarcoma virus	8	71
c-myc	Avian meylocytomatosis virus	8	15, 92
c-abl	Abelson murine leukemia virus	9	34
c-ras^H	Harvey murine sarcoma virus	11	19, 56
c-ras^K	Kirsten murine sarcoma virus	12	81
c-fes	Feline sarcoma virus	15	16, 34
c-src	Rous sarcoma virus	20	82
c-sis	Simian sarcoma virus	22	17, 91

oncogene v-myc (15,92). The human c-myc oncogene has been shown
to locate on chromosome 8 by somatic cell hybrid method and also
confirmed by the use of a human CHO-K1 cell hybrid containing a
single human chromosome 8 (15). Thus, the translocation of the
cellular c-myc gene from chromosomes 8 to the variable region of
the immunoglobulin heavy chain gene complex may result in the
alteration or activation of the c-myc gene and lead to the
development of Burkitt lymphoma.

In addition, other non-African forms of Burkitt lymphoma were
found to involve translocations between chromosomes 8 and 2, or 8
and 22 (44). Gene mapping studies have assigned the gene complex
encoding the kappa light chain of the immunoglobulin to chromosome
2 (53,55) and the gene for the lambda light chain to chromosome 22
(55). Thus, it is possible that these forms of Burkitt lymphoma
also involve translocation of the c-myc gene from chromosome 8 to
the chromosomal sites where the immunoglobulin light chain gene
complexes reside.

Similarly, mouse plasmacytomas involve consistent
translocations between chromosomes 15 and 12 (44). Previously,
the mouse immunoglobulin heavy chain gene complex was mapped to
mouse chromosome 12 using cell hybrid method (21). Thus it was
not surprising to find that the mouse c-myc gene is not only
located on chromosome 15, but is also translocated to the heavy
chain region of chromosome 12 in plasmacytoma (13,33,92).
Intensive research activities are currently being carried out

attempting to elucidate the role of the c-myc gene in eliciting
these malignancies (1,22,54).

Two other oncogenes have been localized to chromosomal
regions involved in consistent translocations. Chronic
myelogenous leukemia (CML) involves consistent translocations
between chromosomes 9 and 22 (9q34/22q11). It has been found that
the oncogene c-abl, located at 9q34, has been translocated from
chromosome 9 to chromosome 22 in CML cells (34). This is also the
first demonstration of the reciprocal nature in these
translocations. The cellular oncogene c-mos has been mapped to
the long arm of chromosome 8, 8q22 (71), a region which is
involved in the consistent translocation 8q22/21q11 in acute
myeloblast leukemia (79). These mapping findings have established
links between translocations and oncogenes and further elucidation
of the genetic mechanisms underlying these hematological diseases
should be actively pursued.

It is interesting to point out that the two murine sarcoma
virus oncogenes, the Harvey strain c-ras^H and the Kirsten strain
c-ras^K, have been mapped to human chromosomes 11 (19,56) and 12
(81), respectively. The c-ras^H gene has been found to be
homologous to the transforming gene of human bladder carcinoma and
the c-ras^K gene homologous to the transforming gene of lung/colon
carcinoma (20,68,83). Human chromosomes 11 and 12 are similar in
morphology and in banding patterns, and possess respectively in
their short arms the isozymes lactate dehydrogenase A and B (58).
It would be interesting to explore the possibility that
chromosomes 11 and 12 were derived from duplicate chromosomes in
their evolutionary history.

There are other non-hematological cancers involving
consistent deletions or translocations of specific chromosomal
segments. In addition to retinoblastoma and Wilms tumor mentioned
earlier (76,97), small cell lung carcinoma involves deletions in
the short arm of chromosome 3, including 3p14-p23 (95).
Chromosome 3 has also been shown to be consistently involved in
translocations found in renal cell carcinoma (11,70). Partial
deletion in the short arm of chromosome 1, including 1pter-p32,
has been found to be a consistent chromosomal abnormality in human
neuroblastoma (8). Thus, detailed gene mapping studies in these
crucial chromosomal regions may lead to identification of specific
genes or DNA sequences responsible for the cancerous growth in
these malignancies.

GENETIC FINE STRUCTURE MAPPING IN HUMAN CANCERS

More than 30% of the human genome is composed of repetitive
sequences, a large portion of which are interspersed with unique

sequences throughout the genome (85). For example, the Alu repeated sequence family consists of 300,000 copies per haploid human genome (37). Another type of repetitive sequence contains only several thousand copies or less (2,26,49,84), and has been found to exhibit distinct bands when hybridized to DNA blots prepared from cell hybrids containing only a single human chromosome (26,49). These multiple bands represent different regions in the chromosome. Thus, such repetitive sequence probes can be used as genetic markers for specific segments of a particular chromosome.

Two repetitive sequence probes have been developed and used as multiple site markers on chromosomes 12 (49) and 11 (26) respectively. Using a cloned 2.2 kb repetitive sequence isolated from a DNA library of human chromosome 12, Law et al. (49) demonstrated that it can form multiple bands when hybridized to the digested DNA from a cell hybrid containing a single human chromosome 12. Characteristic band patterns were also found with other single human chromosomes. Regional mapping of some of these bands on chromosome 12 has been achieved by using cell hybrids containing various deletions of chromosome 12 (49,51).

Another repetitive sequence probe was derived from human β-globin gene complex on chromosome 11. When this probe was hybridized to digested chromosome 11 DNA, 19 discrete bands were identified (26). Using a series of cell hybrids containing different terminal deletions of chromosome 11 (38), these bands were assigned to eleven specific regions of the chromosome (26).

These studies demonstrate the use of repetitive sequence probes in identifying multiple sites on a chromosome. The power of this approach can be increased by applying different restriction enzymes and additional repetitive sequence probes. Using this approach, a detailed fine structure map showing specific restriction fragments in the chromosome can be constructed which should be useful in detecting possible DNA sequence alterations in cancers or other genetic diseases with consistent chromosomal abnormalities. An increasingly refined map for a particular part of the chromosome can be established by using repetitive sequence probes with higher copy numbers.

CONCLUSIONS

Human gene mapping has experienced explosive growth during the past 10-15 years. This remarkable progress has resulted mainly from the development of an entirely new approach in gene mapping, namely the somatic cell hybrid method. More than 420 genes have been successfully mapped to different parts of the human genome. By combining somatic cell genetics with recombinant

DNA techniques, a variety of oncogenes have been identified in the human genome. Using the gene mapping procedures described here, at least 9 oncogenes have been successfully localized to specific human chromosomal sites (Table 1). These achievements have furnished the basis for a converging cellular genetic concept in carcinogenesis. The involvement of the c-myc oncogene in Burkitt lymphoma has provided molecular approaches to the study of the etiology of cancer. More and more oncogenes will be identified, cloned, characterized and mapped. Fine structure analysis of oncogenes and their flanking sequences will increase our understanding of the regulatory processes involved in the various abnormal growth behaviors. Genetic mapping of these genes in the human genome will elucidate the relationships between the gene location and its function, and their roles in the etiology of cancer. Hopefully, manipulating the turning on and off of gene activities may offer promise in the molecular therapy of certain cancers.

ACKNOWLEDGEMENTS

This is contribution number 439 of the Eleanor Roosevelt Institute for Cancer Research. I wish to thank Dr. Martha Liao Law for critical reading of the manuscript. The work from the author's laboratory was supported by NIH grants GM26631 and HD02080.

REFERENCES

1. Adams, J.M., S. Gerondakia, E. Webb, L.M. Corcoran, and S. Cory (1983) Cellular myc oncogene is altered by chromosome translocation to an immunoglobulin locus in murine plasmacytomas and is rearranged similarly in human Burkitt lymphoma. Proc. Natl. Acad. Sci. USA 80:1982-1986.
2. Adams, J.W., R.E. Kaufman, P.J. Kretschmer, M. Harrison, and A.W. Nienhuis (1980) A family of long reiterated DNA sequences, one copy of which is next to the human beta globin gene. Nucleic Acids Res. 8:6113-6128.
3. Alhadeff, B., M. Velivasakis, and M. Siniscalco (1977) Simultaneous identification of chromatid replication and of human chromosomes in metaphases of man-mouse somatic cell hybrids. Cytogenet. Cell Genet. 19:236-239.
4. Barski, G., S. Sorieul, F. Cornefert (1960) Production dans des cultures in vitro de deux souches cellulaires en association de dellules de caractere "hybride". C.R. Acad. Sci. 251:1825-1827.
5. Bishop, J.M. (1981) Enemies within: The genesis of retrovirus oncogenes. Cell 23:5-6.
6. Bishop, J.M. (1982) Oncogenes. Sci. Amer. 246:81-92.

7. Bobrow, M., and J. Cross (1974) Differential staining of human and mouse chromosomes in interspecific cell hybrids. Nature 251:77-79.

8. Brodeur, G.M., A.A. Green, F.A. Hayes, K.J. Williams, D.L. Williams, and A.A. Tsiatis (1981) Cytogenetic features of human neuroblastomas and cell lines. Cancer Res. 41:4678-4686.

9. Caspersson, T., L. Zech, and C. Johansson (1970) Differential banding of alkylating fluorochromes in human chromosomes. Exp. Cell Res. 60:315-319.

10. Chu, E.H.Y., and H.V. Malling (1968) Mammalian cell genetics. II. Mutational chemical induction of specific locus mutations in Chinese hamster cells in vitro. Proc. Natl. Acad. Sci. USA 61:1306-1312.

11. Cohen, A.J., F.P. Li, S. Berg, D.J. Marchetto, S. Tsai, S.C. Jacobs, and R.S. Brown (1979) Hereditary renal-cell carcinoma associated with a chromosomal translocation. New Engl. J. Med. 301:592-595.

12. Cooper, G.M. (1982) Cellular transforming genes. Science 218:801-806.

13. Crews, S., R. Barth, L. Hood, J. Prehn, and K. Calame (1982) Mouse c-myc oncogene is located on chromosome 15 and translocated to chromosome 12 in plasmacytomas. Science 218:1319-1321.

14. Croce, C.M., M. Sahnder, J. Martinis, L. Cicurel, G.G. D'Ancona, T.W. Dolby, and H.L. Koprowski (1979) Chromosomal location of the genes for human immunoglobulin heavy chains. Proc. Natl. Acad. Sci. USA 76:3416-3419.

15. Dalla-Favera, R., M. Bregni, J. Erikson, D. Patterson, R.C. Gallo, and C.M. Croce (1982) Human c-myc onc gene is located on the region of chromosome 8 that is translocated in Burkitt lymphoma cells. Proc. Natl. Acad. Sci. USA 79:7824-7827.

16. Dalla-Favera, R., G. Franchini, S. Martinotti, F. Wong-Staal, R.C. Gallo, and C.M. Croce (1982) Chromosomal assignment of the human homologues of feline sarcoma virus and avian myeloblastosis virus onc genes. Proc. Natl. Acad. Sci. USA 79:4714-4717.

17. Dalla-Favera, R., R.C. Gallo, A. Giallongo, and C.M. Croce (1982) Chromosomal localization of the human homolog (c-sis) of the simian sarcoma virus onc gene. Science 218:686-688.

18. Davidson, R.L. (1974) Gene expression in somatic cell hybrids. Ann. Rev. Genet. 8:195-218.

19. De Martinville, B., B.J. Fiacalone, C. Shih, R.A. Weinberg, and U. Francke (1983) Oncogene from human EJ bladder carcinoma is located on the short arm of chromosome 11. Science 219:498-501.

20. Der, C.J., T.G. Krontiris, and G.M. Cooper (1982) Transforming genes of human bladder and lung carcinoma cell lines are

homologous to the ras genes of Harvey and Kirsten sarcoma viruses. Proc. Natl. Acad. Sci. USA 79:3637-3640.

21. D'Eustachio, P., D. Pravtcheva, K. Marcu, and F.H. Ruddle (1980) Chromosomal location of the structural gene cluster encoding murine immunoglobulin heavy chains. J. Exp. Med. 151:1545-1550.

22. Erikson, J., A. Ar-Rushdi, H.L. Drwinga, P.C. Nowell, and C.M. Croce (1983) Transcriptional activation of the translocated c-myc oncogene in Burkitt lymphoma. Proc. Natl. Acad. Sci. USA 80:820-824.

23. Erikson, J., J. Finan, P.C. Nowell, and C.M. Croce (1982) Translocation of immunoglobulin V_H genes in Burkitt lymphoma. Proc. Natl. Acad. Sci. USA 79:5611-5615.

24. Ferguson-Smith, M.A., and D.A. Aitken (1981) The contribution of chromosome aberrations to the precision of human gene mapping. Cytogenet. Cell Genet. 32:24-42.

25. Friend, K.K., S. Chen, and F.H. Ruddle (1976) Differential staining of interspecific chromosomes in somatic cell hybrids by alkaline Giemsa stain. Somat. Cell Genet. 2:183-188.

26. Gusella, J.F., C. Jones, F.T. Kao, D. Housman, and T.T. Puck (1982) Genetic fine-structure mapping in human chromosome 11 by use of repetitive DNA sequences. Proc. Natl. Acad. Sci. USA 79:7804-7808.

27. Gusella, J.F., C. Keys, A. Varsanyi-Breiner, F.T. Kao, C. Jones, T.T. Puck, and D. Housman (1980) Isolation and localization of DNA segments from specific human chromosomes. Proc. Natl. Acad. Sci. USA 77:2829-2833.

28. Gusella, J., A. Varsanyi-Breiner, F.T. Kao, C. Jones, T.T. Puck, C. Keys, S. Orkin, and D. Housman (1979) Precise location of human β-globin gene complex on chromosome 11. Proc. Natl. Acad. Sci. USA 76:5239-5243.

29. Haldane, J.B.S., and C.A.B. Smith (1947) A new estimate of the linkage between the genes for colour-blindness and haemophilia in man. Ann. Eugen. 14:10-31.

30. Harper, M.E., A. Ullrich, and G.F. Saunders (1981) Localization of the human insulin gene to the distal end of the short arm of chromosome 11. Proc. Natl. Acad. Sci. USA 78:4458-4460.

31. Harris, H., and D.A. Hopkinson (1976) Handbook of Enzyme Electrophoresis in Human Genetics, North-Holland, Amsterdam, The Netherlands.

32. Harris, H., and J.F. Watkins (1965) Hybrid cells derived from mouse and man: Artificial heterokaryons of mammalian cells from different species. Nature 205:640-646.

33. Harris, L.J., P. D'Eustachio, F.H. Ruddle, and K.B. Marcu (1982) DNA sequence associated with chromosome translocations in mouse plasmacytomas. Proc. Natl. Acad. Sci. USA 79:6622-6626.

34. Heisterkamp, N., J. Groffen, J.R. Stephenson, N.K. Spurr, P.N. Goodfellow, E. Solomon, B. Carritt, and W.F. Bodmer

(1982) Chromosomal localization of human cellular homologues of two viral oncogenes. Nature 299:747-749.

35. Hobart, M.J., T.H. Rabbitts, P.N. Goodfellow, E. Solomon, S. Chambers, N. Spurr, and S. Povey (1981) Immunoglobulin heavy chain genes in humans are located on chromosome 14. Ann. Hum. Genet. 45:331-335.

36. Human Gene Mapping 6, Oslo Conference (1981) Cytogenet. Cell Genet. 32:1-341.

37. Jelinek, W.R., T.P. Tommey, L. Leinwand, C.H. Duncan, P.A. Biro, P.V. Choudary, S.M. Weissman, C.M. Rubin, C.M. Houck, P.L. Deininger, and C.W. Schmid (1980) Ubiquitous, interspersed repeated sequences in mammalian genomes. Proc. Natl. Acad. Sci. USA 77:1398-1402.

38. Jones, C., and F.T. Kao (1978) Regional mapping of the gene for human lysosomal acid phosphatase (ACP$_2$) using a hybrid clone panel containing segments of human chromosome 11. Hum. Genet. 45:1-10.

39. Kao, F.T., J.A. Hartz, M.L. Law, and J.N. Davidson (1982) Isolation and chromosomal localization of unique DNA sequences from a human genomic library. Proc. Natl. Acad. Sci. USA 79:865-869.

40. Kao, F.T., R.T. Johnson, and T.T. Puck (1969) Complementation analysis of virus-fused Chinese hamster cells with nutritional markers. Science 164:312-314.

41. Kao, F.T., C. Jones, and T.T. Puck (1976) Genetics of somatic mammalian cells: Genetic, immunologic, and biochemical analysis with Chinese hamster cell hybrids containing selected human chromosomes. Proc. Natl. Acad. Sci. USA 73:193-197.

42. Kao, F.T., C. Jones, and T.T. Puck (1977) Genetics of cell-surface antigens: Regional mapping of three components of the human cell-surface antigen complex, A$_L$, on chromosome 11. Somat. Cell Genet. 3:421-429.

43. Kao, F.T., and T.T. Puck (1968) Genetics of somatic mammalian cells. VII. Induction and isolation of nutritional mutants in Chinese hamster cells. Proc. Natl. Acad. Sci. USA 60:1275-1281.

44. Klein, G. (1981) The role of gene dosage and genetic transpositions in carcinogenesis. Nature 294:313-318.

45. Klobutcher, L.A., and F.H. Ruddle (1979) Phenotype stabilization and integration of transferred material in chromosome-mediated gene transfer. Nature 280:657-670.

46. Kohler, G., and C. Milstein (1975) Continuous culture of fused cells secreting antibody of predefined specificity. Nature 256:495-497.

47. Krontiris, T., and G.M. Cooper (1981) Transforming activity of human tumor DNAs. Proc. Natl. Acad. Sci. USA 78:1181-1184.

48. Langer-Safer, P.R., M. Levine, and D.C. Ward (1982) Immunological method for mapping genes on Drosophila polytene chromosomes. Proc. Natl. Acad. Sci. USA 79:4381-4385.

49. Law, M.L., J.N. Davidson, and F.T. Kao (1982) Isolation of a human repetitive sequence and its application to regional chromosome mapping. Proc. Natl. Acad. Sci. USA 79:7390-7394.

50. Law, M.L., and F.T. Kao (1978) Induced segregation of human syntenic genes by 5-bromodeoxyuridine and near-visible light. Somat. Cell Genet. 4:465-476.

51. Law, M.L., and F.T. Kao (1979) Regional assignment of human genes TPI$_1$, GAPDH, LDHB, SHMT, and PEPB on chromosome 12. Cytogenet. Cell Genet. 24:102-114.

52. Littlefield, J.W. (1964) Selection of hybrids from matings of fibroblasts in vitro and their presumed recombinants. Science 145:709-710.

53. Malcolm, S., P. Barton, C. Murphy, M.A. Ferguson-Smith, D.L. Bentley, and T.H. Rabbitts (1982) Localization of human immunoglobulin k light chain variable region genes to the short arm of chromosome 2 by in situ hybridization. Proc. Natl. Acad. Sci. USA 79:4957-4961.

54. Marcu, K.B., L.J. Harris, L.W. Stanton, J. Erikson, R. Watt, and C.M. Croce (1983) Transcriptionally active c-myc oncogene is contained within NIARD, a DNA sequence associated with chromosome translocations in B-cell neoplasia. Proc. Natl. Acad. Sci. USA 80:519-523.

55. McBride, O.W., P.A. Heiter, G.F. Hollis, D. Swan, M.C. Otey, and P. Leder (1982) Chromosomal location of human kappa and lambda immunoglobulin light chain constant region genes. J. Exp. Med. 155:1480-1490.

56. McBride, O.W., D.C. Sean, E. Santose, M. Barbacid, S.R. Tronick, and S.A. Aaronson (1982) Localization of the normal allele of T24 human bladder carcinoma oncogene to chromosome 11. Nature 300:773-774.

57. McKusick, V.A. (1978) Mendelian Inheritance in Man, 5th edition, The Johns Hopkins University Press, Baltimore, Maryland.

58. McKusick, V.A. (1980) The anatomy of the human genome. J. Hered. 71:370-391.

59. McKusick, V.A., and F.H. Ruddle (1977) The status of the gene map of the human chromosomes. Science 196:390-405.

60. Meisler, M.H., L. Wanner, F.T. Kao, and C. Jones (1981) Localization of the uroporphyrinogen I synthase locus to human chromosome region 11q13ter and interconversion of isozymes. Cytogenet. Cell Genet. 31:124-128.

61. Grossman, L., and K. Moldave, Eds. (1980) Methods in Enzymology, Vol. 65, Academic Press, New York.

62. Migeon, B.R., and C.S. Miller (1968) Human-mouse somatic cell hybrids with single human chromosome (Group E): Link with thymidine kinase activity. Science 162:1005-1006.

63. Miller, O.J., P.W. Allderdice, D.A. Miller, W.R. Breg, and B.R. Migeon (1971) Human thymidine kinase gene locus: Assignment to chromosome 17 in a hybrid of man and mouse cells. Science 173:244-245.

64. Moore, E.E., C. Jones, F.T. Kao, and D.C. Oates (1977) Synteny between glycinamide ribonucleotide synthetase and superoxide dismutase (soluble). Amer. J. Hum. Genet. 29:389-396.
65. Morton, N.E. (1955) Sequential tests for the detection of linkage. Amer. J. Hum. Genet. 7:277-318.
66. Okada, Y. (1958) The fusion of Ehrlich's tumor cells caused by HVJ virus in vitro. Biken J. 1:103-110.
67. Owerbach, D., G.I. Bell, W.J. Rutter, and T.B. Shows (1980) The insulin gene is located on chromosome 11 in humans. Nature 286:82-84.
68. Parada, L.F., C.J. Tabin, C. Shih, and R.A. Weinberg (1982) Human EJ bladder carcinoma oncogene is homologue of Harvey sarcoma virus ras gene. Nature 297:474-478.
69. Pardue, M.L., and J.G. Gall (1970) Chromosomal localization of mouse satellite DNA. Science 168:1356-1358.
70. Pathak, S., L.C. Strong, R.E. Ferrell, and A. Trindade (1982) Familial renal cell carcinoma with a 3;11 chromosome translocation limited to tumor cells. Science 217:939-941.
71. Prakash, K., O.W. McBride, D.C. Swan, S.G. Devare, S.R. Tronick, and S.A. Aaronson (1982) Molecular cloning and chromosomal mapping of a human locus related to the transforming gene of Moloney murine sarcoma virus. Proc. Natl. Acad. Sci. USA 79:5210-5214.
72. Puck, T.T. (1972) The Mammalian Cell as a Microorganism: Genetic and Biochemical Studies In Vitro, Holden-Day, San Francisco, California.
73. Puck, T.T., and F.T. Kao (1982) Somatic cell genetics and its application to medicine. Ann. Rev. Genet. 16:255-271.
74. Puck, T.T., and P.I. Marcus (1955) Rapid method for viable cell titration and clone production with HeLa cells on tissue culture: The use of X-irradiated cells to supply conditioning factors. Proc. Natl. Acad. Sci. USA 41:432-437.
75. Renwick, J.H. (1969) Progress in mapping human autosomes. Brit. Med. Bull. 25:65-73.
76. Riccardi, V.M., E. Sujansky, A.C. Smith, and U. Francke (1978) Chromosomal imbalance in the aniridia-Wilms' tumor association: 11p interstitial deletion. Pediatrics 61:604-610.
77. Ringertz, N.R., and R.E. Savage (1976) Cell Hybrids, Academic Press, New York.
78. Rowley, J.D. (1982) Human oncogene locations and chromosome aberrations. Nature 301:290-291.
79. Rowley, J.D. (1982) Identification of the constant chromosome regions involved in human hematologic malignant disease. Science 216:749-751.
80. Ruddle, F.H. (1981) A new era in mammalian gene mapping: Somatic cell genetics and recombinant DNA methodologies. Nature 294:115-120.

81. Sakaguchi, A.Y., S.L. Naylor, T.B. Shows, J.J. Toole, M. McCoy, and R.A. Winberg (1983) Human c-Ki-ras2 proto-oncogene on chromosome 12. Science 219:1081-1083.
82. Sakaguchi, A.Y., S.L. Naylor, R.A. Weinberg, and T.B. Shows (1982) Organization of human proto-oncogenes. Amer. J. Hum. Genet. 34:175A.
83. Santos, E., S.R. Tronick, S.A. Aaronson, S. Pulciani, and M. Barbacid (1982) T24 human bladder carcinoma oncogene is an activated form of the normal human homologue of Balb- and Harvey-MSV transforming genes. Nature 298:343-347.
84. Schmeckpeper, B.J., F.W. Huntington, and K.D. Smith (1981) Isolation and characterization of cloned human DNA fragments carrying reiterated sequences common to both autosomes and the X chromosome. Nucleic Acids Res. 91:1853-1872.
85. Schmid, C.W., and W.R. Jelinek (1982) The Alu family of dispersed repetitive sequences. Science 216:1065-1070.
86. Shih, C., B. Shilo, M.P. Goldfarb, A. Dannenberg, and R.A. Weinberg (1979) Passage of phenotype of chemically transformed cells via transfection of DNA and chromatin. Proc. Natl. Acad. Sci. USA 76:5714-5718.
87. Shih, C., and R.A. Weinberg (1982) Isolation of a transforming sequence from a human bladder carcinoma cell line. Cell 29:161-169.
88. Shows, T.B., and A.Y. Sakaguchi (1980) Gene transfer and gene mapping in mammalian cells in culture. In Vitro 16:55-76.
89. Sorieul, S., and B. Ephrussi (1961) Karyological demonstration of hybridization of mammalian cells in vitro. Nature 190:653-654.
90. Southern, E.M. (1975) Detection of specific sequences among DNA fragments separated by gel electrophoresis. J. Mol. Biol. 98:503-517.
91. Swan, D.C., O.W. McBride, K.C. Robbins, D.A. Keithley, E.P. Reddy, and S.A. Aaronson (1982) Chromosomal mapping of the simian sarcoma virus onc gene analogue in human cells. Proc. Natl. Acad. Sci. USA 79:4691-4695.
92. Taub, R., I. Kirsch, C. Morton, G. Lenoir, D. Swan, S. Tronick, S. Aaronson, and P. Leder (1982) Translocation of the c-myc gene into the immunoglobulin heavy chain locus in human Burkitt lymphoma and murine plasmacytoma cells. Proc. Natl. Acad. Sci. USA 79:7837-7841.
93. Tjio, J.H., and A. Levan (1956) The chromosome number of man. Hereditas 42:1-6.
94. Weiss, M.C., and H. Green (1967) Human-mouse hybrid cell lines containing partial complements of human chromosomes and functioning human genes. Proc. Natl. Acad. Sci. USA 58:1104-1111.
95. Whang-Peng, J., C.S. Kao-Shan, E.C. Lee, P.A. Bunn, D.N. Carney, A.F. Gazdar, and J.D. Minna (1982) Specific chromosome effect associated with human small-cell lung cancer: Deletion 3p(14-23). Science 215:181-182.

96. Wigler, M., A. Pellicer, S. Silverstein, and R. Axel (1979) Biochemical transfer of single-copy eucaryotic genes using total cellular DNA as donor. Cell 14:725-731.
97. Wilson, M.G., A.J. Ebbin, J.W. Towner, and W.H. Spencer (1977) Chromosomal anomalies in patients with retinoblastoma. Clin. Genet. 12:1-8.
98. Yunis, J.J. (1976) High resolution of human chromosomes. Science 191:1268-1270.

RELATIONSHIP BETWEEN MUTAGENESIS AND CARCINOGENESIS

Claes Ramel

Wallenberg Laboratory
University of Stockholm
S-106 91 Stockholm, Sweden

SUMMARY

The sequence of events in cancer induction is usually thought
to include at least three steps - initiation, promotion and
progression. Several lines of evidence speak in favor of a
mutational origin of cancer, particularly at the initiation level.
This evidence is mostly indirect, but recent analyses of oncogenes
have given the theory more direct support. These investigations
have indicated that base substitutions as well as chromosomal
rearrangements are involved. Other genetic mechanisms of cancer
initiation have also been suggested, such as hypomethylation of
DNA bases and transposition of DNA segments. However the lack of
response both of prokaryotic and eukaryotic transposable elements
to carcinogenic agents makes it doubtful whether transpositions
are involved in chemically induced cancer. Although experimental
data clearly indicate that alterations of DNA are involved in
cancer induction, there are also experimental observations on
nuclear transplantations, on transmission of induced cancer
properties to the offspring, and on the frequency of neoplastic
transformations, which are not readily explained by a mutational
origin of cancer. Although this circumstance calls for some
cautiousness in excluding a nongenetic origin of cancer in some
cases, there are nevertheless strong reasons to believe that some
kind of somatic mutation events constitute the predominant
mechanism for cancer initiation. At the practical screening level
the empirical correlation between results from animal cancer tests
and short term mutagenicity assays has given a firm foundation for
the use of mutagenicity screening for prediction of
carcinogenicity.

INTRODUCTION

Transformation of normal cells to cancer cells implies the acquisition of a permanent trait which is transmitted from cell generation to cell generation. The simplest and most obvious explanation would be that cancer evolves as a consequence of a mutagenic change. Under this hypothesis, which was put forward several decades ago, carcinogenic chemicals could be expected to be mutagenic and vice versa. However, when this hypothesis was tested experimentally during the 1950s it was found that the correlation between mutagenic and carcinogenic properties of chemicals was very poor indeed. Since it was also found that tumors could be induced in mammals by certain viruses, whose mechanism of action was not recognized at the time, it became difficult to maintain the somatic mutation hypothesis of carcinogenesis. The picture was changed, however, by the finding that most chemical carcinogens are not carcinogenic per se but require a metabolic transformation and activation in the mammalian body into carcinogenic forms. Thus the lack of mutagenic effect by known carcinogens was explained by the fact that the test systems used for mutagenicity screening did not exhibit the necessary biotransformation. Furthermore, when the viral interaction with eukaryotic chromosomes and DNA was revealed, the carcinogenic properties of oncogenic viruses became conceivable and logical in relation to the somatic mutation hypothesis. Since the 1960s the hypothesis of a mutational origin of cancer has gained strong experimental support.

Evidence for a Mutational Origin of Cancer

Several lines of evidence in favor of the somatic mutation hypothesis for cancer are given below:

1. Cancer property is an irreversible and transmissable cellular characteristic in accordance with a genetically determined trait.

2. Tumors have been shown by immunological methods to be monoclonal, i.e., they emanate from one transformed cell (25). If the reverse had been true--that tumors are formed from several transformed cells--a mutational origin would have been difficult to maintain.

3. Known carcinogenic and mutagenic chemicals share the property of being electrophilic and bind to nucleophilic centers of macromolecules, including DNA (24).

4. Experimental evidence, taking into consideration mammalian biotransformation, has shown that most carcinogens are also mutagenic.

5. In the human disease Xeroderma pigmentosum increase of tumors is associated with a genetically-determined lack of excision repair of DNA (6). In another system using fish, tumor induction by UV was counteracted by visible light, which cleaves UV-induced thymine dimers through photoreactivation, suggesting the involvement of DNA repair in tumorigenesis (37).

6. Specific cellular and homologous viral genes, oncogenes, have been shown to be involved in tumor formation. Recent investigations have given direct evidence that certain cancer forms are attributable to defined alterations in DNA in or in the vicinity of oncogenes.

Oncogenes

The study of oncogenes has rapidly become a central issue in the attempt to understand the mechanism of carcinogenicity and particularly during the last year there has been a quite dramatic development within this area of research (for a review see 7,19). Acute transforming RNA viruses (retroviruses) contain specific genes, oncogenes or transforming genes, which can cause neoplastic transformation of cells. These oncogenes are homologous to normal cellular genes. The viral oncogenic sequences lack the introns of the cellular counterparts and they evidently emanate from cellular genes which have been picked up by the retroviruses. Around 20 such retrovirus oncogenes have been recognized, some of them occurring in several independent viruses (for a review see 7). These oncogenes can be reintegrated from retroviruses into new sites of cellular DNA ("processed genes").

Cellular oncogenes, on the other hand, show a high evolutionary stability and must have a normal function in the cell, presumably in connection with cellular differentiation. The oncogenic property of these genes is acquired through activation. This activation can occur by different genetic means, resulting in either an increased level of gene expression or in an alteration of the gene product through a mutational change of the oncogene in question.

The transforming property of tumor DNA in general and the oncogenes in particular has been subjected to extensive analyses by means of so-called transfection assays in combination with the use of restriction enzymes. The transfection assay is essentially equivalent to the classic genetic transformation of bacteria with DNA, as first reported by Avery et al. (1). Recipient cells are exposed to donor DNA isolated, for instance, from tumor cells. The commonly used cell line for this purpose is a mouse fibroblast cell line, NIH 3T3, in which a stable integration of DNA occurs with a high frequency (39). This NIH cell line is a permanent cell line, which, as with many such cell lines, balances between a

normal state and a state of oncogenic transformation. It is thus
"half" transformed.

While transfection with DNA from normal nontransformed human
or animal cells does not give rise to transformation of NIH cells,
similar transfection experiments with DNA from tumor cells often
gives rise to a high frequency of transformation. This
transformation property implies a permanent genetic change which
is transmitted between cell generations of the NIH cells.

Activation of Oncogenes by Point Mutation

Recently three research groups (34,41,42) have independently
shown that the transforming property of cells from the two related
human bladder tumor cell lines EJ and T24 depends on a single base
substitution from guanine to thymine in an oncogene. This
oncogene is closely related to viral oncogenes of Harvey sarcoma
virus (v-$\underline{ras}^H$) and the cellular homolog in human DNA (c-$\underline{ras}_1^H$). The
gene product of the ras oncogene is a protein located at the inner
cell membrane with a molecular weight of 21000 (p 21). The
transversion of the basepair GC to TA in the oncogene of the human
bladder cells has implied a corresponding alteration of the p 21
gene product by one amino acid, that is from glycine to valine.
It is likely that this simple amino acid alteration is responsible
for the transforming property of the NIH cells transfected with
DNA from the bladder tumor cell lines. The pronounced effect of
this amino acid substitution may stem from the fact that glycine
is the only unbranched amino acid and is able to take part in
bending and folding of the protein. An exchange of this amino
acid will confer a configuration or restriction to the protein in
this respect (41). The importance of this region is illustrated
by the fact that in the rat the v-$\underline{ras}^H$ oncogene deviates from the
normal rat cellular oncogene by coding for arginine instead of
glycine in exactly the same position of p 21 (41) and an
equivalent change of glycine to serine is indicated with the
closely related v-$\underline{ras}^{Ki}$ oncogene (43) (see Fig. 1).

It should be emphasized that DNA from normal bladder cells
does not exhibit the ability to transform cells in transfection
assays and the base substitution found in the bladder carcinoma
cells is not just a normal genetic polymorphism but seems to be
directly linked to the transforming property. These results on
human bladder carcinoma constitute the first direct evidence that
a simple mutational change of one basepair in DNA is responsible
for carcinogenic transformation of cells.

The picture is, however, more complicated. Both experimental
data and epidemiological considerations strongly point to the fact
that carcinogenesis is a multistep process. The NIH cell line

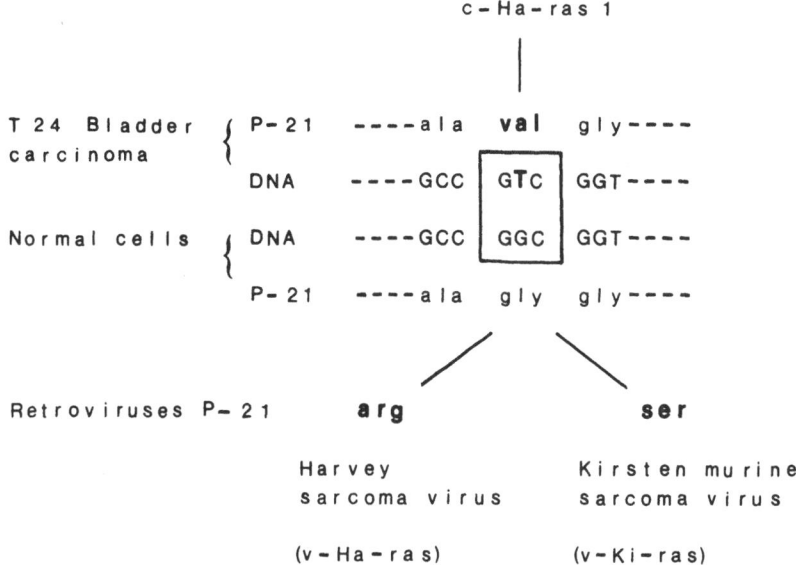

Figure 1. Activation of the cellular and viral ras oncogene
family through point mutations and substitutions of the
twelfth amino acid (glycerine) in the gene product p 21.

used in these transfection experiments is already partly
transformed and presumably the transfected DNA from the bladder
carcinoma cells furnishes only one of the steps required in the
sequence of events leading to transformation. The NIH cells are
themselves responsible for the remaining step or steps.
Experimental data give some support for this interpretation.

The involvement of two events in one transformation is
indicated by B cell lymphoma induced by avian lymphoid leukosis
virus (LLV). In 80% of LLV lymphomas one step implies a viral
activation by means of the insertion of the viral transcriptional
promoter in the vicinity of the cellular oncogene c-myc. However
no integration of a viral DNA sequence was detected in NIH cells
transformed by B cell lymphoma DNA. A cellular DNA sequence not
of a viral origin must have been responsible for the transforming
property of the transfected DNA (12). LLV does not contain any
viral oncogene (11) and it also deviates from retroviruses, which
carry oncogenes, by being only weakly oncogenic and having a
longer latency period.

An important step towards an understanding of the genetic
changes leading to cancer has recently been provided by
transfection experiments with normal diploid fibroblasts. Such
cells cannot be transformed by single activated oncogenes, but

transformation is acquired by combined transfections. Thus these
cells will transform if they are given the ras-1 oncogene,
carrying the base substitution mutation described above, in
combination with either a virally promoted myc oncogene (20) or
genes coding for antigens from a DNA virus (20,36). Normal
fibroblasts can also be transformed by the ras-1 oncogene,
provided they are first made "immortal" by carcinogen treatment
(26). These results indicate that neoplastic transformation
involves at least two separate mutational events.

Chromosomal Rearrangements and Oncogenes

Cancers of blood cells generally involve translocation of
chromosomes and it has been suggested that these rearrangements
lead to an altered expression of cellular oncogenes (19). There
are several lines of evidence in support of this view. Oncogenes
have been located at the break points of translocations, such as
the ones associated with myeloblastic leukemia, Burkitt's lymphoma
and chronic myelogenous leukemia (the Philadelphia chromosome)
(35). It has been shown that antibody genes are located at the
break points of such translocations. The translocations bring the
oncogenes close to the highly active transcription promoter of the
antibody genes and the data indicate that the increased expression
of such oncogenes is responsible for the carcinogenic property of
cells carrying these translocations.

An artificial activation of an oncogene by means of an
increased gene expression has been acquired by combining a segment
of the human oncogene c-ras^H, with the long terminal repeat of the
viral counterpart v-ras^H, which functions as a transcriptional
promoter. The resulting gene exhibits an increased production of
the p 21 protein and causes oncogenic transformation (9). A
similar activation through the incorporation of a viral promoter
in the vicinity of an oncogene is the B lymphoma induced by avian
lymphoid leukosis virus (LLV) as mentioned above.

In the case of translocated c-myc oncogene, which is
involved in Burkitt's lymphoma as well as mouse plasmacytomas, it
has been found that the break point of the translocation is
located in the repeated DNA sequence of the heavy chain coding
segment of the antibody gene which, in turn, is involved in the
normal antibody rearrangements of DNA. It is therefore possible
that these translocations are in fact to some extent the
consequence of the gene rearrangements normally performed by
antibody genes.

Although several investigations have observed an increased
expression of oncogenes in connection with translocations, there
are also data showing a shorter than normal transcript of c-myc
oncogene involved with the translocation. This indicates an

intragenic rearrangement of the c-myc oncogene as well (38). That a qualitative change of the gene product can give rise to transformation is in accordance with the base substitution of the oncogene in the bladder carcinoma cells as discussed above.

Genetic Mechanisms of Initiation in Cancer

The correlation between carcinogenicity and mutagenicity has mostly been considered at the level of initiation of cancer, while promotion and progression of tumors are usually supposed to proceed by non-genetic mechanisms. It should however be realized that the division of carcinogenicity into discrete steps is based on operational definitions of these steps, but in reality clearcut boundaries between these steps may not occur. Furthermore, experimental data indicate that more than three steps are often involved in tumor formation and the initiation and promotion steps can be further subdivided.

The experimental data on oncogenes point to the fact that the transforming property does not result from one single genetic change but must involve some additional alterations. This fact makes it difficult to assign the genetic changes reported in oncogenes and leading to oncogenic transformation to one of the conventional steps in carcinogenicity. It is, however, likely that the genetic alterations of oncogenes concern the early stage of carcinogenicity and therefore essentially correspond to the initiation step.

The recent experimental data on oncogenes strongly support the notion that mutational alteration of the genetic material is involved in carcinogenicity. As discussed above, the results identify two types of mutational changes behind oncogenic transformation: point mutations and translocations. Two crucial questions can be raised in this connection - firstly, whether other genetic changes can be involved and secondly, whether non-genetic (epigenetic) mechanisms may also be operating in cancer initiation.

There is little direct experimental evidence that genetic mechanisms other than point mutations and translocations are involved in cancer initiation. Solid tumors seem, however, to involve deletions rather than translocations, but no data on the functional relationship between the deletion and the carcinogenicity, for instance concerning oncogenes, are yet available (22).

Retroviruses cause cancer by insertion of viral DNA sequences and reinsertion of processed oncogenes. A somewhat similar insertion of DNA sequences can occur by means of transposition of cellular elements, particularly movable

transposable elements. Such elements seem to have a wide
distribution in both lower and higher organisms and could possibly
play a role in carcinogenicity (4). The insertion of such an
element, resembling bacterial insertion sequence (IS element) has
been suggested for the activation of the c-mos oncogene in a
chemically induced myeloma in mouse (33). However, the presence
of a viral transcriptional promoting sequence, a long terminal
repeat, has been traced and offers an alternative explanation of a
viral activation of this oncogene (27). It should be pointed out
that transposition of transposable elements in higher organisms,
i.e. Drosophila (16, and Jernelöv and Ramel, unpublished) and IS
elements in bacteria are remarkably unaffected by treatment with
known mutagenic and carcinogenic agents. This fact speaks against
chemically induced cancer being caused by such transpositions,
although it does not rule out a connection between transposable
elements and spontaneous cancer.

It should finally be pointed out that cancer is linked to the
differentiation process and methylation of DNA is probably
involved in the regulation of transcription and cellular
differentiation. Holliday and Pugh (15) presented a model in
which methylation of DNA adjacent to structural genes determines
the transcription and differentiation, including a method for
"counting" of the number of cell divisions in different tissues.
Holliday (14) has extended this model to including the induction
of cancer by means of a loss of methylation in connection with
repair of DNA. A hypomethylation of DNA in accordance with this
hypothesis has in fact been reported in some human cancers (8).

Can Cancer be Initiated by Non-Genetic Mechanisms?

Although much experimental data strongly suggest that
mutational changes of DNA are involved in cancer initiation, this
does not rule out the possibility that some cancer forms may be
caused by epigenetic mechanisms. It is reasonable to assume that
the origin of cancer is related to the process of differentiation
in multicellular organisms and this process is probably an
epigenetic event. Initiation of cancer could, in accordance with
this view, occur as a result of an alteration of an epigenetic
control of differentiation. Some support for such a mechanism has
been suggested by transplantation experiments in both animals and
plants. Transplantation of nucleic from frog renal carcinomas
into enucleated frog eggs has thus given rise to normal tadpoles
(10). Reversion of malignancy to non-malignancy has also been
shown after transplantation of several types of tumors (30). It
may also be mentioned that transfection experiments with NIH cells
and DNA from various tumors give rise to transformation only in
about 50 percent of the cases (7).

There are other experimental results which are difficult to explain on the basis of mutational events. This applies to the increase of cancer in the offspring and further generations after treatment of male and female germ cells with carcinogenic agents. Nomura (28) has reported from large scale experiments that treatment of male mice with X rays gave an increased cancer incidence in the F_2 and F_3 offspring of such a high magnitude that it can only be explained by one of three possibilities: a very large number of genetic sites being involved; a remarkably high mutation frequency, or a mechanism other than conventional mutations. It may be pointed out that a similar unexpectedly high mutation frequency has been indicated for polygenic mutations (31). This may point to a relationship between this kind of "cancer mutation" and quantitative traits, controlled by polygenes.

More difficult to explain on the basis of mutations are the results reported by Nomura on urethane. Urethane treatment of male mice gave rise to an increased incidence of cancer in the offspring, in spite of the fact that urethane itself is usually negative in mutagenicity assays, for instance the Ames Salmonella test, and it was negative in dominant lethal tests after treatment of male mice in experiments by Nomura. However, Nomura found an increased incidence of sex-linked recessive lethals in Drosophila after treatment with urethane.

Although none of these results proves an epigenetic origin of human cancers, they do at least call for some cautiousness before one can accept a genetic mechanism behind initiation of cancer to the exclusion of alternative mechanisms.

Mechanism of Promotion

The second major step in carcinogenesis after initiation is tumor promotion. Promotion is usually demonstrated in skin painting experiments on mice (2). After application of an initiating carcinogen, repeated application of a promoting agent enhances the formation of tumors. Some chemicals can accomplish both these steps (complete carcinogens). The classic promoters are phorbol esters, which do not act as initiators, do not bind to DNA and are not active in most mutagenicity tests such as the Ames Salmonella/microsomal assays. Promoters, unlike initiators, have therefore been assumed to function by non-genetic mechanisms. The phorbol esters are highly active biologically and nearly a hundred different effects on biological systems have been reported. This bewildering number of biological effects has inevitably caused confusion, although certain promising lines of investigation have evolved, particularly concerning effects on cell membranes which are connected with several properties of promotion such as cell-cell communication, anchorage independence and lipid peroxidation.

Although phorbol esters and many other promoters do not seem
to act directly on the genetic material, some experimental
evidence nevertheless points to the possibility of an indirect
genetic action. Promoters like phorbol esters stimulate cell
proliferation and DNA synthesis (44). It is conceivable that the
capacity for error-free repair of DNA can be exhausted by this
effect and give rise to error-prone repair.

Of particular importance is the finding that promoters,
including phorbol esters, causes an oxidative burst and the
production of superoxide radicals and hydrogen peroxide by
polymorphonuclear leukocytes (PMN) (5,40). The radicals formed
can cause damage to DNA in these as well as in neighboring cells.
Birnboim (3) has shown that promoters may induce a high frequency
of strand breaks in DNA of human white cells. There was a
correlation between the tumor-promoting effect and the DNA damage.
The role of radicals in this process is indicated by the fact that
the DNA damage is prevented by superoxide dismutase and catalase,
enzymes which function as protections against radicals.

Phorbol esters have also been shown to cause aneuploidy in
yeast (29) and sister chromatid exchange (18). These effects have
been suggested as a mechanism for changing the expression of
recessive mutations produced by initiating agents. Recently it
has been shown that conversion of benign papillomas to maligant
carcinomas in mouse skin tumor assay is unaffected by the non-
mutagenic promoter TPA, but increased by initiators (mutagens)
(13).

It is thus possible that some effects on the genetic material
are also involved at some stage of tumor promotion.

Short-Term Mutagenicity Tests for Detection of Carcinogenic Chemicals

The vast majority of human cancers clearly are caused by
environmental factors and therefore supposedly are to a large
extent avoidable. The recognition of such carcinogenic factors
constitutes a major biological problem. The rapidly increasing
use of synthetic chemicals in addition to the exposure to natural
chemicals of unknown carcinogenic potential has made the search
for carcinogenic chemicals a formidable task. Conventional animal
cancer tests are too expensive, too time-consuming and often also
too insensitive for the large-scale testing required.
Epidemiological data of course are of high informative value, when
sufficient and reliable material can be obtained, but those cases
are indeed few. And, furthermore, particularly with respect to
the long latency period of cancer, by the time such data are
available the damage to the population is already done.

The correlation between carcinogenicity and mutagenicity has, however, made it possible to extrapolate data from mutagenicity tests to carcinogenicity. A large number of various short term mutagenicity tests have been applied for this purpose and a few particularly useful ones have reached a more general international adoption. The correlation between the most widely used test, the Ames Salmonella/microsomal assay, and carcinogenicity has usually been about 85 percent. The addition of other short-term tests has increased this correlation by decreasing false negative results, but at the same time the chance of false positive results must increase. The combination of test systems is, however, not primarily a mathematical or statistical problem but a biological one. The sequence of events from the time of exposure to a carcinogen to the actual development of tumors is a complex process involving major levels of complications--such as the biotransformation and induction of genetic alterations. These processes cannot be imitated by a single short-term test.

Considering biotransformation, it is a well-known fact that most carcinogens are not carcinogenic or mutagenic per se but require an activation, which is primarily performed by the mixed function oxidase system, occurring in many tissues but centralized in the liver cells. The use of liver microsomes for this metabolic activation in in vitro test systems such as the Salmonella/microsomal test is therefore highly appropriate. It must be realized, however, that the metabolic conversion of procarcinogens does not involve only this activation process but also deactivations by means of conjugation and other mechanisms. The balance between the activation and deactivation is, furthermore, often different in different tissues. The use of liver microsomes in mutagenicity testing tends to overemphasize the activation part of the process and therefore makes the system more responsive to indirect carcinogens. While conjugation with glutathione constitutes a deactivation pathway for most chemicals, it functions as an efficient activation to potent carcinogens with a few chemicals, for instance 1,2-dichlorethane (32).

The construction of short-term test systems for mutagenicity has focused on the testing of electrophilic compounds and metabolites, which act as mutagens by covalent binding to DNA. It is, however, likely than an indirect action of chemicals by means of lipid peroxidation and the formation of oxygen radicals play an important role for both initiation and promotion of cancer. New Salmonella strains, which are sensitive to peroxides, have been synthesized by Ames' group (21) and will probably be of great value for the detection of mutagenic and carcinogenic chemicals acting by means of radical formations.

At the level of mutations the test systems are constructed to respond to more or less specific lesions--point mutations, various

chromosome aberrations, transpositions, deviating chromosome
numbers etc. The analyses of oncogenes have made it clear that
genetic changes involved in cancer are not of a uniform type but
include point mutations, translocations, deletions, and probably
other types of lesions as well. This fact must be taken into
account when using short-term mutagenicity tests for the screening
of carcinogenicity.

It should finally be emphasized that in many cases the
sensitivity of the mutagenicity test systems has been increased by
manipulating the DNA repair, particularly by the elimination of
error-free repair. In vivo the repair system varies between
different species, tissues and individuals and quantitative
extrapolation from in vitro tests is therefore difficult.

CONCLUSION

In conclusion, it can be stated that the short-term
mutagenicity tests have been constructed with the aim of obtaining
a maximum sensitivity towards both the metabolic activation and
the genetic end points. Such a high susceptibility for
carcinogenic chemicals may, however, not be unrealistic in vivo in
some tissues or under some specific circumstances. From a
qualitative point of view, an extrapolation from mutagenicity
tests to carcinogenicity is therefore justified.

When it comes to the quantitative relationship between
mutagenicity and carcinogenicity and the capacity of mutagenicity
tests to predict carcinogenicity potency the situation is
different. A correlation between mutagenicity and carcinogenicity
potency was reported by Meselson and Russel (23) for a sample of
various chemicals and a high correlation was reported for 26
nitrosamines in a mammalian cell assay with rat hepatocytes as a
metabolizing system (17). In many other cases, however, the
correlation has been poor. A close correlation between
mutagenicity and carcinogenicity in this respect cannot be
expected if one considers the fact that cancer is at least a two-
step event involving initiation and promotion. Unless the
mechanisms of initiation and promotion are the same, which is
clearly not the case, a quantitative correlation cannot be
expected between mutagenicity and carcinogenicity, as the
mutagenicity assays primarily reflect the initiation but not the
promotion step.

PERSPECTIVES

The hypothesis that carcinogenicity involves a mutagenic
event has essentially been based on empirical associations between

mutagenicity and carcinogenicity. Recent developments within the field of oncogenes have, however, given a powerful and more direct support for the concept that there is also a causal relationship between mutations and cancer. This also demonstrates that the development and use of mutagenicity assays to predict carcinogenicity is justified and that further development along these lines should be given high priority. The prospect for the application of a more accurate mutagenicity testing protocol in order to predict carcinogenicity will to a large extent depend on further elucidation of the mechanisms of cancer induction. The exciting new techniques of analyzing the molecular mechanisms behind oncogene activation by means of transfection assays and DNA sequence analysis are indeed promising in this respect. At this time there is a very rapid development within this area and much important information can be expected in the near future.

REFERENCES

1. Avery, O.T., C.M. MacLeod, and M. McCarty (1944) Studies on the chemical nature of the substance inducing transformation of pneumococcal types. I. Induction of transformation by a deoxyribonucleic acid fraction from pneumococcus type III. J. Exp. Med. 79:137-158.
2. Berenblum, I. (1975) Sequential aspects of chemical carcinogenesis: Skin. In Cancer: A Comprehensive Treatise, F. F. Becker, Ed., Plenum Press, Vol. 1, pp. 323-344.
3. Birnboim, H.C. (1981) DNA strand breakage in human leukocytes exposed to a tumor promoter, phorbol myristate acetate. Science 215:1247-1249.
4. Cairns, J. (1981) The origin of human cancer. Nature 289:359-357.
5. Cerutti, P. (1978) Repairable damage in DNA. In DNA Repair Mechanisms, P.H. Hanawalt, E. Friedberg and C. F. Fox, Eds., Academic Press, pp 1-14.
6. Cleaver, J.E. and D. Bootsma (1975) Xeroderma pigmentosum - Biochemical and genetic characteristics. Ann. Rev. Genet. 9:19-38.
7. Cooper, G.M. (1982) Cellular transforming genes. Science 218:801-806.
8. Feinberg, A.P. and B. Vogelstein (1983) Hypomethylation distinguishes genes of some human cancers from their normal counterparts. Nature 301:89-92.
9. DeFeo, D., M.A. Gonda, H.A. Young, E.H. Chang, D.R. Lowy, E.M. Scolnick and R.W. Ellis (1981) Analysis of two divergent rat genomic clones homologous to the transforming gene of Harvey murine sarcoma virus. Proc. Natl. Acad. Sci. USA 78:3328-3332.
10. Gurdon, J.B. (1974) The Control of Gene Expression in Animal Development. Harvard Univ. Press.

11. Hanafusa, H.   (1977)   Cell transformation by RNA viruses.   In
     Comprehensive Virology, Vol. 10, H. Fraenkel-Conrat and
     R.R. Wagner, Eds., Plenum Press, pp. 401-483.

12. Hayward, W.S., B.G. Neel and S.M. Astrin   (1981)   Activation
     of a cellular onc gene by promoter insertion in ALV-induced
     lymphoid leukosis. Nature 290: 475-480.

13. Hennings, H., R. Shores, M.L. Wenk, E.F. Spangler, R. Tarone
     and S.H. Yuspa (1983) Malignant conversion of mouse skin
     tumours is increased by tumour initiators and unaffected by
     tumour promoters. Nature 304:67-69.

14. Holliday, R.   (1979)   A new theory of carcinogenesis. Brit.
     J. Cancer 40:513-522.

15. Holliday, R. and J.E. Pugh   (1975)   DNA modification
     mechanisms and gene activity during development. Science
     187:226-232.

16. Ising, G. and C. Ramel   (1976)   Transposition of an X-
     chromosome segment in Drosophila.   In The Genetics and
     Biology of Drosophila, M. Ashburner and E. Novitski, Eds.,
     Academic Press, Vol. 1b, pp. 947-954.

17. Jones, C.A., P.J. Marlino, W. Lijinsky and E. Huberman (1981)
     The relationship between carcinogenicity and mutagenicity of
     nitrosamines in a hepatocyte mediated mutagenicity assay.
     Carcinogenesis 2:1075-1077.

18. Kinsella, A.R. and M. Radman   (1978)   Tumor promoter induces
     sister chromatid exchanges: relevance to mechanism of
     carcinogenesis. Proc. Natl. Acad. Sci. USA 75:6149-6153.

19. Klein, G.   (1981)   The role of gene dosage and genetic
     transposition in carcinogenesis. Nature 294:313.

20. Land, H., L.F. Parada and R.A. Weinberg   (1983)   Tumorigenic
     conversion of primary embryo fibroblasts requires at least
     two cooperating oncogenes. Nature 304:648-651.

21. Levin, D.E., M. Hollstein, M.F. Christman, E.A. Schwiers, and
     B.A. Ames (1982) A new Salmonella tester strain (TA102) with
     A-T base pairs at the site of mutation detects oxidative
     mutagens. Proc. Natl. Acad. Sci. USA 79:7445-7449.

22. Marx, J.L.   (1982)   The case of the misplaced gene. Science
     218:983-985.

23. Meselson, M. and K. Russel   (1977)   Comparisons of
     carcinogenic and mutagenic potency. In Origins and Human
     Cancer, H.H. Hiatt, J.D. Watson and J.A. Weinstein, Eds.,
     Cold Spring Harbor, Vol. 4, pp. 1473-1481.

24. Miller, E.C. and C.M. Miller   (1971)   The mutagenicity of
     chemical carcinogens: correlations, problems and
     interpretations. In Chemical Mutagens: Principles and
     Methods for Their Detection, Vol. 1, A. Hollaender, Ed.,
     Plenum Press, pp. 83-119.

25. Möller, G. and E. Möller   (1975)   Consideration of some
     current concepts in cancer research. J. Natl. Cancer
     Inst. 55:755-759.

26. Newbold, R.F. and R.W. Overell (1983) Fibroblast immortality is a prerequisite for transformation by EJ c-Ha-ras oncogene. Nature 304:648-651.

27. Newmark, P. (1983) What has moved into c-mos? Nature 301:196.

28. Nomura, T. (1982) Parental exposure to X rays and chemicals induces heritable tumours and anomalies in mice. Nature 296:575-577.

29. Parry, J.M., E.M. Parry and J.C. Barrett (1981) Tumour promoters induce mitotic aneuploidy in yeast. Nature 294:363-365.

30. Pierce, G.B. (1970) Differentiation of normal and malignant cells. Fed. Proc. 29:1248-1254.

31. Ramel, C. (1982) Polygenic effects and genetic changes affecting quantitative traits. Mutation Res. 114:107-116.

32. Rannug, W., A. Sundvall and C. Ramel (1978) The mutagenic effect of 1,2-dichloroethane on Salmonella typhimurium. I. Activation through conjugation with glutathione in vitro. Chem.-Biol. Interactions 20:1-16.

33. Rechari, G., D. Givol and E. Canaani (1982) Activation of cellular oncogene by DNA rearrangement: possible involvement of an IS-like element. Nature 300:607-611.

34. Reddy, E.P., R.K. Reynolds, E. Santos and M. Barbacid (1982) A point mutation is responsible for the acquisition of transforming properties by the T24 human bladder carcinoma oncogene. Nature 300:149-152.

35. Rowley, J.D. (1983) Human oncogene locations and chromosome aberrations. Nature 301:290-291.

36. Ruley, H.E. (1983) Adenovirus early region 1A enables viral and cellular transformation genes to transform primary cells in culture. Nature 304:648-651.

37. Setlow, R.B. and R.W. Hart (1975) Direct evidence that changed DNA results in neoplastic transformation - A fish story. In Proc. Fifth Int. Congr. Radiation Res., O.F. Nygaard, H.I. Adler and W.K. Sinclair, Eds., Academic Press, pp 879-884.

38. Shen-Ong, G.L.C., E.J. Keath, S.P. Piccoli and M.D. Cole (1982) Novel myc oncogene RNA from abortive immunoglobulin-gene recombination in mouse plasmacytomas. Cell 31:443-452.

39. Shih, C., B.-Z. Shilo, M.P. Goldfarb, A. Dannenberg and R.A. Weinberg (1979) Passage of phenotypes of chemically transformed cells via transfection of DNA chromatin. Proc. Natl. Acad. Sci. USA 76:5714-5718.

40. Slaga, T.J., A.J.P. Klein-Szanto, L.L. Triplett, L.P. Yotti and J.E. Trosko (1981) Skin tumor-promoting activity of benzoyl peroxide, a widely used free radical-generating compound. Science 213:1023-1025.

41. Tabin, C.J., S.M. Bradley, C.I. Bargmann, R.A. Weinberg, A.G. Papageorge, E.M. Scolnick, R. Dahr, D.R. Lowy and

E.H. Chang   (1982)   Mechanism of activation of a human
oncogene.   Nature 300:143-149.

42.   Taparowsky, E., Y. Suard, O. Fasano, K. Schimiza, M. Goldfarb
and M. Wigler   (1982)   Mechanism of activation of a human
oncogene.   Nature 300:762-765.

43.   Tsuchida, N., T. Ryder and E. Ohtsubo   (1982)   Activation of
the T24 bladder carcinoma transforming gene is limited to a
single amino acid change.   Science 217:937-939.

44.   Yuspa, S.H., U. Lichti, T. Ben, E. Patterson, H. Hennings,
T.J. Slaga, N. Colburn and W. Kelsey   (1976)   Phorbol esters
stimulate DNA synthesis and ornithine decarboxylase activity
in mouse epidermal cell cultures.   Nature 262:402-404.

THEORETICAL BASIS OF MUTAGENESIS

F.E. Würgler

Institute of Toxicology
Swiss Federal Institute of Technology
and University of Zürich, Switzerland

SUMMARY

Mutations are changes in the hereditary material which are propagated through successive generations in cells and whole organisms. Mutations can occur spontaneously or under the influence of physical and chemical mutagenic agents. Depending on the type and size of the change in the genome we distinguish: point or gene mutations (base pair substitutions, frameshift mutations, small deletions), chromosome mutations or aberrations (e.g., large deletions, translocations), aneuploidy (trisomy, monosomy, etc.) and genome mutations (polyploidy). All these types of mutations are known to occur in man. Their phenotypic consequences can, even within one class of mutation, vary from lethality, malformations, and metabolic disorders to mild or even hardly detectable changes. Estimates suggest that a minimum of 10-11% of all live born will manifest at birth, or during development, or later in adulthood, a very wide range of serious genetic defects.

The mechanisms by which mutations are induced by mutagens are only poorly understood. A number of model systems (e.g., bacteria, fungi, insects, mammalian cells in culture) are suitable to study particular aspects of mutagenesis in prokaryotic and eukaryotic (nucleosomal) chromosomes.

In wild-type cells the induced frequency of a given type of mutation is the result of a complex interaction between normal DNA replication, error-free and error-prone DNA repair activities, and recombinational events. In some cases, in which the genetic control of the crucial steps is known (e.g., the rec-lex-dependent error-prone repair in E. coli), detailed and experimentally

testable models could be worked out. In other cases (e.g., the
induction of sister chromatid exchanges or chromosomal aberrations
in eukaryotes) the models are less detailed. Some reasons for
this are: the low number of suitable mutants available; multiple
types of mutations induced by one particular mutagen; and
difficulty in identifying the premutational lesion(s),
particularly because of the insensitivity of the available
biochemical assays. The complexity of the chemical mutagenesis is
also reflected in the observation that differnt mechanisms can
lead to the same type of mutation, e.g., the induction of
chromosome aberrations is dependent on the normal S-phase DNA-
synthesis with some agents but not with others. In addition, not
only primary DNA damage (such as strand breaks, covalently bound
mutagens, intercalations) but also non-DNA damage (e.g., effects
of metals on the enzymes of DNA-synthesis, or interaction of
spindle poisons with the polymerization of microtubules, or
nucleotide pool imbalances) can lead to mutations.

A discussion of some models of the productions of different
types of mutation illustrates our present knowledge on basic
mutation mechanisms in pro- and eukaryotes. The importance of the
knowledge of these mechanisms for genetic toxicology has to be
stressed. We can learn more about the types of lesions for which
we should look in mutagenicity screening tests, and we can
approach the construction of better defined and more sensitive
test systems and test batteries.

INTRODUCTION

Mutations can be detected as a sudden, heritable change.
The term "mutation" was coined in 1901 by de Vries based on
his studies with the plant Oenothera (for definitions, detailed
descriptions and explanations of the use of genetic terms see
138). The basic idea was that a mutation is a simple change of a
hereditary unit (gene). Due to this change (of unknown nature)
the originally present gene is transformed into its mutant form.
The original state of the gene is also called the allele a1 of the
particular gene and a mutation changed it into allele a2.

Although this is true for many of the mutations we shall
discuss, in the particular case of Oenothera the experimentally
observed phenomenon was misleading. Subsequent cytological
studies revealed a very special type of chromosome segregation in
combination with recombinational events which lead to the frequent
phenotypic expression of pre-existing (recessive) mutations. Even
if the basic idea of mutation from wild type to a mutated allele
was an extremely successful idea, the historical paradox is
remarkable. A central term of modern genetics was initially
defined based on a nonconclusive experiment, but was later applied

to the intuitionally correctly assumed basic phenomenon and developed into the starting point of a whole branch of modern genetics: mutation research. Today, we are much more aware of the problems connected with generalizations which are based on individual experiments. We still face the difficult problems inherent in the exercises to extrapolate from mutational changes in prokaryotic DNA to the comparable changes in eukaryotic DNA, and in particular human DNA. Only recently has it become possible to sequence DNA from different sources.

Although basic research in this area is primarily concerned with the study of the signals needed and used to regulate different processes of DNA metabolism such as replication. Transcription and recombination, more and more mutant DNA sequences are becoming known. Hopefully, the analysis of the basic mutation mechanisms will help us to understand similarities and discrepancies between the different systems in which mutations are studied and will lead us to reasonable extrapolations of data from experimental test systems to the human genome. This is particularly important if we are to try to estimate the impact of our present day environment (air, water, food) and our life style (eating, drinking, smoking, etc.) on the stability of the genetic material in germ cells as well as somatic cells.

The Organization of the Genetic Material

We distinguish three types of organisms:

Viruses. These are organisms which need living cells for their reproduction. A special type of viruses are the bacteriophages, the viruses of bacteria. They are often simply called "phages." Most viruses have a protein coat which protects the genetic material, DNA or RNA, from external influences.

Prokaryotes. These are the bacteria and the blue algae. They have a nuclear equivalent made up of double stranded DNA and not a true cell nucleus.

Eukaryotes. These are organisms with real cell nuclei which are enclosed in a nuclear envelope. Broadly speaking plants, animals, and humans are eukaryotic organisms.

Phage Chromosomes

Among the phages studied in most detail are the T-phages which can multiply in the bacterium Escherichia coli. The chromosome of a T-phage consists of a naked DNA double-strand molecule. The DNA is not associated with proteins. Of special interest is the arrangement of the genes on chromosomes of different phages. The chromosomes of the phage T2 and T4 are longer than one complete phage-genome. The chromosomes of these phages are terminal redundant and circular permutated. For details see Hayes (60).

Bacterial Chromosomes

Bacteria have naked ring chromosomes of double stranded DNA. Usually the DNA is attached at one or more sites to the cell membrane in the vicinity of the nucleoid region which contains the DNA but very few plasma components. Electron micrographs of E. coli chromosomes (26) show the condensed packing of DNA in loops (10-80 per chromosome) and supercoils. It could be shown that these areas of condensed chromosomes contain the DNA, the nascent mRNA chains and the enzyme for the RNA synthesis (RNA polymerase), but no ribosomes. This leads to the conclusion that in bacteria the protein synthesis does not take place in the immediate neighborhood of the coding DNA.

Chromosomes of Eukaryotes

The major amount of DNA of eukaryotes is contained in the chromatin of the cell nucleus. The extranuclear DNA is naked, circular double-stranded DNA in the mitochrondria and chloroplasts. This DNA resembles bacterial chromosomes, but will not be discussed any further.

During the nuclear cycle (mitotic and meiotic) the chromatin is replicated and then condensed into chromosomes in order to be distributed to daughter cells. Within every living species the condensed elements, the chromosomes, which become visible at the time of cell division are characteristic with respect to number, size, shape and the position of the centromere. During the anaphase, in which the replicated chromosomes are distributed to the daughter cells, spindle fibers are attached to the centromeres. They are, therefore, also called spindle fiber attachment sites (SFA). The total of the chromosomes of an eukaryotic cell is called the karyotype.

In animals the karyotype in the two sexes is usually different. In mammals, including man, they differ in the sex chromosomes (female XX, male XY). The autosomes (the non-sex-specific chromosomes) are constant within a species, but differ from species to species. The normal human karyotype consists of 46 chromosomes, 22 pairs of homologous autosomes and one pair of sex chromosomes. This is a diploid karyotype made up to two haploid chromosomes sets. Modern banding techniques (168, 189, 190) allow the identification of every single chromosome and even small sections of individual chromosomes.

Today it is assumed that every chromosome contains a single, continuous doublestranded DNA molecule (23, 72, 128, 187). The

DNA molecule is associated with basic proteins (histones) and acid proteins (non-histone). The arrangement of DNA, histones and non-histones is characteristic for all eukaryotic chromosomes. There are four histones (lysine rich histones H1, H2A and H2B; arginine rich histone H4). The histone genes are very conservative in evolution because they have a fundamental and universal function in chromosome structure. The non-histone proteins, on the other hand, are numerous. They are heterogeneous and include enzymes responsible for DNA metabolism such as DNA polymerases for replication and RNA polymerases for transcription. There are usually more than 100 different non-histone proteins present in chromatin.

The histones play a central role in the packing of chromatin by which, for example, in condensated chromosomes the DNA is packed to about one hundredth of its lengths. Of the multilevel folding the lowest levels are best understood. The nucleosome structures were discovered in electron micrographs of spread chromatin (for a review see 105). A nucleosome consists of two each of the following histones: H2A, H2B, H3 and H4. This histone core which has the shape of a short cylinder is associated with 140 DNA base pairs. The DNA is wound outside the nucleosome so that a DNA supercoil results. Nucleosomes are connected by DNA linkers and are thought to form a 25-30 nm thick chromatin fiber (45, 140, 170). Histone H1, which is present by about one H1 molecule per nucleosome, plays a fundamental role in stabilizing the nucleosomes and in maintaining the higher order structures of the chromatin fibers (19, 81, 169, 170). It is still unclear how the chromatin fiber is folded and organized to form metaphase chromosomes. Several models have been proposed. In the "folded fiber" model, Du Praw (34-36) proposes that the 25 nm fiber is repeatedly folded back on itself to make up the body of the chromatid. In another model, a protein backbone ("scaffold") in the axis of the chromosome is postulated to which the DNA would be attached to form loops of chromatin fibers (126, 163). Such loops were observed to be radially oriented (4, 103). In a third model the attachment points for such loops are not necessarily located along the axis of the chromosomes, but are thought to be distributed all over the chromosome ("chromomere loop" model: 24, 123). A decision as to which model best describes the actual structure of the condensated metaphase chromosomes has not been reached (81, 123).

TYPES OF MUTATION

As a basis for a classification of mutations we may choose the type and the size of the change in the genetic material. We will restrict our discussion to those cases in which the genetic information is contained in double stranded DNA.

Table 1. Types of Mutation

Class	Types	Description
Point or gene mutation	Base pair substitution	Transition
		Transversion
	Frameshift	Addition
		Deletion
	Block mutation	e.g., deletion of several adjacent base pairs
Chromosome mutations or chromosome aberrations	Deletion	Loss of a large DNA (chromosome) section
	Duplication	A given chromosome section is contained twice in a haploid chromosome set
	Transposition	A chromosome segment is inserted into or attached to a new chromosome region
	Reciprocal translocation	An exchange between terminal segments of nonhomologous chromosomes
Aneuploidy	Monosomy	Only one copy of one particular chromosome exists in an otherwise diploid genome
	Trisomy	Three copies of a particular chromosome exist in otherwise diploid genome
Polyploidy or genome mutations	Triploid	A cell nucleus contains three haploid chromosome sets
	Tetraploid	A cell nucleus contains four haploid chromosome sets

We first look only at so-called forward mutations. These are changes in the genome from the normal (wild type, the type predominating in wild populations) genetic information to a new, mutant information.

Table 1 gives a rough overview of the most important types of mutation known to occur in prokaryotic and eukaryotic organisms. Historically two main classes of mutations were distinguished: (i) those visible in the light microscope (chromosome aberrations, aneuploidy and genome mutations), and (ii) those not visible in the light microscope (point mutations). Even if the definition of the borderline cases poses some problems today, this simple "morphological" criterion turned out to be of basic importance: visibility in the light microscope was found to correlate also with certain aspects of the mechanisms leading to particular types of mutation. As we will learn in this chapter point mutations, chromosome aberrations and aneuploidy are mutations resulting from three distinctly different mutational processes.

Among the simple base pair substitutions we have transitions and transversions (49). In a transition the purine in the original base pair is replaced by the other purine.   Two types of transitions are possible: AT to GC and GC to AT. In a transversion the original base pair is replaced in such a way that the purine is replaced by a pyrimidine and the pyrimidine by a purine. The new bases again form a complementary base pair. Transitions and transversions may result in a change of the coding property of the altered DNA section.

Frameshift mutations consist, in the simplest case, of changes which are one base pair in size. They may be additions (insertion of one or more adjacent base pairs) or deletions (loss of one or more adjacent base pairs). The stable hereditary change will interfere neither with DNA replication nor with transcription. The genetic consequences will show up only at the time of translation on the ribosome. The altered reading frame of the triplet code which starts at the mutant site will lead to a misinterpretation of the base sequence.

The simple one base pair deletion (frame shift) mutation is the smallest of a whole continuous series of deletion mutations. As the size of the deleted section increases block mutations will occur or, in eukaryotes, chromosomal deletions which are visible in the light microscope and which may involve anything from a few to a great many genes.

The larger intragenic deletions (block mutations) may, operationally, be separated from the smaller frameshift mutations of the deletion type by their nonrevertability (see below).

Chromosome aberrations are, in the broadest sense, all types of changes in the chromosome structure and chromosome number. A standard karyotype is used for comparison and identification of the type of change. Some of the most important groups are listed in Table 1. The group of the chromosome mutations contains all structural changes involving the gain, loss or relocation of chromosome segments. The group of aneuploidy contains genotypes having one or more whole chromosomes of a euploid complement absent from or in addition to that complement. A complete chromosome set is called euploid. Genome mutations are changes in the number of whole chromosome sets. A polyploid cell may have thre (triploid), four (tetraploid), five (pentaploid), or more complete chromosome sets instead of two (diploid) or one (haploid).

CONSEQUENCES OF MUTATIONS

Any given type of mutation may lead to a wide spectrum of consequences. The severity of the phenotypic change occasionally reflects the extent of a change in the genome. Phenotypic consequences may vary from lethality to severe metabolic disturbances, functional inabilities of cells, organs or organisms, or to mild or hardly detectable changes. In some situations silent mutations without any phenotypic consequence may occur.

If we first look at the point or gene mutations we can expect quite different consequences on the cellular level if we assume that a mutation occured in a structural gene (e.g., one coding for an enzyme), in a RNA gene (e.g., coding for a tRNA or a ribosomal RNA) or a regulatory gene. One of the decisive factors in a expressibility of newly arisen mutations is the uniqueness or redundancy of the genetic information. In the case of a diploid cell with two copies of most of the unique genes, a mutation is only phenotypically expressed if the new allele is codominant, or if both copies of the gene are expressed independently (e.g., co-dominance as in the case of hemoglobin genes), or if only one copy of the gene is present (e.g. in lower mammals and man the X-linked genes in males).

Base Pair Substitutions

The basic consequences of a base-pair substitution is usually a change in the primary polypeptide structure. The type of change can be deduced from the knowledge of the genetic code:

(i)   A triplet coding for a particular amino acid may be changed into a new triplet now coding for another amino acid: missense mutation. An example is the mutation from human hemoglobin A to the sickle cell hemoglobin S where at position 7 of the beta chain glutamic acid is replaced by valine (for details see 174). There is a whole group of transitions which can often not be recognized phenotypically, e.g., among non-polar amino acids (alanine, glycine, isoleucine, leucine, threonine, valine), which do not significantly alter the tertiary folding of the corresponding enzyme (109).

Interesting cases, from an evolutionary point of view, may be found among proteins from more or less related species for which the amino acid sequences were determined systematically, e.g., the cytochrome c or alpha and beta chains of hemoglobins.

(ii)   In some cases a triplet coding for an amino acid may be changed into a stop codon due to an exchange of a coding base (nonsense mutation). An example is the ATC triplet (amber) which might result from a transversion taking place in the first position of a GTC triplet (coding for glutamic acid). An amber mutation will, during transcription of the messenger RNA, lead to a premature termination of the polypeptide chain as soon as this codon is read on the ribosome. This phenomenon has been studied for a number of amber sites known to occur in the gene for the head protein in different strains of T4 phages. The further away from the origin of the gene the amber mutation occurred, the longer was the polypeptide fragment synthesized (144). This experiment proved the co-linearity of the nucleotide sequence (genetic map) and the amino acid sequence.

(iii)   Due to the degeneracy of the code, the possibility exists that a triplet coding for a particular amino acid mutates to a synonymous triplet coding for the same amino acid. This will result in a silent mutation without a phenotypic effect. It is possible that silent mutations may lead to a phenotypic expression after some additional cellular events. One can imagine a crossing-over event between particular synonymous triplets which may lead to a missense mutation. Such a situation may be encountered in a heterozygote for GCA and TCC. These two triplets are coding for arginine. An intra-codon crossing-over may lead to GCC and TCA. The codon GCC still codes for arginine, but TCA will code for serine.

Frameshift Mutations

With both types of frameshift mutations, i.e., deletions and additions, the mutant site displaces the starting point for the interpretation of triplets in the messenger RNA. The mRNA is misread distal to the mutation.   Codons which result after a frameshift mutation fall into three categories: (i) sense codons, which are translated the same as before; (ii) missense codons, which code for a different amino acid; (iii) nonsense codons, which doe for no amino acid.

In general we expect that after a certain number of wrong and occasionally correct amino acids a nonsense codon will appear by chance and stop the synthesis of the polypeptide.   We therefore expect that a large fraction of frameshift mutations leads to nonfunctional polypeptide products and thus eliminates the biological function of the mutant gene product.   A special case is represented by those mutants where three or a multiple of three pase pairs are changed.   Here, additional amino acids may be added to the polypeptide or one or more amino acids may be deleted.   But distal to this change the original amino acid sequence is retained.   From an evolutionary point of view, these are very interesting mutations.

Block Mutations

With the term "block-mutations" we designate a particular group of deletion and insertion mutations. We can easily define the deletion mutations because these are larger than the frameshift deletion mutations. They are operationally separated from them by genetic tests: they are mutants which do not revert (see below) to the original form and do not recombine in genetic crosses with two or more point mutations which do revert and recombine.   As opposed to chromosome deletions, the block mutations cannot be identified in the light microscope.   The genetic consequences of deletion mutations are primarily due to the loss of genetic information.   The function of one or several genes may be lost. The consequences of large insertions, e.g., of a transposon, is the interruption of the genetic information. Block mutations may be inherited as recessive mutations.

In E. coli a limited number fo specific DNA sequences can insert themselves into different sites in the bacterial genome to shut off gene activity.   To date, repeated examples of four such insertion sequences (IS) have been documented: These have been labeled IS1 and IS4. They contain, respectively, about 800, 1300, 1200 and 1400 base pairs. Insertion can occur at many sites but is not random, and there are "hot spots" for IS induced mutations. The presence of some IS elements can lead to a high frequency of spontaneous deletions, which begin at the site of insertion. Interestingly enough, similar phenomena have been found in maize and Drosophila (for a review see 18). The study of the insertion sequences has become a very active field of research. For example, in Drosophila melanogaster more than 50 families of IS are present in the genome without obvious phenotypic effects (28) and have been shown to be associated with high frequencies of spontaneous deletions involving the two X chromosome gene loci, yellow body color (y) and while eye color (w). It has been postulated that other types of spontaneous chromosomal aberrations, translocations, inversions etc., that occur infrequently, may also be intimately associated with IS. The numerical abundance of IS in the Drosophila genome also argues for their role in spontaneous genetic events (58).

Russell (141) pointed out in 1964 how the coat-color mosiacism in mice exhibited by certain alleles of the pearl, albino, and aguti loci would be explained in terms of the controlling elements of maize. Experimental evidence to this end awaited the demonstration by Jenkins et al. (71) that a specific ustable allele, the dilute (d) coat-color mutation of DBA/2J mice, was associated with the (variable) site of integration of a ecotropic MuLV genome. In a recent general review, Whitney and Lamourex (182) systematically developed the parallelism between the phenotypic characteristics of the presence of transposons in maize and Drosophila and the behavior of the three mouse alleles mentioned earlier.

Chromosome Aberrations

The different types of chromosome aberrations can have quite different genetic consequences.

Deletion. The genetic consequences of deletions are primarily due to the loss of genetic information and secondarily to quantitative changes in the genotype as well as changes of the gene balance (138). Depending upon the size of the deleted chromosome segment, a deletion may act as a recessive or dominant mutation, often with a lethal phenotype.

In human chromosomes deletions containing up to 20 genes may be invisible in the light microscope (78). In the very special case of the salivary gland chromosomes of Drosophila one-band deletions containing only one or a few genes can be seen (89). A classical example of the genetic consequences of a deletion is the "cri du chat" syndrome in man, resulting from a heterozygous deletion of the short arm of chromosome 5. This was the first deletion syndrome discovered in man my Lejeune et al. (90) based on cytogenetic studies.

Duplication. The genetic consequences of duplications depend on the type and amount of genetic information involved and the change in the gene balance effected by them. If homo- and heterozygous, they may cause an increase or decrease in the viability of their carriers and, in extreme cases, they may act as lethals. In evolution small duplications may provide a basis for the mutational differentiation of genetic material (122).

The classical case of a duplication with a dominant phenotypic expression is the small eye mutation (bar eye, Bar) on the X-chromosome of <u>Drosophila melanogaster</u> (14). The consequences of up to 8 copies of the Bar region within an otherwise diploid genome could be studied (136). Extensive studies showed that the phenotypic expression is the result of a position effect. It was recently shown in <u>Drosophila kikkawai</u> (6) that long tandem duplications can also occasionally occur.

In humans, duplications are not very common, but they can result as secondary aberrations from the meiotic chromosome segregation in translocation heterozygotes or inversion heterozygotes (with crossing-over within a pericentric inversion) which leads to complementary duplication-deficiencies.

<u>Inversion</u>. Provided that no detrimental changes occur at the break points, inversions are without phenotypic consequences in somatic cells in homo- as well as in heterozygous condition. Complications arise in meiosis in conjunction with heterozygous inversions, i.e., if a structurally normal homologous chromosome is present as the homologous partner of the inversion chromosome. The meiotic pairing behavior of the inversion chromosome and of its structurally normal partner depends on the length of the inversion chromosome. The meiotic pairing behavior of the inversion chromosome and of its structurally normal partner depends on the length of the inversion and the longitudinal relationship of the inverted and uninverted chromosome segments. If the inversion is long, chromosome pairing involves the formation of a retograde loop in the homologous, structurally unchanged chromosome; the inversion chromosome affixes itself to its partner in such a way that homologous loci pair with one another. If the inverted segment is so short that loop formation is not possible, either the inversion segment remains unpaired or pairing of nonhomologous segments can take place. If the inversion segment is very long, it can pair without loop formation and the uninverted terminal segments then remain unpaired.

Crossing-over and chiasma formation within and outside the inverted segment gives rise to secondary structural changes (duplications and deletions), depending on (i) the inversion-type, (ii) the number of chiasmata, and (iii) the localization of the chiasmata. Such structural changes in turn lead to meiotic products with unbalanced sets of chromosomes.

In the case of a heterozygous, paracentric inversion, a chiasma within the loop results in the formation of a dicentric chromatid and an acentric fragment in which the distal chromosome segments lacking in the dicentric chromatid are doubled.   The chromatids not involved in crossing-over remain unchanged, i.e., one possesses the gene order of the inverted chromosome, the other that of the normal partner.   During anaphase I, the dicentric chromatid is stretched between the centromeres, which has migrated to the opposite poles, and thus forms a chromatid bridge.   If two chiasmata are formed within the inversion loop, a double bridge and two fragments may be produced in anaphase.

In the case of heterozygous, pericentric inverstions, chiasma formation does not result in the production of a dicentric chromatids, fragments or loop-univalents.   However, the crossing-over chromatids differ owing to the formation of single or diagonal chiasmata in the inverstion loop through duplication and deletion of the terminal segments.

In the germ line inversions have two important effects: (i) aberrant meiotic products, which are inviable, may be formed and will lead to a reduced fertility of inversion heterozygotes.   (ii) Inversions often affect crossing-over frequencies.   In this context a distinction is made between (a) the effects within the inversion bivalents (a reduced crossing-over frequency within the inversion and outside near the break points), and (b) interchromosomal effects (an increased crossing-over frequency in the non-inverted heterologous bivalents, the so-called "Schultz-Redfield effect").   Inversions are of evolutionary importance; since because of the extensive crossing-over reduction within heterozygous inversions balanced complexes ("super genes") may be built up.   If the gene complexes prove to be adaptable to certain environmental conditions the inversion heterozygotes carrying them can be the starting point for the development of new local populations.

Translocation.   In many organisms, translocations result in specific patterns of genetic segregation, such as unusual linkages and reduced meiotic recombination, reduced viability of a fraction of the offspring, familial patterns of trisomy, etc.

It is not possible to present here the manifold possibilities of different types of translocations (details are found in 138). Two important types of translocation will be discussed: (i) Reciprocal symmetrical translocations (interchanges).   These result from an exchange of terminal segments between nonhomologous chromosomes.   In individuals heterozygous for a symmetrical reciprocal translocation (structural hybrids), four chromosomes share a partial homology but no two are identical.   Consequently, chromosome pairing at prophase of the first division of meiosis results in a cross configuration involving the four chromosomes. In the mouse, for example, such cross configuration are visible in cytological preparations of spermatogonial prophases (3, 53).   The subsequent behavior of such a cross configuration depends upon the frequency and location of the chiasmata and the mode of centromere orientation. Depending on whether homologous or non-homologous centromeres reach the same spindle pole, gametes which are characterized by duplications and deletions. The latter are genetically unbalanced and give rise to gametic sterility in plants and to zygotic sterility in animals. The percentage of sterility is usually about 50% (between 30% and 70%). This phenomenon is called "semisterility". In mice, the semisterility may be used to score for translocations (52). (ii) Whole-arm translocations. These are the translocations in which whole (or nearly whole) chromosome arms are transposed or interchanged. For example, from two metacentric chromosomes (A.B and C.D) two new chromosomes (A.C and B.D) are formed. In humans, this type of translocation can be found in cases with familial Down's syndrome. Whole arm exchanges can occur between chromosome 21 and chromosome 13 or 14 (174). The small product consisting of the two short arms of 21 and either 13 or 14, respectively, is lost without any phenotypical consequences, and only the new chromosome with the long arm of 21 and the long arm of, for example, chromosome 14 is transmitted to the progeny. A remarkable consequence is that humans with absolutely inconspicuous phenotype may have a diploid chromosome set of only 45 chromosomes. Upon segregation of the chromosomes from such a genotype heterozygous for a 21/14 chromosome, trisomies and nullisomies for the long arm of 21 are formed. The nullisomies are lethal and the trisomies exhibit Down's syndrome.

Aneuploidy

The phenotype resulting from a gain or loss of one or more whole chromosomes from an euploid chromosome complement is unpredictable. A well-known case in which all possible types of trisomies (2n + 1) are viable and distinguishable based on the phenotype is the plant Datura stramonium (12a).

In humans, monosomies seem to be lethal whereas trisomies for very small chromosomes, e.g., 22, 21, survive until birth or may reach adulthood, trisomes for medium sized chromosomes, e.g., 13 (Patau syndrome), 18 (Edwards syndrome), may be born out but have a very short life expectancy, and those for large chromosomes, e.g., 1, 2 etc., are lethal in early embryonic or foetal life. An exception are the sex-chromosomes. This more complex situation will not be discussed here; see Vogel and Motulsky (174) for details.

The genetic consequences of aneuploidy are loosely correlated with the size of the fraction of the karyotype involved and result primarily from quantitative changes of the genetic balance. Depending on the severity of the resultant changes aneuploidy

might appear as a recessive or dominant, visible or lethal
mutation.

Polyploidy

In higher plants, polyploidy in the form of allopolyploidy
(containing unlike chromosome sets) has been of major significance
in evolution. Among lower plants and animals, polyploidy is quite
rare and is virtually confined to some hermaphroditic groups and
to forms reproducing by parthenogenesis in animals. The main
barriers to evoutionary polyploidy in the animal kingdom are
(181): (i) the almost universal presence of sex chromosome
mechanisms in the bisexual groups; (ii) the prevalence of
obligatory cross-fertilization mechanisms in the hermaphroditic
groups. These facts prevent the establishment of polyploidy
because a newly arisen tetraploid will encounter diploid mates and
will produce sterile triploid offspring. In humans, triploids are
found among abortuses and therefore are non-viable. Intra-
individual somatic polyploidy via endomitosis and
endoreduplication results when the exact coupling of chromosome
duplication and mitosis, controlling diploidy, is relaxed by
disturbances or upsets of the spindle function. Such
endomitotically-arisen polyploid cells may be found in the human
liver. Those cells with an enlarged nucleus have also an
increased volume of cytoplasm. Due to this nucleocytoplasmic
ratio (63) polyploid cells are usually larger than the diploid
ones.

RELATION OF MUTATIONS TO CARCINOGENESIS

There are four lines of evidence indicating that
carcinogenesis in humans is related to genetic phenomena,
including mutagenesis:

(i)   There exist hereditary tumors of childhood (e.g.,
        retinoblastoma, 106) which follow simple Mendelian
        genetics. The possibility that certain deletions e.g., in
        chromosome 13 (retinoblastoma) or in the short arm of
        chromosome 11 (Wilms tumor), are related to carcinogenesis
        is under intense study (78).

(ii)   A decreased DNA repair capacity as in the case of Xeroderma
        pigmentosum leads to an increased tumor incidence.

(iii)   The data on human and experimental animal carcinogenicity
        for 54 chemicals, groups of chemicals, and industrial
        processes were evaluated (69). On the basis of evidence
        from human studies 18 of the 54 chemicals and industrial
        products are human carcinogens. A further 18 chemicals are

probably carcinogenic for humans, although the data are considered not adequate to establish a causal association.

(iv)   In experimental test systems, the large majority of the known chemical carcinogens which act as initiators lead to covalent DNA binding (101) and act as mutagens (66). These substances are classified as genotoxic carcinogens.

Since this aspect is presented elsewhere in this Workshop, it will not be discussed any further.

BACK MUTATIONS

A back mutation is a heritable change in a mutant gene resulting in a revertant which has regained the active enzyme or the function that was lost due to the so-called "forward mutation". A true back mutation restores the original nucleotide sequence which had been changed by the forward mutation. The observation of a wild-type phenotype in a mutant line does not necessarily mean that this reversion has occurred by an actual reversal of the original mutational event, that is, by true back mutation. True back mutations can be simulated by suppressor mutations, gene conversions, etc. (57).

Base Pair Substitutions

Since we know that forward mutations can result from either of two types of transitions, from AT to GC or from GC to AT we expect that true back mutations should be possible. For example a forward mutation from AT to GC and a back mutation at the same position in the polynucleotide chain from GC to AT will restore the original base sequence. In fact, every new transition at a transition site will be a back mutation. chemical mutagens, such as 5-bromodeoxyuridine or 2-aminopurine, are base analogs capable to revert mutations which they have induced due to their ability to induce two-way transitions (49).

With transversions, the same sequence of forward and back mutations can be assumed. Theoretically, their frequencies per mutational event should be lower than with transitions because "wrong" transversions, not restoring the initial base sequence, are possible.

Frameshift Mutations

Let us assume that by recombination we can combine two frameshift mutations of opposite sign (an addition and a deletion of the same size, e.g., one base pair) within a small region of the same gene. The mRNA from this double mutant will be correctly interpreted from the start to the first mutational site.   From here until the second mutational site wrong triplets are present but distal to the second mutation the original reading frame is restored.   Provided that no stop codon appears in the section between the two mutational sites, and that the small sequence of wrong amino acids in the peptide chain produced at the ribosome does not eliminate the functional activity of the otherwise correct peptide, we will see a "reversion" phenotypically. With frameshift mutations a true back mutation, restoring exactly the original base sequence, is expected to be very rare. But we can get reversions by any second frameshift mutation in the neighborhood of a preexisting one if the two mutations together restore the original reading frame and the still mutated base sequence between the two frameshifts does not contain a stop codon.

Specificity

Some chemical mutagens are specific inducers of transitions (e.g., 5-bromouracil or 2-aminopurine, see below). With these at hand, one can test for the presence of transitions among certain groups of forward mutations.   This is possible because it is impossible to revert a frameshift mutation by a transition. Freese (49) has reported such experiments which indicate that 2-aminopurine-induced mutations are easily reverted with 2-aminopurine but that no revertants are obtained if proflavin-induced mutations are treated with 2-aminopurine.   This is explained by the fact that 2-aminopurine induced base pair substitutions whereas proflavin induced frameshift mutations.   In the experimental data the situation turns out not to be exactly alternative, because the mutants analyzed are "contaminated" with some spontaneously arisen mutants which can either be substitutions or frameshifts.

Block Mutations

With larger deletions of several base pairs or more it becomes less and less probable that a compensating back mutation of the frameshift type can restore the reading frame and at the same time eliminate the functional deficiency of the polypeptide. Therefore, the larger deletions can be separated operationally from the frameshift mutations of the deletion type by their nonrevertability by spontaneous events or under the influence of a mutagenic treatment.

Chromosome Aberrations

Chromosome aberrations cannot be reverted by the same mechanisms by which they were created. As an exception, true revertants can be obtained in meiotic cells homozygous for a tandem duplication. Due to asymmetric crossing-over in the improperly paired duplicated region, a chromosome with one piece and a chromosome with three pieces originally duplicated on both chromosomes can appear. This phenomenon makes tandem duplications, as in the case of the Bar region in Drosophila, comparatively unstable.

Aneuploidy

Aneuploidy in viable and fertile individuals can be reverted in a number of cases due to the segregation of chromosomes during meiosis. From a karyotype with a trisomy (2n+1) for one chromosome, haploid (n) and disomic (n+1) gametes may be formed. Theoretically both types should occur with equal frequency. The different types may occur among the progeny in frequencies differing from the gametic ratio due to differential viability etc. An example from human genetics is the inheritance of the Trisomy 21 (Down's syndrome). The syndrome can be transmitted only by female patients because males are infertile. The theoretical expectation of 50% children with Down's syndrome among the progeny of such females is not found: among 23 children only 9 showed the Down's syndrome. This probably reflects, at least partly, the higher foetal mortality of trisomics as compared to normal karyotypes (185).

Polyploidy

Polyploidy is not revertable except in particular cases where segregation and fertilization (or parthenogenesis) result in a diploid configuration.

MECHANISMS LEADING TO POINT MUTATIONS

Most of the studies concerning gene mutation mechanisms were carried out in microbial systems. Our knowledge of the role of "mutagenic"DNA repair results particularly from studies with E. coli. In this chapter we will, therefore, primarily talk about the ideas concerning mutagenesis in prokaryotic systems and point to the situation in eukaryotes only where specific pieces of evidence are available. We try to extrapolate our knowledge of prokaryotes to eukaryotes, or, to express it more cautiously we use the mechanisms found in prokaryotic systems as models or working hypotheses for the analysis of the phenomena in eukaryotes, which often turn out to be more complicated.

Fidelity of DNA Replication

DNA replication is catalysed by multi-enzyme complexes in vivo (79). It is one of the central genetic processes. The fidelity expected from physico-chemical knowledge of the stability of non-Watson-Crick base pairs and the occurrence of alternate tautomeric forms of the bases would predict 10^{-4} to 10^{-5} errors per base pair replicated (171). In vivo studies revealed a much higher fidelity. In order of 10^{-8} to 10^{-10} per base pair replicated in T4 bacteriophage and E. coli, respectively (30).

The recognition of Watson-Crick base-pairing plays a central role in establishing this high fidelity. Polymerases tend to select the correct base or nucleotide triphosphate. Evidence has been presented that several mutator genes in E. coli are in fact concerned with the mismatch correction (131). The full mechanism for DNA replication in E. coli is complex and appears from in vivo studies to require at least thirteen components (5, 183). There is obviously considerable scope for mutations in any of these components to reduce the accuracy of polymerization. For the T4 bacteriophage, genetic analyses revealed that alterations in any of the seven major DNA replication proteins can increase mutation rates (113, 114). In prokaryotes, such as E. coli and its bacteriophage T4, a special enzyme activity has been found connected with the DNA polymerase: the 3'-5' proofreading exonuclease activity (17). This activity of the polymerase enables it to remove a wrong base it has just inserted. This back-tracking has been called editing or proofreading and is significantly more likely to occur when an incorrect base has been inserted. Very elegant experimental evidence for the importance of this process has been obtained using mutants of T4 which are altered in DNA polymerase. Both mutator and antimutator DNA polymerases may be altered in 3'-5' exonuclease activity, which is revealed as a turnover of deoxyribonucleotide triphosphate during DNA replication (11, 115). 3'-5' Exonuclease activity also occurs in certain microbial eukaryotic DNA polymerases (Ustilago maydis, 8 and yeast, 22). On the other hand, purified DNA polymerases from higher eukaryotes lack any associated exonuclease activity (20, 98) but the possibility exists that such an activity might be found in association with the enzyme (98, 131). In addition to the selection of bases by the polymerase and the 3'-5' exonuclease activity, a third system was discovered in E. coli which reduces the frequency of spontaneous mutations: the mismatch correction. The mutants uvrD and uvrE are deficient in this process. The uvrD gene has recently been cloned with a phage vector (120) and the gene product has been identified as a 75'000 MW polypeptide which possesses DNA-dependent ATPase activity (121). The size and chromatographic properties of this protein are consistent with those of DNA-dependent ATPase I. Richet et al. (137) have suggested that DNA-dependent ATPase I and DNA helicase II are the same protein. Mismatch correction is possible because the cell is able to recognize the newly synthesized DNA strand which is undermethylated (129). DNA of E. coli contains two naturally methylated bases, N^6-methyl adenine and 5-methyl cytosine. These

methyl groups are added enzymatically fairly slowly after the DNA
has been replicated. In human cells, only 5-methyl cytosine has
been found (86). It is likely that the methylation enzymes
recognize specific sequences so that a total of only about 1% of
adenine residues and 0.4% of cytosine residues are methylated.
Methylation of cytosine, and a small proportion of adenine, is
concerned with restriction and modification, the process whereby
E. coli can recognize and destroy foreign DNA, unless specific
sequences are methylated. In appears, however, that most
methylation of adenine is not concerned with this process. Still
it must be important since dam mutants of E. coli are partly
deficient in methylation of adenine and show a range of altered
properties (110). Experiments of Radman et al (132) suggest a
role of methylation of adenine in the strand recognition process.
They found that mismatch correction in reconstructed
heteroduplexes of lambda bacteriophate with one strand from a dam
and one from a dam$^+$ host occurs with high efficiency in favor of
the fully methylated strand.

Manganese (Mn^{2+}) has been shown to be mutagenic in vivo (125,
139). If in reconstructed DNA replicating systems in vitro the
ordinary Mg^{2+} is replaced by Mn^{2+}, the error frequency is
increased (112). Since a strong competition effect of Mg^{2+} is
absent it is unlikely that manganese exerts its effect via its
complex with the nucleotide triphosphates to which the binding
constants for both, Mg^{2+} and Mn^{2+}, are very similar. The
mutagenic activity of manganese is likely to be due largely to its
interaction with the DNA polymerase or with other replication
proteins. This was shown for in vitro systems containing
polymerases with 3'-5' exonuclease activity, e.g., T4 (64), as
well as those containing polymerases without this activity (157).

Other, nearly unexplored mutational mechanisms could occur
(32). Deoxyribonucleoside triphosphate imbalance during DNA
replication is expected to be mutagenic. Thymine deprivation
appears to be a good example (158). As a further possibility, it
was shown that certain amino acid analogues are mutagenic (164,
165), probably because they produce defective enzymes of DNA
metabolism. It also seems likely that the direct chemical
modification (e.g., alkylation) of enzymes of DNA metabolism is
sometimes mutagenic, but evidence for such effects is wholly
lacking at present (32).

The genetic effects of deoxyribonucleotide pool imbalances
have recently been studied in some detail (for a review see 80).
In vivo DNA precursor imbalances can have profound genetic
consequences in prokaryotes and eukaryotes. In vitro studies have
demonstrated that the fidelity of DNA replication is dependent on
correct balances of deoxyribonucleoside triphosphates during DNA
synthesis. These findings suggest that intracellular

concentrations of DNA precursors may be regulated in order to minimize the frequency of genetic change. In addition, they indicate the existence of important non-DNA targets for the induction of mutation, recombination, chromosome aberrations, etc.

Historically, the first mechanism by which chemically induced mutations could be explained was the mispairing model of tautomeric forms. The migration of a single proton within the coding (hydrogen bonding) face of a base can alter the amino forms of cytosine and adenine into their imino forms, or the keto forms of thymine and guanine into their enol forms, thus producing mutations of the transition type. This was suggested for the normal base pairs (177) and for base analogue-induced mutagenesis (47, 48). For the situation with natural bases the extremely high fidelity of DNA replication made it impossible to prove this theoretically very appealing scheme. In the case of 5-bromouracil (BU), 5-bromodeoxyuridine (BUdR) and 2-aminopurine (2-AP), the error frequencies were orders of magnitude higher, and for both base analogues the back mutation tests indicated the presence of two-way-transitions. After BUdR and 2-AP mutagenesis the altered bases in the mutant single strand were determined and proved the tautomeric shift model (67, 167).

The coding properties of a base may also then direct the DNA replication complex to incorporate an incorrect base into the newly synthesizes strand. For example, hydroxylamine (H_2NOH) can specifically react with cytosine, producing a derivative which is N-hydroxylated in the 4 position. The derivative will exist in two tautomeric forms. In one form it will form hydrogen bonds to adenine in place of the original guanine. Therefore, if transforming DNA or DNA viruses are treated with high concentrations of hydroxylamine (at a low pH of approximately 6), the induced mutagenic DNA alterations of cytosine will give rise to specific base pair transitions (GC to AT). For a detailed discussio n and references see (31). Recently Shugar et al. (154) have presented evidence that in a number of instances hydroxylamine can effect transitions other than the normally expected C to T. It is, therefore, unlikely that hydroxylamine-induced mutations result exclusively from simple Watson-Crick base pair transitions. Another interesting case is the deamination of bases by nitrous acid. At a low pH of about 3 free nitrous acid can be produced from sodium nitrite. It can oxydatively deaminate adenine, guanine and cytosine to hypoxanthine, xanthine and uracil, respectively. There is experimental evidence that most of the mutagenic effect of nitrous acid results from transitions induced by the deamination products of adenine and cytosine. In order to avoid the impression of an oversimplification it should be added that nitrous acid can also induce interstrand cross-links, presumably via the formation of diazonium complexes (9, 102) and in this way induce deletion mutations (166).

Interaction of alkylating agents with DNA can lead to a variety of alkylation products at nearly 20 different sites, some of which will lead to incorrect Watson-Crick base-pairing (155). Miscoding has been shown for DNA and/or RNA synthesis for O^6-methylguanine (1, 54), O^4-alkylthymine (1, 87, 143), and for O^2-alkyluracil and O^4-alkyluracil (156). Other alkylated bases which do not show miscoding are O^2-methylthymine (143) or 7-methylguanine (1, 61,100).

Of particular interest are the consequences of bulky adducts such as those resulting from well known genotoxic carcinogens. Two carefully studied compounds are the 7,8-dihydro-9,10-epoxyde of benzo(a)pyrene (BPDE) and the 2-acetylaminofluorene (AAF). They have recently been reviewed by Grunberger and Weinstein (59). Several studies have documented that the BPDE is the major metabolic intermediate in the covalent binding of this polycyclic aromatic hydrocarbon to cellular nucleic acids. It has also been established that in both RNA and DNA the major nucleoside adduct results from linkage of the 10 position of BPDE to the 2-amino (N^2) group of guanine.

Examination of the model of double stranded DNA with Watson-Crick geometry indicates that this N^2 group of guanine is relatively exposed in the minor groove of the helix. It is likely, therefore, that BPDE can bind covalently to the N^2 residue with little distortion of the native DNA conformation. The S1 nuclease digestion and formaldehyde denaturation data do suggest that there may be a slight destabilization of the helix at sites of BPDE-modified guanine residues. This may be due to interference in the usual hydrogen bonding of the N^2 group of guanine with cytosine, perhaps secondary to a slight distortion of the helix to fully accomodate the bulky BPDE residue in the minor groove. This has been called the "minor groove model."

In cellular systems the effects of benzo(a)pyrene are complicated by the fact that different isomeres of BPDE are formed and can react with DNA. It is expected that N^6 groups of adenine and N^4 groups of cytosine may be attacked. There is experimental evidence that BPDE modification of adenine residues produces a marked destabilization of the A-T base pair. Recent studies suggest that BPDE may also attack the N^7 position of guanine, but this adduct is readily lost from DNA due to labilization of the glycoside bond and depurination. Depurination can lead to spontaneous or enzymically mediated chain scission and thus to genetic consequences.

The BPDE modified DNA has decreased template activity. The interference with progression of the replicating fork during DNA synthesis in vivo may explain the action of benzo(a)pyrene and other polycyclic aromatic hydrocarbons as frameshift mutagens in

Salmonella typhimurium and as inducers of phage production and SOS error-prone DNA repair in E. coli.

The covalent attachment of another bulky carcinogen, AAF, present major steric problems and can be associated with major distortions in the native conformation of DNA. Based on studies with AAF-modified dimers the so-called "base displacement model" was developed. This model also proved to be valid for AAF-modified DNA molecules. As the major product obtained from hydrolysates of AAF-modified DNA, N-(deoxyguanosine-8-yl) AAF was identified. The base displacement model predicts that in a Watson-Crick DNA configuration the altered guanine is displaced out of the helix in the syn configuration and that the fluorene rings are stacked inside the helix with the rings of the neighboring bases. This produces a localized denaturation of about 12 base pairs around the site of the AAF-modified guanine.

The genetic consequences of AAF-modified guanine will result from the fact that the displaced base will no longer participate in base pairing. This has been elegantly shown with modified RNA. It was observed that AAF covalently bound to a guanine residue in a GAA or AAG codon resulted in the inactivation of that codon. No miscoding was obtained. The effect was different if a guanine residue in a GAAA or AAAG tetramer was modified. In this case the carcinogen led to an approximately 50% reduction of the recognition of the AAA codon by the lysyl-tRNA. Similar effects may be expected for transcription and replication of AAF-modified DNA.

The gross change in configuration of the AAF adduct to the C^8 makes adducts easily recognizable for the DNA excision repair system. It is rapidly removed from rat liver DNA in vivo with a half-life of about 7 days. On the other hand, the guanine-N^7 adduct, a minor product of AAF-modified DNA, seems to occupy the minor groove of the DNA helix, thus leading only to minor distortions. Probably because it is not easily seen by the excision repair this adduct remains presistently bound to DNA. Taken together, this suggests that the DNA excision repair enzyme system preferentially recognizes and excises lesions associated with major distortions of the DNA helix. Assuming that this type of repair operates with high fidelity, persistent carcinogen substituents, such as adducts at the N^2 position of guanine, might be more significant in terms of mutagenesis and carcinogenesis. The potency of a chemical might be a function of two factors: (a) ability to bind to DNA and alter its template function and (b) ability to bind in a form that does not produce a conformational distortion which is readily recognized and excised by DNA repair systems.

In one case of a bulky adduct, namely Aflatoxin Bl, the probability of leading to a mutation was determined. It was estimated (159) that one in 27 to one in 37 adducts formed lead to a observable mutation. How these figures obtained for an extremely potent mutagen relate to other mutagens and carcinogens remains to be determined.

A further important lesion resulting in a reduced fidelity of DNA replication are apurinic sites. It has been estimated (95) that depurination of normal DNA in vivo has a rate constant in the order of 1.8×10^{-9}. Alkylation (107, 108) and specific glycosidases (77, 93) increase the depurination rate constant of those altered bases to $0.08 - 1.4 \times 10^{-3}$, respectively. The rate of depurination is greater for methylated than for ethylated bases (84, 161). The effect of depurination on polynucleotide templates on the fidelity of DNA synthesis in vitro has been determined by Sherman and Loeb (152). They found an increased error rate with E. coli DNA polymerase I, avian myeloblastosis virus DNA polymerase and human placenta DNA polymerase-beta. Kinetic studies and nearest-neighbor studies suggest that the incorporation of non-complementary nucleotides occurs randomly as single base substitutions and that any of the four deoxyribonucleotides can be incorporated opposite apurinic sites. In E. coli, as well as in cultured human fibroblasts and HeLa cells, it could be shown that, among other possibilities, a very specific repair pathway exists: insertases are able to fill apurinic sites directly with the correct missing purine base (27, 97).

Error-prone repair is a third important mechanism which can lead to mutations of the base pair substitution type. If a DNA damage, e.g., a strand break or a cross-link, prevents the DNA from acting as a template for replication, this damage must be repaired in order for the cell to survive. Although most repair mechanisms are highly accurate, a minor error-prone repair pathway has been found in E. coli (184) and elsewhere (76).

The basic observation of Witkin (184) was that lexA strains of E. coli are immutable by UV. It transpired in subsequent studies that mutagenesis depends on the presence of recA$^+$ and lexA$^+$ gene products. They also play a central role in the phenomenon of UV-reactivation of Weigle-reactivation (178, 179) and have a large number of pleiotropic effects (for a discussion see 56). Radman (129) developed the stimulating hypothesis that Weigle-reactivation and bacterial mutagenesis are two manifestations of an inducible last-ditch repair process (SOS repair) which operates by permitting polymerization past noncoding lesions in the DNA. Recently, Green (56) has speculated that it is simplest to consider base substitution mutagenesis by agents such as UV as arising through suppression of the 3'-5' exonuclease

activity by DNA polymerase III. This exonuclease suppression is
under strict control and may be induced following DNA damage,
although a constitutive level of this error-prone repair is likely
(15).

Frameshift Mutagenesis

Our knowledge of mutagenesis by frameshift is less precise
than our knowledge of mutagenesis by base pair substitution. This
is unfortunate, since many polycyclic hydrocarbons and aromatic
amines are carcinogens and specific frameshift mutagens in
bacterial tests (70, 104). Bacterial studies indicate that there
exist two classes of frameshift mutagens. The first type is the
group of the classic intercalating agents such as acridines. They
can intercalate into the DNA double helix (29, 92, 124) and some
mutagenic activity may be observed without covalent binding (191).
Acridine-type mutagens might also act by stabilizing extrahelical
bases by simple stacking (32, 92). The second type, to which
belong many carcinogenic aromatic amines and polycyclic
hydrocarbons, seems to need covalent binding to DNA in order to be
mutagenic. These adducts cause far greater distortions of the DNA
than, for instance, intercalated acridines or UV-induced
pyrimidine dimers (25). Three types of mechanisms for frameshift
mutagens have been proposed: (i) looping out of one or a few
bases in a repetitive sequence during DNA repair (162); (ii)
incorrectly aligned repetitive sequences during recombination (91,
150); and (iii) looping out of bases during DNA replication (180).
The first two hypotheses seem to be most useful for the
explanation of the mutagenic effect of intercalating agents
without covalent binding, whereas the last mechanism seems to
apply for the induction of frameshift mutations by large
covalently bound mutagens such as polycyclic hydrocarbons and
aromatic amines.

Block Mutations

These can be of two types: large deletions or large
insertions. The latter group may result from the integration of
DNA pieces into the parental DNA, e.g., insertion sequences,
episomes, viruses such as phages in bacteria or tumor viruses in
mammalian cells. In chemical mutagenesis the induction of
deletions is of importance. The mechanisms leading to this type
of mutation have yet to be studied in detail. The simplest
hypothesis would be to adapt the Campbell model (21) which assumes
illegitimate exchanges between sections within the same chromosome
during recombination. The problem remains, however, that deletion
mutagenesis can occur in recombination deficient bacteria. Other
possible mechanisms might be deletion formation in connection with
the transposition of insertion sequences, or simply the action of
ligase randomly joining the free ends of DNA (56).

Structural Changes of Chromosomes

There is little doubt that the basic unit structure involved
in chromosome breakage and exchange is a single duplex of DNA
(40). Two types of aberrations can be observed: chromosome-type
and chromatid-type. The chromosome-type aberrations are those in
which breakage and exchange occur in G1 prior to chromosome
replication. The chromatid-type aberrations are those in which
the unit of the chromosome that is broken or exchanged is the half
chromosome or chromatid, and this kind of aberration is produced
during S or G_2. It has been learned that at least two mechanisms
must exist that lead to chromosome aberrations: (i) a mechanism
whereby aberrations are induced independently of DNA synthesis, by
agents such as ionizing radiation (42, 148), bleomycin (173),
streptonigrin (74) and 8-ethoxycaffeine (147); (ii) a mechanism
which is dependent on DNA synthesis. The majority of chemical
mutagens, as well as UV light, produce lesions in G1 or G2
chromosomes that only develop into aberrations when the cell
proceeds through a DNA replication phase. This leads to chromatid
aberrations; and (iii) a third exchange mechanism does not lead to
aberrations but to sister chromatid exchanges (SCEs). SCEs are
known to be formed at the time of DNA synthesis from lesions which
may be induced in cells at all stages of the cell cycle (186).

Irrespective of the primary damage induced by a mutagen, the
ultimate lesions responsible for aberration formation seem to be
DNA strand breaks. There is evidence (116) that X-ray-induced
aberrations can arise in a manner envisaged by the classic
"breakage first hypothesis" (145). A general model has been
developed by Bender et al. (10). It is based on the following
basic features: (i) a G1 chromosome contains one DNA double
strand; (ii) DNA is the primary target; (iii) aberrations arise
from polynucleotide strand breaks and recombination between broken
ends; and (iv) double strand breaks can be directly induced or
mediated through enzymatic repair processes or normal DNA
synthesis.

The strongest experimental support for the crucial role of
DNA double strand breaks in the formation of aberrations has been
presented by Natarajan and coworkers (116, 117). In these studies
Neurospora endonuclease (specific for cleaving single stranded
DNA) was introduced into mutagenized Chinese hamster ovary (CHO)
cells. Single strand breaks resulting from X-ray, MMS, or
bleomycin treatment of G2 cells were converted into double strand
breaks and led to an increased frequency of DNA synthesis-
independent aberrations and chromatid breaks. It was shown
earlier (75) that two lesions are necessary for exchange
formation. As the cells progressed from early G2 into late G2/
early prophase, the increase in exchanges disappeared, whereas
chromatid breaks were increased. Most probably this resulted from

changes in the organization of the chromosome, such as condensation, in the nuclei preparing for mitosis. Not only the temporal coexistence of two breaks, but also their spatial vicinity are prerequisites for aberration formation. A rejoining distance of 0.1 - 0.3 μ was estimated by Wolff et al. (188) within which breaks can lead to exchanges. Steffensen (160) discussed the variability of this distance in relation to cellular activity, temperature and other conditions. The actual exchange might well be performed by some kind of recombinational repair. There is good evidence for the existence of recombinational repair in bacterial systems (68) and some evidence of it in mammalian cells (111). A comparative model for the suspected differences between the recombination repair processes in the two cell types has been worked out by Green (56).

Double strand breaks are easily induced by densely ionizing radiation (high LET radiation such as alpha particles or protons from fast neutrons). With X-rays, a single energy loss event of 50 eV produces a single strand break, and two such independently induced single strand breaks, if less than five bases apart, can result in a complete scission across both polynucleotide strands (172). There is evidence that two, and in some cases one, energy loss events can result in a chromatid exchange (118). This fits nicely with the facts (i) that with X-rays, exchange aberrations usually involve two electron tracks and increase in proportion to the square of dose, whereas with densely ionizing radiations that produce double stranded breaks, they increase linearly with dose, and (ii) that densely ionizing radiations are more efficient per unit of energy absorbed, i.e., they have a high RBE (38, 39).

MMS is inefficient in inducing S-phase independent chromosome aberrations, despite the fact that apurinic sites and single strand breaks are found as secondary lesions. This indicates that in normal cells not many single strand breaks are converted into double strand ones by normal cellular processes (116).

There is one interesting observation that non-DNA alkylations can play a role in the formation of DNA double strand breaks and their consequences. Sega and Owens (149) propose the following mechanism for the action of EMS in mouse germ cells which leads to chromosomal damage: After protamine synthesis, but before the cysteine disulfide bonds are formed, EMS could ethylate the sulfhydryl groups and thus effectively block normal disulfide bond formation. This in turn would prevent normal chromatin condensation in the sperm nucleus, leading to stresses in the chromatin structure in the chromatin could develop away from the actual site of ethylation so that the break might occur at some distance from where the ethyl group is bound. W. R. Lee (personal communication) has also suggested that a chromosome break might be more likely to occur in a region of chromatin where a single

strand DNA break has already occurred as a result of depurination
of an ethylated base and hydrolysis of the phosphodiester bond.
Stresses in the chromatin resulting from ethylation of cysteine
residues in the protamine might then be sufficient to rupture the
other DNA strand and produce a chromosome break.

The S-phase dependent formation of chromosome aberrations
seems to be initiated by a wide variety of lesions, e.g., cross-
links (146), pyrimidine dimers (40), simple alkylations, single
strand breaks (117).   The list of chemical agents that induce
chromosome aberrations is very extensive and includes base
analogs, intercalating agents, agents which interfere with DNA
replication etc. (40). Unfortunately the identification of
lesions involved in the formation of aberrations does not mean
that the mechanism of aberration production is known.   Basically,
aberrations must be formed as a consequence of the cell's attempt
to rectify DNA damage, or to circumvent lesions during DNA
replication.   Evans (40) emphasizes three important points: (i)
many of the lesions that ultimately give rise to aberrations may
affect bases, or phosphodiester linkages, on only one strand of
the DNA; (ii) whole duplexes of DNA are exchanged or broken; and
(iii) the lesions develop into aberrations during the period when
the normal process of semiconservative replication takes place.
Based on our knowledge of bacterial repair enzymes, it is
relatively easy to envisage models which take into account the
first two points.   There is at the moment no experimental evidence
leading to one particular model that takes into account the
discontinuous replication, possible exchanges between Okasaki
fragments (79), formation of hybrid DNA between strands from
different parental strands, etc.

To illustrate in more detail the state of our knowledge on
the formation of chromosomal structural changes we will briefly
discuss the effects of cross-linking agents.   They are very
attractive mutagens because (a) they are very potent inducers of
chromosome aberrations and especially SCEs (127), and (b)
mammalian cell lines are known with different sensitivities to
this class of chemical mutagens.   For example, for the sulfur
mustard it is known (46, 83, 85) that basically three types of
lesions are induced in double stranded DNA: (i) inter-strand
cross-links;   (ii)   intra-strand   cross-links;   and (iii)
monofunctional alkylations.

Scott (146) has studied the consequences of these different
lesions in the chromosomes of two cell lines from rat tumors
(Yoshida sarcoma) which exhibit a pronounced differential
sensitivity to the lethal effects of bifunctional alkylating
agents.   In spite of the fact that sulfur mustard treatment gave
equal alkylation on the DNA of the two cell lines, much more
chromosome damage was observed with the sensitive cells (YS).   The

aberrations were produced during the time of DNA synthesis and were only chromatid-type aberrations (42, 148). Since both cell types have normal excision repair capacity, they must have entered the S-phase with an equal number of lesions in the chromatin. The difference in the aberration frequency must have therefore resulted from different consequences of lesions present during DNA synthesis. It can be assumed that such lesions may lead to the formation of gaps in the newly synthesized (nascent) DNA opposite the lesions. The gaps may then be filled up by a gap-filling process called postreplication repair. This repair process can be initiated by non-toxic doses of caffeine, which leads to an enhanced frequency of the S-phase-dependent chromosome aberrations. Whereas the frequencies of chromosome aberrations in the resistant line (YR) were enhanced by caffeine (to about the level found in the sensitive line), a caffeine treatment was without effect on the frequencies in the sensitive (YS) cells. This indicates that the postreplication repair normally avoids the formation of chromosome aberrations, and that gaps in the newly synthesized DNA are somehow involved in the production of chromosome aberrations. This is not true for the SCE formation, because the SCE frequencies induced in the two cell lines are equal. A possible interpretation would be to assume different lesions leading to aberrations and SCEs (146). Since both sister chromatids are involved in an exchange, the important lesion might be inter-strand cross-link. The actual exchange might take place in connection with the replication by-pass mechanism suggested by Shafer (151) whereby replication can continue past a cross-link as a result of SCE formation. In chromosome aberrations usually only one chromatid is involved. This might reflect an initial damage in only one strand, probably an intra-strand cross-link. Such cross-links, especially in the absence of an efficient postreplication repair system, may result in gaps and subsequent breaks in the damaged parental strand leading to double strand breaks. Two such double strand breaks resulting from two independent intra-strand cross-links within the limits of an exchange site will then lead to a chromosome aberration such as a chromatid interchange. It is clear, that the healing of a double strand break does not lead to the aberration, but that some sort of misrepair has to take place (40).

Aneuploidy

Aneuploidy with additional or missing chromosomes may result from mitotic or meiotic nondisjunction, i.e., from an irregular distribution of sister-chromatids or homologous chromsomes to the poles, as inferred from cytological observations. Spontaneous nondisjunction is under genetic control, as exemplified by the increased nondisjunction in asynaptic mutants (e.g., in Drosophila melanogaster, 7). In Drosophila oogenesis a special type of aberration-mediated segregation of chromosomes was discovered.

Whether this phenomenon is of general importance or a specialty of
Drosophila remains to be determined. In humans, nondisjunction
frequencies increase with parental age (174).

Mitotic poisons (e.g., colchicine, colcemide, etc.) may
induce nondisjunction by interfering with tubulin assembly when
microtubuli are forming the spindle fibers (13, 50). Colcemide
fed to Drosophila females induces aneuploidy in oocytes (171a).
Other compounds which can induce aneuploidy as well as chromosome
aberrations and mitotic inhibition and which also act as
teratogens are the Vinca alcaloids such as Vincristine and
Vinblastine (51). A number of organic mercury compounds have been
shown to induce aneuploidy in somatic cells as well as in germ
cells (16, 133, 134, 135). The induction of aneuploidy can be
studied in normal human lymphocyte cultures and lymphoid cell
lines (62) or in germ cells of mice (142). The possibility of
chemically induced aneuploidy in man has been discussed with
respect to children of parents treated with colchicine for gout
(43, 44) and with respect to a possible genetic disposition for
increased nondisjunction (65). Today these problems are still
open to discussion.

Polyploidy

The same agents which may induce aneuploidy can be used to
induce polyploidy, a technique routinely used in plant genetics
(37). Spontaneous polyploidization may result in various
biological phenomena such as polyspermy, disturbances of polar
body formation in female meiosis, endomitosis etc.

MUTATION AVOIDANCE MECHANISMS

As far as we know, all living organisms have developed
mutation avoidance mechanisms in order to survive and to keep the
mutation frequency at a reasonable level (30). The fidelity of
DNA replication, as well as numerous DNA repair systems, determine
the frequency of spontaneous as well as induced mutations. The
situation as it is known today for E. coli (56) seems to be
characteristic for prokaryotes and, with some modifications, also
for eukaryotes. It now appears that in a normal repair proficient
strain of E. coli, there is a most elegant coordination of repair
processes which minimizes the chance of a UV-induced pyrimidine
dimer giving rise to a mutation. In the first instance, the dimer
may be photoreactivated or excised, both efficient and accurate
processes. If the dose of UV is high, DNA replication becomes
stalled temporarily and restarts at the origin of replication (12,
73). This allows additional excision repair to occur. If a dimer
is replicated, a gap is formed in the newly synthesized DNA that
can be filled accurately by recombination repair. If gaps in

daughter strands overlap, thus preventing recombination repair, reannealing and postreplication excision occurs. It is only when all these processes fail that error-prone repair will be called into operation. In a repair proficient strain, this may well not be activated for a residuum of repair resistant damage, but for the rare cases where conventional repair goes wrong.

Neel (119) has reviewed the data available on comparative mutation rates in Drosophila, mice and humans, as derived from the study of both biochemical and morphological traits. There is a notable similarlity in these rates, despite the large differences in average generation time, mean body temperature and number of cell divisions intervening between fertilization of the egg and production of functional gametes. This suggests the evolution of superior genetic (DNA) repair strategies in long-lived humans. Evidence raises the possibility of higher mutation rates in mostly tropical dwelling, tribal human populations than in temperate-dwelling civilized groups.

REFERENCES

1. Abbott, P. J. and R. Saffhill (1977) DNA synthesis with methylated poly (dA-dT): possible role of 0^4-methylthymidine as a promutagenic base. Nucleic Acids Res. 4:761-769.
2. Abbott, P. J. and R. Saffhill (1979) DNA synthesis with methylated poly (dC-dG) templates: evidence for a competitive nature to miscoding by 0^6-methylguanine. Biochim. Biophys. Acta 562:51-61.
3. Adler, I. D. (1978) The cytogenetic heritable translocation test. Biol. Zentralblatt 97:441-451.
4. Adolph, K. W. (1981) A serial sectioning study of the structure of human mitotic chromosomes. Eur. J. Cell Biol. 24:146-153.
5. Albert, B. and R. Sternglanz (1977) Recent excitement in the DNA replication problem. Nature(Lond) 269:655-661.
6. Baimai, V. and S. Kitthawee (1981) A spontaneous tandem duplication in a Drosophila chromosome. Experientia 37:345-346.
7. Baker, B. S., J. B. Boyd, A. T. Carpenter, M. M. Green, T. D. Nguyen, P. Ripoll and P. D. Smith (1976) Genetic controls of meiotic recombination and somatic DNA metabolism in Drosophila melanogaster. Proc. Natl. Acad. Sci. USA 73:4140-4144.
8. Banks, G. R. and G. T. Yarranton (1976) DNA polymerase from Ustilago maydis. 2. Properties of associated deoxyribonuclease activity. Eur. J. Biochem. 62:143-150.
9. Becker, E. F., B. K. Zimmerman and E. P. Geiduschek (1964) Structure and function of cross-linked DNA. 1. Reversible

denaturation and Bacillus subtilis transformation. J. Mol. Biol. 8:377–391.

10. Bender, M. A., H. G. Griggs and J. S. Bedford (1874) Mechanisms of chromosomal aberration production. III. Chemicals and ionizing radiation. Mutation Res. 23:197–212.

11. Bessman, M. J., N. Muzyczka, M. F. Goodman and R. L. Schnaar (1974) Studies on the biochemical basis of spontaneous mutation. II. The incorporation of a base and its analogue into DNA by wild-type, mutator and antimutator DNA polymerases. J. Mol. Biol. 88:409–421.

12. Billen, D. (1969) Replication of the bacterial chromosome: location of new initiation sites after irradiation. J. Bacteriol. 97:1169 (1969).

13. Borisy, G. G. and E. W. Taylor (1967) The mechanism of action of colchicine. J. Cell. Biol. 34:525–535.

14. Bridges, C. B. (1936) The Bar "gene" a duplication. Science 38: 210–211.

15. Bridges, B. A. and R. P. Mottershead (1978) Mutagenic DNA repair in Escherichia coli. VII. Constitutive and inducible manifestations. Mutation Res. 52:151–159.

16. Bruhin, A. (1955) Über die polyploidisierende Wirkung eines Samenbeizmittels. Phytopathol. Z. 23:381–394.

17. Brutlag, D. and A. Kornberg (1972) Enzymatic synthesis of deoxynucleic acid. XXXVI. A proofreading function for the 3'-5' exonuclease activity in deoxyribonucleic acid polymerase. J. Biol. Chem. 247:241–248.

18. Bukhari, A. I., J. A. Shapiro and S. L. Adhya (1977) DNA Insertion Elements, Plasmids and Episomes. Cold Spring Harbor Laboratory, Cold Spring Harbor, New York, pp. 1–782.

19. Butler, P. J. G. and J. O. Thomas (1980) Changes in chromatin folding in solution. J. Mol. Biol. 140:505–529.

20. Buttula, N. and L. A. Loeb (1976) On the fidelity of DNA replication. J. Biol. Chem. 251:982–986.

21. Campbell, A. M. (1962) Episomes. Adv. Genet. 11:101–145.

22. Chang, L. M. S. (1977) DNA polymerase from bakers-yeast. J. Biol. Chem. 252: 1873–1880.

23. Comings, D. E. (1972) Structure and function of chromatin. Adv. Hum. Genet. 3:237–431.

24. Comings, D. E. (1978) Mechanisms of chromosome banding and implication for chromosome structure. Ann. Rev. Genet. 12:25–46.

25. Daune, M. P. and R. P. P. Fuchs (1977) Structural modification of DNA after covalent binding of a carcinogen. Colloq. Int. CNRS, 256:83.

26. Delius, H. and W. Worcel (1974) Electron microscopic visualization of the folded chromosome of Escherichia coli. J. Mol. Biol. 82:107–109.

27. Deutsch, W. A. and S. Linn (1979) DNA binding activity from cultured human fibroblasts that is specific for

partially depurinated DNA and that inserts purines into apurinic sites. Proc. Natl. Acad. Sci. USA 76:141-144.

28. Dowsett, A. P. and M. W. Young (1982) Differing levels of dispersed repetitive DNA among closely related species of Drosophila. Proc. Natl. Acad. Sci. USA 79:4570-4574.

29. Drake,, J. W. (1969a) Mutagenic mechanisms. Ann. Rev. Genet. 3:247-268

30. Drake J. W. (1969b) Comparative rates of spontaneous mutation. Nature 221:1132.

31. Drake, J. W. (1970) The Molecular Basis of Mutation. Holden-Day, San Francisco, pp. 1-273.

32. Drake, J. W. (1977) Fundamental mutagenic mechanisms and their significance for environmental mutagenesis, in: Progress in Genetic Toxicology, D. Scott, B. A. Brdiges and F. H. Sobels, (Eds.), Elsevier/North Holland Press, Amsterdam-New York-Oxford, pp. 43-55.

33. Drake, J. W. and R. H. Baltz (1976) The biochemistry of mutagenesis. Ann. Rev. Biochem. 45:11-37.

34. DuPraw, E. J. (1965) Macromolecular organization of nuclei and chromosomes - a folded fibre model based on whole-mount electron microscopy. Nature 206:338-343.

35. DuPraw, E. J. (1966) Evidence for a "folded-fibre" organization in human chromsomes. Nature 209:577-581.

36. DuPraw, E. J. (1968) Cell and Molecular Biology. Academic Press, New York-London.

37. Eigsti, O. J. and P. Dustin (1954) Colchicine in Agriculture, Medicine, Biology and Chemistry. The Iowa State College Press, Ames.

38. Evans, H. J. (1962) Chromosome aberrations induced by ionizing radiations. Int. Rev. cytol. 13:221-321.

39. Evans, H. J. (1974) Effects of ionizing radiation on mammalian chromosomes in: Chromsomes and Cancer, J. German (Ed.), John Wiley, New York, pp. 191-238.

40. Evans, H. J. (1977) Molecular mechanisms in the induction of chromosome aberrations, in: Progress in Genetic Toxicology, D. Scott, B. A. Bridges and F. H. Sobels (Eds.), Elsevier/North Holland, Amsterdam, New York, Oxford, pp. 57-74.

41. Evans, H. J. (1980) How effects of chemicals might differ from those of radiations in giving rise to geneticl ill-health in man, in: Progress in Environmental Mutagenesis, M. Alacevic (Ed.), Elsevier/North Holland, Amsterdam, New York, Oxford, pp. 3-21.

42. Evans, H. J. and D. Scott (1964) Influence of DNA synthesis on the production of chromatid aberrations by X-rays and maleic hydrazide in Vicia faba. Genetics 49:17-38.

43. Ferreira, N. R. and A. Buoniconti (1968) Trisomy after colchicine therapy. Lancet ii: 1304.

44. Ferreira, N. R. and O. Frota-Pessoa (1969) Trisomy after cholchicine therapy. Lancet i:1160-1161.

45. Finch, J. T. and A. Klug· (1976) Solenoidal model for superstructure in chromatin. Proc. Natl. Acad. Sci. USA 73:1897–1901.

46. Flamm, W. G., N. J. Bernheim and L. Fishbein (1970) On the existence of intrastrand cross-links in DNA alkylated with sulfur mustard. Biochim. Biophys. Acta 224:657–659.

47. Freese, E. (1959a) The difference between spontaneous and base-analogue induced mutations of phage T4. Proc. Natl. Acad. Sci. USA 45: 622–633.

48. Freese, E. (1959b) The specific mutagenic effect of base analogues on phage T4. J. Mol. Biol. 1:87–95.

49. Freese, E. (1963) Molecular mechanisms of mutations, in: Molecular Genetics, J. H. Taylor (Ed.), Plenum Press, New York, pp. 207–269.

50. Gaulden, M. E. and J. G. Carlson (1951) Cytological effects of colchicine in the grasshopper neuroblast in vitro with special reference to the origin of the spindle. Exp. Cell. Res. 2:416.

51. Gebhart, E., G. Schwanitz and G. Hartwich (1969) Zytogenetische Wirkung von Vincristin auf menschliche Leukozyten in vivo und in vitro. Med. Klin. 51:2366–2371.

52. Generoso, W. M., W. L. Russell, S. W. Huff, S. K. Stout and D. G. Gosslee (1974) Effects of dose on the induction of dominant-lethal mutations and heritable translocations with ethyl methanesulfonate in male mice. Genetics 77:741–752.

53. Generoso, W. M., K. T. Cain, S. W. Huff and D. G. Gosslee (1978) Heritable translocation test in mice, in: Chemical Mutagens – Principles and Methods for Their Detection, A. Hollaender and F. J. deSerres (Eds.), Plenum Press, New York, Vol. 5, pp. 55–77.

54. Gerchman, L. L. and D. B. Ludlum (1973) The properties of O^6-methylguanine in templates for RNA polymrase. Biochim. Biophys. Acta 308:310–316.

55. Gersch, N. F. and D. O. Jordan (1965) Interaction of DNA with aminoacridines. J. Mol. Biol. 13:138–156.

56. Green, M. H. L. (1979) Mutagenic consequences of chemical reaction with DNA, in: Chemical Carcinogens and DNA, P. L. Grover (Ed.), CRC Press, Boca Raton, Florida, Vol. II, pp. 95–132.

57. Green, M. M. (1959) Reverse mutation in Drosophila and the status of the particulate gene. Genetica 29:1–38.

58. Green, M. M. (1982) Genetic instability in Drosophila melanogaster: Deletion induction by insertion sequences. Proc. Natl. Acad. Sci. USA 79: 5367–5369.

59. Grunberger, D. and I. B. Weinstein (1979) Conformational changes in nucleic acids modified by chemical carcinogens, in: Chemical Carcinogens and DNA, P. L. Grover (Ed.), CRC Press, Boca Raton, Florida, Vol. II, pp. 59–93.

60. Hayes, W. (1968) The Genetics of Bacteria and Their Viruses, 2nd Edition, Blackwell Scientific Publications, Oxford, Edinburgh.

61. Hendler, S., E. Furer and P. R. Srinivasan (1970) Synthesis and chemical properties of monomers and polymers containing 7 methylguanine and an investigation of their substrate or template properties for bacterial DNA or RNA polymerases. Biochemistry 9:4141-4153.

62. Henrich, R. T., T. Nogawa and A. Morishima (1980) In vitro induction of segregational errors of chromosomes by natural cannabinoids in normal human lymphocytes. Environ. Mutagen. 2:139-147.

63. Hertwig, R. (1903) Ueber die Korrelation von Zell- und Kerngrösse und ihre Bedeutung für die geschlechtliche Differenzierung und die Teilung der Zelle. Biol. Zbl. 23:108.

64. Hibner, U. and B. M. Albers (1980) Fidelity of DNA replication catalysed in vitro on a natural DNA template by the T4 bacteriophage multi-enzyme complex. Nature 285:300-305.

65. Hoefnagel, D. (1969) Trisomy after colchicine therapy. Lancet i:1160.

66. Hollstein, M., J. McCann, F. A. Angelosanto and W. W. Nichols (1979) Short-term tests for carcinogens and mutagens. Mutation Res. 65:133-226.

67. Howard, B. D. and J. Tessman (1964) Identification of the altered bases in mutated single stranded DNA. II. In vivo mutagenesis by 5-BUdR and 2-AP. J. Mol. Biol. 9:364-371.

68. Howard-Flanders, P. (1973) DNA repair and recombination. Brit. Med. Bull. 29:226-235.

69. IARC (1979) Monographs on the Evaluation of the Carcinogenic Risk of Chemicals to Humans. International Agency for Research on Cancer, Lyon, IARC Monographs, Suppl. 1, pp. 1-71.

70. Isono, K. and J. Yourno (1974) Chemical carcinogens as frameshift mutations: Salmonella DNA sequence sensitive to mutagenesis by polycyclic carcinogens. Proc. Natl. Acad. Sci. USA 71:1612-1617.

71. Jenkins, N. A., N. G. Copeland, B. A. Taylor and B. K. Lee (1981) Dilute (d) coat color mutation of DBA/2J mice is associated with the site of integration of an ecotropic MuLV genome. Nature 293:370-374.

72. Kavenoff, R. and B. H. Zimm (1973) Chromosome-sized DNA molecules from Drosophila. Chromosoma 41:1-27.

73. Kelner, A. (1953) Growth, respiration and nucleic acid synthesis in UV-irradiated and in photo-reactivated Escherichia coli. J. Bacteriol. 65:252.

74. Kihlman, B. A. (1964) The production of chromosomal aberrations by streptonigrin in Vicia faba. Mutation Res. 1:54-62.

75. Kihlman, B. A., A. T. Natarajan and H. C. Andersson (1978) Use of 5-bromodeoxyuridine-labelling technique for exploring mechanisms involved in the formation of chromosomal aberrations. I. G2 experiments with root tips of Vicia faba. Mutation Res. 55:85-120.

76. Kimball, R. F. (1978) The relation of repair phenomena to mutation induction in bacteria. Mutation Res. 55:85-120.

77. Kirtikar, D. M. and D. A. Goldthwait (1974) The enzymatic release of O^6-methylguanine and 3-methyladenine from DNA reaction with the carcinogen N-methyl-N-nitrosourea. Proc. Natl. Acad. Sci. USA 71:2022-2026.

78. Kolata, G. B. (1980) Genes and cancer: The story of Wilms tumor. Science 207:970-971.

79. Kornberg, A. (1979) Aspects of DNA replication. Cold Spring Harbor Symp. Quant. Biol. 43:1-9.

80. Kunz, B. A. (1982) Genetic effects of deoxyribonucleotide pool imbalances. Environ. Mutagen. 4:695-725.

81. Labhart, P., T. Koller and H. Wunderli (1982) Involvement of higher order chromatin structures in metaphase chromosome organization. Cell 30:115-121.

82. Laurence, D. J. R. (1963) Chain breakage of deoxyribonucleic acid following treatment with low doses of sulphur mustand. Proc. Royal Soc. A 271:520-530.

83. Lawley, P. D. (1966) Some effects of chemical mutagens and carcinogens on nucleic acids. Progr. Nucl. Acid. Res. Mol. Biol. 5:89-131.

84. Lawley, P. D. and P. Brookes (1963) Further studies on the aklylation of nucleic acids and their constituent nucleotides. Biochem. J. 89:127-138.

85. Lawley, P. D., J. H. Lethbridge, P. A. Edwards and K. V. Shooter (1969) Inactivation of bacteriophage T7 by mono and bifunctional sulphur mustard in relation to cross-linking and depurination of bacteriophage DNA. J. Mol. Biol. 38:181-198.

86. Lawley, P. D., A. R. Crathorn, S. A. Shah and B. A. Smith (1972) Biomethylation of deoxyribonucleic acid in cultured human tumor cells (HeLa). Methylated bases other than 5-methylcytosine not detected. Biochem. J. 128: 133-138.

87. Lawley, P. D., D. J. Orr, P. B. Farmer and M. Jarman (1973) Reaction products from N-methyl-N-nitrosourea and DNA containing thymidine residues. Synthesis and identification of a new methylation product, O^4-methylthymidine. Biochem. J. 135:193-201.

88. Ledda, G. M., A. Columbano, P. M. Rao, S. Rajalakshmi, S. Sarma, D. S. R. Sarma (1980) In vivo replication of carcinogen-modified rat liver DNA: increased susceptibility of O^6-methylguanine compared to N^7-methylguanine in replicated DNA to S-1-nuclease. Biochem. Biophys. Res. Comm. 95:816-821.

89. Lefevre, G. (1976) A photographic representation and interpretation of the polytene chromosomes of Drosophila melanogaster salivary glands, in: The Genetics and Biology of Drosophila, M. Ashburner and E. Novitski (Eds.), Academic Press, New York, Vol. 1a, pp. 32-67.

90. Lejeune, J., J. Lafourcade, R. Berger, J. Vialatte, M. Roeswillwald, P. Seringe, and R. Turpin (1963) Trois cas de deletion partielle du bras court d'un chromosome 5. C.R. Acad. Sci. (Paris) 257:3098-3102.

91. Lerman, L. S. (1963) The structure of the DNA-acridine complex. Proc. Natl. Acad. Sci. USA 49:94-102.

92. Lerman, L. S. (1966) Acridine mutagens and DNA structure. J. Cell. Comp. Physiol. 64(Suppl. 1):1-18.

93. Lindahl, T. (1976) New class of enzymes acting on damaged DNA. Nature 256:64-66.

94. Lindahl, T. and A. Andersson (1972) Rate of chain breakage at apurinic sites in double stranded DNA. Biochemistry 11:3618-3623.

95. Lindahl, T. and B. Nyberg (1972) Rate of depurination of native deoxyribonucleic acid. Biochemistry 11:3610-3618.

96. Liquori, A. M., B. DeLerma, F. Ascoli, C. Botre, and M. Trasciatti (1962) Interaction between DNA and polycyclic aromtic hydrocarbons. J. Mol. Biol. 5:521-526.

97. Livneh, Z., D. Elad, and J. Sperling (1979) Enzymatic insertion of purine bases into depurinated DNA in vitro. Proc. Natl. Acad. Sci. USA 76:1089-1093.

98. Loeb, L. A. (1974) Eukaryotic DNA polymerases, in: The Enzymes, P. Boyer (Ed.), Academic Press, New York, Vol. 10, pp. 173-209.

99. Loveless, A. (1969) Possible relevance of 0^6-alkylation of deoxyguanosine to mutagenicity and carcinogenicity of nitrosamines and nitrosamides. Nature 223:206-207.

100. Ludlum, D. B. (1970) The properties of 7 methylguanine containing templates for RNA polymerase. J. Biol. Chem. 245:477-482.

101. Lutz, W. (1979) In vivo covalent binding of organic chemicals to DNA as a quantitative indicator in the process of chemical carcinogenesis. Mutation Res. 65:289-356.

102. Luzzati, D. (1962) The action of nitrous acid on transforming desoxyribonucleic acids. Biochem. Biophys. Res. Comm. 9:508-516.

103. Marsden, M. P. F. and U. K. Laemmli (1979) Metaphase chromosome structure: evidence for a radial loop model. Cell 17:849-858.

104. McCann, J., E. Choi, E. Yamasaki, and B. N. Ames (1975) Detection of carcinogens as mutagens in the Salmonella/ microsome test: Assay of 300 chemicals. Proc. Natl. Acad. Sci. USA 72: 5135-5139.

105. McGhee, J. D. and G. Felsenfeld (1980) Nucleosome structure. Ann. Rev. Biochem. 49:1115-1156.

106. McKusick, V. A. (1978) Mendelian Inheritance in Man, (5th edition), The Johns Hopkins University Press, Baltimore, pp. 1-975.

107. Margison, G. P., and P. J. O'Connor (1973) Biological implications of the instability of N-glycosidic bond of 3-metnyldeoxyadeonsine in DNA. Biochim. Biophys. Acta 331:349-356.

108. Margison, G. P., M. J. Capps, P. J. O'Connor, and A. A. Craig (1973) Loss of 7 methylguanine from rat liver DNA after methylation in vivo with methyl methanesulfonate or dimethyl nitrosamine. Chem. Biol. Interact. 6:119-124.

109. Margoliash, E., and E. L. Smith (1956) Structural and functional aspects of cytochrome c in relation to evolution, in: Evolving Genes and Proteins, V. Bryson and H. J. Vogel (Eds.), Academic Press, New York, pp. 221-242.

110. Marinus, M. G., and N. R. Morris (1975) Pleiotropic effects of a DNA adenine methylation mutant (dam-3) in Escherichia coli. Mutation Res. 28:15-26.

111. Meneghini, R. (1976) Gaps in DNA synthesis by ultraviolet light irradiated W 138 human cells. Biochim. Biophys. Acta 425:419-427.

112. Mildvan, A. S., and L. A. Loeb (1979) The role of metal ions in the mechanisms of DNA and RNA polymerases. CRC Crit. Rev. Biochem. 6:219-244.

113. Mufti, S. (1979) Mutator effects of alleles of phage T4 genes 32, 41, 44, and 45 in the presence of an antimutator polyerase. Virology 94:1-9.

114. Mutfi, S., and H. Bernstein (1974) The DNA-delay mutants of bacteriophage T4. J. Virol. 14:860-871.

115. Muzyczka, N., R. L. Poland, and M. J. Bessman (1972) Studies on the biochemical basis of spontaneous mutation. I. A comparison of the deoxyribonucleic acid polymerase of mutator, antimutator and wild-type strains of bacteriophage T4. J. Biol. Chem. 247:7116-7122.

116. Natarajan, A. T., and G. Obe (1978) Molecular mechanisms involved in the production of chromosomal aberrations. I. Utilization of Neurospora endonuclease for the study of aberration production in G2 stage of the cell cycle. Mutation Res. 52:137-149.

117. Natarajan, A. T., G. Obe, A. A. van Zeeland, F. Palitti, M. Meijers and E. A. M. Verdegaal-Immerzeel (1980) Molecular mechanisms involved in the production of chromosomal aberrations. II. Utilization of Neurospora endonuclease for the study of aberration production by X-rays in G1 and G2 stages of the cell cycle. Mutation Res. 69:293-305.

118. Neary, G. T., J. R. K. Savage, and H. J. Evans (1964) Chromatid aberrations in Tradescantia pollen tubes induced by monochromatid X-rays of quantum energy 3 and 1.5 keV. Int. J. Radiat. Biol. 8:1-19.

119. Neel, J. V. (1983) Frequency of spontaneous and induced "point" mutations in higher eukaryotes. J. Heredity 74:2-15.

120. Oeda, K., T. Horiuchi, and M. Sekiguchi (1981) Molecular cloning of the uvr D gene of Escherichia coli that controls ultraviolet sensitivity and spontaneous mutation frequency. Mol. Gen. Genet. 101:227-244.

121. Oeda, K., T. Horiuchi, and M. Sekiguchi (1982) The uvr D gene of E. coli encodes a DNA-dependent ATPase. Nature 298:98-100.

122. Ohno, S. (1970) Evolution by Gene Duplication, Springer-Verlag, Heidelberg, New York, pp. 1-160.

123. Okada, T. A., and D. E. Comings (1980) A search for protein cores in chromosomes: is the scaffold an artifact? Am. J. Hum. Genet. 32:814-832.

124. Orgel, L. E. (1965) The chemical basis of mutation. Advan. Enzymol. 27:289-346.

125. Orgel, A., and L. E. Orgel (1965) Induction of mutations in bacteriophage T4 with divalent manganese. J. Mol. Biol. 14:453-457.

126. Paulson, J. R., and J. K. Laemmli (1977) The structure of histone-depleted metaphase chromosomes. Cell 12:817-828.

127. Perry, P., and H. J. Evans (1975) Cytological detection of mutagen-carcinogen exposure by sister chromatid exchange. Nature 258:121-125.

128. Prescott, D. M. (1970) The structure and replication of eucaryotic chromosomes, in: Advances in Cell Biology, D. M. Prescott, L. Goldstein and E. McConkey (Eds.), Appleton-Century-Crofts, New York, pp. 57-117.

129. Radman, M. (1975) SOS repair hypothesis: Phenomenology of an inducible DNA repair which is accompanied by mutagenesis, in: Molecular Mechanisms for the Repair of DNA, Part A, P. C. Hanawalt and R. B. Setlow (Eds.), Plenum Press, New York, pp. 355-367.

130. Radman, M., S. Spadari, and G. Villani (1978a) Mutagenesis and cell transformation by ultraviolet radiation: Many hypotheses for few results. Natl. Cancer Inst. Monogr. 50:121-127.

131. Radman, M., S. Boiteux, O. Doubleday, G. Villani, and S. Spadari (1978b) Making and correcting errors in DNA synthesis: in vitro and semi-in vivo studies of mutagenesis. J. Supramol. Struct 12(Suppl):14.

132. Radman, M., G. Villani, S. Boiteux, A. R. Kinsetta, B. W. Glickman, and S. Spadari (1979) Replication fidelity: mechanisms of mutation avoidance and mutation fixation. Cold Spring Harbor Symp. Quant. Biol. 43: 937-946.

133. Ramel, C. (1969) Genetic effects of organic mercury compounds. I. Cytological inviestigations on Allium roots. Hereditas 61:208-230.

134. Ramel, C. (1972) Genetic effects, in: Mercury in the Environment, an Epidemiological and Toxicological Appraisal, L. Fridberg and I. Vostal (Eds.), CRC Press, Boca Raton, Florida, pp. 169–181.

135. Ramel C., and J. Magnusson (1969) Genetic effects of organic mercury compounds. II. Chromosome segregation in Drosophila melanogaster. Hereditas 61:231–254.

136. Rapoport, J. A. (1940) Multiple linear repetitions of chromosome blocks and their evolutionary significance. J. Gen. Biol. (Moscow) 1:235–270.

137. Richet, E., R. Kern, M. Kobiyama, and M. Hirota (1980) Isolation of DNA-dependent ATPase I mutants of E. coli, in: Mechanistic Studies of DNA Replication and Genetic Recombination, B. Alberts (Ed.), Academic Press, New York, pp. 606–608.

138. Reiger, R., A. Michaelis, and M. M. Green (1968) A Gloccary of Genetics and Cytogenetics, Springer-Verlag, New York, pp. 1–647.

139. Ripley, L. S. (1975) Transversion mutagenesis in bacteriophage T4. Molec. Gen. Genet. 141:23–40.

140. Ris, H., and D. F. Kubai (1970) Chromosome structure. Ann. Rev. Genet. 4:263–294.

141. Russell, L. B. (1964) Genetic and functional mosaicism in the mouse, in: Role of Chromosomes in Development, M. Locke (Ed.), 23rd Symp. Soc. Develop. and Growth, Academic Press, New York.

142. Russell, L. B. (1976) Numerical sex-chromosome anomalies in mammals: Their spontaneous occurrence and use in mutagenesis studies, in: Chemical Mutagens - Principles and Methods for Their Detection, A. Hollaender (Ed.), Plenum Press, New York, Vol. 4, 55–91.

143. Saffhill, R., and P. J. Abbott (1978) Formation of O^2-methylthymidine in poly(dA-dT) on methylation with N-methyl-N-nitrosourea and dimethyl sulfate. Evidence that O^2-methylthymidine does not miscode during DNA synthesis. Nucleic Acid Res. 5:1971–1978.

144. Sarabhai, A., A. D. W. Stretton, S. Brenner, and A. Bolle (1964) Colinearity of the gene with the polypeptide chain. Nature 201:13–17.

145. Sax, K. (1940) An analysis of X-ray induced chromosome aberrations in Tradescantia. Genetics 25:41–68.

146. Scott, D. (1980) Molecular mechanisms of chromosome structural changes, in: Progress in Evironmental Mutagenesis, M. Alacevic (Ed.), Elsevier/North-Holland, Amsterdam, New York, Oxford, pp. 101–113.

147. Scott, D., and H. J. Evans (1964) On the non-requirement for deoxyribonucleic acid synthesis in the production of chromosome aberrations by 8-ethoxycaffeine. Mutation Res. 1:146–156.

148. Scott, D., and H. J. Evans (1967) X-ray-induced chromosomal aberrations in Vicia faba: Changes in response during the cell cycle. Mutation Res. 4: 579-599.

149. Sega, G. A., and J. G. Owens (1978) Ethylation of DNA and protamine by ethyl methanesulfonate in the germ cells of male mice and the relevancy of these molecular targets to the induction of dominant lethals. Mutation Res. 52:87-106.

150. Sesnowitz-Horn, S., and E. A. Adelberg (1968) Proflavin treatment of Escherichia coli: Generation of frameshift mutations. Cold Spring Harbor Symp. Quant. Biol. 33:393-402.

151. Shafer, D. A. (1977) Replication bypass model of sister chromatid exchanges and implications for Bloom's syndrome and Fanconi's anemia. Hum. Genet. 39:177-190.

152. Sherman, C. W., and L. A. Loeb (1977) Depurination decreases fidelity of DNA synthesis in vitro. Nature 270:537-538.

153. Sherman, C. W., and L. A. Loeb (1979) Effects of depurination on the fidelity of DNA synthesis. J. Mol. Biol. 128:197-218.

154. Shugar, D., C. P. Huber, and G. I. Birnbaum (1976) Mechanism of hydroxylamine mutagenesis. Crystal structure and confirmation of $1,5$-dimethyl-N^4-hydroxy-cytosine. Biochim. Biophys. Acta 447:274-284.

155. Singer, B. (1977) Sites in nucleic acids reacting with alkylating agents of differing carcinogenicity or mutagenicity. J. Toxicol. Environm. Health 2:1279-1295.

156. Singer, B., H. Fraenkel-Conrat, and J. T. Kusmiereck (1978) Preparation and template activities of polynucleotides containing O^2- and O^4-alkyl uridine. Proc. Natl. Acad. Sci. USA 75:1722-1726.

157. Sirover, M. A., and L. A. Loeb (1980) On the fidelity of DNA synthesis: Effects of steroids and intercalating agents. Chem.-Biol. Interactions 30:1-8.

158. Smith, M. D., R. R. Green, L. S. Ripley, and J. W. Drake (1973) Thyminless mutagenesis in bacteriophage T4. Genetics 74: 393-403.

159. Stark, A. A., J. M. Essigmann, A. L. Demain, T. R. Skopek, and G. N. Wogan (1979) Aflatoxin B1 mutagenesis, DNA binding, and adduct formation in Salmonella typhimurium. Proc. Natl. Acad. Sci. USA 76:1343-1347.

160. Steffensen, D. M. (1962) A variable distance for chromosome exchange dependent on cellular activity, temperature and other conditions. Radiation Res. 16:590.

161. Strauss, B., and T. Hill (1970) The intermediate in the degradation of DNA alkylated with a monofunctional alkylating agent. Biochim. Biophys. Acta 213:14-25.

162. Streisinger, G., Y. Okada, J. Emrich, I. Newton, A. Tsugita, E. Terzaghi, and M. Inouye (1966) Frameshift mutations and

the genetic code. Cold Spring Harbor Symp. Quant. Biol. 31:77–84.

163. Stubblefield, E., and W. Wray (1971) Architecture of the Chinese hamster metaphase chromosome. Chromosoma 32: 262–294.

164. Talmud, P. J., and D. Lewis (1974a) Mutagenicity of amnio-acid analogs in Coprinus lagopus. Genet. Res. 23:47–61.

165. Talmud, P. J., and D. Lewis (1974b) Mutagenicity of amino-acid analogs in eukaryotes. Nature 249:563–564.

166. Tessman, I. (1962) The induction of large deletions by nitrous acid. J. Mol. Biol. 5:442–445.

167. Tessman, J., R. K. Poddar, and S. Kumar (1964) Identification of the altered bases in mutant single stranded DNA. I. In vitro mutagenesis by hydroxylamine, ethyl methanesulfonate and nitrous acid. J. Mol. Biol. 9:352–363.

168. Therman, E. (1980) Human Chromosomes, Structure, Behavior, Effects, Springer–Verlag, New York, pp. 1–235.

169. Thoma, F., and T. Koller (1981) Unravelled nucleosomes, nucleosome beads and higher order structures of chromatin: influence of non–histone components and histone H1. J. Mol. Biol. 149:709–733.

170. Thoma, R., T. Koller, and A. Klug (1979) Involvement of histone H1 in the organization of the nucleosome and of the salt–dependent superstructures of chromatin. J. Cell. Biol. 83:403–427.

171. Topal, M. D., and J. R. Fresco (1967) Complementary base pairing and the origin of substitution mutations. Nature 263:285–289.

172. Van der Schans, G. P. (1969) On the production of breaks in DNA by gamma–rays. Int. J. Radiat. Biol. 16:58.

173. Vig, B. K., and R. Lewis (1978) Genetic toxicology of bleomycin. Mutation Res. 55: 121–145.

174. Vogel, F., and A. G. Motulsky (1979) Human Genetics: Problems and Approaches, Springer–Verlag, Heidelberg, New York, pp. 1–700.

175. de Vries, H. (1901) Die Mutationstheorie, Bd. 1, Veit & Comp., Leipzig.

176. Waring, M. J. (1965) Complex formation between ethidium bromide and nucleic acids. J. Mol. Biol. 13:269–282.

177. Watson, J. D., and F. H. C. Crick (1953) The structure of DNA. Cold Spring Harbor Symp. Quant. Biol. 18:123–131.

178. Weigle, J. J. (1953) Induction of mutations in bacterial virus. Proc. Natl. Acad. Sci. USA 39:628–636.

179. Weigle, J. J., and R. Dulbecco (1953) Induction of mutations in bacteriophage T3 by ultraviolet light. Experientia 9:372–373.

180. Weinstein, I. B., and D. Grunberger (1974) Structural and functional changes in nucleic acids modified by chemical carcinogens, in: Chemical Carcinogenesis, Part A, P. O. P.

T'so and J. A. DiPaolo (Eds.), Marcel Dekker, New York, pp. 217-235.

181. White, M. J. D. (1973) The Chromosomes, 6th Edition, Chapman and Hall, London.

182. Whitney, J. B. III, and M. L. Lamoreaux (1981) Transposable elements controlling genetic instabilities in mammals. J. Hered. 73:12-18.

183. Wickner, S., and J. Hurwitz (1974) Conversion of ϕ X174 viral DNA to double-stranded form by purified Escherichia coli proteins. Proc. Natl. Acad. Sci. USA 71:4120-1424.

184. Witkin, E. M. (1967) Mutation-proof and mutation-prone modes of survival in derivatives of Escherichia coli B differing in sensitivity to ultraviolet light. Brookhaven Symp. Biol. 20:17-55.

185. Witkowski, R., and F. H. Herrmann (1976) Einführung in die Klinische Genetik, Vieweg, Braunschweig, pp. 1-218.

186. Wolff, S. (1977) Sister chromatid exchange. Ann. Rev. Genet. 11:183-201.

187. Wolff, S., and P. Perry (1975) Insights on chromosome structure from sister chromatid exchange ratios and the lack of both isolabelling and heterolabelling as determined by the FPG technique. Exp. Cell Res. 93:23-30.

188. Wolff, S., K. C. Atwood, M. L. Randolph, and H. E. Luippold (1958) Factors limiting the number of radiation induced chromosome exchanges. I. Distance: Evidence from non-interaction of X-ray and neutron-induced breaks. J. Biophys. Biochem. Cytol. 4:365-372.

189. Yunis, J. J. (1976) High resolution of human chromosomes. Science 191:1268-1270.

190. Yunis, J. J. (1980) Nomenclature for high resolution human chromosomes. Cancer Genet. Cytogenet. 2:221-229.

191. Zampieri, A., and J. Greenberg (1965) Mutagenesis by acridine orange and proflavin in Escherichia coli strains S. Mutation Res. 2:552-556.

DNA REPAIR PROCESSES

Tomas Lindahl

Imperial Cancer Research Fund
Mill Hill Laboratories
London, NW7 1AD, United Kingdom

SUMMARY

Several different mechanisms for correction of damaged or
altered nucleotide residues in DNA are described. In a few cases,
direct enzymatic reversal of potentially mutagenic alterations in
DNA may occur. A more common and general pathway of repair
involves excision of the damaged moiety. This may be initiated by
one of several DNA glycosylases, a set of small enzymes which
specifically cleave a base-deoxyribose bond at an altered site.
Alternatively, a complex nuclease may recognize a distorted region
in DNA and liberate an oligonucleotide containing the damaged
residue. Repair of certain important lethal lesions such as
inter-strand cross-links can occur by a similar pathway. Damage
to DNA near replicating regions poses special problems, but the
resulting daughter-strand gaps can be corrected by a recombination
process. Several of the above repair pathways are inducible in
lower organisms, but it is at present unclear if inducible DNA
repair occurs in human cells.

INTRODUCTION

Mutagenic alterations in the cellular DNA occur as a
consequence of exposure to a wide variety of environmental
chemicals and ionizing or ultraviolet radiation. In addition, DNA
is susceptible to slow spontaneous hydrolysis at neutral pH, and
potentially dangerous lesions continuously arise in this way as a
background of mutagenic damage. Living cells possess several
different DNA repair systems to correct or minimize such
chromosomal alterations, and these repair functions are of crucial

importance in preventing unacceptably high frequencies of
mutagenesis and carcinogenesis.

NATURE OF MUTAGENIC DAMAGE

Many different kinds of alterations in the covalent structure
of DNA can occur as a consequence of genomic damage. It would
clearly simplify the consideration of various mutagenic lesions,
and their repair, if a simple scheme of classification of
different kinds of damage into a few groups could be constructed.
A number of such schemes have been proposed recently, but none is
entirely successful. While there is general agreement that
single-site mutations in DNA arise as a consequence of damage to
the nitrogenous bases rather than to the sugar residues or
phosphodiester bonds, the large variety of lesions that can be
obtained by exposure to many different group-specific reagents
have to some extent defied easy classification, and the effect and
repair of a particular kind of lesion may often have to be
regarded as an individual problem. However, it is sometimes
useful to classify DNA base lesions roughly in two groups, i.e.
those causing major or minor helix distortion (2). An example of
the former type is a pyrimidine dimer, and of the latter an
apurinic site. An alternative system of classification is to
arrange different types of DNA damage into groups determined by
the particular type of enzymic event which initiates the repair
process (8). The rationale for this scheme is that knowledge of
the properties and specificities of the various DNA repair enzymes
has made great progress in recent years, and several such enzymes
may now be used as probes to identify certain types of damage.
Thus, all lesions initially attacked by the uvr endonuclease would
fall into one group, which is similar (but not identical) to the
class of lesions causing major helix distortion in Cerutti's
scheme (2). Additional important aspects of DNA lesions depend on
whether the mutagen reacts at a single site or at two adjacent
sites, causing the formation of a cross-link in DNA. Moreover,
many important mutagens for mammalian cells require metabolic
activation by microsomal mixed function oxidases in order to be
converted to a form that will react directly with DNA.

Alkylating agents form the largest group of chemical
mutagens, and many of the important environmental carcinogens
belong to this group. These agents bind covalently at
nucleophilic centers in DNA. Simple methylating and ethylating
agents, which do not require metabolic activation, have often been
used for model studies on mutagenesis and carcinogenesis. Such
alkylating agents may be subdivided into two groups, depending on
whether they act primarily by a bimolecular mechanism with
formation of a transition complex with the nucleophile (S_N2) or by
a unimolecular mechanism (S_N1) in which the formation of a

reactive ion is the rate-limiting step. The S_N2 reagents show a strong preference for the most nucleophilic sites, while S_N1 reagents attack all nucleophiles and consequently yield a broader variety of products. The S_N1 agents are strongly mutagenic.

Several research groups have characterized the spectrum of alkylation products in DNA, and the major products as well as the significant minor products are known in detail. The most important contributions to this area of research have been those by P.D. Lawley and co-workers (7) and of B. Singer (17). The most abundant DNA lesions are 7-methylguanine and 3-methyladenine. A relatively minor lesion of great biological significance is O^6-methylguanine (12). The alkylation of guanine at this position freezes the residue in an erroneous tautomeric form, which base-pairs with thymine instead of cytosine. O^6-Methylguanine in DNA is induced efficiently by S_N1 reagents, that is, by agents such as methylnitrosourea and methylnitrosoguanidine. In consequence, such agents are strongly mutagenic and carcinogenic, although a cellular DNA repair process to some extent alleviates the effect.

More complex alkylating agents, such as the clinically-used chloroethylnitrosoureas, react at the same sites in DNA as the methylating agents. However, these compounds also generate cross-links in the DNA, because the chloroethyl mono-adducts initially formed in the DNA are chemically reactive and can bind at a second site. For steric reasons large alkylating agents such as the metabolically activated form of benzopyrene tend to bind preferentially at other sites, i.e. at the N^2 position of guanine and at the C^8 position of purines, in addition to the N^7 position of guanine. The latter position is the preferred reaction site not only of methylating and ethylating agents, but also of bulkier agents such as nitrogen mustards and the liver carcinogen aflatoxin (17).

Ultraviolet irradiation has often been used to introduce DNA damage in a controlled fashion in studies on DNA repair processes. Sunlight is an important mutagen; what is more, it induces a single type of stable and easily measured lesion, i.e. the pyrimidine dimer. Recently, DNA sequencing studies have indicated that in addition to the classic form of cyclobutane pyrimidine dimers, non-cyclobutane dimers (called 6-4 dimers) also occur, and the latter lesions may account for a sizeable part of the mutagenic effect of ultraviolet light (10). Besides this complication, mutagenesis by bulky lesions such as pyrimidine dimers requires the presence of certain proteins, the umuC and umuD gene products, in the E. coli model system (4). The reason for this is not yet understood, but it is believed that such proteins allow a DNA polymerase to replicate over the damaged area, inserting erroneous nucleotides into DNA. Without the accessory factors, DNA replication would presumably just stop at

the pyrimidine dimer, resulting in a cell killing effect but no
mutations. In contrast, small DNA lesions such as O^6-
methylguanine are misread by DNA polymerases even in in vitro
systems without accessory proteins.

The mutagenic base lesions caused by ionizing radiation have
been much less studied than those introduced by ultraviolet light.
This is largely because many different lesions are introduced
(18), which are difficult to analyze separately. In addition, DNA
chain breaks are also caused by X-rays, and these may generate
chromosomal sequence rearrangements which contribute to the
biological effect. Much work has been done on DNA single-strand
breaks introduced by exposure to ionizing radiation, but this
easily-repairable type of lesion is probably of little biological
significance with regard to the mutagenic and carcinogenic effects
of radiation exposure.

In addition to these various types of DNA alterations caused
by external agents, the spontaneous hydrolysis of DNA generates
apurinic sites and deaminated bases (8). The latter, in
particular, are premutagenic lesions. This background damage is
very efficiently repaired, however, and such repair processes are
of course essential to prevent high spontaneous mutation rates in
living cells.

EXCISION-REPAIR

The most common and important form of DNA repair involves the
enzymatic removal of damaged DNA bases, followed by their
replacement with the correct, undamaged counterparts. This mode
of repair, termed excision-repair, was discovered in 1964 by
P. Howard-Flanders (1), R.B. Setlow (16), and their coworkers.
The initial incision of the DNA at the damaged sites can occur in
two different ways; either a nuclease cleaves a phosphodiester
bond, adjacent to the damaged site, or a DNA glycosylase cleaves
the base-sugar bond of the damage nucleotide. In the latter case,
an endonuclease specific for apurinic and apyrimidinic sites in
DNA generates a chain break at the site where an altered base has
been liberated. In either case, a short patch is excised, and
then filled in by a DNA polymerase in a repair replication
process, followed by sealing of the last phosphodiester bond by
DNA ligase (Figs. 1 and 2).

The specificity of the repair process is governed by the
enzyme that initially recognizes (or does not recognize) a certain
lesion. In the case of pyrimidine dimers, a large E. coli enzyme
comprises 3 different subunits; it requires ATP as a cofactor, and
is needed for specific incision of ultraviolet-irradiated DNA
(15). These subunits are the products of the uvrA, uvrB, and uvrC

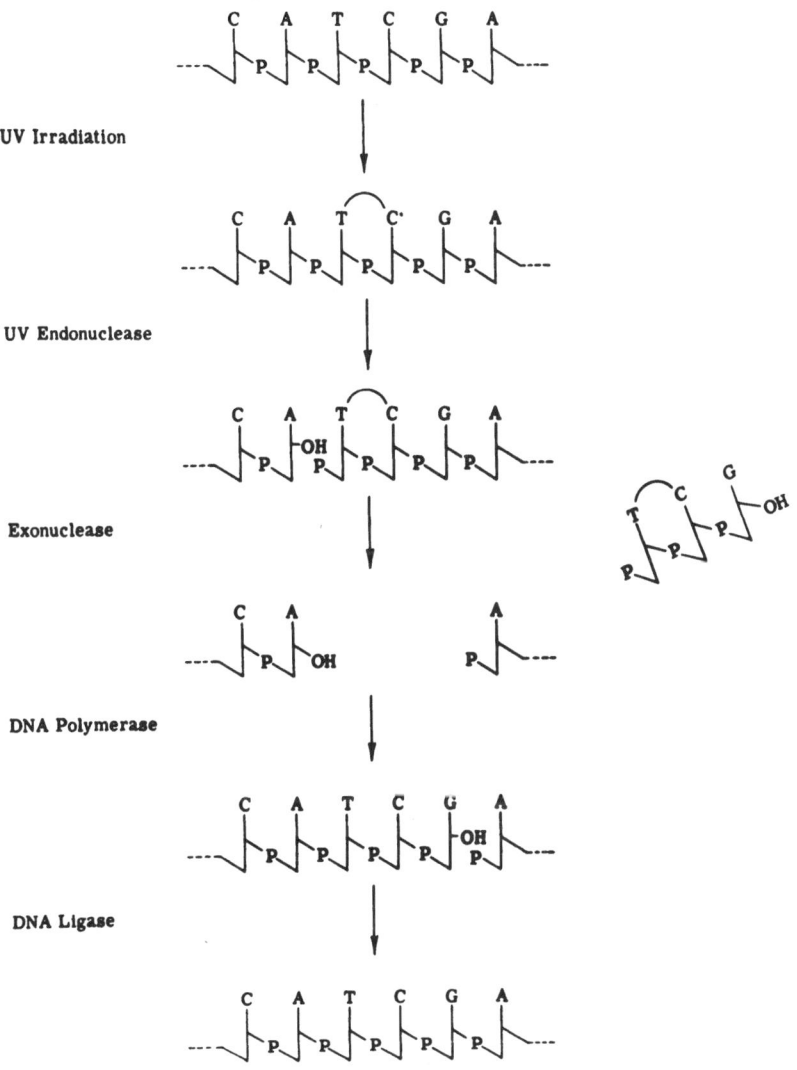

Figure 1. Excision-repair of a pyrimidine dimer in DNA. The
      initial repair event involves cleavage of a phosphodiester
      bond adjacent to the altered residues by a damage-specific
      endonuclease. The UV endonuclease and exonuclease steps may
      occur as a single concerted step, catalyzed by the same
      enzyme. The complementary strand in the double-helical DNA
      is not shown.

genes of E. coli, and all three genes have now been cloned by
recombinant DNA techniques onto small, multicopy plasmids (14).
It is thought that the enzyme incises DNA at two sites in the same
strand, about 13 nucleotides apart and on each side of the dimer,

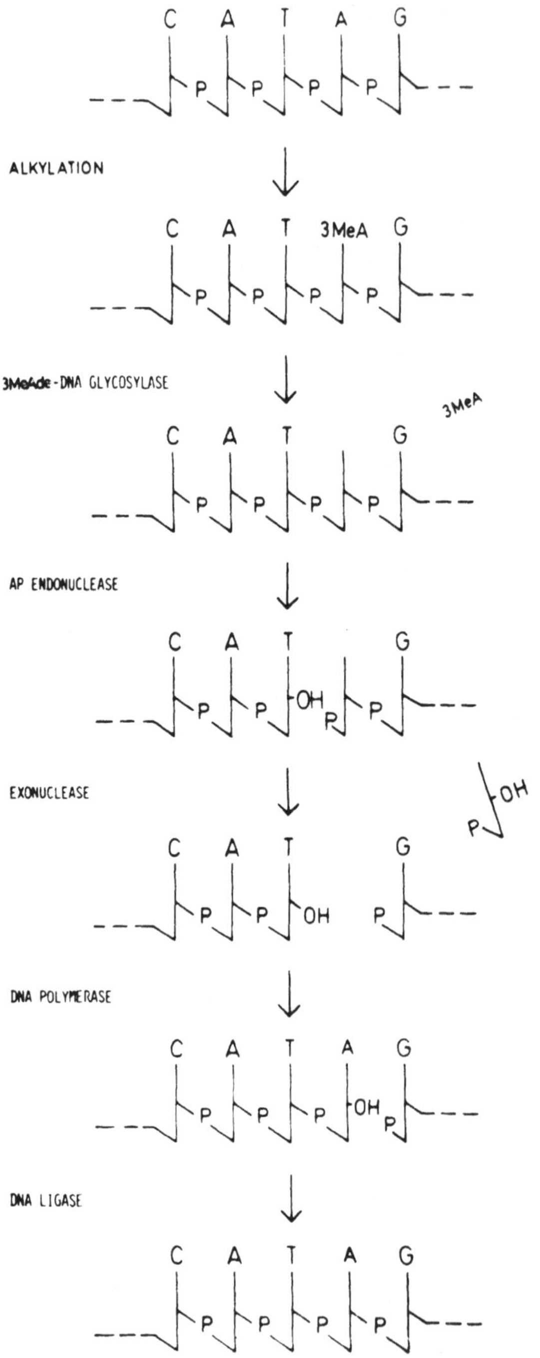

Figure 2.   Excision-repair of an alkylated adenine residue in DNA.
The initial reapir event involves cleavage of the base-sugar
bond of the damaged nucleotide residue by a DNA glycosylase.
The complementary strand in the double-helical DNA is not
shown.

to generate a small gap in the DNA. Thus, a separate excision step catalyzed by an exonuclease might not be necessary (see Fig. 1). The uvr endonuclease clearly recognizes a major distortion in the DNA helix, associated with a change in the covalent DNA structure, because the same enzyme can incise DNA at several chemically dissimilar but bulky lesions, such as psoralen base adducts and benzpyrene diol epoxide base adducts.

Human cells possess a corresponding, complex endonuclease for incision of DNA at pyrimidine dimers. An inherited disorder, xeroderma pigmentosum, appears to result from a deficiency in this enzyme (5). Interestingly, xeroderma pigmentosum patients exhibit a vastly increased frequency of ultraviolet light-induced skin cancer. The biochemistry of the human enzyme is at present not as well understood as that of the E. coli enzyme.

In contrast to the broad substrate specificity of the uvr enzyme, DNA glycosylases are small proteins (molecular weight 20,000 to 30,000), with narrow substrate specificities. Thus, several different enzymes of this class have been found, and each one seems to recognize only one, or a small group of lesions. Both E. coli and mammalian cells have distinct DNA glycosylases for the removal of the following abnormal or fragmented DNA bases: uracil, hypoxanthine, 3-methyladenine, 3-methylguanine, formamidopyrimidine, and urea (9). In addition, certain ultraviolet-resistant organisms such as Micrococcus luteus and phage T4-infected E. coli possess a special pyrimidine dimer-DNA glycosylase for unhooking pyrimidine dimers, thereby allowing their excision (6). In conjunction with the DNA glycosylases, endonucleases for apurinic and apyrimidinic sites (AP endonucleases) are required, and several such activities have been described in E. coli and in mammalian cells (9).

REVERSAL OF DAMAGE

The first DNA repair process discovered was the photoreactivation of pyrimidine dimers. An enzyme, photolyase, can directly monomerize a dimer in a light-dependent reaction, leading to increased survival after radiation exposure. The phenomenon of photoreactivation of DNA has been well documented in bacteria and fungi, but it is doubtful if it occurs in mammalian cells. The light dependence of the enzymatic reaction is believed to depend on an absorbing co-enzyme, which has been tentatively identified as a riboflavin derivative.

An entirely different type of direct reversal of DNA damage occurs for the highly mutagenic alkylation product O^6-methylguanine. In this case, a methyltransferase of molecular weight about 20,000 catalyzes the transfer of the methyl group

from the alkylated guanine residue in DNA to one of its own
cysteine residues. Thus, unmethylated guanine is regenerated in
DNA, while S-methylcysteine is formed in the protein. The
methylated enzyme is trapped as a dead-end complex as a result of
the reaction. This means that each repair event consumes an
enzyme molecule. The biochemistry of this unusual reaction is
exactly the same both in $\underline{E}$. $\underline{coli}$ and in mammalian (including
human) cells (9).

Recently, many human tumor cell lines and tumor virus-
transformed cells have been found to be deficient in the O^6-
methylguanine-DNA methyltransferase. Such cells, called Mer⁻ or
Mex⁻, are more sensitive to killing by simple alkylating agents
than normal cells. In this regard, it is not known if cellular
transformation occurs more readily in rare Mer⁻ variant cells in a
normal cell population, leading to an overrepresentation of this
phenotype among the transformed cells, or if growth of certain
types of tumor cells in tissue culture favors selection of the
Mer⁻ phenotype.

REPAIR OF DAMAGE IN NEWLY REPLICATED DNA

When a DNA replication fork reaches a pyrimidine dimer, DNA
synthesis cannot proceed over the dimer-containing region, but
resumes at some point further along the template. In consequence
a gap will be present in the newly synthesized DNA chain opposite
the dimer in the parental chain. This lesion may be repaired by a
recombination event, employing the intact information in the other
newly synthesized daughter DNA molecule. The process of postreplication repair was discovered by W.D. Rupp and P. Howard-
Flanders in 1967 (13). It provides an alternative DNA repair
pathway to excision-repair. An obligatory requirement for postreplication repair is the protein product of the $\underline{recA}^+$ gene (11).
This interesting protein binds to DNA and facilitates the
occurrence of recombinational events between homologous sequences.

In cells that have received large doses of DNA damage, the
$\underline{recA}^+$ protein is induced to high levels. At the same time,
several other proteins are induced, at least some of which (such
as the products of the $\underline{uvrA}^+$ and $\underline{uvrB}^+$ genes) are enzymes active
in DNA repair. Consequently, the repair capacity of the cell is
improved. The mechanism of induction is known in considerable
detail: the $\underline{recA}^+$ protein, when bound to single-stranded DNA,
cleaves and inactivates a common repressor protein, which is the
product of the $\underline{lexA}^+$ gene. The induction of these repair
functions has been termed the SOS response, because the extra
repair capacity obtained is error-prone and therefore mutagenic.
The latter phenomenon depends critically on two of the induced SOS
functions, called $\underline{umuC}^+$ and $\underline{umuD}^+$ (4). These two proteins, which

are synthesized from the same operon, somehow permit DNA replication to proceed over a damaged template. However, because the template cannot be read properly, incorrect deoxynucleoside triphosphates are to some extent used as precursors during synthesis.

It is at present unclear if higher cells have an inducible, error-prone DNA repair system. Much more work is required in this area. So far, it has been shown that some lower eukaryotic cells have a protein similar to the recA$^+$ protein of E. coli, but it is not known if there is a counterpart in human cells.

An inducible repair system specific for higher cells, and different from those described above, involves the enzymatic synthesis of the unusual polymer poly(ADP-ribose) in cell nuclei, using NAD$^+$ as precursor. The presence of this polymer in increased amounts facilitates the repair of strand breaks in the DNA of mammalian cells, possibly by enhancing the activity of a DNA ligase (3).

In conclusion, several different DNA repair pathways occur in living cells. A defect in any of these processes may be expected to result in increased susceptibility of the individual to environmental mutagens and carcinogens. Studies of the spontaneous and induced mutation frequencies in cells with specific DNA repair defects will help to elucidate and evaluate the effects of mutagens on living cells.

REFERENCES

1. Boyce, R.P., and P. Howard-Flanders (1964) Release of ultraviolet light-induced thymine dimers from DNA in E. coli K-12. Proc. Natl. Acad. Sci. USA 51:293-300.
2. Cerutti, P.A. (1975) Repairable damage in DNA: Overview, In Molecular Mechanisms for Repair of DNA, P.C. Hanawalt, and R.B. Setlow, Eds., Plenum Press, New York, pp. 3-12.
3. Creissen, D., and S. Shall (1982) Regulation of DNA ligase activity by poly(ADP-ribose). Nature 296:271-272.
4. Elledge, S.J., and G.C. Walker (1983) Proteins required for ultraviolet light and chemical mutagenesis. J. Mol. Biol. 164:175-192.
5. Friedberg, E.C. (1981) Xeroderma pigmentosum - a human model of defective DNA repair, In Chromosome Damage and Repair, E. Seeberg, and K. Kleppe, Eds., Plenum Press, New York, pp. 313-320.
6. Grafstrom, R., N. Shaper, L. Grossman, and W. Haseltine (1981) Incision of pyrimidine dimer containing DNA by small molecular weight enzymes, In Chromosome Damage and Repair,

E. Seeberg, and K. Kleppe, Eds., Plenum Press, New York, pp. 85-90.

7. Lawley, P.D. (1966) Effects of some chemical mutagens and carcinogens on nucleic acids. Progr. Nucleic Acid Res. Mol. Biol. 5:89-131.

8. Lindahl, T. (1979) DNA glycosylases, endonucleases for apurinic/apyrimidinic sites, and base excision-repair. Progr. Nucleic Acid Res. Mol. Biol. 22:135-192.

9. Lindahl, T. (1982) DNA repair enzymes. Ann. Rev. Biochem. 51:61-87.

10. Lippke, J.A., L.K. Gordon, D.E. Brash, and W.A. Haseltine (1981) Distribution of UV light-induced damage in a defined sequence of human DNA: Detection of alkaline-sensitive lesions at pyrimidine nucleoside-cytidine sequences. Proc. Natl. Acad. Sci. USA 78:3388-3392.

11. Little, J.W., and D.W. Mount (1982) The SOS regulatory system of E. coli. Cell 29:11-22.

12. Medcalf, A.S.C., and P.D. Lawley (1980) Time course of O^6-methylguanine removal from DNA of methylnitrosourea-treated human fibroblasts. Nature 289:796-798.

13. Rupp, W.D., and P. Howard-Flanders (1968) Discontinuities in the DNA synthesized in an excision-defective strain of E. coli following ultraviolet irradiation. J. Mol. Biol. 31:291-304.

14. Sancar, A., R.P. Wharton, S. Seltzer, B. Kacinski, N. Clarke, and W.D. Rupp (1981) Identification of the uvrA gene product. J. Mol. Biol. 148:45-62.

15. Seeberg, E. (1978) Reconstitution of an E. coli repair endonuclease activity from the separated $uvrA^+$ and $uvrB^+$/ $uvrC^+$ gene products. Proc. Natl. Acad. Sci. USA 75:2569-2573.

16. Setlow, R.B., and W.L. Carrier (1964) The disappearance of thymine dimers from DNA: An error-correcting mechanism. Proc. Natl. Acad. Sci. USA 51:226-231.

17. Singer, B., and J.T. Kusmierek (1982) Chemical mutagenesis. Ann. Rev. Biochem. 51:655-693.

18. Teoule, R. (1978) Radiation-induced degradation of the base component in DNA and related substances - final products, In Effects of Ionizing Radiation on DNA, J. Hutterman, W. Kohnlein, and R. Teoule, Eds., Springer-Verlag, Berlin, pp. 171-251.

THE ROLE OF DNA REPAIR IN MUTATION INDUCTION

F.E. Würgler
Institute of Toxicology
Swiss Federal Institute of Technology
and University of Zürich, Switzerland

SUMMARY

Mutations are changes in coding or regulating DNA sequences, or else, represent rearrangements of the genetic material within the genome. To explain the action of physical and chemical mutagens, it is generally assumed that most of these agents interact with DNA. Some chemicals interact directly with DNA, but many carcinogens need metabolic activation before a reactive metabolite can react as the ultimate carcinogen with DNA. Some mutagens, however, act in a rather indirect way, e.g., by changing nucleotide pools or other components of DNA metabolism, or via oxygen radicals. The primary DNA defects caused by the damaging agents can be modified by intracellular processes such as DNA repair. They may either lead to the removal of premutational lesions or may allow the cell to cope with the adverse consequences of replication on damaged templates. We assume that all living organisms are capable of repairing damaged DNA. Experimental studies on the biological and genetical significance of these pathways have concentrated so far primarily on the analysis of a few suitable models such as E. coli, yeast, Drosophila, and mice or mammalian and human cells in culture. In some mutants which were found to be defective in certain modes of DNA repair, mutagen-induced killing, mutation and recombination are enhanced over wild-type responses. Other repair defective mutants exist that are sensitive to the lethal effects of mutagens but in which induced mutagensis is suppressed or absent. Hence, it seems that DNA repair systems represent causative as well as ameliorative factors in mutagenesis. We distinguish the error-free modes of repair from the error-prone or mutagenic modes which are involved in the production of mutations. It remains to be

determined biochemically how distinct error-prone repair or
misrepair may be from misreplication.

INTRODUCTION

Over the past two decades it has been established that many
mutagens are capable of producing, either directly, or following
metabolic activation, a variety of DNA structural lesions (133).
The lesions can be classified in a number of ways, but from a
biological point of view (56), it is convenient to recognize three
categories: (a) those that are potentially lethal, (b) those that
are potentially mutagenic, and (c) those that are potentially
recombinogenic. These lesions are not necessarily different
chemical entities and, for example, ultraviolet light (UV)-induced
pyrimidine dimers can produce all three biological effects.

Killing, mutation and recombination may all be enhanced over
wild-type responses in mutant strains defective in certain modes
of DNA repair. This is taken to indicate that many of the lesions
capable of stimulating these effects may be removed by repair in
wild-type strains before they have had an opportunity to provoke
genetic effects. On the other hand, mutants also exist that are
sensitive to the lethal effects of mutagens but in which induced
mutagenesis is suppressed or absent.

These results have led to the view, first enunciated for UV
mutagenesis in Escherichia coli (130, 131) and subsequently
applied to other organisms, that DNA repair systems are involved
as causative as well as ameliorative factors in mutagenesis. From
an operational point of view, repair systems that are involved in
the production of mutations are said to be error-prone or
mutagenic. All other modes of repair are said to be error-free.
These observations form the basis for the so-called DNA damage-
repair concept, which has been used to account qualitatively for
the existence of strains both sensitive and refractory to the
effects of mutagens and, qualitatively, for the dose-response
relations for cell killing and mutagenesis (for a review see 56).
However, it is important to remember that, unlike the
biochemically defined process of error-free excision repair (55),
error-prone repair is still a hypothesis for which direct
biochemical evidence has only recently started to emerge.

If we now first concentrate on the processes responsible for
mutation fixation of gene-mutations we might use a simple initial
concept for a systematic analysis of mutagen-induced mutagenesis.
We will use some selected sets of experimental evidence to
estimate the relative importance of the two major pathways leading
from different initial lesions to various genetic endpoints.

AN INITIAL CONCEPT

Based on mechanistic studies with a variety of test systems it emerged that two major pathways for mutation fixation have to be envisaged.

a)   Misreplications: In this case a mutagen-induced change, e.g., a damage on the template, a disturbance in the nucleotide pool etc., leads finally to a novel base sequence in the newly-synthesized strand during the normal scheduled DNA synthesis. Provided that the novel base sequence is not corrected during or after synthesis a heritable change has resulted. It is plausible to assume that in addition to base pair substitutions resulting from miscoding bases, frameshift mutations and possibly larger insertion and deletion mutations can also result if certain loop configurations are present in the template strand at the time the replication fork passes by.

b)   Misrepair: In this case mutation fixation may result from errors in DNA synthesis unrelated to the S-phase synthesis, or from metabolic activities which come into action if the normal replication apparatus is not capable of coping with severe distrubances.

EXPERIMENTAL OBSERVATIONS

Bacteriophage T4

A large body of information on misreplication has been obtained for the E. coli phage T4. Numerous studies have demonstrated the central role for the DNA polymerase of bacteriophage in the fidelity of DNA replication. A number of mutant polymerase alleles have substantial effects on mutation rates. Although mutant alleles of other genes of DNA metabolism can also increase mutation frequencies, their effects are generally smaller (100) for example, one temperature-sensitive DNA polymerase allele can increase the mutation rate per round of replication of an rII ochre codon approximately 100-fold (99), whereas less than tenfold increases are produced by mutant alleles in other DNA-metabolism genes (e.g., 129). Additional support for a central role of the DNA polymerase in determining mutation rates stems from the isolation of mutant polymerase alleles that decrease mutation rates. These "antimutator" polymerases sharply reduce the frequency of spontaneous A:T to G:C transition mutations (31). Although mutant DNA polymerases may increase mutation rates by a variety of mechanisms that need not be related to those for spontaneous mutation, a polymerase can decrease spontaneous rates only if it reduces the number of errors normally made or reduces the conversion of spontaneous mutation

intermediates to mutation (100). Hence, the antimutator effect
strongly argues for the importance of the polymerase in defining
spontaneous A:T-site transition rates.

Certain T4 DNA polymerase mutants strongly influence the
frequency and specificity of frameshift mutagenesis, if tested for
reversion frequencies of frameshift mutations in the T4 rII genes.
Most polymerase mutants increase frameshift frequencies, but a few
alleles which are antimutators with respect to the base
substitution mutations decrease the frequencies of certain
frameshifts while increasing the frequencies of others. The
various patterns of enhanced or decreased frameshift mutation
frequencies suggest that T4 DNA polymerase is likely to play a
variety of roles in the metabolic events leading to frameshift
mutations. Differences in frameshift frequencies at similar DNA
sequences (sequences with six adjacent A:T pairs) within the rII
genes, the influence of mutant polymerase alleles on these
frequencies, and the presence or absence of the dinucleotide
sequence associated with initiation of Okasaki pieces at
frameshift sites led to the suggestion that the discontinuities
associated with discontinuous DNA replication may contribute to
spontaneous frameshift mutation frequencies in T4 (101). Base
analogue mutagenesis would be expected to be a simple case to
stress the role of induced mispairing. But it turned out that
already during the normal replication, incorporation accuracies
are very complex. For example, one mutant allele of DNA
polymerase which displays a strong antimutator effect on
spontaneous A:T to G:C transition mutation is at the same time a
weak mutator for spontaneous G:C to A:T transitions (31) and for
transversions (98).

Surprisingly important is the template substrate orientation
of misincorporation which also influences the relative fidelity of
wild-type and antimutator polymerases (99). The relative
frequency of 2-aminopurine (AP):cytosine base pairs differs by an
order of magnitude when AP occupies the template position compared
to when cytosine occupies that position.

The complexities should be even more pronounced with
mutagenesis induced by alkylating agents which lead to a variety
of premutational lesions in DNA. A good means of determining the
relative importance of misreplication versus misrepair would be
with an agent known to induce a relatively minor fraction of
miscoding lelsions. Methyl methanesulfonte (MMS) with its low
Swann-Scott constant is known to induce only a minor fraction of
its alkylations at the O^6 position of guanine which results in a
miscoding alkylated base (110, 122). After technical difficulties
were overcome it was possible to study MMS-induced mutagenesis in
T4 phage (30).

MMS-induces reverse rII mutations from a wild-type background in bacteriophage T4. About 56% are base pair substitutions, about 30% are frameshift mutations, and the remainder is a miscellaneous set of rapidly-reverting or leaky mutants of unknown composition, but deletions were not detected (30). In wild-type phage the three genes w^+, px^+, and y^+ constitute the so-called WXY system which is involved in DNA repair. MMS-induced forward mutation is sharply reduced by the mutations px and y, which also reduce UV, photodynamic and gamma-ray mutagenesis and increase killing by all of these agents. Thus, many of the induced mutations arise via the T4 WXY system. The induction of G:C to A:T transitions was detected even in a px or y background using sensitive reversion tests, and the few forward rII mutations that were induced from this background also behaved like transition mutations. Thus, some MMS-induced mutations arise independently of the WXY system, perhaps as a result of the rather weak ability of MMS to alkylate the O^6 position of guanine (64) and leads to mispairing. It should be noted that by no means all T4 DNA repair genes have been examined for effects upon MMS mutagenesis (30). Candidates for interesting effects include "mms", which is sensitive to MMS lethality (32) and uvs79, which is sensitive to UV (26). MMS induced mutagenesis in T4 provides us with an example in which the majority of induced mutations result from error prone repair, that is, misrepair, and only a tiny fraction from simple misreplication.

Escherichia coli

In the bacterium E. coli DNA replication as well as a variety of DNA repair activities were studied in great detail. The fidelity in vivo estimated from the frequency of spontaneous base-pair substitutions (29) is much greater than expected from physochemical knowledge of the stability of non-Watson-Crick base pairs (57, 120). This is a strong indication that powerful error avoidance mechanisms must be involved in the DNA replication process.

In addition to the classic case in which recA, lexA and umuC were shown to be essential for mutation fixation, several other repair defect were studied with respect to alteration of chemical mutagenesis. Many of these studies were undertaken with the aim of increasing the detection capacity of E. coli mutagenicity test systems (79). Basically, there are two possibilities for increasing the sensitivity: (a) error-free pathways are eliminated and in that way more premutational damages chanelled into replication on damaged templates or error-prone repair, (b) the error-proneness of repair pathways may be increased, e.g., by the introduction of plasmids such as pKM101 (124). A classic demonstration of increased mutagenesis is found in those cases in which excision repair deficient strains are used (uvrA, uvrB). In such strains a large variety of chemicals show increased mutagenic activity: several alkylating agents, including epoxides, alkane sulfonates, dialkylsulfates, lactones, sultones, nitrosamines, nitrosamides, organophosphates, captan, folpet, nitrogen mustard derivatives, ethyleneimino derivatives, nitroso compounds, 8-methoxypsoralen (in the dark), hycanthone methanesulfonate, nitro-o-phenylenedimine, cysteine adducts of chlorinated cyclohexanes, halogenated cyclohexanes (79). There are sporadic cases in which a mutagenic activity in detected only in cells with uvrB background, e.g., the photodynamic activity of chloropromazine (Ellenberger, unpubl.). On the other hand, some chemicals show a decrease of mutagenic activity in uvrB backgrounds, e.g., acridine orange under white light irradiation, and some others such as methylene blue, malonaldehyde and mitomycin C will not be detectable in such strains (62, 86).

Analogous shifts in detectable mutation spectra have been described upon introduction of dam^- mutations. The dam^+ gene is responsible for a methylation instructed error-avoidance mechanism which is responsible for a factor 10^4 to 10^5 increased DNA replicating fidelity in vivo (43, 80). Hypermutability in dam background has been observed for the base analogues AP, 6-N-hydroxyaminopurine, 5-bromouracil (44, 95), the intercalating agent 9-aminoacridine and the phenylating agent methylphenylnitrosamine. Some enhancement is seen with ethyl methanesulfonate and normal mutagenicity with UV, gamma-rays and mitomycin C. An unexpected suppression of mutagenicity is observed with the methylating agents methylnitronitrosoguanidine (MNNG) and MMS. A further interesting possibility for modifying the mutagenic response of E. coli cells to chemical mutagens is the introduction of the mutagenesis-enhancing plasmid pKM101. Because of its ability to enhance chemical mutagenesis, pKM101 was incorporated into the Ames Salmonella tester strains for the detection of carcinogens as mutagens (74), and it has played a major role in the success of the system (73, 125). The regulation and function of the pKM101 plasmid is currently being studied by Walker and coworkers (126) in E. coli. They discovered that damage to DNA of E. coli results in the increased expression of a set of at least ten (din = damage induced) genes whose products play biological roles in processes as diverse as chemical mutagenesis and filamentous growth. One of the most interesting genes of this group is the umcC gene. Its product plays a key, but as yet undefined, role in UV and chemical mutagenesis. An analogous function seems to be encoded by the pKM101 plasmid and has been located to about a 2000 base-pair region on the plasmid. It is termed muc (mutagenesis: UV and chemical).

Experimental evidence indicates that the pKM101 may act by providing an analogue of the umcC gene product. It seems that the muc region codes for this analogue. The pKM101 plasmid makes cells more susceptible to mutagenesis by a variety of chemical agents (28, 83). It increases the susceptibility of the cells not only to base substitution but also to frameshift mutagenesis with the appropriate mutagenic compounds. The picutre becomes much less clear if one realises that pKM101 seems to increase only certain types of base pair substitutions and frameshifts, but not others (80, 119, 126).

In Vitro Studies

The in vivo studies clearly indicate a complex genetic control of mutagenesis. For the elucidation of particular mutagenic events in vitro studies seem to be more promising. According to our basic model, two different approaches may be needed to study misreplication and misrepair. We have to assume that premutational lesions exist which have basically different consequences:

(a) lesions produced by some mutagen, (e.g., MNNG, NaHSO$_3$, simple alkylating agents) can alter the coding properties of the DNA bases directly and are expected to lead to mutations via misrepairing during replication;

(b) others, such as the UV-induced pyrimidine dimers and those produced by bulky adducts (e.g., Aflatoxins, polycyclic aromatic hydrocarbons) act as blocks to DNA replication. This latter type of lesion will require the active involvement of an error-prone repair process for the generation of mutations.

Misreplication: Only recently the detailed study of molecular mechanisms was started using in vitro studies. A first problem is the miscoding by natural bases, base analogues and carcinogen-modified bases, for example the model alkylation product: 0-6-alkyl guanine or the base analogue 2-aminopurine. There are two ideas discussed with respect to the achievement of a high fidelity of DNA polymerase:

(a) One general class of models (128 and references therein) postulates that the polymerase either accepts or rejects candidate deoxyribonucleotides for insertion into DNA according to an enzyme mechanism that can be thought to as base selective or literate; that is, the enzyme is able to read each DNA template base. A literate polymerase would alter its properties to decrease insertion rates for deoxyribonucleotides that are noncomplementary at each DNA template site. It could do this perhaps by some allosteric mechanism.

(b)   In contrast with the literate polymerase model a simple alternative has been proposed (19, 38) in which polymerases control fidelity not by altering their properties in response to the template but simply by exploiting differences in base-pair free energies.

In this model, phosphodiester-bound formation is catalized by the same rate for any nucleotide substrate resident at the polymerase active site independently of the template base. Fidelity is determined by the ratio of residence of "sticking" times for complementary nucleotides compared with noncomplementary nucleotides (38). A reduced residence time for a wrong nucletotide would result from its lower stability, i.e., weaker base pairing (hydrogen bonding and base stacking) relative to that of the proper base pairing partners.

(c)   Watanabe and Goodman (128) have recently tested the second model by comparing the kinetics of the formation of 2-aminopurine:cytosine and 2-aminopurine:thymidine base pairs. This comparison gives optimal expected differences because in the common tautomeric states the $2Ap^{\cdot}T$ pair forms two hydrogen bounds where as 2 Ap:C forms just one. In addition they used human DNA polymerase which lacks detectable 3'-endonuclease activity so that the results cannot be biased by such a correction mechanism. They found a ratio between the Michaelis constants of Km(C)/Km(T) of 25 which agrees well with the misinsertion ratio of 21.7. This result supports the model in which base-pairing stability and NOT polymerase active site selection or base exclusion determine the mutation frequencies of miscoding bases.

The polymerase is also viewed as having the same nonselective properties during nucleotide excision whenever a 3'-exonuclease proofreading activity is present. The excision probability for a newly-formed base pair is determined by its probability of being melted out, which depends on the base-pair stability. Therefore, for any given allele of an DNA polymerase misinsertion and proofreading reflect only the physio-chemical properties of the miscoding base. On the other hand, different alleles of the same polymerase and polymerases from different sources may copy a given DNA template with different degrees of accuracy due to mutational differences among the enzyme molecules. For example, Singer, et al. (111) studied in vitro discrimination of replicates (E. coli DNA polymerase I and cucumber RNA-dependent RNA polymerase) acting on the same carcinogen-modified polynucleotide templates. They found that both qualitatively and quantitatively, replication errors resulting from carcinogen-modified bases ($3-N^{4}$ ethenodeoxycytidine and 3-

methyldeoxy cytidine) are less frequent than errors in transcription of the same deoxypolynucleotides.

Error Prone Repair. The UV-induced pyrimidine dimers has been studied as a model lesion which needs bypass of blocked DNA replication. Rabkin et al. (94) published an in vitro system which as they suppose, mimics a possible model of operation of SOS repair.

In E. coli, replication of UV-damaged cells that are excision-deficient results in gaps in the newly-synthesized DNA, most of which is processed error free by a recombination mechanism. In cases in which recombination is impossible, such as on single-strand DNA or at overlapping daughter-strand gaps, de novo DNA synthesis directed by the SOS system must in some way add bases opposite the lestion and allow filling of the gap left by replication. The in vivo enzymology of such a process is completely unknown, but the involvement of a DNA polymerase with relaxed fidelity is a stimulating hypothesis.

DNA synthesis on UV-irradiated templates usually terminates one nucleotide before the first (3') pyrimidine in a pyrimidine dimer when DNA synthesis is catalyzed by any one of three prokaryotic and two enkaryotic enzymes (81). This termination is independent of the operation of a proofreading exonuclease activity. If in a reaction mixture of UV irradiated DNA and the E. coli polymerase I (Klenow fragment pol I) the usual Mg^{2+} is substituted by Mn^{2+} the synthesis appears to terminate at the site of the lesion (81, 82). Obviously Mn^{2+} results in a relaxed nature of the polymerase pol I specificity in its presence.

In their studies, Rabkin et al (94) started for a first step of the experiment with a single-stranded UV-irradiated phage ϕX174 DNA and allowed the polymerase I in the presence of magnesium to synthesize the complementary strand up to the position immediately before the pyrimidine dimer. These radioactive labelled molecules are isolated and used as substante in a second reaction with pol I in the presence of manganese (Mn^{2+}) and specific nucleotides. In this way the insertion of nucleotides opposite the 3'pyrimidine of the dimer. Sequence studies of the final products reveal a preference for purine nucleotides in particular adenine, over pyrimidine nucleotides opposite the dimer. This is most probably not due to residual pairing because the pol I shows also a preferential incorporation of purines opposite to a-basic sites where no base pairing is possible.(144).

A direct quantitative extrapolation to the in vivo situation might not be possible because in damaged cells, nucleotide pool bias may play a role in determining mutagenicity and viability (77,145).   Induction of the SOS functions by thymine deprivation of E. coli results within 30 minutes in a 4- to 5-fold increase in dATP. Overall, the tendency to insert adenine opposite the pyrimidine dimers means that a large proportion of potential mutations will be "lost" due to the insertion of the "correct" base, thymine being the most frequent pyrimidine in a dimer. In addition, transitions should be more frequent than transversions because of the preference for purine insertion opposite dimers. However, the observation that pyrimidines are sometimes incorporated opposite dimers means that a significant proportion of UV-induced mutations should be transversions. This prediction is substantiated by the data available. Whereas in E. coli MNNG leading to mispairing results in about 400 times more transitions than transversions, the preference for transitions after UVirradiation is only about 4 (for details see 94).

Yeast

In the yeast Saccharomyces cerevisiae a large number of strains have been isolated based on their hypersensitivity to certain lethal (and mutagenic) agents, particularly UV light (23), ionizing radiation, MMS, and/or nitrous acid (139). To date, at least 95 mutations are known that confer sensitivity to radiation and/or chemicals; however, the allelic relations of all of these mutations have not yet been established (56).

In attempts to analyze the interrelationship between different loci, double mutants were most frequently studied. In principle, three different situations may arise: (a) the two genes may be epistatic, i.e., the double mutant may be no more sensitive than its most sensitive single-mutant parent; (b) the two genes may have an additive effect on each other, i.e., the killing produced by a given dose in the double mutant could be equal to the sum of the killing in the wild type plus the incremental killing of the two parents over wild type; and (c) the two genes may have a synergistic effect on each other, i.e., the interaction would be more than additive. Epistasis implies that the gene products mediate steps in the same repair pathway or that they are part of a multimeric repair complex; additivity imples that the gene products act independently on different substrates, and synergism implies that the genes control steps in two repair systems that compete for the same lesions. Three epistasis groups have been identified, and loci in these groups affect at least three modes of DNA repair as well as other phenotypic characteristics of the cells. As a matter of convenience the three groups are named after a prominent locus in each.

RAD3 Group. Loci of the RAD3 group are sensitive to UV killing and generally show enhanced UV mutagenesis. At least nine of these loci are known to control error-free excision of pyrimidine dimers. A few seem to be defective in the initial incision step to dimer repair because they are defective in the production of single-strand DNA breaks during post-UV incubation, but extracts are capable of excising dimers from preincised UV-irradiated DNA. The mutant rad3 is unable to remove psoralen plus 360nm light-induced DNA interstrand cross-links or monoadducts (Jachymczyk and von Borstel, unpulbished).

RAD52 Group. The RAD52 group is sensitive primarily to X-rays (39), although loci in this group also control a so-called minor pathway for repair of UV damage (24). The majority of the mutants have been shown to depress UV-induced and chemically-induced mutation. The mutant rad51 is defective in repair of double-stand breaks, but is capable to repairing induced single-strand breaks (85) and can remove DNA interstrand cross-links induced by psoralen plus 360nm light (Jachymczyk and von Borstel, unpublished). Three other mutants were shown to be able to repair DNA double-strand breaks. Cells homozygous for mating type are sensitive to X-rays but not to UV (36), are sensitive to MMS, display reduced frequencies of spontaneous and UV-induced recombination. Morrison et al. (84) have suggested that an error-free repair process is present in a/alpha diploids but is absent from strains homozygous for mating type.

RAD6 Group. Loci of the RAD6 group influence sensitivity to both UV and X-rays and control error-prone repair; some of these mutants may also affect error-free repair (56). The RAD6 epistasis group may subdivided (76):

(a) The mutant rad6 can excise thymine dimers (93), although dimer excision may be inefficient (53), and it is unable to repair DNA strand breaks induced by MMS (17, 59).

(b) Loci, such as rad18 concerned only with rad6-dependent, error-free repair. The mutant rad18 accumulates single- and double-strand DNA breaks during post-X-irradiation incubation (85) and may be partially defective in dimer excision (7).

(c) Genes that function in an error-free or error-prone mode depending on the nature of the lesions attached.

(d) Genes concerned exclusively with error-prone repair which convey only a moderate degree of radiation sensitivity.

There is, at least with respect to UV damage, good evidence that all epistasis groups have been discovered (24, 56). However, the epistasis groups are not mutually exclusive because it appears that some alleles belong simultaneously to two groups.

It was argued initially that each group corresponds to a distinct "biochemical pathway" for repair and, certainly, genes coding for enzymes involved in sequential repair reactions would be expected to be epistatic to one another. However, the fact that the RAD3 group contains nine loci controlling dimer excision (and probably more are involved) suggests that a larger number of gene products are required for the process than might be expected on strict enzymological grounds. Thus, it is possible that the epistasis groups may control the formation of macromolecular complexes that are required to mediate the three major manifestations of DNA metabolism, namely, excision repair, normal replication (possibly including some steps of precursor synthesis), and recombination. If such "metabolic complexes" exist in yeast, then it is not surprising that there should be a substantial number of loci epistatic to one another. If these complexes share any common proteins, or in some way interact with or overlap one another, then it is also not surprising that some loci might exist that appear to belong to two epistasis groups. Although at present we have no biochemical evidence for the existence of such complexes in yeast, enzyme complexes associated with DNA precursor synthesis have been isolated from bacteriophage-T4-infected bacteria (72) and Ehrlich ascites carcinoma cells (109). Similarly, DNA replication complexes have been detected in adenovirus-infected HeLa nuclei (1) and in Chinese hamster embryo fibroblasts (96). Finally, several enzymes involved in excision repair in E. coli are believed to function as a unit (108). Thus, it would not be surprising to find such complexes in yeast.

The yeast data may be used to emphasize two points: (a) The existence of longlasting physiological changes in mutagen exposed cells and (b) the phenomenon of untargeted mutagenesis. Within a mutagenized genome an induced mutation may appear at the position at which the original DNA strand was damaged by the mutagen. This is called targeted mutagenesis. As described by Witkin and Wermundsen (132) high frequencies of untargeted mutations, events that occur in lesion-free stretches of DNA, may be found. The occurrence of untargeted mutagenesis has been clearly established in experiments with bacteriophages (e.g., 62), but the relative proportions of targeted and untargeted events have not been previously estimated. Lawrence et al. (65) attempted to do this in mating experiments with yeast.

An excision-deficient haploid strain carrying a nonrevertible deletion of the whole CYC1 locus is mated with an excision-deficient haploid of the opposite mating type that contains the UV-reversible ochre allele, cyc1-91. Estimates of the total number of diploid clones formed, and of the proportion of CYC1 revertants among them, can be obtained by plating the mating mixture on suitable selective media. Irradiation of the deletion, but not the cyc1-91 parent, gives an estimate of the frequency of untargeted mutations, while the reciprocal treatment gives an estimate of the total frequency of both types of events. Data from strains in which excision was blocked by two mutations (rad 1-2, rad 2-5) give estimates that approach 40%.

The fact that mating takes 5 hours indicates that the cellular response to UV which leads to untargeted mutagenesis is a long-lasting effect. The mechanism by which untargeted mutagenesis occurs has not yet been clarified. UV induced lesions in the deletion genome might give rise to untargeted mutations by one or both of two general mechanisms: (a) by recombination between homologues and the introduction of lesions into the unirradiated chromosome, followed by some sort of long patch error-prone repair, or (b) by the induction (or release) of a diffusable agent that reduces replication fidelity or otherwise promotes mutation.

Data from experiments with excision deficient strains that carry the rad52-1 allele, and are, therefore, also deficient with respect to induced and spontaneous recombination (104) show that untargeted mutagenesis does not depend to any detectable extent on recombination. In consequence, it remains an aim for future experiments to identify the suspected diffusible agent. Overall, the very extensive studies conducted with yeast indicate the existence of a complex network of error-free and error-prone repair activities which may be modulated by a large variety of factors.

Drosophila

Systematic studies of DNA repair and related phenomena in Drosophila melanogaster became possible after P.D. Smith isolated the first X-chromosomal mutant with larval hypersensitivity to alkylating agents (113). In addition to the sophisticated genetics, the availability of suitable mutants was a prerequisite to initiate studies of DNA metabolism in this eukaryote which represents an attractive intermediate between microorganisms and mammals. Drosophila allows biochemical analyses with cultured embryonic cells as well as genetic in vivo studies with both germ cells and somatic cells.

A classic genetic analysis recently identified 29 independent loci which are required for resistance to mutagen exposure (116). They have been mapped on all of the major chromosomes; loci are present on the X chromosome as well as on chromosomes 2 and 3. Two main classes of mutation affecting DNA metabolism have been identified in Drosophila, mutagen-sensitive (mus) mutants exhibiting larval hypersensitivity to mutagens, and meiotic (mei) mutants exhibiting aberrant chromosome behavior during meiosis (2 , 10, 12, 114, 118). A subset of the meiotic mutants was found to be highly mutagen-sensitive. Genetic and cytological analyses

have shown that many of the mus and mei genes function in both
somatic and meiotic cells and affect recombinational events,
chromosome stability and mutation frequency monitoring several
different genetic endpoints (3, 5, 22, 41, 116, 134).

By exploiting repair assays developed for mammalian cell
cultures, Boyd and coworkers (11) have identified seven genes
essential for normal excision repair. These genes can be divided
into two classes based on the strength of the exicision defect
expressed by their mutant alleles. The class with strong excision
deficiencies mei-9 and mus(2)201 lack the resynthesis step of
excision repair, and retain all detectable pyrimidine dimers
during post-UV incubation (13, 54, 89). Since no single strand
DNA breaks are detected which are normally associated with the
initial incision step of excision repair, these mutants are
blocked at or near the first step of the excision process (54).
From alkylation mutagenesis experiments with repair-deficient E.
coli stains, Smith et al. (115) noted that the mei-9 strain
exhibits pleiotropic mutant phenotypes very similar to those
displayed by the uvr D mutant. By analogy with these studies,
they speculate that mei-9, like uvr D, is deficient in a DNA
unwinding protein. Among the mutants with partial excision
deficiencies two also influence the incision step, the remaining
three mutants are being further studied to determine whether they
block a post-incision step of excision repair.

The term postreplication repair refers to the capacity of a
cell to synthesize DNA on a damaged template. Operationally this
function is monitored by measuring the molecular weight of pulse
labelled DNA after mutagen treatment. Mutants at four loci
strongly disrupt this process and six are associated with moderate
deficiencies (8, 9). The reassortment of pyrimidine dimers in the
parental strands to newly synthesized strands as it occurs during
postreplication repair in E. coli (40) was studied in Drosophila
mutants (11). Since even at the maximum apparent frequency of
dimer transfer is only one sixth of that observed in bacteria,
recombination is unlikely to serve as the basis for most
postreplication repair in Drosophila. To this extent the
Drosophila and mammalian mechanisms are similar (66). Brown and
Boyd (16) were able to divide the strong postreplication repair
deficient mutants into two classes: two mutants block DNA
synthesis at pyrimidine dimers whereas in two other mutants
synthesis proceeds past dimers leaving gaps for alkali labile
sites. Postreplication repair defective mutants at five loci
disrupt meiosis in unmutagenized females (12, 51a). In one mei-41
mutant, electron microscopy revelaed a reduction in the number of
recombination nodules obserbed at the pachytene stage of meiosis
(140), and the morhology of the residual nodules is abnormal. The
freqency of exchanges between sister chromatids is apparently
elevated in the germ line of a mus(3)302 mutant (138). A striking

correlation has been found between postreplication repair defects
and increases in a form of somatic mutation which may involve
mobilization of a transpositional element (37). Furthermore, two
postreplication   repair   deficient   mutants   exhibit   somatic
chromosome instability. Of particular interest are the mei-41
mutants which have been shown to lose 15% of their cells every
cycle due to lethal chromosome breaks. Fine structure mapping of
seven mei-41 alleles has revealed that this locus is one of the
largest known in Drosophila (J. M. Mason, personal communication).

Previous demonstrations of photorepair in Drosophila have
been extended by purification of a highly stable photoreactivating
enzyme from tissue culture cells (6). Boyd et al. (11) recently
identified a strain which is deficient in the photoreversal of UV-
induced pyrimidine dimers.

An interesting aspect is the inducibility of repair. Data
obtained in split-dose X-ray experiments with somatic cells are
interpreted as evidence for a X-ray-inducible DNA repair system
(60). Brown and Boyd (16) identified some postreplication repair
functions (precursor incorporation; synthesis of large single-
stranded DNA) that may be enhanced in response to UV-radiation.
Harris (see 11) detected the apparent induction of an early
cellular response to UV irradiation: increased accessibility of
DNA damage to repair enzymes. The proportion of pyrimidine dimers
accessible to an endonuclease increases during the first 90 min by
about 60%. Since this phenomenon is abolished by including
novobiocin or courmermycin during incubation, a topoisomerase may
mediate this increased chromatin accessibility.

The study of repair functions, their pleiotropic expression
and the specific effects of particular alleles on spontaneous and
induced mutation frequencies give at least an indirect insight
into the role of DNA repair functions in mutagenesis in the
complex eukaryotic genome of Drosophila. Direct exposure of males
carrying repair defects gives variable results in postmeiotic
stages of spermatogenesis (90, 121). A study by Smith et al.
(115) indicates that in particular in the spermatogonia of adult
males the normally observed low frequency of alkylation-induced
mutations results from an efficient excision of premutational
lesions. If excision repair is blocked by a mei-9 mutation,
nonexcised lesions lead to a 4-8 fold increased mutability. This
indicates that in spermatogonia paternal repair systems and other
activities of DNA metabolism determine mutation frequencies. In
connection with the meiotic divisions the genes responsible for
these activities are turned off, the enzyme activities disappear
(possibly in a gradual manner that needs to be analyzed in more
detail (48, 121)) and finally the mature sperm is incapable of
repairing premutational lesions. Putting mutagenized spermatozoa
into oocytes with different DNA repair capabilities indicates that

in this situation enzyme activities encoded by the maternal genome
in the oocyte determine the extent of repair and mutations
(fixation due to the premutational lesions in the paternal
chromosomes) (46, 48, 134). For example, cross links in paternal
chromosomes introduced into mus(1)101 oocytes are, in contrast to
the situation in wild type oocytes, non-mutagenic. The observed
increased frequency of dominant lethals indicates that, due to the
inability of the mus(1)101 oocytes to repair cross links, they act
as lethal lesions and in that way all premutational lesions are
eliminated (49). In contrast to this, mutation frequencies are
increased if EMS, MMS or MNU exposed spermatozoa are introduced
into excision defective mei-9 oocytes. Here, at least a fraction
of the lesions present are not excised in the usual manner and
channeled into mutagenic pathways (miscoding at DNA
replication?) (49, 121).

Interestingly enough the functions encoded by many of the DNA
repair genes are active in germ cells as well as in somatic cells
(3, 4, 42). In particular the mei-9 mutants have been shown to
increase spontaneous as well as induced mutational events in
somatic cells (37, 135).

Recently a somatic mutation and recombination test using wing
disc cells of mutagen exposed larvae was developed (50, 51). This
assay, which is at present under extensive validation, might turn
out to be faster, cheaper and more flexible than the standard
Drosophila tests used at present. Graf synthesized a stock which
allows the testing of chemically-induced mutagenesis in wing disc
cells in excision proficient and excision defective larvae
simultaneously (50). Initial studies indicate that for certain
types of chemicals (e.g., MMS, formaldehyde, safrole) the excision
defect increases the detection capacity in this promising short
term assay (121).

Mammals and Man

Repair and mutagenesis in mammalian cells will be discussed
here only briefly because details are to be found in other
chapters of this volume (e.g., 18). Mammalian and human cells in
culture are available as primary low passage cells or as
established aneuploid cell lines. Experimental techniques allow
the study of mutagen-induced toxic effects (reduced survival;
lethal hits) as well as mutagenic effects (dominant, codominant
and recessive gene mutations; sister chromatid exchanges;
chromosome and chromatid aberrations; aneuploidy). For an
overview of available techniques and of problems related to
metabolic activation of pro-mutagens and pro-carcinogens see
Howard-Flanders (58). An extremely strong stimulus for
experimental studies resulted from the disovery of the involvement
of a DNA repair defect in the human hereditary disease Xeroderma

pigmentosum (XP) (20, 67). The intrest and experimental potential increased with the identification of additional DNA repair defects in humans (14, 21). The mutants available allow critical investigation of the role of DNA repair processes in mutagenesis by comparing mutation frequencies induced in repair-competent and repair-deficient cell lines.

The (simplified) basic concept assumes that there exist error-free and error-prone processes in DNA metabolism. If a mutation eliminates an error-free pathway, then more premutational lesions than normal would be processed by the error-prone process and hypermutability is observed. On the other hand, if the error-prone pathway is blocked a cell line might appear non-mutable.

The best example of this approach may be found in studies on the XP cell lines (25). Cells cultured from classic XP patients are more sensitive than normal to both the lethal and mutagenic effects of UV. However, on the basis of mutations per lethal hit, classical XP cells and normal cells do not differ. This observation implies that the thymine dimer excision process, deficient in classical XP, operates in a largely error-free manner (75). However, cells from XP variant patients show a considerable increase over normal in the number of mutations per lethal event after UV (71, 87), suggesting that the postreplication repair system, deficient in XP variants, normally operates to correct potentially mutagenic lesions in DNA daughter strands resulting from the presence of photoproducts that escape pre-replicative excision. The hyper-mutability but near normal sensitivity of XP variant cells to the lethal effects of UV would then require these cells to repair daughter strand damage by a second, error-prone, route.

The removal, or modification, of altered DNA bases or chemical adducts from DNA by repair processes has been demonstrated in cultured cells treated with a variety of different chemical mutants (102). However, it is necessary to be able to identify those lesions with a significant mutagenic potential in order to comment on the role of DNA repair in chemical mutagenesis (25, 58). Such comment has been possible in the case of the mutagen benzo(a)pyrene-diolepoxide. The excision repair defect in XP human cells renders them sensitive to this epoxide. Recent studies (136) show that the mutagenicity of this agent in XP cells correlates with the failure to excise an N^2-deoxyguanosine adduct from DNA. This implies that this adduct is the principal premutational lesion and, in addition, that its excision in normal cells is largely error-free.

Recently the techniques of molecular biology have been adopted to study the molecular mechanisms of repair and mutagenesis in mammalian cells. An impressive example are the studies of O^6-methylguanine by Mitra and coworkers (78). The O^6-methylguanine (m6G) produced in nuclear DNA by simple alkylating agents is believed to be a critical factor in the ultimate expression of mutagenesis and/or carcinogenesis by such agents (61, 70, 106). Mammalian cells as well as bacteria are capable of removing O^6-methylguanine from DNA by the action of a DNA-guanine-O^6-methyltransferase. The protein reacts stoichiometrically (103) with m6G in DNA, resulting in transfer of the methyl group to a cysteine thiol group of the protein causing its inactivation (68, 69). Similar protein activities have been found in E. coli (91), in rodent and human liver fractions and in cultured human cells (35, 88, 92, 123, 137).

The capacity for removal of m6G from DNA varies widely among different mammalian cells. Extracts of human tumor cells classified as proficient (Mer$^+$) in the ability to reactivate MNNG-damaged adenovirus (27) have been found to contain m6G-DNA-methyltransferase activities equivalent to 30,000 to 220,000 molecules per cell, whereas extracts of deficient (Mer$^-$) cell lines contain little or no activity (35, 137). The similar classification of human lymphoblastoid lines as Mex$^+$ or Mex$^-$ is based on the ability of the cells to remove m6G produced in their DNA by MNNG treatment (112). Both CHO and V79 Chinese hamster cell lines are reportedly deficient in m6G removal (45, 127).

Whereas in bacteria a single subtoxic dose of MNNG can induce the methyltransferase activity (in E. coli 100 x, in B. subtilis 10 x; 52, 78, 69, 103), mammalian cells are non-inducible and show only a constitutive activity (34). Using synthetic polydeoxynucleotides containing O^6-methylguanine as the only modified base the base-pairing properties of O^6-methylguanine in template DNA during in vivo DNA replication can be studied (117). O^6-methylguanine paired with either thymine or cytosine but with a much higher preference for thymine. Contrary to theoretical predictions that the m5dG:dT pair should be comparable to the dA:dT pair, the presence of m6G in the template inhibited DNA synthesis. Different DNA polymerases behaved differently in m6G-directed deoxynucleotide turnover and incorporation (117). Physiological factors other than DNA repair are also likely to influence the mutability of cells, e.g., the presence of natural clastogenic substances (33), or changes in the availability of DNA precursors for repair or replication (77). Consequently, changes in the mutability of a cell strain should not be interpreted as a direct DNA-repair effect without sound biochemical evidence. Nevertheless, quantitative mutagenesis experiments with somatic cell lines, having characterized defects in DNA-repair or metabolism, have great potential for elucidating mutagenic mechanisms. The de novo isolation and characterization of DNA repair mutants from established cell lines (105, 107), should help to widen the narrow view of the whole problem mostly based on

viable human diseases.   In addition, such mutant cell lines will
be of great help in elucidating the mechanisms of mutagenesis.

REFERENCES

1. Abboud, M. M., and M. S. Horowitz (1979) The DNA
    polymerases associated with the adenovirus type 2
    replication complex: Effect of 2'-3' dideoxythymidine-5'-
    triphosphate on viral DNA-synthesis. Nucleic Acids
    Res. 6:1025.
2. Baker, B. S., and A. T. C. Carpenter (1972) Genetic
    analysis of sex chromosomal meiotic mutants in Drosophila
    malanogaster. Genetics 71:255-286.
3. Baker, B. S., and D. A. Smith (1979) The effect of
    mutagen-sensitive mutants of Drosophila melanogaster in
    nonmutagenized cells. Genetics 92:833-847.
4. Baker, B. S., A. T. C. Carpenter, and P. Ripoll (1978) The
    utilization during mitotic cell division of loci controlling
    meiotic recombination and disjunction in Drosophila
    melanogaster. Genetics 90:531-578.
5. Baker, B. S., J. B. Boyd, A. T. C. Carpenter, M. M. Green,
    T. D. Nguyen, P. Ripoll, and P. D. Smith (1976) Genetic
    controls of meiotic recombination and somatic DNA metabolism
    in Drosophila melanogaster. Proc. Natl. Acad. Sci. USA
    73:4140-4144.
6. Beck, L. A., and R. M. Sutherland (1979) Purification of a
    photoreactivating enzyme from Drosophila melanogaster. Am.
    Soc. Photobiol. 7:130 (Meeting Abstract).
7. Boram, W. R., and H. Roman (1976) Recombination in
    Saccharomyces cerevisiae: A DNA-repair mutation associated
    with elevated mitotic gene conversion. Proc. Natl. Acad.
    Sci. USA 73:2828.
8. Boyd, J. B., and R. B. Setlow (1976) Characterization of
    postreplication repair in mutagensensitive strains of
    Drosophila melanogaster. Genetics 84:507-526.
9. Boyd, J. B., and K. E. S. Shaw (1982) Postreplication
    repair defects in mutants of Drosophila melanogaster. Mol.
    Gen. Genet. 186:289-294.
10. Boyd, J. B., M. D. Golino, T. D. Nguyen, and M. M. Green
    (1976) Isolation and characterization of X-linked mutants
    of Drosophila melanogaster which are sensitive to mutagens.
    Genetics 84:485-506.
11. Boyd, J. B., P. V. Harris, J. M. Presley, and M. Narachi
    (1983) Drosophila melanogaster: A model eukaryote for the
    study of DNA repair. In press.
12. Boyd, J. B., M. D. Golino, K. E. S. Shaw, C. J. Osgood, and
    M. M. Green (1981) Third chromosomal mutagen-sensitive
    mutants of Drosophila melanogaster. Genetics 97:607-623.

13. Boyd, J. B., R. D. Synder, P. V. Harris, J. M. Presely,
    S. F. Boyd and P. D. Smith (1982) Identification of a
    second locus in Drosophila melanogaster requried for
    excision repair. Genetics 100:239–257.

14. Bridges, B. A. (1982) Some DNA repair-deficient human
    syndromes and their implications for human health, in:
    Environmental Mutagens and Carcinogens, T. Sugimura,
    S. Kondo and T. Takebe (Eds.), Alan R. Liss, Inc. New York,
    pp. 47–57.

15. Brown, T. C., and J. B. Boyd (1981a) Abnormal recovery of
    DNA replication in ultraviolet-irradiated cell cultures of
    Drosophila melanogaster which are defective in DNA repair.
    Mol. Gen. Genet 183:363–368.

16. Brown, T. C., and J. B. Boyd (1981b) Postreplication
    repair-defective mutants of Drosophila melanogaster fall
    into two classes. Mol. Gen. Genet. 183:356–362.

17. Chlebowitz, E., and W. J. Jachymczyk (1979) Repair of MMS-
    induced double strand breaks in haploid cells of
    Saccharomyces cerevisiae, which requires the presence of a
    duplicate genome. Mol. Gen. Genet. 167:279.

18. Chu, E. H. Y., I.-C. Li, and J. Fu (1984) Mutagenesis in
    cultured mammalian cells: Problems and prospects. This
    volume.

19. Clayton, L. K., M. F. Goodman, E. W. Branscomb, and D. I.
    Galas (1979) Error induction and correction by mutant and
    wild type T4 DNA polymerases. J. Biol. Chem. 254:1902–1913.

20. Cleaver, J. E. (1968) Defective repair replication of DNA
    in xeroderma pigmentosum. Nature 218:652–656.

21. Cleaver, J. E. (1982) DNA damage, DNA repair and DNA
    replication in short-term tests for exposure to mutagens,
    in: Chemical Mutagenesis, Human Population Monitoring and
    Genetic Risk Assessment, Progress in Mutation Research,
    K. C. Bora, G. R. Duglas and E. R. Nestmann (Eds.), Elsevier
    Biomedical Press, Amsterdam, Oxford, New York, Vol. 3,
    pp. 111–124.

22. Cooper, S. F., and S. Zimmering (1981) A genetic study of
    the effects of the repair-deficient mei-9[a] mutation in
    Drosophila on spontaneous and X-ray-induced paternal sex
    chromosome loss. Mutation Res. 80:281–287.

23. Cox, B. S., and J. M. Parry (1968) The isolation, genetics
    and survival characteristics of ultraviolet light-sensitive
    mutants in yeast. Mutation Res. 6:37–55.

24. Cox, B. S., and J. C. Game (1974) Repair systems in
    Saccharomyces. Mutation Res. 26:257–264.

25. Cox, R. (1982) Mechanisms of mutagenesis in cultured
    mammalian cells, in: Environmental Mutagens and Carcinogens,
    T. Sugimura, S. Kondo, and H. Takebe (Eds.), University of
    Tokyo Press, Tokyo, pp. 157–166.

26. Cupido, M., O. Schreij-Visser, and P. Van der Ree (1982) A bacteriophage T4 mutant defective in both DNA replication and replication repair. Mutation Res. 93:285-295.

27. Day, R. S., C. H. J. Ziolkowski, D. A. Scudiero, S. A. Meyer, and M. R. Mattern (1980) Human tumor cell strains defective in the repair of alkylation damage. Carcinogenesis 1:21-32.

28. Drabble, W. T., and B. A. D. Stocker (1968) R (transmissible drug-resistance) factors in S. typhimurium: Pattern of transduction by phage P22 and ultraviolet protection effect. J. Gen. Microbiol. 53:109.

29. Drake, J. W. (1969) Spontaneous mutation. Nature 221:1128-1132.

30. Drake, J. W. (1982) Methyl methanesulfonate mutagenesis in bacteriophage T4. Genetics 102:639-651.

31. Drake, J. W., E. F. Allen, S. A. Forsberg, R. M. Preparata, and E. O. Greening (1969) The genetic control of mutation rates in bacteriophage T4. Nature 221:1128-1132.

32. Ebisuzaki, K., C. L. Dewey, and M. T. Behme (1975) Pathways of DNA repair in T4 phage. I. Methyl methanesulfonate sensitive mutant. Virology 64:330-338.

33. Emerit, I., and P. A. Cerutti (1981) Clastogenic activity from Bloom syndrome fibroblast cultures. Proc. Natl. Acad. Sci. USA 78:1868-1872.

34. Foote, R. S., and S. Mitra (1984) Lack of induction of O^6-methylguanine-DNA methyltransferase in mammalian cells treated with N-methyl-N'-nitro-N-nitrosoguanidine. Carcinogenesis 5:277-281.

35. Foote, R. S., B. C. Pal, and S. Mitra (1983) Quantitation of O^6-methylguanine-DNA methyltransferase in HeLa cells. Mutation Res. 119:221-228.

36. Friis, J., and H. Roman (1968) The effect of the mating type alleles on intragenic recombination in yeast. Genetics 59:33-36.

37. Fujikawa, K., and S. Kondo (1982) Frequency patterns of somatic eye-color mutations induced in repair-deficient strains of Drosophila melanogaster by alkylating agents (MMS, ENNG and MNNG). Japan. J. Genet. 57:661 (Abstract).

38. Galas, D., and E. W. Branscomb (1978) Enzymatic determinants of DNA polymerase accuracy. Theory of coliphage T4 polymerase mechanisms. J. Mol. Biol. 124:653-687.

39. Game, J. C., and R. K. Mortimer (1974) A genetic study of X-ray sensitive mutants in yeast. Mutation Res. 24:281-292.

40. Ganesan, A. K. (1974) Persistence of pyrimidine dimers during post-replication repair in ultraviolet light-irradiated Escherichia coli K12. J. Mol. Biol. 87:103.

41. Gatti, M. (1979) Genetic control of chromosome breakage and rejoining in Drosophila melanogaster: Spontaneous

chromosome aberrations in X-linked mutants defective in DNA metabolism. Proc. Natl. Acad. Sci. USA 76:1377–1381.

42. Gatti, M., S. Pimpinelli, and B. S. Baker (1980) Relationship among chromatid interchanges, sister chromatid exchanges, and meiotic recombination in Drosophila melanogaster. Proc. Natl. Acad. Sci. USA 77:1575–1579.

43. Glickman, B. W., and M. Radman (1980) Escherichia coli mutator mutants deficient in methylation-instructed DNA mismatch correction. Proc. Natl. Acad. Sci. USA 77:1063–1067.

44. Glickman, B., P. Van den Elsen, and M. Radman (1978) Induction mutagenesis in dam⁻ mutants of Escherichia coli: A role for 6-methyladenine residues in mutation avoidance. Mol. Gen. Genet. 163:307–312.

45. Goth-Goldstein, R. (1980) Inability of Chinese hamster ovary cells to excise 0^6-methylguanine. Cancer Res. 40:2623–2624.

46. Graf, U., and F. E. Würgler (1982) DNA repair dependent mutagenesis in Drosophila melanogaster, in: Advances in Genetics, Development and Evolution of Drosophila, S. Lakovaara (Ed.), Plenum Press, New York, pp. 85–99.

47. Graf. U., and F. E. Würgler (1983) Excision repair defect sensitizes a novel Drosophila mutagenicity test. Experientia 39:685 (Abstract).

48. Graf, U., M. M. Green, and F. E. Würgler (1979) Mutagen-sensitive mutants in Drosophila melanogaster. Effects on premutational damage. Mutation Res. 63:101–112.

49. Graf, U., A. Kägi, and F. E. Würgler (1982) Mutagenesis in spermatozoa of Drosophila melanogaster by cross-linking agents depends on the mus(1)101⁺ gene product in the oocyte. Mutation Res. 95:237–249.

50. Graf, U., H. Juon, A. J. Katz, H. J. Frei, and F. E. Würgler (1983) A pilot study on a new Drosophila spot test. Mutation Res. 120:233–239.

51. Graf, U., F. E. Würgler, A. J. Katz, H. Frei, H. Juon, C. B. Hall, and P. G. Kale (1984) Somatic mutation and recombination test in Drosophila melanogaster. Environ. Mutagen. 6:153–188.

51a. Green, M. M. (1982) Genetic instability in Drosophila melanogaster: Deletion induction by insertion sequences. Proc. Natl. Acad. Sci. USA 79: 5367–5369.

52. Hadden, C. T., R. S. Foote, and S. Mitra (1983) Adaptive response of Bacillus subtilis to N-methyl-N'-nitro-N-nitrosoguanidine. J. Bacteriol. 153:756–762.

53. Haladus, E., and J. Zuk (1979) Mitotic recombination in rad6-1 mutant of Saccharomyces cerevisiae. Stud. Biophys. 76:61.

54. Harris, P. V., and J. B. Boyd (1980) Excision repair in Drosophila, analysis of strand breaks appearing in DNA of

mei-9 mutants following mutagen treatment. Biochim. Biophys. Acta 610:116-129.

55. Haseltine, W. A., L. K. Gordon, C. P. Lindan, R. H. Grafstrom, N. L. Shaper, and L. Grossman (1980) Cleavage of pyrimidine dimers in specific DNA sequences by a pyrimidine dimer DNA-glycosylase of M. luteus. Nature 285:634-641.

56. Haynes, R. H., and B. A. Kunz (1981) DNA repair and mutagenesis in yeast, in: The Molecular Biology of the Yeast Saccharomyces: Life Cycle and Inheritance, J. N. Strathern, E. W. Jones and J. R. Broach (Eds.), Cold Spring Harbor Laboratory, New York, pp. 371-413.

57. Hibner, U., and B. M. Alberts (1980) Fidelity of DNA replication catalyzed in vitro on a natural DNA template by the T4 bacteriophage multi-enzyme complex. Nature 285:300-305.

58. Howard-Flanders, P. (1981) Mutagenesis in mammalian cells. Mutation Res. 86:307-327.

59. Jachymczyk, W. J., E. Chlebowicz, Z. Swietlinska, and J. Zuk (1977) Alkaline sucrose sedimentation studies of MMS-induced DNA single strand breakage and rejoining in the wild-type and UV-sensitive mutants of Saccharomyces cerevisiae. Mutation Res. 43:1-10.

60. Kennison, J. A., and P. Ripoll (1981) Spontaneous mitotic recombination and evidence for a X-ray-inducible system for the repair of DNA damage in Drosophila melanogaster. Genetics 98:91-103.

61. Kleihues, P., and G. P. Margison (1974) Carcinogenicity of N-methyl-N-nitrosourea: possible role of excision repair of O^6-methylguanine from DNA. J. Natl. Cancer Inst. 53L1839-1841.

62. Kondo, S. (1973) Evidence that mutations are induced by errors in repair and replication. Genetics 73:109-122.

63. Kondo, S., and H. Ichikawa (1973) Evidence that pretreatment of Escherichia coli cells with N-methyl-N'-nitro-nitrosoguanidine enchances mutability of subsequently infecting phage. Mol. Gen. Genet. 126:319.

64. Lawley, P. D., and S. A. Shah (1972) Reaction of alkylating mutagens and carcinogens with nucleic acids: detection and estimation of a small extent of methylation of o^6 of guanine in DNA by methyl methane sulfonate in vitro. Chem. Biol. Interact. 5:286-288.

65. Lawrence, C. W., R. Christensen, and A. Schwartz (1982) Mechanisms of UV mutagenesis in yeast, in: Molecular and Cellular Mechanisms of Mutagenesis, J. F. Lemontt and W. M. Generoso (Eds.), Plenum Press, New York and London, pp. 109-120.

66. Lehmann, A. R. (1978) Replicative bypass mechanisms in mammalian cells, in: DNA Repair Mechanisms, P. C. Hanawalt,

E. C. Friedberg and C. F. Fox (Eds.), Academic Press, New York, pp. 485–488.

67. Lehmann, A. R., S. Kirk-Bell, C. F. Arlett, P. H. M. Lohman, E. A. de Weerd-Kastelein and D. Bootsma (1975) Xeroderma pigmentosum cells with normal levels of excision repair have a defect in DNA synthesis after UV irradiation. Proc. Natl. Acad. Sci. USA 72:219–223.

68. Lindahl, T., E. Demple, and P. Robins (1982) Suicide inactivation of the E. coli O^6-methylguanine-DNA transferase. EMBO J. 1:1359–1363.

69. Lindahl, T., B. Rydberg, T. Hjelmgren, M. Olsson, and A. Jacobsson (1981) Cellular defense mechanisms against alkylation of DNA, in: Molecular and Cellular Mechanisms of Mutagenesis, J. F. Lemontt and W. M. Generoso (Eds.), Plenum Press, New York, pp. 89–107.

70. Loveless, A. (1969) Possible relevance of O^6-alkylation of deoxyguanosine to mutagenicity and carcinogenicity of nitrosamines and nitrosamides. Nature 223:206–207.

71. Maher, V. M., L. M. Quelett, R. D. Curren, and J. J. McCormick (1976) Frequency of ultraviolet light-induced mutations is higher in xeroderma pigmentosum variant cells than in normal cells. Nature 261:593–595.

72. Matthews, C. K., T. W. North, and G. P. V. Reddy (1979) Multienzyme complexes in DNA precursor biosynthesis. Adv. Enzyme Regul. 17:133.

73. McCann, J., and B. N. Ames (1976) Detection of carcinogens as mutagens in the Salmonella/microsome test: Assay of 300 chemicals: Discussion. Proc. Natl. Acad. Sci. USA 73:950–954.

74. McCann, J., N. E. Spigarn, J. Kobori and B. N. Ames (1975) Detection of carcinogens as mutagens: Bacterial tester strains with R factor plasmids. Proc. Natl. Acad. Sci. USA 72:979–983.

75. McCormick, J. J., and V. M. Maher (1978) in: DNA Repair Mechanisms, P. C. Hanawalt, E. C. Friedberg, and C. F. Fox (Eds.), Academic Press, New York, pp. 739–749.

76. McKee, R., and C. W. Lawrence (1979) Genetic analysis of gamma-ray mutagenesis in yeast. I. Reversion in radiation sensitive strains. Genetics 93:361–373.

77. Meuth, M., N. L'Heureux-Huard, and M. Trudel (1979) Characterization of a mutator gene in Chinese hamster ovary cells. Proc. Natl. Acad. Sci. USA 76:6505–6509.

78. Mitra, S., B. C. Pal, and R. S. Foote (1982) O^6-Methylguanine-DNA-methyltransferase in wild-type and ada mutants of Escherichia coli. J. Bacteriol. 152:534–537.

79. Mohn, G. R., and J. Ellenberger (1980) Appreciation of the value of different bacterial test systems for detecting and for ranking chemical mutagens. Arch. Toxicol. 46:45–60.

80. Mohn, G. R., N. Guijt, and B. W. Glickman (1980) Influence of DNA adenine methylation (dam) mutations and of plasmid

pKM101 on the spontaneous and chemically induced mutability of certain genes in Escherichia coli K-12. Mutation Res. 74:255-265.

81. Moore, P. D., K. K. Bose, S. D., Rabkin, and B. S. Strauss (1981) Sites of termination of in vitro DNA synthesis on ultraviolet- and N-acetylaminofluorene-treated ϕX templates by prokaryotic and eukaryotic DNA polymerases. Proc. Natl. Acad. Sci. USA 78:110-114.

82. Moore, P. D., S. D. Rabkin, and B. S. Strauss (1982) In vitro replication of mutagen-damaged DNA: Site of termination, in: . Molecular and Cellular Mechanisms of Mutagenesis, J. F. Lemontt and W. M. Generoso (Eds.), Plenum Press, New York and London, pp. 179-197.

83. Mortelmans, K. E., and B. A. D. Stocker (1976) Ultraviolet light protection, enhancement of ultraviolet light mutagenesis, and mutator effect of plasmid R46 in Salmonella typhimurium. J. Bacteriol. 128:271.

84. Morrison, D. P., S. K. Quah, and P. J. Hastings (1980) Expression in dipolids of the mutator phenotype of some mutator mutants of Saccharomyces cerevisiae. Can. J. Genet. Cytol. 22:51.

85. Mowat, M., and P. J. Hastings (1979) Repair of gamma-ray induced DNA strand breaks in radiation sensitive (rad) mutants of Saccharomyces cerevisiae. Can. J. Genet. Cytol. 21:574.

86. Mukai, F. H., and B. D. Goldstein (1976) Mutagenicity of malonaldehyde, a decomposition product of peroxydised polysaturated fatty acids. Science 191:868-869.

87. Myhr, B. C., D. Turnbull, and J. A. DiPaolo (1979) Ultraviolet mutagenesis of normal and xeroderma pigmentosum variant human fibroblasts. Mutation Res. 62:341-353.

88. Myrnes, B., K. E. Giercksky, and H. Krokan (1982) Repair of O^6-methylguanine residues in DNA takes place by a similar mechanism in extracts from HeLa cells, human liver, and rat liver. J. Cell Biochem. 20:381-392.

89. Nguyen, T. D., and J. B. Boyd (1977) The meiotic mei-9 mutants of Drosophila melanogaster are deficient in repair replication of DNA. Mol. Gen. Genet. 158:141-147.

90. Nix, C. E., C. McKinley, and J. L. Epler (1980) Examination of the effect of mei-9 and mei-41 alleles on mutation induction in Drosophila males. Genetics 94:s77. Abstract.

91. Olsson, M., and T. Lindahl (1980) Repair of alkylated DNA in Escherichia coli. J. Biol. Chem. 255:10569-10571.

92. Pegg, A. E., M. Roberfroid, C. von Bahr, R. S. Foote, S. Mitra, H. Bresil, A. Likhachev, and R. Montesano (1982) Removal of O^6-methylguanine from DNA by human liver fractions. Proc. Natl. Acad. Sci. USA 79:5162-5165.

93. Prakash, L. (1977) Repair of pyrimidine dimers in radiation sensitive mutants rad3, rad4, rad6, rad9 of Saccharomyces cerevisiae. Mutation Res. 45:13-20.

94. Rabkin, S. D., P. D. Moore and B. S. Strauss (1983) In vitro bypass of UV-induced lesions by Escherichia coli DNA polymerase I: Specificity of nucleotide incorporation. Proc. Natl. Acad. Sci. USA 80:1541-1545.

95. Radman, M., G. Villani, S. Bioteux, A. R. Kinsella, B. W. Glickman, and S. Spadari (1979) Replication fidelity: mechanisms of mutation avoidance and mutation fixation. Cold Spring Harbor Symp. Quant. Biol. 43:937-946.

96. Reddy, G. P. V., and A. B. Pardee (1980) Multienzyme complex for metabolic channeling in mammalian DNA-replication. Proc. Natl. Acad. Sci. USA 77:3312-3316.

97. Reynolds, R. J., J. D. Love, and E. C. Friedberg (1981) Molecular mechanisms of pyrimidine dimer excision in Saccharomyces cerevisiae: Excision of dimers in cell extracts. J. Bacteriol. 147:705.

98. Ripley, L. S. (1975) Transversion mutagenesis in Bacteriophage T4. Mol. Gen. Genet. 141:23-40.

99. Ripley, L. S. (1981) Influence of diverse gene 43 DNA polymerases on the incorporation and replication in vivo of 2-aminopurine at A:T base-pairs in bacteriophage T4. J. Mol. Biol. 150:197-216.

100. Ripley, L. S. (1982) The infidelity of DNA polymerase, in: Induced Mutagenesis: Molecular Mechanisms and Their Implication for Environmental Protection, C. W. Lawrence, L. Prakash, and F. Sherman (Eds.), Plenum Press, New York and London, pp. 85-116.

101. Ripley, L. S., and N. B. Shoemaker (1983) A major role for bacteriophage T4 DNA polymerase in frameshift mutagenesis. Genetics 103:353-366.

102. Roberts, J. J. (1978) The repair of DNA modified by cytotoxic, mutagenic, and carcinogenic chemicals, in: Advances in Radiation Biology, J. T. Lett and H. Adler (Eds.), Vol. 7, pp. 211-436.

103. Robins, P., and J. Cairns (1979) Quantitation of the adaptive response to alkylating agents. Nature 280:74-76.

104. Saeki, T., I. Machida, and S. Nakai (1980) Genetic control of diploid recovery after X-irradiation in the yeast Saccharomyces cerevisiae. Mutation Res. 73:251-265.

105. Sato, K., N. Hieda-Shiomi, and H. Hama-Inaba (1983) X-ray-sensitive mutant mouse cells with various sensitivities to chemical mutagens. Mutation Res. 121:281-285.

106. Schendel, P. F., and P. E. Robins (1978) Repair of 0^6-methylguanine in adapted Escherichia coli. Proc. Natl. Acad. Sci. USA 75:6017-6020.

107. Schultz, R. A., C. Chang, and J. E. Trosko (1981) The mutation studies of mutagen-sensitive and DNA repair mutants of Chinese hamster fibroblasts. Environ. Mutagen. 3:141-150.

108. Seeberg, E. (1978) Reconstitution of an Escherichia coli repair endonuclease activity from the separated uvrA,

uvrB$^+$, uvrC$^+$ gene products. Proc. Natl. Acad. Sci. USA 75:2569-2573.

109. Shoaf, T., and M. E. Jones (1973) Uridylic acid synthesis in Ehrlich ascites carcinoma cells. Properties, subcellular distribution and nature of enzyme complexes of the six biosynthetic enzymes. Biochemistry 12:4039.

110. Singer, B. (1982) Mutagenesis from a chemical perspective: Nucleic acid reactions, repair, translation, and transcription, in: Molecular and Cellular Mechanisms of Mutagenesis, J. F. Lemontt and W. M. Generoso (Eds.), Plenum Press, New York and London, pp. 1-42.

111. Singer, B., J. T. Kusmierer, and H. Frankel-Conrat (1983) In vitro discrimination of replicases acting on carcinogen-modified polynucleotide templates. Proc. Natl. Acad. Sci. USA 80:969-972.

112. Sklar, R., and B. Strauss (1981) Removal of 0^6-methylguanine from DNA of normal and xeroderma pigmentosum-derived lymphoblastoid lines. Nature 289:417-420.

113. Smith, P. D. (1973) Mutagen sensitivity of Drosophila melanogaster. I. Isolation and preliminary characterization of a methylmethanesulfonate-sensitive strain. Mutation Res. 20:215-220.

114. Smith, P. D. (1976) Mutagen sensitivity of Drosophila melanogaster. III. X-linked loci governing sensitivity to methylmethanesulfonate. Mol. Gen. Genet. 149:73-85.

115. Smith, P. D., C. F. Baumen, and R. L. Dusenbery (1983) Mutagen sensitivity of Drosophila melanogaster. VI. Evidence from the excision-defective mei-9^{A1T} mutant for the timing of DNA-repair activity during spermatogenesis. Mutation Res. 108:175-184.

116. Smith, P. D., R. L. Dusenbery, S. F. Cooper, and C. F. Beumen (1982) Examining the mechanism of mutagenesis in DNA repair-deficient strains of Drosophila melanogaster, in: Environmental Mutagens and Carcinogens, T. Sugimura, S. Kondo and H. Takebe (Eds.), Alan R. Liss, Inc., New York, pp. 147-155.

117. Snow, E. T., R. S. Foote, and S. Mitra Base-pairing properties of 0^6-methylguanine in template DNA during in vitro DNA replication. J. Biol. Chem. (in press).

118. Synder, R. D., and P. D. Smith (1982) Mutagen sensitivity of Drosophila melanogaster. V. Identification of second chromosomal mutagen sensitive strains. Mol. Gen. Genet. 188:249-255.

119. Todd, P. A., C. Monti-Bragadin, and B. W. Glickman (1979). MMS mutagenesis in strains of Escherichia coli carrying the mutagenic enhancing plasmid R46: phenotypic analysis of arg$^+$ revertants. Mutation Res. 62:227-237.

120. Topal, M. D., and J. R. Fresco (1976) Complementary base pairing and the origin of substitution mutations. Nature 263:285-289.

121. Vogel, E. W. (1982) Dependence of mutagenesis in Drosophila males on metabolism and germ cell stage, in: Environmental Mutagens and Carcinogens, T. Sugimura, S. Kondo, and H. Takebe (Eds.), Alan R. Liss, Inc., New York, pp. 183–194.

122. Vogel, E. W., and A. T. Natarajan (1982) The relation between reaction kinetics and mutagenic action of monofunctional alkylating agents in higher eukaryotic systems: Interspecies comparisons, in: Chemical Mutagens, Principles and Methods for Their Detection, F. J. deSerres and A. Hollaender (Eds.), Plenum Press, New York and London, Vol. 7, pp. 295–336.

123. Waldstein, E. A., F.-H. Cao, and R. B. Setlow (1982) Adaptive increase of O^6-methylguanine-acceptor protein in HeLa cells following N-methyl-N'-nitro-N-nitrosoguanidine treatment. Nucleic Acid Res. 10:4595–4604.

124. Walker, G. C. (1977) Plasmid (pKM101)-mediated enhancement of repair and mutagenesis. Dependence on chromosomal genes in Escherichia coli K-12. Mol. Gen. Genet. 152:93–103.

125. Walker, G. C. (1981) Molecular principles underlying the Ames Salmonella/microsome test: Elements and design of short term mutagenicity test, in: In Vitro Toxicity Testing, A. Kolber (Ed.), Plenum Press, New York and London.

126. Walker, G. C., C. J. Kenyon, A. Bagg, S. J. Elledge, K. L. Perry, and W. G. Shanabruch (1982) Regulation and function of Escherichia coli genes induced by DNA damage, in: Molecular and Cellular Mechanisms of Mutagenesis, J. F. Lemontt and W. M. Generoso (Eds.), Plenum Press, New York and London, pp. 43–63.

127. Warren, W., A. R. Crathorn, and K. V. Shooter (1979) The stability of methylated purines and of methylphosphotriesters in the DNA of V79 cells after treatment with N-methyl-N-nitrosourea. Biochim. Biophys. Acta 563:82–88.

128. Watanabe, S. M., and M. F. Goodman (1978) Mutator and antimutator phenotypes of suppressed amber mutants in genes 32, 41, 44, 45, and 62 of bacteriophage T4. J. Virol. 25:73–77.

129. Watanabe, S. M., and M. F. Goodman (1982) Kinetic measurement of 2-aminopurine·cytosine and 2-aminopurine·thymine base pairs as a test of DNA polymerase fidelity mechanisms. Proc. Natl. Acad. Sci. USA 79:6429–6433.

130. Witkin, E. M. (1967) Mutation-proof and mutation-prone modes of survival in derivatives of Escherichia coli B differing in sensitivity to ultraviolet light. Brookhaven Symp. Biol. 20:17–55.

131. Witkin, E. M. (1969) Ultraviolet-induced mutation and DNA repair. Ann. Rev. Genet. 3:525–552.

132. Witkin, E. M., and I. E. Wermundsen (1979) Targeted and untargeted mutagenesis by various inducers of SOS function in Escherichia coli. Cold Spring Harbor Symp. Quant. Biol. 43:881.

133. Wright, A. S. (1980) The role of metabolism in chemical mutagenesis and chemical carcinogenesis. Mutation Res. 75:215-241.

134. Würgler, F. E., and U. Graf (1980) Mutation induction in repar-deficient strains of Drosophila melanogaster, in: DNA Repair and Mutagenesis in Eukaryotes, W. M. Generoso, M. D. Shelby and F. J. deSerres (Eds.), Plenum Press, New York and London, pp. 223-240.

135. Würgler, F. E., U. Graf, and H. Frei Somatic mutation and recombination test in wings of Drosophila melanogaster, in: Collaborative Study of Short-Term Tests for Carcinogens, J. Ashby, F. J. deSerres, M. Draper, M. Ishidate, Jr., B. Margolin, B. Matter and M. D. Shelby (Eds.), Elsevier North-Holland, in press.

136. Yang, L. L., V. M. Maher, and J. J. McCormick (1980) Error-free exision of the cytotoxic, mutagenic N_2-deoxy-guanosine DNA adduct formed in human fibroblasts by 7 beta, 8 alpha-dihydroxy-9 alpha, 10 alpha-epoxy-7,8,9,10-tetrahydro benzo(a)pyrene. Proc. Natl. Acad. Sci. USA 77:5933-5937.

137. Yarosh, D. B., R. S. Foote, S. Mitra, and R. S. Day (1983) Repair of O^6-methylguanine in DNA by demethylation is lacking in Mer⁻ human tumor cell strains. Carcinogenesis 4:199-205.

138. Zimmering, S. (1982) Preliminary data suggesting that females of the repair-deficient strain designated s t mus302 are strong potentiators of chromosome loss induced by procarbazine and diethylnitrosamine (DEN) in the male genome of Drosophila. Mutation Res. 103:141-144.

139. Zimmerman, F. K. (1968) Sensitivity to methyl methanesulfonate and nitrous acid of ultraviolet light sensitive mutants of Saccharomyces cerevisiae. Mol. Gen. Genet. 102:247-256.

140. Carpenter, A.T.C. (1979) Recombination nodules and syn-aptonemal complex in recombination defective females of Drosophila melanogaster. Chromosoma (Berlin) 75:259-292.

141. Cox, R. (1980) Comparative mutagenesis in cultured mammalian cells, in: Progress in Environmental Mutagenes M. Alacevic (Ed.) Elsevier/North Holland Biomedical Press, Amersterdam, New York, Oxford, pp. 33-46.

142. Generoso, W.M., M.D. Shelby and F.J. DeSerres (1980) (Eds.) DNA Repair and Mutagenesis in Eurkaryotes, Plenum Press, New York and London p. 458.

143. Meuth, M. (1981) Role of deoxynucleotide triphosphate pools in the cytotoxic and mutagenic effects of DNA alkylating agents. Somat. Cell Genet. 7:89-102.

144. Strauss, B.S., S. Rabkin, D. Sagher and P. Moore (1982. The
      role of DNA polymerase in base substitution mutagenesis
      on non-instructional templates. Biochimie 64:829-838.
145. Weinberg, T., B. Ullman and D.W. Martin (1981) Mutator
      phenotypes in mammalian cell mutants with distinct bio-
      chemical defects and abnormal deoxyribonucleoside triphos-
      phate pools. Proc. Natl. Acad. Sci. USA 78:2447=2451.

USE OF THE <u>BACILLUS</u> <u>SUBTILIS</u> REC-ASSAY IN

ENVIRONMENTAL MUTAGEN STUDIES

Tsuneo Kada, Yoshito Sadaie, Yutaka Sakamoto[1] and
Koichi Hirano[2]

Department of Induced Mutation
National Institute of Genetics
Mishima, Shizuoka-ken 411, Japan

[1]Central Research Division
Takeda Chemical Industries Ltd.
Osaka 532, Japan

[2]Laboratories of Animal Science and
Toxicology, Sankyo Co. Ltd.
Fukuroi, Shizuoka-ken 437, Japan

SUMMARY

Since certain types of DNA damage are subject to cellular
recombination repair, recombinationless bacteria are usually more
sensitive than the wild type. Agents showing increased lethal
activity on Rec^- in comparison with Rec^+ cells may have caused
cellular DNA damage. Because of the simplicity of the procedures,
positive chemicals are very easily selected from a great number of
samples. <u>Bacillus</u> <u>subtilis</u> (gram-positive) is useful in this
assay because its cellular membrane is more permeable to chemicals
than that of some gram-negative enteric bacteria.

Rec-assay procedures were used with fruits in studies on
environmental mutagens. A number of new chemical mutagens were
detected. They included food additives, mycotoxins, methyl
compounds, etc. For known mutagens or carcinogens that are shown
to be positive in the rec-assay, the rec-assay procedure was
adopted to screen environmental factors that suppress or
inactivate mutagenic activities. We define these agents as
desmutagens and are attempting to use them for the prevention of
environmental carcinogenesis.

INTRODUCTION

Research conducted in the past ten years in many laboratories
has shown that a number of known carcinogens are also mutagens in
microorganisms (1,36). It is also likely that the mutagens
detected by the microbial method could also be mutagens in mammals
or in man, if these chemicals were resistant to in vivo metabolic
degradation and transported to sexual organs, thus inducing
heritable mutations. Because mutagenesis experiments in microbial
systems are relatively easy and inexpensive, it is reasonable to
test mutagens first in microorganisms and further to check their
genotoxic potentialities in mammals. The microbial procedures can
also be used efficiently to investigate environmental carcinogens
whose mutagenicity is evident, thus providing basic information
about the genetic mechanism of carcinogens in the environment.

We developed the Bacillus subtilis rec-assay system for the
efficient detection of mutagens and carcinogens (3,12,17). Growth
of bacteria that are genetically deficient in the DNA repair
capacity is more inhibited than that of wild cells. Since damage
induced in DNA by various types of mutagens is subject to
recombination repair, the rec-assay procedure is able to detect,
in a minimum amount of time, any agent that damages DNA. The
specificity of each mutagen can be determined, as a second step,
by other microbial mutagenicity assays.

GENETIC BACKGROUND

A Marburg strain 15 (argA15 trp-3) was first obtained from
the laboratory of Y. Ikeda and H. Saito (4). After single-colony
purification from this strain, a strain designated H17 (or NIG17)
(argA15 trp-3) was obtained which had a stable recipient property
for transformation. After treating cells of strain H17 with N-
methyl-N'-nitro-N-nitrosoguanidine, a γ-ray-sensitive strain was
isolated and named M45 (or NIG45) (31).

The strain M45 showed high sensitivities to γ-rays,
ultraviolet light (UV) and mitomycin C. Deficiency in genetic
recombination of this strain was shown by experiments on its
capacity for transformation, SPO2 transfection, and PBS1 phage
transduction, as well as its radiation and drug sensitivities and
its Hcr$^+$ capacity for UV-exposed phage M2. Mapping studies
revealed that the mutation rec-45 was found to be tightly linked
to recE. The mutation rec-45 reduced the frequency of
recombination involved in both transformation and PBS1
transduction. SPO2 lysogens of the strain M45 were not inducible,
indicating involvement of a rec-45 gene product in the development
of SPO2 prophage to a vegetative form. The UV-induced DNA

degradation in vegetative cells was more pronounced in the rec-45 strain than in the wild strain.

Recently de Vos et al. (39) showed that the cell with recM45, a mutation at the recE gene, has a modified protein with a molecular weight of 45,000 and an isoelectric point higher than that of the wild strain. The high sensitivity of a recE strain to mitomycin C was restored by the introduction of a plasmid carrying a recA gene of E. coli. It was shown that this recA gene was expressed in the cells of B. subtilis, thus producing the recA protein. These findings suggest that the M45 strain of B. subtilis is lacking a function of recA (as defined in E. coli) and that the rec-assay positiveness is related to the recA function.

EXPERIMENTAL PROCEDURES OF THE REC-ASSAY

(1) Procurement, Selection and Stock of Bacteria

Slants or spore suspensions of Rec^+ and Rec^- bacteria will be mailed on request by Drs. T. Kada and Y. Sadaie (National Institute of Genetics, Mishima, Sizuoka-ken 411, Japan). These cultures may be used immediately in experiments. However, it is recommended to select good stocks. Cells taken from slants or spore suspensions are diluted appropriately and spread on broth agar to obtain visible isolated colonies. From the grown colonies, certain numbers (usually 10 or more) of pure independent liquid cultures are made (for example, use 3 ml of liquid broth). Each culture is streaked radially on the dry surface of broth agar by means of small pipettes and a paper disk (diameter 16 mm) impregnated with 20 µl of Mitomycin C solution (10 µg/ml) (Fig. 1). Portions of each culture are stocked either by streaking on broth agar slants, or by freezing with glycerin supplement (to 3 ml culture add 1 ml of 50% glycerin, stock at -40 – -80°C). The stocks giving unfavorable results (for example, reduced sensitivities to Mitomycin C) are discarded and only good stocks are used for future experiments.

(2) Preparation of Test Samples

The best stable positive controls are mitomycin C (for assays without metabolic activation) and 2-aminoanthracene (for assays with metabolic activation). The standard results with these drugs are shown in Table 1.

To do assays of the chemicals whose DNA-damaging activities are unknown, it is recommended to start with the highest concentrations. When a sample at first shows too much killing effect, it should be assayed at lower concentrations in the next

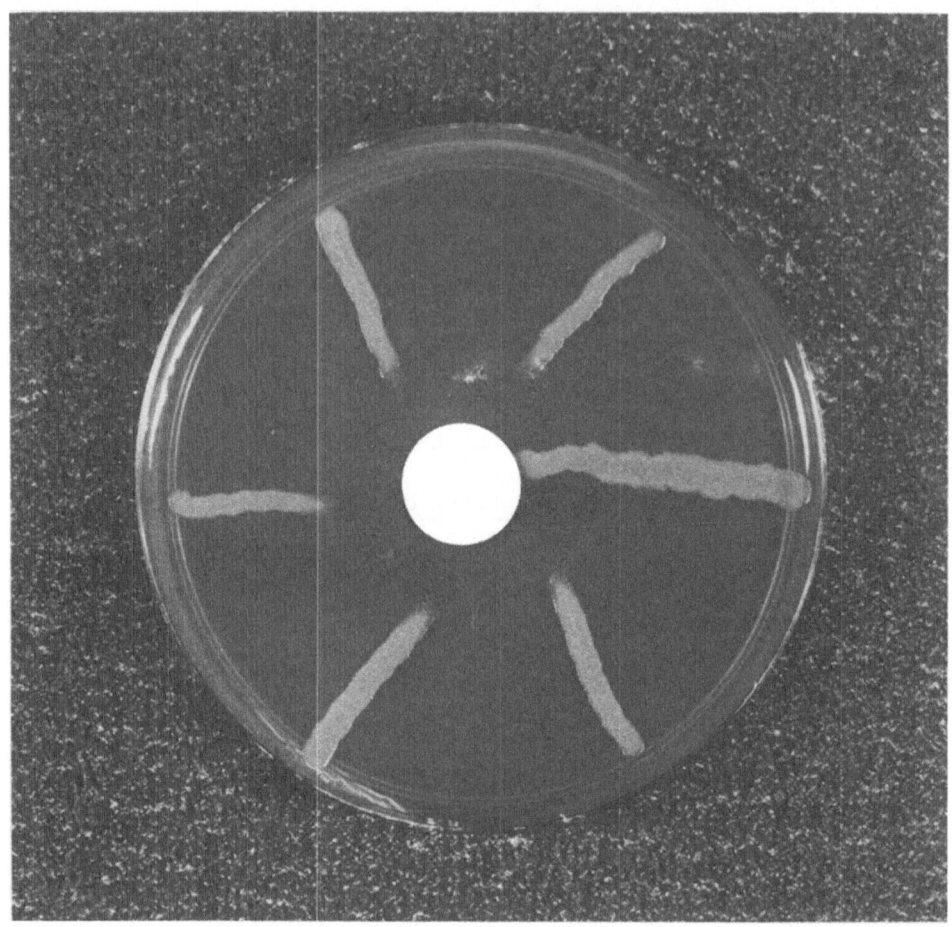

Figure 1.   Selection of "good" Rec⁻ subclones by radial
      streakings.   Mitomycin C (0.2 µg per disk) was used as a
      positive control.

steps.   A characteristic of the repair test of diffusing samples
in agar is that meaningful results are obtainable at any
concentration.

   It is not necessary to be careful about the sterility of the
sample, since the results of the rec-assay will usually come out
before the growth of any contaminating microorganisms from outside
the disk is detected.   The following solvents are currently used
without any interference with the results: water, DMSO, ethyl
acetate, acetone and ethyl alcohol.   However, it

Table 1. Results of the B. subtilis rec-assay of typical chemical
        mutagens and non-mutagens without metabolic activation using
        freshly grown vegative cells of H17 Rec$^+$ and M45 Rec$^-$.

Drug	Dose (μg/disk)	mm of inhibition zone	
		H17 Rec$^+$	M45 Rec$^-$
Mitomycin C	0.1	0	9
AF2	0.02	0	6
Captan	20	1	10.5
4NQO	0.02	0	1
Benzotrichloride	2.6 μM	0	5
$K_2Cr_2O_7$	(50 μl of 0.005 M)	3	20
V_2O_5	(50 μl of 0.5 M)	0	5
Aflatoxin B_1	10	0	3
Kanamycin	20	14	16

is recommended, in carrying out the solvent control experiments,
to check the sensitivities of each solvent to Rec$^+$ and Rec$^-$
bacteria.

(3) Rapid Streak Method (Without Metabolic Activation) (17)

Usually 3 ml of liquid broth (wet meat extract 10 g,
polypepton dry powder 10 g, NaCl 5 g, water 1000 ml; pH adjusted
to 7.0) is inoculated with a small amount of cells or spores and
the test tube is incubated overnight (for about 16 hours) at 37°C
and shaken. The test tube containing the overnight fresh culture
can be kept in an ice-water mixture for several hours before use
without any change in the results.

Suspensions of vegetative cells of two strains H17 Rec$^+$ and
M45 Rec$^-$ are streaked radially on the "dry" surface of broth agar
by means of small pipettes, and a paper disk (diameter 16 mm)
containing the test solution is placed on the starting points of
the streaks. The plates are incubated at 37° C for about 20 hrs,
after which the lengths of the inhibition are measured.
Photographs of typical plates are shown in Fig. 2. Results
obtained with typical mutagens are shown in Table 1.

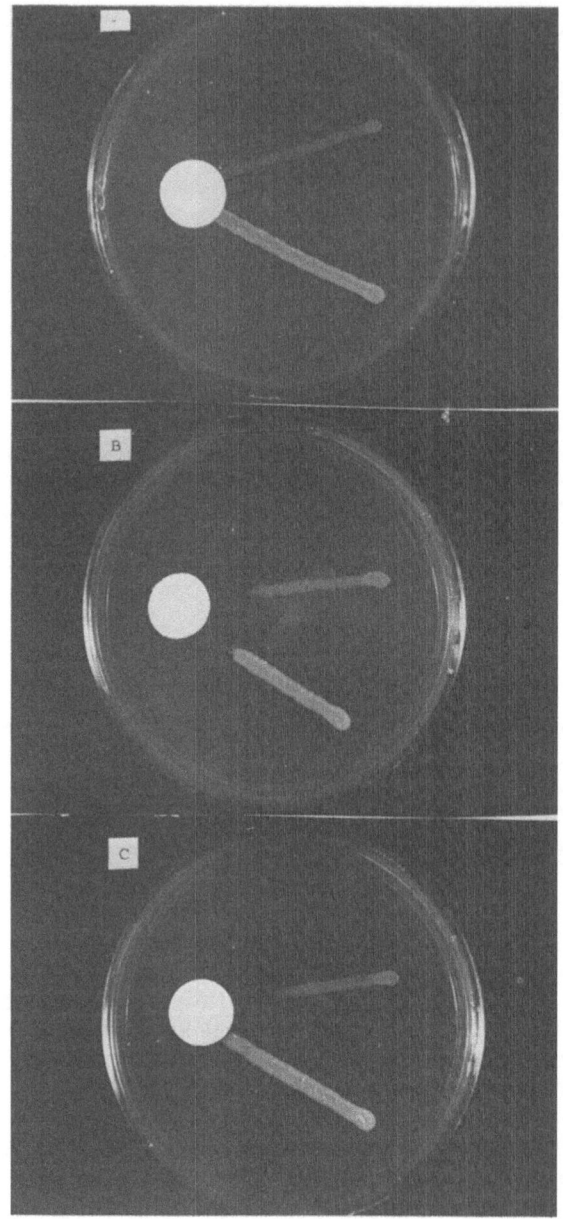

Figure 2. Photographs of plates in the rec-assay of chemicals.
           A, no drug control; B, Kanamycin (KM); C, 4-nitroquinoline-N-
           oxide (4NQO) (Kada et al, 1972).

In place of the fresh vegetative cells, frozen stocks or spores can be used for streaking. The higher sensitivities are obtained by using frozen stocks or spores rather than using fresh vegetative cells. To prepare frozen stocks, the 3 ml culture is supplemented with 1 ml of 50% glycerol (50 g of glycerol are filled up with distilled water to 100 ml and autoclaved) and kept between -40 and -80°C (the final concentration of glycerol being 12.5%). The stocks are unfrozen and streaked. The highest sensitivities can be obtained by streaking the spore suspensions (10^4 - 10^5 spores per 1 ml in distilled water). See the following section for procedures in the preparation of spores.

(4) Preparation of Rat Liver Microsomal Fraction (S9)

The method is essentially based on that currently carried out for the Salmonella reversion assay (1,25,41). For the spore rec-assay, we do not prepare the S9 mixture. Instead, we incorporate the 9,000 xg supernatant solution into the agar plate without co-factors. In brief, rat liver previously treated with one or two chemical inducers of drug metabolizing enzymes (PCB, β-naphtoflavone plus phenobarbitol) is removed and homogenized; the supernatant fraction obtained by 9,000 xg centrifugation is pooled and stocked at -80°C.

(5) Spore Rec-Assay (With or Without Metabolic Activation) (3,12)

Preparation of spores. An overnight culture of M45 or H17 is spread on modified Schaeffer's agar containing 16 g Difco nutrient broth, 2 g KCl, 0.5 g $MgSO_4$ · $7H_2O$, 19.8 mg $MnCl_2$ · $4H_2O$, 278 μg $FeSO_4$ · $7H_2O$, 236 mg $Ca(NO_3)_2$ · $4H_2O$ and 1 g glucose, and solidified with 15 g Difco agar/liter. After incubation at 37°C for 3 days for the strain H17 or for 5 days for the strain M45, cells are scraped up, washed once with minimal medium (MM), resuspended in MM, and treated with lysozyme (final concentration: 2 mg/ml) at 37°C for 30 min. Then, sodium dodecyl sulfate is added to the mixture (final concentration: 1%) and incubated for another 30 min. The spores are then washed five times by centrifugation with distilled water and resuspended in distilled water for storage at 4°C.

Spores can be prepared also in liquid culture. Bacteria are fully grown in the above modified Schaeffer's medium (without agar) and the culture is continued with active aeration for 2-3 days at 37°C. Formation of spores is checked by sampling a small portion of the culture under microscope.

The number of spores is determined by diluting the stock solution with sterile water, and by spreading 0.1 ml of it on the surface of broth agar and counting the growth colonies. It is convenient to adjust the concentration to about 2 x 10^7 per ml for the stock. Spores are quite stable in a water suspension at 4°C; no modification of the drug sensitivities takes place after at least two years stock of both Rec⁺ and Rec⁻ spores.

(6) Preparation of a Cofactor Solution

A cofactor solution containing 20 mg glucose-6-phosphate sodium salt (64.1 μmol) and 40 mg NADP disodium salt (50.6 μmol) per ml is prepared on the day of experiments by dissolving the salts in a buffer solution (8 μM $MgCl_2$; 33 μM KCl; 100 mM phosphate buffer, pH 7.2). Usually no sterilization process is necessary if there are no expected heavy contaminations. Paper disks impregnated with 20 μl of the above cofactor solution can be maintained as a stable stock in a freezer (below −20°C).

Spore rec-assay procedure. An appropriate volume (0.1 − 0.3 ml) of 9,000 xg supernatant of S9 and spore of H17 (Rec⁺) or M45 (Rec⁻) strains (each 0.1 ml of a suspension of 2×10^7 spores per ml) are placed into an empty Petri dish (90 mm diameter), then 10 ml of molten broth medium (autoclaved with 0.8% Difco agar and kept at 43°C) is added and mixed well (Fig. 3). When the agar

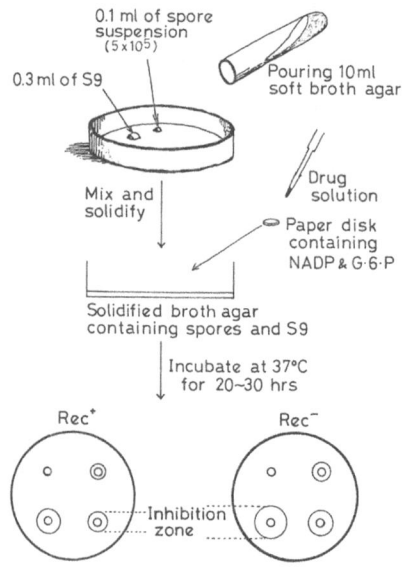

Figure 3.   Procedure of the spore rec-assay
(Hirano et al., 1982).

Table 2. Results of the B. subtilis spore rec-assay with
metabolic activation on some typical mutagens. Standard rat
liver homogenate (S9) was used.

| Drug | Dose (μg/disk) | mm of inhibition zone | |
		H17 Rec[+]	M45 Rec[−]
Aflatoxin B$_1$	0.1	3	15.0
2-Aminofluorene	200	6.3	13.0
2-Aminoanthracene	200	0	6.4
Sterigmatocystin	10	0	4.0
Trp-P-1	1.2	0	4.0
Trp-P-2	3.0	0	5.6
DMN	1000	0	8.0

medium is well solidified (it is recommended to keep the plates at
4℃ for 30 min before use), a paper disk (diameter 8 mm; thickness
1 mm) is impregnated successively with 20 μl of the cofactor
solution and 20 μl of the chemical solution. After 20 hour's
incubation at 37°C, the lengths of inhibition zones appearing
around the disk are measured and the values of the Rec[+] strain are
compared with those of the Rec[−] strains (Table 2). The lengths of
clear inhibition zones are measured after incubation of the plates
20 hrs. The inhibition zones become clearer by additional for
overnight incubation (2 days in total). We usually place 4 disks
in a plate: one positive control and three samples of increasing
doses. Figure 4 shows typical results of the spore rec-assays.

For the assays of samples that do not require metabolic
activation, use of S9 and the co-factor is omitted in the above
procedure. Each sample is impregnated in a paper disk without the
co-factor and placed on the spore agar, not containing the S9.

Spore agar plates without S9 can be used for several days
after preparation, if stocked at 4°C for several hours.

(7) Differential Killing Assays in Liquid (With and Without
Metabolic Activation) (16,32)

Preparation of bacteria. Five ml of nutrient broth
(polypepton powder 10 g, yeast extract powder 2 g, sodium chloride

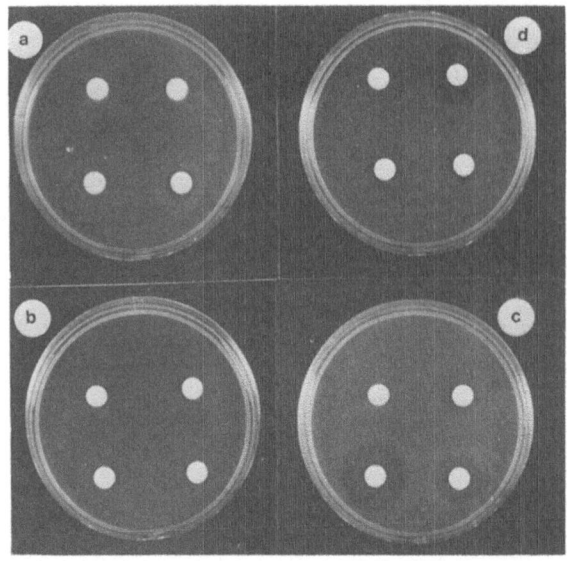

Figure 4.   Typical plates of the spore rec-assay.   Plates: (A)
Rec[+], without S-9; (B) Rec[-], without S-9; (C) Rec[+], with S-9;
(D) Rec[-], with S-9; Samples, counterclockwise, from top
left: no drug; mytomycin C (0.02 µg); Trp-P2 (37.5) and
Trp-Pl (6.5 µg).   (Kada et al., 1980).

2 g, dissolved in 1,000 ml of distilled water; pH adjusted to 7.2)
are inoculated with 0.1 ml of the seed bacterial suspension (H17
Rec[+] or M45 Rec[-]).   They are shaken at 37°C for 16 hours.   The
bacterial titers of fresh cultures are about 1.5×10^9 for H17
Rec[+] and about 4×10^8 for M45 Rec[-].

   Plating.   The nutrient soft agar containing 6 g of agar
powder in 1,000 ml of liquid nutrient broth is prepared.   4.5 ml
portions are distriubted into test tubes and kept at 48°C before
use.   The important feature is that these nutrient agar plates
must be supplemented with 0.2% of glucose and 1 mM of Mg[++].   Non-
supplemented nutrient agar will often cause false positive
effects.

   Test chemicals.   Test chemicals are usually dissolved in DMSO
or distilled water and more than 3 doses are first prepared for
each chemical.   The upper limit concentration of each test
chemical solution is that of saturated solution.   AF2 (1 ng/tube)
and DAPA (5 µg/tube) are used as positive controls in experiments
without metabolic activation.   For those requiring on S9 mix, 2-
aminoanthracene (2AA, 20 µg/tube) is used as positive control.

Kanamycin sulfate (0.3 μg/tube) is used for a negative control giving uniform killing effects both on Rec$^+$ and Rec$^-$ cells.

A test sample solution (usually 0.1 ml), 0.5 ml of the S9-mix and 0.2 ml of the bacterial suspension (freshly grown in liquid nutrient broth) are added to a sterile test tube, mixed and incubated by shaking at 37°C for 30 minutes. To this pre-incubated mixture, 3 ml of nutrient molten soft agar is added; and the mixture is poured on nutrient broth agar plate supplemented with glucose (0.2%), MgSO$_4$ · 7H$_2$O (1 mM) and agar (1.5%). When the soft agar overlay has hardened, an additional 4.5 ml soft agar is overlaid. Plates are incubated at 37°C for 40 hours and grown colonies are counted. Results of typical chemicals are shown in Fig. 5.

Scoring and analysis of the results. Colonies grown on plates are counted.

H$_o$ and M$_o$: Mean numbers of the colonies of H17 Rec$^+$ and M45 Rec$^-$ on the plates belonging to the control (solvent) plates, respectively.

H$_i$ and M$_i$: Mean numbers of the colonies of H17 Rec$^+$ and M45 Rec$^-$ on the plates belonging to the test group, respectively.

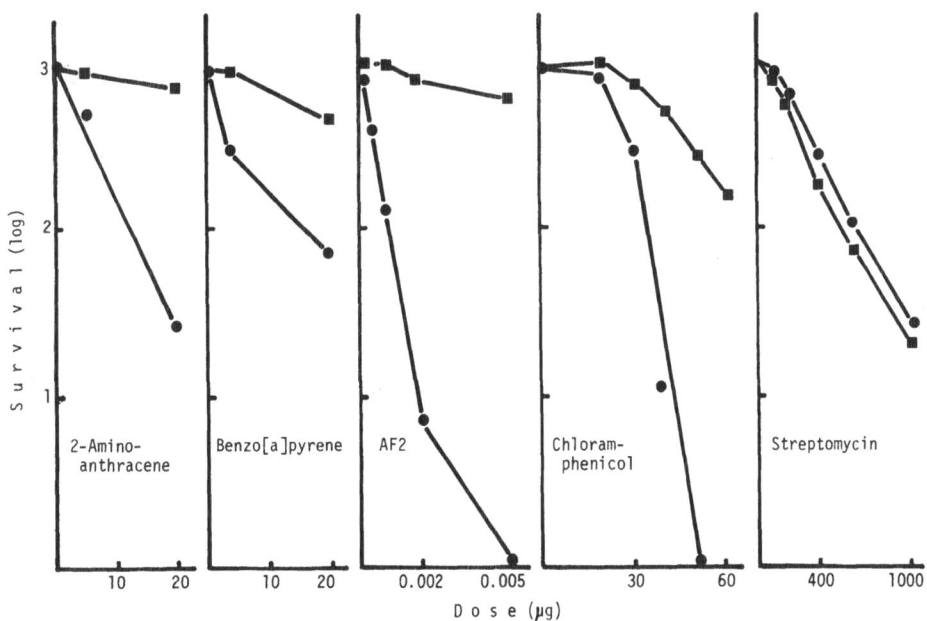

Figure 5. Results of differential killing assay in liquid of typical chemical samples (Sakamoto, 1981).

The ratio R_i is calculated for each sample as follows:

$$R_i = (M_i/M_o)(H_i/H_o)$$

if the R_i is less than 0.5 and the dose–response effects are observed, the test sample is concluded as positive in the present assay.

QUANTITATIVE ANALYSIS OF RESULTS

First, it should be pointed out that the assay procedure using agar plates is applicable to chemicals that diffuse in agar medium. A small number of chemicals are not diffusable or diffuse with difficulty in agar. Assay of these samples should be done with the differential killing method where formation of the diffusion gradient in the agar is not required.

To realize quantitative relationships between drug concentration and the length of growth inhibition zones of $\underline{Rec}^+$ and $\underline{Rec}^-$ cells, the depth of the agar medium in plates should not be large (not greater than 2 mm). As shown in Fig. 6, the chemical placed on the surface of the agar medium diffuses in two directions, horizontally and vertically. If the agar is thin enough, the vertical diffusion is negligible and the chemical diffuses mostly horizontally. (In this case, the concentration C of the drug at a distance r from the center is proportional to $1/r^2$.) Since the growth inhibition is exactly proportional to the drug concentration, the length of the growth inhibition zone is proportional to the logarithmic value of the drug concentration.

As shown in Fig. 6, dose–effects curves for $\underline{Rec}^+$ and $\underline{Rec}^-$ cells by a positive chemical are usually represented by two straight lines. When each line is extrapolated to zero inhibition, the values (X and Y) of the concentration giving the minimum inhibition are obtained. The ratio X/Y indicates an index of DNA–damage by the test chemical.

When lengths of inhibition zones produced by negative samples such as streptomycin in $\underline{Rec}^+$ and $\underline{Rec}^-$ strains are plotted (Fig. 7), they produce a straight line whose angle is slightly lower than 45°. Positive samples give deviated points as indicated in Figure 7 for samples such as Mitomycin C or AF2.

It is noted that the above dose–effects straight lines are usually parallel which means that the difference in length of the inhibition zones for $\underline{Rec}^+$ and $\underline{Rec}^-$ cells is constant for any drug doses over the X dose. For many years, the results of the rec-assay using the simplest streak procedure were shown as the

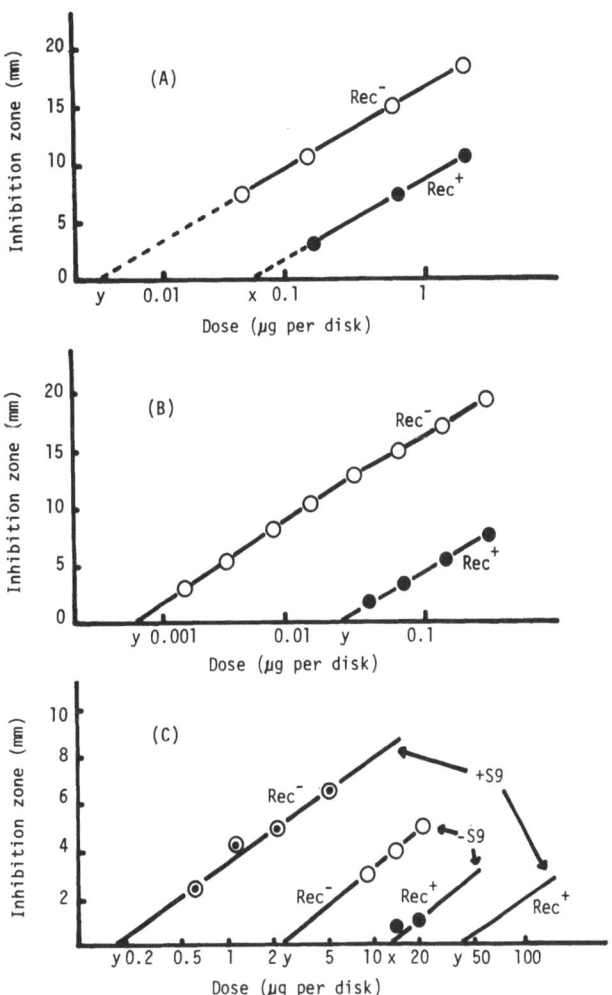

Figure 6. Relationships between mutagen doses and their
    inhibition zones.

(A) Mitomycin C   Streak method with vegative cells without
                  metabolic activation

(B) Mitomycin C   Spore method without metabolic activation

(C) Trp-P-1   Spore method with metabolic activation

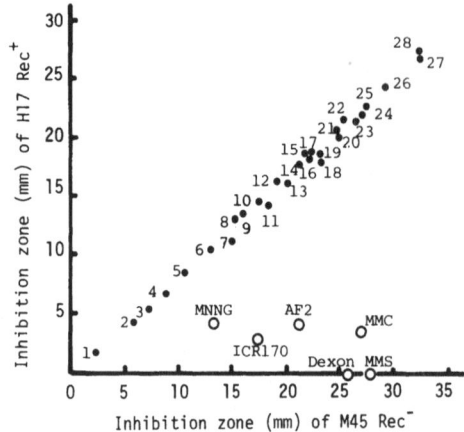

Figure 7. Relationship between inhibition zones of Rec$^+$ and Rec$^-$
    strains by different chemicals. (O) Rec-assay-positive
    samples; (●) 28 negative samples: 1, Kolistin; 2, polymyxin;
    3, fradiomycin (7.5 µg); 4, fradiomycin (15 µg); 5,
    fradiomycin (30 µg); 6, fradiomycin (60 µg); 7, lincomycin;
    8, fradiomycin (commercial disk); 9, kanamycin (20 µg); 10,
    kanamycin (40 µg); 11, oxytetracycline; 12, kanamycin (80
    µg); 13, paromomycin; 14, kanamycin (160 µg); 15, demethyl
    chlortetracycline; 16, methacycline; 17, aminodeoxy
    kanamycin; 18, tetracycline, 19, gentamicin; 20,
    oleandomycin; 21, novobiocin; 22, kanamycin (commercial
    disk); 23, josamycin; 24, leucomycin; 25, sulbenicillin; 26,
    penicillin, 27, benzyl penicillin; 28, calbenicillin (Kada et
    al, 1980).

difference between two lengths of inhibition zones for Rec$^+$ and
Rec$^-$ (Table 1). That this value represents certain quantitative
aspects is supported by the above reasoning, in the case when the
drug dose produces some minimum inhibition for Rec$^+$ cells (in
Fig. 7: the dose over X) and the agar medium in the plates is
thin enough (we usually use 10 ml of agar medium for plates having
a diameter of 9 cm).

NATURE OF POSITIVE SAMPLES

        The rec-assay has been applied for several years in testing
more than one thousand chemicals (8,9,21-23). Certain positive
samples are classified and listed in Table 3. The following are
characteristic features of the results:

Table 3. Examples of chemical carcinogens that
gave positive results in the rec-assay.

AF2	Hydrazine SO_4
2-AAF	Hexa-met-phosphoramide
4-Aminoazobenzene	* (−)-Luteoskyrin
4-Aminobiphenyl	Mitomycin C
o-Aminoazotoluene	MOCA
Auramine	Methylazoxymethanolacetate
3-Aminotriazole	α-Naphtylamine
Butylbuthanol-nitrosamine	β-Naphtylamine 4
N-n-Butyl-N-nitrosourea	4NQO
N-n-Butyl-N-nitrosourethane	n-Nitromorpholine
*Berylium (BeSO$_4$)	*Patuline
Benzidine	*Penicillic acid
Benzo(α)pyrene	1, 3-Propane sultone
Chromium ($K_2Cr_2O_7$)	β-Propiolactone
*Cortisone acetate	Quinoline
Chloroform	*Safrole
Cyclophosphamide	*Thioacetamide
n-Dibutylnitrosamine	*Thiourea
4-Dimethylaminostilbenc	*Tween 60
*Diethylstilbestrol	o-Tolunidine HCl
Dimethylnitrosamine	Trp-P1
1,2-Dimethylhydrazine	Trp-P2
Dimethylanthracene	Urethane
Ethylenethiourea	*Vanadium (VOCl$_2$)
Epichlorohydrin	

* Non genotoxic chemicals that are negative in the reversion assays.

(1) When typical carcinogens are assayed, about 85% of them
are detected by the rec-assay. They include carcinogens
that are negative in other microbial assays such as Ames'
reversion assays of Salmonella. In contrast, there are
some carcinogens giving positive responses in the Ames
assay which are negative in the rec-assay. In an
international programme of blind assay (9), it was
indicated that the rec-assay has the highest
detectability of known carcinogens, that is 85%, and this
value increased to 96% if combined with the Ames assay.

(2) The existence of carcinogens that gave negative results in reverse mutation assays using Ames Salmonella strains but positive results in the rec-assay can be explained as follows. Reverse mutations take place only by specific modifications of base pairs at the site of genes where auxotrophic characters are determined. It is likely that DNA damages induced by chemicals and requiring recombination for their repair have wider spectra than those responsible for reverse mutations. Because the rec-assay needs only about one tenth of the time and expense required for bacterial reversion assays, we save much time if we use this assay for initial screening. The positive samples can then be checked using other systems to determine their specificities as mutagens. These procedures have been adopted for food additives (6,7,38), pesticides (15,33,34), metals (19), crude drugs (26), etc.

(3) Past studies on the detection of chemical carcinogens revealed that mutagenic specificities of certain rec-assay positive carcinogens (chemicals with asterisks in Table 3) are not determined in any kind of bacterial systems or that these chemicals are probably not acting directly on DNA. They often include skin irritants, tumor promoters and allergens (18). The implications of these findings are under examination.

USE OF THE REC-ASSAY IN PREVENTION OF ENVIRONMENTAL CARCINOGENS

Since experiments using the rec-assay are so easy and simple, it is also very efficient to use it to screen for environmental desmutagens and anticarcinogens. For example, we have screened vegetable desmutagens acting on mutagenic carcinogens formed by pyrolysis of amino acids and protein (35). Though Trp-P-1, a tryptophan pyrolysate, usually requires S9 activation to express its mutagenicity, it is positive in the spore rec-assay of Bacillus subtilis without metabolic activation. When 0.2 µg of Trp-P-1 is mixed with juices prepared from different kinds of vegetables and fruits and placed on spore plates, we can easily detect vegetable samples that are irreversibly reducing the mutagenicity of pyrolysate amino acid mutagens such as cabbage, burdock, etc. (5,14,27). Similar assays were successful to find desmutagens acting on mutagenic reaction products formed from two food additives, sorbic acid and sodium nitrite (2,10,28-30). Recent studies on desmutagens and antimutagens have been summarized (11,13).

The rec-assays were applied to the detection of different kinds of chemicals and the results have been evaluated (9,12,20-24,34,37,40).

ACKNOWLEDGEMENT

The present study was supported in part by research grants from the Japanese Ministry of Education, the Japanese Ministry of Welfare, the Science and Technology Agency of Japan, the Takamatsunomiya Cancer Research Foundation, and the Nissan Science Foundation.

REFERENCES

1. Ames, B.N., W.E. Durston, E. Yamasaki, and F.D. Lee (1973) Carcinogens are mutagens: A simple test system combining liver homogenates for activation and bacteria for detection. Proc. Natl. Acad. Sci. USA 70:2281-2285.
2. Hayatsu, H., K.C. Chung, T. Kada, and T. Nakajima (1975) Generation of mutagenic compound(s) by a reaction between sorbic acid and nitrite. Mutation Res. 30:417-419.
3. Hirano, K., T. Hagiwara, Y. Ohta, H. Matsumoto, and T. Kada (1982) Rec-assay with spores of Bacillus subtilis with and without metabolic activation. Mutation Res. 97:339-347.
4. Ikeda, Y., H. Saito, K. Miura, J. Takagi, and H. Aoki (1965) DNA base composition, susceptibility to bacteriophages, and interspecific transformation as criteria for classification in the genus Bacillus. J. Gen. Appl. Microbiol. 11:181-190.
5. Inoue, T., K. Morita, and T. Kada (1981) Purification and properties of a plant desmutagenic factor for the mutagenic principle of tryptophan pyrolysate. Agric. Biol. Chem. 45:345-353.
6. Kada, T. (1973) Escherichia coli mutagenicity of furylfuramide. Japan J. Genetics 48:301-305.
7. Kada, T. (1975) Mutagenicity and carcinogenicity screening of food additives by the rec-assay and reversion procedures, In Screening Tests in Chemical Carcinogenesis, R. Montesano, H. Bartsch, and L. Tomatis, Eds., IARC Scientific Publication No.12, pp. 105-112.
8. Kada, T. (1981) Mutagenicity of selected chemicals in the rec-assay in Bacillus subtilis, In Comparative Chemical Mutagenesis, R.J. de Serres, and M.D. Shelby, Eds., Plenum Publishing Corporation, pp. 19-26.
9. Kada, T. (1981) The DNA-damaging activity of 42 coded compounds in the rec-assay, In Evaluation of Short-Term Tests for Carcinogens, F.J. de Serres, and J. Ashby, Eds., Elsevier North-Holland, Vol. 1, pp. 175-182.

10. Kada, T. (1981) Recent research on environment mutagens. Nippon Nôgeikagaku Kaishi 55:597–605.

11. Kada, T. (1982) Mechanisms and genetic implications of environmental antimutagens, In Environmental Mutagens and Carcinogens, T. Sugimura, S. Kondo, and H. Takabe, Eds., University of Tokyo Press, Tokyo and Alan R. Liss, Inc., New York, pp. 355–359.

12. Kada, T., K. Hirano, and Y. Shirasu (1980) Screening of environmental chemical mutagens by the rec–assay system with Bacillus subtilis, In Chemical Mutagens, F.J. de Serres, and A. Hollaender, Eds., Plenum Publishing Corporation, Vol. 6, pp. 149–173.

13. Kada, T., T. Inoue, and M. Namiki (1982) Environmental desmutagens and antimutagens, In Environmental Mutagenesis, Carcinogenesis, and Plant Biology, E.J. Klekowski, Jr., Ed., Praeger, Vol. 1, pp. 133–152.

14. Kada, T., K. Morita, and T. Inoue (1978) Anti–mutagenic action of vegetable factor(s) on the mutagenic principle of tryptophan pyrolysate. Mutation Res. 53:351–353.

15. Kada, T., M. Moriya, and Y. Shirasu (1974) Screening of pesticides for DNA interactions by "rec–assay" and mutagenesis testing, and frameshift mutagens detected. Mutation Res. 26:243–248.

16. Kada, T., Y. Sadaie, and Y. Sakamoto. Bacillus subtilis repair test, In Handbook of Mutagenicity Test Procedures, B.J. Kilbey, Ed., Elsevier Ltd., New York, in press.

17. Kada, T., K. Tutikawa, and Y. Sadaie (1972) In vitro and host–mediated "rec–assay" procedures for screening chemical mutagens; and phloxine, a mutagenic red dye detected. Mutation Res. 16:165–174.

18. Kada, T., and S. Watanabe (1983) Bacillus subtilis rec–assay with and without metabolic activation: Improvements and applications, In In Vitro Toxicity Testing of Environmental Agents, A.R. Kolber, T.K. Wong, L.D. Grant, R.S. DeWoskin, and T.J. Hughes, Eds., Plenum Publishing Corporation, New York, pp. 41–60.

19. Kanematsu, N., M. Hara, and T. Kada (1980) Rec assay and mutagenicity studies on metal compounds. Mutation Res. 77:109–116.

20. Karube, I., T. Matsunaga, T. Nakahara, S. Suzuki, and T. Kada (1981) Preliminary screening of mutagens with a microbial sensor. Anal. Chem. 53:1024–1026.

21. Kawachi, T., T. Komatsu, T. Kada, M. Ishidate, M. Sasaki, T. Sugiyama, and Y. Tazima (1980) Results of recent studies on the relevance of various short–term screening tests in Japan, In The Predictive Value of Short–Term Screening Tests in Carcinogenicity Evaluation, G.W. Williams et al., Eds., Elsevier/North Holland Biomedical Press, pp. 253–267.

22. Kawachi, T., T. Yahagi, T. Kada, Y. Tazima, M. Ishidate, M. Sasaki, and T. Sugimura (1980) Cooperative programme on

short-term assays for carcinogenicity in Japan, In <u>Molecular</u>
<u>and Cellular Aspects of Carcinogen Screening Test</u>,
R. Montesano, H. Bartsch, and L. Tomatis, Eds., Lyon (IARC
Sci. Publ. No. 27), pp. 323-330.

23. Leifer, Z., T. Kada, M. Mandel, E. Zeiger, R. Stafford, and
    H.S. Rosenkranz (1981) An evaluation of tests using DNA
    repair-deficient bacteria for predicting genotoxicity and
    carcinogenicity. <u>Mutation Res</u>. 87:211-297.

24. Matsui, S. (1980) Evaluation of a <u>Bacillus subtilis</u> rec-assay
    for the detection of mutagens which may occur in water
    environments. <u>Water Res</u>. 14:1613-1619.

25. Matsushima, T., M. Sawamura, K. Hara, and T. Sugimura (1976) A
    safe substitute for polychlorinated biphenyls as an inducer
    of metabolic activation system, In <u>In Vitro Metabolic</u>
    <u>Activation in Mutagenesis Testing</u>, F.J. de Serres et al.,
    Eds., Elsevier/North-Holland, Amsterdam, pp. 85-88.

26. Morimoto, I., F. Watanabe, T. Osawa, T. Okitsu, and T. Kada
    (1982) Mutagenicity screening of crude drugs with <u>Bacillus</u>
    <u>subtilis</u> rec-assay and <u>Salmonella</u>/microsome reversion assay.
    <u>Mutation Res</u>. 97:81-102.

27. Morita, K., M. Hara, and T. Kada (1978) Studies on natural
    desmutagens: Screening for vegetable and fruit factors active
    in inactivation of mutagenic pyrolysis products from amino
    acids. <u>Agric. Biol. Chem</u>. 42:1235-1238.

28. Namiki, M., and T. Kada (1975) Formation of ethylnitrolic acid
    by the reaction of sorbic acid with sodium nitrite. <u>Agric.</u>
    <u>Biol. Chem</u>. 39:1335-1336.

29. Namiki, M., S. Udaka, T. Osawa, K. Tsuji, and T. Kada (1980)
    Formation of mutagens by sorbic acid--nitrite reaction:
    Effects of reaction conditions on biological activities.
    <u>Mutation Res</u>. 73:21-28.

30. Osawa, T., H. Ishibashi, M. Namiki, and T. Kada (1980)
    Desmutagenic actions of ascorbic acid and cysteine on a new
    pyrrole mutagen formed by the reaction between food
    additives; sorbic acid and sodium nitrite. <u>Biochem. Biophys.</u>
    <u>Res. Commun</u>. 95:835-841.

31. Sadaie, Y., and T. Kada (1976) Recombination-deficient mutants
    of <u>Bacillus subtilis</u>. <u>J. Bacteriol</u>. 125:489-500.

32. Sakamoto, Y., O. Nagayabu, K.S. Yamamoto, and Y. Kikuchi
    (1981) Rec-assay with <u>B. subtilis</u> by survival colony
    counting. <u>Third Inter. Conf. Environ. Mutagens</u>, 3P27.

33. Shirasu, Y., M. Moriya, K. Kato, A. Furuhashi, and T. Kada
    (1976) Mutagenicity screening of pesticides in the microbial
    system. <u>Mutation Res</u>. 40:19-30.

34. Shirasu, Y., M. Moriya, K. Kato, F. Lienard, H. Tezuka,
    S. Teramoto, and T. Kada (1977) Mutagenicity screening on
    pesticides and modification products: A basis of
    carcinogenicity evaluation, In <u>Origins of Human Cancer</u>, Cold
    Spring Harbor Laboratory, pp. 267-285.

35. Sugimura, T. (1982) Mutagens, carcinogens, and tumor promoters in our daily food. Cancer 49:1970-1984.
36. Sugimura, T., S. Sato, M. Nagao, T. Yahagi, T. Matsushima, Y. Seino, M. Takeuchi, and T. Kawachi (1976) Overlapping of carcinogens and mutagens, In Fundamentals in Cancer Prevention, P.N. Magee et al., Eds., University of Tokyo Press, Tokyo/University Park Press, Baltimore, pp. 191-215.
37. Suter, W., and I. Jaeger (1982) Comparative evaluation of different pairs of DNA repair-deficient and DNA repair-proficient bacterial tester strains for rapid detection of chemical mutagens and carcinogens. Mutation Res. 97:1-18.
38. Tazima, Y., T. Kada, and A. Murakami (1975) Mutagenicity of nitrofuran derivatives, including furylfuramide, a food preservative. Mutation Res. 32:55-80.
39. de Vos, W.M., and G. Venema (1982) Transformation of Bacillus subtilis competent cells: Identification of a protein involved in recombination. Mol. Gen. Genet. 187:439-445.
40. Yasuo, K., S. Fujimoto, M. Katoh, Y. Kikuchi, and T. Kada (1978) Mutagenicity of benzotrichloride and related compounds. Mutation Res. 58:143-150.
41. Yoshikawa, K., T. Nohmi, R. Miyata, M. Ishidate Jr., T. Kawachi, and T. Kada (1982) Differences in liver homogenates from Donryu, Fischer, Sprague-Dawley and Wistar strains of rat in the drug-metabolizing enzyme assay and the Salmonella/hepatic S9 activation test. Mutation Res. 96:167-186.

MUTAGENESIS IN YEASTS

Nicola Loprieno

Istituto di Biochimica
Biofisica e Genetica
Università di Pisa, Italy

SUMMARY

Among lower eukaryotes, the genetics of the yeasts
Saccharomyces cerevisiae and Schizosaccharomyces pombe are the
most extensively studied. They are highly suitable for use in
assaying genetic effects of environmental chemicals, as their
genic and chromosomal structures show strong similarities to those
of higher organisms. S. cerevisiae strains are suitable for
routine use only for the measurement of genetic endpoints that are
detectable in haploids. The advantages of yeasts in mutagenicity
studies are:the variety of genetic end-points that may be assayed
(gene-mutation; mitotic recombination and gene-conversion;
aneuploidy); the low cost and the limited requirements of
technical expertise and laboratory facilities; and the existence
of extensive background information.

Both forward and reverse mutations can be detected. The
former type can be analyzed in S. pombe strain ade 6-60, rad
10-198, h-, whereas for reverse mutations, the XV185-14 strain of
S. cerevisiae, which contains a variety of different molecular
revertible sites in 5 genes, is used. In both strains base-pair
substitution and insertion-deletion mutations are inducible and
can be easily scored. D7 strain of S. cerevisiae can detect
reverse-mutations induced in a diploid structure. D4 and JD1
diploid strains of S. cerevisiae are usually utilized for mitotic
gene-conversion induction, while D7 diploid strain is used to
detect mitotic gene-conversion and mitotic crossing-overs. D6
diploid strain of S. cerevisiae had been used for induction of
chromosome aneuploidy, a genetic end-point of great relevance for
the identification of environmental mutagens.

Several in vitro and in vivo methodologies have been
developed for the two species of yeast. The former may include
the use of exogenous as well as endogenous metabolic activation
systems. Host mediated assay techniques have been shown to be
useful for testing the in vivo formation of mutagens from
unreactive precursors.

The evaluation made by the GENE-TOX Program has indicated
that up to 1981 some 521 chemicals were tested on yeasts, of which
257 gave a positive response. Since that time, about 200 more
chemicals have been tested.

INTRODUCTION

Fungi and yeasts have been employed extensively in mutation
studies since the discovery of chemical mutagens, and provide
highly suitable media for rapid mutagenicity screening. The
species generally used are the yeasts Saccharomyces cerevisiae and
Schizosaccharomyces pombe and the moulds Neurospora crassa and
Aspergillus nidulans. With these organisms it is possible to
evaluate the induction of: (a) forward mutations for base changes;
(b) reverse mutations for base substitutions or frameshift; (c)
chromosomal aberrations; (d) inter- and intragenic recombination
or gene-conversion; (e) chromosome non-disjunction; and (f)
mutations in mitochondrial DNA. In these eukaryotic
microorganisms, the role of a number of DNA repair mechanisms
involved in the formation of mutations is well understood. The
cell stages can be controlled during treatment with mutagens, and
the presence of P-450 cytochrome system under certain growth
condition may eliminate the need for the exogenous microsomal
activation system.

In this paper the use of yeast mutagenicity systems and other
fungal systems employed in mutagenicity tests will be discussed.
Extensive reviews on all these organisms are available in the
literature (1-7).

GENETIC END-POINTS IN YEASTS

Yeast mutagenicity tests allow detection of several
definitive changes in the genetic apparatus of these organisms
produced by a chemical agent. These end-points include point
mutation in chromosomal and mitochondrial genes, recombination
between and within genes, and chromosomal aneuploidy during
mitosis and meiosis.

(A) Point Mutation

A variety of forward mutation systems have been used with
yeast cultures. However, the most extensively used system has
been the one involving the induction of recessive alleles of the
genes for adenine biosynthesis. The system involves the use of
cultures carrying defective mutations of the genes of ade 1 and
ade 2 of S. cerevisiae and ade 6 and ade 7 of S. pombe, the
presence of which results in the production of red or red/purple
pigmented yeast colonies due to the presence of an intracellular
pigment (1,2,7).

Forward mutations are detected in such strains by the
induction of mutations at 5 other genes that control the
production of the red/purple pigment in the entry phase of the
adenine synthetic pathway. Such mutations result in the
production of doubly defective colonies which may be visually
observed as white colonies or sectors. The system is illustrated
in Table 1.

The spontaneous mutation frequency at the five white loci is
in the order of 10^{-4}. Strong mutagens can be detected by scoring
2500–5000 colonies, whereas weak mutagens need many more
(20,000–30,000 colonies). With this assay, mutations and
lethality are scored in the same medium. Dose–response
relationships are usually obtained in order to make a quantitative
assessment of the mutagenicity of a chemical.

Examples of results obtained with this method are presented
in Table 2. These data were subjected to a regression analysis in

Table 1.

STRAIN : *S. pombe* P1 . Genotype : *ade 6-60/rad 10-198/h-*

ade 6-60	Forward mutations produced by BS and FS mechanism	*ade 6-60/ade X* [*]
(red/purple colonies)		(White colonies)

(*) *ade X* = new mutation induced in one out of five loci controlling the
early phases of the adenine pathway (*ade 1, ade 3, ade 4, ade 5, ade 9*):
these genes are located in different DNA segments of the three chromosomes of the yeast

Table 2.   Forward mutations induced in vitro by a
series of alkene oxides on growing cells of
<u>s</u>. <u>pombe</u> (taken from Migliore <u>et</u> <u>al</u>., 9)

Compound structure	Buffer				S9		
	Dose (mM)	Survival (%)	Number of mutants / Number of colonies	Mutation freq.[a]	Survival (%)	Number of mutants / Number of colonies	Mutation freq
Ethylene oxide $CH_2 - CH_2$ (O)	0	100	5/77 665 (2)	0.66 ± 0.59	100	5/94 999 (2)	0.59 ± 0.22
	0.5	74.78	15/66 079 (2)	1.89 ± 1.00	100	28/82 943 (2)	3.32 ± 0.96
	1.5	99.19	35/82 266 (2)	4.17 ± 0.75	76.64	44/62 475 (2)	7.15 ± 0.24
	5	80.3	148/96 740 (2)	18.77 ± 0.72	100	121/100 950 (2)	14.33 ± 7.62
	15	35.14	127/22 695 (2)	66.21 ± 29.44	42.87	115/22 948 (2)	50.28 ± 1.76
Propylene oxide $CH_3 - CH - CH_2$ (O)	0	100	5/143 003 (4)	0.37 ± 0.13	100	6/160 038 (3)	0.41 ± 0.39
	3	71.7	44/90 814 (3)	4.87 ± 0.18	93.84	42/151 305 (3)	2.79 ± 0.12
	10	96.3	75/122 797 (3)	5.95 ± 1.25	86.65	112/121 514 (3)	8.42 ± 1.81
	30	73.52	205/59 643 (4)	34.07 ± 8.68	68.99	207/39 967 (2)	24.05 ± 2.19
Epichlorohydrin $CH_2 - CH-CH_2Cl$ (O)	0	100	7/102 644 (3)	0.67 ± 0.21	100	22/196 574 (5)	1.25 ± 1.46
	0.2	98.2	8/59 901 (2)	1.45 ± 0.86	100	16/80 934 (2)	1.87 ± 2.26
	0.4	77.83	24/67 830 (3)	3.67 ± 1.51	97.5	11/58 425 (2)	2.09 ± 2.10
	0.8	78.9	51/105 763 (4)	4.78 ± 0.89	94.1	41/71 728 (3)	5.67 ± 0.64
	1.6	86.4	133/91 593 (3)	14.50 ± 5.80			
	3.2	44.14	148/72 405 (5)	21.01 ± 4.50	56.0	107/80 344 (4)	13.01 ± 1.47
	6.4				35.5	29/21 019 (2)	14.10 ± 2.55
Glycidol $CH_2 - CH-CH_2OH$ (O)	0	100	15/149 556 (3)	0.99 ± 0.16	100	4/102 607 (2)	0.36 ± 0.16
	0.01	100	5/113 030 (2)	0.43 ± 0.36	100	10/85 755 (2)	1.36 ± 0.73
	0.1	100	14/113 326 (2)	1.28 ± 0.25	71.01	17/267 780 (2)	0.65 ± 0.16
	0.3	100	19/85 520 (2)	1.87 ± 1.72	100	14/129 176 (2)	1.32 ± 1.03
	1	100	33/106 036 (2)	3.31 ± 0.22	87.34	13/44 508 (2)	3.02 ± 0.95
	10	38.30	82/26 786 (2)	29.84 ± 2.64	78.23	130/62 511 (2)	20.13 ± 2.69

Compound structure	Buffer				S9		
	Dose (mM)	Survival (%)	Number of mutants / Number of colonies	Mutation freq. [a]	Survival (%)	Number of mutants / Number of colonies	Mutation freq. [a]
1.1.1-Trichloropropylene oxide CH$_2$—CH—CCl$_3$	0	100	9/200 375 (5)	0.48 ± 0.12	100	9/181 742 (4)	0.50 ± 0.21
	0.1	83.00	12/154 172 (3)	0.78 ± 0.19	93.06	11/133 917 (3)	1.12 ± 1.33
	0.3	90.63	15/131 438 (3)	0.62 ± 0.24			
	1	79.09	19/175 058 (4)	1.04 ± 0.61	82.41	20/162 148 (3)	1.44 ± 1.43
	3	74.31	32/104 895 (4)	3.72 ± 3.18	81.07	25/141 506 (3)	2.11 ± 1.27
1.2-Epoxybutane C$_2$H$_5$—CH—CH$_2$	0	100	18/133 073 (4)	1.07 ± 0.56	100	11/153 599 (4)	0.71 ± 0.32
	0.4	100	10/47 974 (2)	2.02 ± 0.22	100	13/58 794 (2)	2.04 ± 0.99
	0.8	100	5/25 301 (2)	1.79 ± 0.44	100	11/103 853 (2)	1.34 ± 0.69
	1.6	77.95	29/138 514 (3)	2.36 ± 0.70	100	20/73 960 (2)	2.80 ± 0.73
	3.2	87.70	34/60 270 (2)	5.52 ± 0.85			
	6.4	95.80	53/57 035 (2)	8.41 ± 3.86	100	22/60 524 (2)	3.85 ± 1.29
	12.8				100	47/51 562 (2)	8.44 ± 2.85
2.3-Epoxybutane CH$_3$—CH—CH—CH$_3$	0	100	6/133 529 (4)	0.41 ± 0.14	100	4/149 393 (3)	0.31 ± 0.22
	1	88.93	5/113 672 (3)	0.46 ± 0.15	88.99	9/139 750 (2)	0.68 ± 0.42
	3	92.5	4/90 370 (2)	0.47 ± 0.13	61.89	6/118 265 (3)	0.60 ± 0.41
	10	80.4	5/70 815 (2)	0.74 · 0.14	55.56	8/89 496 (2)	0.97 ± 0.28
	30	76.10	4/54 450 (2)	0.68 · 0.16	61.72	22/109 224 (3)	1.64 ± 1.05
	100	82.45	13/51 024 (2)	2.63 ± 0.80	45.65	66/72 268 (2)	8.11 ± 2.59

In parentheses, number of experiments pooled.
[a] Mutants / 10^4 survivors.

order to evaluate dose–effect relationships and calculate the specific activities (Table 3 and Fig. 1).

This mutagenicity test has also been employed in in vivo assays (the so-called host-mediated assays) for investigating the formation of mutagenic metabolites from non-mutagenic precursors.

Table 3. Linear dose–effect relationships and specific activities
of the epoxides. Evaluated either in absence (–) or in
presence (+) of mouse liver extracts (S9)

Compound	S9	Equations [a]	r [b]	Specific activity [c]
Ethylene oxide	–	$y = -1.18 + 4.44x$	0.93 ***	0.148
	+	$y = 0.86 + 3.24x$	0.99 ***	0.108
Propylene oxide	–	$y = -0.57 + 1.12x$	0.95 ***	0.037
	+	$y = 0.45 + 0.79x$	0.99 ***	0.026
Glycidol	–	$y = 0.76 + 2.90x$	0.996 ***	0.097
	+	$y = 0.79 + 1.94x$	0.99 ***	0.065
Epichlorohydrin	–	$y = 0.81 + 6.53x$	0.93 ***	0.218
	+	$y = 2.39 + 2.34x$	0.89 ***	0.078
1,1,1-Trichloropropylene oxide	–	$y = 0.38 + 1.07x$	0.68 ***	0.036
	+	$y = 0.82 + 0.45x$	0.50 *	0.015
1,2-Epoxybutane	–	$y = 1.06 + 1.17x$	0.83 ***	0.039
	+	$y = 1.14 + 0.55x$	0.91 ***	0.018
2,3-Epoxybutane	–	$y = 0.39 + 0.02x$	0.93 ***	0.0007
	+	$y = 0.19 + 0.09x$	0.94 ***	0.0098

[a] y, mutants/10^4 survivors; x, dose of the compound as mM.
[b] r, correlation coefficient. * $P < 0.05$; ** $P < 0.01$; *** $P < 0.001$.
[c] Specific activity, mutation frequency $\times 10^4 \times locus^{-1} \times mM^{-1} \times h^{-1}$

Examples, taken from the studies by Barale et al. (10,11), are
shown in Tables 4 and 5 and Fig. 2.

N-nitroso–dimethylamine, a well known pro–mutagen and procarcinogen, is mutagenic in this assay system when Aminopyrine is
administered to animals fed with $NaNO_2$. Furthermore, the
formation of a mutagenic metabolite can be quantitatively
determined as shown in Table 5. The mechanism of this in vivo
reaction is shown in Fig. 3. The host–mediated assay is also
useful in studying compounds that suppress mutagenic activity, as
shown in Figs. 4, 5 and 6. These studies are particularly
relevant in the investigations of factors which can modulate the
formation in vivo of mutagenic agents as happens with the human
diet.

Fifty seven chemicals tested in S. pombe have been included
in the GENE–TOX report (1). In addition, 72 chemicals, including
pesticides, drugs, cosmetic ingredients and industrial chemistry
intermediates have been tested in this system (9–14).

The most widely used yeast strains for the detection of
reverse mutation is the haploid strain of S. cerevisiae XV 185–14C
of the genotype a/ade 2-1/arg 4-17/lys1-1/trp 5-48/his 1-7/hom
3-10 developed by Von Borstel et al. (15). This strain contains
ochre nonsense mutations that are revertible by base substitution-
inducing mutagens, ochre suppressor mutations in t–RNA loci, a

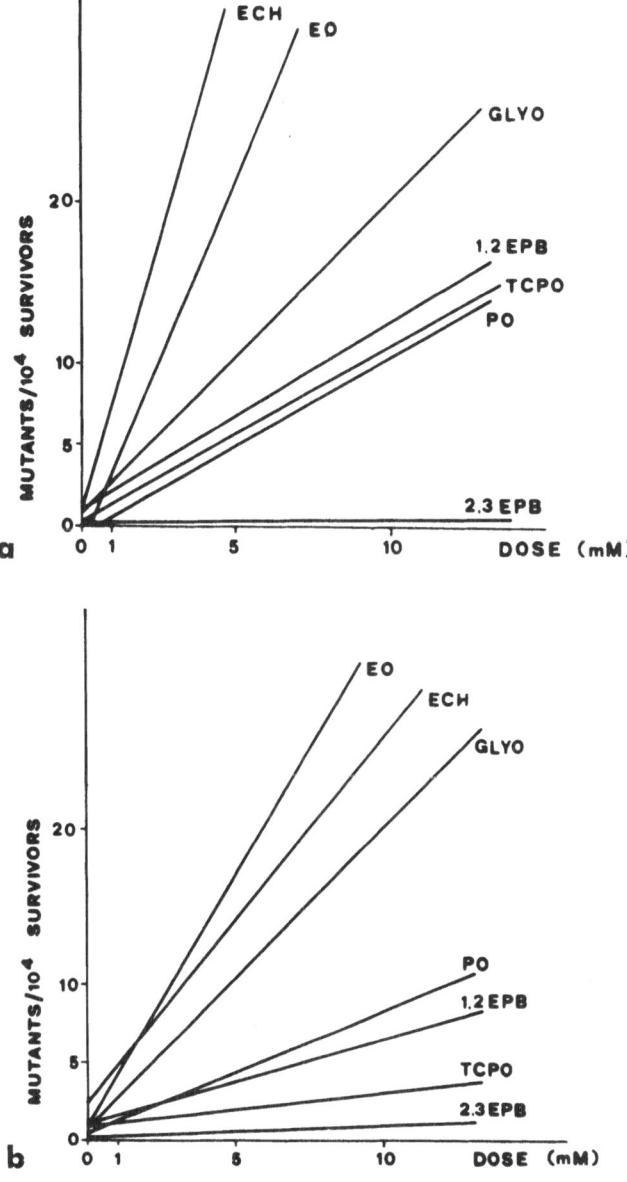

Figure 1. Linear dose-response relationships obtained by testing
      the seven epoxides in the S. pombe assay. a) In buffer. b)
      In presence of PB-pretreated mouse liver S9.

missense mutation (his 1-7) reverted by a second site mutation,
and a frameshift alteration (hom 3-10). In a diploid strain of S.
cerevisiae (strain D7) the ilv 1-92 mutation in a homoallelic
condition allows this strain to be used in point-mutation assay.

Table 4. Mutagenic activity of dimethylnitrosamine on yeast S. pombe (forward mutation) in the host-mediated assay in intravenous injection of yeast cells. Animals treated with 13.5 µmoles/kg.

Incubation time (min)	Mutation frequency $X10^{-4} \pm S.D.$
Control, 0 min	2.21 ± 1.45 [2]
Control, 150 min	0.82 ± 0.14 [3]
Treated, 15 min	1.20 ± 0.11 [3]
Treated, 30 min	4.20 ± 1.78 [3]
Treated, 60 min	7.23 ± 2.25 [6]
Treated, 150 min	17.36 ± 1.78 [3]

Regression analysis:
$y = 1.17 + 0.107x$
$y = M.F. \times 10^{-4}$
x = min of treatment.
$R = 0.99389$***

(B) Genetic Recombination

In eukaryotic cells recombinational events occur primarily during meiosis. In yeast genetic exchanges between homologous chromosomal regions occur also during the mitotic division (mitotic recombination) with a frequency lower than those occurring during meiosis.

Mitotic recombinational events between different genes (mitotic crossing over) or between different sites within the same gene (mitotic gene conversion) can be detected in yeast. Mitotic recombination and mitotic gene conversion are produced by genetically active chemicals in a "non-specific way" - i.e., they represent the cell's response to a wide spectrum of genetic

Table 5. Mutagenicity of aminopyrine in vivo in presence of nitrite. Mice-mediated assay on the yeast S. pombe (forward mutation) incubated for 5 h (intrasanguineous assay).

Aminopynne (µmoles kg)	NaNO$_2$ (µmoles kg)	Number of animals[a]	Mutation frequency $X10^{-4} \pm S.D.$
0	1449	3	0.90 ± 0.22
21.6	1449	3	8.88 ± 3.00
108	1449	3	21.52 ± 3.77
432	1449	3	37.18 ± 4.53

[a] Treatment by gavage.

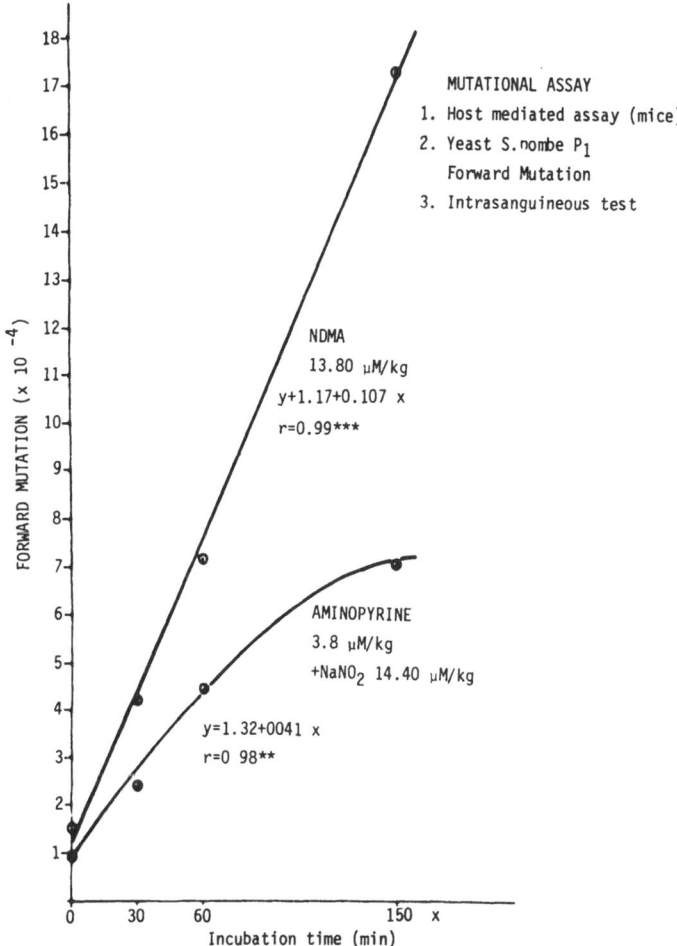

Figure 2. Comparisons of mutagenic activities of NDMA (13.50 μmoles/kg) and simultaneous gastric intubation of amniopyrine (3.8 μmoles/kg) and NaNO₂ (14.40 μmoles/kg).

damage. Therefore increased mitotic recombination and/or mitotic gene conversion are good indications that the chemical agent producing such effects has genotoxic properties.

In a battery of mutagenicity tests, the tests for mitotic recombination and mitotic gene-conversion have been designated as "indicator tests" by N. Loprieno (1983). Examples are shown in Figures 7 and 8.

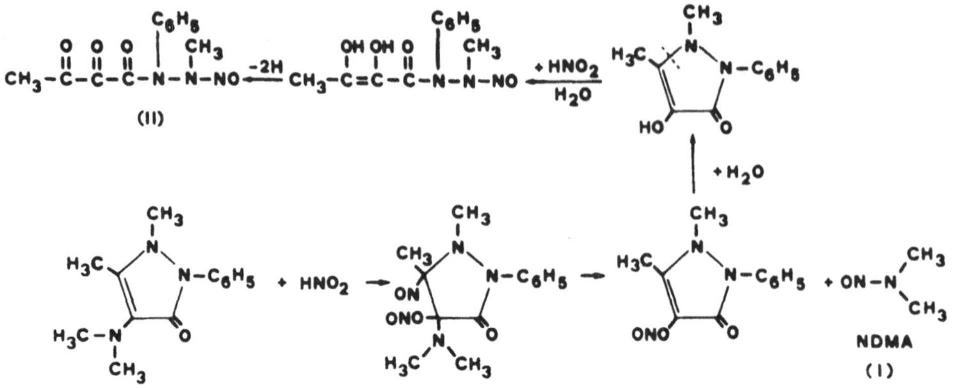

AMINOPYRINE NITROSATION

$(pK_a = 5.04 ; pH = 2.0; K(M^{-2} sec^{-1}) = 80)$
(Op.)

(I) = N - NITROSODIMETHYLAMINE
(II) = I - DIKETOBUTYRYL - I - PHENYL - 2 - METHYL - 2 - NITROSOHYDRAZIDE

Figure 3. The fate of amniopyrine plus $NaNO_2$ in acidic conditions. First, AP is nitrosated to a nitroso compound which spontaneously releases NDMA. The pyrine ring is then broken and further nitrosated to produce a second nitroso compound: 1-diketobutyryl-1-phenyl-2-methyl-2-nitrosohydraside.

A number of suitable strains of S. cerevisiae have been used in genotoxicity testing. Mitotic gene conversion may be assayed using the diploid strain D4, whose genotype (ade 2-2, ade 2-1/trp 5-12, trp 5-27) contains two heteroalleles at the AD2 and TRP5 loci. A genotoxic chemical agent may produce prototrophic colonies carrying one wild type allele, which allows for growth on selective medium lacking either tryptophane or adenine, as a result of the mitotic gene conversion.

A particularly convenient multipurpose strain of S. cerevisiae is D7 (16). It carries a set of genetic markers which allow the simultaneous assay of mitotic crossing-over, gene conversion and point mutation. The genotype of D7 is as follows:

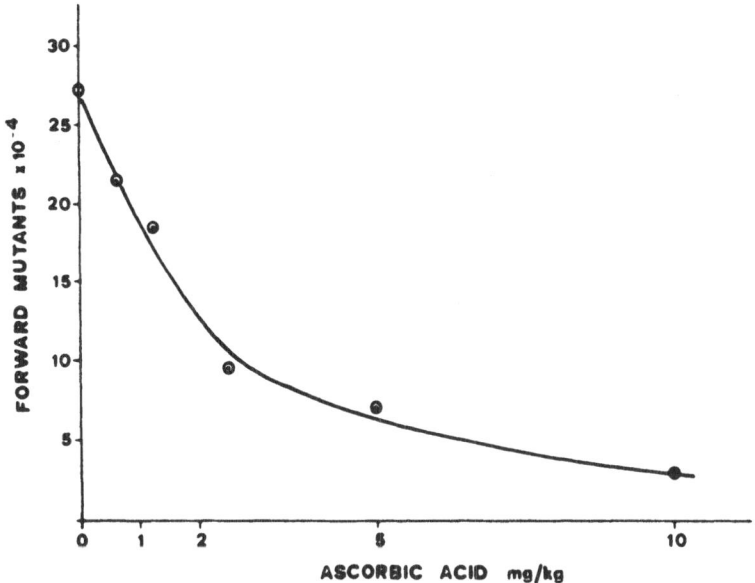

Figure 4. Suppression of mutagenic compound formation in vivo
     after AP and NaNO$_2$ feeding by simultaneous intubation of mice
     with various amounts of ascorbic acid.

$$\text{Chromosome III } \underline{a}; \text{ Chromosome VIII } \frac{\text{trp 5-12 , cyn}_2^r}{\text{trp 5-27 , CHY}_2^r};$$

$$\text{Chromosome V } \frac{\text{ilv 1-92}}{\text{ilv 1-92}}; \text{ Chromosome XV } \frac{\text{ade 2-40}}{\text{ade 2-119}}.$$

The heteroalleles, trp 5-12 and trp 5-27, of the TRP 5 locus
are used in the scoring for induced mitotic gene conversion as in
the D4 strain. ade 2-40 is a recessive allele of ADE-2 locus
which produces deep red colonies, whereas ade 2-119 is a leaky
allele causing accumulation of only a small amount of pigment,
thus producing pink colonies. The two ade 2 alleles complement
each other to give rise to white adenine independent colonies.
Mitotic crossing-over in D7 may give rise to the production of
cells homoallelic for the ade 2 mutations and thus lead to the
observation of both red and pink reciprocal products. Mitotic
crossing-over may also be assayed in D7 by the use of the
recessive cycloheximide resistant cyh$_2^r$ allele on chromosome VII.
Induction of base-substitution mutation at the homoallelic

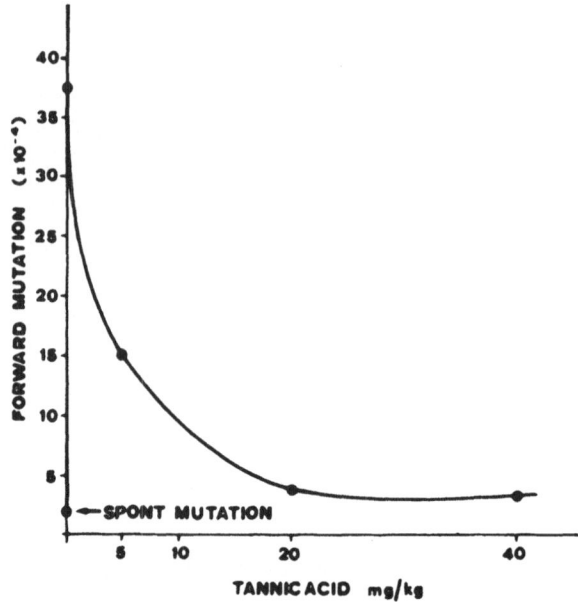

Figure 5.   Suppression of mutagenic compound formation in vivo
after AP and NaNO$_2$ feeding mice by simultaneous intubation
with different amounts of tannic acid.

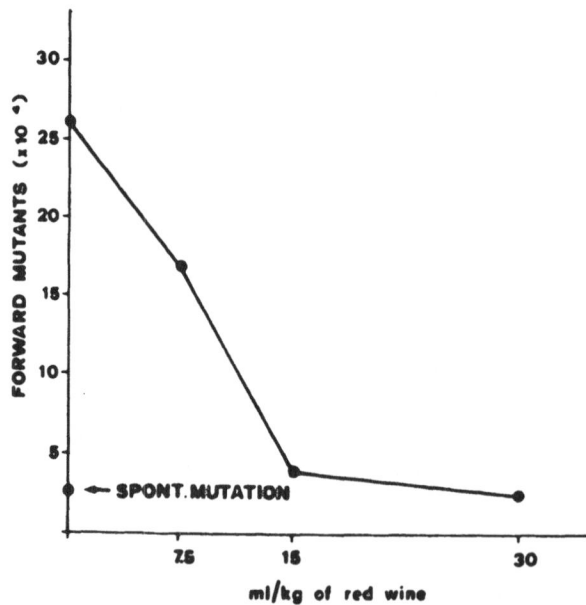

Figure 6.   The ability of a common diet component such as red wine
to suppress mutagenic activity in vivo, of AP and NaNO$_2$ was
evaluated by simultaneous intubation of mice with various
amounts of wine.   Each mouse received the same final intubation
volume (1 ml).

ilv 1-92 markers may be analyzed by scoring prototrophic colonies
which grow on selective minimal medium lacking isoleucine.

(C) Aneuploidy

Abnormal chromosome segregations leading to the production of
numerical chromosome aberrations may be detected in yeast by
genetic means using appropriate yeast strains. Suitable strains
of S. cerevisiae are available which are capable of detecting and
quantifying the induction of monosomy (chromosome loss) during
mitotic cell division from 2n to the 2n-1 conditions and the
production of disomy (chromosomal gain) and diploidization in
spores produced during meiotic cell division (sporulation). The
D6 strain described by Parry and Zimmermann (17) has been used for
the assay of this genetic end-point.

Tables 6 and 7 show the genotype of S. cerevisiae (a) and S.
pombe (b) strains used in mutagenicity studies.

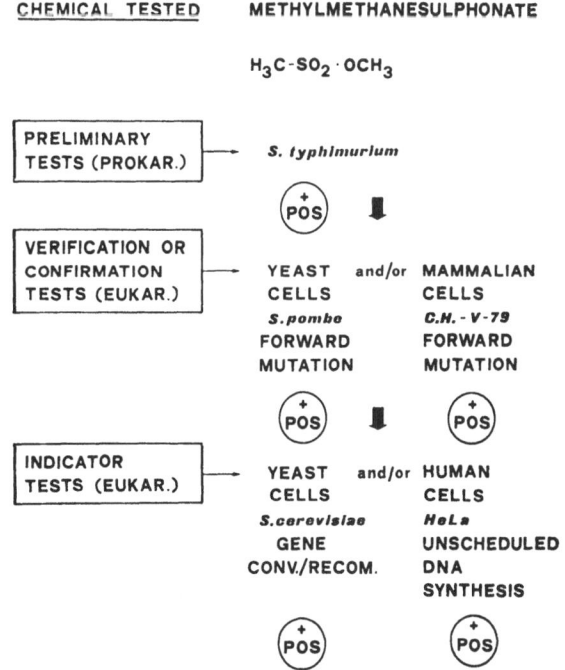

Figure 7. Sequential approach for evaluating chemicals for
their potential for in vitro gene mutations induction.

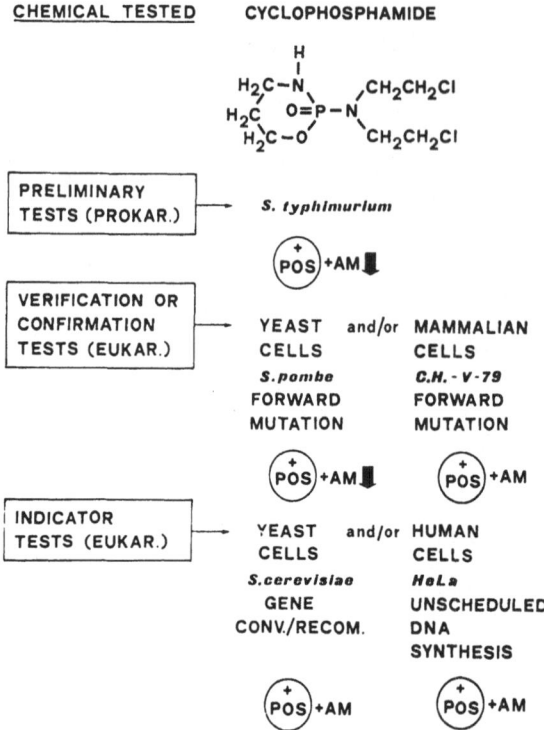

Figure 8.   Sequential approach for evaluating chemicals for
       their potential for in vitro gene mutations induction.

INTERPRETATION OF THE RESULTS OBTAINED WITH YEAST MUTAGENICITY
TESTS

1.   A positive response in yeast mutation assays is
indicative of the ability of a chemical to induce point mutations
in eukaryotic DNA.

2.   A positive response in assays for mitotic recombination
indicates the potential of a chemical to produce unspecific DNA
alterations and interaction in a eukaryotic diploid cell.

3.   A positive response in assays for chromosomal aneuploidy
indicates the potential of a chemical to produce changes in
chromosome number in eukaryotic cells.   However, it is possible
that at least a percentage of such chemicals may be specific for
the fungi and their activity requires confirmation in a mammalian
germ cell assay.

Table 6. Strains and genotypes of *Saccharomyces cerevisiae* Commonly used for assaying chemical carcinogens

STRAIN	GENOTYPE	GENETIC ENDPOINT	REFERENCES
D3	CYH4 ade2 his8 / + + +	MITOTIC RECOMBINATION (DIPLOID)	ZIMMERMANN ET AL. 1966
D4	a gal2 adz2-2 trp5-12 leu1 / α + ade2-1 trp5-27 +	MITOTIC GENE CONVERSION (DIPLOID)	ZIMMERMANN & SCHWAIER, 1967
D5	a trp1 ade2-40 MAL1 + / α + ade2-119 + MAL4	MITOTIC RECOMBINATION, GENE CONVERSION AND MUTATION (DIPLOID)	ZIMMERMANN, 1973
D6	a his4 tifade3 leu1 trp5 cyh2 met13 ade2-40 / α + + + + + + ade2-40	MITOTIC NONDISJUNCTION (DIPLOID)	PARRY & ZIMMERMANN, 1976
D7	a ade2-40 trp5-12 ilv1-92 / α ade2-119 trp5-27 ilv1-92	INDUCED MITOTIC CROSSING-OVER, GENE CONVERSION AND REVERSE MUTATIONS (DIPLOID)	ZIMMERMANN ET AL. 1975
6117	α + + cyh2 met13 tyr3 lys5 ade5-7 ade2-1 / a leu1 trp5 + + + + + ade2-1 ; ura4 can1 / + +	MITOTIC GENE CONVERSION (DIPLOID)	SORA ET AL., 1979B
JD1	a his4ABC + + + trp5U6 / a his4C ade2-1 ser7 his8 trp5-U9	MITOTIC GENE CONVERSION (DIPLOID)	DAVIES ET AL., 1975
XV185-14C	a trp5-48 arg4-17 lys1-1 ade2-1 his1-7 hom3-10	REVERSION OF OCHRE, MISSENSE AND PUTATIVE FRAMESHIFT (HAPLOID)	QUAH & VON BORSTEL, CITED IN SHAHIN & VON BORSTEL, 1977, 1978
XV1000-1A	a ade2-912 art4-17 his1-7 hom3-10 trp5-48	FORWARD MUTATION (HAPLOID)	VON BORSTEL, UNPUBLISHED
6126/16c	a his4-1 ade1-10 arg4-17 tyr7-1 trp1-1	REVERSION OF OCHRE, AMBER, MISSENSE AND PUTATIVE FRAMESHIFT (HAPLOID)	SORA ET AL., 1979B
X 4-8C	a ade2-1 his1-7 hom3-10 lys1-1 trp5-48 rad2	REVERSION OF OCHRE, MISSENSE AND FRAMESHIFT (HAPLOID)	LARIMER ET AL., 1978

Table 7. Strains and genotypes of *Schizosaccharomyces pombe* commonly used for assaying chemical carcinogens[A]

STRAIN	GENOTYPE	GENETIC ENDPOINT	REFERENCE
P1	SP-198 ade6-60 rad10-198 h-	FORWARD MUTATION AT 5 LOCI (HAPLOID)	LOPRIENO, 1973
P2	ade7-50 / ade7-150	MITOTIC GENE CONVERSION (DIPLOID)	LOPRIENO, 1978
P3	ade7 / +	MITOTIC RECOMBINATION (DIPLOID)	LOPRIENO, 1978
DC	met4	REVERSE AND FORWARD MUTATION AT SUPPRESSOR LOCI	LOPRIENO & CLARKE, 1965

[A] FROM LOPRIENO ET AL., 1974

REFERENCES

1. Barale, R. et al. (1981) A mutagenicity methodology for
    assessing the formation of N-Dimethylnitrosamine in vivo.
    Mutation Res. 85:57-70.
2. Barale, R. et al. (1983) The intragastric host-mediated assay
    for the assessment of the formation of direct mutagens in
    vivo. Mutation Res. 113:21-32.
3. Gutz, H. et al. (1974) Schizosaccharomyces pombe, In Handbook
    of Genetics, R.C. King, Ed., Plenum Press, New York,
    pp. 395-446.
4. Loprieno, N. (1981) Mutagenicity of selected chemicals in
    yeast: Mutation-induction at specific loci, In Comparative
    Chemical Mutagenesis, F.J. de Serres, and M.D. Shelby, Eds.,
    Plenum Press, New York and London, pp. 139-150.
5. Loprieno, N. (1981) Screening of coded-carcinogenic-non
    carcinogenic chemicals by a forward-mutation system with the
    yeast Schizosaccharomyces pombe, In Evaluation of Short-Term
    Tests for Carcinogens, F.J. de Serres, and J. Ashby, Eds.,
    Elsevier/North-Holland, New York, pp. 424-433.
6. Loprieno, N. et al. (1980) Report 4. Mutagenesis assays with
    yeasts and moulds. IARC Monographs, Suppl. 2:135-155.
7. Loprieno, N. et al. (1982) Mutagenic studies on the hair dye
    2-(2',4'-diaminophenoxy)ethanol with different genetic
    systems. Mutation Res. 102:331-346.
8. Loprieno, N. et al. (1983). Testing of chemicals for
    mutagenic activity with Schizosaccharomyces pombe. A report
    of the U.S. Environmental Protection Agency Gene-Tox Program.
    Mutation Res. 115:215-223.
9. Mehta, R.D., and R.C. Borstel (1981) Mutagenic activity of 42
    encoded compounds in the haploid yeast reversion assay,
    strain XV 184-146, In Evaluation of Short-Term Tests for
    Carcinogens, F.J. de Serres, and J. Ashby, Eds., Elsevier/
    North-Holland, New York, pp. 414-423.
10. Migliore, L. et al. (1982) Mutagenic action of structurally
    related alkene oxides on Schizosaccharomyces pombe. The
    influence in vitro of mouse-liver metabolizing system.
    Mutation Res. 102:425-437.
11. Parry, J.M., and F.K. Zimmermann (1976) The detection of
    monosomic colonies produced by mitotic non-disjunction in the
    yeast Saccharomyces cerevisiae. Mutation Res. 36:49-66.
12. Rossi, A.M. et al. (1983) In vivo and in vitro mutagenicity
    studies of a possible carcinogen, trichloroethylene, and its
    two stabilizers, Epichlorohydrin and 1,2-Epoxybutane.
    Teratogen., Carcinogen., and Mutagen. 34:75-87.
13. Zimmermann, F.K. et al. (1975) A yeast strain for the
    simultaneous detection of induced crossing-over, mitotic gene
    conversion and reverse mutation. Mutation Res. 28:381-388.
14. Zimmermann, F.K. et al. Testing of chemicals for mutagenic
    activity with Saccharomyces cerevisiae. A report of
    U.S. Environmental Protection Agency Gene-Tox Program.

A COMPARISON OF GENOTOXIC ACTIVITY IN SOMATIC TISSUE

AND IN GERM CELLS OF DROSOPHILA MELANOGASTER

Ekkehart W. Vogel

Department of Radiation Genetics and
Chemical Mutagenesis, Sylvius Laboratories
State University of Leiden
Wassenaarseweg 72, 2333 AL Leiden
The Netherlands

SUMMARY

Studies on the mutagenicity in germ line assays of
representatives from practically all classes of procarcinogens/
promutagens, altogether about 80 chemicals, indicated that
Drosophila melanogaster has the enzymatic potential for converting
a wide array of procarcinogens into genetically active species.
Biochemical analysis of the enzymes involved in their
biotransformation proved the presence and the inducibility in
Drosophila of several types of cytochrome P-450 and of other
components of the xenobiotics-metabolizing enzymes, such as aryl
hydrocarbon hydroxylase (AHH) activity.

The considerable AHH-activity present in microsomal
preparations stands in marked contrast to the low mutagenic
effectiveness generally observed when testing carcinogenic
aromatic amines and polycyclic hydrocarbons in the recessive
lethal assay. Thus one crucial question concerning the action of
these large two classes of procarcinogens is how their genetic
effectiveness in Drosophila can be enhanced. Since it is not a
priori possible to conclude that gonadal tissue will always be the
proper target for chemicals requiring metabolic activation,
somatic mutation assays may provide a useful alternative for
testing purposes, as opposed to germ cell assays.

In an attempt to study the usefulness of this approach, we
compared the genotoxic effectiveness of a number of model mutagens
in male and female germ cells with their activity in somatic

tissue of females (white/white^{CO} test) by treating first-instar
larvae. By using the same protocol, MMS, bleomycin, 9,10-
dimethylanthracene and 7,12-dimethylbenz(a)anthracene were quite
efficient in producing somatic recombination and mutations in
female larvae, as opposed to only marginal effects in male and/or
female germ cells. These findings raise the possibility that a
high proportion of mutations were induced in premeiotic cells, and
that those carrying a lethal mutation were eliminated by germinal
selection. Calibration studies with 17 additional reference
mutagens also support the conclusion that somatic mutation assays
might be particularly valuable as a complement to recessive lethal
tests on heritable genetic damage.

The broad spectrum of heritable and somatic genetic
alterations which can be determined in Drosophila suggests a model
function for this system which may prove to be of increasing value
in the future. By determining the mutagenic profile of a given
genotoxic reagent as completely as possible, employing the nine
parameters currently available for such an integrated analysis in
Drosophila, it should be feasible to link certain types of DNA
alkylation adducts with various genetic end points, as has already
been demonstrated for a series of monofunctional alkylating
agents, and for hexamethylphosphoramide (HMPA). Thus, analysis of
multiple genetic parameters in Drosophila provides a means of
gaining insight into the relationship between chemical interaction
pattern and mutagenic mechanism.

INTRODUCTION

A large number of chemical mutagens of diverse molecular
structures and physical properties have been found to induce
heritable genetic damage in Drosophila melanogaster, and most of
these are carcinogenic to mammals. Stocks are available or can be
constructed to compare the induction in germ cells of gene
mutations, deletions, and of almost all possible types of
chromosomal rearrangements. The study of such relationships has
been a major area of research that is closely related to the
molecular mechanisms of chemical mutagenesis.

Comparative investigations on the reliability of these
different end points, however, clearly revealed that the X-linked
recessive lethal test (SLRL) is by far the most sensitive and
reliable assay in Drosophila to screen chemicals for their ability
to cause heritable genetic damage (43). One of the major reasons
is that the phenomenon of recessive lethality can have different
origins, i.e. recessive lethals comprise point mutations
(intragenic changes), deletions affecting more than one gene, and
both small and large rearrangements (2). Thus, a mutagen which
produces only gene mutations would not be detected in a test for

chromosomal rearrangements, but would still be picked up in the recessive lethal assay.

There now seems little doubt that most genotoxic agents undergo metabolic activation within cells to intermediates that react covalently with cellular macromolecules and in particular with DNA. Gradually, through the identification as mutagens in the SLRL of very diverse groups of indirect-acting carcinogens, and through the discovery of the various elements of the microsomal mixed-function oxidase system, definite proof has been obtained that Drosophila melanogaster is capable of activating a wide array of procarcinogens: nitrosamines, aryldialkyltriazenes, azo- and azoxyalkanes, pyrrolizidine alkaloids, aflatoxins, haloalkanes, halo-olefins, oxazaphosphorines, hydrazines, aromatic amines and polycyclic hydrocarbons (Fig. 1). It is not intended to review here in detail the mutagenic activity of the various classes of procarcinogens, since the different aspects of this work on approximately 80 procarcinogens have been discussed earlier (39,42,47).

The extensive amount of information that has accumulated for Drosophila has helped to bring into focus not only the possibilities this system has to offer for attacking relevant problems in chemical mutagenesis, but also those problems which remain to be examined further. This certainty holds true for the

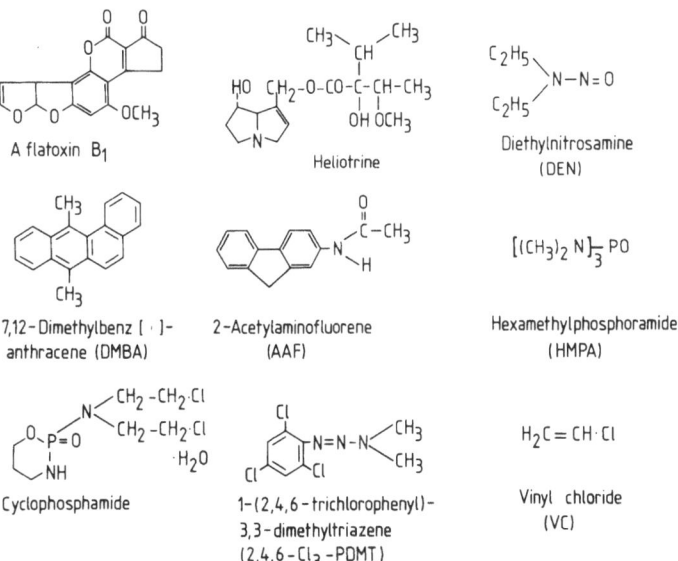

Figure 1. Structures of some procarcinogens which are mutagenic in the Drosophila recessive lethal assay.

difficulties encountered with the detection of certain
carcinogenic aromatic amines and polycyclic hydrocarbons in the
SLRL. The recessive lethal test, when compared with the fast
systems utilizing unicellular organisms or mammalian cells in
culture, is a relatively tedious and time-consuming method. For
this reason, and also in view of the possible differences in the
capacity of somatic versus gonadal tissues to metabolize
procarcinogens, several research groups are currently
investigating the possibilities of somatic assays to function as
complements to recessive lethal assays on germ cells. The aims of
the current chapter are as follows:

(1) to summarize the current knowledge on xenobiotics-
    metabolizing enzymes in Drosophila, with special
    reference to the class of carcinogenic aromatic amines
    and polycyclic hydrocarbons;

(2) to describe what is known at present about the utility of
    somatic mutation assays versus the conventional germ cell
    tests; and

(3) to discuss briefly a concept bearing on quantitative
    aspects of chemical mutagenesis.

I. Xenobiotics-Metabolizing Enzymes

Enzyme analysis. Although initially, most of the important
findings in chemical mutagenesis came from experiments on
Drosophila (the first known chemical mutagen, mustard gas, was
discovered in a recessive lethal assay carried out in 1940-1941
(3), and although the mutagenicity of procarcinogens (some
pyrrolizidine alkaloids) was already reported in 1959 (11),
analysis of the xenobiotics-metabolizing enzymes was not conducted
until the mid 1970s. Over the last years, numerous studies have
shown the presence of considerable aryl hydrocarbon hydroxylase
(AHH) activity and other elements of the xenobiotics-metabolizing
enzymes in Drosophila melanogaster (5,6,20). From spectral
analysis it was deduced that, in microsomal preparations of
Drosophila, the cytochrome P-450 and b_5 are present. Moreover,
microsomes appeared to contain AHH and epoxide hydratase
activities, and post-microsomal supernatants were able to
conjugate appropriate substrates with reduced glutathione and
phosphate (4,5,6). Furthermore, studies on the enzyme
distribution pattern in larvae and adult flies have established
that the spectral and enzyme features of the larval cytochrome
differ considerably from those present in adult flies (48,51).

Recently, evidence for inducibility of cytochrome P-450,
benzo(a)pyrene hydroxylation, p-nitroanisole demethylation (pNA)
and aminopyrene (AP) demethylation was shown in different strains

of flies that were pretreated with Aroclor 1254 (AC),
phenobarbital (PB), polychlorinated biphenyls or butylated
hydroxytoluene (BHT) (18,19,28,40,48). The work of Zijlstra (51)
has revealed considerable differences in some enzyme activities
between insecticide-resistant (IR) and insecticide-sensitive (WT)
strains. pNA demethylation and AP demethylation were
substantially higher in all IR populations, while no correlation
could be found between their increased insecticide resistance and
BP hydroxylating capacity or P-450 content of the microsomes.
This finding may mean that the use of IR strains in mutagenicity
testing may not improve the sensitivity of the SLRL method. Since
there is no information available on how the various forms of
cytochrome P-450 may alter activation and deactivation processes
in Drosophila, a shift in the ratio of P-450s might have
unpredictable consequences, leading either to an increase or
decrease of mutagenic effectiveness of the compound under
investigation, or simply leaving the situation unchanged.

   Effects of enzyme modifiers on procarcinogen mutagenesis.
From the above spectrum of effects, we can infer that
procarcinogen mutagenesis may be altered by the application of
inducers and inhibitors of cytochrome P-450 and its adjunct
enzymes. In fact, Magnusson and Ramel (29) and Magnusson et
al. (28) found that pretreatment of adult males with PB caused an
increase in the mutagenic effects of vinyl chloride. In similar
experiments, the frequency of recessive lethals was about doubled
in a white strain when pretreatment with PB was carried out before
exposure to styrene or styrene oxide (14). In experiments carried
out in our department, pretreatment of male larvae and cotreatment
of the hatching males with PB had no influence on the mutagenicity
of several carcinogens, namely cyclophosphamide,
dimethylnitrosamine, and 2,4,6,-Cl_3-PDMT (48). The effect of 3-MC
applications was a twofold or threefold increase in the frequency
of recessive lethals after exposure to cyclophosphamide or 2,4,6-
Cl_3-PDMT. Aroclor pretreatment also resulted in a threefold
enhancement in the mutagenic response of 2,4,6-Cl_3-PDMT treated
males. However, application of inducers of microsomal enzymes did
not lead to any changes in their mutagenic effectiveness of a
series of carcinogenic aromatic amines and polycyclic hydrocarbons
(49). In conclusion, enzyme induction generally failed to enhance
the mutagenic effectiveness of weakly active procarcinogens in
Drosophila.

   Another important question relevant to chemical mutagenesis
in Drosophila is the possible effect on genotoxic activity of
enzyme inhibition. When male larvae from strain Berlin K are
pretreated with l-phenylimidazole, an inhibitor of cytochrome
P-450, a potentiating effect on mutagenesis of 1,2-dibromoethane
is observed (J.A. Zijlstra; personal communication). Similarly, a
drastic increase in recessive lethal frequency was found in Berlin

K males which, before exposure to cyclophosphamide, had been
pretreated with iproniazide, an inhibitor blocking N-oxidation.
Moreover, enzyme inhibition can have a strong potentiating effect
on mutagenesis of direct-acting carcinogens. In adult feeding
experiments, cotreatment with 1-phenylimidazole or iproniazide led
to up to a 5-fold increase in the yields of recessive lethals
produced by nor-nitrogenmustard or methyl p-toluenesulfonate
(J.A. Zijlstra; personal communication). And since both direct-
acting chemicals were hardly mutagenic in the absence of those
enzyme inhibitors, this "first-pass" effect may mean that an
efficient detoxification system is located in the digestive tract
of Drosophila.

II. Results of Mutagenicity Testing in Drosophila

Induction of heritable genetic damage. Evaluation of the
intrinsic activity of chemicals to cause heritable genetic damage
is best performed in the recessive lethal assay which has been
calibrated against a wide array of direct-acting agents and of
mutagenic procarcinogens. To date, 421 compounds have been tested
in this system, according to a report of the US GENE-TOX PROGRAM
(24). 198 compounds were found to be active and 46 inactive at
the highest concentration tested. A third group as large as 177
compounds was not classified as either positive or negative,
because a very rigid criterion was used, namely a test of at least
7,000 chromosomes in both the control and the treated groups (per
one dose level), with an average spontaneous frequency of 0.2%.
The fulfillment of this criterion would enable the detection of a
doubling of the recessive lethal frequency (24). The inherent
program with this approach is that flexibility is diminished, and
that too much weight may be put on the one experimental condition
selected by assuming that this will provide optimal and reliable
results.

The alternative procedure, which seems more realistic in view
of the complex metabolic system of Drosophila, would be to use a
flexible protocol in mutagenicity testing. Reliance should not be
placed on a large number of chromosomes tested at only one dose,
but instead a variety of experimental conditions (e.g. injection
vs. feeding; larvae vs. adult treatment) may be used to identify
optimal experimental conditions for the chemical under
investigation. A good example of the applicability of the second
approach is the demonstration by Zijlstra et al. (51) that 7,12-
dimethylbenzanthracene (DMBA), methyl p-toluenesulfonate and nor-
nitrogenmustard are strongly mutagenic, weakly mutagenic or
inactive in recessive lethal assays, depending on the route of
administration selected.

Another crucial aspect concerns the question of adult
treatment versus larvae feeding in relation to metabolism. It has

become clear from recent studies on the enzyme distribution patterns in larvae and adult flies that the spectral and enzymic features of the larval cytochrome P-450 differ considerably from those present in adult flies (48,50). Most of the studies with indirect-acting mutagenic carcinogens in Drosophila have concentrated on adult males, but the available information about their possible effects in larvae is scanty. Clark (12) has recently used the X-linked recessive lethal test to compare mutagenicity of some naturally occurring mutagens in adult versus larval stages. Increased levels of mutagenicity observed in some cases (heliotrine, bracken fern extracts) in third-instar larvae let Clark (12) suggest that both larval and adult stages should be tested in screening programs based on the recessive lethal method. The germ cells present in larval testis are spermatogonial and spermatocyte stages (1,27). Thus, a difficulty associated with larval feeding experiments might be clusters of mutants of common origin. The occurrence of such clusters necessitates special statistical treatment of the data (15,32). Moreover, from cell-stage comparisons in adult males, after treatment of the entire male germ-cell cycle, it has become apparent that highest yields of X-chromosomal recessive lethals are mostly obtained from cells that have already passed the meiotic divisions. The reason that has been advanced to account for these large changes in sensitivity within the male germ cell cycle is known under the term "germinal selection" introduced by Muller (33). It refers to the fact that X-linked deletions that include a gene required for development of a spermatogonial cell into a spermatozoon are removed by germinal selection. Thus, the possibility of strong segregational elimination during meiosis of cells carrying a mutation should be taken into consideration as a potential disadvantage when testing chemicals in Drosophila larvae.

Genetic damage in somatic tissue. The recessive lethal assay is, when compared with the fast systems utilizing bacteria or lower eukaryotes, a relatively tedious and time-consuming method. For this reason, it is applicable in practice for only a small number of chemicals. In view of this intrinsic limitation of the recessive lethal test as a means of screening large numbers of chemicals, several investigators have recently drawn attention to the potential use of somatic mutation assays.

A system developed by Rasmusson (36) has shown promise for screening purposes, although evaluation of this method with a wider variety of mutagens is needed. The method is based on the scoring of somatic mutations in an unstable duplication of the white locus, leading to red sectors against a background of yellow eyes. The instability is caused by the insertion of a piece of DNA into the regulatory part of the white locus. This unstable strain shows a resolving power for X-rays which is somewhat higher than that for X-linked recessive lethals. A series of chemical mutagens, including both direct-acting and promutagens which require metabolic activation, gave positive results with this system (16,34). Another comparable system involving a stable duplication of the white locus was considerably less sensitive.

There are numerous somatic assay systems in which chromosomes carry suitable markers for determining genetic damage in somatic tissue (10,17). A somatic mutation that is used in our laboratory, devised by Becker (8), is based on the determination of twin mosaic spots and single mosaic spots in the eyes of females which had been treated during their larval development (9,30,31). These clones may arise from many types of genetic alterations, i.e., as a consequence of somatic crossing-over, somatic nondisjunction, chromosome loss, deletions or forward mutation.

Genetic damage in larval germ cells versus somatic tissue. It will be recalled that two distinct levels of biological action may determine differences in mutational response between adult and larval stages, i.e., metabolic factors and germinal selection. Since the aspect of segregational elimination is irrelevant when dealing with somatic tissue as the target, one is again compelled to ask whether somatic mutation assays may be valuable as a complement to recessive lethal assays, or may even replace them. In view of the possible usefulness of larval tissue of Drosophila for mutagen screening, a more systematic comparison has been made of the responses of gonadal tissue versus somatic cells, after treatment of first-instar larvae with a series of reference mutagens.

For the determination of recessive lethals induced in male or female larvae, a set of five strains was selected that was originally devised by Dr. W.R. Lee (23,25) for the analysis of delayed mutations. Four of these strains are needed to produce the two genotypes (A/B male; C/D female) used in these larval feeding experiments, as shown in Figure 2. Thus, induction of recessive lethals could be measured in the B chromosome in male larvae and in the D chromosome in female larvae. The advantage of this arrangement is that pre-existing lethals do not cause a problem after treatment of female larvae because they receive their D chromosome from their father. In fact, we have not observed any spontaneous clusters in the control runs. It is also seen in Figure 2 that this scheme enables the simultaneous detection of recessive lethals produced in paternal and maternal X-chromosomes. For the tests, F_1 (B/D) virgins (the males die or are sterile) are pair-mated with males (E/F) from another stock carrying an attached X-Y^L chromosome (F). In the F_2, absence of males carrying a copy of the treated B chromosome of D chromosomes indicates a lethal mutation on the paternal or maternal X-chromosome, respectively. And since both types of males lack the

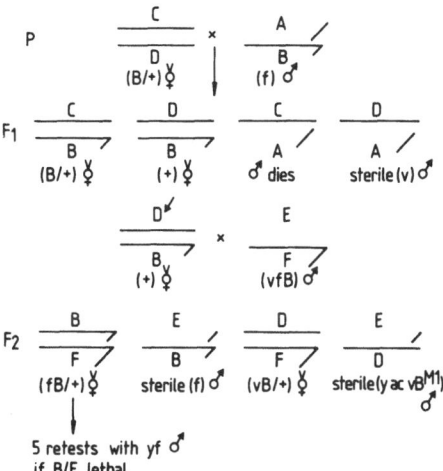

Figure 2. Mating scheme for detecting X-linked recessive lethal
     mutations in male and female larvae (slightly modified
     version of the scheme by Lee (23). Phenotypes are given in
     parentheses; each chromosome is designated by a letter with
     genotype as follows: (A) $Y^{L} \cdot sc^{S1}$, y^{+} ac bb_{8}; (B_{8}= treated
     parental X) $X \cdot Y^{S}$, In(1)EN2 , f; (C) In(1)sc^{8}, sc^{8} l B; (D =
     treated maternal X) In(1) dl - 49 + B^{M1}; (E) Y^{S}; (F) $X \cdot Y^{L}$, In
     dl - 49, v f B; .

long arm of the Y, the females produced in F_{2} are virgins,
allowing any further verification (retesting) of suspected
lethals. For a description of the genetic markers and of the
mutant stocks used in these experiments, see Lindsley and Grell
(26).

     Using the same experimental protocol as for recessive
lethals, parallel studies were performed with the w/w^{co} system of
Becker (Figure 3). Females of the genotype w^{co} sn^{2}; se h were
crossed with w; se h males in vials and were permitted to lay eggs
for four days on food supplemented with different concentrations
of chemicals that are stable under physicological conditions
(method A). With more labile chemicals, with MMS or MNU for
instance, a somewhat different protocol (method B) was used. The
flies were allowed to oviposit in vials for 41 h and 7 hours later
the developing cultures were treated by dropping 0.2 ml mutagen
solution on the surface of the food.

     The data on the induction of recessive lethals in male and
female larvae and that on somatic effects in female larvae are
summarized in Tables 1, 2 and 3, respectively. Considering first

Table 1. Percentages of induced X-linked recessive lethals after treatment of male and female larvae.

Chemical	Method	Concentration	Brooding (in days)	Tests	% induced lethals ♂	% induced lethals ♀	Activity ♂	Activity ♀
MMS	B	5.0 mM	0-2	1079	0.41	0.24	(+)	i
			2-4	963	0.87	0.15	+	i
ENU	B	1.0 mM	0-2	521	7.20	10.80	+	+
			2-4	572	6.70	6.20	+	+
Bleomycin	B	0.005%	0-2	787	0.50	0	(+)	-
			2-4	679	0.47	0	(+)	-
DMBA	A	1.0/2.0 mM	0-2	1565	0.25	1.60	i	+
			2-4	1194	0.40	1.00	(+)?	+
DA	A	1.0/4.0 mM	0-2	2322	1.60	0.16	+	i
			2-4	1704	0.55	0.30	(+)	i

i, inconclusive; (+), weakly mutagenic; +, mutagenic; - no activity detectable

Table 2. Chemically induced mosaic twin spots (TS) and mosaic light spots (ML) in w^{CO} $sn^2/w +$; se h/se h females treated at larvae.

Chemical	Method	Concentration	Eyes tested	TS %	ML %	TS:ML (induced)	Activity
MMS	B	5.0 mM	1820	5.27	6.10	0.89	+
ENU	B	1.0 mM	1424	1.19	3.79	0.32	+
Bleomycin	B	0.005%	1872	2.35	0.85	4.19	+
		0.010%	1320	4.17	1.52	3.37	+
DMBA	A	1.0/2.0 mM	3108	0.84	3.57	0.23	+
DA	A	1.0/4.0 mM	1318	0.79	1.37	0.58	+
Controls	-	-	5854	0.09	0.31	-	

Table 3.   Chemically induced mosaic twin spots (TS) and mosaic light spots (ML) in w^{co} sn^2/w +; se h/se h females treated as larvae.

Chemical	Method	Concentration	Eyes tested	TS %	ML %	TS:ML (induced)	Activity
EMS, ethylmethanesulfonate	B	20.0 mM	1098	2.91	7.10	0.42	+
DES, diethylsulfate	B	40.0 mM	514	1.17	1.95	0.66	+
MNU, N-methyl-N-nitrosourea	B	1.0/2.0 mM	522	3.26	5.36	0.63	+
MNNG, N-methyl-N'-nitro-N-nitrosoguanidine	B	5.0 mM	256	4.69	4.69	1.05	+
DMN, dimethylnitrosamine	B	5.0 mM	432	0.69	1.85	0.39	+
	A	1.0 mM	290	0.34	2.41	0.12	+
ENNG, N-ethyl-N'-nitro-N-nitrosoguanidine	B	10.0 mM	1030	0.58	2.91	0.19	+
DEN, diethylnitrosamine	B	10.0 mM	1048	0.19	1.53	0.08	+
	A	1.0 mM	1398	0.50	2.07	0.23	+
MEN, methylethylnitrosamine	A	1.0 mM	1480	0.95	2.77	0.35	+
DPN, dipropylnitrosamine	A	1.0 mM	538	0.37	1.12	(0.35)	+
DNPT, dinitrosopentamethylenetetramine	A	1.6 mM	1182	0.42	1.27	0.34	+
HMPA, hexamethylphosphoramide	B	0.6 mM	784	0.26	0.51	(0.75)	(+)[1]
Cyclophosphamide	A	0.5 mM	1236	0	0.65	-	(+)[1]
Procarbazine	A	2.0 mM	1186	0.25	2.61	0.07	+
AAF, 2-acetylaminofluorene	A	0.5/1.0 mM	730	0.68	0.27	-	(+)[1]
3-MC, 3-methylcholanthrene	A	2.0/4.0	1062	0.38	0.56	(1.10)	(+)[1]
NA, 2-naphthylamine	A	1.5	754	0.26	0.79	(0.31)	(+)[1]
DBE, 1,2-dibromoethane	A	0.35	514	0.78	7.00	0.10	+
Controls	-	-	5854	0.09	0.31	-	-

1) further testing required to confirm tentative conclusion (+)
   (+), weakly mutagenic; +, mutagenic

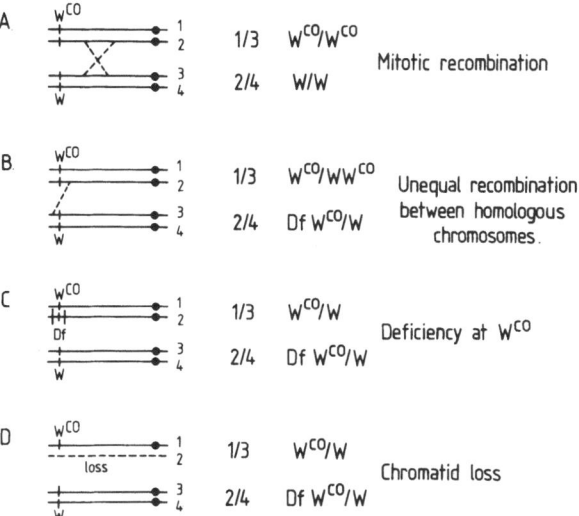

Figure 3. Some instances of genetic alterations in somatic eye tissue of female w^{CO}/w larvae.

the results on recessive lethals, it can be seen that ENU (N-ethyl-N-nitrosourea), showing strong mutagenic effectiveness, was the only reference mutagen readily detectable in both sexes, whereas the identification as mutagens of MMS (methylmethanesulfonate) and BL (bleomycin) caused severe problems (Table 1). Both mutagens were either ineffective or only showed marginal activity at rather high dosages. DMBA (7,12-dimethylbenz(a)anthracene) was more active in larval testis, DA (9,10-dimethylbenzanthracene), more so in the female germ line. Except for ENU, which predominantly induces point mutations, and with DA for which a few clusters were found, no large clusters of mutants of common origin were observed. (It is outside the scope of this paper to analyze the actual data exhaustively; these results will be published in full length elsewhere). These observations raise the possibility that most of the mutations were induced in spermatocyte and oocyte stages, and that gonial cell-stages are either resistant to mutation induction, or that those carrying a recessive lethal mutation were eliminated during meiosis.

The pattern of effects described for the recessive lethal assay is quite different from the markedly high response of somatic tissue to these five chemicals (Table 2). The most striking difference between both assays is the high response of somatic tissues to BL and MMS, as opposed to the marginal effects in male and female germ cells. Thus, a drastic difference in

detection capacity will be noted between germ line assay and
somatic mutation test. However, quantitative comparisons cannot
be made, because the eye disc at the time of response consisted of
at least 20 cells per female. This will somewhat reduce the
difference between germinal and somatic changes. Another
interesting aspect of this analysis concerns the proportion of
mosaic twin spots:mosaic single light spots. Twin spots are
assumed to reflect somatic recombination events and chromosomal
aberrations, single light spots may predominantly be due to
mutations or deletions (8). It is seen in Table 2 that the
proportion of twin spots (TS) versus light spots (ML) varied
considerably as a function of the type of DNA damage introduced.
BL and MMS, reagents that are known to be efficient inducers of
chromosome breaks, produced the highest TS:ML ratios (Table 2).
Thus, we are inclined to the view that strong segregational
elimination during meiosis may be responsible for the low yield of
heritable mutations seen with these reagents.

In view of the encouraging results in these pilot studies
with the w/w^{co} system, it seemed of interest to further validate
its efficiency with a somewhat larger number of reference
mutagens. Inspection of Table 3 will reveal that a wide array of
mutagenic carcinogens are active in this assay. In essence, the
recent studies encourage us to believe that somatic mutation
assays represent a very useful adjunct to recessive lethal assays
on heritable mutations in adult flies, keeping in mind the
existing differences in enzyme activities between larval stages
and adult flies, as well as the potential limitations associated
with the testing of premeiotic cell-stages.

Assessment of genotoxic profiles in Drosophila . Chemically
induced genetic alterations arise from a multiplicity of causes,
and represent a number of different mutational DNA lesions. While
there exist for the first screening phase, i.e., the qualitative
identification and characterization of the mutagenic activity, a
number of well-established genetic bioassays, we are faced with
enormous problems when investigating what data generated in short-
term bioassays mean in terms of potential genetic risk for
mammals, ultimately for man. What is the predictive value and the
real significance of in vitro mutagenesis in relation to the whole
mammal? The special difficulties that confront us can best be
visualized if we consider the fact that the kinds of genetic
damage and the proportion with which they occur can vary
considerably, depending upon dose, cell type, repair, genetic end
point measured, etc. Since the actual mutagen dose at the genetic
target is mostly unknown, there is often no way of relating the
patterns of genetic effects observed in different detection
systems. It is therefore hardly surprising that it is still a
major unresolved problem whether general features and common
characteristics can be established for genotoxic effects of

chemicals that are valid beyond the particular species, strain, cells stage and set of experimental conditions under which the correlations were established and thus, at a later stage, permit interspecies comparison and extrapolations.

In order to tackle these questions, there is certainly an urgent need to gain more insight into the basic associations between the structure of a chemical mutagen, its mode of interaction with DNA, the modifying (secondary) factors determining the expression of genetic damage and, as far as possible, all of its genotoxic properties. This type of information is required not only for any quantitative risk evaluation, but also for establishing possible relationships existing between "mutagenic and carcinogenic potency".

At the present time, Drosophila constitutes one of the very few eukaryotic in vivo systems which would enable us to detect, in the same population of treated cells, practically all the known changes capable of occurring in the genetic material (Table 4). This suggests a model function for Drosophila which may prove of increasing value in the future.

Following the concept outlined here, a series of 11 alkylating agents (AAs) with well-defined physico-chemical properties and reaction patterns have been scrutinized by our group in the past years (44,45). AAs are particularly well suited for approaches aimed at identifying relations between primary DNA damage and the resulting genetic consequences, because detailed information is available on the actual distribution at various DNA reactive sites (21,22,35,37,38). But the most compelling argument for choosing this class of mutagens is that the relative distribution in DNA of the same type of alkyl group can be manipulated, depending on the alkylating substrate selected. There is as yet no other class of known mutagens which would enable us to produce a similar distribution of DNA adducts.

In brief, our previous studies (41,44-46) have revealed that there is for monofunctional AAs a general, direct association between chromosome-breaking efficiency, the occurrence of delayed genetic damage and N-alkylation in DNA; and a general inverse association between the latter parameter and the ability of AAs to induce point mutations. Conversely, those AAs acting more extensively at the oxygen atoms in DNA, while being less active with regard to the production of breaks, are more potent as inducers of point mutations. Table 5 summarizes the mutagenic profiles of three representative of AAs: MMS, EMS and ENU. Recently, Bartsch et al. (7) have demonstrated for ten AAs a positive relationship between carcinogenicity in rodents and the initial ratios of $7:0^6$ alkyl-guanine formed or expected after their reaction with double-stranded DNA in vitro. These authors

Table 4. Analysis of multiple biological end points for the determination
          of genotoxic profiles of mutagens in D. melanogaster.

BIOLOGICAL END POINT	ASPECT INVESTIGATED
In germ cells	
1. Mutation induction relative to cytotoxicty	primary interaction with DNA
2. The ratio of chromosomal aberrations versus that of recessive lethal mutations	primary interaction with DNA
3. Effect of storage on the formation of chromosomal aberrations	expression time – secondary factors
4. The ability of the mutagen to cause delayed mutations, expressed by the ration of F_2-lethals : F_3-lethals	primary interaction with DNA
5. The ratio in postmeiotic versus premeiotic cells of X-chromosomal to IInd chromosomal recessive lethal mutations	germinal selection – secondary effects
6. Modification of mutagenesis by DNA repair processes	role of excision and postreplication repair – secondary effects
7. The proportion of multi-locus deletions among X-linked recessive lethals	initial interaction with DNA and influence of expression time
8. Mutation induction relative to primary interaction with DNA	initial interaction with DNA – molecular dosimetry
In somatic cells	
9. The proportion of recombination/chromosomal breakage in relation to mutations	primary interaction with DNA

Table 5. The genotoxic profile of HMPA (hexamethylphosphoramide) in comparison to MMS, EMS and ENU

Parameter	MMS	EMS	ENU	HMPA
1. Storage effect on induction of chromosomal aberrations	strong	strong	weak	not detectable
2. Induction of 2-3 translocations relative to recessive lethals	high	moderate	low	low
3. Ratio F_2-lethals : F_3-lethals	high	moderate	moderate/low	low
4. Hypermutability with the excision-repair deficient mutant mei-9^{LI} :				
(a) recessive lethals	strong	moderate	not detectable	not detectable
(b) ring-X loss	moderate	moderate	moderate	decrease
5. Induction of mosaic twin spots relative to single spots in somatic tissue	high	moderate	low	low
6. Reaction in DNA with				
(a) N-sites in purines	high	moderate	low	low/not measurable
(b) O-sites in purines	low	moderate	high	low/not measurable

suggested that alkylation of guanine at position O^6 (or at other O-atoms of DNA bases) may be a critical DNA-base modification which determines the overall carcinogenicity of these alkylating agents in rodents. If one relates this with the Drosophila data (44 and Table 5), it becomes immediately apparent that the strong point mutagens in Drosophila (ENU, ENNG, DEN) are also strong initiators of tumors in rodents while, in turn, the reagents more efficient in the production of chromosomal breakage in Drosophila (MMS, EMS) are weak carcinogens. However, while these data suggest a major role of O^6 alkylation for point mutations in carcinogenesis, they do not rule out a role for chromosomal aberrations and other mutagenic endpoints as well.

The basic concept behind the approach described here has been to use multiple genetic end points to unravel some of the relevant primary DNA lesions responsible for a given genotoxic event, by choosing model substrates with well-defined reaction mechanisms. This, in turn, raises the question of whether mutagenic profiles might serve as a basis for predicting the presence (or absence) of certain DNA-interaction patterns for mutagens with unknown DNA reaction products. A mutagen which may fulfill model function in this respect is HMPA (hexamethylphosphoramide). The mutational pattern of HMPA, in many respects, differs from that seen for the class of monofunctional AAs (Table 5): (i) there is no storage effect on HMPA-induced translocations (41); (ii) the ability of HMPA to cause delayed mutations is low; (iii) absence of excision-repair has no influence on mutation induction by HMPA; and (iv) this mutagen causes extraordinarily high frequencies of ring-X losses (which in this case are likely to result from the induced SCEs) as opposed to only marginal frequencies of partial chromosome losses and 2-3 translocations (Vogel, unpublished observations). HMPA was, therefore, to be anticipated to attack DNA at sites other than the N-positions in guanine. In fact, in two experiments in which Drosophila DNA was reacted in vivo with (^{14}C) HMPA, by far the major part of radioactive label was at a position of unidentified products, possibly the pyrimidines or the phosphate group, whereas HMPA did not bind to a measurable extent with the O^6 and the N-7 positions of guanine (van Zeeland, personal communication). In other words, the negative correlation to be seen here is absence of the storage phenomenon for a mutagen that does not seem to interact measurably with the positions in guanine usually attacked by AAs (Table 5). This observation, in turn, reinforces the concept of the specific significance of N-7 and N-3 alkyl guanine for the formation of those chromosomal abnormalities occurring with a delay. Clearly, exceptional cases like HMPA can lead to a better understanding of mutation at the molecular level.

ACKNOWLEDGEMENTS

This work was funded by the National Institute of Environmental Health Sciences (USA), Contract No. 1027-07/08, and the "Stichting Koningin Wilhelmina Fonds" (The Netherlands), Contract No. S.G. 81.90. Part of this investigation also received support from the Association Contract No. 139-77-1 ENV N between the European Communities (Environmental Research Programme) and the State University of Leiden. The technical assistance of Ms. Ineke Bogerd, Ms. Henny de Gunst and Mrs. Wanda Ten Cate Hoedemakers-Gelder is greatly appreciated.

REFERENCES

1. Alderson, T., and M. Pelecanos (1964) The mutagenic activity of diethyl sulphate in Drosophila melanogaster. II. The sensitivity of the immature (larval) and adult testis. Mutation Res. 1:182-192.
2. Auerbach, C. (1962) Mutation. An Introduction to Research on Mutagenesis. Part I: Methods. Oliver and Boyd, Edinburgh and London.
3. Auerbach, C., and J.M. Robson (1947) Tests of chemical substances for mutagenic action. Proc. R. Soc. Edinburgh Section B: 62B, 284-291.
4. Baars, A.J., M. Jansen, and D.D. Breimer (1979) Xenobiotica-metabolizing enzymes in Drosophila melanogaster. Activities of epoxide hydratase and glutathione S-transferase compared with similar activities in rat liver. Mutation Res. 62:279-291.
5. Baars, A.J., J.A. Zijlstra, E. Vogel, and D.D. Breimer (1977) The occurrence of cytochrome P-450 and aryl hydrocarbon hydroxylase activity in Drosophila melanogaster microsomes, and the importance of this metabolizing capacity for the screening of carcinogenic and mutagenic properties of foreign compounds. Mutation Res. 44:257-268.
6. Baars, A.J., J.A. Zijlstra, M. Jansen, E. Vogel, and D.D. Breimer (1980) Biotransformation and spectral interaction of xenobiotics with subcellular fractions from Drosophila melanogaster. Arch. Toxicol. Suppl. 4:54-58.
7. Bartsch, H., B. Terracini, C. Malaveille, L. Tomatis, J. Wahrendorf, G. Brun, and B. Dodet (1983) Quantitative comparisons of carcinogenicity, mutagenicity and electrophilicity of ten direct-acting alkylating agents and of initial O^6:7-alkylguanine ratio in DNA with carcinogenic potency in rodents. Mutation Res. 110:181-219.
8. Becker, H.J. (1966) Genetic and variegation mosaics in the eye of Drosophila. Curr. Topics Developm. Biol. 1:155-171.

9. Becker, H.J. (1975) X-ray- and TEM-induced mitotic
    recombination in Drosophila melanogaster: Unequal and sister-
    strand recombination. Molec. gen. Genet. 138:11-24.
10. Brink, N.G. (1982) Somatic and teratogenic effects induced by
    heliotrine in Drosophila. Mutation Res. 104:105-111.
11. Clark, A.M. (1959) Mutagenic activity of the alkaloid
    heliotrine in Drosophila. Nature (London) 183:731.
12. Clark, A.M. (1982) The use of larval stages of Drosophila in
    screening for some naturally occurring mutagens. Mutation
    Res. 92:89-97.
13. Demopoulos, N.A., N.D. Stamatis, and G. Yannopoulos (1980)
    Induction of somatic and male crossing-over by bleomycin in
    Drosophila melanogaster. Mutation Res. 78:347-351.
14. Donner, M., M. Sorsa, and H. Vaino (1979) Recessive lethals
    induced by styrene oxide in Drosophila melanogaster.
    Mutation Res. 67:373-376.
15. Engels, W.R. (1979) The estimation of mutation rates when
    premeiotic events are involved. Environ. Mutagen. 1:37-43.
16. Fahmy, J.M., and O.G. Fahmy (1980) Altered control of gene
    activity in the soma by carcinogens. Mutation
    Res. 72:165-172.
17. Graf, U, H. Juon, A.J. Katz, H.Y. Frei and F.E. Wügler (1983)
    A pilot study on a new Drosophila spot test. Mutation
    Res. 120:233-239.
18. Hällström, I., A. Blanck, R. Graftström, U. Rannug, and A.
    Sundvall (1980), In Biochemistry, Biophysics and Regulation
    of Cytochrome P-450, J.-A. Gustafsson et al., Eds., Elsevier/
    North Holland Biomedical Press, pp. 109-112.
19. Hällström, I., and R. Graftström (1981) The metabolism of
    drugs and carcinogens in isolated subcellular fractions of
    Drosophila melanogaster. II. Enzyme induction and metabolism
    of benzo(a)pyrene. Chem.-Biol. Interact. 34:145-159.
20. Hodgson, E. (1974) Comparative studies of cytochrome P-450 and
    its interaction with pesticides, In Survival in Toxic
    Environments, M.A.Q. Khan, and J.P. Berderka, Jr., Eds.,
    Academic Press, New York, pp. 213-260.
21. Lawley, P.D., D.J. Orr, and M. Jarman (1975) Isolation and
    identification of products from alkylation of nucleic acids:
    Ethyl- and isopropyl-purines. Biochem. J. 145:73-84.
22. Lawley, P.D., and S.A. Shah (1972) Reaction of alkylating
    mutagens and carcinogens with nucleic acids: Detection and
    estimation of a small extent of methylation at O-6 of guanine
    in DNA by MMS in vitro. Chem.-Biol. Interact. 5:286-288.
23. Lee, W.R. (1976) Chemical mutagenesis, In The Genetics and
    Biology of Drosophila, A. Ashburner, and E. Novitski, Eds.,
    Academic Press, London-New York-San Francisco, pp. 1299-1341.
24. Lee, W.R., S. Abrahamson, R. Valencia, E.S. Von Halle,
    F.E. Würgler, and S. Zimmering (1983) The sex-linked
    recessive lethal test for mutagenesis in Drosophila

melanogaster: A report of the US EPA GENE-TOX Program. Mutation Res. 123:183-279.

25. Lee, W.R., G.A. Sega, and E.S. Benson (1972) Transmutation of carbon-14 within DNA of Drosophila melanogaster spermatozoa. Mutation Res. 16:195-201.

26. Lindsley, D.L., and E.H. Grell (1968) Genetic variations of Drosophila melanogaster. Carnegie Inst. Wash. Publ. 627, 472 pp.

27. Lindsley, D.L., and K.T. Tokuyasu (1980) Spermatogenesis, In The Genetics and Biology of Drosophila, Vol. 2d, M. Ashburner, and T.R.F. Wright, Eds., Academic Press, London, pp. 226-287.

28. Magnusson, J., I. Hällström, and C. Ramel (1979) Studies on metabolic activation of vinyl chloride in Drosophila melanogaster after pretreatment with phenobarbital and polychlorinated biphenyls. Chemico.-Biolog. Interactions 24:287-298.

29. Magnusson, J., and C. Ramel (1978) Mutagenic effects of vinyl chloride on Drosophila melanogaster with and without pretreatment with sodium phenobarbiturate. Mutation Res. 57:307-312.

30. Mollet, P., and W. Weileman (1976) Characteristics of a new mutagenicity test. Induction of somatic recombination and mutation in Drosophila by different chemicals. Mutation Res. 38:131-132.

31. Mollet, P., and F.E. Würgler (1974) Detection of somatic recombination and mutation in Drosophila. A method for testing genetic activity of chemical compounds. Mutation Res. 25:421-424.

32. Muller, H.J. (1952) The standard error of the frequency of mutants some of which are of common origin. Genetics 37:608.

33. Muller, H.J. (1954) The manner of production of mutations by radiation, In Radiation Biology, vol. 1, A. Hollaender, Ed., McGraw Hill, New York, pp. 475-626.

34. Nylander, P.O., H. Olofsson, B. Rasmusson, and H. Svahlin (1978) The use of Drosophila melanogaster, I. Effects of benzene and 1,2-dichloroethane. Mutation Res. 57:163-167.

35. Pegg, A.E., and J.W. Nicoll (1976) Nitrosamine carcinogenesis: The importance of the persistence in DNA of alkylated bases in the organotropism of tumor induction, In Screening Tests in Chemical Carcinogenesis, Sci. Publ., No. 12, R. Montesano, H. Bartsch, and L. Tomatis, Eds., International Agency for Research on Cancer, Lyon, pp. 571-592.

36. Rasmuson, B., H. Svahlin, A. Rasmuson, I. Montelli, and H. Olofsson (1978) The use of a mutationally unstable X-chromosome in Drosophila for mutagenicity testing. Mutation Res. 54:33-38.

37. Singer, B. (1976) All oxygens in nucleic acids react with carcinogenic ethylating agents. Nature (London) 264:333-339.

38. Sun, L., and B. Singer (1975) The specificity of different
    classes of ethylating agents toward various sites of Hela
    cell DNA in vitro and in vivo. Biochemistry 14:1795-1802.
39. Vogel, E. (1975) Some aspects of the detection of potential
    mutagenic agents in Drosophila. Mutation Res. 29:241-250.
40. Vogel, E. (1981) Recent achievements with Drosophila as an
    assay system for carcinogens, In Short-Term Tests for
    Chemical Carcinogens, H.F. Stich, and R.H.C. San, Eds.,
    Springer Verlag, New York-Heidelberg-Berlin, pp. 379-398.
41. Vogel, E.W. (1983) Approaches to comparative mutagenesis in
    higher eukaryotes: Significance of DNA modifications with
    alkylating agents in Drosophila melanogaster, In Cellular
    Systems on Toxicity Testing, G.M. Williams, V. Dunkel, and
    V.A. Ray, Eds., Ann. N. Y. Acad. Sci. 407:208-220.
42. Vogel, E., W.G.H. Blijleven, P.M. Klapwijk, and J.A. Zijlstra
    (1980) Some current perspectives of the application of
    Drosophila in the evaluation of carcinogens, In The
    Predictive Value of Short-Term Screening Tests,
    G.M. Williams, R. Kroes, H.W. Waaijers, and K.W. van de Poll,
    Eds., Elsevier/North Holland Biomedical Press, Amsterdam,
    pp. 125-147.
43. Vogel, E., and B. Leigh (1975) Concentration-effect studies
    with MMS, TEB, 2,4,6-triCl-PDMT, and DEN on the induction of
    dominant and recessive lethals, chromosome loss and
    translocations in Drosophila sperm. Mutation
    Res. 29:383-396.
44. Vogel, E., and A.T. Natarajan (1979) The relation between
    reaction kinetics and mutagenic action of mono-functional
    alkylating agents in higher eukaryotic systems. I. Recessive
    lethal mutations and translocations in Drosophila. Mutation
    Res. 62:51-100.
45. Vogel, E., and A.T. Natarajan (1979) The relation between
    reaction kinetics and mutagenic action of mono-functional
    alkylating agents in higher eukaryotic systems. II. Total
    and partial sex-chromosome loss in Drosophila. Mutation
    Res. 62:101-123.
46. Vogel, E., and A.T. Natarajan (1982) The relation between
    reaction kinetics and mutagenic action of mono-functional
    alkylating agents in higher eukaryotic systems. III.
    Interspecies comparisons, In Chemical Mutagens, vol. 7,
    A. Hollaender, and F.J. de Serres, Eds., Plenum Press, New
    York, pp. 235-236.
47. Vogel, E., and F.H. Sobels (1976) The function of Drosophila
    in genetic toxicology testing, In Chemical Mutagens, vol. 4,
    A. Hollaender, Ed., Plenum Press, New York, pp. 91-142.
48. Vogel, E., J.A. Zijlstra, W.G.H. Blijleven, and D.D. Breimer
    (1983) Metabolic activation and mutagenic properties of
    procarcinogens in Drosophila, In In Vitro Toxicity Testing of
    Environmental Agents, Current and Future Possibilities,

A.R. Kolber et al., Eds., Plenum Press, New York, pp. 215-233.

49. Vogel, E.W., J.A. Zijlstra, and W.G.H. Blijlevel (1983) Mutagenic activity of selected aromatic amines and polycyclic hydrocarbons in Drosophila melanogaster, Mutation Res. 107:53-77.

50. Zijlstra, J.A., E. Vogel, and D.D. Breimer (1979) Occurrence and inducibility of cytochrome P-450 and mixed-function oxidase activities in microsomes from Drosophila melanogaster larvae. Mutation Res. 64:151-152.

51. Zijlstra, J.A., E.W. Vogel, and D.D. Breimer (1983) Strain-differences and inducibility of microsomal oxidative enzymes in Drosophila melanogaster flies. Chem.-Biol. Interactions (in press).

52. Zijlstra, J.A., E.W. Vogel, and D.D. Breimer (1983) Mutagenicity of 7,12-dimethylbenz(a)anthracene and some aromatic mutagens in Drosophila melanogaster. Mutation Res. (in press).

THE P-FACTOR: A TRANSPOSABLE ELEMENT IN DROSOPHILA[1]

James F. Crow

Genetics Laboratory
University of Wisconsin
Madison, WI   53706

SUMMARY

A number of seemingly unrelated phenomena, including high mutability, chromosome breakage, temperature-sensitive sterility, and male crossing-over, appear in certain inter-strain crosses in Drosophila melanogaster. The collection of symptoms has been called "hybrid dysgenesis". The explanation is a 2907 bp transposable element, designated P. The P element is highly mobile in an appropriate maternally-derived cellular environment, designated M, but hardly at all otherwise. The various manifestations of hybrid dysgenesis are thought to be consequences of P factor transposition.

A complete P factor has a 31-base reverse repeat at each end and two internal protein-coding sequences. Incomplete P factors having the repeated ends but lacking the coding regions are transposable, but only when at least one complete P factor is present in the nucleus. A specific, extremely highly mutable state of the singed bristle locus is thought to be due to the insertion of one or more incomplete P factors. Mutability at this locus has been employed as an indicator of the presence of a complete P factor in experiments showing DNA transformation by injecting P-containing DNA into embryos.

Three features of the P system distinguish it from other transposable elements. One is the extremely high rate of transposition. Second is complete genetic control over

---

[1] Paper Number 2667 from the Laboratory of Genetics.

transposition by choice of the appropriate maternal contribution. Third, the phenomenon is confined almost entirely to germ cells. These properties make the system particularly useful for a variety of genetic experiments, for example those using P factors as vehicles for cloning and DNA transformation.

The role of transposable elements as contributors to the spontaneous mutation rate is yet to be determined.

INTRODUCTION

High mutation rates in Drosophila have been reported many times (1,11,23,31,35,38,39,41,59), and probably have been observed but not reported many more times. These often involve a few specific loci, frequently include both gene and chromosomal mutations, and sometimes are associated with strain or species crosses (54). Typically the high mutability property disappeared before it could be fully analyzed, and there was sometimes evidence of non-Mendelian inheritance. Green (24) has reviewed the earlier literature. Similar phenomena have been reported in several plant species.

Recently a number of seemingly unrelated abnormalities have been reported in certain strain crosses in Drosophila melanogaster. These include high mutability, chromosome breakage, sterility, male recombination, and distorted segregation ratios (28,60,61). The first to suggest that these have a common etiology were Kidwell, Kidwell, and Sved (32), who introduced the name "Hybrid Dysgenesis".

I shall review the genetic and molecular analysis that has led to the identification of a transposable element, the P-factor, as the cause. The emphasis will be on the work of my colleagues William Engels, Michael Simmons, and Christine Preston, but I should like to emphasize that one reason for the rapid progress in understanding this system has been the cooperation and exchange of materials among workers in several laboratories, and these will be identified in the course of this review. The first to suggest that a transposable element might be involved was M. M. Green (25).

Drosophila strains can be divided into two sharply contrasting types, M (for maternal contributing) and P (for paternal contributing). Hybrid dysgenesis occurs in the progeny when M females are mated to P males. The reciprocal mating and matings within a type do not produce dysgenic effects, or produce them only in very low frequencies. Old laboratory strains are invariably M, while flies from nature are typically P.

This system has another striking property: the effects are all confined to cells of the germ line. Somatic tissues are essentially normal.

This article is a brief review of (i) the genetic and cytological analysis suggesting the existence of a transposable element, (ii) molecular identification and characterization of the P elements, (iii) the analysis of the striking difference between reciprocal crosses, and (iv) some possible implications of this and other systems for evolution and human mutagenesis.

A more extensive review has been written by Engels (19). Several reviews of transposable elements are available (5,7,9,45,55) including a book edited by Shapiro (50). Temin and Engels have discussed evolutionary implications (56).

GENETIC ANALYSIS OF HYBRID DYSGENESIS

Three phenotypic manifestations of hybrid dysgenesis have proven to be the most suitable for genetic analysis: (i) female sterility, (ii) mutability at the singed (sn) locus, and (iii) chromosome rearrangements. I shall discuss them briefly in turn and show that they have the same rules of inheritance, suggesting a common underlying mechanism.

Female Sterility

An attractive feature of sterility as a phenotype for study is that it is manifest in the individual concerned, not its progeny. Furthermore, hybrid dysgenesis-induced sterility is sharply temperature-dependent; at 29° C there is up to 100% sterility, while at 24° or below fertility is normal. This means that one can identify sterility by using high temperatures while perpetuating the desired genotypes at low temperatures. Female sterility is easier to assess than male sterility, especially with some labor-saving innovations (20), so most experimental analysis has been done with females.

The female progeny of M female X P male and P female X M male matings are chromosomally identical, yet the former are sterile (at 29°) while the latter are fully fertile. Clearly some extra-chromosomal influence is playing a role. However, the mother's being M is not sufficient, otherwise M X M matings would produce sterile progeny. There must be some joint influence of M maternal inheritance with P chromosomes.

Engels and Preston (16), by using known mutant genes as chromosome markers, showed that the P property is polygenic. There are usually several P factors on each chromosome and the

larger the number of such elements the greater is the degree of
sterility (provided, of course, that the mother was M). The data
are consistent with the assumption that each P chromosome has an
independent probability of producing sterility. This suggests the
possibility that the P factors are all alike and represent some
element that is repeated several times in the genome. It is not
necessary for the P chromosome to be transmitted from the father
for it to be effective (16). Male sterility is less fully
analyzed, but generally conforms to the same rules of inheritance.

The sterility in either sex is caused by the absence of germ
cells and resembles the phenotype caused by mutations or
environmental treatments that destroy early germ cells. The
temperature-sensitive period begins at the age where germ cells
are just beginning to divide rapidly. The sterile flies are
otherwise normal (e.g., in sexual structures not derived from germ
cells and in mating behavior). The evidence all points to the
phenotype's being caused by destruction of early germ cells (20).

Engels (16,18) suggested the word "cytotype" to designate the
cellular environment in which the P factors occur. The results of
the experiments can be summarized by saying that sterility depends
on the individual's having the P genotype and an M cytotype. The
cytotype is ordinarily transmitted extra-chromosomally by the
mother. I shall return to cytotype later, after consideration of
other phenotypes and the nature of P factors.

Mutability

High mutability has usually been recognized by increased
frequency of lethals or of classical mutant phenotypes
(6,24,51,52). A locus where high mutability has been observed
repeatedly is the gene for singed bristles (25). One allele, sn^w
(singed-weak) arose in a dysgenic hybrid and was spectacularly
mutable (15). In M female X P male matings sn^w mutates to two
relatively stable forms, sn^+ (normal) and sn^e (extreme singed).
The two stable forms occur in approximately equal numbers and
together may constitute over 50% of the offspring. Such a high
rate of mutation means that the progeny of a single fly are
sufficient to establish whether the parent is normal or mutable.
This makes it feasible to study mutability as a phenotype without
the large numbers of progeny usually required.

The pattern of mutability is shown in Figure 1. High
mutability requires that the mother be of the M cytotype and that
chromosomes carry one or more P factors. Since the singed locus
is on the X chromosome, it is convenient to study mutability in
males mated to compound X females so that the hemizygous sons
receive their X chromosome from their father. The cytotype can be
controlled by mating to compound X females from an M or P strain.

With M cytotype and P chromosomes:

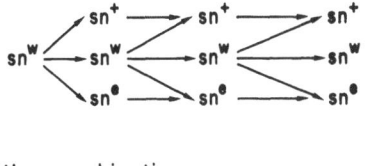

Other combinations:

$$sn^W \longrightarrow sn^W \longrightarrow sn^W$$

Figure 1. The pattern of mutability of singed-weak in dysgenic
matings (upper). The lower part illustrates the stability in
other matings.

If the cytotype is M the sn^W allele is highly unstable (40-60%
mutants); if it is P the allele is nearly stable ($\leq$1% mutants).
In dysgenic matings other loci are also mutable, but the rates are
much less extreme (15,18).

Genetic analysis (17) shows that the rules of inheritance of
mutability are the same as those for sterility, strongly
suggesting that the underlying phenomenon is the same. There is
one difference, however. The sn^W allele by itself is not mutable.
At least one additional P factor must be present. The P element
at the singed locus must differ from the others; so P factors are
not identical after all.

Chromosome Breakage

Chromosome breakage is one of the most striking
manifestations of the hybrid dysgenesis phenomenon (2,60,61).
Screening for chromosome breakage is greatly facilitated if a fly
with a chromosome break can easily be identified by its external
appearance. Such a phenotype was discovered by Engels and Preston
(21). In one of the P strains a chromosome break at a specific
site on the X chromosome causes a characteristic and conspicuous
phenotype, heldup wings (hdp). This suggests that there may be a
P factor at this site, but regardless of the mechanism it provides
a simple screen for chromosome breaks. A fly with heldup wings is
easily detected among a large number of normal flies. Using this
method, Engels and Preston identified hundreds of flies with
chromosome rearrangements. The position of the breaks was
determined by standard salivary-gland chromosome methods.

In addition to 649 inversions, there were 96 more complicated
rearrangements and one translocation, but the experimental

procedures favored the detection of inversions on the X chromosome
(Engels, personal communication). Figure 2 shows the positions of
the breaks in two strains. Always one break was at the heldup
locus because this phenotype was used for screening. The breaks
were not located uniformly throughout the chromosome, but tended
to be at a few hot spots. This immediately suggests that these
spots are the locations of P factors, and that these factors
somehow cause localized chromosome breakage. That the breakage
hotspots are restricted to P chromosomes was demonstrated as
follows. When parts of chromosomes from an M strain were
introduced by crossing over into the P chromosome, the hot spots
in this region disappeared (21). Figure 2 illustrates another
property of the breakage hot spots: their locations are not the
same in different P chromosomes.

The specificity of the breakage sites can be illustrated
further by doing the reverse experiment. Starting with flies
having heldup wings, and therefore having one or more inversions,
Engels and Preston screened for normal progeny. In these, as
expected, the heldup region had reverted to its normal sequence.
In many cases there was an exact re-inversion involving both
break-points, so that the original sequence was restored.

As with sterility and mutability, a high rate of chromosome
breakage occurs only in dysgenic crosses; that is, when there are
P-containing chromosomes and the mother is of M cytotype.
Furthermore, the frequencies of sterility, mutability, and
chromosome breakage are highly correlated (19). Thus, all three
phenotypic manifestations of hybrid dysgenesis appear to depend on
the same underlying mechanism, somehow mediated by P factors in
the proper cellular environment. There is one difference,
however; only sterility has the striking temperature dependence,

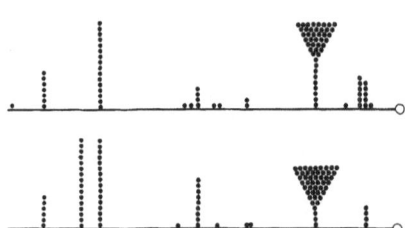

Figure 2. The distrution of breaks along the X chromosome induced
by hybrid dysgenesis. These are mainly inversions with one
break at the heldup region (the large cluster). Notice that
the lower chromosome has a hot spot near the left end that is
missing in the upper chromosome. Whether there is an extra
hot spot at the right end of the upper chromosome is
uncertain because of the difficulty of precise location in
and near the heterochromatin.

suggesting that the temperature effect is secondary to the underlying mechanism.

Comparison of several P strains, most of them from the same wild population, showed that each contained a number of P factors. But, as mentioned before, they were not at the same positions in different lines (see Fig. 2). This provides a strong hint that the factors are moveable, and more recently, actual transpositions have been documented (19).

The hypothesis suggested by all the genetic studies is that sterility, mutability, chromosome breakage, and other aspects of hybrid dysgenesis are consequences of P factors in M cytotype and are somehow associated with transposition events.

MOLECULAR NATURE OF THE P FACTORS

Often a mutation produced under the influence of the P-M system turned out to be unstable under dysgenic conditions. This suggests that the mutant involves the transposition of a P factor into the gene, causing the mutant phenotype, and that further mutation to wild type or another allele involve excision or modification of the inserted P factor. If such an insertion mutant should occur in a locus that had been cloned, this would open up the chance to clone the P factor with all the attendant molecular possibilities that this offers.

Simmons and Lim (51) in their study of mutants produced by the P-M system found some at the classical white eye locus. This locus had already been cloned (4), so the P-induced mutants at this locus were readily cloned and studied by molecular methods. The P-induced mutants at the white locus differed from the normal allele in having kilobase-size insertions (3,46). Four independent mutants of this sort have been obtained. Three of them have an insertion at the same site, arguing for some site-specificity of P-induced insertions. DNA prepared from these inserted elements was shown to be homologous to the locations of the chromosome breakage hot spots by in situ hybridization in salivary gland chromosomes, thus establishing that the breaks occur at sites where there is an inserted P element (19). No such homology is found in M chromosomes. There is also homology with the singed-weak mutant, but not with the normal allele at this locus.

O'Hare et al. (42) have determined the nucleotide sequence of several P factors, including the one from the heldup region. About 1/3 are 2907 bases long while the remaining 2/3 are shorter and of variable length. All of these include a 31-base reverse repeat at each end, along with a direct repeat of 8 bases of host chromosome DNA immediately outside the P element at each end. Inside the 2907 base element are two large open reading frames with start codons, in the same direction. (Actually there are three open reading frames, but one has no ATG start sequence, suggesting that two of these are a single gene with an intervening sequence.) Except for the host-DNA repeats, the large P factors are identical. The smaller pieces all have deletions, although the 31-base repeats at the ends are intact.

Earlier I mentioned that $\underline{sn}^W$ is not mutable unless at least one additional P factor is present. The location of the additional P factors is immaterial. This argues that one of the genes in the complete P factor codes for a trans-acting product (transposase?), which is necessary for the transposition events that produce the high mutability at the <u>singed</u> locus.

The conclusion suggested by all these (and other) observations is that transposition of a P factor requires at least the terminal repeats plus some product that can be produced only by a complete factor. The various manifestations of hybrid dysgenesis are then consequences of transposition.

WHAT IS CYTOTYPE?

Dysgenic effects occur in flies that have one or more P factors and whose mothers were of the M type. That the M cytotype is not simply a delayed expression of the mother's genotype is shown by the matings in Figure 3. The two classes of females in the third generation have the same genotype and their mothers had the same genotype; yet one class is sterile (at high temperatures) and the other fertile. The difference is that the maternal grandmothers of the sterile class were M whereas those of the fertile class were P. So the cytotype property, M or P, persists through two generations, suggesting that cytotype depends on some sort of self-reproducing property.

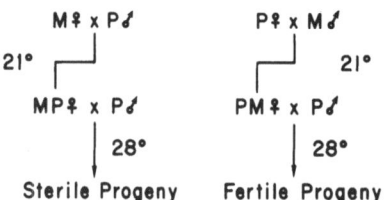

Figure 3. Pattern of matings illustrate the maternal transmission of cytotype. The sterile and fertile progeny differ only in whether the maternal grandmother was M or P.

However, the cytotype does not persist indefinitely in the presence of chromosomes of the opposite type. If female hybrids from M female X P male matings are mated to P males, the female progeny again mated to P males, and the process repeated for several generations, after 4-6 generations the cytotype becomes P despite the fact that the extra-chromosomal, maternal ancestry is entirely M. In the reverse situation of repeated backcross to M males, the cytotype becomes M after several generations, although the maternal ancestry is entirely P. After a sufficient number of backcross generations, the cytotype comes to agree with the chromosomal type. Engels has shown that the probability of having P cytotype depends on the mother's cytotype and the number of P factors, and has provided a rough quantitative picture of the relationship (19). If there are very many P factors the fly has a high probability of being P even if the mother was M. Conversely, if there are very few P factors there is a high probability of an M fly even if the mother was P. In M X P hybrids with an intermediate level the cytotype is regularly that of the mother.

The breeding experiments suggest that the P strain has a property lacking in the M strain. This is confirmed by molecular evidence for the physical reality of P factors and by the fact that there is no DNA in M strains that is homologous to P factors. This suggests that all the dysgenic phenomena are the result of P factors and that M is simply their absence.

The nature of cytotype still is not understood. There are a number of reasonable hypotheses, recently summarized by Engels (19). I shall mention only one. Assume that one of the two genes in the P factor codes for a transposase. This seems reasonable because some trans-acting product of a normal P factor can render the incomplete P factor at the singed locus mutable. This is also consistent with what is known in bacterial transposons. Assume further that the P factor can generate a plasmid that is capable of slow self-replication and that produces a transposase repressor when in the plasmid form. Perhaps coding for the repressor is the function of the second gene. In M female X P male hybrids the repressor is not initially present, not being transmitted by the sperm, and because of its limited self-replicating ability does not build up rapidly in the hybrid. Thus the hybrids show the dysgenic effects of the unrepressed P factor. On the other hand, when the mother is P the repressor is present and transposition does not occur (or occurs at very slow rates), even when P factors are present.

This hypothesis may or may not turn out to be correct. It has the merit of being heuristic and it is a convenient way to organize a number of heterogeneous observations.

DISCUSSION

Some Applications of P Factors

    One of the most exciting uses of P factors is the
demonstration of DNA transformation by Spradling and Rubin (53),
who took advantage of the fact that singed-weak is not mutable
unless additional P factors are present. They used cloned P
factors from the heldup region to prepare DNA which was injected
into embryos of M cytotype carrying the sn^w mutant. When the
embryos developed into adults it was easy to determine whether the
injected P factors had been incorporated by examining the progeny
for sn^+ and sn^e phenotypes. About half of the injected embryos
that developed into fertile adults had incorporated the P factor.
The incorporated factors were located at various sites in the
genome, as expected. Rubin and Spradling (47) also used this
system to repair the effects of the mutant rosy by injecting DNA
from the normal allele which had been incorporated into a P
element. The normal alleles, which were incorporated at several
chromosomal sites, supplied the enzyme missing in the homozygous
mutant flies.

    P factors have also been used for gene cloning. A gene that
has mutated because of a P insertion can be readily cloned and
identified, and this procedure (3) has now been used for several
genes, for example RNA polymerase II (49).

Special Features of the Hybrid Dysgenesis System

    The P factor is one of a rapidly growing number of
transposable elements that have been identified in Drosophila and
elsewhere. The P-M system has several properties that are of
special interest.

    1. P transposition occurs at an extremely high rate, on the
order of one per chromosome arm per generation. In addition, this
high rate is found only in progeny from dysgenic matings; the rate
within the P cytotype is two orders of magnitude lower. The
complete genetic control over the process opens up a number of
experimental possibilities.

    2. The transposition process is associated with conspicuous
phenotypes (sterility and changes at the singed and heldup loci).
This means that the powerful technology of traditional Drosophila
genetics can be combined with molecular approaches in a two-
pronged attack.

    3. The phenomenon is restricted to the germ line; somatic
cells are essentially normal. The nature of this restriction is
entirely unknown, but it offers experimental possibilities (for

example a search for mutants or environmental treatments that break down the distinction between germ and soma). There is an obvious evolutionary advantage to confining transpositions to the germ line: it permits the element to spread through the population while doing minimal damage to the survival of its host. But the mechanisms, both developmental and evolutionary, are completely unknown.

Comparison with Other Transposable Elements

It has been estimated that transposable elements make up from 10 to 20 percent of the Drosophila euchromatin. They can be grouped into several families, within each of which there is considerable DNA homology but between which there is little or none (45). Two types have been found in all Drosophila strains that have been examined. The copia-like elements are characterized by having direct repeats at their ends. The foldback elements, in contrast, have inverted repeats (58). They are also responsible for instability at some gene loci (34). These and the other widely distributed elements in Drosophila differ from P factors in not having any such high rate of transposition as occurs in dysgenic matings.

The first system in Drosophila where specific transposition could be identified and followed over many generations was that of Ising and Ramel (29,30), who showed that a chromosome region from the distal end of the X chromosome could be moved around the genome. More than 150 such events were documented and the positions located. A system that resembles the P factor in causing a high frequency of mutations and transpositions is the L factor (36,37). The I-R system (5,6,44) is similar to the P system in having striking reciprocal cross differences and in producing sterility as one of the phenotypes; the sterility, however, is of a different sort.

The P factor thus far appears to be confined to D. melanogaster. However, D. ananassae has properties that are similar in several ways (26,27,40). The phenotypic manifestations are male crossing over and high mutability, and there are reciprocal cross differences indicating some sort of extrachromosomal influence.

Evolutionary Implications

Clearly, systems that move DNA around the genome, producing mutations and chromosome rearrangements in the process, can have important effects on the organisms that possess them. If transpositions are frequent, the effects can be profound, as in dysgenic matings in the P-M system. The large number of pseudogenes of various ages that have been found in mammals argue strongly that mechanisms for DNA transposition have existed for a long time. Some of these appear to have involved RNA intermediates. How important such events have been in evolution, for example by gene duplication, remains to be seen. They could well be responsible for dispersed repetitive sequences. On the other hand, the gene order is conserved over long evolutionary periods. Those changes that are observed are mainly the consequences of standard cytogenetic events, such as inversions and translocations. An important questions is: With all the mechanisms that exist for moving genes around, why is the genome so stable?

It has often been suggested that a high mutation rate, perhaps in response to some environmental crisis, provides an increase in genetic variability that is important for survival of the species. It is also argued that conventional neo-Darwinian assumptions are inadequate to account for speciation. A generation ago, mass mutation and systemic mutations were invoked by Goldschmidt (22). Neel (41) suggested that mutator genes were the "pacemakers of evolution", and the same idea (in the same words) has been suggested in connection with P factors (57). The novelty and spectacular properties of these systems invite free-wheeling speculation (12).

I prefer the opposite view, that a very high mutation rate (whether or not it is in response to an environmental stress) is not likely to be evolutionarily advantageous. The conventional neo-Darwinian view seems to be more reasonable. For one thing, in any large, sexually reproducing population the amount of standing genetic variability is enormously larger than that which arises by mutation in a few generations. For another, standing variation is to some extent preselected by the worst mutants, having been eliminated through natural selection. Segregation and recombination of existing variants seem to me to be a much better way to keep up with environmental changes than mass mutation. It is very difficult to see how there can be efficient adjustment of mutation rates in sexual species to strike an optimum balance between a rate high enough to keep up with a changing environment and low enough not to incur a large load of deleterious mutations (33).

On the other hand, in asexual species there is likely to be a much closer connection between mutation rate and evolutionary rate. A gene that changes the mutation rate remains tied to the genes whose mutation rates it is affecting, rather than being separated by segregation and recombination. In fact, chemostat experiments have shown that bacterial strains with higher mutation rates can win out over rival strains with lower rates (8,10). Transposons may give a sexual population the possibility of selection for or against high mutability of the transposon itself and of loci linked to it, but whether this is of any significance to other than the transposon remains to be seen.

"Selfish DNA" (43,48) would be expected to spread more rapidly in biparental than in uniparental species. Similarly, a transposon that could direct a copy of itself to the homologous chromosome would maximize its capacity to spread in the population. In this context it is interesting that in one system there seems to be a preferential spread of mutability from a chromosome to its homolog (37).

Implications for Human Mutagenesis

For the purposes of this workshop, the important question about transposons is the possible damage such systems may do to the human germ plasm by increasing the burden from gene and chromosome mutations. What fraction of the spontaneous mutation rate in man is due to such events is of course unknown. The Drosophila evidence suggests that it may be appreciable. A substantial fraction of the spontaneous mutant genes that have been analyzed by molecular methods have turned out to have insertions; but the numbers are too small, and perhaps the sample of loci too unrepresentative, for any conclusions to be reached at present.

In Drosophila it seems likely that some episodes of high mutability in natural populations can be explained by the presence of both M and P types in the same area. If any such thing occurs in the human population, it would be one more reason for suspecting that mutation rates are highly heterogeneous and that a few individuals contribute disproportionately to the mutational burden of future generations.

A related question is the possible influence of environmental chemicals on the frequency of gene and chromosome mutations induced by transpositions. Such limited studies as have been done have suggested that this form of mutability is not significantly influenced by ordinary mutagens. For example, tests of ethylnitrosourea and methylmethanesulfonate in the P-M system showed little or no effect, although combinations of repair-deficient mutants enhanced the mutability (13,14). However, it is likely that if environmental chemicals do influence the rate of transposition, they will not be the same substances as cause other mutations, for the molecular mechanisms may be quite different.

ACKNOWLEDGEMENT

This article is substantially improved because of the ministrations of Bill Engels, Mike Simmons, and Christine Preston.

I am also indebted to them for permission to discuss their unpublished results.

REFERENCES

1. Berg, R.L. (1974) A simultaneous mutability rise at the singed locus in two out of three Drosophila melanogaster populations studied in 1973. Dros. Inf. Serv. 51:100-102.
2. Berg, R., W.R. Engels, and R.A. Kreber (1980) Site-specific X-chromosome rearrangements from hybrid dysgenesis in Drosophila melanogaster. Science 207:606-611.
3. Bingham, P.M., M.G. Kidwell, and G.M. Rubin (1982) The molecular basis of P-M hybrid dysgenesis: The role of the P element, a P strain-specific transposon family. Cell 29:995-1004.
4. Bingham, P.M., R. Levis, and G.M. Rubin (1981) The cloning of the DNA sequences from the white locus of Drosophila melanogaster using a novel and general method. Cell 25:693-704.
5. Bregliano, J.C., and M.G. Kidwell (1983) Hybrid dysgenesis determinants, In Mobile Genetic Elements, J.A. Shapiro, Ed., Academic Press, New York, pp. 363-410.
6. Bregliano, J.C., G. Picard, A. Bucheton, A. Pelisson, J.M. Lavige, and P. L'Heritier (1980) Hybrid dysgenesis in Drosophila melanogaster: The inducer-reactive system. Science 207:606-611.
7. Calos, M.P., and J.H. Miller (1980) Transposable elements. Cell 20:579-595.
8. Chao, L., C. Vargas, B.B. Spear, and E.C. Cox (1983) Transposable elements as mutator genes in evolution. Nature 303:633-635.
9. Cold Spring Harbor Symposia on Quantitative Biology (1981) Vol. 45. Cold Spring Harbor Laboratory, New York.
10. Cox, E.C., and T.C. Gibson (1974) Selection for high mutation rates in chemostats. Genetics 77:169-184.
11. Demerec, M. (1937) Frequency of spontaneous mutations in certain stocks of Drosophila melanogaster. Genetics 22:469-478.
12. Echols, H. (1981) SOS functions, cancer, and inducible evolution. Cell 25:1-2.
13. Eeken, J.C.J., and F.H. Sobels (1981) Modification of MR activity in repair-deficient strains of Drosophila melanogaster. Mut. Res. 83:191-200.
14. Eeken, J.C.J., and F.H. Sobels (1983) Modification of insertion mutation in Drosophila by ENU, MMS, or deficient DNA-repair. Envir. Mut. 5:456.
15. Engels, W.R. (1979) Extrachromosomal control of mutability in Drosophila melanogaster. Proc. Natl. Acad. Sci. USA 76:4011-4015.

16. Engels, W.R. (1979) Hybrid dysgenesis in Drosophila melanogaster: Rules of inheritance of female sterility. Genet. Res. 33:219-234.

17. Engels, W.R. (1981) Germline hypermutability in Drosophila and its relation to hybrid dysgenesis and cytotype. Genetics 98:565-587.

18. Engels, W.R. (1981) Hybrid dysgenesis in Drosophila and the stochastic loss hypothesis. Cold Spring Harbor Symp. Quant. Biol. 45:561-565.

19. Engels, W.R. (1983) The P family of transposable elements in Drosophila. Ann. Rev. Genet. 17: in press.

20. Engels, W.R., and C.R. Preston (1979) Hybrid dysgenesis in Drosophila melanogaster: The biology of male and female sterility. Genetics 92:161-174.

21. Engels, W.R., and C.R. Preston (1981) Identifying P factors in Drosophila by means of chromosome breakage hotspots. Cell 26:421-428.

22. Goldschmidt, R. (1940) The Material Basis of Evolution. Yale University Press, New Haven.

23. Green, M.M. (1970) The genetics of a mutator gene in Drosophila melanogaster. Mut. Res. 10:353-363.

24. Green, M.M. (1976) Mutable and mutator loci, In The Genetics and Biology of Drosophila, 1b, M. Ashburner and E. Novitski, Eds., Academic Press, New York, pp. 929-946.

25. Green, M.M. (1977) Genetic instability in Drosophila melanogaster: De novo induction of putative insertion mutations. Proc. Natl. Acad. Sci. USA 74:3490-3493.

26. Hinton, C.W. (1981) Nucleocytoplasmic relations in a mutator-suppressor system of Drosophila ananassae. Genetics 98:77-90.

27. Hinton, C.W. (1983) Relations between factors controlling crossing over and mutability in males of Drosophila ananassae. Genetics 104:95-112.

28. Hiraizumi, Y. (1971) Spontaneous recombination in Drosophila melanogaster males. Proc. Natl. Acad. Sci. USA 68:268-270.

29. Ising, G., and K. Block (1981) Derivation-dependent distribution of insertion sites for a Drosophila transposon. Cold Spring Harbor Symp. Quant. Biol. 45:527-544.

30. Ising, G., and C. Ramel (1973) A white-suppressor behaving as an episome in Drosophila melanogaster. Genetics 74:s123.

31. Ives, P.T. (1950) The importance of mutation rate genes in evolution. Evolution 4:236-252.

32. Kidwell, M.G., J.F. Kidwell, and J.A. Sved (1977) Hybrid dysgenesis in Drosophila melanogaster: A syndrome of aberrant traits including mutation, sterility, and male recombination. Genetics 86:813-833.

33. Leigh, E. (1973) The evolution of mutation rates. Genetics 73(suppl):1-18.

34. Levis, R., M. Collins, and G.M. Rubin (1982) FB elements are the common basis for the instability of the w^{DzL} and w^c Drosophila mutations. Cell 30:551-565.

35. Levitan, M. (1963) A maternal factor which breaks paternal chromosomes. Nature 200:437-438.

36. Lim, J.K. (1979) Site-specific instability in Drosophila melanogaster: The origin of the mutation and cytogenetic evidence for site specificity. Genetics 93:681-701.

37. Lim, J.K., M.J. Simmons, J.D. Raymond, N.M. Cox, R.F. Doll, and T.P. Culbert (1983) Homologue destabilization by a putative transposable element in Drosophila melanogaster. Proc. Natl. Acad. Sci. USA, in press.

38. Mampell, K. (1943) High mutation frequency in Drosophila pseudoobscura race B. Proc. Natl. Acad. Sci. USA 29:137-144.

39. Minamori, S., and K. Ito (1971) Extrachromosomal element delta in Drosophila melanogaster. VI. Induction of recurrent lethal mutations in definite regions of second chromosomes. Mut. Res. 13:361-369.

40. Moriwaki, D., and Y.N. Tobari (1975) Drosophila ananassae, In Handbook of Genetics, Plenum Press, New York, Vol. 3, pp. 513-535.

41. Neel, J.V. (1942) A study of a case of high mutation rate in Drosophila melanogaster. Genetics 27:519-536.

42. O'Hare, K., and G.M. Rubin (1983) Structures of P transposable elements of Drosophila melanogaster and their sites of insertion and excision. Cell 34:25-35.

43. Orgel, L.E., and F.H.C. Crick (1980) Selfish DNA: The ultimate parasite. Nature 285:604-607.

44. Picard, G., J.C. Bregliano, A. Bucheton, J.M. Lavige, A. Pelisson, and M.G. Kidwell (1978) Non-Mendelian female sterility and hybrid dysgenesis in Drosophila melanogaster. Genet. Res. 32:275-287.

45. Rubin, G.M. (1983) Dispersed repetitive DNAs in Drosophila, In Mobile Genetic Elements, J.A. Shapiro, Ed., Academic Press, New York, pp. 329-361.

46. Rubin, G.M., M.G. Kidwell, and P.M. Bingham (1982) The molecular basis of P-M hybrid dysgenesis: The nature of induced mutations. Cell 29:987-994.

47. Rubin, G.M., and A.C. Spradling (1982) Genetic transformation of Drosophila with transposable element vectors. Science 218:348-353.

48. Sapienza, C., and W.F. Doolittle (1981) Genes are things you have whether you want them or not. Cold Spring Harbor Symp. Quant. Biol. 45:177-182.

49. Searles, L.L., R.S. Jokerst, P.M. Bingham, R.A. Voelker, and A.L. Greenleaf (1983) Molecular cloning of sequences from a Drosophila RNA polymerase locus by P element transposon tagging. Cell 31:585-592.

50. Shapiro, J.A., Ed. (1983) Mobile Genetic Elements, Academic Press, New York.

51. Simmons, M.J., and J.K. Lim (1980) Site specificity of mutations arising in dysgenic hybrids of Drosophila melanogaster. Proc. Natl. Acad. Sci. USA 77:6042-6046.

52. Slatko, B.E., and Y. Hiraizumi (1973) Mutation induction in the male recombination strains of Drosophila melanogaster. Genetics 75:643-649.

53. Spradling, A.C., and G.M. Rubin (1982) Transposition of cloned P elements into Drosophila germ line chromosomes. Science 218:341-347.

54. Sturtevant, A.H. (1937) Essays on evolution. I. On the effects of selection on the mutation rate. Quart. Rev. Biol. 12:464-467.

55. Sved, J.A. (1979) The hybrid dysgenesis syndrome in Drosophila melanogaster. BioScience 29:659-664.

56. Temin, H.M., and W.R. Engels (1983) Movable genetic elements and evolution, In Evolution Prospects in the 1980's, J.W. Possard, Ed., J. Wiley and Sons, New York, in press.

57. Thompson, J.N., and R.C. Woodruff (1978) Mutator genes -- pacemakers of evolution. Nature 274:317-321.

58. Truett, M.A., R.S. Jones, and S.S. Potter (1981) Unusual structure of the FB family of transposable elements in Drosophila. Cell 24:753-763.

59. Voelker, R.A. (1974) The genetics and cytology of a mutator factor in Drosophila melanogaster. Mut. Res. 22:265-276.

60. Yamaguchi, O., R. Cardellino, and T. Mukai (1976) High rates of spontaneous chromosome aberrations in Drosophila melanogaster. Genetics 83:409-442.

61. Yamaguchi, O., and T. Mukai (1974) Variation of spontaneous occurrence of chromosomal aberrations in the second chromosome of Drosophila melanogaster. Genetics 78:1209-1221.

SILKWORM GENETICS AND CHEMICAL MUTAGENESIS

Y. Tazima

National Institute of Genetics
Mishima, JAPAN

SUMMARY

The domesticated silkworm (Bombyx mori L.), although not as frequently used as Drosophila for screening mutagenic substances, is also well suited for mutation studies, and the information obtained from such research could be applied to the sericultural industry. The most outstanding feature of silkworm genetics is the richness of genetic traits that manifest themselves at early developmental stages, such as the egg or newly hatched larva. This, in combination with the ease of egg collection, furnishes a great advantage in the detection of mutants and analysis of the mutants in the progeny.

In this paper, methods developed for mutation detection in the silkworm will be introduced, with comments on their merits and demerits, with special reference to their capability and specificity in mutation detection. For details, the reader may refer to my article (5).

INTRODUCTION

Several convenient systems have been developed for screening environmental mutagens, using bacteria or cultured mammalian cells. Most of these systems deal with mutagenic events that occur in somatic cells. From the genetic viewpoint, however, we need systems that can be applied for the detection of mutagenicity of agents that affect germ cells in the gonad and for the analysis of mutations transmitted to the offspring. Systems adequate for these requirements are rather limited, and Drosophila has been frequently used in a number of laboratories. We have, however, developed for this purpose a convenient detection system using the domesticated silkworm, <u>Bombyx</u> <u>mori</u> L.

Genetically, <u>Bombyx</u> <u>mori</u> contrasts with <u>Drosophila</u> <u>melanogaster</u> in several respects. The sex chromosome formula of <u>Bombyx</u> is female heterogametic, ZW-ZZ, while that of <u>Drosophila</u> is male heterogametic, XX-XY. In both insects, recombination of linked genes occurs only in the homogametic sex; namely, it occurs in the male <u>Bombyx</u> and in the female <u>Drosophila</u>. The chromosome of <u>Bombyx</u> is known to be holokinetic, while in <u>Drosophila</u> it is monokinetic. The germ cells develop almost simultaneously during development of the silkworm (Fig. 1), whereas they mature successively in <u>Drosophila</u>. The feeding habit is also quite different in the two species. These contrasting characteristics sometimes cause rather different modes of mutation response between the two. For instance, chromosome fragments, once produced, are likely to be lost in <u>Drosophila</u>, whereas they are

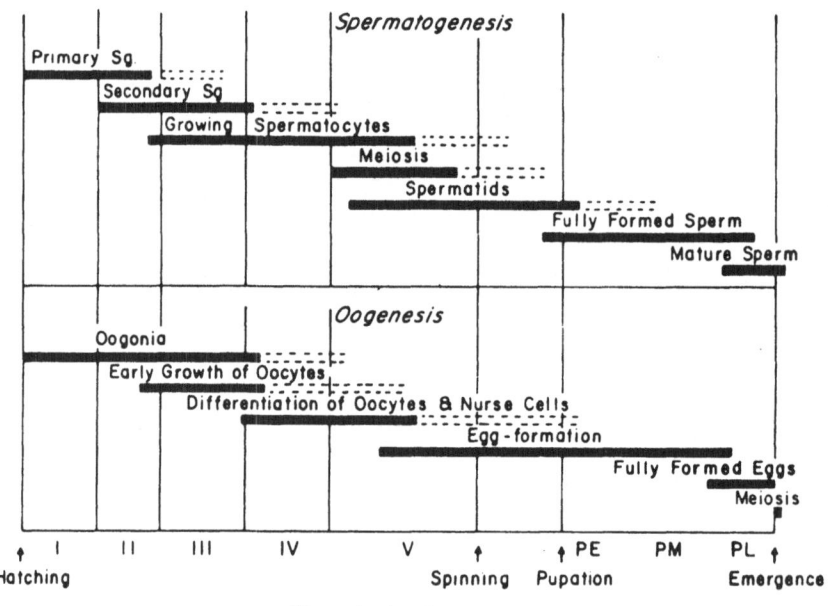

Figure 1. Schematic illustration of gametogenesis of the silkworm in relation to the developmental stages in both sexes. Duration represented by a dashed line indicates a period during which differentiation of the same cell type continues but utilizable gametes are not produced, because of the shortness of time. (I, II, ...) First, second, ... larval instars. Meiosis denotes meiotic metaphases I and II.

seldom eliminated in Bombyx; thus, the pairing and disjunctional behavior of chromosomes is quite different in the two species. Furthermore, the metabolic activation capacity of xenobiotics appears to differ between the two organisms (3).

These contrasting characteristics make Bombyx a suitable subject in the study of mutagenesis, even if it lacks several advantages of Drosophila. In Bombyx, a suitable sex-linked recessive lethal method is not yet available. Instead, however, a specific locus method is conveniently used, and the work to be reviewed in this article was mostly performed using this method.

Several methods have been used in the silkworm for the quantitative assessment of mutations induced by radiation and/or chemical agents. These are: the specific locus method; the recessive lethal method; sex-linked recessive lethal method; the dominant lethal method; and the method for detection of nondisjunction. The recessive lethal method assesses lethal mutations occurring over a whole genome and is used for the study of both spontaneously accumulated and freshly induced lethals in the silkworm. The system, however, is rather complicated and tedious and will not be dealt with in this article. The reader may refer to the reviews (4,5) for further details.

THE SPECIFIC LOCUS METHOD

This method essentially consists of mating treated homozygous dominant individuals with a test strain homozygous for known recessive marker loci. Mutations are detected in F_1 as recessives at marked loci, and mutation frequencies are given directly by their proportion to the total number observed. The method is simple and accurate, but it requires a large number of F_1 individuals for observation to obtain a reasonable number of mutants. This would usually be a great disadvantage. However, with the silkworm, this does not pose any particular problem, when the marker genes used are concerned with those which manifest themselves in early developmental stages, such as the egg or newly hatched larva. One female moth produces approximately 300 to 500 eggs, so that 100,000 eggs (or larvae) can be obtained from only 200 to 350 females.

The marker genes that have been most conveniently used in the silkworm are pink (pe, 5 - 0.0) and red (re, 5 - 31.5). Both are easily detected a few days after oviposition, when wild-type eggs turn from light yellow to dark brown. Other markers that have been used are body-color mutants of newly hatched larvae, chocolate (ch, 13 - 9.6) and sex-linked chocolate (sch, 1 - 21.5). Several mutants with translucent skins, os (1 - 0.0), od (1 - 49.6), oc (5 - 40.8), ok (5 - 4.7) etc., can also be utilized as

Table 1. Comparison of Mutation Response to MMC among
strains after treatment of spermatids in
mature larvae[a].

Strain	Dose (μg/larva)	Number of observed eggs	Observed mutation frequency ($\times 10^{-5}$)				Induced mutation (frequency/locus/μg)
			pe	pe, +[b]	re	re, +	
Aojuku	0	54,898	7.3	9.0	1.8	1.8	—
	10	51,867	21.2	250.6	7.7	108.0	18.4
C108	0	42,323	11.8	4.7	2.4	2.4	—
	10	40.128	17.4	149.5	2.5	69.7	10.9
Sekko	0	58,016	13.8	3.4	0	1.7	—
	10	32,634	33.7	465.8	6.1	306.4	39.7
Ascoli	0	46,982	14.9	14.9	2.1	10.6	—
	10	33,844	38.4	186.1	11.8	70.9	13.3
rb	0	36,157	16.6	5.5	0	2.8	—
	10	9,258	10.8	302.4	0	183.6	23.6

[a] Injection: 0.025 ml, MMC 5 μg/insect.
[b] pe, + denotes mosaic for pe and +.

markers, but these require raising F_1 larvae at least up to the
third instar.

    The simplest way of administering chemicals is by injection.
Several examples are given in Table 1.   In these experiments
methyl methanesulfonate (MMs) in physiological saline solution was
injected into the body cavity of several strains.   The injection
was performed into males at the mature larval stage, when germ
cells were at the spermatid stage.   Mutation-bearing individuals
were detected either as the whole or the partial (mosaic).   It can
be noted that the induced mutation frequencies are very high.
This may depend on the characteristics of the marker genes, which
manifest themselves at the early developmental stage.

    The induced mutation frequency varies with the development
stage of germ cells, as already known in other organisms.
Mutation frequency also varies with the strain (Table 1).

THE NATURE AND TRANSMISSIBILITY OF INDUCED MUTANTS

    The silkworm is not suitable for the study of mutation
mechanisms at the molecular level, and identification of gene
mutation in the strict sense is rather difficult.

After testing the complementation pattern between induced
egg-color mutants, we reached the conclusion that chemically
induced specific locus mutations of the silkworm comprise not only
point mutations and small deficiencies but also a sizable fraction
of gross chromosomal changes.

Since these mutations manifest themselves at the egg stage,
they are detected at a .very early stage of the development. The
silkworm is unique in this respect; in Drosophila and the mouse,
most mutations are observed in the adult. Therefore, direct
comparison of mutation frequencies of the silkworm with those
organisms is not reasonable. In this regard, the further fate of
these observed mutants was investigated and their transmission
rate assessed in comparison with that of non-mutant individuals.
The materials used for this purpose were radiation-induced egg-
color mutants.

The parameters used for comparison were the hatchability (A),
the survival up to late pupa or moth (B), and the rate of fertile
female moths (C). By multiplying these parameters, the rate of
transmissability was obtained as shown in Table 2. The
transmission rate thus calculated was very low in comparison to
that of non-mutant type individuals, giving 0.16 for pe and 0.05
for re or roughly 0.1 on an average. The low transmission rates
indicate that of those observed mutants at very high frequencies
only a small fraction of mutation was transmitted to the next
generation.

IMPROVEMENT OF SENSITIVE TEST SYSTEM

For screening environmental mutagenic substances, it is
essential to use a highly sensitive method that can detect agents
of weak mutagenicity. It is known in Drosophila, as well as in
the silkworm, that oocytes are extraordinarily sensitive to the

Table 2. Relative transmissibility of radiation-induced
          pe and re mutants compared to non-mutants of
          the same treatment group.

	Hatch-ability(%) (A)	Survival up to late pupa or moth (B)	% fertilized-egg-laying moths (C)	Rate of total transmission[†] A × B × C	Relative total transmissibility
Non-mutants	97.7	86.6	97.3	82.3	1.000
pe mutants	26.3	64.4	80.9	13.7	0.166
re mutants	18.1	33.1	77.3	4.6	0.056

† The low transmissibility of recovered mutants may chiefly be attributed to gross
chromosomal changes involving deficiencies at marked loci.

mutagenic action of radiation around the stage of meiotic
metaphase. Therefore, efforts have been made to find the most
effective stage for treating oocytes. The meiotic metaphase takes
place in this insect immediately after oviposition. The deposited
egg is covered with a thick chorion. However, when eggs were
soaked in 0.01 M KNO_2 aqueous solution, almost all were killed
when the pH was 4.2, indicating that substances of low molecular
weight can pass through the chorion. Hence, the penetrability
through chorion was assessed with MMC. The chemical was injected
into the hemocoel of female pupae one day before emergence. An
almost linear relationship was obtained between the administered
dose and the observed mutation frequencies, but the frequencies
were not as high as expected from other experiments in which
injection was done at earlier stages. Perhaps penetration of the
chemical might have been interrupted by the chorion to an
appreciable extent.

Injection was performed at various developmental stages of
the egg cells from one to five days before emergence (Table 3).
Mutation frequency in females injected three days before emergence
was about ten times and in those injected five days before
emergence it was more than 100 times as effective as in those
injected one day before emergence. The effectiveness appeared to
be negatively correlated with chorion formation, which starts
about five days before and is completed one day before emergence.
Higher mutation frequencies obtained by the injection at earlier
stages seemed to indicate that a large amount of the chemical was
incorporated into the egg plasm. This view was confirmed by using
a short-lived mutagen, diethyl sulfate.

Table 3. Change in mutation frequency with
stage of injection into female pupae
(treated ++♀ x pe re♂, 723 rb).

Injected stage[1]	Injection (% × ml)	No. of eggs obsd.	No. of mutants	Mut. freq. (10^{-5})/ μg/locus
Control[2]	0.025	26,682	9	12.8
1 day before	.02 × 0.025[3]	19,509	10	25.7
3 days before	.02 × 0.025	14,122	87	308.1
5 days before	.02 × 0.025	5,910	355	3003.4

[1] Stage of injection is indicated in days before emergence.
[2] Saline only was injected 3 days before emergence.
[3] Injected dose per head was 5 μg.

COMPARISON OF MUTAGENIC POTENTIALITY BETWEEN CHEMICAL MUTAGENS

The relationship between the treatment dosage and the frequency of induced mutations has been investigated with a number of chemicals in the silkworm, but the number of tested points of different concentration was not sufficient for each compound to determine whether or not the threshold existed at the low dose range. Murakami (1) carried out a fairly large scale experiment, in which he tested at least five different dose levels for each mutagen. Most compounds exhibited a linear relation with the increasing dose within the test range. But a minor tranquilizer, diazepoxide, showed clearly a sigmoid curve rising very slowly at a dose range lower than 100 μg/head. Furthermore, even a known strong mutagen, aflatoxin B_1, exhibited a nearly sigmoid curve at a dose range lower than 0.1 μg/head for pe locus mutation. In these low dose experiments, as many as 194,285 F_1 individuals were used for the observation at each point. From these results we may conclude that most mutagens respond in a similar way to those observed for diazepoxide: namely, there is a threshold range.

Since the reaction curve is not linear, it is hardly possible, in the strictest sense, to compare the mutagenic potentialities of different compounds. But the comparison may be permitted if we express the potentiality on μg/head basis. Hence Murakami plotted mutation frequencies obtained per μg/head basis on a linear scale (Fig. 2). In this figure mutation potentialities are indicated for both pe and re loci separately. They vary widely over a range of five orders of magnitude depending on the compounds. The range is two orders of magnitude narrower than that reported for Salmonella.

PECULIAR MUTATION RESPONSE

During the course of our experiment, we noticed a peculiar mutation response of the silkworm. To at least two known mutagens/carcinogens the silkworm response was quite different from those reported for other organisms. For instance, 4-nitroquinoline-1-oxide is a strong mutagen, whose mutagenicity has been reported for various organisms from bacteria to the mouse. However, mutagenicity has not been detected so far in the silkworm, even though several trials have been made. N-dimethylnitrosamine (DMNA) is also a known mutagen and carcinogen. The mutagenic potential is not as high, but its mutagenicity has been reported in many organisms, and is clearly positive in Drosophila. However, after several efforts we have not been able to confirm the mutagenicity of this compound in the silkworm. Murakami (2) discovered recently that N-diethylnitrosamine, in contrast to the negative effect of DMNA, showed positive mutagenicity in the silkworm. He attributed this finding to the

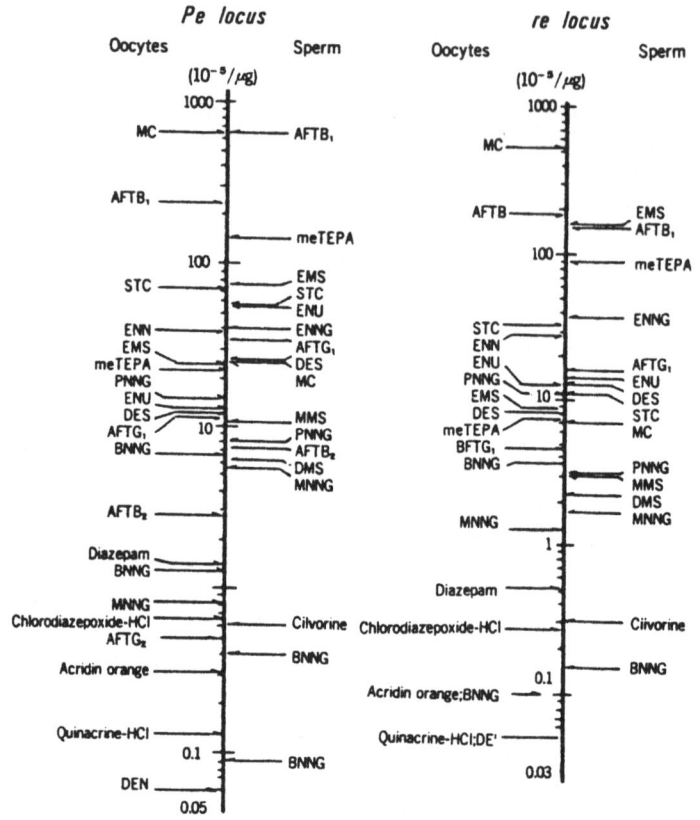

Figure 2. Relative mutagenic effectiveness of several
environmental chemicals in pupal germ-cells of
the silkworm.

different repairability between methylated and ethylated sites.
But it may also be explained by assuming a strong killing effect
of methylated sites. Further studies are necessary to elucidate
these results.

REFERENCES

1. Murakami, A. (1982) Studies on genetic hazard of environmental
    mutagens with silkworm germ cells, In Report of Special Study
    Group on Environmental Science, B. 124 R20-5, pp. 64-77.
2. Murakami, A. (1982) Comparison of mutagenic activity between
    methylating and ethylating agents in the silkworm. Abst.
    54th Ann. Meet. of Genet. Soc. Japan, p. 35.

3. Tazima, Y. (1964) The Genetics of the Silkworm, Logos Press,
    London, pp. 1-253.
4. Tazima, Y. (1978) Mutagenicity testing of environmental
    chemicals, In The Silkworm, Y. Tazima, Ed., Kodansha, Tokyo,
    pp. 247-268.
5. Tazima, Y. (1980) Chemical mutagenesis in the silkworm, In
    Chemical Mutagens, F.J. de Serres, and A. Hollaender, Eds.,
    Plenum Press, New York, pp. 203-238.

MUTATION INDUCTION AND DETECTION IN ARABIDOPSIS

G.P. Rédei[*], Gregoria N. Acedo[*] and S.S. Sandhu[**]

[*]Department of Agronomy, University of Missouri
Columbia, MO 65211

[**]Environmental Protection Agency
Research Triangle Park, NC 27711

SUMMARY

Mutation is an infrequent event, and the stability of the various gene loci is quite different. Additional complexity appears when mutability is studied in different taxonomic groups displaying a variety of genome organization. Higher plants permit the study of mutation in three types of genomes: nuclear, plastidic and mitochondrial. In Arabidopsis the size of the nuclear genome is about 1.3×10^{12} dalton whereas the chloroplast and the mitochondrial genomes are estimated to be 4 to 5 orders of magnitude smaller. The nuclear genome of Arabidopsis is much less redundant than that of other higher plants, thus mutation can be detected at an estimated 19,688 loci during embryonic development and the total number of loci with visible mutations appears to be 27,875. This chapter outlines the methods of culture of intact plants and isolated tissues as they are most suitable for mutation studies. An analysis of the development of the "germline" is presented, and the calculation of mutation rate in the various assay systems is outlined. Theoretical considerations and practical applications for planning mutation experiments are described. Special attention is given to analysis of mutations within the plastome. The chloroplast genome contains numerous copies per cell yet with the availability of a mutator locus in the nuclear genome, extremely high frequency of plastome mutation can be detected.

INTRODUCTION

The first systematic study of mutation began with plants, and
in 1901 Hugo de Vries (9) summarized his observations in two
monumental volumes. It took, however, another quarter of a
century to learn how to induce mutations experimentally (39).
Today, it is know that hereditary alterations can be produced in
plants not only by ionizing or other types of radiation and by a
large variety of chemicals (3), but also by biological factors
such as mutator genes and "controlling elements" (31). Recently,
means were developed for the introduction of foreign genes into
the plant genome by bacterial plasmids (7) and other prokaryotic
vectors (12).

The purpose of mutation studies may be either to obtain
mutants in order to investigate genetic or other biological
problems, or to study the mechanism of the mutation process
itself. The progress in understanding the basic features of
mutation makes possible the application of the principles to
societal and economic problems. The most widely used short-term
assays of environmental pollutants (carcinogens) are mutation
tests (8). Between 1969 and 1975 the number of agriculturally or
horticulturally useful varieties produced by induced mutation had
doubled (24), and in the Netherlands nearly all of the
commercially grown white and yellow varieties of chrysanthemums
are radiation-induced mutants (4).

Whatever the purpose of mutation studies, we depend on the
efficiency of the techniques to achieve success. Some of these
essential techniques will be outlined below.

ARABIDOPSIS AS A TOOL IN MUTATION STUDIES

In research, as in any other human activity, the work must be
carried out with maximum efficiency to have clear-cut results at
the lowest cost. Therefore, we use a simple yet representative
model. Biological principles are generally applicable and valid
across specific boundaries, therefore if we understand well a
relatively simple system, the information obtained can be used for
conducting work with other species.

Arabidopsis thaliana is a member of the crucifer family, and
it is thus a higher plant. The chromosome number (x = 5) is one
of the smallest among the angiosperms and so is its DNA content
(about 2×10^6 kbp/genome). The various features of the plant had
been summarized earlier (30), but a few of its advantageous
properties may be worth repeating here. The lifecycle of the
plant is very short; up to 8 generations can be raised annually.

The seed output is very high; a single plant may produce over 50,000 seeds. Its size is quite variable but one can raise to maturity 300 or more plants in a Petri dish (9 cm in diameter) or other similar vessel (Figure 1); in a 12 cm pot we can classify a seedling population of 1000 individuals.

This plant grows very well in aseptic culture, and in a standard test tube (15 x 160 mm) over 100 seeds may be produced. Callus and cell cultures are quite easy to establish, and the regeneration of normal plants from cultured cells can be routinely accomplished. Sometimes from a single callus several seedlings develop on an appropriate culture medium (1).

Auxotrophic mutants in the thiamine pathway can easily be obtained (18,27) and they can be utilized for genetic studies

Figure 1. Arabidopsis plants grown at high density in a 9 cm Petri dish containing "Promix" commercial medium, a mixture of peat moss, perlite and other unspecified material.

requiring high sensitivity. Unusual mutator loci affecting the plastome genetic material are known (29,34,36). Arabidopsis is an excellent host to Agrobacterium tumefaciens and cauliflower mosaic virus. These two systems are considered the most promising vectors for molecular genetic engineering.

CONVENTIONAL METHODS OF MUTATION INDUCTION

In our laboratory, we generally treat mature seed (24 hrs imbibed) with the mutagen. Because of the small size (about 20 mg/1000 seeds), the seed is placed in cloth bags tied with thread, in order to prevent losses during handling. Treatment with chemical mutagens is generally followed by 4 hours washing in running water.

Planning of Mutation Experiments

The "germline" of the plant. Although plants do not have a definite germline as it is known in animals, the sexual organs and the gametes develop from a limited number of cells present in the mature embryo. A mature seed of Arabidopsis (excluding the testa) contains about 6000-7000 diploid cells, usually only about two of these contribute to the formation of the inflorescence (gametes), and only about a dozen cells form the initials of the vegetative rosette of the plants (33). In other species these number may be different. As the seeds germinate and the seedlings grow the original number of the cells in the germline increases.

The number of cells in the germline can be determined by genetic means in hermaphroditic plants. If a mutation occurs in the single cell germline of the plant, that cell becomes heterozygous and its offspring will display 3:1 segregation. This segregation is frequently detectable in the fruits borne by the heterozygous individual (Fig. 2). In the majority of plants, mutation at a very large number of loci affects the production of leaf pigments, and pigment-deficient embryos can be easily identified in Arabidopsis by opening the fruits before the seed coat becomes brown and opaque. If the germline was represented by a single cell at the time of mutation, all the siliquae show both green and chlorophyll-deficient embryos by about 10 to 12 days after anthesis. Thus in a single culture, mutation induction and identification becomes possible. This fact enabled Gregor Mendel to classify segregation in peas heterozygous for cotyledon color (yellow and green) and cotyledon structure (smooth and wrinkled), saved him precious time and facilitated large-scale studies. The same principle was applied first for the detection of mutation in Arabidopsis by Müller (25).

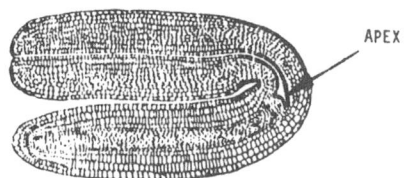

APEX

THE APICAL MERISTEM OF THE EMBRYO IN THE
MATURE SEED CONTAINS THE GERMLINE REPRE-
SENTED BY TWO DIPLOID CELLS

IF MUTATION OCCURS IN ONE OF THESE CELLS
THE DEVELOPING PLANT IS A GENETIC CHIMERA

MEIOSIS RESULTS IN SEGREGATION OF THE GENES AND FERTILIZATION
TAKES PLACE BY INDEPENDENT ASSORTMENT OF THE GAMETES; THE EM-
BRYOS WITHIN THE FRUIT DEVELOPED FROM THE HETEROZYGOUS SECTOR
OF THE CHIMERA DISPLAY MENDELIAN SEGREGATION BECAUSE THEY REP-
RESENT THE SECOND GENERATION

THE POOLED PROGENY OF THE TWO SECTORS IS EXPECTED TO SEGREGATE IN THE PROPORTION OF 7:1

Figure 2. The developmental path of the germline
in Arabidopsis

If the germline contains two cells at the time of the
mutation, half of the fruits are expected to show segregation and
half of them are expected to contain only wild-type embryos. If
we harvest all the fruits of each plant and sow separately the
pooled seeds of each individual, we can expect to obtain--in case
of recessive mutation--an approximately 7:1 segregation for
dominant:recessive phenotypes. This 7:1 ratio is the outcome of
the 4:0 (progeny of the nonmutant cells of the germline), and the

3:1 segregation (progeny of the heterozygous cell of the germline), in the pooled seed output of the individual (Fig. 2).

If the germline contains 4, 8 or 16 cells at the time of the mutation, the segregation ratios in the pooled progenies may become 15:1, 31:1, and 63:1, respectively. These wider ratios are observed because only 1 in 4, 1 in 8 or 1 in 16 cells, respectively, is heterozygous (Fig. 3).

These theoretical ratios are rarely manifested exactly because some of the mutant genes (especially the deletions and other aberrations of the chromosomes) fail to transmit according to mathematically predictable chance. The mutation may affect the function of the gametes or may interfere with the normal development of the embryo or with the germination of the seed. Therefore, it is not uncommon that the mutant class is reduced in number, and when the expected proportion of the mutants is 1/4, actually it may become 1/10 or even less. If we use only those progenies which contain point mutations the segregation ratio provides, however, a reliable estimate of the number of cells in the germline at a particular developmental stage.

This number of cells in the germline we called the genetically effective cell number (GECN) (19) because, in effect, only these cells contribute to the generative progeny. In an actual experiment involving approximately 500 ethyl

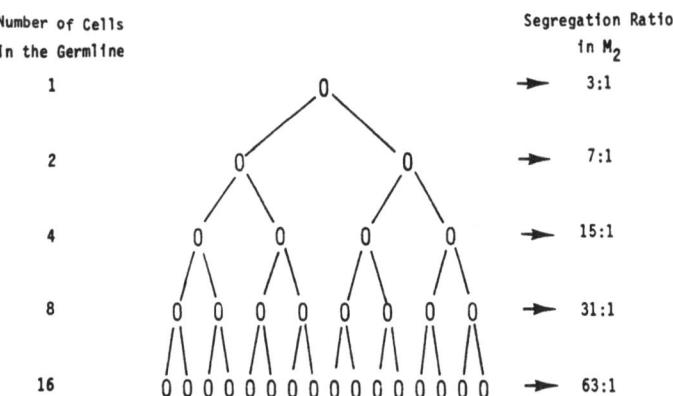

Figure 3. As the number of germline cells increases, mutations occurring at any particular cell will be represented in a proportionally smaller sector of the inflorescence and consequently in a smaller fraction of the progeny. In the segregation ratio column the proportions are based on the assumption that each germline cell gives rise to 4 offspring (or a multiple of this number).

methanesulfonate-treated mature seeds and their M_2 progenies, the range of segregation ratios varied, yet the median (the most frequent class) coincided with the 7:1 class, i.e. the GECN appeared to be 2 (Fig. 4).

Determination of critical population size. If we wish to obtain a particular type of mutant or a certain number of mutant classes, we must know the size of the experiment that will be sufficient to provide the type of mutants sought. However, if we have never seen a particular mutant, we have no firm basis for predicting how frequently it may occur, yet general experience indicates that the induced mutability at many loci is within the range of 10^{-4} to 10^{-5}. For our calculation, as an example, we use the higher frequency because at the lower range an ordinary electronic pocket calculator does not permit sufficient accuracy at its 10 digit capability.

For the determination of the required GECN, we must know (i) how many seeds are needed--at an expected mutation frequency of 1 x 10^{-4}--so that at least one mutation of the desired kind would

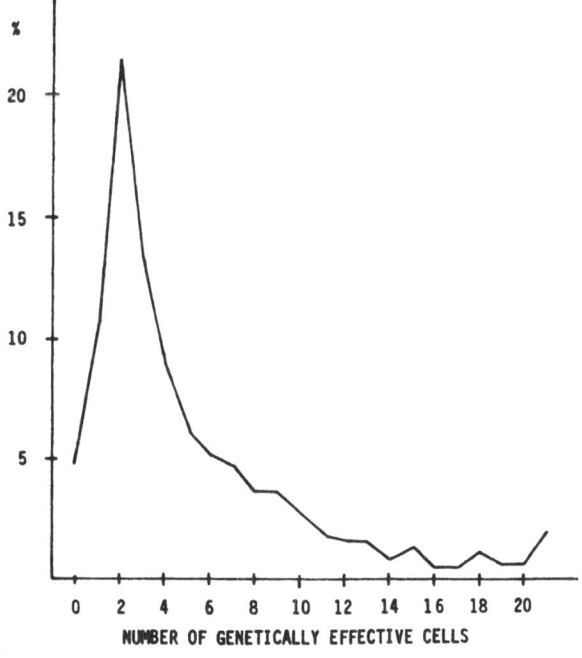

Figure 4. The frequency distribution of the genetically effective cell number based on actual segregation ratios among the populalion of 486 plants exposed to 0.3% ethyl methanesulfonate at the seed stage. No corrections were made for reduced transmission of the affected chromosomes.

occur, and also (ii) what is the probability of our finding such a mutation should it occur.

For an answer to the first question we must solve the general equation of

$$f^n = P$$

where f is the fraction of the GECN in which no mutations are expected, and n is the total number of germline cells which must be used in order that the failure-rate of recovering at least one heterozygous mutant cell would not exceed P. In other words, we determine what population of cells assures us that not all the germline cells will remain unaltered, i.e., at least one heterozygous cell will occur.

An example: we chose the P to be 1/100, the f = 9999/10000 and the value of n is sought. Then $f^n = P = (9999/10000)^n$, hence n = (log 1 - log 100)/(log 9999 - log 10000) = -2/-0.0000434 ≈ 46,083. In case the genetically effective cell number per seed is 1 we need 46,083 seeds to have 0.99 probability for the induction of at least one mutation of the desired type. (If the GECN = 2, obviously, 46,083/2 seeds are enough.) This number suffices, however, only if 100% of the mutations can be recovered.

If the mutant allele segregates in the second generation in a normal Mendelian fashion then the wild type and recessive individuals will occur in the proportion of 3:1 among the descendants of that mutant germline cell. Thus if a single individual is tested in the progeny of each heterozygous GEC (genetically effective cell), there is 0.25 chance that it will be homozygous recessive. If we test a larger number of individuals among the segregants, the chances increase for identifying one individual with the recessive phenotype. The actual probability can be determined again exactly the same way as we have shown above ($P = 1 - f^n$), e.g. if we test 16 individuals in the segregating progeny the probability of finding at least one recessive is $P = 1 - (3/4)^{16} = 0.989977$. In case the segregating allele is dominant and only 2 individuals are tested the $P = 1 - (1/4)^2 = 0.9375$. Table 1 lists the minimal probabilities for various population sizes. In a similar manner we can determine the probability of finding at least one recessive individual in the segregating progeny of plants containing a GECN of 2. Thus e.g. $P = 1 - (7/8)^{16} = 0.881933$.

Obviously, if the required minimum M_1 population size is 10,000 but the efficiency of recovery of a mutant in M_2 is only 0.25, then we must use 4 x 10,000 GECN in the M_1 generation (and not 10,000).

Table 1.   The probability of recovering at least one mutant
in the progeny of a diploid germline cell heterozygous
for a recessive of dominant mutation.

	Number of Individuals Tested in the M_2 Generation										
	1	2	4	6	8	10	12	14	16	24	48
Recessive	0.250	0.437	0.683	0.822	0.899	0.943	0.968	0.982	0.989	0.998	0.999999
Dominant	0.750	0.937	0.996	0.999	0.9999						

The efficiency of mutant isolation. We have seen (Table 1)
that the efficiency of recovery of the mutant is much greater if
the size of the individual families is large. If each family in
M_2 is represented by a single individual, there is only 0.25
chance of finding a recessive individual, whereas if the number of
individuals in each heterozygous family is 48 the probability of
recovering at least one mutant increases to 0.999999.

Obviously, it would be desirable to recover all the mutations
induced but the experimental means are not unlimited, and we have
to find the conditions under which the largest number of different
mutants can be isolated. Thus, for example, if we are limited to
testing only a total of 576 individuals heterozygous for a
different recessive gene, our goals can be pursued in different
ways. We can test 576 families, each with a single individual, or
we can screen for mutants only in 24 families, each represented by
24 offspring (24 x 24 = 576). In the former case the probability
of recovering a recessive individual is only 0.25, in the latter
that chance increases to 0.998, at least. Following the first
alternative, we can expect to recover 576 x 0.25 = 144 different
mutants in M_2, choosing the latter we can expect 24 x 0.998 =
23.95 different mutants, albeit the total number of mutants in the
latter case may also be 144 (but each will occur in about 6
identical copies). At other family sizes the efficiency of
recovery will be in between these two extremes (Table 2). The
calculations indicate that by using large number of families but
each with small number of individuals, the effectiveness of
recovering different mutants may increase by a factor of about 6.

Arabidopsis has about two diploid cells in the germline at
the stage of the mature seeds. In an experiment we used 0.2%
ethyl methanesulfonate for the induction of mutation and searched
for different mutants in populations comprising approximately 1000
individuals. The 1000 plants in each series were derived from 42
to 1000 families. The design of the experiment and the results
are tabulated (Table 3). Although the frequency of mutation
identified per family was lowest when only a single individual was
tested per family, the total number of different mutants was the
largest in the two series where only 1 or 2 individuals were

Table 2. In the event that each family represents the
descendants of a heterozygous individual, the
size of the family determines the probability
of recovering independent mutations.

Number of Individuals in Each Family	Number of Families	Expected Independent Recessive Mutations Recovered	
		Number	Percent
1	576	144.00	100.00
4	144	98.35	68.35
8	72	64.73	44.95
16	36	35.60	24.72
24	24	23.95	16.63

tested per family, just as expected on the basis of theoretical
considerations.

In the numerous but small family series we failed to recover
most of the mutations induced, nevertheless we obtained more types
of mutants than in the series where only few families with large
number of offspring were tested.

Cost Effectiveness of Mutation Experiments

Theoretical considerations and actual experiments have shown
that more different mutants can be isolated by using a relatively
large number of M_2 progenies with only a minimal number of
individuals in each family. Obviously therefore the choice of the
proper balance between the size of the M_1 and M_2 generations is
not a scientific but an economic decision. This economic decision
will, however, determine the success of the scientific efforts.

Table 3. Mutation experiment with Arabidopsis in
which approximately identical numbers of
seeds (1000) were tested at different
family sizes (GECN$\cong$2).

Number of Families Tested	Number of Seeds Planted of Each M_2 Family	Number of Mutations		Frequency of Mutations per Family	Mutation Rate Per Genome
		Observed	Expected		
42	24	11	11.5	0.262	0.0655
125	8	23	22.7	0.184	0.0460
250	4	30	28.6	0.120	0.0300
500	2	40	32.5	0.080	0.0200
1000	1	36	34.7	0.036	0.0090

Frequently, the cost of producing an M_1 individual involves less labor (or other investment) because fewer special efforts (or screening procedures) are required than with an M_2.

No matter what the cost of production is of an experimental plant, large numbers of individuals in M_2 contribute significantly to the cost of a mutation (Table 4). The most economical M_1:M_2 ratios are underlined in the body of this table. According to these calculations, there seems to be no advantage in testing more than 4 individuals per germline cell progeny even if the cost of raising an M_1 individual exceeds four times that of the cost of each M_2 individual. In view of this finding it is quite surprising that most of the mutation experiments reported in the literature involve M_2 family sizes far exceeding the most economical ones.

Reduction of the cost of mutant isolation. One of the major costs of producing the M_1 generation arises from keeping the individual pedigrees separate. Most of this expense can be eliminated by planting and harvesting large numbers of progenies in bulk. The seed of Arabidopsis is small, and when it is suspended in a viscous agar solution it can be spread onto the surface of the culture dishes with the aid of a separatory funnel. This way several 100,000s of seeds can be planted reasonably uniformly within an hour. Smaller quantities of seeds are

Table 4. Cost effectiveness of mutation experiments. In the calculations the cost of production of a plant was assumed to vary between 1 and 4 in both M_1 and M_2 generations. The cost of a mutation is determined by the sum of the cost of the M_1 and M_2 generations and divided by the theoretically expected probability of detection depending on the number of individuals in the M_2. The figures in the body of the table represent the calculated cost of an identified mutation.

Number of Individuals in the M_2 Families →	1	2	4	8	12	16	24
Size of $M_1 + M_2$ →	24+24	12+24	6+24	3+24	2+24	1.5+24	1+24
Expected No. of Mutations Recovered →	6	5.244	4.098	2.697	1.936	1.484	1
Cost of M_1:M_2							
1 : 1	8	6.865	7.321	10.011	13.430	17.183	25
1 : 2	12	11.442	13.177	18.910	25.826	33.356	49
1 : 3	16	16.018	19.034	27.809	38.223	49.528	73
1 : 4	20	20.595	24.890	36.707	50.620	65.701	97
2 : 1	12	9.153	8.785	11.123	14.463	18.194	26
3 : 1	16	11.442	10.249	12.236	15.496	19.205	27
4 : 1	20	13.783	11.713	13.384	16.529	20.216	28

frequently placed in vials containing only water and distributed with Pasteur pipets.

In our laboratory, the seed is generally harvested in bulk by threshing the fruits over paper sheets. The harvested material is stored in open vials or beakers in an air-conditioned room in order to assure longevity. Seeds kept at a combined temperature (F) and relative humidity value below 100 germinate well for 4 to 5 years or longer. (If the room temperature is 65° F, the relative humidity must remain below 35%.)

From the bulked seed representative samples can be easily withdrawn by volume, and the relative contribution of each family can be determined on the basis of the Poisson distribution. Thus, if we know that the mass-produced M_1 generation involved 1000 individuals and approximately 2000 seeds are withdrawn from the bulk, theoretically only 13.5% of the individuals will be missed and 86.5% of the M_1 families will be represented by at least one seed. In case we wish to withdraw an average of 4 seeds per family, about 95% of the M_1 individuals will be represented albeit only about 56.7% of the families will be represented by exactly 4 seeds. The lack of representation of a small fraction of the families will be compensated for partly by the families which were favored by the samplings, and thus have a higher chance to display at least one of the mutants expected to be there. A bulk-planted experiment is shown in Figure 5.

Reversion Assays

In Arabidopsis conditional lethal thiamine auxotrophs are available at at least four loci and these can effectively be screened for revertants (Fig. 6). Large quantities of seed can be readily obtained from the homozygous recessive stocks by feeding the mutants with the vitamin by periodic spraying of the pots. Usually, the mutants perish on soil within about 2 weeks after emergence whereas the revertants—sometimes after a short period of log—continue to grow because the mutant sector can supply the readily diffusable vitamin to the entire plant (Fig. 7). For these types of experiments, it is indispensable to use genetically marked stocks, so the revertants may be distinguished from the contaminants (Fig. 8) by some conspicuous morphological traits.

Besides the genetic markers, we can take advantage of organization of the germline for the safe identification of the true revertants. If back mutation occurred at the two cell stage of the germline, in the progeny of the revertant, segregation for the selected marker would result in a ratio of 5 auxotrophs and 3 prototrophs. If the reversion took place at later stages of the development of the plants, the second generation of the revertants should display an even larger proportion of the auxotrophs

Figure 5. Mutants can be identified in populations
planted at a density of one plant per 0.5 cm^2.

relative to the prototrophs. By the combination of these
criteria, even if only a single revertant is obtained, its nature
can be definitely determined. Thus, the precision of
identification of revertants in these diploid plants exceeds that
of any haploid organism.

Arabidopsis populations of several millions of individuals
can be screened without difficulty. The efficiency of the
screening for reversion can be doubled by using autotetraploid
auxotrophs.

Selective isolation of dominant and recessive forward
mutations. Any compound which interferes with the germination of
the seeds (such as herbicides) can be used as a selective agent
for the isolation of resistant mutants. In a 500 ml Erlenmeyer
flask five million or more seeds can be readily germinated in a
liquid medium. By selecting a critical concentration of the
poison, only the mutants are expected to germinate. After a
period of about 2 weeks the seed suspension can be filtered
through an appropriate mesh hardwarecloth and only the resistant
mutant plants--which germinated--are expected to be trapped on the
filter. After this mechanical separation, the seedlings can be
transferred to culture media for seed production.

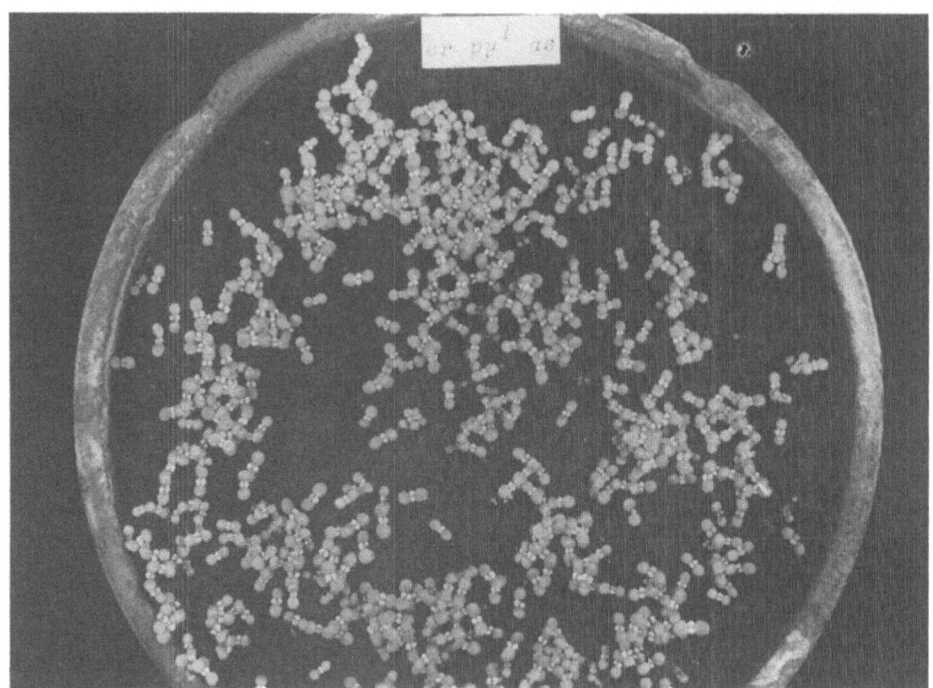

Figure 6. For reversion experiments genetically marked material
must be used. These seedlings are auxotrophic for 2-methyl,
4-amino, 5-hyroxymethyl pyrimidine and carry the flanking
morphological markers er and as, 5 and 9 m.u. from the py
locus, respectively. In such an experiment where revertants
are selected among auxotrophs in a 12 cm in. diameter pot,
1000 seedlings can easily be screened because only the
revertants can survive.

Determination of Mutation Rates

In haploid organisms or in haploid cells of diploids the rate
of mutation is generally expressed as the mutant fraction of the
survivors. Such a calculation may frequently lead to an
overestimation of the actual induced mutation rate because of the
differences in the survival of the treated and the control
populations. For example, if in a haploid organism revertants of
auxotrophs are screened, such a procedure may lead to erroneous
estimates. Obviously in the control the survival is better, thus
if in the concurrent control 10 mutants are found per 100
surviving cells and in the treated series also 10 mutants are
observed among the 50 survivors the rate of mutation in the
treated appears to be twice the rate of the control. In this

Figure 7.   The well-grown revertant plant is conspicuous
among the bleached and dying sibs.

case, however the mutation rate may be the same and only the
survival rate be different. If we express the mutation on the
basis of the survivors in the treated series, no increase of
mutation rate in the treated would be evident.

     In Arabidopsis we generally express the mutation rate of both
treated and control series on the basis of the survivors in the
treated populations.

     In diploids--especially in plants--the calculation of
mutation rate is carried out in a number of different ways (28).
Most of these procedures are undesirable because mutation rate
must always be expressed on a genome basis. Such a calculation is
possible, however, only if we know the number of the genetically
effective cells in the germline at the time of the mutation. Thus
in Arabidopsis, if at the time of the induction of mutation there
are two diploid cells in the germline, the number of mutational

Figure 8.   This growing plant among the dead auxotrophs is a
            contaminant because it is not only a prototroph but fails to
            display the two markers <u>er</u> and <u>as</u>.

events is divided by the product of the number of survivors and
the number of genomes (4 in this example). By using this
procedure, we are permitted to compare the rate of mutation at a
gene controlling a specific biochemical step in organisms as
diverse as <u>Arabidopsis</u> and prokaryotes. Such a comparative study
has indicated that the mutation rate for thiamine auxotrophy in
<u>Penicillium</u> and <u>Arabidopsis</u> is practically the same. This, of
course, should be expected because in plants and fungi the
biosynthetic path of thiamine appears to be the same, and
presumably the size of the enzymes and the size of the genes are
also similar and only a difference in the genetic repair system
would be responsible for the difference in the mutation rates if
such should be encountered.

"Fluctuation Test" and Determination of the Differences in Mutation Rates

When the rate of mutation is very low, it is sometimes difficult to determine whether the mutants observed were induced by a particular treatment or if the experiment just selected and enriched the mutants present before treatment. This problem may require more attention when the isolation is carried out in bulked progenies as outlined above, rather than by the pedigree method.

Bacterial geneticists distinguish between preexisting and induced mutations by the statistical procedures of the fluctuation test (20). According to this test, a large number of cells maintained in bulk culture are seeded on plates containing a substance of unknown mutagenic properties and in a parallel experiment several individually propagated cultures (initiated from small inocula) are exposed to the same unknown agent in a manner identical to the bulk series. If the mutations are uniformly distributed on the plates in both series, then there is a possibility (although not a proof) that all the mutations arose as a consequence of the exposure. When, however, in the second series there is a much larger variance among the plates than in the first set of plates, the treatment obviously did not induce the mutations, but only selected the mutants which pre-existed in the population. The larger variance of the second series indicates that spontaneous mutations occurred at random. On the plates seeded with preexisting mutants a large number of mutant colonies arose, whereas the progenies picked from the vessels free of spontaneous mutations displayed no mutants at all on the plates containing the agent under study for mutagenic properties. In the first series the uniform distribution of the mutants is expected because the individual plates represent a random sample of the homogeneous original bulk population.

Keeping in mind the lessons of these bacterial tests, we carried out a "fluctuation test" for ascertaining the apparent low mutagenicity of ascorbate. In each of 120 pots we grew to maturity an average of 325 plants that were exposed to 0.2% sodium ascorbate at the seed stage. Similarly in 121 pots we grew to maturity untreated control plants at approximately identical population density. Both untreated and control series were harvested in bulk by pot, and then an average of about 2 seeds per individual were withdrawn from the seed lots and tested pot by pot for mutations (Table 5).

A total of 89 mutants were observed in the treated pots, and 41 were found in the untreated control. If phenotypically identical mutants were found in the same pot-progeny, these replicates were considered as results of the same mutational event although this could not be proven. After this correction for

Table 5.   Distribution of mutants in an experiment where
the seeds of <u>Arabidopsis</u> were treated with
ascorbate, and the data of the
concurrent untreated control.

Number of Mutants/Pot	0	1	2	3	4	5	6	
TREATED SERIES								Total
No. of Pots	64	35	14	3	3	1	0	120
Expected by Poisson	57.14	42.40	15.73	3.89	0.72	0.11	0.01	120
CONCURRENT CONTROL								Total
No. of Pots	89	23	9	0	0	0		121
Expected by Poisson	86.21	29.23	4.95	0.56	0.05			121

independent mutational events, the mean and the variance of the two series (treated [T] and control [C]) became $M_T = 0.666$, $V_T = 0.639$ and $M_C = 0.289$, $V_C = 0.255$, respectively. The calculations indicated a rather uniform distribution of the mutational events in both series and in accordance with the Poisson series because the variance and the mean were similar. Thus, the increase in mutation frequency in the treated series could be safely attributed to the treatment.

According to the t test the difference between the frequency of mutations is significant at the 0.001 level.

The mutation rates on genome basis in the two series (treated and control) were 80/336,000 = 0.000238, and 35/338,000 = 0.000103, respectively. The actual mutation rates must have been higher because the sampling procedure used permitted a study of only 63.2% (according to the cumulative terms of the Poisson series) of the descendants of the two germline cells. Furthermore, by testing a single offspring from each germline cell only about 25% of the mutations were actually detected. These biases affected, presumably, both series to the same extent, therefore did not much change the relative values. The absolute values of the mutation rates were expected to be about five times higher than the calculated ones.

The statistical significance of the difference between frequencies can be determined also by calculating the standard errors of the two frequencies according to the formula

$$s_f = \sqrt{p(1-p)/n}$$

where p stands for the proportions observed. According to the normal distribution, standard errors ±1s, ±2s, and ±3s define 68.26, 95.45 and 99.73% of the populations around the means, respectively. Since the two observed mutation frequencies plus-minus their 3s do not overlap, there is more than 0.9973 probability that the two frequencies are not identical.

Alternatively, the critical ranges of the mutation frequencies may be determined by the values of the formula

$$\sum_{r=x_1}^{x_1+x_2} \binom{x_1+x_2}{r} p^r (1-p)^{x_1+x_2-r} \leq \propto$$

where x_1 and x_2 are the number of mutants in two series to be compared, p = the relative size of population with x_1 mutations, r is the critical value for each level of probability ($\propto$) that x_1 must exceed in order that the difference between the mutation rates should be considered statistically significant at the level of $\propto$. This computation is somewhat cumbersome but Kastenbaum and Bowman (14) constructed tables for the critical values. Applying this procedure of calculation to the data provided above: x_1 = 80, x_2 = 35, p = 120/(120+121) = 0.4979 and N = 80 + 35 = 115. In case the critical value for $\propto$ = 0.01 is 70 or 71, i.e. less than the observed x_1 = 80, the null hypothesis that the two frequencies are identical is not supported by the data. Thus, the difference is assured at the 0.01 level of significance.

Estimation of the Occurrence of Undetected Mutations

Some mutagenic agents have very low potency yet in very large populations they may cause significant number of genetic alterations. We may need, therefore, to know what is the possible number of mutations which escaped detection in a certain size of experiment. Such information may be extremely important before we classify a particular agent as genetically neutral if very large numbers of individuals are routinely exposed to it (e.g. as a food additive or drug).

The theoretical considerations are as follows: each biological event is expected to occur with a certain average number (with equal tails around the mean if the distribution is "normal"). We can observe, however, only the event which actually is visible and we cannot directly ascertain the other half of the distribution, i.e. the nonoccurring cases. These non-events are expected to have the same frequency as the events (at binomial

distribution), and thus the probability of their occurrence is
half of the total probability. Actually, we are dealing here with
a sampling process, and we assume that at low frequency only half
of all events were observed in the samples of limited size. (When
the events follow the Poisson distribution, the situation is more
complicated. Also, we make the assumption that the penetrance
[transmission] of the mutation as well as the expressivity of the
gene(s) are 100%.)

With all these stipulations and restrictions in mind, we will
consider a population of n genomes and take the frequency of the
mutant genomes as (1 - q) and if we chose the level of probability
as 0.999, we can derive the value of q as follows:

$$(1 - q)^n = (1 - P)/2 \text{ hence}$$

$$q = 1 - \sqrt[n]{(1 - P)/2},$$

and if we find no mutations at all among 1000 genomes, after
substitutions we get

$$q \approx 1 - \sqrt[1000]{(0.001)/2}$$

$$\approx 1 - 992428 \approx 0.007572$$

Expressing this estimate in another way, we can say that even if
no mutations could be observed in the trial, there is a high
probability that 7 mutations escaped our detection in that
population of 1000. Also, there is the same chance that in a
genome population of 1,000,000 the number of mutations could be as
high as 1,000,000 x 0.007572 i.e. 7,572, a very large number
indeed.

RAPID ASSAY OF MUTAGENICITY

By November 1977 the American Chemical Abstracts Service
(CAS) registered 4,039,907 distinct chemicals and at that time the
list was growing at a weekly rate of about 6000. Currently,
approximately 60,000 to 70,000 chemicals are used on a large scale
by industry, medicine and agriculture (22). Many of them pose a
potential health hazard to living creatures, including man. There
is no known assay system that would identify the genetic and
carcinogenic hazards with absolute certainty (32).

Arabidopsis mutation assays can be used for studying these
potentially hazardous agents by an embryo test. We routinely

expose the seeds to the compound, and identify mutations in the
fruits of the plants which developed from the treated seeds
(Figs. 2 and 9).

For these assays we generally grow at least 200 plants in a
tall Petri dish and before the seed coat turns brown, two fruits
at opposite sides of each plant are opened with the aid of sharp
forceps and the number of fruits containing pale or white embryos
are recorded. This way each plant provides information on
mutation in an average of two cells (4 genomes). Each test
includes a concurrent untreated control of approximately identical
size.

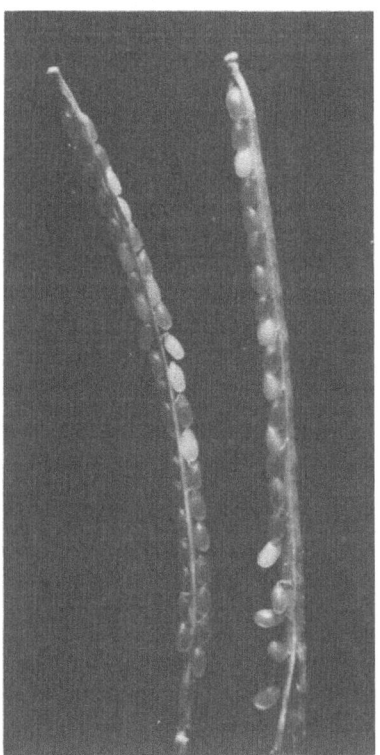

Figure 9. Detection of mutation in the fruits of Arabidopsis
        segregating for white cotyledons. The immature siliqua was
        opened by sharp forcepts at the base and the carpels removed
        to examine the rows of embryos. Usually during this handling
        some of the embryos fall from the replum and therefore these
        gaps appear on the photograph.

This assay requires only one generation because the segregating embryos within the fruits of the M_1 plants represent the M_2 offspring. Such an assay generally can be completed within four weeks. Although this period is much longer than that required by several other types of short-term mutagen assays, the efficiency compares favorably with the other tests because a very large number of loci are responding.

ESTIMATION OF GENE NUMBER IN ARABIDOPSIS

The DNA content of the genome does not permit at this time a valid estimate of the number of genes because a variable fraction of it is highly reiterated and does not seem to be expressed as the common structural genes.

A more reasonable inference regarding gene number can be obtained on the basis of mRNA complexity. Even this approach has its shortcomings because of the large variation in transcription at different developmental stages and the substantial deviations in the estimates of the different laboratories (37). Kamalay and Goldberg (13) estimated on molecular basis the gene number of tobacco to be about 60,000, whereas Kiper, Bartels and Köchel (15) found an acting gene number of about 10,000 in parsley leaves and about 14,000 in barley leaves.

The number of mutable genes can be estimated also on the basis of overall mutation frequencies if an estimate is available on the average rate of mutation per locus. The overall rate of mutation can be easily obtained in experiments with high doses of efficient mutagens. We have found the ethyl methanesulfonate (0.3%, 15 hours exposure) induced mutation frequency for alterations expressed at the embryo stage of Arabidopsis to be 0.47. An additional 0.12 fraction of mutations were detected at the later stages of development. Thus, approximately 0.59 fraction of the exposed genomes expressed one or another type of mutation. Since mutations are distributed according to the Poisson series, the average frequency of mutation can be extrapolated from the size of the zero mutation class (e^{-u}) that appeared to be $1 - 0.59 = 0.41$ in this particular study. From this figure the average frequency of mutation is $u = -\ln 0.41 \approx 0.892$.

The average frequency of mutation at four thiamine loci based on numerous independent mutations appeared to be 3.2×10^{-5}. Thus, the total number of genes that mutated in this experiment can be estimated as $0.892/(3.2 \times 10^{-5} = 27,875$. Also, we can see that the majority of the genes are expressed already at the embryo stage and less than one third of the loci expressed mutations during later stages of development.

These estimates of gene number in Arabidopsis are obviously loaded with some error. We do not know how representative is the mutability of these four thiamine loci relative to the rest of the gene pool. Certainly, there are significant differences in stability among these four genes. Another source of error is that when we determined overall mutation rates, we did not examine how many of the mutations might be allelic. This latter source of error may not be very substantial, however, because even if we suppose that some of the loci were highly mutable, e.g. mutated by a frequency of 10^{-2}, the repeated occurrence of mutants was expected only at a rate of 10^{-4} and thus affecting only the fourth decimal in the estimate. Even if we observed 100 such mutable loci, the biases of our estimates would be affected by relatively small fractions.

A MUTATOR SYSTEM IN ARABIDOPSIS

The cells of higher plants contain three different types of genomes. The bulk of the genetic material is in the nucleus but important genes are located in the plastids and mitochondria. These organelles have much smaller genomes than the nucleus but they occur in large numbers, each containing several presumably similar copies of genetic information. Detection of new mutations in the plastids and mitochondria is more difficult because the mutant phenotype can generally be observed only after some cycles of sorting out.

According to Röbbelen (35) spontaneous mutation of the plastids of Arabidopsis occurs only in a 0.0001 fraction of the plant population. If we conservatively estimate that a plant contains 10,000,000 cells and an average of 40 plastids per cell, the spontaneous rate of mutation may be as low as about 1×10^{-11}.

The chm nuclear gene (29,34), however, greatly enhances the rate of mutation in the plastids (Figure 10). On some leaves an uncountable number of sectors may occur as the combined result of forward and reverse mutation and sorting out (Fig. 11).

Under the electron microscope, 5 to 6 different types of plastids may be seen in single cells (Fig. 12). In the presence of the chm nuclear gene no plants are found that would fail to display mutant sectors. The chm gene had been located in linkage group 3 approximately 15 map units from gl (glabrous).

By using the sectorial plants as females and outcrossing them to chm[+] gl[+] males the mutator chromosome can be easily removed and the cytoplasmic alterations stabilized. Generally each cell contains several plastome mutations but in the absence of the mutator the sorting out may be quite rapid. Although the

Figure 10. Variegated plant homozygous for the plastid mutator
         allele chm³.

Figure 11. Arabidopsis leaves homozygous for some of the chm
         alleles display both sharp sectors and transition areas of
         active sorting out. In the 3rd leaf from left presumed
         revertant cell colonies are visible.

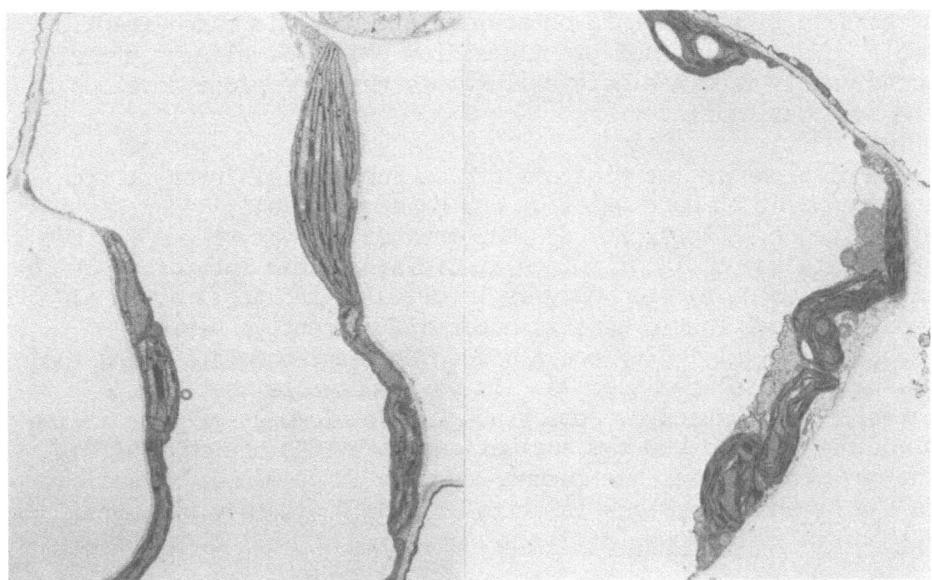

Figure 12.   (Left) A single mature leaf cells with several
    different types of plastids induced by the cloroplast mutator
    nuclear gene.   (Right) Mutant cell containing the same type
    of abnormal albeit green and functional plastids.

literature frequently interprets organellar sorting out by some
stochastic models (5), according to our observations fairly
complete sorting out could occur in several sectors during the
life of a plant even if the mutator had been just introduced and
became homozygous.

    According to the predictions of Michaelis (23), ten times as
many cell generations would be required as the number of sorting
units to reach the homoplastidic state.   On the basis of such a
process, assuming 30 plastids per cell, the complete sorting out
would take more than 2^{300} cell divisions, i.e. approximately 2×10^{90} cells.   Since a small plant may contain no more than 10^6
cells, it would take over 80 plant generations to achieve
homogeneous sectors.   (In view of the fact that each plastid
contains numerous copies of the ring DNA the complete sorting out
should take a much longer time.)   As a matter of fact, however,
homogeneous sectors occur within less than a generation.   The
recovery of homoplastidic mutant sectors in the germline may be
more difficult, because many, but not all, plastome mutations
interfere with the development of the photosynthetic apparatus
and/or cause male or female sterility.

Some of these problems can now be overcome by maintaining the sectors in tissue culture on appropriately supplemented aseptic media. Regeneration of plantlets from callus is also practical. Occasionally from a single callus more than one plant develops to the seedbearing stage.

Mutations in the plastome are of substantial interest from the viewpoint of pure genetics and have potentially great significance in application. The mutator activity needs further study with the tools of molecular biology. This mutator function may be based on altered enzymes involved in the replication or repair of the plastid genetic material. Recently, several laboratories (11,16,26) found gross DNA rearrangements among the plastomes of related species. It is conceivable that the mutability is caused by some transposable elements present in the plastome (17) and the chm nuclear gene provides the control for the movement whereas the mutant may make it possible, in analogy to the Ds and Ac systems in maize and the P elements of hybrid dysgenesis in Drosophila (21,38).

Recently it was found that the resistance of plants to some herbicides containing as active ingredient dichlorophenyl dimethylurea (diuron) or some of the triazines (atrazine) is determined by their binding to a 32 kd peptide of the thylakoid membrane. The PSII-3 photogene has been precisely mapped to the chloroplast DNA by Herrmann (see in 31). In the atrazine-resistant mutant of plants this specific binding site of the herbicide is lost and this trait is maternally inherited (2,6). Similar maternal inheritance has been observed for resistance to the diuron type compounds (10).

In addition, there is the possibility of inducing other types of mutants involved in the photosynthetic system. Furthermore, about 50% of the cellular proteins are contained within the plastids and a good fraction of them are presumably coded by the plastid DNA. This type of a mutator system may thus open up a new approach for breeding agronomically valuable crops.

ACKNOWLEDGEMENT

Contribution from the Missouri Agricultural Experimental Station. Journal Series Number 9368.

REFERENCES

1. Acedo, G.N. (1983) Laboratory Experiments in Plant Tissue Culture, University of Missouri, Columbia, MO.

2. Arntzen, C.J., K. Pfister, and K.E. Steinack (1982) The
    mechanism of triazine resistance: Alterations in the
    cloroplast site of action, In Herbicide Resistance in Plants,
    La Baron, H., and J. Gressel, Eds., Wiley, New York,
    pp. 185-214.
3. Auerbach, D. (1976) Mutation Research, Halsted Press, New
    York.
4. Broetjes, C., P. Koene, and J.W. Van Veen (1980) A mutant of a
    mutant of a mutant of a..: Radiation-induced mutation in
    mutation-breeding programme with Chrysanthemum morifolium
    Ram. Euphytica 29:525-530.
5. Chapman, R.W., J.C. Stephens, R.A. Lansman, and J.C. Avise
    (1982) Models of mitochondrial DNA transmissions genetics and
    evolution in higher eucaryotes. Genet. Res., Cambridge
    40:41-57.
6. Darr, S., V.S. Machado, and C.J. Arntzen (1981) Uniparental
    inheritance of a chloroplast photosystem II polypeptide
    controlling herbicide binding. Biochim. Biophys. Acta
    634:219-228.
7. De Greve, H., J. Leemans, J.P. Harnalsteens, L. Thia-Toong,
    M. De Beuckeleer, L. Willmitzer, L. Otten, M. Van Montagu,
    and J. Schell (1982) Regeneration of normal fertile plants
    that express octopine synthase, from tobacco crown galls
    after deletion of tumour-controlling functions. Nature
    300:752-755.
8. de Serres, F.J., and J. Ashby, Eds. (1981) Evaluation of
    Short-Term Tests for Carcinogens, Elsevier/North Holland, New
    York.
9. De Vries, H. (1901) Die Mutationstheorie, Veit, Leipzig,
    Germany.
10. Galloway, R.E., and L. Mets (1982) Non-Mendelian inheritance
    of 3-(3,4-dichlorophenyl)-1, 1-dimethylurea-resistant
    thyaloid membrane properties in Chlamydomonas. Plant
    Physiol. 70:1673-1677.
11. Gordon, K.H.J., E.J. Crouse, H.J. Bohnert, and R.G. Herrmann
    (1982) Physical mapping of the differences in chloroplast DNA
    of five wild-type plastomes in Oenothera subsection
    Euoenothera. Theor. Appl. Genet. 61:373-381.
12. Howell, S.H. (1982) Plant molecular vehicles: Potential
    vectors for introducing foreign DNA into plants. Annu. Rev.
    Plant Phys. 33:609-650.
13. Kamalay, J.C., and R.B. Goldberg (1980) Regulation of
    structural gene expression in tobacco. Cell 19:935-946.
14. Kastenbaum, M.A., and K.O. Bowman (1966) The minimum
    significant number of successes in a binomial sample. Oak
    Ridge Natl. Lab. ORNL-3909, pp. 1-69.
15. Kiper, M., D. Bartels, and H. Köchel (1979) Gene number
    estimates in plant tissues and cells, In Plant Systematics
    and Evolution, Genome and Chromatin: Organization, Evolution,

Function, Nagl, W., V. Hemleben, and F. Ehrendorfer, Eds., Springer, New York, pp. 129-140.

16. Kung, S.D., Y.S. Zhu, and G.F. Shen (1982) Nicotiana chloroplast genome III. Chloroplast DNA evolution. Theor. Appl. Genet. 61:73-79.

17. Lewin, R. (1983) Promiscuous DNA leaps all barriers. Science 219:478-479.

18. Li, S.L., and G.P. Redei (1969a) Thiamine mutants of the crucifer Arabidopsis. Biochem. Genet. 3:163-170.

19. Li, S.L., and G.P. Redei (1969b) Estimation of mutation rates in autogamous diploids. Radiat. Bot. 9:125-131.

20. Luria, S.E., and M. Delbrück (1943) Mutations of bacteria from virus sensitivity to virus resistance. Genetics 28:491-511.

21. Marx, J.L. (1983) A transposable element of maize emerges. Science 219:829-830.

22. Maugh, T.H., II (1978) Chemicals: How many are there? Science 199:162.

23. Michaelis, P. (1955) Über Gesetmässigkeiten der Plasmon-Umkombination und über eine Methode zur Trennung einer Plastiden-, Chondriosomen-, resp., Sphaerosomen-(Mikrosomen)- und einer Zytoplasma-Vererbung. Cytologia 20:315-338.

24. Micke, A. (1975) Induced mutations in plant breeding. Can. J. Plant Sci. 55:865.

25. Müller, A.J. (1963) Embryonentest zum Nachweis recessiver Latalfaktoren bei Arabidopsis thaliana. Biol. Zbl. 83:133-163.

26. Orihara, Y., and K. Tsunewaki (1982) Molecular basis of the genetic diversity of the cytoplasm in Triticum and Aegilops I. Diversity of the chloroplast and its lineage revealed by the restriction pattern of ct-DNAs. Japan J. Genet. 57:371-396.

27. Redei, G.P. (1965) Genetic blocks in the thiamine synthesis of the angiosperm Arabidopsis. Amer. J. Bot. 52:834-841.

28. Redei, G.P. (1970) Arabidopsis thaliana (L.) Heynh. A review of the genetics and biology. Bibliogr. Genet. 20:1-151.

29. Redei, G.P. (1973) Extrachromosomal mutability determined by a nuclear gene locus in Arabidopsis. Mutation Res. 18:149-162.

30. Redei, G.P. (1975) Arabidopsis as a genetic tool. Annual Rev. Genet. 9:111-127.

31. Redei, G.P. (1982) Genetics, Macmillan, New York.

32. Redei, G.P., G.N. Acedo, and S.S. Sandhu (1983) Sensitivity, specificity and accuracy of the Arabidopsis assay in the identification of carcinogens. (in this volume, pp. 691-710).

33. Redei, G.P., and S.L. Li (1969b) Effects of X-rays and ethyl methanesulfonate on the chlorophyll b locus in the soma and on the thiamine loci in the germline of Arabidopsis. Genetics 61:453-459.

34. Rédei, G.P., and S.B. Plurad (1973) Hereditary structural alterations of plastids induced by a nuclear mutator gene in Arabidopsis. Protoplasma 77:361–380.
35. Röbbelen, G. (1962) Plastomemutationen nach Röntgen-bestrahlung von Arabidopsis thaliana (L.) Heynh. Zeitschr. Vererb.-Lehre 93:25–34.
36. Röbbelen, G. (1966) Chloroplastendifferenzierung nach geninduzierter Plastommutation bei Arabidopsis thaliana (L.) Heynh. Z. Pflanzenphysiol. 55:387–403.
37. Solignac, M., and J. Genermont (1982) Les estimations du nombre de genes et la complexification des genomes. Ann. Biol. 21(3):210–273.
38. Spradling, A.C., and G.M. Rubin (1982) Transposition of cloned P elements into Drosophila germ line chromosomes. Science 218:341–347.
39. Stadler, L.J. (1928) Mutations in barley induced by X-rays and radium. Science 68:186–187.

MUTAGENESIS STUDIES WITH CULTURED MAMMALIAN CELLS:

PROBLEMS AND PROSPECTS[1]

Ernest H. Y. Chu, I-Chian Li and Jiliang Fu[2]

Department of Human Genetics
University of Michigan Medical School
Ann Arbor, Michigan 48109-0010, U.S.A.

SUMMARY

Environmental mutagenesis is thought to play an important role in hereditary and somatic diseases in man. Numerous studies have been made in recent years assaying the mutagenic response of cultured mammalian cells, with the hope of better understanding the effects of environmental agents on human populations. Different test systems have been developed to detect either gene or chromosome mutations at the cellular and molecular levels. Although some of the heritable variations could be epigenetic, direct evidence is available demonstrating that true point mutations indeed occur in mammalian cells. It is now possible to identify specific molecular alterations in mammalian cell mutants arising in vitro.

The development of methods for quantifying single gene mutations in cultured mammalian cells has been considered in this presentation, with special reference to (a) the choice of cell material, (b) development of selective markers, (c) the various factors that affect the expression of mutations, and (d) the application of appropriate statistical methods for the estimation of mutation rates. The role of DNA replication and repair in the mutational process can be illustrated by a concomitant change in mutability, mutagen-sensitivity and other metabolic and genetic alterations in mutant cells that are defective in DNA metabolism.

[1]Supported by research grants GM 20608 and CA 26803 from National Institutes of Health, United States Public Health Service.
[2]Visiting Scholar from the School of Public Health, Sichuan Medical College, Chengdu, Sichuan, China.

315

Recent technical advances have permitted analysis of somatic
cell mutants at the protein and nucleic acid levels. Amino acid
sequence analysis of certain enzymes and studies on DNA structure
changes in cell mutants arising in vitro are now feasible.
Furthermore, methods have been developed to screen for various
classes of electrophoretic mutants occurring at more than 40 well
defined gene loci among cell clones isolated after chemical or
physical mutagenesis. Unlike the selective procedures which detect
single-locus mutations leading to the loss of a vital function,
this multilocus approach has permitted determinations of the
frequencies of mutations which result in variant proteins less
drastically modified from the wild type. Finally, human somatic
cell mutation rates, both spontaneous and induced, at various loci
may be estimated by changes in protein products visualized by two-
dimensional polyacrylamide gel electrophoresis. The results
obtained may afford us the opportunity to compare the mutation
rates of functionally vital and selectively neutral markers, as
well as those in human somatic and germ cells using the same panel
of protein markers.

I. INTRODUCTION

There is an increasing awareness that environmental
mutagenesis plays an important role in hereditary and somatic
diseases in man. Although the impact of mutations in the germline
on health-related problems is well known (cf. 100), many non-
geneticists have been extremely skeptical towards the role of
somatic mutations in human diseases. However, recent
demonstrations (107,112,149) that an oncogene in human bladder
cancer differs from its normal counterpart by a single nucleotide
base change adds further credence to the belief that somatic
mutation can play a significant role in human diseases such as
cancer.

It has been argued that a substantial percentage of
malignancies may result from chemical and physical mutagens
existing in the environment (16). Similarly, atherosclerotic
lesions may result from events related to cellular transformation,
which could also arise from environmentally based mutagenesis
(11,141). To identify the potentially hazardous substances, to
estimate human risks and to minimize human exposure to these
substances, various test systems for carcinogenic or mutagenic
potential of environmental agents have been developed. Because
every conceivable assay system has drawbacks, new techniques are
constantly being evaluated. Mammalian cell culture is in some
respect closer to a revertant system for measurement of mutation in
man than bacteria, yeast or Drosophila. At the same time, it
offers the advantage of ease of handling, low cost and rapidity of
assay, as compared to whole animal studies (35).

Mammalian cell culture is also a convenient system for the study on the mechanisms of human mutagenesis. The role of DNA replication and repair in mutagenesis can be analyzed by the use of cell strains developed from human subjects with hereditary disorders, by isolation of mutant cell lines defective in DNA metabolism, and by experimental modification of these fundamental cellular processes with extrinsic agents. Finally, if the rates of spontaneous and induced mutations in cultured human somatic cells can be determined accurately, they can be compared with the human somatic mutation rates in vivo as well as the human germinal mutation rates (102). If a "correction factor" can be established with confidence, the cultured mammalian cells can serve to estimate the genetic load in human populations.

In this paper we will examine the characteristics of mammalian cells in culture in relation to their use in environmental mutagenesis studies. Since several recent books and reviews on the subject have been published (13,27,33,35,54,58,59,99), we shall attempt to minimize duplication and to complement the earlier reviews with newer materials and perhaps different emphasis.

II. NATURE OF SOMATIC VARIATION IN VITRO

Mutation is generally defined as a permanent hereditary alteration in the genetic material. Short of a direct demonstration of a change in the nucleotide sequence and in the absence of sexual phenomenon in somatic cells, heritable phenotypic variation observed in these cells could be interpreted as the result of either genetic alteration or modulation of gene expression. The latter contention has received some recent experimental support from studies on DNA methylation and gene expression. For instance, treatment of human-mouse somatic cell hybrids with 5-azacytidine results in reactivation of genetically inactive human genes on the X chromosome (98). Harris (51) showed that brief exposure of 5-bromodeoxyuridine-tolerant, thymidine kinase deficient (TK$^-$) Chinese hamster cells to 5-azacytidine resulted in a massive conversion to HAT$^+$ state (ability to grow in a medium containing hypoxanthine, aminopterin, and thymidine), suggesting that the induction of revertants might have resulted from changes in DNA methylation patterns. Similarly, 5-azacytidine treatment has been used to reactivate the expression of HPRT (hypoxanthine guanine phosphoribosyltransferase) on the inactive gene in Chinese hamster cells (44). However, the correlation between DNA methylation and genetic inactivation observed in cells and cell hybrids in vitro may not represent accurately the mammalian organism. Wolf and Migeon (158) used cloned DNA fragments from human X chromosome as probes and examined the restriction enzyme digestion pattern of DNA from placentas and cultured skin fibroblasts. They found that DNA methylation of the

X chromosome changes with replication, is not correlated with the
number of X chromosomes or transcriptional activity, and is less
stable and more prevalent than the human X chromosome present in
interspecific cell hybrids. They further showed that 5-azacytidine
did not cause allelic reactivation of the two heterozygous X-linked
genes in normal female fibroblasts in culture. This is in contrast
to the findings, mentioned earlier, in established cell lines or
cell hybrids. Thus, DNA methylation may explain only in part the
observed, heritable changes in the phenotypes of cultured somatic
cells. Furthermore, this type of modification is not genetic
because no alteration in the basic genetic information is involved.

On the other hand, a large body of experimental evidence
reviewed elsewhere (18,37,54,99,125) suggests a genetic basis for
mammalian somatic cell variation. A number of criteria has been
applied to assess the mutational origin of phenotypic variation
observed in cultured mammalian cells (26,31,125). These criteria
include: (a) random occurrence, (b) retention of stable phenotype
in the absence of selection, (c) induction with mutagens in a dose-
dependent manner, (d) mutagenic specificity, (e) conditional
lethality, (f) interallelic complementation, (g) changes in the
activity and physicochemical or immunological properties of
specific gene products, (h) changes in amino acid sequence of the
gene product and (i) changes in nucleotide sequence of the gene.
To be sure, some of the supporting evidence is circumstantial and
indirect, and not all criteria have been met in every instance of
the observed somatic variation. However, evidence is rapidly
accumulating to support the view that many phenotypic variants
isolated from cultured mammalian cell populations indeed arose from
gene mutation.

III. DEVELOPMENT OF METHODS FOR MUTATION INDUCTION IN CULTURED
MAMMALIAN CELLS

A. Choice of Cell Material

Numerous cell lines, markers and selective systems are now
available; some of the most frequently used are listed in Table 1.
Primary low passage fibroblasts are obtained by cell dissociation
of embryo, newborn or adult animal and human tissues. They are
diploid and grow through 20 or 30 cell doublings, after which the
cells grow more slowly, becoming senescent at 50 or more divisions.
Plating efficiencies are sometimes low, but low passage cells tend
to retain more completely the enzyme activities characteristic of
tissue of origin and are useful for metabolic studies (60). In
addition, Syrian and Chinese embryo fibroblasts serve as the same
target cells for studies on mutagenesis and cell transformation
(9,62). Human fibroblasts also express genetic disorders in
culture, particularly those affecting DNA repair (39,64,67,89,90).

Table 1. Cultures of Mammalian Cells and
          Selective Markers for Mutation Assays[1]

Cell Type	Selective Marker[2]
Low passage cells	
Syrian hamster embryo cells	AG, OUA
Human skin fibroblasts (normal)	AG, OUA, DAP, Dip
Human fetal lung fibroblasts (normal)	OUA
Human skin fibroblasts (Xeroderma pigmentosum)	AG
Established aneuploid cell lines	
Chinese hamster lung fibroblasts, V79	AG, OUA, TG, DAP
Chinese hamster ovary fibroblasts, CHO	Emt, Tri, Dip, TG
Mouse lymphoma cells, L5178Y	AG, BrdUrd, TFTrd
Mouse fibroblasts, 10 T 1/2	AG
Human fibroblasts, chemically transformed	TG, OUA

[1]Modified from 54.

[2]Abbreviations: AG, 8-azaguanine; BrdUrd, 5-bromodeoxyuridine;
DAP, 2,6-diaminopurine; Dip, diphtheria toxin; Emt,
emetine; OUA, ouabain; TFTrd, trifluorothymidine; TG, 6-
thioguanine; Tri, trichodermin.

Established cell lines have been obtained from normal tissues
of rodents and other animals. Chinese hamster ovary (CHO) and lung
(V79) cell lines with near-diploid but altered karyotypes show
rapid cell growth and high plating efficiency. They have been used
widely in studies on the actions of radiations and mutagenic
chemicals. Another frequently used cell line is the mouse lymphoma
L5178Y, which is highly adapted to cell culture conditions, can be
cloned in soft agar, and has good growth characteristics (72). The
mouse embryonic fibroblast 10 T 1/2 cell line remains sensitive to
cell density inhibition in spite of transformation to a permanent
line. The BHK cell line consists of fibroblasts from Syrian
hamster kidney and has the advantage of retaining a capacity for
enzymatic activation of carcinogenic hydrocarbons (132,150).

There is evidence (113) that the repair of DNA damage differs
both in process and extent between human and rodent cells.
Clearly, our knowledge of the similarities and differences in DNA
replication and repair of human and other mammalian cells is
lacking because of the dearth of human cell mutants for these
fundamental processes. Therefore, studies of mutagenesis with
human diploid fibroblasts, lymphoblastoid cell lines or other

established near-diploid cell lines should reduce the extrapolation to man.

B. Selectable Genetic Markers

Drug-resistant markers are probably the easiest to obtain by direct selection in the presence of a cytotoxic agent. Different classes of chemicals have been used, including the analogs of purine and pyrimidine bases and nucleosides, isotopically labelled metabolic precursors, folic acid antagonists, antibiotics, mitotic poisons and membrane-active drugs. The underlying rationale is that cells resistant to an antimetabolite or to an agent either are defective in certain steps of metabolic pathways or produce abnormal proteins that fail to bind to the extrinsic selective agent. There may be, however, more than one mechanism of resistance for a specific drug. In addition, there are instances in which permeation of the drug with the cell may be impaired. In some cases selective procedures are available for both forward change to drug resistance and reverse change to drug sensitivity (e.g., 6-thioguanine resistance); in other cases back selection is not possible (e.g., ouabain resistance).

Of the 120 published papers containing primary data on mutagenesis by chemicals in V79 cells, mutations (resistance to purine analogs) affecting the HPRT locus were measured in 73% of the papers; mutations (resistance to ouabain) affecting the membrane-associated, sodium-potassium-activated adenosine triphosphatase (Na^+/K^+ ATPase) in 24%; and the rest were reports on auxotrophic and prototrophic mutations, temperature sensitive mutations, and mutations measured by resistance to various chemicals (13). Drug resistance also appears to be the most widely used selective marker for environmental mutagenesis using other cell types (Table 1).

On the basis of classic studies with corn and _Drosophila_, the rates of spontaneous mutation may differ among different loci. Evidence has been presented (94) that in certain mutator strains of CHO cells the spontaneous mutation rates were 5- to 50-fold higher than in parental cells for two markers (resistance to 6-thioguanine or to ouabain) but the rates for two other markers (reversion of proline auxotrophy to prototrophy and forward mutation to emetine resistance) were unaffected. As shown by several investigators (3,21,135), certain agents such as X- and gamma-rays fail to induce ouabain resistant mutants in mammalian cells, although they are capable of inducing 6-thioguanine-resistant mutants in the same cell type. Therefore, it appears that mutation rates may be site specific for mammalian cells as well. From both the theoretical and practical points of view, one of the important tasks is to develop additional selectable genetic markers for comparative

mutagenesis among loci and for more accurate assessment of
mutagenic potential of environmental chemicals.

C. Expression of Mutations

Genetic complementation studies with somatic cell hybrids
formed between independently isolated mutants of different
(e.g. HPRT$^-$ X TK$^-$, 81) or similar function (e.g. gly$^-$ A X gly$^-$ B,
69) indicate that nearly all the mutants isolated in various
mammalian cell lines so far examined behave like recessive
characters. However, there are a few markers, such as resistance
to ouabain (3,6,92), vinblastine sulfate (50) or alpha-amanitin
(84) which appear to be dominant or codominant. In theory,
dominant mutations should be readily recognized in diploid
mammalian cells under appropriate selective conditions.

X-linked recessive mutations may be expressed because of the
functional haploidy of this chromosome in male mammals. X-linked
recessive mutations in somatic cells from female mammals can also
be expressed in clones in which the X chromosome bearing the normal
allele is genetically inactivated (87). The gene for HPRT is
located on the X chromosome of human and rodent cells. The
phenotypes of cells with low or undetectable HPRT activity is
recessive in Chinese hamster cells containing more than one active
X-chromosome (18,23,57), although regulatory gene mutations that
suppress all active HPRT genes could be dominant (66).

The expression of autosomal recessive mutations in diploid
cells could be due to homozygosis as a result of mutations of both
alleles of the gene. This is a highly improbable, if not
impossible, event. In normal human fibroblasts, the frequency of
chemically induced 2,6-diaminopurine resistant mutants (affecting
the expression of the adenine phosphoribosyltransferase locus) in
homozygous strains was approximately the square of the induced
frequency in heterozygous strains, as would be expected for the
occurrence of two independent mutational hits (130). Clive et
al. (33) constructed mouse lymphoma cell lines heterozygous at the
thymidine kinase locus (TK$^{+/-}$) in which homozygous TK$^{-/-}$ were
obtained at much higher frequencies than could be obtained in wild
type TK$^{+/+}$ lines. However, in aneuploid cell lines such as CHO and
V79 there are extensive chromosomal structural rearrangements
(68,134) so that effective hemizygosity for many loci exists as a
result of either deletion or inactivation (45,129).

Similarly, the expression of autosomal recessive character
could be the result of mitotic recombination followed by
segregation, changing from a preexisting heterozygous state to the
homozygous recessive state. Such mitotic recombinational events in
cultured cells have not yet been investigated due to the lack of
identifiable linked markers.

D. Spectrum of Induced Mutations

Even though the change is genetic, an alteration in the phenotype of the cultured cells could be the result of a genic or chromosomal event. For instance, Cox and Masson (36) have shown that x-ray-induced mutations to 6-TG resistance in cultured human diploid fibroblasts consist mostly of chromosomal deletions or rearrangements. Similar results were obtained in radiation-induced human lymphocyte antigen (HLA) variants in a lymphoblastoid cell line (71). Hozier and his associates (55) have made cytogenetic analysis of L5178Y/$TK^{+/-} \to TK^{-/-}$ mouse lymphoma mutagenesis assay system. Mutant $TK^{-/-}$ colonies form a bimodal frequency distribution of colony sizes for most mutagenic or carcinogenic test substances. Large-colony $TK^{-/-}$ mutants with normal growth kinetics appear karyologically identical within and among clones and with the $TK^{+/-}$ parental cells. In contrast, most slow-growing small colony $TK^{-/-}$ mutants have a readily recognizable chromosome rearrangement involving the mouse chromosome 11 on which the TK locus is located.

In defining a mutagen screening system, many factors must be examined to distinguish between genetic and epigenetic events and certain basic criteria must be met if a true mutational assay system is to be claimed. The types of mutation that are measured, whether genic, chromosomal or both, should be delineated. In fact, for screening of mutagenicity of environmental chemicals, forward mutations which encompass a broad spectrum of mutational events are preferable to reverse mutations. On the other hand, reverse mutations can help to identify the mutagenic specificity of chemicals (Table 2, references 24,30,91).

IV. FACTORS AFFECTING MUTANT RECOVERY

A. Genetic and Metabolic Heterogeneity of Cells

Mammalian cells cultivated in vitro undergo spontaneous changes. Although normal human fibroblast strains and lymphoblastoid cell lines remain essentially at the euploid state, karyotypic alterations have been known to occur. Similarly, fibroblast strains derived from rat, Chinese hamster and Syrian hamster tissues are fairly stable during early periods of in vitro cultivation, but fibroblasts from the embryonic or adult tissues of the mouse show spontaneous chromosome changes soon after explantation (28). It is well known that established mammalian cell lines of both normal and neoplastic origin exhibit extensive chromosome changes in both number and structure. It is conceivable that mammalian cell populations maintained in continuous growth in vitro are heterogenous with respect to their genetic and metabolic state. For mutagenesis studies, these sources of variation can

Table 2. Revertibility of Azaguanine-Resistant Mutants
in Chinese Hamster Cells[1]

Mutant origin[2]	Reversion Mechanism[3]				No. of mutants
	BPS	+/-	SP	NON	
Spontaneous	8	1	6	10	25
MMS induced	8	0	7	3	18
EMS induced	4	0	6	9	19
ICR-170 induced	0	1	15	12	28
N-AcO-AAF induced	2	4	7	12	25
X-ray induced	9	12	10	29	60
				Total =	195

[1]Data taken from 24, 30 and 91.

[2]MMS, methyl methanesulfonate; EMS, ethyl
methanesulfonate; ICR-170, acridine half mustard; N-
AcO-AAF, N-acetoxy-2-acetylaminofluorene.

[3]BPS, base-pair substitution; +/-, base-pair insertion or
deletion; SP, reverts only spontaneously (reversion
mechanism unknown); NON, non-revertable: no reverse
mutations obtained in all the revertibility tests.

best be controlled by careful attention to variables that could
affect the response of cells to mutagens. These include: storage
of cells in a frozen state; frequent recloning of cells;
standardization of cell culture conditions; and periodic
examination of the growth rate, cell and colony morphology and the
karyotype. It is important to maintain, as much as feasible,
uniformity of the biological material to ensure reproducible
experimental results.

B. Phenotypic Expression

When cells are exposed to mutagenic agents, newly induced
lesions may not immediately display their altered phenotype but may
require a period of time prior to selection for maximum expression;
this phenomenon has been called phenotypic delay, as illustrated in
experiments on 8-AG or 6-TG resistance (1,29,38,90,103,137). The
phenomenon appears to be the culmination of a series of unknown
processes leading to the eventual recognition of the mutant.

Forward mutations, such as HPRT⁻ or TK⁻ which represent a loss of
function, could be the result of a spectrum of mutational events.
The lesion, such as a loss or change of a base, could be repaired
accurately thus producing no mutation, or misrepaired leading to a
change (mutation) in the normal nucleotide sequence. In theory,
two rounds of DNA replications are needed for "error replication"
to occur resulting in an altered base pair. Furthermore, cell
divisions are necessary to allow dilution or degradation of the
parental gene product(s) in the cell carrying the mutated gene.
There also may be a need to accumulate the altered gene product in
the mutant cell.

The length of this period can vary among different cell lines
and different genetic loci. For instance, in the case of 6-TG
resistance (HPRT⁻) of CHO cells induced by ethyl methanesulfonate
(EMS), O´Neill and Hsie (105) have demonstrated a lag period of 7
to 9 days for maximum expression. The expression time for the
HPRT⁻ phenotype in V79 cells is between 4 and 6 days (13,101,120),
but that in mouse lymphoma L5178Y cells is 14 days (115). The
expression time for ouabain resistance for the Chinese hamster V79
cell line is 2 days (19,20). However, variations up to 45 hours in
optimum expression time have been reported with UV light,
alkylating agents and hydrocarbon carcinogens (3,21). Usually the
yields of Oua^r mutants decline after the optimum expression time
due to a relative selective disadvantage of Oua^r cells as compared
to Oua^s cells in nonselective medium.

Li (77) has devised a simple technique for growth of CHO cells
as unattached cultures. The optimum expression time for 6-TG
resistance was shorter for unattached cells (6 days) than attached
cells (9 days). However, different new mutants may have different
growth rates and different selective advantage, relative to each
other and to the parental cells. Some mutants may have divided
faster than the parental cells and be over-represented; others may
be slow growers which may be missed in the enumeration of mutant
colonies at a fixed time. Still other mutants may not thrive at
all. Using the respreading technique, the mutation frequency is
therefore only an average estimate.

C. Cell Density (Metabolic Cooperation)

New HPRT⁻ mutants may be lost if too many cells are placed in
the culture dishes and the density of cells is great enough to
permit metabolic cooperation during the period of selecting for
resistance to guanine or hypoxanthine analogs (29,133,143,145).
This phenomenon is due to a transfer of metabolite, such as 6-
thioguanosine monophosphate, through the gap junctions on the cell
membrane from the dying parental cells to mutants, killing the
latter in the process. For CHO cells, for example, reconstruction

experiments have determined the maximum cell density ($2 \times 10^5/9$ cm dish) that would not affect HPRT$^-$ mutant frequency (105).

The problem of metabolic cooperation may be less serious in L5178Y mouse lymphoma cells grown in suspension due to the small number of gap junctions on the cell surface. Metabolic cooperation between HPRT$^+$ and HPRT$^-$ cells may be reduced or abolished in the presence of a tumor promoter (159).

For certain membrane markers, such as resistance to ouabain or abrin (77), there was no demonstrable metabolic cooperation. Arlett et al. (3) observed a decline of yield of ouabain-resistant mutants at high cell densities, but this may be related to starvation from overcrowding. For still other markers under different circumstances, cross-feeding in crowded cultures may actually increase rather than decrease the mutation frequency. Harris (49) showed that the incidence of spontaneous variation to puromycin resistance in pig kidney cells increased in proportion to population density. Large colonies or aggregates of sensitive cells grew progressively in puromycin at concentrations which destroyed the same populations completely when present in dispersed form. Harris concluded that the response to puromycin is conditioned by cellular interactions as well as by genetic susceptibility, although the genetic nature of puromycin resistance in these cells was not demonstrated.

D. The Type and Concentration of the Selective Agent

In our first study of chemical mutagenesis in cultured mammalian cells, we used 8-AG to select resistant variants (29). We observed that mutant yield depends on the concentration of the selective agent: the yield decreases with increasing concentrations of 8-AG. It was subsequently shown (133) that an unidentified serum factor reduces the toxicity of 8-AG at 37° C. Several measures to overcome this source of error with 8-AG selection have been adopted (see review: 13), including (a) the use of dialyzed serum in the medium to reduce the exogenous sources of native purines, (b) the use of high concentrations (> 30 μg/ml) of 8-AG, (3) the use of other purine analogs such as 6-TG or 8-mercaptopurine, both of which have a greater affinity for HPRT than does 8-AG (18,142). However, 6-TG has the potential disadvantage that it is toxic to some classes of HPRT$^-$ (8-AG^r) mutants, and may, therefore, reduce the total yield of mutants (25,42,119). Dialyzed serum has the disadvantage that it reduces the yield of some classes of mutants (111), lowers cloning efficiency and decreases the overall mutant frequency (88,109). Thus, two facts have become clear that should be borne in mind in quantitative mutagenesis studies: (a) the use of a particular selective agent at a fixed concentration may ensure more reproducible results, and (b) the mutant yield so obtained is only a minimal estimate.

E. Back Mutation and Preexisting Mutants

Mutant HPRT$^-_+$(8-AG^r) V79 cells that are HAT^s have been induced to revert to HPRT$^+$ phenotypes by various chemical mutagens (17,24,61,108). Revertant frequencies increased with the dose of mutagen (17,52,108). However, spontaneous reversion frequencies up to 4 x 10^{-3} revertants/colony-forming units were observed with some mutants (108). In addition, induced reversion frequencies ($\approx$500 x 10^{-6} survivors) were high for reversion of structural gene mutations (17,40,108). Furthermore, mutant levels of HPRT were found in revertants that remained 8-AG^s (40). These problems show that complete characterization of HPRT$^+$ revertant of an HPRT$^-$ mutant has not yet been reported. More recently, it was discovered that HPRT$^-$ V79 cells grown in HAT medium gave rise to survivors which were mostly due to amplification of the HPRT gene and only infrequently to true reversion at the locus (R. Fenwick, personal communication).

On the other hand, Hodgkiss et al. (52) have measured in V79 cells reversion in 6-TG^r clones by alkylating agents. These workers obtained spontaneous reversion frequencies of $\leq$ 2 revertants/10^7 survivors and induced frequencies of $\approx$4 revertants/10^5 survivors, which are much lower than forward mutation frequencies at the HPRT locus.

During the course of mutagenicity testing it is generally assumed that there was no back mutation, but this possibility needs to be verified. If it does indeed occur, the frequency of back mutation should be accounted for in an estimate of spontaneous and induced forward mutations.

The presence of pre-existing mutants of spontaneous origin may be troublesome in the evaluation of a weak mutagen. There are experimental procedures, however, such as recloning or pre-growth in counter-selective media, to eliminate or reduce the number of pre-existing mutants in cell populations prior to mutagenicity testing.

F. Effects of Mutagens at Low Doses

At very low doses of chemicals, the linearity of dose-effect curves has not yet been settled. In both CHO and L5178Y cell systems, dose-effect relationships within a certain range of concentrations have been obtained for many chemicals, such as ethyl methanesulfonate and ethyl nitrosourea (33,56). However, it is debatable whether a linear extrapolation to lower doses is either reasonable or desirable.

A number of biological factors may play important roles during the entry of a chemical (or its derivatives) into the cell and the

nucleus, before its interaction with the genetic material. The
question in testing is to find the lowest effective dose that can
induce mutation, not the theoretical threshold. The practical
problem is the chronic exposure to low doses, rather than the
demonstration of mutation induction after a single exposure at a
very low dose. It seems that a dose-curve using a series of doses
is more informative than the determination of a "doubling dose" of
the spontaneous frequency.

G. Metabolic Activation of Test Chemicals

     Some chemical mutagens and carcinogens are active without
enzymatic changes, whereas others such as polycyclic hydrocarbons,
nitrosamines, and aromatic amines require metabolic activation to
convert them to the reactive derivatives, the ultimate carcinogens
(cf. 97). Many cell types, including CHO, V79 and L5178Y cells,
are not capable of activating procarcinogens or promutagens.
Metabolic activation of chemicals can be achieved by adding a liver
microsomal homogenate to a culture (10,95) or by plating the cells
on a feeder layer of X-irradiated primary human or rodent cells
capable of metabolizing polycyclic hydrocarbons to the active form
(58,60,63,73). Tong and Williams (140) have developed a system in
which adult rat liver epithelial cells are used to detect genotoxic
compounds by mutations at the HPRT locus.

     The subcellular enzyme fractions, such as rat liver microsomal
S-9 fractions, are simpler to use and have been shown to activate a
wide variety of procarcinogens. However, it has been shown that
the profile of the metabolites and DNA adducts formed after
metabolism of various potent carcinogens are different from that of
intact cells. This indicates that the use of sub-cellular
fractions does not accurately simulate in vivo conditions. It is
also known that for certain classes of chemicals metabolic
activation or inactivation is species specific. If it can be
metabolically activated in vitro will it also be activated in
humans? The major dilemma here is that, given a chemical about
which little is known, it is difficult to ensure that the
activation system used will detect its possible mutagenic
potential. Furthermore, the concentration of the S9 fraction used
in the cell-mutagenesis experiments may affect cell viability, thus
introducing a further source of biological variability to
quantitative mutagenesis.

V. ESTIMATION OF MUTATION RATES

     Mutation rates (μ) in cultured mammalian somatic cell
populations have generally been estimated by applying the
fluctuation test of Luria and Delbrück (86). The test is a
"stochastic", or random, approach. Mutation is a rare event and

its occurrence may be regarded as following a Poisson distribution. The first term of the distribution, i.e., the probability that no mutation has occurred, or P_o, can be used to determine m, the number of mutations in a cell population after a certain period of growth. As $P_o = e^{-m}$, or $m = -\ln P_o$, mutation rate is calculated as $\mu = (m \cdot \ln 2)/(N_t - N_o)$, where N_o and N_t are, respectively, the sizes of the initial and the final cell populations of a culture. μ is expressed as the probability of mutations per cell per division cycle.

In addition to the P_o method, there are other methods that can be used to calculate μ from the result of a fluctuation test. These methods have been described in Lea and Coulson's paper (76) on the distribution of mutant numbers in parallel cultures. The calculations are tedious and must refer to the tables and formulae given by the authors. Furthermore, each method has a larger variance of m than that derived from the P_o method. For these reasons, the P_o method has been the most popular method for the calculation of mutation rate. It should be noted, however, that μ estimated by the P_o method is biased, even though the bias is about two orders of magnitude smaller than μ (80). There are indications that μ estimated by other methods is also biased (5). Hence, when comparing mutation rates of different cell lines or experiments, the difference of two μ values cannot be assumed to distribute normally and the Student t test cannot be used for equality testing. To fulfill the need, we (80) have proposed a test statistic and have applied it to experimental results for a test of equality of mutation rates in different cell lines.

Another approach to estimate μ is the measurement of the increase of the "mutant fraction" in a large cell population during a certain period of logarithmic growth. This is a "deterministic" approach, because in order to obtain an increase in mutant fraction with time, at least one mutation must occur during every generation of population doubling. To make this possible, a large cell population must be maintained, usually at least ten times greater than the reciprocal of the mutation rate (85). Consequently, the experimental scale becomes such that it is sometimes technically difficult. The method was originally designed for bacterial systems by Shapiro (118) and Newcombe (104), and further developed by Stocker (131) and Armitage (4,5). To our knowledge, it has not yet been applied to the mammalian cell system, due probably to the difficulty of experimentation and the complications involved in mutation rate calculations. Armitage (5) offered a complicated formula for the analysis of the regression relationship between the multiple mutant fractions (MF) and the number of cell generations (g). We (unpublished data) have simplified the formula by applying the power series expansion of Taylor and by taking the first term to make $MF = \mu \cdot g \cdot \ln 2$. Thus, if MF is plotted against g, a straight regression line is obtained whose slope is mutation rate.

The mutation rate is expressed as number of mutations per cell per generation. A statistical method is available for the equality test of μ obtained from either two cell lines or two experiments. It is based on the existing statistics for the comparison of two linear regression coefficients. As compared to the fluctuation analysis, the mutant fraction method is more laborious in experimentation, but is straightforward, using our simplified formula, in the calculation of μ and equality testing. However, before the deterministic approach can be routinely used for the estimation of mutation rates in mammalian cell systems, preliminary studies should be made e.g., (a) to determine the most appropriate experimental design for obtaining a reliable rate estimate, (b) to assess the effect of different growth rates of the normal and mutant cells on the estimation of μ, and (c) to convert the unit of mutation rate from one approach to the other, because the time units used for the two approaches are different.

In Section VII.C below, we shall describe the application of the two-dimensional polyacrylamide gel electrophoresis technology to the study of somatic mutations in human cells. The deterministic approach appears to us to be the only feasible way to estimate the rates of somatic mutations at multiple loci, as measured by this technique. On each gel, a large array of cellular proteins is assayed under the identical nonselective conditions. Let the average mutation rate for a total of S loci be μ mutations per cell per generation per locus. After one generation of cell growth when the population size N becomes 2N, the number of mutations occurred would be N · μ · S, which is also the number of mutant cells generated. Since a number of proteins as specified by S loci is surveyed simultaneously, cells possessing a mutation at any one locus can be regarded as mutants. In this sense, mutation rate can be viewed as having been increased S fold, or lambda = μ · S mutations/cell/generation/S loci. Because mutations in this situation are probably neutral in nature, no selective disadvantage against them can be assumed and the growth rates of the various mutants are expected to be equal to that of the normal cells. The regression analysis of the increase of the proportion of mutant cells for several generations is the same as that for the single locus with equal growth rates, except that now the slope of the linear regression line is equal to gamma. Thus μ = slope/S.

VI. GENETIC PREDISPOSITION TO MUTAGENESIS

Mutation is much more than a chemical or configurational change in DNA; it is a biological process deeply enmeshed in the structural and biological complexities of the cell and modulated by both internal and external factors. The manifestation of a primary DNA lesion to a substantial alteration in gene expression follows a complex process, which acts as a sieve that allows only a

proportion of the DNA lesions to proceed toward the final product
of a mutant clone. Many of the enzymes of DNA replication and
repair contribute to the fate of primary lesions, and the
specificity of these enzymes contributes to the diversity of the
mutational response.

The use of mutants of bacteria has aided us in linking various
DNA repair, recombination and replication to mutagenesis. Advances
made with various mutagen-sensitive and DNA repair defective
mutants of yeast, Drosophila and rodent cells have given
investigators new insights into the many complex ways by which
genes can influence the basic processes needed to maintain the
fidelity of DNA replication and repair of damaged DNA (cf. 41).

A. The Role of DNA Repair in Mutagenesis

DNA repair is controlled by many genes. The best known
example in man is the hereditary disorder xeroderma pigmentosum
(XP) in which affected individuals suffer from a greatly increased
sensitivity of the skin to UV- light. XP mutations affect the
frequency of UV and chemically-induced mutations. This has been
demonstrated using low passage human skin fibroblasts from
individuals affected by this disorder (89,90). The increased
mutation rates observed in XP cells show that genetically
controlled repair processes act on premutational damage in DNA
molecules. However, the number of mutant cell lines from animals,
including mice and men, is very limited, making it difficult to
apply genetic methods for investigating mutagenesis to mammalian
cells.

More recently, development of in vitro techniques has allowed
a wide variety of indirect and direct means to isolate mutagen-
sensitive and DNA repair defective mutants of mouse, Chinese
hamster and human cells (15,22,75,116,121,122,138,139). Results of
comparative studies indicate that not only are mammalian DNA repair
mechanisms different in many respects from those of bacteria or
lower organisms, but also that some human cells studied are quite
different from mouse or Chinese hamster cells in their DNA excision
repair (113). Therefore, there is an urgent need to isolate human
mutants and to use these, in combination with the rodent mutants,
for cross-species comparisons. This will help to aid cross-species
predictions of potential mutagenic effects of environmental
chemicals, where studies on the whole organism can be made in
animals, but not in humans.

B. Fidelity of DNA Replication

Mechanisms other than DNA repair can influence mutagenesis in
mammalian cells. For instance, Bloom's syndrome has been shown
recently to be hereditary "mutator" mutants (45,146). The defect,

which was thought to be a DNA repair defect (117), might in fact affect DNA replication (47,48,70).

Studies on various rodent cell lines also indicate that mutagenesis can be influenced by perturbations of DNA replication either directly or indirectly. Liu et al. (82) have shown that a V79 cell mutant selected for its resistance to aphidicolin, a specific inhibitor of DNA polymerase alpha, is characterized by slow growth, UV sensitivity, and hypersensitivity to UV-induced mutations. More recently, the purified DNA polymerase alpha from the mutant, as compared to that from the parental cells, has been shown to be more resistant to aphidicolin and to have a decrease in the K_m for dCTP (83). Their results also indicate that an altered DNA polymerase alpha may be intrinsically mutagenic during normal semiconservative replication as well as during UV-induced repair synthesis.

When the pools of DNA precursors become unbalanced, DNA synthesis is increasingly error-prone. For instance, the fidelity of ϕX174 replication in vitro can be altered by varying the 2´- deoxyribonucleotide triphosphate (dNTP) levels (148). Mutation rate is increased in thymine-starved bacteria (14,128) and yeast (7,8,74). Similarly in mammalian cells, imbalances in intracellular DNA precursor pools as a result of either an excessive supply of exogenous nucleosides (12,110,111), or a genetic alteration (93-95,147), have led to a significant elevation of mutations.

In our laboratory, we have isolated mutant clones from V79 cells which show structural gene mutations either at the CTP synthetase (32) or the dTMP synthesis locus (79). In both mutant types, altered dNTP pools and elevated rates of spontaneous mutations have been demonstrated (80). Conceivably, these and other hypermutable cell strains will be very useful in increasing the sensitivity for detecting certain classes of mutagens. It is of particular importance to obtain human cell mutants defective either in DNA repair or DNA replication for such purposes.

VII. ANALYSIS OF MAMMALIAN CELL MUTATIONS AT THE MOLECULAR LEVEL

Thus far, we have discussed the assessment of somatic mutations of cultured mammalian cells at the phenotypic level, based on the selection for a few traits such as drug resistance, nutritional requirement and antigenic variation. Although there is increasing evidence that somatic variation observed in cultured mammalian cells may be largely due to alterations in the genetic material, the nature of specific changes at the molecular level in mutant cells arising in vitro has been inferred but not demonstrated. However, technical advances in recent years now permit analysis of somatic cell mutants at the protein or nucleic acid levels. In the following, we shall discuss four investigative techniques that may render the cultured mammalian cells amenable to molecular analysis.

A. Amino Acid Sequence Analysis of HPRT

The enzyme HPRT catalyzes the formation of purine nucleotides, IMP and GMP from 5-phosphoribosyl-1-pyrophosphate and the respective purine bases. In man, an inherited deficiency of this enzyme is associated with two distinct clinical syndromes: a virtually complete deficiency of HPRT activity has been described in patients with Lesch-Nyhan syndrome, while a partial enzyme deficiency is found in some male patients who present with hyperuricemia and a severe form of gout.

Human HPRT exists, in its native state, as a tetramer of identical subunits, encoded by a single X-linked gene locus (53). The complete amino acid sequence of HPRT from human erythrocytes has been defined (153). Each subunit of the enzyme is 217 amino acids long with a molecular weight equal to 24,470. The enzyme undergoes two post-translational modifications: acetylation of the NH_2 terminal alanine (153) and deamination of asparagine 106 (154). Jolly et al. (65) reported the isolation and preliminary characterization of cloned cDNA sequences of the human HPRT gene. The amino acid sequence predicted from the nucleotide sequence of the cDNA is in complete agreement with that defined by protein sequencing except that the NH_2 terminal methionine encoded by the initiator codon is absent in the mature enzyme.

Wilson and his coworkers (152,155) have purified HPRT from lymphoblastoid cells and erythrocytes and found the existence of 5 unique structural variants of human HPRT. More recently, the molecular basis for HPRT deficiency in one patient with a severe form of gout has been defined (156). This enzyme variant, called $HPRT_{London}$, has a serine to leucine substitution at position 109. This substitution can be explained by a single nucleotide change in the codon for serine (UCA $\rightarrow$ UUA). $HPRT_{Toronto}$ is an electrophoretic variant of the human enzyme that was isolated from a male patient with a partial deficiency of enzyme activity and a severe form of gout. Sequence analysis of a single peptide in the variant enzyme revealed an arginine to glycine substitution at position 50 (codon for Arg_{50} changed from CGA to GGA (157). It is, therefore, entirely possible that the molecular defects of HPRT mutants, spontaneous and induced, that arise in cell cultures can be identified in this manner.

Due to the degeneracy of the genetic code, information on a specific amino acid substitution of a peptide does not always permit inference to be made to a corresponding change in the nucleotide sequence. However, there is the possibility of finding mutational "hot spots" in the coding sequence of the gene. If this turns out to be the case, further analysis of the nucleotide sequence of selected regions by means of recombinant DNA technology would be more realistic in view of the size ($\leq$ 30 kilobases) of the human HPRT gene (65).

B. Multilocus Somatic Mutations Assayed by One-Dimensional Gel Electrophoresis

The number of selective mutation systems generally used to calculate mutation frequencies and assess the genetic risk of environmental agents is limited. The selective phenotypes studied usually involve a mutational event which results in the loss of enzyme activity that is vital to the survival of the cell. It is possible that mutational events resulting in variant protein less drastically modified may be inducible at much higher frequency. For instance, mouse myeloma antigen binding variants which could be attributable to point mutation causing single amino acid substitutions have been shown to arise spontaneously at a rate of 10^{-4}/cell/generation (34).

Using one dimensional gel electrophoresis, Siciliano and his coworkers (123,124) have randomly isolated hundreds of clones of CHO cells with or without prior mutagenesis and screened for different classes of mutagenic events at over 40 well defined loci. The average spontaneous frequency of these classes of electrophoretic mutants was on the order of 10^{-4}. The mean induced mutation frequency following ethyl methanesulfonate treatment or ultraviolet irradiation was 4.4 x 10^{-3} (2) and 0.7 x 10^{-3} (124), respectively. Furthermore, there is a wide range in the susceptibility of gene loci to mutagens. Thus, this multilocus mutation assay system not only provides a sensitive and effective method for detecting a broad range of mutational events in mammalian somatic cells but also has the potential for assaying somatic mutations, both in vitro and in vivo, in animals and man.

C. Multilocus Somatic Mutations Assayed by Two-Dimensional Gel Electrophoresis

Genetic markers suitable for quantitative cell mutagenesis are few in number. Even the most extensively studied HPRT system is not free from technical limitations. The mutation rates thus far reported vary widely among genetic loci, cell types and laboratories. At the University of Michigan the two-dimensional polyacrylamide gel electrophoresis (114) and computer-assisted pattern recognition technology (126,127) are being applied and developed for an analysis of human germinal mutation rates. In our laboratory, we are applying this technical development to assay somatic mutations occurring in cultured human somatic cells. A

broad array of cellular proteins, some of which have already been
identified, are being assayed under the identical non-selective
conditions. It will then become possible to compute mutation rates
for a large variety of loci. The technique should be applicable to
different cell types, including human lymphoblastoid cell lines,
fibroblasts and transformed cell lines. It may permit a direct
comparison of mutation rates obtained from parallel studies on the
same panel of protein markers with leukocytes in the study of human
germinal mutations (102). The technique, when fully developed,
should also be useful for a determination of mutation rates in
hypermutable human cell strains and for mutagenicity testing of
environmental agents.

D. DNA Alterations in Mammalian Cell Mutants

A few mammalian genes have now been cloned and, where
applicable, may be used to probe their mutant alleles in the
mammalian genome. For instance, Graf and Chasin (43), using a
cloned cDNA probe, screened for DNA sequence changes at the
dihydrofolate reductase locus in gamma-ray-induced CHO cells
lacking the enzyme activity. Two of nine mutants they screened
displayed an altered restriction fragment pattern, suggesting the
occurrence of DNA deletions or rearrangements. Similarly, a
bacterial plasmid containing the hamster adenine
phosphoribosyltransferase (APRT) genomic sequence was used to
examine EMS-induced APRT-deficient mutants of CHO cells for
possible alterations in the APRT gene structure (96). Base pair
changes as detected by loss of restriction enzyme sites were found,
but no major internal gene rearrangements were detected.

Such studies, however significant and interesting, will not
provide information on the influence of chromosomal location on
either mutation frequency or the types of mutation induced. An
approach which might satisfy these aspects of mutation analysis is
to introduce a unique, defined and selectable gene into an
appropriate mammalian cell by DNA-transfer techniques. Recently,
the recombinant DNA molecule pSV2-gpt, which contains the bacterial
gene coding for xanthine-guanine phosphoribosyltransferase (XGPRT)
activity, was introduced into a hamster cell line lacking the
equivalent mammalian enzyme (HPRT) (136). Hamster cell sublines
which were found with stable expression of XGPRT activity were used
to study spontaneous and mutagen-induced mutations of the
integrated pSV2-gpt DNA sequence. It is to be expected that within
the next few years mutation studies at the DNA level in mammalian
systems will continue to expand and flourish. The ability to study
changes in defined sequences at various sites in the genome of
mutant cells will ultimately lead to an understanding of the
molecular nature of mutations as they affect the specific cellular
functions.

REFERENCES

1. Abbondandolo, A. (1977) Prospects for evaluating genetic
    damage in mammalian cells in culture. Mutation
    Res. 42:279-298.
2. Adair, G.M., M.J. Siciliano, and R.M. Humphrey (1981)
    Induction of electrophoretic mutants at multiple gene loci in
    CHO cells: Differential mutability of gene loci and apparent
    "hot spots" for mutation. J. Cell. Biol. 91:382a.
3. Arlett, C.F., D. Turnbull, S.A. Harcourt, A.R. Lehmann, and
    C.M. Colella (1975) A comparison of the 8-azaguanine and
    ouabain-resistance systems for the selection of induced mutant
    Chinese hamster cells. Mutation Res. 33:261-278.
4. Armitage, P. (1952) The statistical theory of bacterial
    populations subject to mutation. J. Roy. Statist. Soc. B
    14:1-40.
5. Armitage, P. (1953) Statistical concepts in the theory of
    bacterial mutation. J. Hygiene 51:162-184.
6. Baker, R.M., D.M. Brunette, R. Mankovitz, L.H. Thompson,
    G.F. Whitmore, L. Siminovitch, and J.E. Till (1974) Ouabain-
    resistant mutants of mouse and hamster cells in culture. Cell
    1:9-21.
7. Barclay, B.J., and J.G. Little (1978) Genetic damage during
    thymidylate starvation in Saccharomyces cerevisiae. Molec.
    Gen. Genet. 160:33-40.
8. Barclay, B.J., B.A. Kunz, J.G. Little, and R.H. Haynes (1982)
    Genetic and biochemical consequences of thymidylate stress.
    Canad. J. Biochem. 60:172-194.
9. Barrett, C.J., N.E. Bias, and P.O.P. Ts´o (1978) A mammalian
    cellular system for the concomitant study of neoplastic
    transformation and somatic mutation. Mutation
    Res. 50:121-136.
10. Bartsch, H., C. Malareille, A.-M. Camus, G. Martel-Planche,
    G. Brun, A. Hautefeuille, N. Sabadie, A. Barbin, T. Kuroki,
    C. Drevon, C. Piccoli, and R. Montesano (1980) Validation and
    comparative studies on 180 chemicals with S. typhimurium
    strains and V79 Chinese hamster cells in the presence of
    various metabolizing systems. Mutation Res. 76:1-50.
11. Benditt, E.P., and J.M. Benditt (1973) Evidence for a
    monoclonal origin of human atherosclerotic plaque. Proc.
    Natl. Acad. Sci. USA 70:1753-1756.
12. Bradley, M.O., and N.A. Sharkey (1978) Mutagenicity of
    thymidine to cultured Chinese hamster cells. Nature
    274:607-608.
13. Bradley, M.O., B. Bhuyan, M.C. Francis, R. Langenbach,
    A. Peterson, and E. Huberman (1981) Mutagenesis by chemical
    agents in V79 Chinese hamster cells: A review and analysis of
    the literature. Mutation Res. 87:81-142.
14. Bresler, S.E., M.I. Mosevitsky, and L.G. Vyacheslavov (1973)
    Mutations as possible replication errors in bacteria growing

under conditions of thymine deficiency. Mutation Res. 19:281-293.

15. Busch, D.B., J.E. Cleaver, and D.A. Glaser (1980) Large scale isolation of UV-sensitive clones of CHO cells. Somat. Cell Genet. 6:407-418.

16. Cairns, J. (1978) Cancer, Science and Society, Freeman, San Francisco, California.

17. Caskey, C.T., A.L. Beaudet, D.J. Roufa, and F.D. Gillin (1974) Characterization of 8-azaguanine resistant Chinese hamster fibroblast mutants, In Molecular and Environmental Aspects of Mutagenesis, L. Prakash, F. Sherman, M.W. Miller, C.W. Lawrence, and H.W. Taber, Eds., C.C. Thomas, Springfield, Illinois, pp. 196-209.

18. Caskey, C.T., and G.D. Kruh (1979) The HPRT locus. Cell 16:1-9.

19. Chang, C.-C., M. Castellazzi, T.W. Glover, and J.E. Trosko (1978a) Effects of harman and norharman on spontaneous and ultraviolet light-induced mutagenesis in cultured Chinese hamster cells. Cancer Res. 38:4527-4533.

20. Chang, C.-C., S.M. D'Ambrosio, R. Schultz, J.E. Trosko, and R.B. Setlow (1978b) Modification of UV-induced mutation frequencies in Chinese hamster cells by dose fractionation, cycloheximide and caffeine treatments. Mutation Res. 52:231-245.

21. Chang, C.-C., J.E. Trosko, and T. Akera (1978c) Characterization of ultraviolet light-induced ouabain-resistant mutations in Chinese hamster cells. Mutation Res. 51:85-98.

22. Chang, C.-C., J.A. Boezi, S.T. Warren, C.L.K. Sabourin, P.K. Liu, L. Glatzer, and J.E. Trosko (1981) Isolation and characterization of a UV-sensitive hypermutable aphidicolin-resistant Chinese hamster cell line. Somat. Cell Genet. 7:235-253.

23. Chasin, L.A. (1973) The effect of ploidy on chemical mutagenesis in cultured Chinese hamster cells. J. Cell Physiol. 82:299-308.

24. Chu, E.H.Y. (1971a) Mammalian cell genetics III. Characterization of X-ray-induced forward mutations in Chinese hamster cell cultures. Mutation Res. 11:23-34.

25. Chu, E.H.Y. (1971b) Induction and analysis of gene mutations in mammalian cell cultures, In Environmental Chemical Mutagens, A. Hollaender, Ed., Plenum Publ. Corp., New York, pp. 411-444.

26. Chu, E.H.Y. (1974) Induction and analysis of gene mutations in cultured mammalian somatic cells. Genetics 78:115-132.

27. Chu, E.H.Y. (1983) Mutation systems in cultured mammalian cells. Ann. N.Y. Acad. Sci. 407:221-230.

28. Chu, E.H.Y., and V.C. Monesi (1960) Analysis of X-ray induced chromosome aberrations in mouse somatic cells in vitro. Genetics 45:981.

29. Chu, E.H.Y., and H.V. Malling (1968) Mammalian cell genetics
     II. Chemical induction of specific locus mutations in Chinese
     hamster cells in vitro. Proc. Natl. Acad. Sci. USA
     61:1306-1312.
30. Chu, E.H.Y., P.A. Brimer, C.K. Schenley, T. Ho, and
     H.V. Malling (1974) Reversion studies of chemically-induced
     mutants in Chinese hamster cells, In Molecular and
     Environmental Aspects of Mutagenesis, L. Prakash, F. Sherman,
     M.W. Miller, C.W. Lawrence, and H.W. Taber, Eds., Charles
     C. Thomas, Publ., Springfield, Illinois, pp. 178-195.
31. Chu, E.H.Y., N.C. Sun, and C.C. Chang (1975) Genetic markers
     associated with hamster chromosomes, In Mammalian Cells:
     Problems and Probes, C.R. Richmond, D.F. Peterson,
     P.P. Mullaney, and E.C. Anderson, Eds., National Technical
     Information Service, U.S. Dept. Commerce, Springfield,
     Virginia, pp. 228-238.
32. Chu, E.H.Y., J.D. McLaren, I.-C. Li, and B. Lamb (1982)
     Pleiotropic CTP synthetase mutants of Chinese hamster cells.
     Genetics 100:s12-s13.
33. Clive, D., K.D. Johnson, J.F.S. Spector, A.G. Batson, and
     M.M.M. Brown (1979) Validation and characterization of the
     L5178Y TK$^{+/-}$ mouse lymphoma assay system. Mutation Res.
     59:61-108.
34. Cook, W.D., S. Rudikoff, A.M. Giusti, and M.D. Scharff (1982)
     Somatic mutation in a cultured mouse myeloma cell affects
     antigen binding. Proc. Natl. Acad. Sci. USA 79:1240-1244.
35. Cox, R. (1982) Mechanisms of mutagenesis in cultured mammalian
     cells, In Environmental Mutagens and Carcinogens, T. Sugimura,
     S. Kondo, and H. Takebe, Eds., A.R. Liss, Inc., New York,
     pp. 157-166.
36. Cox, R., and W.K. Masson (1978) Do radiation-induced
     thioguanine-resistant mutants of cultured mammalian cells
     arise by HGPRT gene mutation or X-chromosome rearrangement?
     Nature 276:629-630.
37. DeMars, R. (1974) Resistance of cultured human fibroblasts and
     other cells to purine and pyrimidine analogues in relation to
     mutagenesis detection. Mutation Res. 24:335-364.
38. Duncan, M.E., and P. Brookes (1973) The induction of
     azaguanine-resistant mutants in cultured Chinese hamster cells
     by reactive derivatives of carcinogenic hydrocarbons.
     Mutation Res. 21:107-118.
39. Feldman, G., J. Remsen, K. Shinohara, and P. Cerrutti (1978)
     Excisability and persistence of benzo(a)pyrene DNA adducts in
     epithelioid human lung cells. Nature 274:796-798.
40. Fox, M., and J.M. Boyle (1976) Factors affecting the growth of
     Chinese hamster cells in HAT selection media. Mutation Res.
     35:445-464.
41. Generoso, W.M., M.D. Shelby, and F.J. de Serres (Editors)
     (1980) DNA Repair and Mutagenesis in Eukaryotes, Plenum
     Publ. Corp., New York.

42. Gillen, F.D., D.J. Roufa, A.L. Beaudet, and C.T. Caskey (1973) 8-azaguanine resistance in mammalian cells. I. Hypoxanthine guanine phosphoribosyltransferase. Genetics 73:320-324.

43. Graf, L.H., and L.A. Chasin (1982) Direct demonstration of genetic alterations at the dihydrofolate reductase locus by gamma irradiation. Mol. Cell. Biol. 2:93-96.

44. Grant, S.G., and R.G. Worton (1982) 5-azacytidine-induced reactivation of HPRT on the inactive X chromosome in diploid Chinese hamster cells. Amer. J. Hum. Genet. 34:171A.

45. Gupta, R.S., D.Y.H. Chan, and L. Siminovitch (1978) Evidence for functional hemizygosity at the Emt R locus in CHO cells through segregation analysis. Cell 14:1007-1013.

46. Gupta, R.S., and S. Goldstein (1980) Diphtheria toxin resistance in human fibroblast cell strains from normal and cancer-prone individuals. Mutation Res. 73:331-338.

47. Hand, R., and J. German (1975) A retarded rate of DNA chain growth in Bloom´s syndrome. Proc. Natl. Acad. Sci. USA 72:758-762.

48. Hand, R., and J. German (1977) Bloom´s syndrome: DNA replication in cultured fibroblasts and lymphocytes. Hum. Genet. 38:297-306.

49. Harris, M. (1967) Phenotypic expression of drug resistance in cell cultures. J. Natl. Cancer Inst. 38:185-192.

50. Harris, M. (1973) Phenotypic expression of drug resistance in hybrid cells. J. Natl. Cancer Inst. 50:423-429.

51. Harris, M. (1982) Induction of thymidine kinase in enzyme-deficient Chinese hamster cells. Cell 29:483-492.

52. Hodgkiss, R.J., J. Brennand, and M. Fox (1980) Reversion of 6-thioguanine resistant Chinese hamster cell lines: Agent specificity and evidence for the repair of promutagenic lesions. Carcinogenesis 1:175-187.

53. Holden, J.A., and W.N. Kelley (1978) Human hypoxanthine-guanine phosphoribosyltransferase: Evidence for a tetrameric structure. J. Biol. Chem. 253:4459-4463.

54. Howard-Flanders, P. (1981) Mutagenesis in mammalian cells. Mutation Res. 86:307-327.

55. Hozier, J., J. Sawyer, M. Moore, B. Howard, and D. Clive (1981) Cytogenetic analysis of the L5178/TK$^{+/-}$ → TK$^{-/-}$ mouse lymphoma mutagenesis assay system. Mutation Res. 84:169-181.

56. Hsie, A.W., P.A. Brimer, T.J. Mitchell, and D.G. Gosslee (1975) The dose-response relationship for ethyl methanesulfonate-induced mutations at the hypoxanthine-guanine phosphoribosyl transferase locus in Chinese hamster ovary cells. Somat. Cell Genet. 1:247-261.

57. Hsie, A.W., P.A. Brimer, R. Machnoff, and M.H. Hsie (1977) Further evidence for the genetic origin of mutations in mammalian somatic cells: The effects of ploidy level and selection stringency on dose-dependent chemical mutagenesis to purine analogue resistance in Chinese hamster ovary cells. Mutation Res. 45:271-282.

58. Hsie, A.W., J.P. O'Neill, and V.K. McElheny (Editors) (1979) Banbury Report 2: Mammalian Cell Mutagenesis - The maturation of test systems, Cold Spring Harbor Laboratory, Cold Spring Harbor, New York.

59. Hsie, A.W., D.A. Casciano, D.B. Couch, D.F. Krahn, J.P. O'Neill, and B.L. Whitfield (1981) The use of Chinese hamster ovary cells to quantify specific locus mutations and to determine mutagenicity of chemicals. A report of the GENE-TOX program. Mutation Res. 86:193-214.

60. Hsu, I.C., G.D. Stoner, H. Autrup, B.F. Trump, J.K. Selkirk, and C.C. Harris (1978) Human bronchus-mediated mutagenesis of mammalian cells by carcinogenic polynuclear aromatic hydrocarbons. Proc. Natl. Acad. Sci. USA 75:2003-2007.

61. Huberman, E., L. Aspiras, C. Heidelberger, P.L. Grover, and P. Sims (1971) Mutagenicity to mammalian cells of epoxides and other derivatives of polycyclic hydrocarbons. Proc. Natl. Acad. Sci. USA 68:3195-3199.

62. Huberman, E., R. Mager, and L. Sachs (1976) Mutagenesis and transformation of normal cells by chemical carcinogens. Nature 264:360-361.

63. Huberman, E., and L. Sachs (1976) Mutability of different genetic loci in mammalian cells by metabolically activated carcinogenic polycyclic hydrocarbons. Proc. Natl. Acad. Sci. USA 73:188-192.

64. Jacobs, L., and R. DeMars (1978) Quantification of chemical mutagenesis in diploid human fibroblasts: Induction of azaguanine-resistant mutants by N-methyl-N-nitrosoguanidine. Mutation Res. 53:29-53.

65. Jolly, D.J., A.C. Esty, H.U. Bernard, and T. Friedmann (1982) Isolation of a genomic clone partially encoding human hypoxanthine phosphoribosyltransferase. Proc. Natl. Acad. Sci. USA 79:5038-5041.

66. Kadouri, A., J.J. Kunce, and K.G. Lark (1978) Evidence for dominant mutations reducing HGPRT activity. Nature 274:256-259.

67. Kakunaga, T. (1977) The transformation of human diploid cells by chemical carcinogens, In Origin of Human Cancer, Cold Spring Harbor Symp. Quant. Biol. 43:1537-1548.

68. Kao, F.-T., and Puck, T.T. (1968) Genetics of somatic mammalian cells. VII. Induction and isolation of nutritional mutants in Chinese hamster cells. Proc. Natl. Acad. Sci. USA 60:1275-1281.

69. Kao, F.-T., R.T. Johnson, and T.T. Puck (1969) Complementation analysis on virus-fused Chinese hamster cells with nutritional markers. Science 164:312-314.

70. Kapp, L.N. (1982) DNA fork displacement rates in Bloom's fibroblasts. Biochim. Biophys. Acta 696:226-227.

71. Kavathas, P., F.H. Bach, and R. DeMars (1980) Gamma ray-induced loss of expression of HLA and glyoxalase I alleles in

lymphoblastoid cells. Proc. Natl. Acad. Sci. USA
77:4251-4255.

72. Knaap, A.G.A.C., and J.W.I.M. Simons (1975) A mutational assay
    system for L5178Y mouse lymphoma cells, using hypoxanthine-
    guanine phosphoribosyl transferase (HGPRT) deficiency as
    marker. The occurrence of a long expression time for
    mutations induced by X-rays and EMS. Mutation Res.
    30:97-110.

73. Krahn, D.F., and C. Heidelberger (1977) Liver homogenate-
    mediated mutagenesis in Chinese hamster V79 cells by
    polycyclic aromatic hydrocarbons and aflatoxins. Mutation
    Res. 46:27-44.

74. Kunz, B.A., and R.H. Haynes (1982) DNA repair and the genetic
    effects of thymidylate stress in yeast. Mutation Res.
    93:353-375.

75. Kuroki, T., and S.Y. Miyashita (1976) Isolation of UV-
    sensitive clones from mouse cell lines by Lederberg style
    replica plating. J. Cell. Physiol. 90:79-90.

76. Lea, D.A., and C.A. Coulson (1949) The distribution of the
    numbers of mutants in bacterial populations. J. Genet.
    49:264-285.

77. Li, A.P. (1981) Simplification on the CHO/HGPRT mutation assay
    through the growth of Chinese hamster ovary cells as
    unattached cultures. Mutation Res. 85:165-175.

78. Li, I.-C., D.A. Blake, I.J. Goldstein, and E.H.Y. Chu (1980)
    Modification of cell membrane in variants of Chinese hamster
    cells resistant to abrin. Exper. Cell Res. 129:351-360.

79. Li, I.-C., and E.H.Y. Chu (1982) Direct selection of mammalian
    cell mutants deficient in thymidylate synthetase. J. Cell.
    Biol. 95:447a.

80. Li, I.-C., J. Fu, Y.-T. Hung, and E.H.Y. Chu (1983) Estimation
    of mutation rates in cultured mammalian cells. Mutation
    Res. 111:253-262.

81. Littlefield, J.W. (1964) The selection of hybrid mouse
    fibroblasts. Cold Spring Harbor Symp. Quant. Biol.
    29:161-166.

82. Liu, P.K., C.-C. Chang, and J.E. Trosko (1982) Association of
    mutator activity with UV sensitivity in an aphidicolin-
    resistant mutant of Chinese hamster V79 cells. Mutation Res.
    106:317-332.

83. Liu, P.K., C.-C. Chang, J.E. Trosko, D.K. Dube, G.M. Martin,
    and L.A. Loeb (1983) Mammalian mutator mutant with an
    aphidicolin-resistant DNA polymerase alpha. Proc. Natl. Acad.
    Sci. USA 80:797-801.

84. Lobban, P.E., and L. Siminovitch (1975) alpha-Amanitin
    resistance: A dominant mutation in CHO cells. Cell 4:167-172.

85. Luria, S.E. (1966) Mutations of bacteria and of bacteriophage,
    In Phage and the Origins of Molecular Biology, Cairns, J.,
    G.S. Stent, and J.D. Watson, Eds., Cold Spring Harbor
    Laboratory, New York, pp. 173-179.

86. Luria, S.E., and M. Delbruck (1943) Mutations of bacteria from virus sensitivity to virus resistance. Genetics 28:491-511.
87. Lyon, M. (1962) Sex chromatin and gene action in the mammalian X-chromosome. Am. J. Hum. Genet. 14:135-148.
88. McMillan, S., and M. Fox (1979) Failure of caffeine to influence induced mutation frequencies and the independence of cell killing and mutation induction in V79 Chinese hamster cells. Mutation Res. 60:91-107.
89. Maher, V.M., L.M. Quellette, R.D. Curren, and J.J. McCormick (1976) Frequency of ultraviolet light-induced mutations is higher in Xeroderma pigmentosum variant cells than in normal human cells. Nature 261:593-595.
90. Maher, V.M., J.J. McCormick, P.L. Grover, and P. Sims (1977) Effect of DNA repair on the cytotoxicity and mutagenicity of polycyclic hydrocarbon derivatives in normal and Xeroderma pigmentosum fibroblasts. Mutation Res. 44:313-326.
91. Malling, H.V., and E.H.Y. Chu (1974) Development of mutational model systems for study of carcinogenesis, In Chemical Carcinogenesis, Part B, P.O.P. Ts´o, and J.A. DiPaolo, Eds., Marcell Dekker, Inc., New York, pp. 545-563.
92. Mankovitz, R., M. Buchwald, and R.M. Baker (1974) Isolation of ouabain-resistant human diploid fibroblasts. Cell 3:221-226.
93. Meuth, M. (1981) Role of deoxynucleoside triphosphate pools in the cytotoxic and mutagenic effects of DNA alkylating agents. Somat. Cell Genet. 7:89-102.
94. Meuth, M., N. L´Hcureux-Huard, and M. Trudel (1979a) Characterization of a mutator gene in Chinese hamster ovary cells. Proc. Natl. Acad. Sci. USA 76:6505-6509.
95. Meuth, M., M. Trudel, and L. Siminovitch (1979b) Selection of Chinese hamster cells auxotrophic for thymidine by 1-beta-D-arabinofuranosyl cytosine. Somat. Cell Genet. 5:303-318.
96. Meuth, M., and J.E. Arrand (1982) Alterations of gene structure in ethyl methanesulfonate-induced mutants of mammalian cells. Mol. Cell. Biol. 2:1459-1462.
97. Miller, E.C. (1978) Some current perspectives on chemical carcinogenesis in humans and experiments on animals. Cancer Res. 38:1479-1496.
98. Mohandas, T., R.S. Sparkes, and L.J. Shapiro (1981) Reactivation of an inactive human X chromosome: Evidence for X inactivation by DNA methylation. Science 211:393-396.
99. Morrow, J. (1983) Eukaryotic Cell Genetics, Academic Press, New York.
100. Motulsky, A.G. (1983) Impact of genetic disease in man. (This Workshop).
101. Myhr, B.C., and J.A. DiPaolo (1975) Requirement for cell dispersion prior to selection of induced azaguanine-resistant colonies of Chinese hamster cells. Genetics 80:157-169.
102. Neel, J.V. (1983) Frequency of spontaneous and induced "point" mutations in higher eukaryotes. J. Hered. 74:2-15.

103. Newbold, R.F., C.B. Wigley, M.H. Thompson, and P. Brookes (1977) Cell-mediated mutagenesis in cultured Chinese hamster cells by carcinogenic polycyclic hydrocarbons: Nature and extent of the associated hydrocarbon DNA reaction. Mutation Res. 43:101-116.

104. Newcombe, H.B. (1948) Delayed phenotypic expression of spontaneous mutation in Escherichia coli. Genetics 33:447-476.

105. O´Neill, J.P., P.A. Brimer, R. Machanoff, G.P. Hirsch, and A.W. Hsie (1977) A quantitative assay of mutation induction at the hypoxanthine-guanine phosphoribosyl transferase locus in Chinese hamster ovary cells (CHO/HGPRT system): Development and definition of the system. Mutation Res. 45:91-101.

106. O´Neill, J.P., and A.W. Hsie (1979) Phenotypic expression time of mutagen-induced 6-thioguanine resistance in Chinese hamster ovary cells (CHO/HGPRT system). Mutation Res. 59:109-118.

107. Papageorge, A.G., E.M. Scolnick, R. Dhar, D.R. Lowry, and E.H. Chang (1982) Mechanism of activation of a human oncogene. Nature 300:143-149.

108. Peterson, A.R., H. Peterson, and C. Heidelberger (1975) Reversion of the 8-azaguanine resistant phenotype of variant Chinese hamster cells treated with alkylating agents and 5-bromo-2´-deoxyuridine. Mutation Res. 29:127-137.

109. Peterson, A.R., D.F. Krahn, H. Peterson, C. Heidelberger, B.K. Bhuyan, and L.H. Li (1976) The influence of serum components on the growth and mutation of Chinese hamster cells in medium containing 8-azaguanine. Mutation Res. 36:345-356.

110. Peterson, A.R., J.R. Landolph, H. Peterson, and C. Heidelberger (1978) Mutagenesis of Chinese hamster cells is facilitated by thymidine and deoxycytidine. Nature 276:508-510.

111. Peterson, A.R., and H. Peterson (1979) Facilitation by pyrimidine deoxyribonucleosides and hypoxanthine of mutagenic and cytotoxic effects of monofunctional alkylating agents in Chinese hamster cells. Mutation Res. 61:319-331.

112. Reddy, E.P., R.K. Reynolds, E. Santos, and M. Barbacid (1982) A point mutation is responsible for the acquisition of transforming properties by the T24 human bladder carcinoma oncogene. Nature 300:149-152.

113. Regan, J.D., and R.B. Setlow (1973) Repair of chemical damage to human DNA, In Chemical Mutagens: Principles and Methods for Their Detection, A. Hollaender, Ed., Plenum Press, New York, Vol. 3, pp. 151-170.

114. Rosenblum, B.B., S.M. Hanash, N. Yew, and J.V. Neel (1982) Two-dimensional electrophoretic analysis of erythrocyte membranes. Clin. Chem. 28:925-931.

115. Sato, K., and N. Heida (1980) Mutation induction in a mouse lymphoma cell mutant sensitive to 4-nitroquinoline-1-oxide and ultraviolet radiation. Mutation Res. 71:233-241.

116. Schultz, R.A., J.E. Trosko, and C.-C. Chang (1981) Isolation and partial characterization of mutagen-sensitive and DNA repair mutants of Chinese hamster fibroblasts. Environ. Mutagenesis 3:53-64.
117. Setlow, R.B. (1978) Repair deficient human disorder and cancer. Nature 271: 713-717.
118. Shapiro, A. (1946) The kinetics of growth and mutation in bacteria. Cold Spring Harbor Symp. Quant. Biol. 11:228-234.
119. Sharp, J.D., N.E. Capecchi, and M.R. Capecchi (1973) Altered enzymes in drug-resistant variants of mammalian tissue culture cells. Proc. Natl. Acad. Sci. USA 70:3145-3149.
120. Shaw, E.I., and A.W. Hsie (1978) Conditions necessary for quantifying ethyl methanesulfonate-induced mutations to purine-analogue resistance in Chinese hamster V79 cells. Mutation Res. 51:237-254.
121. Shiomi, T., and K. Sato (1979) Isolation of UV-sensitive variants of human FL cells by a viral suicide method. Somat. Cell Genet. 5:193-201.
122. Shiomi, I., N. Hieda-Shiomi, and K. Sato (1982) A novel mutant of mouse lymphoma cells sensitive to alkylating agents and caffeine. Mutation Res. 103:61-69.
123. Siciliano, M.J., J. Siciliano, and R.M. Humphrey (1978) Electrophoretic shift mutations in Chinese hamster ovary cells: Evidence for genetic diploidy. Proc. Natl. Acad. Sci. USA 75:1919-1923.
124. Siciliano, M.J., B.F. White, and R.M. Humphrey (1983) Electrophoretically detectable mutations induced in CHO cells by varying doses of ultraviolet radiation. Mutation Res. 107:167-176.
125. Siminovitch, L. (1976) On the nature of heritable variation in cultured somatic cells. Cell 7:1-11.
126. Skolnick, M.M. (1982) An approach to completely automatic comparison of two-dimensional electrophoresis gels. Clin. Chem. 28:979-986.
127. Skolnick, M.M., S.R. Sternberg, and J.V. Neel (1982) Computer programs for adapting two-dimensional gels to the study of mutation. Clin. Chem. 28:969-978.
128. Smith, M.D., R.R. Green, L.S. Ripley, and J.W. Drake (1973) Thymineless mutagenesis in bacteriophage T4. Genetics 74:393-403.
129. Stallings, R.L., M.J. Siciliano, G.M. Adair, and R.M. Humphrey (1982) Structural and functional hemi- and dizygous Chinese hamster chromosome 2 gene loci in CHO cells. Somat. Cell Genet. 8:413-422.
130. Steglich, C., and R. DeMars (1982) Mutations causing deficiency of APRT in fibroblasts cultured from human heterozygotes for mutant APRT alleles. Somat. Cell Genet. 8:115-141.

131. Stocker, B.A.D. (1949) Measurement of rate of mutation of flagellar antigenic phage in Salmonella typhimurium. J. Hygiene 47:398-413.

132. Stoker, M., and I.A. Macpherson (1964) Syrian hamster fibroblast cell line BHK21 and its derivatives. Nature 203:1355-1357.

133. Subak-Sharpe, H., R.R. Burk, and J.D. Pitts (1969) Metabolic cooperation between biochemically marked mammalian cells in tissue culture. J. Cell Sci. 4:353-367.

134. Thacker, J. (1981) The chromosomes of a V79 Chinese hamster line and a mutant subline lacking HPRT activity. Cytogenet. Cell Genet. 29:16-25.

135. Thacker, J., M.A. Stephens, and A. Stretch (1978) Mutation to ouabain resistance in Chinese hamster cells: Induction by ethyl methanesulfonate and lack of induction by ionizing radiation. Mutation Res. 51:255-270.

136. Thacker, J., P.G. Debenham, A. Stretch, and M.B.T. Webb (1983) The use of a cloned bacterial gene to study mutation in mammalian cells. Mutation Res. 111:9-23.

137. Thompson, L.H., and R.M. Baker (1973) Isolation of mutants of cultured mammalian cells, In Methods in Cell Biology, D.M. Prescott, Ed., Academic Press, New York, Vol. 6, pp. 209-281.

138. Thompson, L.H., J.S. Rubin, J.E. Cleaver, G.F. Whitmore, and K. Brookman (1980) A screening method for isolating DNA repair-deficient mutants of CHO cells. Somat. Cell Genet. 6:391-405.

139. Thompson, L.H., D.B. Busch, K. Brookman, C.L. Mooney, and D.A. Glaser (1981) Genetic diversity of UV-sensitive DNA repair mutants of Chinese hamster ovary cells. Proc. Natl. Acad. Sci. USA 78:3734-3737.

140. Tong, C., and G.M. Williams (1980) Definition of conditions for the detection of genotoxic chemicals in the adult rat-liver epithelial cell/hypoxanthine-guanine phosphoribosyl transferase (ARL/HGPRT) mutagenesis assay. Mutation Res. 74:19.

141. Trosko, J.E., and C.C. Chang (1980) An integrative hypothesis linking cancer, diabetes and atherosclerosis. The role of mutations and epigenetic changes. Med. Hypotheses 6:455-468.

142. van Diggelen, O.P., T.F. Donahue, and S.-I. Shin (1979) Basis for differential cellular sensitivity to 8-azaguanine and 6-thioguanine. J. Cell. Physiol. 98:59-72.

143. van Zeeland, A.A., M.C.E. van Diggelen, and J.W.I.M. Simons (1972) The role of metabolic cooperation in selection of hypoxanthine-guanine phosphoribosyltransferase (HG-PRT)-deficient mutants from diploid mammalian cell strains. Mutation Res. 14:355-363.

144. van Zeeland, A.A., and J.W.I.M. Simons (1975) The effect of calf serum on the toxicity of 8-azaguanine. Mutation Res. 27:135-138.

145. van Zeeland, A.A., and J.W.I.M. Simons (1976) Linear dose-response relationships after prolonged expression times in V79 Chinese hamster cells. Mutation Res. 35:129–138.

146. Warren, S.T., R.A. Schultz, C.-C. Chang, M.H. Wade, and J.E. Trosko (1981) Elevated spontaneous mutation rate in Bloom syndrome fibroblasts. Proc. Natl. Acad. Sci. USA 78:3133–3137.

147. Weinburg, G., Ullman, B., and Martin, D.W., Jr. (1981) Mutator phenotypes in mammalian cell mutants with distinct biochemical defects and abnormal deoxyribonucleoside triphosphate pools. Proc. Natl. Acad. Sci. USA 78:2447–2451.

148. Weymouth, L.A., and L.A. Loeb (1978) Mutagenesis during in vitro DNA synthesis. Proc. Natl. Acad. Sci. USA 75:1924–1928.

149. Wierenga, R.K., and W.G.J. Hol (1983) Predicted nucleotide-binding properties of p21 protein and its cancer-associated variant. Nature 302:842–844.

150. Wigley, C.B., R.F. Newbold, J. Ames, and P. Brookes (1979) Cell-mediated mutagenesis in cultured Chinese hamster cells by polycyclic hydrocarbons: Mutagenicity and DNA reaction related to carcinogenicity in a series of compounds. Intern. J. Cancer 23:691–696.

151. Wilson, J.M., B.W. Baugher, L. Landa, and W.N. Kelley (1981) Human hypoxanthine-guanine phosphoribosyltransferase: Purification and characterization of mutant forms of the enzyme. J. Biol. Chem. 256:10306–10312.

152. Wilson, J.M., B.W. Baugher, P.M. Mattes, P.E. Daddona, and W.N. Kelley (1982a) Human hypoxanthine-guanine phosphoribosyltransferase. Demonstration of structural variants in lymphoblastoid cells derived from patients with a deficiency of the enzyme. J. Clin. Invest. 69:706–715.

153. Wilson, J.M., G.E. Tarr, W.C. Mahoney, and W.N. Kelley (1982b) Human hypoxanthine-guanine phosphoribosyl-transferase: Complete amino acid sequence of the erythrocyte enzyme. J. Biol. Chem. 257:10978–10985.

154. Wilson, J.M., L.E. Landa, R. Kobayashi, and W.N. Kelley (1982c) Human hypoxanthine-guanine phosphoribosyl-transferase: Tryptic peptides and posttranslational modifications of the erythrocyte enzymes. J. Biol. Chem. 257:14830–14834.

155. Wilson, J.M., and W.N. Kelley (1983) Molecular basis of hypoxanthine-guanine phosphoribosyl transferase deficiency in a patient with Lesch-Nyhan syndrome. J. Clin. Invest. 71:1331–1335.

156. Wilson, J.M., G.E. Tarr, and W.N. Kelley (1983a) Human hypoxanthine (guanine) phosphoribosyltransferase: An amino acid substitution in a mutant form of the enzyme isolated from a patient with gout. Proc. Natl. Acad. Sci. USA 80:870–873.

157. Wilson, J.M., R. Kobayashi, I.H. Fox, and W.N. Kelley (1983b) Molecular abnormality in a mutant form of the enzyme hypoxanthine-guanine phosphoribosyltransferase (HPRT$_{Toronto}$). J. Biol. Chem. 258:6458–6460.

158. Wolf, S.F., and B.R. Migeon (1982) Studies of X chromosome DNA methylation in normal human cells. <u>Nature</u> 295:667–671.

159. Yotti, L.P., C.C. Chang, and J.E. Trosko (1979) Elimination of metabolic cooperation in Chinese hamster cells by a tumor promoter. <u>Science</u> 206:1089–1091.

THE CONTROL OF CELL TRANSFORMATION, MUTAGENESIS, AND DIFFEREN-TIATION BY CHEMICALS THAT INITIATE OR PROMOTE TUMOR FORMATION[1]

Eliezer Huberman

Division of Biological and Medical Research
Argonne National Laboratory
Argonne, Illinois

SUMMARY

Transformation of normal cells to malignant cells in vitro by chemical agents has been achieved in a number of fibroblastic and epithelial cell systems. The commonly used assays for cell transformation employ either diploid Syrian hamster embryo fibroblasts or cells from some permanent mouse fibroblast lines. Under culture conditions these cells have an oriented pattern of growth, are cell density (or contact) inhibited, and are not tumorigenic when implanted in host animals. During cell transformation, these "normal" or control cells are converted into cells that have a hereditary random pattern of cell growth and can form tumors in appropriate hosts. The random pattern of growth of the transformed cells in either a colony or a focus is used as an end point for quantitatively determining cell transformation.

Carcinogenesis (and in vitro cell transformation) is at present believed to be a multistage process in which the initial step is due to a "mutation-like event" caused by the activated carcinogen. Using an in vitro mammalian cell-mediated mutagenicity assay, we were able to establish a relationship between the degree of mutagenesis in this assay, using Chinese hamster V79 cells as target cells, and the degree of activity of carcinogens (initiators) in experimental animals. This assay has also provided the means to study the organ specificity of chemical carcinogens. It is possible, however, that some environmental

---

[1] This work is supported by the United States Department of Energy under Contract No. W-31-109-ENG-38.

agents may act not as primary tumor initiating agents, but as
tumor promoters, at a later stage of the carcinogenic process.
Tumor promoters, including phorbol diesters and teleocidin, which
also enhance cell transformation in vitro, usually do not bind to
DNA and because they are devoid of mutagenic activity cannot be
detected with mutagenicity assays. However, these agents can
induce various effects, including a modulation of differentiation
processes, in a number of cell types. In some human melanoma and
leukemia cells in culture these tumor promoters, after binding to
specific cellular receptors, induce the treated cells to
differentiate into cells with characteristics of mature cells.
This property of the tumor promoter to induce cell differentiation
suggests that tumor promotion may, among other things, involve the
expression of "mutated tumor genes" in a process similar to gene
expression during cell differentiation. This paper discusses the
usefulness of in vitro cell transformation, differentiation, and
mutagenesis systems for studies of the mechanisms of
carcinogenesis.

INTRODUCTION

Clinical observations, epidemiological retrospective surveys,
and studies with experimental animals provide evidence that some
chemicals in our environment are responsible for a significant
proportion of human cancers. Furthermore, these types of studies
suggest that certain human cancers are caused by the interaction
of multiple factors and by a multistep process (19,28). From an
operational point of view, carcinogenesis can be divided into
three sequential processes: initiation, promotion, and progression
(3,10,21,22,25,60). The first two involve the steps that lead to
the transformation of a normal cell into a malignant cell, whereas
progression covers the processes whereby this transformed cell
develops into a malignant tumor.

A model in which initiation and promotion are clearly defined
in the mouse skin two-stage carcinogenesis assay was developed by
Berenblum (3,10,25). In this model, the skin is primed first with
a single application of an "initiator", which is a carcinogen
given at a low dose that in itself does not cause the induction of
skin tumors. This is followed either immediately or at a later
time (even up to a year) by repeated applications of a "promoting
agent", which in itself is either inactive or weakly active as a
carcinogen (10). Such treatments result in the production of
multiple tumors on the mouse skin. Changing the sequence of
treatment, i.e., starting the process with promotion followed by
initiation, does not lead to an increased tumor yield. Nor is the
process effective when the skin receives a limited number of
applications of the tumor promoter (10). These and other
experiments imply that initiation causes an irreversible change

that remains in the memory of the initiated cell, whereas promotion is apparently reversible in its early stage (10,66,70). Studies in the last ten years have revealed that a similar two-stage process can be detected in a number of other organs, including the liver, lung, colon, and bladder (2,21,59,70).

An understanding of the mechanisms underlying initiation and promotion is needed to clarify the processes that control the conversion of a normal cell into a malignant cell. To avoid the complexity of studying such mechanisms in whole animals, a number of in vitro cell systems have been developed. These systems, in addition to allowing study of the nature of cellular changes brought about by chemicals, have proved to be useful in identifying chemicals with potential carcinogenicity.

RESULTS AND DISCUSSION

I. Malignant Cell Transformation In Vitro

Transformation of normal cells to malignant cells in vitro by chemical agents has been achieved in a number of fibroblastic and epithelial cell systems (7,8,26,27,30,32,44,45,52,56,67, 70,71,73). The commonly used cell transformation assays employ either diploid Syrian hamster embryo fibroblasts (7,16,27,30,41) or cells from some permanent mouse cell lines which, in vitro, exhibit the property of cell density (contact) inhibition (26,27,45). Normal Syrian hamster embryo cells are diploid cells, have an oriented pattern of growth in culture, exhibit a limited life-span in vitro, lack the ability to grow in soft agar medium, and are not tumorigenic. During cell transformation, these cells acquire a hereditary random pattern of cell growth, can grow continuously in culture, can form colonies in soft agar medium, acquire an aneuploid karyotype, and are able to form tumors in appropriate hosts (7,14,16-18,41). Additional markers that are also used to indicate the acquisition of the transformation state are listed in Table 1. The markers are detected at different times after the initial treatment with carcinogens (4,5,7,41,48), and are expressed as a result either of an inductive or selective process (4,5,37).

The earliest marker observed, the random pattern of growth of the cells in either a colony or a focus, is used as a quantitative end point in determining the process of cell transformation (7,17,27). It has been used to establish a quantitative relationship between cell transformation in vitro and tumor induction in experimental animals by a series of chemical carcinogens (7,27). Use of this end point has also indicated that the process of initiation and promotion can be shown in cell cultures (47,55,70).

Table 1.   Markers Used to Characterize the
Phenotype of Transformed Cells.

1. Tumorigenicity in susceptible hosts

2. Loss of anchorage dependency for growth

    a. Soft agar/agarose

    b. Methyl cellulose

3. Lack of density-dependent inhibition of replication

4. Morphological changes in colonies (criss-cross, piled-up,

    rounded, refractile)

5. Enhanced fibrinolytic activity

6. Chromosomal changes

7. Unlimited life-span

8. Growth at low serum concentrations

9. Reduced requirement for calcium

10. Changes in membrane properties

    a. Altered glycoproteins and glycolipids

    b. Increased rate of transport

    c. Loss of high molecular weight, surface glycoprotein

    d. Increased agglutinability by plant lectins

    e. Increased mobility of membrane proteins

This inherited phenotypic change to a random, disoriented growth pattern, which has been shown to be the consequence of an induction process, may result in alterations in gene expression induced by "mutation-like events" (6,9,26,27,30,31,36,49,54) after the interaction of the carcinogen with cellular DNA (11,26,30,38,56).   In agreement with this suggestion it was shown that, after an appropriate metabolic activation, carcinogens (e.g., nitrosamines and polycyclic aromatic hydrocarbons) can induce mutations in various cultured mammalian cells (9,24,30,35,39,42,49,61,67) including those that are employed in cell transformation studies (6,9,36,49).   Furthermore, the frequency of transformed colonies or foci correlated with the frequency of mutants observed within the same cell system, although the latter was lower.   This correlation led to the suggestion that cell transformation is the result of a mutation in one of a number of similar or different genes (9,36) that control malignancy (oncogenes?) or, alternatively, that the mutation is in "hot spots" present in such a gene or genes (36).   There are, however, other studies in which an epigenetic process is suggested to explain the acquisition of the transformed phenotype (5,6,54).

II. Control of Mutagenesis by Chemical Carcinogens-Initiators

The concept of a multistage process of carcinogenesis implies that chemicals may act at one or more of these stages to initiate or promote tumor formation. It is presumed that initiation of carcinogenesis involves alterations in the genetic machinery of the cells. The alteration may arise from interactions between the chemical and cellular DNA (11,26,30,38,56). Indeed, many chemical carcinogens-initiators are metabolized by the microsomal mixed function oxidase system to electrophilic intermediates (56) which are capable of binding to cellular constituents such as DNA (11,26,38) and, as a consequence, are mutagenic in a variety of biological systems (1,30,31). In order to establish the involvement of mutation-like processes in carcinogenesis-initiation, it is important to show that a relationship exists between the mutagenicity of a chemical and its carcinogenic-initiating effectiveness and, furthermore, that the same metabolites are responsible for both events. Since most chemical carcinogens are highly cell/tissue specific in their effects, this specificity should also be reflected in the mutagenesis assays.

a. Mammalian cell-mediated mutagenesis assays. For the identification and study of chemicals that can act as carcinogens-initiators, we have developed a simple mammalian cell culture system designated the cell-mediated mutagenesis system (shown diagrammatically in Fig. 1). In this system, mutable target cells that cannot generally metabolize many chemicals are cocultivated with intact normal cells that can metabolically activate chemical carcinogens (30,35). Generally, Chinese hamster V79 cells are used as the targets since they display a relatively stable karyotype, high cloning efficiency, and short generation time, but they have little capacity to metabolize xenobiotics (30). In the original assay, primary rodent fibroblasts were used as the source of activating enzymes (30). These cells are irradiated prior to cocultivation with V79 cells so that they will not persist after the period of carcinogen treatment when V79 cells are reseeded to determine the frequency of induced mutations (30). The fibroblast-mediated assay was used to evaluate the mutagenicity of a series of polycyclic hydrocarbons, and a good correlation between the carcinogenic activity of this chemical class and their activity in the mutagenesis system could be demonstrated (24,30,39,42,61,67). In a related study, the metabolites of the polycyclic hydrocarbon, benzo(a)pyrene [B(a)P] were examined for their mutagenic activity toward V79 cells (40). The primary metabolite, B(a)P-7,8-diol, was more mutagenic than B(a)P but the further metabolite of B(a)P-7-8-diol, the B(a)P-diol epoxide I, was the most mutagenic species tested (40). In vitro cell transformation studies established that B(a)P-diol epoxide I was again the most active metabolite of B(a)P (52). These studies

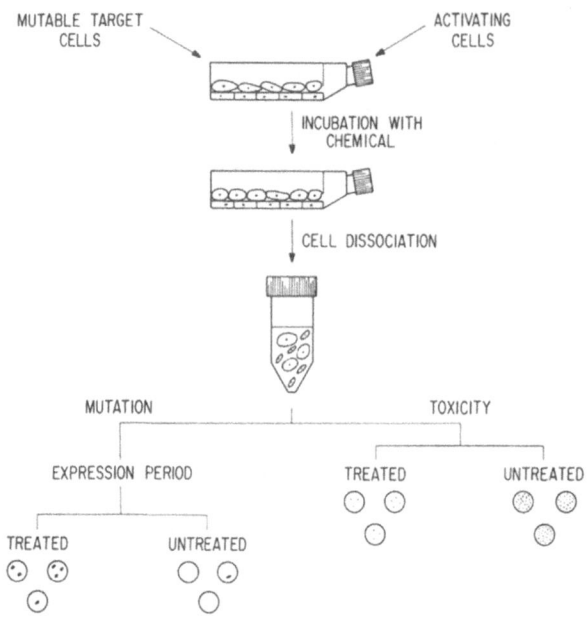

Figure 1. Scheme for the cell-mediated mutagenesis assays.

suggest that the same metabolites are responsible for both the
mutagenic and cell transformation activity of B(a)P.

    The scope of chemical carcinogens that can be activated by
the embryonic fibroblasts is rather limited. In contrast, primary
hepatocytes possess a broader range of metabolizing activities and
hence can metabolize many different classes of carcinogens,
including nitrosamines, aflatoxins, and arylamines. To expand the
efficiency of the cell-mediated mutagenesis, we have substituted
primary hepatocytes for fibroblasts in carcinogen activation (35).

    The hepatocyte-mediated assay proved to be a highly sensitive
system for studying the mutagenicity of nitrosamines (35). We
have used this system to establish a relationship between the
mutagenic activity of a series of 30 nitrosamines and their
carcinogenicity in rats (43). In these studies, the hepatocytes
were prepared from adult male Fischer rats of a genetic stock
similar to that of the rats used in the in vivo experiments. The
mutagenicity of the nitrosamines was assessed from the frequency
of ouabain-resistant or 6-thioguanine-resistant mutants induced in
V79 cells by the activated chemicals. Indices of mutagenic and
carcinogenic activity were ascribed to each nitrosamine (43). The
quantitative relationship between the degree of mutagenicity and
degree of carcinogenicity for this chemical class is shown in
Fig. 2.

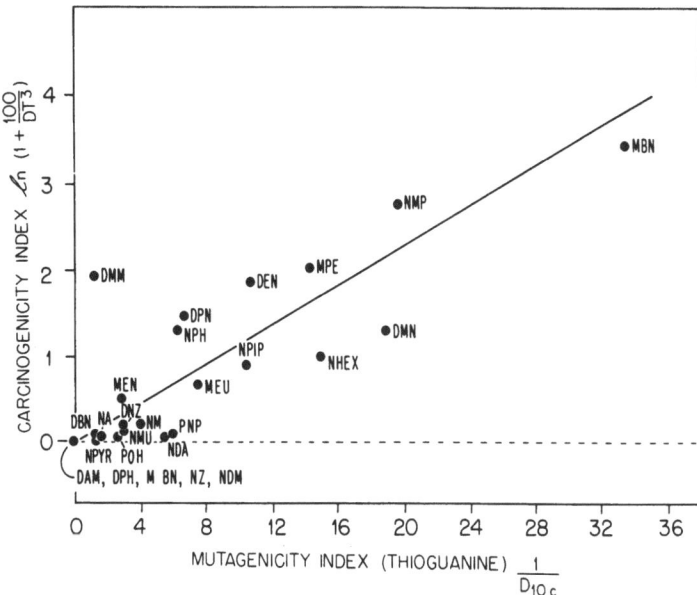

Figure 2. Relationship between carcinogenicity and mutagenicity
for nitrosamines with 6-thioguanine resistance as the genetic
marker. In the Mutagenicity Index, D = dose (nM) yielding a
mutant frequency 10 times higher than the spontaneous mutant
frequency. In the Carcinogenicity Index, D = dose (mol) at
which 50% of rats die from tumors in time T (weeks). DMN,
nitrosodimethylamine; DEN, nitrosodiethylamine; DAM,
nitrosodiallylamine; NMP, notrosomethyl-n-propylamine; MBN,
nitrosomethyyl-n-butylamine; MPE, nitrosomethyl-2-
phenylethylamine; NDA, nitrosomethyldodecylamine; NZ, 1-
nitrosopiperazine; DPH, nitrosodiphenylamine; DPN, nitrosodi-
N propylamine; DBN, nitroso-sodi-N butylamine; POH,
nitrosobis-(2-hydroxypropyl)amine; MEN, nitrosomethyl-
ethylamine; MtBN, nitrosomethyl-tert-butylamine; NMU,
nitrosomethylundecylamine; NA, 1-nitrosoazetidine; NPYR, 1-
nitrosopyrrolidine; NPIP, 1-nitrosopiperidine; PNP, 4-
phenylnitrosopiperidine; NHEX, 1-nitrosohexamethyleneimine;
NPH, 1-nitrosoheptamethyleneimine; NM, 4-nitrosomorpholine;
DMM, 2,6-dimethylnitrosomorpholine; MEU, nitroso-1-
methyl-3,3-diethylurea. From reference 43.

b. Cell specificity in the activation of chemical mutagens
and carcinogens. Many chemical carcinogens exhibit cell
specificity in vivo, but the factors that determine the
susceptibility of a particular cell type to a given carcinogen
remain largely unknown. One possibility could pertain to
differences in the balance between competing metabolic pathways
that control the level and nature of critical intermediates that

might be produced from a procarcinogen, which then interact with
cellular substituents. The nature and quantity of the interaction
between these intermediates and DNA, and the fidelity of
subsequent repair of the DNA, could be further determining steps.

We analyzed such a cell specificity in the cell-mediated
assay using primary cells for carcinogen activation (44,50).
Primary intact cells simulate the in vivo activation of chemicals
better than subcellular fractions (65). The primary cells may be
isolated from various organs, thus facilitating an investigation
of the cell-mediated mutagenic activity of a particular chemical
with each activating cell type and the relationship of the
mutagenic activity to the carcinogenicity of the chemical in the
original organs.

The abilities of rat fibroblasts and hepatocytes to activate
several organ-specific carcinogens were compared (Table 2).
Nitrosodimethylamine and aflatoxin B_1, both of which produce liver
tumors but not fibrosarcomas, were activated into potent mutagens
by hepatocytes but not by fibroblasts. This lack of mutagenicity
in the fibroblast-mediated assay was explained by the absence in
the fibroblasts of the requisite metabolizing capacity for these
compounds.

7,12-Dimethylbenzanthracene produces both liver tumors and
fibrosarcomas in rats. This chemical was mutagenic in both the
hepatocyte- and fibroblast-mediated assays.

B(a)P, which can efficiently induce fibrosarcomas but is
usually not active as a liver carcinogen, was activated at a dose
of 3 µg/ml to a mutagen for V79 cells in the fibroblast- but not
in the hepatocyte-mediated assay (Table 2). Both hepatocytes and
fibroblasts could extensively metabolize B(a)P to products that
bind to the cellular DNA. However, the critical carcinogenic/
mutagenic adduct, which is derived from the diol epoxide
metabolite, was detected only in the fibroblasts and not in the
hepatocytes (Fig. 3). In the hepatocytes, the major adducts were
hydrophilic nucleoside derivatives that are presumably
nonmutagenic. Mutagenesis with B(a)P can be observed when the
metabolizing liver cells are from newborn animals or from animals
pretreated with methylcholanthrene or when high doses of B(a)P are
being used (44).

These results indicate that the different responses of
hepatocytes and fibroblasts to B(a)P are determined by the overall
balance of oxidative and conjugative processes in the cells and,
hence, by the levels of critical active intermediates that
interact with DNA. Demonstration of the overall metabolism of a
carcinogen or the overall binding to DNA in a particular cell type

Table 2. Fibroblast and Hepatocyte-mediated Mutagenesis
of V79 Cells by Organ -specific Carcinogens.

| Chemical | Number of ouabain-resistant mutants/10^6 survivors[a] | |
	Fibroblast Activation	Hepatocyte Activation
Nitrosodimethylamine	1	120
Aflatoxin-B_1	2	41
7,12-Dimethylbenzanthracene	90	30
Benzo(a)pyrene	50	2

Carcinogen concentration was 3 μg/ml.

[a]Data based on published results (44,50).

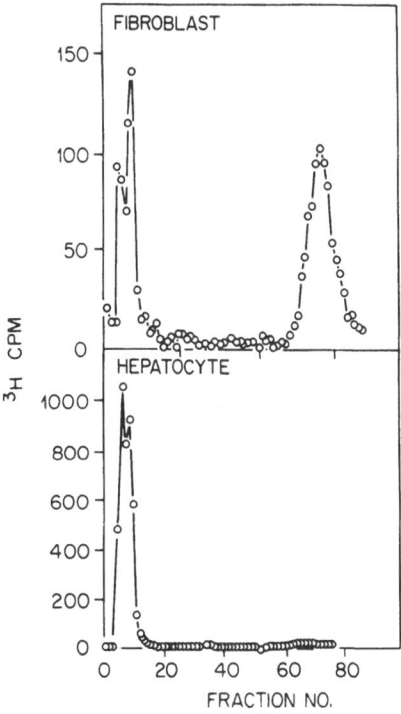

Figure 3. Binding of benzo(a)pyrene to hepatocyte and fibroblast
DNA. Hepatocytes (10^7) or fibroblasts (10^7) were incubated
with ^{3}H-benzy(a)pyrene (1 μg/ml) for 18 h. After extraction of
the cellular DNA, the bound adducts were separated by Sephadex
LH-20 chromatography. The column was eluted with water until
fraction 40, when the solvent was changed to methanol. (From
reference 44, with permission).

is not sufficient to predict the susceptibility of that cell type to the mutagenic/carcinogenic effect of the chemical.

Thus, we have demonstrated that the use of various cell types, such as hepatocytes and fibroblasts, for metabolic activation of carcinogens in co-cultivation with mutable V79 cells provides a sensitive system for the investigation of cell specificity in the activation of carcinogens and mutagens. This approach provides a basis for understanding the significance of the mutational events in the carcinogenic process, as well as a sound estimate of potential carcinogenic risk.

III. The Control of Cell Differentiation in Cultured Human Cells by Tumor Promoting Agents

The transformation of a normal cell into a malignant tumor presumably involves, in its early stages, an irreversible "mutational event of tumor genes" designated as "initiation". In the subsequent development of the tumor, the initiated cell passes through a number of stages during which exposure to various environmental chemicals might be critical. In model systems it has been demonstrated that certain chemicals (e.g., phorbol diesters, teleocidin) (23,25) can promote the formation of tumors on mouse skin that had previously been treated with a low dose of an initiator (e.g., 7,12-dimethyl benz(a)anthracene). Unlike tumor initiators, the chemicals that promote tumor formation are devoid of mutagenic activity. These chemicals may exert their promotional effect by causing the expression of the "mutated tumor genes" in a process similar to gene expression during cell differentiation. Indeed, phorbol diesters and teleocidin have been shown by us and others to act as inducers of cell differentiation in a number of cell types (13,15,32-34,51, 62-64,73). In our cell-differentiation studies with tumor promoters, we employed three different human cell types, HO melanoma cells (34) and the promyelocytic HL-60 (33) and T lymphoid CEM leukemia cells (64). These cells were chosen since they display useful markers of cell differentiation.

In the HO melanoma cells the prototype phorbol diester, phorbol-12-myristate-13-acetate (PMA), at doses as low as 10^{-10} to 10^{-9} M, induces cell differentiation characterized by an inhibition of cell growth, an increased synthesis of melanin, and dendrite-like structures (34). In this system, we have demonstrated a relationship between the tumor-promoting activity of a series of phorbol diesters in vivo and the degree to which these agents induce differentiation in vitro (34). A similar relationship was demonstrated using the HL-60 cells (Fig. 4) (33). In these cells, the phorbol diesters and teleocidin induce a macrophage cell differentiation characterized by an inhibition of cell growth, appearance of morphologically mature cells, increased

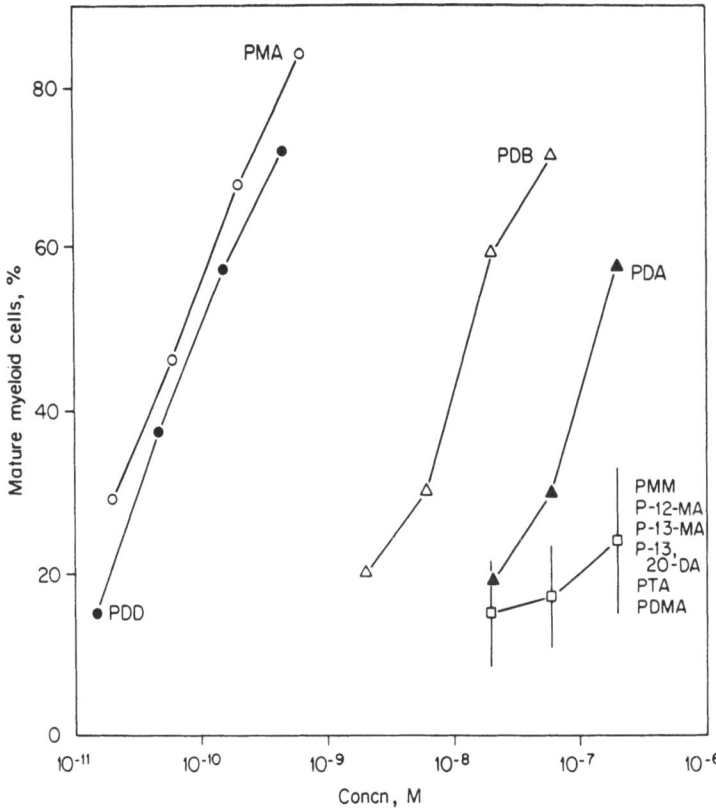

Figure 4. Percent morphologically mature HL-60 cells at 2 days
   after treatment with different phorbol esters. The percent
   of morphologically mature cells was determined after cells
   were stained with Wright-Giemsa stain. The control cultures
   contained about 15% mature cells. o phorbol-12-myristate-13-
   acetate; ● phorbol-12,13-didecanoate; △ phorbol-12,13-
   dibutyrate; ▲ phorbol-12,13-diacetate; □ phorbol-12-
   monomyristate. The results from phorbol-12-monoacetate,
   phorbol-13-monoacetate, phorbol-13,20-diacetate,
   phorbol-12,13,20-triacetate, and phorbol-20-oxo-12-
   myristate-13-acetate were similar to the results with
   phorbol-12-monomyristate. From reference 23, with
   permission.

phagocytic capability, and changes in reactivity of the treated
cells with monoclonal antibodies and in the activity of certain
enzymes (Fig. 5) (Table 3) (12,32,51,63).

   In the CEM cells, PMA causes the cells to express a phenotype
that resembles that of a suppressor T lymphocyte (64). The new
phenotype was characterized in part by the appearance of a
specific antigenic pattern. More specifically, PMA caused the

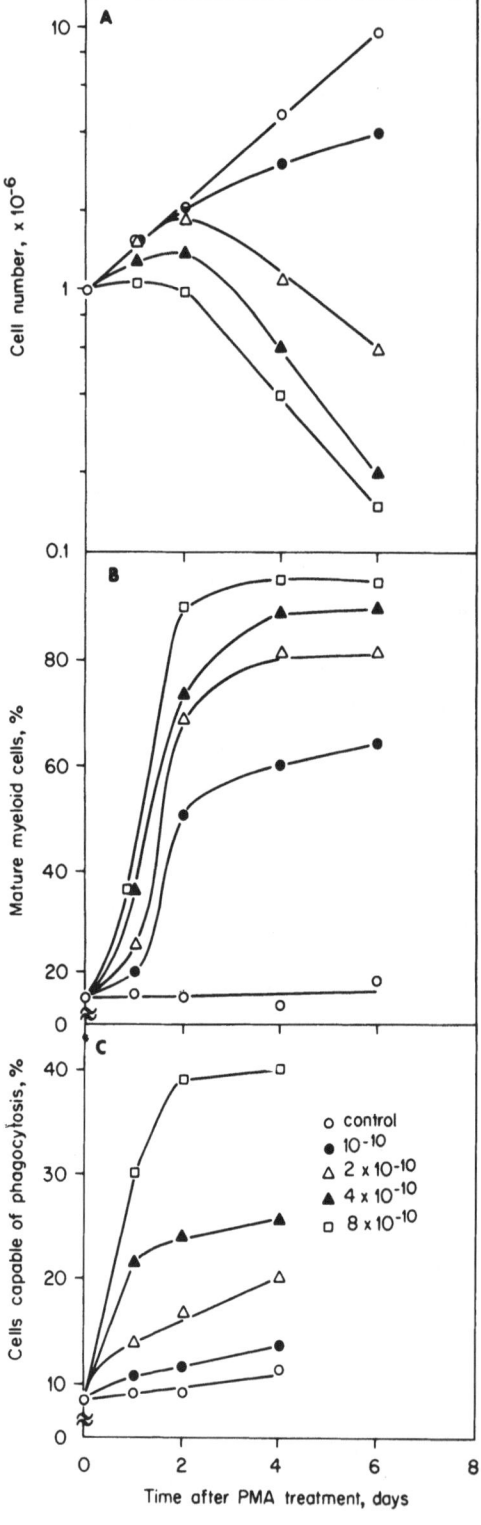

Table 3. The Reactivity of CEM and HL-60 Cells With
Monoclonal Antibodies After Treatment With PMA[a]

| Cell type | Inducer (conc.) | Cell Number, x 10^6 ml | Cells reacting with the monocolonal antibody, % | | | | |
			OKT3	OKT4	OKT6	OKT8	OKM1
CEM	None	3.4	15	52	38	42	< 1
	PMA (16 nM)	1.8	96	8	3	76	< 1
HL-60	None	2.3	< 1	97	< 1	<1	1
	PMA (16 nM)	0.2	< 1	3	< 1	<1	95

[a]A total of 0.3 x 10^6 cells/ml were seeded and analyzed after 4 days of treatment with the inducer. From reference 63.

cells to exhibit increased reactivity with OKT3 (Fig. 6) (Table 3) and OKT8 monoclonal antibodies, which characterize mature suppressor and cytotoxic T lymphocytes, respectively (Table 3) (64). PMA also reduced the reactivity of the treated cells with the OKT4 monoclonal antibody, which detects inducer/helper T lymphocytes, and with the OKT6 antibody, which detects immature

Figure 5. Cell growth (A), percent of morphologically mature cells (B), and phatocytizing cells (C) in HL-60 cells at different times after treatment with different concentrations of phorbol-12-myristate-13-acetate (PMA). The cultures were treated a day after seeding 5 x 10^5 cells in 5 ml of growth medium in 60-mm Petri dishes. Cells were counted to determine cell growth and stained with Wright-Giemsa to determine the numbers of mature cells. The mature cells were composed mainly of myelocytes and metamyelocytes. A small fraction of banded and segmented cells was also observed. Phagocytosis was determined after the HL-60 cells were incubated for 30 min with 4 x 10^6 cells per ml of the diploid S. cerevisiae strain XY 664 and stained with Wright-Giemsa. o, Control. PMA: ●, 10%(-10)M; △ 2 x 10^{-10} M; ▲, 4 x 10^{-10} M; □, 8 x 10^{-10} M. From reference 33, with permission.

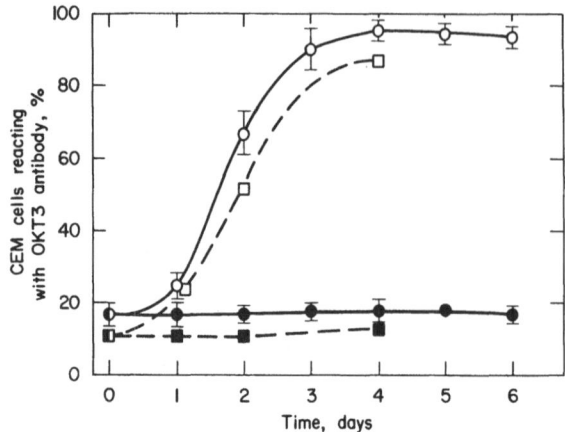

Figure 6. Reactivity of CEM cells with OKT3 monoclonal antibody
    at different times after initiation of treatment with 16 nm
    PMA. Untreated, ●,■. PMA treated, 0,□. Reactivity was
    determined either by counting the percent of cells that
    exhibited immunofluorescence, ●, 0, or by counting the
    fraction of cells that stained with trypan blue after
    antibody- and complement-mediated lympholysis,■,□. The
    results with immunofluorescence are given as the mean + SD of
    3 experiments, each including two independent determinations.
    Each determination involved the analysis of 200 cells. From
    reference 63, with permission.

"common" thymocytes (Table 3) (64). In addition, the treated CEM
cells, in common with mature suppressor T lymphocytes, were able
to suppress [^{3}H]thymidine incorporation into phytohemagluttinin-
activated peripheral blood lymphocytes but, unlike cytotoxic
lymphocytes, were unable to induce a cytotoxic response in a
number of human and rodent target cells (64).

The ability of PMA to induce differentiation of CEM cells
into suppressor cells raises the possibility that tumor promotion
by PMA and related agents in the mouse skin may, among other
things, involve a suppression of the immunity against "initiator"-
induced papillomas and carcinomas. Such a suppression could
result from an increase in the number of mature suppressor T
lymphocytes in the promoted animal.

Despite enormous scientific efforts, the mechanism by which
these chemicals promote tumorigenesis and alter differentiation
processes is poorly understood. Recently, it was found that
phorbol diesters and teleocidin bind to cellular receptors (20)
that seem to be associated with a specific protein kinase (58).
We have reported that in HL-60 cells this binding reaches a

maximum at 20 minutes and is followed by a diminished binding, a "down regulation" (69). This decreased binding of the phorbol diester to the cellular receptors did not occur in an HL-60 cell variant that is resistant to induction of cell differentiation by PMA and teleocidin; however, total binding of the tumor promoter was equivalent to that in HL-60 cells (69). Based on these results, we suggested that binding of phorbol diesters to cellular receptors and the down regulation of this binding are required for the initiation of the biological activity of the phorbol diesters in the HL-60 cells and perhaps other cell types. The interaction of these agents with specific receptors, and perhaps the down regulation of these receptors, may activate the protein kinase and consequently phosphorylate specific proteins. These proteins may then cause the expression of the "mutated tumor genes", either directly or indirectly through a series of changes, including the activation of phospholipid methylation (29) which, in turn, can affect DNA methylation, a process considered to be one of the regulators of gene expression.

In summary, in vitro systems that employ human and rodent cells offer useful tools for studying the control of mutagenesis and cell differentiation, as well as malignant cell transformation.

REFERENCES

1. Ames, B.N., F.D. Lee, and W.E. Durston (1973) An improved bacterial test system for the detection and classification of mutagens and carcinogens. Proc. Natl. Acad. Sci. USA 70:782-786.
2. Armuth, V., and I. Berenblum (1974) Promotion of mammary carcinogenesis and leukemogenic action by phorbol in virgin female Wistar rats. Cancer Res. 34:2704-2707.
3. Berenblum, I. (1969) A re-evaluation of the concept of cocarcinogenesis. Prog. Exp. Tumor Res. 11:21-30.
4. Barrett, J.C., B.D. Crawford, D.L. Grady, L.D. Hester, P.A. Jones, W.F. Benedict, and P.O.P. Ts'o (1977) Temporal acquisition of enhanced fibrinolytic activity by Syrian hamster embryo cells following treatment with benzo(a)pyrene. Cancer Res. 37:3815-3823.
5. Barrett, J.C., and P.O.P. Ts'o (1978) Evidence for the progressive nature of neoplastic transformation in vitro. Proc. Natl. Acad. Sci. USA 75:3761-3765.
6. Barrett, J.C., and P.O.P. Ts'o (1978) The relationship between somatic mutation and neoplastic transformation. Proc. Natl. Acad. Sci. USA 75:3297-3301.

7. Berwald, Y., and L. Sachs (1965) In vitro transformation of
    normal cells to tumor cells by carcinogenic hydrocarbons. J.
    Natl. Cancer Inst. 35:641-661.
8. Borek, C. (1979) Malignant transformation in vitro: Criteria,
    biological markers, and application in environmental
    screening of carcinogens. Radiat. Res. 79:209-232.
9. Bouck, N., and G. diMayorca (1976) Somatic mutation as the
    basis for malignant transformation of BHK cells by chemical
    carcinogens. Nature 264:722-727.
10. Boutwell, R.K. (1974) The function and mechanism of promoters
    of carcinogenesis. Critical Reviews in Toxicology 2:419-443.
11. Brookes, P., and P.D. Lawley (1964) Evidence for the binding
    of polynuclear aromatic hydrocarbons to the nucleic acids of
    mouse skin: Relation between carcinogenic power of
    hydrocarbons and their binding to deoxyribonucleic acid.
    Nature 202:781-784.
12. Cabot, M.C., C.J. Welsh, M.F. Callaham, and E. Huberman (1980)
    Alterations in lipid metabolism induced by 12-0-
    tetradecanoylphorbol-13-acetate in differentiating human
    myeloid leukemia cells. Cancer Res. 40:3674-3679.
13. Cohen, R., M. Pacifici, N. Rubenstein, J. Biehl, and
    H. Holtzer (1977) Effect of a tumor promoter on myogenesis.
    Nature 266:538-540.
14. Colburn, N.H., W.F. Vorder Bruegge, J.R. Bates, R.H. Gray,
    J.D. Rossen, W.H. Kelsey, and T. Shimada (1978) Correlation
    of anchorage-independent growth with tumorigenicity of
    chemically transformed mouse epidermal cells. Cancer Res.
    38:624-634.
15. Diamond, L., T.G. O'Brien, and G. Rovera (1977) Inhibition of
    adipose conversion of 3T3 fibroblasts by tumor promoters.
    Nature 269:247-249.
16. DiPaolo, J.A., R.L. Nelson, and P.J. Donovan (1969) Sarcoma
    producing cell lines derived from clones transformed in vitro
    by benzo[a]pyrene. Science 165:917-918.
17. DiPaolo, J.A., R.L. Nelson, and P.J. Donovan (1971)
    Morphological, oncogenic, and karyological characteristics of
    Syrian hamster embryo cells transformed in vitro by
    carcinogenic polycyclic hydrocarbons. Cancer Res.
    31:1118-1127.
18. DiPaolo, J.A., N.C. Popescu, and R. Nelson (1973) Chromosomal
    binding patterns and in vitro transformation of Syrian
    hamster cells. Cancer Res. 33:3250-3258.
19. Doll, R., and R. Peto (1981) The causes of cancer:
    Quantitative estimates of avoidable risks of cancer in the
    United States today. J. Natl. Cancer Inst. 66:1191-1308.
20. Dunphy, W.G., K.B. Delclos, and P.M. Blumberg (1980)
    Characterization of specific binding of [^{3}H]phorbol 12,13-
    dibutyrate and [^{3}H]phorbol 12-myristate 13-acetate to mouse
    brain. Cancer Res. 40:3635-3641.

21. Farber, E. (1981) Sequential events in chemical carcinogenesis. In Cancer - A Comprehensive Treatise, F.F. Becker, Ed., 2nd Edition, Plenum Press, New York.
22. Foulds, L. (1975) Neoplastic Development, Vol. 2, Academic Press, London.
23. Fugiki, H., M. Suganuma, N. Matsukura, T. Sugimura, and S. Takayama (1982) Teleocidin from Streptomyces is a potent promoter of mouse skin carcinogenesis. Carcinogenesis 3:895-898.
24. Hass, B.S., C.K. McKeown, D.J. Sardella, E. Boger, P.K. Ghoshal, and E. Huberman (1982) Cell-mediated mutagenicity in Chinese hamster V79 cells of dibenzopyrenes and their bay-region fluorine-substituted derivatives. Cancer Res. 42:1646-1649.
25. Hecker, E. (1971) Isolation and characterization of the co-carcinogenic principles from croton oil, In Methods in Cancer Research, Vol. 6, H. Busch, Ed., Academic Press, Inc., New York, pp. 439-484.
26. Heidelberger, C. (1975) Chemical carcinogenesis. Ann. Rev. Biochem. 44:79-122.
27. Heidelberger, C., A.E. Freeman, R.J. Pienta, A. Sivak, J.A. Bertram, B.C. Casto, V.C. Dunkel, M.C. Francis, T. Kakunaga, J.B. Little, and L.M. Schechtman (1982) Cell transformation by chemical agents: A review and analysis of the literature. Mutation Res. 114:283-285.
28. Higginson, V. (1972) The role of geographical pathology in environmental carcinogenesis, In Environment and Cancer, 24th Symposium of Fundamental Cancer Research, Williams and Wilkins, Baltimore, pp. 62-92.
29. Hoffman, D.R., and E. Huberman (1982) The control of phospholipid methylation by phorbol diesters in differentiating human myeloid HL-60 leukemia cells. Carcinogenesis 3:875-880.
30. Huberman, E. (1978) Mutagenesis and cell transformation of mammalian cells in culture by chemical carcinogens. J. Environ. Pathol. Toxicol. 2:29-42.
31. Huberman, E. (1977) Viral antigen induction and mutability of different genetic loci by metabolically activated carcinogenic polycyclic hydrocarbons in cultured mammalian cells, In Origins of Human Cancer, Vol. 4, H.H. Hiatt, J.D. Watson, and J.A. Winsten, Eds., Cold Spring Harbor, New York, pp. 1521-1535.
32. Huberman, E., G.R. Braslawsky, M. Callaham, and H. Fujiki (1982) Induction of differentiation of human promyelocytic leukemia (HL-60) cells by teleocidin and phorbol-12-myristate-13-acetate. Carcinogenesis 3:111-114.
33. Huberman, E., and M.F. Callaham (1979) Induction of terminal differentiation in human promyelocytic leukemia cells by tumor-promoting agents. Proc. Natl. Acad. Sci. USA 76:1293-1297.

34. Huberman, E., C. Heckman, and R. Langenbach (1979) Stimulation of differentiated functions in human melanoma cells by tumor-promoting agents and dimethyl sulfoxide. Cancer Res. 39:2618-2624.

35. Huberman, E., and C. Jones (1980) The use of liver cell cultures in mutagenesis studies. Ann. N.Y. Acad. Sci. 349:264-272.

36. Huberman, E., R. Mager, and L. Sachs (1976) Mutagenesis and transformation of normal cells by chemical carcinogens. Nature 264:360-361.

37. Huberman, E., and L. Sachs (1966) Cell susceptibility to transformation and cytotoxicity by the carcinogenic hydrocarbon benzo[a]pyrene. Proc. Natl. Acad. Sci. USA 56:1123-1129.

38. Huberman, E., and L. Sachs (1976) DNA binding and its relationship to carcinogenesis by different polycyclic hydrocarbons. Int. J. Cancer 19:122-127.

39. Huberman, E., and L. Sachs (1976) Mutability of different genetic loci in mammalian cells by metabolically activated carcinogenic polycyclic hydrocarbons. Proc. Natl. Acad. Sci. USA 73:188-192.

40. Huberman, E., L. Sachs, S.K. Yang, and H.V. Gelboin (1976) Identification of the mutagenic metabolites of benzo(a)pyrene in mammalian cells. Proc. Natl. Acad. Sci. USA 73:607-611.

41. Huberman, E., S. Salzberg, and L. Sachs (1968) The in vitro induction of an increase in cell multiplication and cellular life span by the water-soluble carcinogen dimethylnitrosamine. Proc. Natl. Acad. Sci. USA 59:77-82.

42. Huberman, E., and T.J. Slaga (1979) Mutagenicity and tumor-initiating activity of fluorinated derivatives of 7,12-dimethylbenz(a)anthracene. Cancer Res. 39:411-414.

43. Jones, C.A., P.J. Marlino, W. Lijinsky, and E. Huberman (1981) The relationship between the carcinogenicity and mutagenicity of nitrosamines in a hepatocyte-mediated mutagenicity assay. Carcinogenesis 2:1075-1077.

44. Jones, C.A., R.M. Santella, E. Huberman, J.K. Selkirk, and D. Grunberger (1981) Cell specific activation of benzo(a)pyrene by fibroblasts and hepatocytes. Carcinogenesis 4:1351-1357.

45. Kakunaga, T. (1973) A quantitative system for assay of malignant transformation by chemical carcinogens using a clone derived from Balb/3T3. Int. J. Cancer 12:463-473.

46. Kakunaga, T. (1978) Neoplastic transformation of human diploid fibroblast cells by chemical carcinogens. Proc. Natl. Acad. Sci. USA 75:1334-1338.

47. Kennedy, A.R. (1982) Promotion and other interactions between agents in the induction of transformation in vitro in fibroblast-like cell culture systems, In Mechanisms of Tumor Promotion, Vol. 3, T.J. Slaga, Ed., Tumor Promotion and

Carcinogenesis in Vitro, CRC Press, Inc., Boca Raton, in press.

48. Kuroki, T., and H. Sato (1968) Transformation and neoplastic development in vitro of hamster embryonic cells by 4-nitroquinoline-1-oxide and its derivatives. J. Natl. Cancer Inst. 41:53-71.

49. Landolph, J.R., and C. Heidelberger (1979) Carcinogens produce mutations to ouabain resistance in transformable C3H/10T1/2 Cl 8 mouse fibroblasts. Proc. Natl. Acad. Sci. USA 76:930-934.

50. Langenbach, R., H.J. Freed, D. Raveh, and E. Huberman (1978) Cell specificity in metabolic activation of aflatoxin B$_1$ and benzo(a)pyrene to mutagens for mammalian cells. Nature 276:277-280.

51. Lotem, J., and L. Sachs (1979) Regulation of normal differentiation in mouse and human myeloid leukemic cells by phorbol esters and the mechanism of tumor promotion. Proc. Natl. Acad. Sci. USA 76:5158-5162.

52. Mager, R., E. Huberman, S.K. Yang, H.V. Gelboin, and L. Sachs (1977) Transformation of normal hamster cells by benzo(a)pyrene diol-epoxide. Int. J. Cancer 19:814-817.

53. Marchok, A.C., J.C. Rhoton, and P. Nettesheim (1978) In vitro development of oncogenicity in cell lines established from tracheal epithelium preexposed in vivo to 7,12-dimethylbenz(a)anthracene. Cancer Res. 38:2030-2037.

54. Mondal, S., and C. Heidelberger (1970) In vitro malignant transformation by methylcholanthrene of the progeny of single cells derived from C3H mouse prostate. Proc. Natl. Acad. Sci. USA 65:219-225.

55. Mondal, S., D.W. Brankow, and C. Heidelberger (1976) Two-stage chemical oncogenesis in cultures of C3H/10T1/2 cells. Cancer Res. 36:2254-2260.

56. Miller, J.A. (1970) Carcinogenesis by chemicals: An overview. G.H.A. Clowes Memorial Lecture. Cancer Res. 30:559-576.

57. Montesano, R., L. Saint Vincent, C. Drevon, and L. Tomatis (1975) Production of epithelial and mesenchymal tumours with rat liver cells transformed in vitro. Int. J. Cancer 16:550-558.

58. Niedel, J.E., L.T. Kuhn, and G.R. Vandenbark (1983) Phorbol diester receptor copurifies with protein kinase C. Proc. Natl. Acad. Sci. USA 80:36-40.

59. Peraino, C., E.F. Staffeldt, D.A. Haugen, L.S. Lombard, F.J. Stevens, and R.J.M. Fry (1980) Effect of varying the dietary concentration of phenobarbital on its enhancement of 2-acetylaminofluorene-induced hepatic tumorigenesis. Cancer Res. 40:3268-3273.

60. Pitot, H.C., T. Goldsworthy, and S. Moran (1981) The natural history of carcinogenesis: Implications of experimental carcinogenesis in the genesis of human cancer. J. Supramolec. Struct. Cell Biochem. 17:133-146.

61. Raveh, D., T.J. Slaga, and E. Huberman (1982) Cell mediated mutagenesis and tumor-initiating activity of the ubiquitous polycyclic hydrocarbon, cyclopenta[c,d]pyrene. Carcinogenesis 3:763-766.

62. Rovera, G., T.A. O'Brien, and L. Diamond (1977) Tumor promoters inhibit spontaneous differentiation of Friend erythroleukemia cells in cultures. Proc. Natl. Acad. Sci. USA 74:2894-2898.

63. Rovera, G., D. Santoli, and C. Damsky (1979) Human promyelocytic leukemia cells in culture differentiate into macrophage-like cells when treated with a phorbol diester. Proc. Natl. Acad. Sci. USA 76:2779-2783.

64. Ryffel, B., C.B. Henning, and E. Huberman (1982) Differentiation of human T-lymphoid leukemia cells into cells that have a suppressor phenotype is induced by phorbol 12-myristate 13-acetate. Proc. Natl. Acad. Sci. USA 79:7336-7340.

65. Selkirk, J.K. (1977) Benzo[a]pyrene carcinogenesis: A biochemical selection mechanism. J. Toxicol. Environ. Health 2:1245-1258.

66. Slaga, T.J., S.M. Fischer, K. Nelson, and G.L. Gleason (1980) Studies on the mechanism of skin tumor promoters: Evidence for several stages in promotion. Proc. Natl. Acad. Sci. USA 77:3659-3663.

67. Slaga, T.J., E. Huberman, J.K. Selkirk, R. Harvey, and W.M. Bracken (1978) Carcinogenicity and mutagenicity of benzo(a)anthracene diols and diol-epoxides. Cancer Res. 38:1699-1704.

68. Slaga, T.J., A. Viaje, W.M. Bracken, S.G. Buty, D.R. Miller, S.M. Fischer, C.K. Richter, and J.N. Dumont (1978) In vitro transformation of epidermal cells from newborn mice. Cancer Res. 38:2246-2252.

69. Solanki, V., T.J. Slaga, M. Callaham, and E. Huberman (1981) Down regulation of specific binding of [20-^{3}H]phorbol 12,13-dibutyrate and phorbol ester-induced differentiation of human promyelocytic leukemia cells. Proc. Natl. Acad. Sci. USA 78:1722-1725.

70. Weinstein, I., A. Horowitz, P. Fisher, V. Ivanovic, S. Gattanicelli, and P. Kirschmeier (1982) Mechanisms of multistage carcinogenesis and their relevance to tumor cell heterogeneity, In Tumor Cell Heterogeneity, Academic Press, New York, pp. 261-283.

71. Williams, G.M., J.M. Elliott, and J.H. Weisburger (1973) Carcinoma after malignant conversion in vitro of epithelial-like cells from rat liver following exposure to chemical carcinogens. Cancer Res. 33:606-612.

72. Yamaguchi, N., and I.B. Weinstein (1975) Temperature-sensitive mutants of chemically transformed epithelial cells. Proc. Natl. Acad. Sci. USA 72:214-218.

73. Yamasaki, H., E. Fibach, U. Nudel, I.B. Weinstein,
    R.A. Rifkind, and P.A. Marks (1977) Tumor promoters inhibit
    spontaneous and induced differentiation of murine
    erythroleukemia cells in cultures. Proc. Natl. Acad. Sci.
    USA 74:3451–3455.
74. Yuspa, S.H., H. Hennings, and V. Lichti (1981) Initiator and
    promoter induced specific changes in epidermal function and
    biological potential. J. Supramolec. Struct. Cell. Biochem.
    17:245–257.

DOMINANT-LETHAL MUTATIONS AND HERITABLE TRANSLOCATIONS IN MICE

Walderico M. Generoso

Biology Division
Oak Ridge National Laboratory
Oak Ridge, TN 37830

SUMMARY

Chromosome aberrations are a major component of radiation or chemically induced genetic damage in mammalian germ cells. The types of aberration produced are dependent upon the mutagen used and the germ cell stage treated. For example, in male meiotic and postmeiotic germ cells certain alkylating chemicals induce both dominant-lethal mutations and heritable translocations while others induce primarily dominant-lethal mutations. Production of these two endpoints appears to be determined by the stability of alkylation products with the chromosomes. If the reaction products are intact in the male chromosomes at the time of sperm entry, they may be repaired in fertilized eggs. If repair is not effected and the alkylation products persist to the time of

---

Research sponsored by the Office of Health and Environmental Research, U.S. Department of Energy under contract W-7405-eng-26 with the Union Carbide Corporation.

Abbreviations: Bap, benzo[a]pyrene; IMS, isopropyl methanesulfonate; EMS, ethyl methanesulfonate; MMS, methyl methanesulfonate; ENU, ethylnitrosourea; MNU, methylnitrosourea; TEM, triethylenemelamine; EtO, ethylene oxide.

pronuclear chromosome replication, they lead to chromatid-type
aberrations and eventually to dominant-lethality. The production
of heritable translocations, on the other hand, requires a
transformation of unstable alkylation products into suitable
intermediate lesions. The process by which these lesions are
converted into chromosome exchange within the male genome takes
place after sperm enters the egg but prior to the time of
pronuclear chromosome replication (i.e., chromosome-type). Thus,
dominant-lethal mutations result from both chromatid- and
chromosome-type aberrations while heritable translocations result
primarily from the latter type. DNA target sites associated with
the production of these two endpoints are discussed.

I.   INTRODUCTION

        Chromosome breakage contributes heavily to human genetic
burden. It may result in chromosome loss, which often leads to
lethality but occasionally to viable aneuploids, or to viable
exchanges and inversions. In mice this class of genetic damage is
readily induced in certain germ cells by ionizing radiations and
by numerous chemical mutagens. Thus, transmissable chromosomal
aberrations resulting from induced breaks and rearrangements are
certainly a major concern in the evaluation of genetic risk. How
they are produced in germ cells and what their consequences are in
affected conceptuses are important questions in our attempt to
evaluate genetic risk from exposure to chemical mutagens and to
understand the mechanisms for aberration induction.

        For obvious reasons, aberrations that lead to early lethality
among conceptuses do not contribute to the population's genetic
burden. On the other hand, aberrations that permit survival and
reproduction have untoward effects on the affected individuals as
well as on some of their immediate descendants. In mice,
dominant-lethal mutations and heritable translocations not only
best exemplify these two classes of end effects but are also the
most widely used endpoints of transmitted chromosome breakage
effects in practical testing and hazard evaluation. Thus, this
paper will be restricted to the discussion of these two endpoints
as they are used in practical testing - i.e., response of male
germ cells - with emphasis on possible mechanisms of induction.

II. DOMINANT-LETHAL MUTATIONS

A. Historical Background

        Dominant-lethal mutations are genetic changes in parental
germ cells that cause death of affected first generation progeny.
The earliest information on dominant-lethal effects in mammals

Table 1. Dose-effect of EMS in the induction of dominant-lethal mutations in male mice[a].

Dose (mg/kg)	No. of mated females[b]	No. of pregnant females	Total implants among fertile females (avg)	Living embryos among fertile females (avg)	Dead implants (percent)	Living embryos as percent of controls	
						Among fertile females	Among all mated females
Control	22	22	7.7	7.1	7	-	-
100	25	22	7.5	7.0	7	99	86
150	21	21	7.2	5.9[c]	19	83	82
200	27	25	6.4[d]	3.5	45	49	46
250	29	15[c]	4.5[c]	1.1	71	15	8
300	21	1	2.0	0	100	0	0

[a]From Ref 17.

[b]All matings occurred 6.5–7.5 days after treatment.

[c]$P < 0.01$ for comparison with control.

[d]$P < 0.05$ for comparison with control.

came from experiments with ionizing radiation (see 38 for review).
It was first observed that exposure of male laboratory animals to
X-rays prior to mating during the presterile period resulted in an
increase in abnormal embryos and a reduction in litter size.
These effects were eventually shown to be caused by induced
chromosomal aberrations.

The first observation of dominant-lethal effects induced by a
chemical mutagen was made with the compound TEM. Jackson and Bock
(29) observed that males given either a single dose, or daily
doses for five consecutive days, or daily doses of TEM for a
prolonged period became infertile, even though the males mated
normally and produced spermatozoa during and after treatment.
This "infertility effect", which was also observed in mice, was
found shortly to be the result of early embryonic death (3,8,47).
The demonstration by Cattanach (7) that TEM induced heritable
translocations at the male postmeiotic stages in which embryonic
lethality effects were also associated suggested strongly that
chromosomal aberrations were the cause of the dominant-lethal
response. This cause and effect relationship was later proved
from cytological analyses of first and early cleavage embryos
(4,28,35). Actually, an "infertility effect" similar to that
induced by TEM was reported first for nitrogen mustard by
Falconer, Slizynski and Auerbach (15). However, up to this time
dominant-lethal data for this compound are inconclusive.

B. Expression of Dominant-Lethal Effects

In mice, embryonic death resulting from dominant-lethal
mutations is usually expressed between the two-cell stage and
shortly after implantation. Low levels of dominant-lethal effects
are usually expressed in terms of increased number of deciduomata.
Dominant-lethal mutations that are expressed in embryos which fail
to implant are usually observed when the dominant-lethal effects
are high. In fact, at extreme levels, all embryos in a female may
die so well before implantation that the female shows no sign of
pregnancy. An example of the induced dominant-lethal effects is
shown in Table 1 where induced effects were observed in the
following order as a function of dose: increase in percentage of
dead implants; reduction in the average number of living embryos;
reduction in the average number of total implants; and reduction
in the proportion of pregnant females among mated ones. It is
generally believed that induced dominant-lethal mutations are
almost exclusively chromosome breakage events. The dominant-
lethal test is useful for testing the mutagenicity of chemicals.
It is a simple and quick procedure, and many laboratories
throughout the world are now using this method in practical
testing.

C. Spermatogenesis Stage Differences in Sensitivity to Dominant-Lethal Induction

It is extremely important for anyone who intends to use the dominant-lethal test to have a good understanding of the variety of responses to chemical mutagens at various stages in spermatogenesis. To date, a great number of chemicals had already been studied for induction of dominant-lethal mutations in male mice. Those which are clear-cut inducers induced dominant lethals only at specific stages in spermatogenesis - i.e., none of them induced dominant lethals at all stages of spermatogenesis and the stage or stages affected may vary from one chemical to another. To illustrate stage differences, only a few chemicals will be considered here.

Chemicals differ dramatically in the stages at which dominant lethals can be induced. For instance, isopropyl methanesulfonate (13) and triethylene melamine (8) induce dominant lethals in spermatozoa, spermatids and spermatocytes; ethyl methanesulfonate, methyl methanesulfonate and n-propyl methanesulfonate only in spermatozoa and spermatids (12,13); Myleran in spermatozoa and spermatocytes (11); Mitomycin C in spermatids, spermatocytes and possibly differentiating spermatogonia (10); and 6-mercaptopurine exclusively in late differentiating spermatogonia and possibly early meiotic spermatocyte (36,18). It should be noted that all chemicals mentioned here induced dominant lethals in postmeiotic germ cells in one stage of development or another, with the exception of 6-mercaptopurine. So far, no chemical agent has been shown unequivocally to be effective in inducing dominant lethals in spermatogonial stem cells. This is not to say that no chemicals induce chromosome breakage in spermatogonia stem cells. On the contrary, there is evidence, albeit equivocal, through cytological analysis of spermatogonial metaphases (mixture of differentiating gonia and stem cells), that certain chemicals do break chromosomes of these cells (1,33). If, indeed, chromosome breakage is induced at this germ cell stage, the absence of dominant-lethal effect may be explained by the possibility that cells with chromosome lesion do not make it to the ejaculate - i.e., they are lost sometime during spermatogonial divisions and spermatogenic maturation.

D. Differences Between Stocks of Females in their Yield of Dominant-Lethal Mutations

The oocyte is ovulated with its chromosomes in the metaphase stage of the second meiotic division. It remains in this stage until stimulated by sperm entry to undergo further development. The fertilized eggs of mice can repair certain premutational lesions present in the fertilizing sperm and the yield of dominant-lethal mutations is affected by the strain of females

Table 2. Differences between stock of male mice in response to dominant-lethal effects of EMS.

Treatment	Stock of males[a]	Number of mated females[b]	Number of pregnant females	Number of live embryos (ave)	Dead implants (%)	Dominant lethals[b] (%)
EMS[c]	(C3H × C57BL)F$_1$	39	36	5.4	47	50
(200 mg/kg)	(101 × C3H)F$_1$	37	35	2.8	68	74
	(SEC × C57BL)F$_1$	45	43	5.7	44	47
	T-stock	43	39	7.6	30	30
Control[d]	(C3H × C57BL)F$_1$	23	22	10.2	9	
	(101 × C3H)F$_1$	18	17	10.6	7	
	(SEC × C57BL)F$_1$	21	19	10.8	2	
	T-stock	23	19	11.5	4	

[a]Females used were from (C3H × C57BL)F$_1$ stock.

[b]Calculated using the formula: % D.L. = $[1 - \dfrac{\text{average number of living embryos (experimental)}}{\text{average number of living embryos (control)}}] \times 100.$

All calculations were based on the pooled control average of 10.8 from all stocks.

[c]Treatment to fertilization interval – 6½ to 9½ days. EMS was administered as a single i.p. injection.

[d]Control females were mated 4½ to 7½ days after injection. They were also used as contemporary control for another study.

used to mate with the treated males (21). The ability of
fertilized eggs to carry out repair and permit survival of the
embryo varies from one strain to another depending on the chemical
mutagen used. However, we have found that certain strains of
females consistently gave relatively high dominant-lethal
frequencies for all chemical mutagens we have studied so far.
Because the genotype of the egg has a significant role in the
processing of premutational lesions that are carried in the
chromosomes of mutagen-treated male germ cells, this phenomenon
must be taken into consideration in practical testing. Without
taking this factor into account, negative results may simply mean
that treated males were mated to females from repair-competent
strains.

E. Differences Between Stocks of Males in their Sensitivity to Induction of Dominant-Lethal Mutations

The yield of dominant-lethal mutations may or may not be
affected by the stock of males treated, depending upon the mutagen
used. As shown in Table 2, the yield of EMS-induced dominant
lethals for $(101 \times C3H)F_1$ was about twice as much as that for T-
stock males. A similarly clear-cut difference between these two
stocks exists for IMS, with T-stock males showing lower dominant-
lethal effects than $(101 \times C3H)F_1$. T-stock males, however, are
not always relatively more resistant to mutagenic chemicals. It
can be seen in Table 2 that T-stock males were no more resistant
than $(101 \times C3H)F_1$ males to dominant-lethal induction with TEM.
The explanation for the demonstrably different responses between
stocks of males to dominant-lethal effects of IMS or EMS is not
known but metabolic inactivation and transport processes are
possible factors.

F. General Procedure for Dominant-Lethal Test in Males

From the practical standpoint, the first objective in doing
the dominant-lethal test must be to determine whether or not the
test compound is a mutagen in germ cells. The second objective is
to determine if the test compound is capable of inducing dominant-
lethal mutations when administered via a route similar to the
human route of exposure. Thus, the route of administration one
uses in practical testing should provide the best chance that a
chemical mutagen will be detected, and at the same time it should
be most relevant to the human situation. In many cases one single
route of administration will not serve both objectives. For
example, it may be more relevant to human hazard to administer a
given chemical by gavage or inhalation, but the best chance that
this chemical will be detected as a germ cell mutagen may be when
it is given by intraperitoneal, intravenous, intramuscular or
subcutaneous injection. It is also essential to remember that,
depending upon the chemical, the rate at which it is given may

determine the outcome with respect to its being declared a mutagen or not. A single large dose or a few large doses spaced one day apart may produce higher dominant-lethal effects than lower doses administered over a much longer period of time. Finally, the treatment protocol may also be affected by the nature of the toxicity of the test compound. Some chemicals can be given daily at a relatively high dose without accumulating toxic effects while with others, toxic effects accumulate even at relatively low daily doses. Therefore, depending upon the nature of the test chemical, it may be necessary to use more than one treatment procedure.

1. Treatment of parental males. Generally, three methods of treatment are used in practical dominant-lethal testing: single dose, daily dosing for five consecutive days, and daily dosing for 8 weeks. Whatever method of treatment is used it is generally believed that the test should include the maximum non-toxic dose or concentration (also called maximum tolerance dose, MTD, or maximum tolerated concentration, MTC). The MTD information, if not yet available, should be determined first in a toxicity study, in which immediately after treatment (in case of single dose) or immediately after the last dosing (in case of repeated treatment) or at the last day of exposure (in case of feeding studies) surviving males are to be caged with untreated females in order to find out the effect of treatment on the mating ability of males.

The optimum number of parental males to use in a dominant-lethal experiment depends in part on the ability of the males to breed and on the reproductive performance of the females. When suitable males and females are used, 36 males for each experimental and control group should be adequate.

2. Mating procedure. The mating protocol to be used depends upon the treatment procedure. Males given a single dose and males given five consecutive daily doses should be mated serially for at least 8 weeks. This length of mating period will ensure analysis of the response of the various stages in spermatogenesis. Males that are exposed for 8 consecutive weeks, on the other hand, would have had their germ cells exposed to the chemical throughout the time spermatogenic cells are going through the maturation process. Consequently, these males need to be mated for only one week beginning immediately after the end of exposure.

When libido is not a problem, each treated male is caged with two untreated females. Every morning, females are examined for the presence of vaginal plugs (indication of mating) and each female that copulated is removed and replaced by a virgin female. All mated females are killed for uterine analysis 12 to 15 days after observation of the vaginal plug.

G. Selection of Females for Use in the Dominant-Lethal Test

In addition to the repair phenomenon described in Section II.D, there is another important consideration in choosing the strain of females to use in the dominant-lethal test. This has to do with the various criteria (see Section II.H) for evaluating dominant-lethal effects. All criteria are based upon information obtainable from analysis of the uterus some time after the female had been mated. Of particular importance is the normal incidence of dead implantation because clear-cut increases in the incidence of dead implants is unequivocal evidence that dominant-lethal mutations were induced in treated males. Obviously, it is most desirable to use a strain of females in which the normal incidence of dead implantation is low. It should be emphasized that the incidence of dead implantation can vary greatly, not only with age, but also with the strain of females. Examples of normal incidence of dead implantation in control females from various strains are shown in Table 3. In addition to low frequency of dead implantation, other qualities of desirable strains are large litter size, high proportion of matings (as indicated by vaginal plugs) during the receptive stage of the estrous cycle and high uniformity among females. These characteristics are more likely to be found among hybrid and random-bred stocks than in inbreds.

H. Evaluation of Mutagenicity

Mutagenicity of the test compound is decided by a combination of the following criteria: (i) increase in the frequency of dead implantations; (ii) increase in the number of females with one or

Table 3. Strain Differences in Reproductive
Performance of Female mice[a].

Strain[b]	Number of mated females[c]	Fertile matings (%)	Number of implants (ave)	Number of living embryos (avg)	Dead implants (%)
T-stock	65	94	8.5	6.8	20
(101 × C3H)F$_1$	51	96	7.3	6.9	6
(SEC × C57BL)F$_1$	40	83	9.8	9.3	5

[a]From Ref 16.

[b]These females served as controls of an experiment.

[c]Mating was indicated by the presence of vaginal plug.

more dead implants; (iii) reduction in the average number of
living embryos; (iv) reduction in the average number of
implantations; and (v) reduction in the frequency of fertile
matings. Generally the first three criteria and, in some cases
(when the induction rate is high), also the fourth criterion are
expressed together. The fifth criterion is expressed only when
dominant-lethal induction approaches 100%--i.e., each fertilizing
sperm carries at least one lethal mutation that results in
embryonic death prior to implantation. Apparent sterility of some
treated males may be due to dominant lethality or to physiologic
reasons such as the inability of treated males to mate during the
posttreatment sick phase; thus the value of checking for vaginal
plugs. In experiments where males are serially mated, analysis
may be done on data pooled into successive two-day intervals. In
experiments where males are mated for only one week, data may be
pooled into two groups--first four days and last three days.

III. HERITABLE TRANSLOCATIONS

One of the consequences of chromosome breakage induced in
specific stages in male gametogenesis is the production of
symmetrical reciprocal translocations which can be passed to some
of the first generation progeny. Carriers of these
translocations, referred to as translocation heterozygotes, are
highly viable and generally cannot be distinguished from normal
mice by casual observation of progeny. Because heritable
translocations are, by definition, scored among live progeny, they
provide a definitive and unequivocal measure of chromosome-
breakage effects. Furthermore, they are generally considered to
be the most important endpoint of induced chromosome aberrations
with respect to genetic risk assessment.

A. Historical Background (see 23 and 38 for reviews)

As with dominant-lethal mutations, the initial evidence that
heritable translocations in mice are readily inducible in certain
postspermatogonial stages came from early studies with ionizing
radiation, which showed that among the progeny of irradiated males
there were a number of semisterile animals which transmitted the
semisterile characteristic to about half of their progeny (42-44).
The hypothesis that the cause of semisterility was the presence of
a translocation leading to the production of gametes with
unbalanced chromosome constitution was confirmed a few years later
through genetic (45,46) and cytological (30,31) studies.

In the late 1940s and early 1950s when it was already clear
that, like X-rays, nitrogen mustard compounds have the ability to
induce chromosome rearrangements in several species, attempts were
made to see if this effect could be found in the mouse (2,15).

Although one of these studies (15) indicated the effectiveness of nitrogen mustard, it failed to demonstrate convincingly the induction of heritable translocation when male mice were treated with this chemical. This failure may be attributed to the low number of progeny tested. Unfortunately, no further study with nitrogen mustard has been reported.

In 1957, the first study showing clear-cut evidence of the induction of heritable translocations by a chemical, triethylenemelamine (TEM), was reported (7). Subsequent studies with TEM and other alkylating chemicals showed that heritable translocations are readily induced in male postmeiotic and meiotic stages but not in earlier stages.

B. Effects of Translocations Among Carriers

Approximately one-third of all chemically- or radiation-induced translocations result in male sterility. The nature of these translocations has been studied extensively (5,6,37). One general class of translocations that results in male sterility consists of sex-chromosome - autosome translocations. Another general class, which constitutes the majority of induced cases of F_1 male sterility, involves translocations between autosomes in which at least one of the breaks occurs close to one end of a chromosome (either distal or proximal). Occasionally sterility also ensues when more than one reciprocal translocation is present. The great majority of sterile males have distinctly small testes (about one-third normal size). In most cases, spermatogenesis is blocked at one stage or another; of the few that have sperm in the epididymis, the concentration is markedly lower than normal and the sperm are generally nonmotile with a high frequency of morphological abnormalities, such as bent tails. Generally, females that are heterozygous for either class of translocation are semisterile (19,37).

The remaining translocations are of the partially sterile kind. The degree of partial sterility is dependent upon the proportion in which balanced and unbalanced gametes are represented in the ejaculate, and this in turn is a function of meiotic segregation. The unbalanced sperm, which are produced through adjacent-1 and adjacent-2 segregations and 3-1 missegregation, are capable of fertilization, but they lead to early embryonic lethality observed primarily as resorption moles. The degree of partial sterility may vary from substantially lower to substantially higher than 50%. On average, the percentage of living embryos among normal females mated to partially sterile translocation males is only 43-44% that of normal females mated to normal males (Table 4). This indicates that the percentage of gametes in the ejaculate with unbalanced chromosome constitution is 56-57%. There is evidence that the length of the translocated

Table 4. Average Fertility of Partial Sterile Male
Translocation Heterozygotes Produced from
Postmeiotic Treatment of Male Mice.

Treatment	Class	No. of males tested	No. of implants[b] (avg)	No. of living embryos[b] (avg)	Dead implants (%)
X-ray	Partially sterile	30	9.1	3.9 (43%)[c]	57
	Normal	39	9.6	9.0	6
TEM	Partially sterile	119	9.2	4.2 (44%)[c]	54
	Normal	69	10.0	9.5	5
EMS	Partially sterile	98	9.3	4.5 (44%)[c]	58
	Normal	39	10.7	10.2	5

[a] From Ref 20.

[b] Six pregnancies were analyzed for each partially sterile male and three for each normal male.

[c] Values in parentheses are percentages of the average number of living embryos for normal mice.

chromosome segment has some influence on the proportion of unbalanced gametes in the ejaculate (24). Long translocated segments appear to favor the formation of unbalanced gametes.

Certain translocations may have other adverse effects in addition to infertility. For example, Selby (40) has found three different translocations, each of which was associated with specific skeletal abnormality. We have recently found a translocation that is associated with behavioral abnormality (chromosomal and pathological studies on this translocation stock are in progress).

C. Inducibility of Heritable Translocations at Various Stages in Male Gametogenesis

Like dominant-lethal mutations, induction of heritable translocations is stage-dependent (19). Among chemicals studied so far, only alkylating chemicals have been shown clearly to induce heritable translocations. However, not all alkylating chemicals are effective in inducing heritable translocations. Chemically induced heritable translocations have been recovered only from treated meiotic and postmeiotic germ cells of males. Extensive study on spermatogonia stem cells showed that none of the chemicals studied (TEM, TEPA and cyclophosphamide) significantly induced heritable translocations at this germ cell stage. Ionizing radiations, on the other hand, are clearly effective in inducing heritable translocations in the gonia stem cell (Generoso et al., in press). This striking difference between alkylating chemicals and ionizing radiations is surprising in view of the fact that the chemicals studied, like ionizing radiations, are potent inducers of heritable translocations in male postmeiotic stages. To date, there is no satisfactory explanation for this difference.

D. General Procedure for the Heritable Translocation Test

The heritable translocation test is carried out in the following sequence: (i) treatment of parental males; (ii) mating of treated males and production of first generation progeny; (iii) testing of progeny for translocation heterozygosity; and (iv) statistical analysis of data. Detailed presentations of the basic principles and extent of use of the heritable translocation test were published previously (19,23). Please refer to these publications for details.

1. Treatment of parental males. Because of stage specificity, the treatment and mating procedures must ensure the sampling of the stage most sensitive to the test chemical if this chemical is indeed a mutagen. Unlike in the dominant-lethal test, it is not practical to use the single dose or the five consecutive days regimen followed by a long-term mating schedule, because ensuring adequate sampling of the most sensitive period would require large total numbers of progeny. Thus, the only practical method of treatment for the heritable translocation test is to subject parental males to long-term exposure. It was mentioned earlier that heritable translocations were inducible by chemical mutagens only in meiotic and postmeiotic stages. To ensure treatment of these stages, males need to be exposed continuously for a minimum of 5 weeks. Exposure time may be extended to 8 weeks or longer to allow manifestation of the effect the chemical may induce on germ-cell maturation (e.g., enzymatic alterations) in ways that could increase subsequent sensitivity to chemically-induced chromosomal lesions. If only one dose level is used, this dose should be the MTD. Finally, because the primary use of the heritable translocation data is in genetic risk assessment, the route of administration should be the one that is most relevant to the human situation.

2. Mating procedure. When libido is not a problem, each
treated male is caged with two untreated females for a period of
one week immediately after the end of treatment. As a general
rule, larger effects are expected when males are mated closer to
the end of treatment. At the end of one week, females are
separated from males and caged individually. All male progeny are
weaned and all female progeny are discarded.

3. Testing of male progeny for translocation heterozygosity.
Screening for translocation heterozygotes may be accomplished by
using one of two general procedures. The first method, referred
to as fertility technique, consists of initial testing of the
males for sterility and partial sterility (also referred to as
semisterility) and subsequent cytological analysis of suspect
progeny. The other method bypasses the fertility test; all male
progeny are subjected to cytological analysis (cytological
technique). These procedures were described in detail in
references 19 and 23.

IV. POSSIBLE MECHANISMS FOR CHEMICAL INDUCTION OF DOMINANT-LETHAL
MUTATIONS AND HERITABLE TRANSLOCATIONS

Germ cell stages differ from one another in many biological
properties, including repair competency and interval between S
phases, and it is reasonable to assume that any particular
mutagenesis-related mechanism that operates for one germ cell
stage may not necessarily operate for another. For this reason,
discussion of the mechanisms for induction of dominant-lethal
mutations and heritable translocations is restricted to stages in
which both endpoints are known to be inducible, i.e., meiotic and
postmeiotic male germ cells.

Because dominant-lethal mutations and heritable
translocations are both endpoints of chromosome breakage events
and because ionizing radiations produce them simultaneously in the
same meiotic and postmeiotic germ cell stages, it was natural to
assume a priori that production of these two endpoints involves
the same initial events that randomly result in symmetrical
(heritable translocations) and asymmetrical exchanges and
deletions (dominant-lethal mutations). This general belief was
strengthened by the observations that the alkylating agents
studied in the beginning, i.e., TEM and EMS, not only induced both
endpoints but the respective rates of induction were positively
correlated in dose-response studies (8,9,17,20,34). To date,
however, new information (i) strongly indicates that the chemical
induction of these two endpoints does not necessarily share the
same mechanism and (ii) provides an insight into the possible
molecular events that lead to the production of dominant-lethal
mutations and heritable translocations.

With respect to alkylating chemicals, it is reasonable to assume that alkylation of DNA is the initial step that leads to aberration formation. [It should be pointed out that Sega and Owens (39) associated protamine alkylation to dominant-lethal effects of EMS]. The challenge, however, lies in determining which specific target sites are responsible for the formation of various types of aberrations. This is a difficult problem because there is a multitude of reactive sites in the DNA molecule and it has not been possible to effect binding only on any single target site in mammalian germ cell chromosomes. Thus, at best, interpretation of the most likely adducts responsible for induction of dominant-lethal mutations and heritable translocations comes from association between genetic data and what is known about binding with DNA and about the properties of various adducts.

Contrary to previously-held theories, dominant-lethal mutations and heritable translocations are not always induced at the same relative rates (Table 5). Among compounds studied in our laboratory over the years, IMS, BaP, and ENU were found to be

Table 5. Association Between DNA adducts and Inducibility
of Dominant-Lethal Mutations and
Heritable Translocations.

Mutagen	Induction of		Alkylation at N-7 Guanine
	Dominant Lethals	Heritable Translocations	
BaP	High	Not Detected	-
IMS[b]	High	Low	7.6
ENU[b]	High	Low	11.5
MNU	High	High	67.0
MMS	High	High	83.0
EMS	High	High	65.0
EtO	High	High	90.0
TEM	High	High	-
Cyclophosphamide	High	High	-

[a] Percentage of total alkylation in DNA (14,32,41).

[b] Most of the reaction products are in oxygen of the phosphate backbone. Oxygen in bases are also alkylated to a lesser extent.

effective in inducing dominant-lethal mutations but they induce
very few or no heritable translocations at the same germ cell
stages. This is in contrast to the effects of EMS, MMS, TEM, EtO,
MNU and cyclophosphamide, all of which are effective in inducing
both endpoints. This finding suggests that the primary lesions
produced by IMS, BaP, or ENU that resulted in dominant-lethal
mutations are different from those that resulted in heritable
translocations. In other words, the simple interpretation that
the same mechanism is responsible for the random production of
dominant-lethal mutations and heritable translocations is not
always correct.

It was concluded in previous reports (22,26,27) that the
relative rates at which dominant-lethal mutations and heritable
translocations are produced from chemical treatment of meiotic and
postmeiotic male germ cells depend upon the stability of
alkylation products with DNA. Heritable translocations are
induced by chemicals at a high rate relative to dominant-lethal
mutations when the corresponding alkylation products are converted
into interchanges prior to the first postfertilization chromosomal
division, and chemicals whose reaction products persist to the
time of first chromosomal division, or possibly even to subsequent
early-cleavage divisions, induced mainly the types of aberrations
that lead to dominant lethality. Further, heritable
translocations arise primarily from chromosome-type exchanges
while dominant-lethal mutations arise from both chromosome- and
chromatid-type aberrations. For chemicals like IMS, ENU and BaP
dominant-lethal mutations appear to come primarily from chromatid-
type aberrations. Thus, chromosome- and chromatid-type
aberrations are associated with unstable and stable reaction
products, respectively.. The question, then, is: What are the
corresponding DNA target sites?

We have stated earlier that our interpretation of the likely
adducts responsible for the production of dominant-lethal
mutations and heritable translocation will have to be made from
association between genetic and molecular data. In Table 3, one
can see the compounds for which genetic and DNA binding data are
available. The chemicals that are effective in inducing heritable
translocations (EMS, MMS, MNU and EtO) alkylate primarily the N-7
position of guanine. The chemicals that are ineffective in
inducing heritable translocations but are effective in inducing
dominant-lethal mutations (IMS and ENU) alkylate primarily the
oxygen of the phosphate backbone (forming phosphotriesters) and
oxygen in bases, such as O-6 position in guanine, and there is
very little alkylation of the nitrogen positions. Consistent with
the stability interpretation, N-7 alkyl guanine adducts are not
stable and are lost via hydrolysis resulting in the formation of
apurinic sites. Alkylation products with oxygen of bases and
phosphotriesters, on the other hand, are highly stable. Thus,

heritable translocations are associated with unstable N-7 alkylguanine and, probably, with N-3 alkyladenine as well since the latter is also highly unstable.

There is evidence that the process of chromosome exchange involved in heritable translocations takes place after sperm entry (24) and that it is necessary for the unstable adducts to be transformed into intermediate lesions, perhaps into apurinic sites, before fertilization in order for this process of exchange to take place (27).

The unstable adducts could also lead to dominant lethality through formation of chromosome-type asymmetrical exchanges and deletions, but the stable oxygen alkylations are associated primarily with the production of dominant-lethal mutations. After fertilization the stable adducts in the male genome are either repaired by the egg (see Section II.D) or lead to chromatid-type aberrations.

Finally, it is inherent in our interpretation that the mechanisms for the production of chromosome- and chromatid-type aberrations are not necessarily mutually exclusive. On the contrary, it allows for both classes to be produced at relative rates that may differ from one mutagen to another, depending upon the array of DNA adducts produced.

REFERENCES

1. Adler, I.-D. (1974) Comparative cytogenetic study after treatment of mouse spermatogonia with mitomycin C. Mut. Res. 23:369-379.
2. Auerbach, C.A., and D.S. Falconer (1949) A new mutant in the progeny of mice treated with nitrogen mustard. Nature 163:678-679.
3. Bateman, A.J. (1960) The induction of dominant lethal mutations in rats and mice with triethylenemelamine (TEM). Genet. Res. 1:381-392.
4. Burki, K., and W. Sheridan (1978) Expression of TEM-induced damage to postmeiotic stages of spermatogenesis of the mouse during early embryogenesis. II. Cytological investigations. Mut. Res. 52:107-115.
5. Cacheiro, N.L.A. (1977) Cytological studies of sterility in sons of male mice treated with TEM in postspermatogonial stages. Genetics 86:9-10.
6. Cacheiro, N.L.A., L.B. Russell, and M.S. Swartout (1974) Translocations, the predominant cause of total sterility in sons of mice treated with mutagens. Genetics 76:73-91.
7. Cattanach, B.M. (1957) Induction of translocations in mice by triethylenemelamine. Nature 180:1364-1365.

8. Cattanach, B.M., and R.G. Edwards (1958) The effects of triethylenemelamine on the fertility of male mice. Proc. Roy. Soc. Edinb. B 67:54-64.
9. Cattanach, B.M., C.E. Pollard, and J.H. Jackson (1968) Ethyl methanesulfonate-induced chromosome breakage in the mouse. Mut. Res. 6:297-307.
10. Ehling, U.H. (1971) Comparison of radiation- and chemically induced dominant-lethal mutations in male mice. Mut. Res. 11:35-44.
11. Ehling, U.H., and H.V. Malling (1968) 1,4-Di(methane-sulfonoxy) butane (Myleran) as a mutagenic agent in mice. Genetics 60:174-175.
12. Ehling, U.H., R.B. Cumming, and H.V. Malling (1968) Induction of dominant-lethal mutations by alkylating agents in male mice. Mut. Res. 5:417-428.
13. Ehling, U.H., D.G. Doherty, and H.V. Malling (1972) Differential spermatogenic response of mice to the induction of dominant-lethal mutations by n-propyl methanesulfonate and isopropyl methanesulfonate. Mut. Res. 15:175-184.
14. Ehrenberg, L., K.D. Hiesche, S. Osterman-Golkar, and I. Wennberg (1974) Evaluation of genetic risks of alkylating agents: Tissue doses in the mouse from air contaminated with ethylene oxide. Mut. Res. 24:83-103.
15. Falconer, D.S., B.M. Slizynski, and C. Auerbach (1952) Genetical effects of nitrogen mustard in the house mouse. J. Genet. 51:81-88.
16. Generoso, W.M., and W.L. Russell (1969) Strain and sex variations in the sensitivity of mice to dominant-lethal induction with ethyl methanesulfonate. Mut. Res. 8:589-598.
17. Generoso, W.M., W.L. Russell, S.W. Huff, S.K. Stout, and D.G. Gosslee (1974) Effects of dose on the induction of dominant-lethal mutations and heritable translocations with ethyl methanesulfonate in male mice. Genetics 77:741-752.
18. Generoso, W.M., R.J. Preston, and J.G. Brewen (1975) 6-Mercaptopurine, an inducer of cytogenetic and dominant-lethal effects in premeiotic and early meiotic germ cells of male mice. Mut. Res. 28:437-447.
19. Generoso, W.M., K.T. Cain, S.W. Huff, and D.G. Gosslee (1978a) Heritable translocation test in mice, In Chemical Mutagens - Principles and Methods for Their Detection, Vol. 5, A. Hollaender, F.J. de Serres, Eds., Plenum Press, New York, London, pp. 55-77.
20. Generoso, W.M., K.T. Cain, and S.W. Huff (1978b) Inducibility by chemical mutagens of heritable translocations in male and female germ cells of mice, In Advances in Modern Toxicology, Vol. 5, W.G. Flamm, and M.A. Mehlman, Eds., Hemisphere Publishing Corporation, Washington, D.C., London, pp. 109-129.
21. Generoso, W.M., K.T. Cain, M. Krishna, and S.W. Huff (1979a) Genetic lesions induced by chemicals in spermatozoa and

spermatids of mice are repaired in the egg. Proc. Natl. Acad. Sci. USA 76:435-437.

22. Generoso, W.M., S.W. Huff, and K.T. Cain (1979b) Relative rates at which dominant-lethal mutations and heritable translocations were induced by alkylating chemicals in postmeiotic male germ cells of mice. Genetics 93:163-171.

23. Generoso, W.M., J.B. Bishop, D.G. Gosslee, G.W. Newell, C.J. Sheu, and E. von Halle (1980) Heritable translocation test in mice: A report of the "GENE-TOX" program. Mut. Res. 76:191-215.

24. Generoso, W.M., K.T. Cain, M. Krishna, E.B. Cunningham, and C.S. Hellwig (1981a) Evidence that chromosome rearrangements occur after fertilization following postmeiotic treatment of male mice germ cells with EMS. Mut. Res. 91:137-140.

25. Generoso, W.M., M. Krishna, K.T. Cain, and C.W. Sheu (1981b) Comparison of two methods for detecting translocation heterozygotes in mice. Mut. Res. 81:177-186.

26. Generoso, W.M., K.T. Cain, C.V. Cornett, E.W. Russell, C.S. Hellwig, and C.Y. Horton (1982a) Difference in the ratio of dominant-lethal mutations to heritable translocations produced in mouse spermatids and fully mature sperm after treatment with triethylenemelamine (TEM). Genetics 100:633-640.

27. Generoso, W.M. (1982b) A possible mechanism for chemical induction of chromosome aberrations in male meiotic and postmeiotic germ cells of mice. Cytogenet. Cell Genet. 33:74-80.

28. Hitotsumachi, S., and Y. Kikuchi (1977) Chromosome aberrations and dominant lethality of mouse embryos after paternal treatment with triethylenemelamine. Mut. Res. 42:117-124.

29. Jackson, H., and M. Bock (1955) Effect of triethylene melamine on the fertility of rats. Nature 175:1037-1038.

30. Koller, P.C. (1944) Segmental interchange in mice. Genetics 29:247-263.

31. Koller, P.C., and C.A. Auerbach (1941) Chromosome breakage and sterility in the mouse. Nature 148:501-502.

32. Lawley, P.D., D.J. Orr, and M. Jarman (1975) Isolation and identification of products from alkylation of nucleic acids: Ethyl- and isopropyl-purines. Biochem. J. 145:73-84.

33. Luippold, H.E., P.C. Gooch, and J.G. Brewen (1978) The production of chromosome aberrations in various mammalian cells by triethylenemelamine. Genetics 88:317-326.

34. Matter, B.E., and W.M. Generoso (1974) Effects of dose on the induction of dominant-lethal mutations with triethylenemelamine in male mice. Genetics 77:753-763.

35. Matter, B.E., and I. Jaeger (1975) Premature chromosome condensation, structural chromosome aberrations, and micronucleic in early mouse embryos after treatment of parental postmeiotic germ cells with triethylenemelamine.

Possible mechanism for chemically induced dominant-lethal
mutations. <u>Mut</u>. <u>Res</u>. 33:251-260.

36. Ray, V.A., and M.L. Hyneck (1973) Some primary consideration
in the interpretation of the dominant-lethal assay. <u>Environ</u>.
<u>Health</u> <u>Perspect</u>. 6:27-36.

37. Russell, L.B., and C.S. Montgomery (1969) Comparative studies
on X-autosome translocations in the mouse. I. Origin,
viability, fertility and weight of five (TX; 1)S'. <u>Genetics</u>
63:103-120.

38. Russell, W.L. (1954) Genetic effects of radiation in mammals,
In <u>Radiation</u> <u>Biology</u>, A. Hollaender, Ed., McGraw-Hill, New
York, Vol. 1, pp. 825-859.

39. Sega, G.A., and J.G. Owens (1978) Ethylation of DNA and
protamine by ethyl methanesulfonate in the germ cells of male
mice and the relevancy of these molecular targets to the
induction of dominant lethals. <u>Mut</u>. <u>Res</u>. 52:87-106.

40. Selby, P.B. (1979) Radiation-induced skeletal mutations in
mice: Mutation rate, characteristics, and usefulness in
estimating genetic hazard to humans from radiation, In
<u>Radiation</u> <u>Research</u>, Proceedings of the 6th Intern. Cong. of
Radia. Res., S. Okada, M. Imamura, T. Terashima, and
H. Yamaguchi, Eds., Toppan Printing Co., Tokyo, Japan,
pp. 537-544.

41. Singer, B. (1982) Mutagenesis from a chemical perspective:
Nucleic acid reactions, repair, translation, and
transcription, In Basic Life Sciences, Vol. 20, <u>Molecular</u> <u>and</u>
<u>Cellular</u> <u>Mechanisms</u> <u>of</u> <u>Mutagenesis</u>, J.F. Lemontt, and
W.M. Generoso, Eds., Plenum Press, New York, pp. 1-42.

42. Snell, G.D. (1933) Genetic changes in mice induced by X-rays.
<u>Am</u>. <u>Naturalist</u> 67:24.

43. Snell, G.D. (1934) The production of translocations and
mutations in mice by means of X-rays. <u>Am</u>. <u>Naturalist</u> 68:178.

44. Snell, G.D. (1935) The induction by X-rays of hereditary
changes in mice. <u>Genetics</u> 20:545-567.

45. Snell, G.D. (1941) Linkage studies with induced translocations
in mice. <u>Genetics</u> 26:169.

46. Snell, G.D. (1946) An analysis of translocations in the mouse.
<u>Genetics</u> 31:157-180.

47. Steinberger, E., W.D. Nelson, A. Boccabella, and W.J.. Dixon
(1959) A radiomimetic effect of TEM on reproduction in the
male rat. <u>Endocrinology</u> 65:40-50.

RECESSIVE AND DOMINANT MUTATIONS IN MICE

U.H. Ehling and J. Favor

Institut für Genetik, Gesellschaft für Strahlen-
und Umweltforschung, D-8042 Neuherberg
Federal Republic of Germany

SUMMARY

The specific locus method consists of mating wild-type mice
to those homozygous for seven recessives and scoring in the first
generation offspring for mutations at any of the marked loci.
This method was used in experiments to investigate the effects of
chemicals and radiation on the mutation frequency with respect to
the following factors: differential spermatogenic response;
changes of the mutation spectrum with different doses; comparison
of the mutation rates under different treatment conditions; sex
differences; and especially for radiation the dose rate effect.

The scoring of specific locus mutations can be combined with
the detection of dominant cataract mutations in mice, and the
latter results can be used to estimate the total impact due to
induced mutations on health in the first generation. The risk
estimation for dominant cataracts must fulfill the following
suppositions: (a) The dose-effect curve for the induction of
dominant cataract mutations is linear. (b) Dominant cataract
mutation rates ($0.45 - 0.55 \times 10^{-6}$ mutations/gamete/R or $7.3 -
10.7 \times 10^{-7}$ mutations/gamete/mg/kg of ethylnitrosourea) are
representative for all dominant mutations. (c) The ratio of the
numbers dominant cataract mutations (20) to the total number of
well established dominant mutations (736) in man is the same as in
the mouse. This ratio gives a multiplication factor of 36.8,
which is used to convert the induced mutation rate of dominant
cataracts to the estimation of the overall frequency of dominant
mutations. Similar assumptions are used for the risk estimation
based on dominant skeletal mutations.

1.   INTRODUCTION

Methods for the detection of mutations in mammals have been discussed by Hertwig (44,45), Snell (98,99), Catcheside (10), and Falconer (39). The pioneering studies of Hertwig (45), Brenneke (5), and Schaefer (85) had already indicated that the litters sired during the pre-sterile period of irradiated mice were of reduced size. Since there was no effect on sperm mobility and since the number of fertilized eggs was normal, it was concluded that the reduced litter size was due to death of embryos after fertilization. The observation of various nuclear and chromosomal abnormalities in fertilized ova led to the conclusion that embryonic death was caused by chromosomal abnormalities, induced by irradiation in spermatozoa.

A systematic effort to investigate dominant mutations was conducted during a period extending from the late 1943 until about 1950, at the University of Rochester School of Medicine, under the auspecies of the wartime Manhattan Project and, later, its sucessor, the U.S. Atomic Energy Commission. It was concluded that the incidence of mortality, rare morphological anomalies, visible mutations and mutations affecting fertility taken together are definitely increased by radiation at the rate of at least 1.16 x 10^{-4} per R (15).

The first description of irradiation-induced recessive mutations in mice was published by Hertwig (46). A systematic attempt to study the induction of recessive mutations was initially made by Russell (70) in what is now called the specific-locus test (87). With this method the effect of various biological and physical factors on the radiation-induced mutation frequency at a sample of seven loci have been explored. The physical factors were radiation dose, dose rate, dose fractionation, and radiation quality. The biological factors included sex, cell stage, and interval between irradiation and fertilization. The investigation of mutation induction by chemical mutagens was initiated by Cattanach (11). The induction of mutations by chemical mutagens is more specific than by ionizing radiation. These experiments open new perspectives for the investigation of the mutation process in mammals.

One important aspect of these experiments is the need for estimating the genetic hazards to humans due to the exposure from these agents. To complement the doubling-dose method for estimation of the genetic hazard of radiation or chemical mutagen exposure, Ehling developed a method to systematically screen for dominant mutations affecting the skeleton of mice (36) or causing cataracts (33,49). These data could then be used for the direct estimation of the genetic hazard of radiation or chemical mutagen exposure.

2. METHODS

For the detection of recessive mutations in the mouse three
different multiple recessive tester stocks are available (24).
The multiple recessive tester stock that has been extensively used
in mutation studies was developed by Russell (70) and is described
in 2.1. For dominant mutations two different methods are
available. One method screens systematically for the detection of
mutations affecting the skeleton (2.2); the other method examines
the lens of the mouse to detect mutations causing cataracts (2.3).
Both methods have only been used in studies with mice, but could
also be used for a comparison of the mutation rate in different
mammalian species. Throughout the text, statistical comparisons
of mutation rate data were made using Fisher's exact test or Chi-
square test, depending upon the frequency of the least frequency
class.

2.1. Specific Locus Mutations

A specific locus test is conducted by mating treated mice
that are homozygous wild type at a set of marker loci to untreated
animals that are homozygous recessive at the marker loci. The
resultant offspring are expected to be heterozygous at the marker
loci. In the event of a mutation at one of the marker loci in the
treated wild type animal, the offspring will express the recessive
phenotype characteristic for the locus. The specific locus stocks
extensively used has the following markers: a, non-agouti; b,
brown; c^{ch}, chinchilla; d, dilute; p, pink eyed dilution; s,
piebald; se, short-ear. The coat color is affected by six markers
and one marker affects the size of the external ear (se, short
ear). The d and se loci are closely linked (recombination,
0.16%); a double d-se mutation may represent a deletion involving
both loci.

Genetic changes that are detectable by this method include
lesions both within the marker locus and external to it. Any
alteration that leads to a change in the gene product resulting in
an altered phenotype is scored by the specific locus method.
Intermediate alleles, i.e., mutations that do not cause complete
absence of the gene product, are very likely to be the result of
intragenic changes. The most common gross changes are small
deficiencies involving the locus in question. The length of a
deficiency that is compatible with viability of heterozygotes
probably depends on the content of specific chromosome regions and
may be as great as seven centimorgans (67). The resulting mutant
phenotypes are normally characteristic for the marker locus.
Although this reduces the need for genetic confirmation of the
presumed mutations, genetic tests of allelism are routinely done
and combined with characterization of effects of the mutation on
viability.

The method is simple and fast, essential for experiments of the dimension required to produce conclusive restuls. The specific locus method has been previously described and results of radiation (76,86,93) and chemical (24,28,69) mutagenicity experiments have been reviewed. The specific locus method has provided the bulk of information regarding factors affecting the mutation processes in mammals. These factors include dose, dose rate, dose fractionation, differential spermatogenic response and sex differences.

2.2. Dominant Mutation Tests

Dominant mutations can be measured by comparing first generation descendants from treated and untreated populations, but, for many characters, it is difficult to distinguish between the effects of newly occurring genetic changes and the within-strain variations. These problems have been solved for dominant mutations affecting the skeleton and the lens of the mouse.

2.2.1. Dominant skeletal mutations

When one screens for dominant visible mutations, one either consciously or unconsciously limits the phenotypes screened for possible mutations causing a change from the wild type to a variant phenotype. Dominant skeletal mutation experiments represent the earliest attempt at systematically screening for mutations affecting one body system. Variability due to differences in the ability to recognize phenotypic changes were, therefore, reduced. In the initial experiments, phenotypic variants of cleared and stained skeletons occurring in offspring were identified and classified according to whether the variant phenotype was unique or occurred more than once. The variants were further subdivided into those that were unilateral or bilateral, and whether additional skeletal variations were simultaneously present. A class of phenotypic variants was identified as highly likely to be mutations and respresented variants occurring only once in the experiments (18,19,36). Three dominant mutations affecting the skeleton were found by Ehling (20) to be transmitted to the second and third generation. One of these mutations, a disproportionate micromelia (Dmm), was described in detail recently by Brown et al. (7).

In a subsequent experiment the inheritance of presumed mutations was tested by Selby and Selby (95-97). F_1 offspring derived from irradiated males were allowed to produce an F_2 generation before skeletons were prepared and classified by the criteria developed by Ehling (19). Selby and Selby (95-97) confirmed the conclusion of Ehling (19) that the presumed mutations are true mutations. Recently, Selby (94) developed a sensitive-indicator method for the detection of skeletal variants.

With this method the excess of presumed mutations among offspring of treated parents over the control frequency can be detected.

2.2.2. Dominant cataract mutations

When compared with screening for dominant skeletal mutations, screening for lens phenotypic variants has the advantage of a more rapid examination which can be performed on living animals. Thus, one may screen a large number of F_1 progeny and then subject the suspect phenotypic variant offspring to a genetic cross for confirmation of presumed mutations in the F_2 generation (31,37,41,48,49). Essentially, at weaning age the eyes are biomicroscopically examined with the aid of slit lamp illumination, after dilation of pupils with atropine treatment. Presumed mutants are outcrossed to normal mice and at least 20 F_2 offspring are examined for the presence of the particular phenotypic variation.

A detailed description of a number of radiation-induced dominant cataract mutation in mice was published by Kratochvilova (48). She compared 11 radiation-induced dominant cataracts with 8 dominant cataracts described earlier for the mouse, and concluded that "all 11 cataract mutations differ from each other with respect to their morphological characteristics, degree of severity, and the associated lesions". No similarities were found between the cataracts described in her publication and the dominant cataracts reported in the literature. In humans at least 20 well defined cataracts with dominant inheritance are known (56).

3. GAMETOGENESIS

In both sexes of mammalian species the primordial germ cells are formed at early embryonic stages. In the mouse, as in most species, mitotic division of oogonia to give rise to meiocytes is completed well before birth and the lifetime supply of oocytes is already present at the time of birth. The main bulk of oocytes remain in the diplotene stage of meiosis for months or years, according to the species, before they are ovulated or lost by atresia. The testes of adult mammals, on the other hand, contain germ cells in many different stages of development, ranging from the spermatogonia stem cells to mature spermatozoa. Relative timings of the various stages in mouse and human are shown in Table 1.

Different stages in gametogenesis may be scored for induced mutations depending upon the interval between treatment and fertilization (58). If spermatogenesis is not affected by the treatment, the various gametogenic stages are sampled in the mouse

Table 1. Timing of events in gametogenesis in man and
mouse (51).

	Man	Mouse
Formation of genital ridge	33 days	10 days
Sexual differentiation of gonads	42 days	12 days
Ovary		
Meiosis begins	3 - 7 months	13-16 days
Primordial follicles	3 - 5 months	18 days
Duration of follicular growth	?	35 days
Testis		
Mitosis ceases in gonocytes		14 days
Spermatogonia formed		Birth
Spermatogenesis begins	after birth	after birth
Duration of spermatogenesis	74 days	35 days
Duration of spermiogenesis	23 days	14 days

during the following post-treatment mating intervals: spermatozoa
(1-7 days), spermatids (8-21 days), spermatocytes (22-35 days),
followed by differentiating spermatogonia and gonial stem cells
(Table 2) (59).   In man the duration of spermatogenesis is roughly
twice that of the mouse, namely 74 days (16). The time required
for stem cell division is about 8.5 days in the mouse and 16 days
in man.

The length of time it takes for a resting or immature oocyte
to mature to ovulation has proved more difficult to ascertain.
Oakberg (61) has recently used labelling of the zona pellucida to
measure the time taken for the mouse oocyte to progress from stage
3b to maturity and has obtained an estimate of six weeks.
Pedersen (64), on the other hand, employed [^{3}H] thymidine
labelling of follicle cells and estimated this time to be 19 days.
No estimate is yet available for the growth of human follicles.

Table 2.   Germ cell stages of the male mouse sampled at
           various times after treatment (62).

Interval (days)	Stage sampled	Cellular Processes
1- 7	Epididymal sperm	
8-14	Testicular sperm and late spermatids	Differentiation
15-21	Early spermatids	
22-35	Spermatocytes	Meiotic division
36-42	Type-B spermatogonia	
43-49	Type-A spermatogonia	Mitotic division
50-56	Type-A_s spermatogonia	

4.   RESPONSE TO IONIZING RADIATION AND CHEMICAL MUTAGENS

Ionizing radiation induces specific locus mutations in all
spermatogenic stages of mice. However, the relative sensitivity
to the induction of mutations is different in the various germ
cell stages. In contrast to radiation, chemical mutagens may
induce mutations only in certain spermatogenic stages (24).
Because these differences may reflect a difference in the primary
action of these mutagenic treatments, the induction of mutations
by radiation and chemical mutagens will be discussed separately.

4.1. Radiation Experiments

Most information on germ cell response to ionizing radiation
has been obtained primarily from studies using the specific locus
method (i.e., recessive mutations). A much more limited number of
experiments have also been performed to study the inducibility of
dominant mutations, making it possible for comparisons between
these two classes of mutations.

4.1.1. Germ Cell Stage Sensitivities.

Both morphological and biochemical differentiation occur
during the development of the gametes. These different processes
may influence the yield of mutations observed after mutagenic
treatment.

4.1.2. Specific locus mutations.

In male mice, the primordial germ cells present in fetal and early post-parturition animals are less sensitive to mutation induction by irradiation than the A$_s$ (stem cell) spermatogonia (9,91). The A$_s$-spermatogonia are less sensitive to irradiation-induced mutations than the post-spermatogonial stages which undergo meiosis and cell differentiation (Table 3).

For female mice all germ cells reach the oocyte stage before birth and remain in a resting stage until before ovulation. Differences in the rates of induced mutation have also been observed and correlate with the switch from the oocytes in the arrested state to those which begin to develop further prior to ovulation. Treated oocytes shortly before birth are less sensitive to mutation induction than A$_s$-spermatogonia (9). Maturing oocytes, which give rise to offspring up to six weeks following treatment (61) are more sensitive to mutation induction than either oocytes in fetal animals or A$_s$-spermatogonia (74). Mature oocytes, which give rise to offspring within the first week following treatment, are less sensitive to mutation induction than those stages two to six weeks post-treatment.

Table 3.   Radiation-induced specific locus mutations in
different germ cell stages of the mouse
(Dose rate: 70-90 R/min)

Germ cell stage	Dose (R)	No. of mutations	No. of offspring	Mutations per locus $\times 10^5$	References
primordial (♂)	200	9	31 253	4.1	[9]
primordial (♂)	300	16	55 456	4.1	[91]
post-spermatogonia	300	39	26 458	21.1	[90]
A$_s$ spermatogonia	300	40	65 548	8.7	[72]
primordial (♀)	200	1	30 289	0.5	[9]
oocytes mature	200	2	4 659	6.1	[55]
oocytes > 6 weeks	50	0	92 059	0	[77]
oocytes 1-6 weeks	50	13	166 604	1.1	[77]
	200	33	45 465	10.3	[77]

Arrested oocytes in adult female mice give rise to offspring at intervals of six weeks or greater after treatment. This stage is less sensitive to mutation induction by irradiation although highly sensitive to cell killing. Eighty-four mutations were observed in a total of 319,399 offspring born up to seven weeks after radiation with 30 - 400 rads and only three mutations in 259,683 offspring born more than seven weeks after exposure of the females. The later frequency corresponds to the control frequency in female mice (76). According to W.L. Russell (73), these patterns of germ cell stage sensitivity to irradiation-induced mutations best correlate with DNA repair capability. Those stages which show higher mutation rate are incapable of repair or have a reduced repair capability.

4.1.3. Dominant skeletal mutations.

The frequency of recovered presumed dominant skeletal mutations in offspring derived from males irradiated with 600 R is 2.6 times higher in post-spermatogonial cell stages than spermatogonia (Table 4). This difference is in good agreement with the observation in the specific locus experiments (19).

Table 4. Frequency of dominant mutations affecting the skeleton of mice

Classification	Dose (R)	Interval between dose fractions	Germ cell stage treated	No. of F_1 skeletons examined	Mutations n	Mutations (%)	References
	0			1 739	1	0.06	
	600	0	post-spermatogonia	569	10	1.8	
Presumed	600	0	spermatogonia	754	5	0.7	[19]
Mutations	100+500	24 hours	spermatogonia	277	5	1.8	
	500+500	10 weeks	spermatogonia	131	2	1.5	
Mutations	100+500	24 hours	spermatogonia	2 646	31-37	1.2-1.4	[95]

4.1.4. Dominant cataract mutations.

For different exposure conditions the mutation frequency in post-spermatogonia was 2 - 4 times higher than in spermatogonia (37). A similar range was observed for the frequency of specific locus mutations in the same experiments (Table 5).

4.1.5. Dose Rate Effect

The dose rate (dose per unit time) at which a total dose is delivered may affect the induced mutation rate observed, depending upon the germ cell stage studied.

4.1.5.1. Specific locus mutations. In A_s-spermatogonia the induced mutation rate is reduced when the total dose of irradiation is delivered at a low dose rate as compared to the same total dose delivered at a high dose rate (81). The observed mutation rate following low dose rate irradiation is approximately 1/3 that observed after high dose rate irradiation. An even greater reduction in the induced mutation rate due to low vs. high dose rate is observed for maturing oocytes (offspring within six weeks of treatment) (Table 6). The interpretation of such results has been controversial. Russell (77) and Lyon et al. (55) considered a limited DNA repair capability as the preferred

Table 5.   Recessive and dominant mutations induced in mice
by γ-rays (37)

Dose (R)	Dose rate (R/min)	Germ cell stage treated	Number of F_1 offspring	Number of mutations at 7 specific loci	Mutations per locus $\times 10^5$	Number of dominant cataract mutations	Mutations per gamete $\times 10^5$
0	-	-	103 218	6(A)	0.8	-	-
0	-	-	8 174	2	3.5	0	0
534	53	post-spermatogonia	1 721	3	24.9	1	58.1
600	53	post-spermatogonia	865	3	49.5	1	115.6
455+455	55	post-spermatogonia	272	2	105.0	1	367.6
534	53	spermatogonia	10 212	7	9.8	3	29.4
600	53	spermatogonia	11 095	14	18.0	3	27.0
455+455	55	spermatogonia	5 231	9(B)	24.6	6	114.7

(A) Untreated historical control of the laboratory

(B) A simultaneous d-se mutation included, which is caused by double non-disjunction

Table 6. Dose rate effect in male and female mice

Dose (R)	Dose rate	No. of mutations at 7 specific loci	No. of F$_1$ offspring	Mutations per locus x10^5	References
A$_s$ Spermatogonia					
600	90	111	119 326	13.3	[80]
600	0.8	10	28 059	5.1	[80]
600	0.001	22	53 380	5.9	[80]
Maturing oocytes					
200	90	33	45 465	10.4	[77]
200	72	2	4 659	6.1	[55]
284(A)	0.009	1	14 402	1.0	[77]
283(A)	0.009	2	13 742	2.1	[77]

(A) Effective dose calculated on that portion of a 400 R chronic dose rate

dose received by sensitive maturing oocyte cells after initiation of

treatment [77].

explanation. In germ cells that are capable of DNA repair, a dose rate effect is observed because at a lower dose rate more pre-mutational DNA lesions may be repaired than when the same total dose is delivered at a high dose rate. Abrahamson and Wolff (1), Brewen and Payne (6) and Wolff (103), on the other hand, argued that a portion of the recovered specific locus mutations result from a two-hit phenomenon of the induced pre-mutational DNA lesions. A dose rate effect is observed because at a lower dose rate, DNA pre-mutational lesions are induced over a longer period of time, during which they may dissipate or be repaired, such that the probability of interaction of two pre-mutational lesions is reduced from that when the pre-mutational lesions are induced essentially simultaneously (high dose rate).

4.1.6. Dose Response

4.1.6.1. Specific locus mutations. Data are available for an adequate examination of the dose response of induced mutations for high or low dose rate for the exposure of A$_s$-spermatogonia and maturing oocytes (Table 7). For high dose rate

Table 7. Dose response of specific locus mutations in mice

Dose (R)	Number of mutations	Number of offspring	Mutations per locus $\times 10^5$	References
Untreated Male Control				
0	28	531 500	0.7	[75]
0	11	157 421	1.0	[86]
0	11	169 955	0.9	[37]
A_s Spermatogonia high dose rate (72-90 R/min)				
300	40	65 548	8.7	[75]
600	111	119 326	13.3	[75]
670	12	11 138	15.4	[53]
1000	29	44 649	9.3	[75]
A_s Spermatogonia low dose rate (0.001-0.009 R/min)				
37.5	7	79 364	1.3	[86]
86	6	59 810	1.4	[75]
300	15	49 569	4.3	[75]
600	22	53 380	5.9	[80]
861	12	24 281	7.1	[75]
Untreated Female Control				
0	8(A)	166 826	0.3	[77]
0	0	37 813	0.0	[3]
Maturing oocyte high dose rate (51.5-71.5 R/min)				
200	7	18 867	5.3	[55]
400	7	7 501	13.3	[55]
600	26	9 875	37.6	[55]
Maturing oocyte low dose rate (0.009-0.05 R/min)				
207(B)	1	7 692	1.9	[77]
283(B)	2	13 742	2.1	[77]
284(B)	1	14 402	1.0	[77]
615	1	10 177	1.4	[8]

(A) 3 indepentent mutational events, one was a cluster of
    6 mutations

(B) Effective doses calculated by Russell [77]

exposure of A_s-spermatogonia the shape of the dose response curve is humped, there being a linear increase of the induced mutation rate up to a maximum observed at 670 R (86), followed by a decrease at higher doses (71). A linear regression analysis of the data between 0 and 670 R yielded the equation Y = [8.10 ± 1.19] x 10^{-6} + [2.19 ± 0.19] x 10^{-7} x D (80). The humped shape of the dose response curve in spermatogonia for high dose rate treatment has been interpreted as due to a heterogeneous cell population for sensitivity to irradiation-induced mutations and cell killing. That a humped dose response curve could exist has been theoretically demonstrated when sensitivities to mutation induction and cell killing are either correlated (63) or stochastic (43). By analyzing the length of radiation-induced sterile period as an indication of spermatogonial cell killing, Cattanach (12) has shown the existence of a heterogenous A_s-spermatogonial cell population. The sensitivity to cell killing pattern in these studies showed a transition point at 600 R, as would be expected from the specific locus dose response data.

The dose-response for low dose rate treatment is linear up to and including the highest dose point for which experimental data are available (861 R). The linear regression equation best fit of the data is:

$$Y = [8.10 \pm 1.19] \times 10^{-6} + [0.73 \pm 0.08] \times 10^{-7} \times D \ (80).$$

For maturing oocytes, the linear regression fit of all low dose rate experiments treating the control cluster as one mutational event is:

$$Y = 2.12 \times 10^{-6} + 0.296 \times 10^{-7} \times D \ (77).$$

For offspring conceived within 7 days of oocyte exposure to high dose rate irradiation, a linear regression fit of the dose-response was marginally significant whereas a quadratic fit was better with the formula:

$$Y = 0.21 \times 10^{-5} + (0.39 \pm 1.05) \times 10^{-6} \times D$$
$$+ (6.26 \pm 2.60) \times 10^{-9} \times D^2 \ (55).$$

4.1.6.2. Dominant skeletal mutations. The only data to date on dose-response relationship data to date indicate a linear increase with dose of the frequency of skeletal abnormalities in offspring resulting from post-spermatogonial treatment with X- or neutron-irradiation (36).

4.1.7. Dose Fractionation

4.1.7.1 Specific locus mutations. The effects of fractionating a total dose delivered at a high dose rate on the induced mutation rate depend upon the germ cell stage treated, the size and number of fractionation doses, the total dose and the fractionation interval between doses. Table 8 indicates that 1,000 R delivered in two fractions of 500 R separated by 24 hr caused an enhancement of the specific locus mutation frequency. Similarly, the mutation frequency is enhanced (1.9 times higher) if 600 R is given as 100 R followed 24 hr later by 500 R. The explanation for these fractionation effects is thought to be that

Table 8. Specific locus mutation frequency in the mouse
after single and fractionated radiation exposure

Dose regime	Interval	No. of mutations	No. of offspring	Induced mutations per locus per R $\times 10^8$	References
A_s Spermatogonia high dose rate (90 R/min)					
600 R	0	111	119 326	20.9	[75]
100+500 R	24 hr	42	24 811	39.1	[75]
1000 R	0	29	44 649	8.5	[75]
600+400 R	15 weeks	10	4 904	28.4	[75]
500+500 R	24 hr	39	11 164	49.2	[75]
500+500 R	2 hr	12	14 879	10.7	[75]
5x200 R	1 week	15	10 968	18.8	[75]
Oocytes high dose rate (50-90 R/min)					
200 rad	0	9	34 813	18.5	[52]
20x 10 rad	24 hr (A)	1	17 682	4.0	[52]
20x 10 rad	2 hr (B)	0	21 620	-	[52]
400 R	0	16	12 853	44.5	[75]
200+200 R	24 hr	9	6 086	52.8	[75]

(A) 5 days per week x 4 weeks

(B) 4 times per day x 5 days

the A_s-spermatogonial population surviving 24 hr after the first irradiation is in a stage of the cycle that is especially sensitive to mutation induction (75). This assumption was later supported by Oakberg (60). He demonstrated by labelling experiments that the population of A_s-spermatogonia surviving the 500 R + 500 R exposure differs from that surviving single exposures of 300 R, 600 R or 1 000 R.

In females (Table 8), the exposure of 20 x 10 rads of irradiation induced less specific locus mutations than a single exposure with 200 rads (52). The interpretation of these results is controversial. Abrahamson and Wolff (1) considered the low dose rate effect and decreased effectiveness of lower doses in inducing mutations to be due to a reduction in the proportion of two-hit phenomena resulting in mutations. Russell (77) considered these observed effects to be due to a limited repair capability of the oocyte, such that at low dose rates or low doses a larger proportion of the premutational DNA damage may be repaired.

In spermatogonia treated weekly with repeated small doses of radiation at high dose rate (12 x 50 rad X-rays at 66 - 70 rad/min) the mutation rate was not significantly different from that after single exposure. In contrast, the mutation rate after 12 x 50 rad γ-rays at 0.06 rad/min was typical of low dose-rate irradiation. This observation indicates that the effect of dose rate on the mutation rate does not depend on the continuity or close spacing of the exposure (54).

4.1.7.2. Dominant skeletal mutations. Limited information is available on the effects of fractionating a total radiation dose on the frequency of presumed mutations (Table 4). A high dose rate regime of 100 R + 500 R with a 24 hr fractionation interval to spermatogonia yielded a higher incidence of presumed skeletal mutations than did a single exposure (19). Although a comparative single exposure with 1000 R is not available, it should be mentioned that a 500 R + 500 R high dose rate irradiation with 24 hr fractionation interval spermatogonial treatment yielded a higher incidence of presumed skeletal mutations than did 600 R single exposure (19).

4.1.7.3. Dominant cataract mutations. A 455 R + 455 R spermatogonial exposure with a 24 hr fractionation interval yielded a higher incidence of dominant cataract mutations than did a single exposure to 534 R or 600 R (Table 5).

4.1.8. Spectra of Specific Locus Mutations and Their Viability

Great variability exists in the observed yield of radiation-induced mutations among the seven specific loci screened. This is a reflection of possible differences in the mutability of the different loci or a difference in the ultimate survival and transmission of a mutation at the particular loci (Table 9). The highest yields of mutations are observed at the s locus; the d, b, p and c loci have an intermediate frequency. In contrast, the a and se loci are least mutable. For spermatogonia, the spectrum of recovered mutations has been reported to be independent of the dose rate (73). It has been suggested that post-spermatogonial stage treatment yields a spectrum of recovered mutations different from that observed in spermatogonial treatment experiments. In the former, there is a less pronounced difference among the loci in the mutation rate and a higher frequency of d-se double mutations (73).

The analysis of the spectra of recovered mutations after treatment of maturing oocytes is still limited and may explain discrepancies in two reports. Based on the relative frequencies of d, se, and d-se double mutants, L.B. Russell (66) suggested that the oocyte results resemble those of the post-spermatogonial stage treatment. Lyon et al. (55), on the other hand, reported a spectrum of recovered mutations similar to that of spermatogonial stage treatment.

Just as the mutation rate among loci varies, the relative number of recovered mutations which are homozygous lethal depend on the particular locus. The d and s loci have a very high proportion of recovered mutations which are homozygous lethal. The overall frequency of homozygous lethal mutations recovered in spermatogonia appears to be similar for X-, gamma- or neutron-irradiation (Table 9).

L.B. Russell (68) has shown that, at the albino locus region, a relatively low frequency of homozygous lethals was found in control, a higher frequency of homozygous lethals was found in treated spermatogonia or oocytes, and the frequency was highest in treated post-spermatogonial cell stages. Since the frequency of d-se double mutations also is highest in post-spermatogonial cell stage treatment, the results suggest that there is an increase in the frequency of small deletions when post-spermatogonial cell stages were treated.

Russell and Kelly (79) have observed a small reduction in the frequency of homozygous lethal specific locus mutations after chronic dose rate spermatogonial treatment (14/28). Lyon et al. (53) have indicated a statistically lower frequency of homozygous lethal mutations after low dose rate spermatogonial treatment as compared to high dose rate treatment (5/16). This may result from a difference in the spectrum of recovered mutations, or more likely a reduction in that proportion of mutations recovered after at low dose rate treatment which are deletions (Section 4.1.2).

4.1.9. Dominant cataract mutations.

Viability data indicate seven of the ten dominant cataract mutations recovered after irradiation of either post-spermatogonial or spermatogonia to be homozygous lethal (48), an

Table 9. Mutation spectrum and lethality of specific locus
mutations after treatment of male mice
with different noxa

Treatment	Germ cell stage treated	Number lethal/number tested per locus									Percent lethal	References
		a	b	c	d	se	d+se	p	s	Total		
Control	-	-	0/ 7	0/ 3	0/ 1	-	-	2/ 4	-	2/15	13	[57]
High dose rate												
X-rays	g	2/2	10/18	2/ 6	12/12	1/2	-	6/14	38/38	71/92	77	[86]
X- and γ-rays	g	0/2	1/10	5/11	14/14	0/1	1/1	1/ 9	8/ 9	30/57	53	[86]
Fission neutrons	g	0/2	4/ 4	2/ 6	6/ 6	0/4	2/2	2/ 9	19/24	35/57	56	[86]
γ-rays	g	-	0/ 6	6/ 9	8/ 8	1/2	2/2	0/ 5	12/13	29/45	64	[57]
Procarbazine	pg	-	-	-	-	-	-	0/ 1	2/ 2	2/ 3	67	[34]
MMS	pg	-	3(A)/3	-	-	1/1	1/1	3/ 4	1(B)/3	9/12	75	[35]
TEM	g	1/1	-	-	3/ 3	-	-	0/ 1	-	4/ 5	80	[11]
Procarbazine	g	0/1	0/ 7	0/ 3	5/ 6	0/3	-	0/ 7	2/ 2	7/29	24	[25]
Mitomycin C	g	0/1	0/ 1	0/ 1	3/ 3	-	-	0/ 5	0/ 1	3/12	25	[25]
ENU	g	0/2	0/ 6	0/ 5	13/16	0/5	-	0/21	4/ 4	17/59	29	[35]

g = spermatogonia

pg = post-spermatogonia

(A) Includes one mutation with dominant deleterious effect

(B) Includes one semisterile mutation (translocation carrier)

after radiation treatment. Further, Selby (92) has reported seven
incidence similar to that for specific locus mutations recovered
of eight dominant skeletal mutations recovered after
spermatogonial irradiation to be homozygous lethal.

4.2. Chemical Experiments

Experiments using a chemical treatment were only relatively
recently initiated by Cattanach (11). Critical reviews of all
specific locus experiments using chemical mutagens have been
published (24,28,69). To date, 25 chemical mutagens have been
subjected to a specific locus test. Of these, results for MMS,
procarbazine and ENU are extensive. For ENU, results are also
available on the frequency of induced skeletal variants and
dominant cataract mutations.

4.2.1. Germ Cell Stage Sensitivities

The differential spermatogenic response of mice to the induction of mutations is very likely due to the different metabolic pathways of the test compound and, therefore, to their different effects on the structural and macromolecular changes during spermatogenesis. The interesting relationship between the germ cell stage specific induction of mutations and unscheduled DNA synthesis was discussed by Sega and Sotomayor (89). The importance of the differential spermatogenic response for the risk estimation of the use of antineoplastic drugs was described in detail by Ehling (21,25,30) and recently discussed by Lyon (51) and Searle (88).

4.2.2 Specific locus mutations

The induction of specific locus mutations after i.p. injection of methyl methanesulfonate (MMS) and cyclophosphamide demonstrates the differential spermatogenic response (Table 10). A dose of 20 mg/kg of MMS increased significantly the mutation frequency in the mating interval 5 - 12 days post-treatment (P = 0.009). The highest mutation rates induced by MMS is 27 times higher than the control rate. For cyclophosphamide the induced specific locus mutation rate in spermatozoa (1 - 7 days post-treatment) is 28 times higher than the mutation frequency in spermatogonia. The mutation frequency in spermatogonia is not significantly different from the control frequency (P = 0.56). The reduction of average litter size indicates an induction of dominant lethal mutations. The data prove that the peak sensitivity to the induction of specific locus mutations corresponds well with the sensitivity pattern for the induction of dominant lethal mutations for these two tested compounds (21,24).

In contrast to MMS and cyclophosphamide, other compounds such as procarbazine (34) and triethylenemelamine (TEM) (11) induce mutations in both post-spermatogonial germ cell stages and spermatogonia, whereas ethylnitrosourea (ENU) and mitomycin C induce specific locus mutations mainly or exclusively in spermatogonia (Table 11). However, until all post-spermatogonial cell stages are sampled (Table 10), the lack of mutations in post-spermatogonia should be interpreted cautiously. For example, the ENU post-spermatogonial stage results reported in Table 11 are mainly derived from treated spermatozoa. It should therefore be concluded that ENU is relatively ineffective in inducing mutations in spermatozoa. Indeed, W. L. russell (78) does mention that some specific locus mutations were recovered in post-spermatogonia but a detailed description of the germ cell stages sampled have not yet been reported.

Table 10. Highly effective compounds for the induction of
specific locus mutations in spermatozoa
and spermatids of mice(A)

Compound	Dose (mg/kg)	Mating intervals (days)	Average litter size	No. of offspring	No. of mutations at 7 loci	Frequencies per locus x 10^5	Reference
		1 - 4	6.6	5 638	1	2.5	
		5 - 8	6.6	5 593	2	5.1	
	20	9 - 12	6.7	5 518	2	5.2	
Methyl		13 - 16	7.0	5 785	1	2.5	
methane-		17 - 20	6.9	5 759	0	-	
sulfonate							[35]
(MMS)		1 - 4	6.0	2 902	2	9.8	
		5 - 8	4.0	1 716	3	25.0	
	40	9 - 12	4.1	1 750	3	24.5	
		13 - 16	6.0	2 799	1	5.1	
		17 - 20	6.6	3 072	0	-	
Methyl		1 - 4	6.6	1 426	0	-	
methane-		5 - 8	4.4	898	1	15.9	
sulfonate	4x10(B)	9 - 12	5.6	1 186	1	12.0	[35]
(MMS)		13 - 16	7.7	1 638	0	-	
		17 - 20	7.6	1 692	0	-	
		1 - 7	4.4	1 627	3	26.3	
Cyclophos-		8 - 14	4.8	1 794	3	23.9	
phamide	120	15 - 21	4.9	1 744	1	8.2	[30]
		22 - 42	7.1	2 678	1	5.3	
		≥ 43	7.4	12 573	1	1.1	

(A) See Table 11 for the control rate

(B) 24 hr apart

4.2.3. Dominant cataract mutations.

Only one germ cell stage comparison of sensitivity to induced
dominant mutations is available. The frequency of dominant
cataract mutations recovered after 250 mg/kg ENU spermatogonial
treatment (17/9352) was higher than the same dose treatment of
post-spermatogonia (2/3360). This is not inconsistent with the
above mentioned specific locus mutation data. It is interesting,
however, that the ENU post-spermatogonia treatment group is the
first in which the yield of dominant cataract mutations is greater
than the yield of specific locus mutations (37,41).

Table 11. Highly effective compounds for the induction of
specific locus mutations in spermatogonia

Compound	Dose (mg/kg)	Germ cell stage treated	No. of F_1-offspring	No. of mutations at 7 loci	No. of minimum independent mutational events	Mutations per locus per gamete x 10^5(A)	Reference
Untreated control	-		119 416	8	8	1.0	
Solvent control	-		86 378	11(B,C)	5	1.8	[35]
Combined control	-		205 794	19(B,C)	13	1.3	
	200	post	6 722	0			
	400	spermato-	3 394	1	1	4.2	
	600	gonia	1 930	2	2	14.8	
	800		1 771	0			
Procarbazine							[34]
	200		37 202	5	5	1.9	
	400	spermato-	35 047	10(B)	9	4.1	
	600	gonia	45 413	16	16	5.0	
	800		40 013	7	7	2.5	
	40	post-	5 028	0			
	80	spermato-	4 660	0			
	160	gonia	4 416	0			
Ethylnitro-	250		3 360	0			[35]
sourea (ENU)	40		11 410	3	3	3.8	
	80	spermato-	4 855	8	8	23.5	
	160	gonia	8 658	35(D)	32	57.8	
	250		9 766	64(D,E)	57	93.6	
	0	-	531 500	28	28	0.8	
	25		3 687	0	0	0	
	50		15 204	10	10	9.4	
Ethylnitro-	75	spermato-	3 015	7	5	33.2	[83]
sourea (ENU)	100	gonia	21 235	64	51	43.1	
	150		7 715	32	26	59.3	
	200		2 080	10	7	68.7	
	250		6 547	32	21	69.8	

(A) Calculation based on total number of mutations

(B) Includes a cluster of 2 mutations

(C) Includes a cluster of 6 mutations

(D) Includes 3 clusters of 2 mutations

(E) Includes 2 clusters of 2 mutations and one cluster of 3 mutations

4.2.4. Sex Differences

The results of induced specific locus mutations in female mice are summarized in Table 12. The limited data base indicates

Table 12. Chemically-induced specific locus mutations in female mice

		Interval between treatment and conception				
		Up to 7 weeks		More than seven weeks		
Treatment	Dose	No. of offspring	No. of mutations at 7 loci	No. of offspring	No. of mutations at 7 loci	References
Procar-bazine	400 mg/kg	9 369	0	23 233	2	[32]
	600 mg/kg	16 888	0	31 599	1	
	total	26 257	0	54 832	3	
Mito-mycin C	2 mg/kg	2 847	1	4 956	0	[32]
	4 mg/kg	1 515	0	1 204	0	
	total	4 362	1	6 160	0	
ENU	160 mg/kg	2 728	2	3 325	0	[35]
	250 mg/kg	390	0	1 040	0	
	total	3 118	2	4 365	0	
TEM	2 mg/kg	10 812	2	-	-	[13]

(A) Three independent mutational events, one of which was a cluster of 6. The total number of

mutations observed was 8 in 204 639 control offspring [77]

differences in the mutagenic response of mature and early oocytes for different chemicals. To date more specific locus mutations have been induced in treated mature oocytes than in immature oocytes by radiation, ENU, and mitomycin C. The opposite seems to be true in the case of procarbazine.

Similarly, the sensitivity between male and female germ cells is different for the various chemicals tested. If spermatogonia and oocytes were equally sensitive to the induction of specific locus mutations by procarbazine, one would expect 26 mutants in a total of 81 089 offspring resulting from oocyte treatment with 400 or 600 mg/kg procarbazine. The three mutations observed are significantly below the expected frequency of 26 ($P < 10^{-4}$). This result indicates that oocytes are less sensitive than spermatogonia for the induction of specific locus mutations with procarbazine (Tables 11 and 12). Similarly, ENU is less effective in oocytes than in spermatogonia. The results of mitomycin C are inconclusive. The mutation rate for TEM is similar in spermatogonia and oocytes (13).

4.2.5. Dose-Response

4.2.5.1. Specific locus mutations. Dose-response data have been collected for two chemicals after treatment of A_s-spermatogonia (Table 11). The frequency of procarbazine-induced specific locus mutations in spermatogonia was shown to increase linearly with dose up to a maximum at 600 mg/kg body weight with a subsequent decrease in the observed mutation rate at 800 mg/kg body weight dose treatment (34).

Based on their data in Table 11, Ehling and Neuhäuser-Klaus (31,35) concluded that the dose effect relationship of ENU induced specific locus mutations is linear. Comparing these results with those of Russell et al. (83) in Table 11, the only apparent point where a difference may exist is at a dose of 250 mg/kg. However, in an independent experiment reported by Russell (78), 72 mutations were observed in 10,146 offspring, yielding a mutation rate of 101.4×10^{-5} per locus per gamete. The results from both laboratories are compatible with a linear dose response curve.

Ehling (31) has identified an exceptionally high mutation frequency of ENU in the 250 mg/kg dose group mainly at the d and p loci. The dose-effect relationship for the remaining five loci (a, b, c, se, s) is similar to the dose-effect curve for ENU-induced dominant cataract mutations. If these differences among loci are real, the combined specific locus data fits a linear dose-response relationship but this linear relationship reflects the pooling of loci with differences in their dose response to the induction of mutations by ENU (31).

4.2.5.2. Dominant cataract mutations. Preliminary data at high dose points indicate a non-linear dose response of ENU-induced dominant cataract mutations in spermatogonia (31,40).

4.2.6. Dose Fractionation

A single dose of 40 mg/kg of MMS induced 23.9×10^{-5} specific locus mutations per gamete in the mating interval 5 - 12 days post-treatment, which is 27 times the spontaneous level. Dividing the total dose of 40 mg/kg into four equal fractions of 10 mg/kg given 24 hours apart induced only 13.5×10^{-5} mutations per locus per gamete, which is 15 times the control mutation rate. However, the sample size of these experiments is small and this difference is not statistically significant ($P = 0.21$). The results of these experiments are summarized in Table 10.

In contrast to the induction of mutations in spermatozoa and spermatids by MMS, the yield of specific locus mutations induced by fractionated doses of procarbazine or ENU in A_s-spermatogonia was reduced. The results of these experiments are summarized in Table 13. For procarbazine, 6 x 100 mg/kg with 24 hr fractionation interval treatment regime produced a significantly lower observed mutation rate than that for an unfractionated 600 mg/kg dose. In the only fractionation experiment carried out at a dose higher than 600 mg/kg, 5 x 200 mg/kg with a weekly fractionation interval showed a non-significant increase in the mutation rate as compared to a single 800 mg/kg dose treatment (25). For ENU, the treatment of 10 mg/kg dose in each of 10 successive weeks significantly reduced the observed mutation rate from that observed for a single acute 100 mg/kg dose treatment (82). However, the frequency of mutations in the fractionated treatment group was still significantly higher than the historical control of 28 mutants in 531 000 (P < 0.001). This result indicates that if the fractionation effect is due to repair of premutational lesions, then the DNA repair efficiency of the spermatogonia is incomplete.

The fractionation of the dose affects the yield of mutations differently in specific locus and dominant lethal experiments. In dominant lethal experiments with MMS there was no effect of dose fractionation (27). The following hypotheses could explain these results: (a) The distinct genetic endpoints have different repair mechanisms. (b) There are differences in the type of lesion induced in the different germ cell stages (differentiating spermatids and spermatozoa versus the dividing spermatogonia).

Table 13. Induction of specific locus mutations in
spermatogonia of mice after fractionation
of the dose

Compound	Dose (mg/kg)	Interval between fractions	No. of F$_1$-offspring	No. of mutations at 7 loci	No. of minimum independent mutational events	Mutations per locus per gamete x 10^5(A)	Reference
Control	-	-	531 500	28	28	0.8	
							[82]
	1x100	-	21 235	64	53	43.1	
ENU	1x100	-	3 679	12	12	46.6	
	10x 10	week	19 991	9	8	6.4	

Compound	Dose (mg/kg)	Interval between fractions	No. of F$_1$-offspring	No. of mutations at 7 loci	No. of minimum independent mutational events	Mutations per locus per gamete x 10^5(A)	References
Control	-	-	205 794	19(B)	13	0.9	[35]
	1x600		45 413	16	16	5.0	
Procar-	2x300	day	13 908	3	3	3.1	[25]
bazine	6x100	day	20 621	2	2	1.4	
	5x200	week	18 393	4	4	3.1	
Control	-	-	531 500	28	28	0.8	
							[81]
	1x100	-	21 235	64	53	35.7	
ENU	1x100	-	3 679	12	12	46.6	
	10x 10	week	19 991	9	8	5.7	

(A) Calculation based on total number of mutations

(B) For details of clusters see Table 2

4.2.7. Spectra of Specific Locus Mutations and Their
       Viability

In addition to the quantitative differences in the mutation
induction by chemicals, qualitative differences can also observed.
The spectrum of induced mutations and the viability of induced
specific locus mutations depend upon mutagenic treatment and are
different for post-spermatogonial and spermatogonial treatment.
The results of the qualitative differences are summarized in Table
9.

Like in the radiation experiments, the largest number of
mutations induced in post-spermatogonia occur at the s locus
(5/15). Approximately 7% of the induced mutations (1/15) are
double d-se mutants. The majority of the mutations induced in
post-spermatogonia are lethal in homozygotes (11/15).

In spermatogonial germ cell stages only 7% of the induced
mutations occur at the s locus. The majority of mutations
occurred at the p and d loci. No double mutant has been observed.

For a total of 135 mutants, i.e., 15 from control groups and
120 from different experiments with chemical mutagens, the
viability tests are complete. All mutations as homozygotes that
cause death before maturity have been classified as lethals. The
overall control frequency so far shows 13% (2 out of 15) were
homozygous lethals. In the procarbazine group, 2 of 3 mutations
(67%), and in the MMS group, 9 of 12 mutations (75%), induced in
post-spermatogonial stages were lethal. In contrast to the high
frequency of lethals in post-spermatogonial germ cell stages, only
7 of 29 mutations in the procarbazine group (24%), 3 of 12
mutations in the mitomycin C group (25%), and 17 of 59 in the ENU
group (29%) induced in spermatogonia were lethal in homozygous
condition. TEM-induced mutations have a similar lethal frequency
as radiation-induced specific locus mutations.

The high frequency of homozygous lethal mutations induced in
post-spermatogonia suggests that small deficiencies are the main
cause for mutations induced in these germ cell stages. This
observation explains likewise the correlation between the induced
dominant lethal and specific locus mutations. The relatively low
frequency of homozygous lethal mutations suggests that mutations
induced by ENU, mitomycin C, and procarbazine are mainly due to
base-pair changes.

4.3. Comparison of Radiation- and Chemically-Induced
     Mutations

Three points should be emphasized regarding the induction of
specific locus mutations by radiation and chemical mutagens.

First, the spectrum of specific locus mutations shifts when chemical mutagens are used as compared to radiation (Table 9). This indicates either inherent differences in mutability or differences in the probability of survival and transmission of mutations at the different loci depends upon the mutagenic treatment and, therefore, the induced DNA lesion ultimately expressed as a mutation. Second, the frequency of homozygous lethal specific locus mutations recovered after spermatogonial irradiation treatment is 2/3 while for chemical treatment is only 1/4 (Table 9). Together these results suggest that the DNA lesion from chemical mutagen treatment ultimately resulting in a specific locus mutation is a very small deletion or a simple base pair change. In contrast, radiation may induce primarily small deletions. Although it has been previously the subject of debate (103), the question remains as to what fraction of the radiation induced deletions represent one-hit or two-hit events. Third, an increase in the frequency of homozygous lethal specific locus mutations in treated post-spermatogonial cell stages is seen in chemical experiments. This suggests tht the DNA lesion ultimately resulting in a specific locus mutation may be qualitatively different in treated post-spermatogonial cell stages as compared to spermatogonia.

4.4. Comparison of Dominant and Recessive Mutations

In a combined experiment, dominant cataract mutations and specific locus mutations were scored in the same offspring (Table 5). A total of 15 dominant cataract and 38 specific locus mutations was scored in 29,396 offspring. A comparison of the overall frequency of induced specific locus mutations and cataract mutations in post-spermatogonia and spermatogonia after single exposure showed that there were 2.5-2.7 times more recessive mutations than dominant mutations induced by γ-radiation. Taking into account that in humans 20 well-established dominant cataracts are known (56), it is likely that in this experiment at least three times as many loci coding for dominant cataracts were scored as recessive mutations. Therefore, on a per-locus rate, radiation induced about eight times more recessive mutations than cataract mutations. A similar difference was observed for ENU (37,41). These data indicate the importance of the dominant mutations for the estimation of the genetic risk in the first generation.

5.   EVALUATION AND RISK ESTIMATION

The specific locus method was introduced to investigate the effects that various physical, chemical, and biological factors might have on mutation frequency. The knowledge of these factors is the foundation for the assessment of the genetic hazard in man. The dose-rate effect is important for the estimation of the genetic risk; the differential spermatogenic response for the induction of mutations by chemicals is important for the evaluation of the genotoxic risk. Although mutagenesis studies with mammalian cells in culture are useful for investigating the basic mechanisms under the conditions of the tests, they are unlikely to be reliably predictive of the mutagenic events that occur in the various germ cell stages, and which are transmitted to descendent generations. For this reason, the Committee on Chemical Environmental Mutagens (14) of the National Academy of Sciences, Washington, D.C. classified the specific locus test and the dominant lethal assay as confirmation tests. The investigation of dominant mutations, affecting the lens or the skeleton of the mouse, was introduced by us in order to estimate the genetic risk in the first generation.

In using the data from the mouse to arrive at quantitative estimates of the radiation genetic risks for humans, three general assumptions are made, according to Sankaranarayanan (84):

a) The amount of genetic damage induced by a given type of radiation under a given set of conditions is the same in the germ cells of humans and in those of the test species which serves as a model.

b) The various biological and physical factors affect the magnitude of the damage in similar ways and to similar extents in the mouse and in humans.

c) At low doses and at low dose rates of low LET irradiation, there is a linear relationship between dose and frequency of genetic effects studied.

Unless there is evidence to the contrary, the first and second condition may also be considered valid for chemical mutagens. The validity of extrapolation from one species to another, taking into consideration uptake, transport, metabolism and excretion of chemicals, should, however, be proven by mutagenicity testing. The third assumption, linear dose-effect relationship, cannot be assumed for chemical mutagens, i.e., data on dose-effect relationship are needed for each chemical.

There are two main approaches in making genetic risk estimates. One of these, termed the direct method, expresses risks in terms of expected frequencies of genetic changes induced per unit dose. The other, referred to as the doubling dose method or the indirect method, expresses risks in relation to the observed incidence of genetic disorders now present in man.

5.1. The Doubling Dose Method

The doubling dose can be defined as the dose necessary to
induce as many mutations as occur spontaneously in one generation.
One underlying assumption for the calculation of the doubling dose
is a linear dose-response relationship. Another assumption is the
similarity between spontaneous and induced mutations. If these
conditions are fulfilled the doubling dose can be used to
calculate the individual as well as the population risk.

5.1.1. Individual risk.

Procarbazine is used in combination treatment of Hodgkin's
disease. The induction of specific locus mutations by
procarbazine fulfills the requirements for the doubling dose
approach (34). The doubling dose for procarbazine, based on the
regression coefficient for the induction of mutations in A_s-
spermatogonia, is 110 mg/kg (30). By comparing the therapeutic
dose of 215 mg/kg with the doubling dose of 110 mg/kg, one may
conclude that the procarbazine treatment of a patient with
Hodgkin's disease would induce two times as many mutations as
arise spontaneously, provided man and mouse are equally sensitive.
A fractionated application schedule in man would reduce the
calculated genetic risk (Table 13). In addition, the calculation
is based on the sensitivity of a male patient. The genetic risk
to a female patient would be drastically lower (Table 12).

5.1.2. Population risk.

Two possibilities exist to quantify the population risk. A
quantification is possible if we use the doubling dose in
combination with the current incidence of genetic disorders. This
method has been used by the UNSCEAR-Report (102) to quantify the
genetic risk of radiation. The other way is to calculate the
number of mutations expected based upon an assumed number of human
genes: [The mutation rate in spermatogonia] x [the number of
genes] x [number of births] x [population dose] x [female
sensitivity] gives the number of expected mutations for one
generation. In addition, we have to assume that the specific
locus mutation rate is representative for the whole genome. Both
approaches have great uncertainties. Therefore, the CCEM-Report
(14) suggests to base the risk estimation on dominant mutation
tests. These tests have the advantage of being homologous to
human genetic disorders and that they directly determine the
genetic damage expressed in subsequent generations.

5.2. Direct Estimation of First-Generation Risk

Based on the induction of dominant mutations in mice, Ehling
(22,23) developed a concept for the direct estimation of the risk

of radiation-induced genetic damage to the human population expressed in the first generation. The quantification of the genetic risk is based on the following assumptions:

a) The dose-effect curve for the induction of dominant cataract mutations is linear.

b) Dominant cataract mutation rates are representative for all dominant mutations.

c) The ratio of the number of dominant cataract loci (20) to the total number of well-established dominant loci (736) in man (56) is the same as in the mouse. This ratio gives a multiplication factor of 36.8. The multiplication factor is used to convert the induced mutation rate of dominant cataracts to the estimation of the overall dominant mutation rate.

The rate of radiation-induced dominant cataracts after single exposure (Table 5) is $0.45 - 0.55 \times 10^{-4}$ mutations/gamete/Gy. The rate has to be multiplied by 36.8 for the calculation of the overall frequency for dominant mutations ($17 - 20 \times 10^{-4}$ mutations/gamete/Gy). For a high dose rate exposure with high-intensity radiation in spermatogonia of man with 1 Gy, one can expect 1,700 - 2,000 induced dominant mutations in the first generation in 1 million liveborns (Table 14).

Table 14. Estimated effect of 1 Gy per generation of high intensity exposure of spermatogonia for 1 million liveborns

	Cataracts	Skeletal defects
Mutations/gamete/Gy	$0.45 - 0.55 \times 10^{-4}$	10.1×10^{-4}
Multiplication factor for the overall dominant mutation rate	36.8	4.6
Expected cases of dominant diseases	1700 - 2000	4600

No experimental data are available for the induction of dominant cataract mutations by low dose-rate radiation in male mice, or for the mutation rate in female mice. Therefore, an estimation of the population risk is only possible with reservations. Such an estimation can only be based on the generalization of results obtained with the specific locus method (26). In general, for exposure with low dose-rate we expect only one-third of the frequency with high dose-rate exposure (102).

For the risk due to total population exposure, a sensitivity factor for female mice of 1.4 (77) or 2 (26) is used. Because of the higher DNA content in the germ cells of man, in comparison to those of mice (2), some authors use a factor of 1.2 for the extrapolation from mice to man (26,100). Other authors claim that mice and man are equally sensitive to the induction of mutations by radiation (84,102). Using these additional assumptions, it is possible to calculate the genetic risk of a population after a low dose rate exposure with 1 Gy per 1 million liveborns to be 800 to 1,600 expected cases of dominant diseases in the first generation.

One advantage of the direct estimation of the genetic risk is that the results based on the induction of dominant cataracts can be compared with the data based on dominant hereditary disorders of another system, for example, or induction of dominant skeletal mutations. The rate of radiation-induced dominant skeletal mutations after single exposure is 10.1×10^{-4} mutations/gamete/Gy (Table 4). This rate can be converted to an overall frequency for dominant mutations by multiplication with 4.6. It follows that for high dose-rate exposure with high-intensity radiation of spermatogonia of man with 1 Gy, we can expect 4,600 induced dominant mutations in the first generation in 1 million liveborn individuals (Table 14).

Using similar suppositions to those in the UNSCEAR-Report (101,102), one can calculate the population risk. This calculation is based on the following assumption: 10.1×10^{-4} dominant skeletal mutations/gamete/Gy x 4.6 (multiplication factor for conversion to the overall mutation rate) x 0.3 (correction factor for dose rate) x 1.4 (exposure of female mice) x 10^{6} (offspring) equals 2,000. This estimate of the effect of 1 Gy per generation of low dose, low dose rate, low LET irradiation on a population of one million liveborn individuals of 2,000 cases with autosomal dominant diseases is based on a generalization of the results obtained with the specific locus method. This figure of 2,000 cases is identical with the estimation of Ehling (23) and the UNSCEAR-Report (101).

For an improvement of the risk estimation it is necessary to extend the data base for the induction of dominant mutations in mice, especially the mutation frequency in female mice and in the low dose-rate range of exposure. In addition, it is necessary to emphasize that a mutation with high penetrance has a better chance of being recovered in these experiments than a mutation with low penetrance. Furthermore, it is likely that the multiplication factors for the determination of the overall estimates will increase as more genes with dominant inheritance are described. Therefore, these quantifications underestimate the radiation-induced genetic damage to the first generation.

Similarly to the quantification of the radiation-induced genetic damage of the first generation, the data of Ehling et al. (37) can be used for the estimation of the expected number of dominant mutations in the first generation after ENU-exposure. The problems of risk estimation due to exposure to chemical mutagens were recently discussed in detail (29-32).

6.   PROBLEMS AND PERSPECTIVES

Results and conclusions presented in this and the preceding chapters indicate that estimations of the genetic risks due to radiation exposure or environmental chemical mutagens are based on the results of mammalian genetic studies. The specific locus method has been used for exploring the effect of various biological and physical factors on induced mutation frequency at a sample of seven genes. However, this exploration has been limited to a very specific genetic background of the cross $(101 \times C3H)F_1 \times$ Test Stock mice. Knowing that the induced mutation rate may depend upon the genetic background (50), it is essential to explore these factors with other strains of mice. The development of techniques to study the induction of dominant mutations makes this exploration possible.

The BEIR (4), the UNSCEAR (102) and the CCEM-Report (14) emphasize the importance of dominant mutations for the quantification of the genetic risk. Because of the importance of these test systems, Sankaranarayanan (84) pointed out correctly that the data base for these estimations are very limited. It is essential to establish dose-response curves for the induction of dominant mutations. In addition, "there are no data on the induction of skeletal or cataract mutations in female mice" and "there is no experimentation so far to verify whether the correction factors used to estimate effects at low doses and at low dose rates are in fact valid for these kinds of mutational events". Similarly, it is necessary to establish a data base for the induction of dominant mutations by chemical mutagens. Because it does not seem possible to determine the total burden of chemical mutagens, it is necessary to establish standards for the exposure limits of individual chemical mutagens. These limits can be expressed as a relative increase of the spontaneous mutation rate or as an accepted number of mutations. For example, for a single compound the allowable level of risk for all dominant mutations could be 10^{-6}. This figure is debatable and depends on the number of chemical mutagens a society will permit. Leaving aside the details of these exposure limits for chemical mutagens, it is necessary to discuss the problem and find an acceptable and responsible solution.

A very important aspect of risk estimation is the possible interaction between different chemical mutagens, and between ionizing radiation and chemical mutagens. Some of the main problems involved in studying interaction between radiation and chemicals have recently been reviewed by Glubrecht et al. (42) and Kada et al. (47). It is likely that the action of a mutagenic agent is not direct, and that cellular functions, such as mutators or repair systems, are involved in the mutagenesis processes initiated by an agent. Such cellular functions can be affected by a second agent. In sexually reproducing organisms, the two agents can also act on separate cells (male and female germ cells) which subsequently fuse. However, it has not yet been possible to estimate the impact of the interaction between radiation and chemicals on risk assessment.

Mammalian genetics has many fascinating aspects besides the urgent problems which have to be solved in order to improve the ways of estimating the genetic risk to the human population due to exposure to genotoxic compounds and ionizing radiation. For example, the lens tissue is ideal for the study of various fundamental biological processes. Because of its ability to preserve a complete record of one of the highest forms of morphological and biochemical specialization, not only can differentiation be studied fruitfully, but also some aspects of the process of aging, which is still poorly understood, since the lens never sheds its cells.

The metabolism of the lens is relatively simple. Experiments are in progress to determine the biochemical defects of the recovered mutations. After isoelectric focusing of soluble lens proteins on ultrathin polyacrylamide gels, it could be shown that, in the region of the β- and γ-crystallins, protein bands present in the wild type were not visible in the heterozygotes and homozygotes of one recovered mutant. Activity determinations of various enzymes of lens extracts revealed that the activities of enolase, lactate dehydrogenase, and hexokinase were enhanced in the mutant genotypes as compared to the wild type (65).

The isolated mutations are ideal subjects for studies in the field of developmental genetics. A recessive cataract gene in rabbits (kat-1) is manifested by an opacity of the lens suture. In the homozygous state, the gene disturbs the metabolism of the

lens and interferes with the maintenance of osmotic balance.
Depending on the state of hydration of the animal during the
second month of life, the manifestation of the kat-1 homozygous
phenotype may progress to complete lens opacification. Two groups
of rabbits with sutural cataracts were compared, one receiving a
diet poor in water to cause dehydration of the animals, the other
receiving a diet rich in water. In 84% of all cases in the group
with limited water intake, the cataract remained stationary. In
the other group 85% of all animals developed a total cataract
(17).

Similar experiments to investigate the manifestation of
cataracts in mice will be performed. The systematic investigation
of cataracts in mice should lead to the understanding of the
genesis of hereditary cataracts. More importantly, these studies
could be a model for the elucidation of the manifestation of
cataracts in man.

REFERENCES

1.   Abrahamson, S. and S. Wolff (1976) Re-analysis of radiation-
      induced specific locus mutations in the mouse. Nature
      264:715-719.
2.   Abrahamson, S., M.A. Bender, A.D. Conger, and S. Wolff (1973)
      Uniformity of radiation-induced mutation rates among
      different species. Nature 245:460-462.
3.   Batchelor, A.L., R.J.S. Phillips, and A.G. Searle (1969) The
      ineffectiveness of chronic irradiation with neutrons and
      gamma rays in inducing mutations in female mice. Br. J.
      Radiol. 42:448-451.
4.   BEIR-Report (1980) (Biological Effects of Ionizing
      Radiations), The Effects on Populations of Exposure to Low
      Levels of Ionizing Radiation: 1980. National Academy Press,
      Washington, D.C.
5.   Brenneke, H. (1937) Strahlenschädigung von Mäuse- und
      Rattensperma, beobachtet an der Frühentwicklung der Eier.
      Strahlentherapie 60:214-238.
6.   Brewen, J.G., and H.S. Payne (1979) X-ray stage sensitivity
      of mouse oocytes and its bearing on dose-response curves.
      Genetics 91:149-161.
7.   Brown, K.S., R.E. Cranley, R. Greene, H.K. Kleinman, and
      J.P. Pennypacker (1981) Disproportionate micromelia (Dmm): An
      incomplete dominant mouse dwarfism with abnormal cartilage
      matrix. J. Embryol. exp. Morph. 62:165-182.
8.   Carter, T.C. (1958) Radiation-induced gene mutation in adult
      female and foetal male mice. Br. J. Radiol. 31:407-411.
9.   Carter, T.C., M.F. Lyon, and R.J.S. Phillips (1960) The
      genetic sensitivity to X-rays of mouse foetal gonads. Genet.
      Res., Camb. 1:351-355.

10. Catcheside, D.G. (1947) Genetic effects of radiations. _Brit._ _J._ _Radiology_ 1:s109-s116.
11. Cattanach, B.M. (1966) Chemically induced mutations in mice. _Mutation Res._ 3:346-353.
12. Cattanach, B.M. (1974) Spermatogonial stem cell killing in the mouse following single and fractionated X-ray doses, as assessed by length of sterile period. _Mutation Res._ 25:53-62.
13. Cattanach, B.M. (1982) Induction of specific locus mutations in female mice by triethylenemelamine (TEM). _Mutation Res._ 104:173-176.
14. CCEM-Report (1983) (Committee on Chemical Environmental Mutagens). Identifying and Estimating the Genetic Impact of Chemical Mutagens. National Academy Press, Washington, D.C.
15. Charles, D.R., J.A. Tihen, E.M. Otis, and A.B. Grobman (1960) Genetic effects of chronic X-irradiation exposure in mice, UR-505 AEC Research and Development Report. The University of Rochester Atomic Energy Project, Rochester, New York.
16. Courot, M., M.T. Hochereau-de Reviers and R. Ortavant (1970) Spermatogenesis, In _The_ _Testis_, Vol. 1, A.D. Johnson, W.R. Gomes, and N.L. Vandemark (Eds.). Academic Press, New York, pp. 339-432.
17. Ehling, U. (1957) Untersuchungen zur kausalen Genese erblicher Katarakte beim Kaninchen. _Z._ _menschl._ _Vererb.-u._ _Konstitutionslehre_ 34:77-104.
18. Ehling, U.H. (1964) Frequency of X-ray induced presumed dominant mutations affecting the skeleton of mice. _Genetics_ 50:246.
19. Ehling, U.H. (1966) Dominant mutations affecting the skeleton in offspring of X-irradiated male mice. _Genetics_ 54:1381-1389.
20. Ehling, U.H. (1970) Evaluation of presumed dominant skeletal mutations, In _Chemical_ _Mutagenesis_ _in_ _Mammals_ _and_ _Man_, F. Vogel and G. Röhrborn (Eds.). Springer, Berlin-Heidelberg-New York, pp. 162-166.
21. Ehling, U.H. (1974) Differential spermatogenic response of mice to the induction of mutations by antineoplastic drugs. _Mutation Res._ 26:285-295.
22. Ehling, U.H. (1974) Die Gefährdung der menschlichen Erbanlagen im technischen Zeitalter (Vortrag beim Deutschen Röntgenkongreß 1974 in Baden-Baden). _Fortschr._ _Röntgenstr._ 124:166-171, 1976.
23. Ehling, U.H. (1976) Estimation of the frequency of radiation-induced dominant mutations. ICRP, CI-TG 14, Task Group on Genetically Determined Ill-Health.
24. Ehling, U.H. (1978) Specific-locus mutations in mice, In _Chemical_ _Mutagens_, Vol. 5, A. Hollaender and F.J. de Serres, (Eds.). Plenum Press, New York, pp. 233-256.
25. Ehling, U.H. (1980) Induction of gene mutations in germ cells of the mouse. _Arch._ _Toxicol._ 46:123-138.

26. Ehling, U. (1980) Strahlengenetisches Risiko des Menschen. Umschau 80:754-759.
27. Ehling, U.H. (1981) Genetische Risiken durch Umweltchemikalien, In Umweltrisiko 80, B. Globel, G. Gerber, R. Grillmaier, R. Kunkel, H.-K. Leetz und E. Oberhausen (Hrsg.). Thieme, Stuttgart-New York, pp. 400-411.
28. Ehling, U.H. (1981) Mutagenicity of selected chemicals in induction of specific locus mutations in mice, In Comparative Chemical Mutagenesis, F.J. de Serres, and M.D. Shelby (Eds.). Plenum Press, New York, pp. 729-742.
29. Ehling, U.H. (1982) From hazard identification to risk estimation of mutagens, In Mutagens in Our Environment, M. Sorsa and H. Vainio (Eds.). Alan R. Liss, Inc., New York, pp. 203-218.
30. Ehling, U.H. (1982) Risk estimations based on germ-cell mutations in mice, In Environmental Mutagens and Carcinogens (Proceedings of the 3rd International Conference on Environmental Mutagens), T. Sugimura, S. Kondo, and H. Takebe, (Eds.). University of Tokyo Press, Tokyo/Alan R. Liss, Inc., New York, pp. 709-719.
31. Ehling, U.H. (1983) Cataracts - Indicators for dominant mutations in mice and man, In Utilization of Mammalian Specific Locus Studies in Hazard Evaluation and Estimation of Genetic Risk, F.J. de Serres, and W. Sheridan (Eds.). Plenum Press, New York, pp. 169-190.
32. Ehling, U.H. (in press) In vivo gene mutations in mammals, Proceedings of the Symposium "Critical Evaluation of Mutagenicity Tests". Bundesgesundheitsamt, Berlin.
33. Ehling, U.H., and J. Kratochvilova (1979) Direct estimation of genetic risk from radiation in the first generation. 6th International Congress of Radiation Research, Tokyo, 13.-19.5.1979, p. 180.
34. Ehling, U.H., and A. Neuhäuser (1979) Procarbazine-induced specific locus mutations in male mice. Mutation Res. 59:245-256.
35. Ehling, U.H., and A. Neuhäuser-Klaus (1982) Chemically-induced mutations in mice, Progress Report: May 1982 - November 1982. Commission of the European Communities.
36. Ehling, U.H., and M.L. Randolph (1962) Skeletal abnormalities in the F_1 generation of mice exposed to ionizing radiations. Genetics 47:1543-1555.
37. Ehling, U.H., J. Favor, J. Kratochvilova, and A. Neuhäuser-Klaus (1982) Dominant cataract mutations and specific locus mutations in mice induced by radiation or ethylnitrosourea. Mutation Res. 92:181-192.
38. Ehling, U.H., D. Averbeck, P.A. Cerutti, J. Friedman, H. Greim, A.C. Kolbye, Jr. and M.L. Mendelsohn (1983) Review of the evidence for the presence or absence of thresholds in the induction of genetic effects by genotoxic chemicals. Mutation Res. 123:281-341.

39. Falconer, D.S. (1949) The estimation of mutation rates from incompletely tested gametes, and the detection of mutations in mammals. J. Genetics 49:226-234.

40. Favor, J. (1982) ENU-induced dominant cataract mutations in mice. Environ. Mutagenesis 4:318.

41. Favor, J. (1983) A comparison of the dominant cataract and recessive specific locus mutation rates induced by treatment of male mice with ethylnitrosourea. Mutation Res. 110:367-382.

42. Glubrecht, H., A.R. Gopal-Ayengar, and L. Ehrenberg (1979) Interactions of ionizing radiation and chemicals and mechanisms of action - summary, In Radiation Research (Proceedings of the 6th International Congress of Radiation Research), S. Okada, M. Imamura, T. Terashima, and H. Yamaguchi (Eds.). Japanese Association for Radiation Research, Tokyo, Japan, pp. 708-710.

43. Haynes, R.H., and F. Eckardt (1979) Analysis of dose-response patterns in mutation research. Can. J. Genet. Cytol. 21:277-302.

44. Hertwig, P. (1932) Wie muss man züchten, um bei Säugetieren die natürliche oder experimentelle Mutationsrate festzustellen? Arch. Rassen- u. Gesellschaftsbiol. 27:1-12.

45. Hertwig, P. (1935) Sterilitätserscheinungen bei röntgenbestrahlten Mäusen. Z. indukt. Abstammungs- u. Vererbungslehre 70:517-523.

46. Hertwig, P. (1939) Zwei subletale rezessive Mutationen in der Nachkommenschaft von röntgenbestrahlten Mäusen. Der Erbarzt 4:41-43.

47. Kada, T., T. Inoue, A. Yokoiyama, and L.B. Russell (1979) Combined genetic effects of chemicals and radiation, In Radiation Research (Proceedings of the 6th International Congress of Radiation Research), S. Okada, M. Imamura, T. Terashima, and H. Yamaguchi (Eds.). Japanese Association for Radiation Research, Tokyo, Japan, pp. 711-720.

48. Kratochvilova, J. (1981) Dominant cataract mutations detected in offspring of gamma-irradiated male mice. J. Heredity 72:302-307.

49. Kratochvilova, J., and U.H. Ehling (1979) Dominant cataract mutations induced by γ-irradiation of male mice. Mutation Res. 63:221-223.

50. Laskowski, W. (1981) Biologische Strahlenschäden und ihre Reparatur. Walter de Gruyter, Berlin-New York.

51. Lyon, M.F. (1981) Sensitivity of various germ-cell stages to environmental mutagens. Mutation Res. 87:323-345.

52. Lyon, M.F., and R.J.S. Phillips (1975) Specific locus mutation rates after repeated small radiation doses to mouse oocytes. Mutation Res. 30:375-382.

53. Lyon, M.F., D.G. Papworth, and R.J.S. Phillips (1972) Dose-rate and mutation frequency after irradiation of mouse spermatogonia. Nature (London) New Biol. 238:101-104.

54. Lyon, M.F., R.J.S. Phillips, and H.J. Bailey (1972) Mutagenic effects of repeated small radiation doses to mouse spermatogonia. I. Specific-locus mutation rates. Mutation Res. 15:185-190.

55. Lyon, M.F., R.J.S. Phillips, and G. Fisher (1979) Dose-response curves for radiation-induced gene mutations in mouse oocytes and their interpretation. Mutation Res. 63:161-173.

56. McKusick, V.A. (1978) Mendelian Inheritance in Man, 5th Edition. Johns Hopkins University Press, Baltimore-London.

57. Neuhäuser-Klaus, A. (1983) Personal communication.

58. Oakberg, E.F. (1956) Duration of spermatogenesis in the mouse and timing of stages of the cycle of the seminiferous epithelium. Am. J. Anat. 99:507-516.

59. Oakberg, E.F. (1975) Effects of radiation on the testis, In Handbook of Physiology, Vol. 5, Endocrinology, pp. 233-243.

60. Oakberg, E.F. (1978) Differential spermatogonial stem cell survival and mutation frequency. Mutation Res. 50:327-340.

61. Oakberg, E.F. (1979) Timing of oocyte maturation in the mouse and its relevance to radiation-induced cell killing and mutational sensitivity. Mutation Res. 59:39-48.

62. Oakberg, E.F., and R.L. Diminno (1960) X-ray sensitivity of primary spermatocytes of the mouse. Int. J. Radiat. Biol. 2:196-209.

63. Oftedal, P. (1968) A theoretical study of mutant yield and cell killing after treatment of heterogeneous cell populations. Hereditas 60:177-210.

64. Pedersen, T. (1970) Follicle kinetics in the ovary of the cyclic mouse. Acta Endocrinol. 64:304-323.

65. Pretsch, W., D.J. Charles, and J. Kratochvilova (1980) Untersuchungen an Linsenextrakten von Mäusen mit erblicher Katarakt: Ultradünnschicht-isoelektrische Fokussierung und Enzymanalysen, In Elektrophorese Forum '80, B.J. Radola (Hrsg.). Technische Universität München, pp. 75-80.

66. Russell, L.B. (1971) Definition of functional units in a small chromosomal segment of the mouse and its use in interpreting the nature of radiation-induced mutations. Mutation Res. 11:107-123.

67. Russell, L.B., and B.E. Matter (1980) Whole-mammal mutagenicity tests: Evaluation of five methods. Mutation Res. 75:279-302.

68. Russell, L.B., W.L. Russell, and E.M. Kelly (1979) Analysis of the albino-locus region of the mouse. I. Origin and viability. Genetics 91:127-139.

69. Russell, L.B., P.B. Selby, E. von Halle, W. Sheridan, and L. Valcovic (1981) The mouse specific locus test with agents other than radiations. Interpretation of data and recommendations for future work. Mutation Res. 86:329-354.

70. Russell, W.L. (1951) X-ray-induced mutations in mice, Cold Spring Harbor Symposia Quant. Biol. 16:327-336.

71. Russell, W.L. (1956) Lack of linearity between mutation rate and dose for X-ray-induced mutations in mice. Genetics 41:658-659.

72. Russell, W.L. (1963) The effect of radiation dose rate and fractionation on mutation in mice, In Repair from Genetic Radiation, F.H. Sobels (Ed.). Pergamon Press, Oxford-London-New York-Paris, pp. 205-217.

73. Russell, W.L. (1964) Evidence from mice concerning the nature of the mutation process, In Genetics Today (Proceedings of the XI International Congress of Genetics, The Hague, The Netherlands, September, 1963). Pergamon Press, New York, pp. 257-264.

74. Russell, W.L. (1965) Effect of the interval between irradiation and conception on mutation frequency in female mice. Proc. Natl. Acad. Sci. USA 54:1552-1557.

75. Russell, W.L. (1965) The nature of the dose-rate effect of radiation on mutation in mice. Jap. J. Genet. 40:128-140.

76. Russell, W.L. (1972) The genetic effects of radiation, In Peaceful Uses of Atomic Energy, Vol. 13, International Atomic Energy Agency (IAEA), Vienna, pp. 487-500.

77. Russell, W.L. (1977) Mutation frequencies in female mice and the estimation of genetic hazards of radiation in women. Proc. Natl. Acad. Sci. USA 74:3523-3527.

78. Russell, W.L. (1982) Factors affecting mutagenicity of ethylnitrosourea in the mouse specific locus test and their bearing on risk estimation, In Environmental Mutagens and Carcinogens (Proceedings of the Third International Conference on Environmental Mutagens), T. Sugimura, S. Kondo and H. Takebe (Eds.), University of Tokyo Press, Tokyo/Alan R. Liss, Inc., New York, pp. 59-70.

79. Russell, W.L., and E.M. Kelly (1982) Specific-locus mutation frequencies in mouse stem-cell spermatogonia at very low radiation dose rates. Proc. Natl. Acad. Sci. USA 79:539-541.

80. Russell, W.L., and E.M. Kelly (1982) Mutation frequencies in male mice and the estimation of genetic hazards of radiation in men. Proc. Natl. Acad. Sci. USA 79:542-544.

81. Russell, W.L., L.B. Russell, and E.M. Kelly (1958) Radiation dose rate and mutation frequency. Science 128:1546-1550.

82. Russell, W.L., P.R. Hunsicker, D.A. Carpenter, C.V. Cornett, and G.M. Guinn (1982) Effect of dose fractionation on the ethylnitrosourea induction of specific locus mutations in mouse spermatogonia. Proc. Natl. Acad. Sci. USA 79:3592-3593.

83. Russell, W.L., P.R. Hunsicker, G.D. Raymer, M.H. Steele, K.F. Stelzner, and H.M. Thompson (1982) Dose-response curve for ethylnitrosourea-induced specific locus mutations in mouse spermatogonia. Proc. Natl. Acad. Sci. USA 79:3589-3591.

84. Sankaranarayanan, K. (1982) Genetic Effects of Ionizing Radiation in Multicellular Eukaryotes and the Assessment of

Genetic Radiation Hazards in Man, Elsevier Biomedical Press,
Amsterdam (Quotation p. 311).

85. Schaefer, H. (1939) Die Fertilität von Mäusemännchen nach
Bestrahlung mit 200 r. Z. mikroskop. anat. Forsch.
46:121-152.

86. Searle, A.G. (1974) Mutation induction in mice, In Advances
in Radiation Biology, Vol. 4, J.T. Lett, H.I. Adler, and
M. Zelle (Eds.). Academic Press, New York-London,
pp. 131-207.

87. Searle, A.G. (1975) The specific locus test in the mouse.
Mutation Res. 31:277-290.

88. Searle, A.G. (1982) Germ-cell sensitivity in the mouse: A
comparison of radiation and chemical mutagens, In
Environmental Mutagens and Carcinogens, T. Sugimura,
S. Kondo, and H. Takebe (Eds.). University of Tokyo Press,
Tokyo/Alan R. Liss, Inc., New York, pp. 169-177.

89. Sega, G.A., and R.E. Sotomayor (1982) Unscheduled DNA
synthesis in mammalian germ cells - its potential use in
mutagenicity testing, In Chemical Mutagens, Principles and
Methods for Their Detection, Vol. 7, F.J. de Serres, and A.
Hollaender (Eds.). Plenum Press, New York-London,
pp. 421-445.

90. Sega, G.A., R.E. Sotomayor, and J.G. Owens (1978) A study of
unscheduled DNA synthesis induced by X-rays in the germ cells
of male mice. Mutation Res. 49:239-257.

91. Selby, P.B. (1973) X-ray induced specific locus mutation rate
in newborn male mice. Mutation Res. 18:63-75.

92. Selby, P.B. (1979) Radiation-induced dominant skeletal
mutations in mice: mutation rate, characteristics, and
usefulness in estimating genetic hazards to humans from
radiation, In Proceedings of the Sixth International Congress
of Radiation Research, S. Okada, M. Imamura, T. Terashima and
H. Yamaguchi (Eds.), Toppan Printing Co., Tokyo, Japan,
pp. 537-544.

93. Selby, P.B. (1981) Radiation genetics, In The Mouse in
Biomedical Research, Vol. 1, H.L. Foster, J.D. Small, and
J.G. Fox (Eds.). Academic Press, New York-London,
pp. 263-283.

94. Selby, P.B. (1982) Induced mutations in mice and genetic risk
assessment in humans, In Progress in Mutation Research,
Vol. 3, K.C. Bora et al. (Eds.). Elsevier Biomedical Press,
pp. 275-288.

95. Selby, P.B., and P.R. Selby (1977) Gamma-ray-induced dominant
mutations that cause skeletal abnormalities in mice. I.
Plan, summary of results and discussion. Mutation
Res. 43:357-375.

96. Selby, P.B., and P.R. Selby (1978) Gamma-ray-induced dominant
mutations that cause skeletal abnormalities in mice. II.
Description of proved mutations. Mutation Res. 51:199-236.

97. Selby, P.B., and P.R. Selby (1978) Gamma-ray-induced dominant mutations that cause skeletal abnormalities in mice. III. Description of presumed mutations. Mutation Res. 50:341-351.
98. Snell, G.D. (1935) The induction by X-rays of hereditary changes in mice. Genetics 20:545-567.
99. Snell, G.D. (1945) The detection of mutations. Relative efficiency of various systems of brother-sister inbreeding in mice. J. Heredity 36:275-278.
100. UNSCEAR-Report (1966) (United Nations Scientific Committee on the Effects of Atomic Radiation), Supplement No. 14. United Nations, New York (Quotation p. 8).
101. UNSCEAR-Report (1977) (United Nations Scientific Committee on the Effects of Atomic Radiation), Sources and Effects of Ionizing Radiation. United Nations, New York.
102. UNSCEAR-Report (1982) (United Nations Scientific Committee on the Effects of Atomic Radiation) Ionizing Radiation: Sources and Biological Effects. United Nations, New York.
103. Wolff, S. (1967) Radiation genetics, In Annual Review of Genetics, Vol. 1, H.L. Roman, L.M. Sandler, and G.S. Stent (Eds.). Annual Reviews, Inc., Palo Alto, California, pp. 221-244.

BENZO(A)PYRENE AND 6-NITROBENZO(A)PYRENE METABOLISM IN HUMAN

AND RODENT MICROSOMES AND TISSUE CULTURE

J.K. Selkirk[1], S. Tong[2], G.D. Stoner[3]
A. Nikbakht[1] and B.K. Mansfield[1]

[1]Biology Division, Oak Ridge National Laboratory
Oak Ridge, TN 37830

[2]University of Tennessee-Oak Ridge Graduate
School of Biomedical Sciences, Oak Ridge, TN 37830

[3]Department of Pathology, Medical College of Ohio
Toledo, OH 43699

SUMMARY

Chemical carcinogens occur in a number of unrelated chemical
structures and comprise a unique set of toxic compounds since they
have the common biological endpoint of cancer induction.
Characteristically these chemicals are biochemically inert and
require some degree of metabolic activation to form the reactive
species of the carcinogen molecule. Currently it appears that all
reactive forms are electrophilic reagents which readily bind to
cellular nucleophiles such as DNA, RNA, and protein. Data from
extensive metabolic studies with polycyclic aromatic hydrocarbons
and several other carcinogens indicate a qualitative similarity of

Research sponsored jointly by the National Cancer Institute Grant
    No. R01-CA30355 and the Office of Health and Environmental
    Research, U.S. Department of Energy, under contract W-7405-
    eng-26 with the Union Carbide Corporation.

metabolites formed in both susceptible and resistant species, tissues, and cells. This suggests that there may be divergent processing of the activated carcinogen and/or its immediate precursor that will dictate the probability of the reactive intermediate reaching a critical target site for malignant transformation. Assay of intracellular and extracellular metabolites of benzo(a)pyrene and its relatively inert isomer, benzo(e)pyrene, in epithelial and fibroblast cells display significant biochemical variation and in both conjugation reactions to water-soluble products and the region of the carcinogen molecule where the drug metabolizing enzymes attack. It is critical to the understanding of the mechanism of action of chemical carcinogens to assemble a complete metabolic pathway followed by the parent carcinogen. This includes interspecies variance in terms of processing the molecule from its activated species through formation of its various hydroxylated intermediates and the specificity of the conjugation reactions. Knowledge of these biochemical schemes in concert with the degree of interaction with cellular macromolecules will allow for assembly of the dynamic processing of the carcinogen in both resistant and susceptible tissues and will help explain the diverse activity of the carcinogens.

The first evidence that cancer was involved with environmental contamination was reported in 1776 by Percival Pott, a British physician, who correctly associated skin cancer in chimney sweeps with soot. However, it was not until 150 years later that the first experimental tumor was successfully grown in the laboratory by Yamagiwa and Ichikawa (29) by painting extracts of soot on the ear of a rabbit. Incomplete combustion of carbonaceous material appears to be the major formative process for carcinogenic polycyclic hydrocarbons. Temperatures under 1000°C allow free-radical polymerization of carbon to build various polyaromatic structures (1) in the most energetically stable forms (Fig. 1). Chemicals of this class are major pollutants of our external environment, including air, water and food. Each year more than 1300 tons of carcinogenic benzo(a)pyrene [B(a)P] alone are dispersed into the air by the burning of waste, industrial processes, and fuel exhausts from gasoline and diesel engines (2).

The voluminous data collected in the last decade has revealed the major steps in metabolic activation and detoxification for a number of these important environmental carcinogens including polycyclic hydrocarbons, nitrosamines, and aflatoxins (14). Many of these compounds are "procarcinogens" or "parent compounds", which are fairly inert chemically and are relatively stable outside of the body. However, once ingested and absorbed, they are subject to the natural drug detoxification processes. Intense

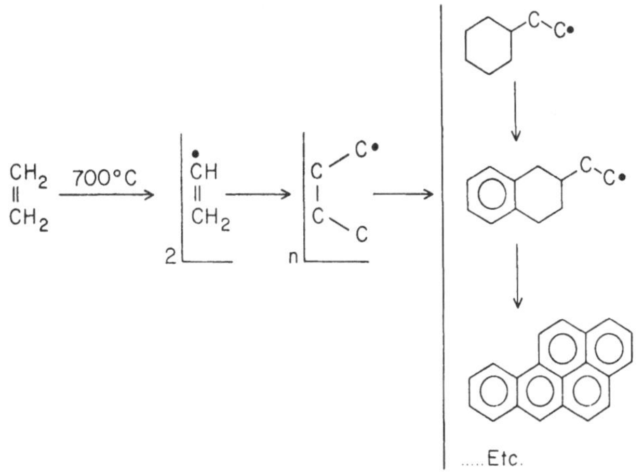

Figure 1. Pyrolytic formation of polycyclic hydrocarbons by free-
    radical polymerization. This reaction is not predominant in
    well-oxygenated, high-temperature combustion, since a more
    complete oxidation to carbon dioxide is favored.

research over the last decade has made it clear that environmental
carcinogens undergo some degree of metabolic activation within the
host tissue before exerting their carcinogenic activity.

The normal biochemical defense against toxic chemicals ·is to
attach functional groups such as hydroxyl, glucuronic acid,
glutathione etc. that tend to make the resulting less harmful
derivatives more water-soluble and readily excretable. However,
chemical carcinogens form highly reactive intermediates on the
path to final detoxification. These chemically unstable
compounds, once formed, can readily attack cellular macromolecules
such as DNA, RNA, and protein, and alkylation of genetic material
by these substances is critically involved with malignant
transformation and mutagenesis.

The divergent molecular structure of chemical carcinogens,
coupled with the wide variation in tissue and species
susceptibility, has enabled investigators to begin comparing the
relative carcinogenic potencies of structurally different chemical
carcinogens. The ultimate aim is to draw this large body of data
into the formulation of a mechanism describing the cause of cancer
by chemicals (18).

Early attempts to explain the carcinogenicity of chemical
types utilized physical chemical calculations (13). These early
hypotheses were based on the assumption that the regions of a
molecule most likely to react in chemical substitution or addition

reactions would most probably react with cellular target sites, causing an irreversible disruption of cell homeostasis. Despite some success for polycyclic hydrocarbons, this concept has not provided a comprehensive generalization even for this molecular type. Indeed, it has become apparent that biochemical activation of such molecules takes place in regions other than those anticipated (23).

To date, the only generalization that can be made with a good degree of certainty is that all known chemical carcinogens are electrophilic (8). It appears that activated forms of carcinogens, irrespective of the parent molecular structure, are electron deficient. They readily attack nucleophilic sites on nucleic acids and proteins. Although the location of such attacks may be random, probability predicts that a finite number of alkylations will occur at critical site(s) for the initiation of tumorigenesis.

The enzyme complex that metabolizes foreign chemicals is the monooxygenase system. It is found in endoplasmic reticulum and, when isolated by ultracentrifugation, comprises the microsomal fraction. This enzyme complex, called cytochrome P-450, is one component of the electron transport system which includes NADH- and NADPH-cytochrome-c-reductase, and cytochrome b_5.

There are several forms of the cytochrome P-450 enzyme which have been differentiated by their relative induction with different chemical types (e.g. steroids, polycyclic hydrocarbons, phenobarbitol), which can exhibit site selectivity when attacking compounds such as B(a)P.

Polycyclic aromatic hydrocarbons (PAH) comprise the most appropriate starting points for a discussion of chemical carcinogenesis, for it was this class of compounds that heralded the concept that the external environment could be directly involved in the development of cancer in man (20). B(a)P has been the most studied of the polycyclic carcinogens and has been the developmental model for the determination of the metabolic activation pathway. B(a)P is metabolized to a series of oxygenated derivatives. The structures seen in Figure 2 represent the major known metabolites found in all eukaryotic systems. They are isolated in their free-oxygenated, nonconjugated forms in cell-free systems (e.g., homogenates, microsomes) (16,17). However, intact cells which contain a full complement of metabolic machinery will further metabolize them to their conjugated excretion products such as glucuronides, glutathiones and sulfates (4,15). Most of these derivatives have been tested for carcinogenic and mutagenic activity in both in vivo (11) and in vitro (7) systems. It is currently felt that the pathway critical for carcinogenesis and mutagenesis by B(a)P is the one found in

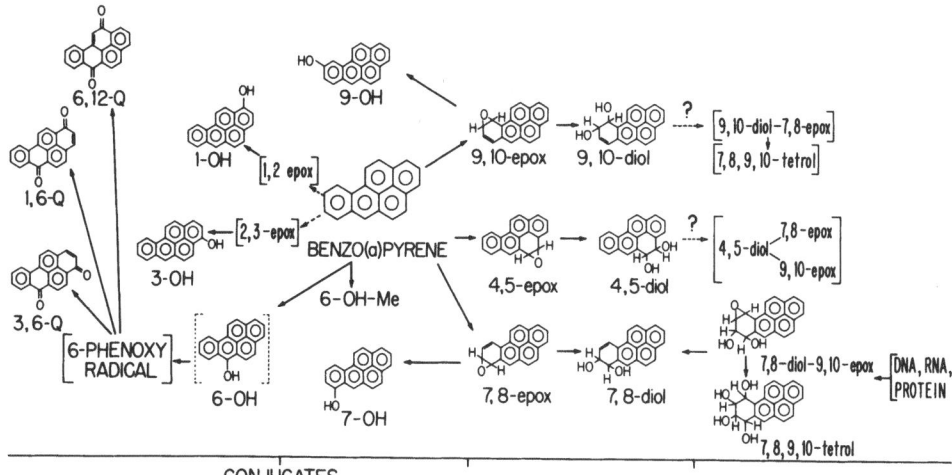

CONJUGATES

Figure 2.   Composite profile of oxygenated derivatives
    metabolically formed by tissue mixed-function oxidase
    complex.   Bracketed structures have not yet been directly
    isolated with the recent exception of 7,8,9,10-tetrol (lower
    right).

the lower righthand corner of Figure 2 where the intermediate 7,8-
epoxide is formed. This becomes hydrated by microsomal epoxide
hydrase and then reactivated by the mixed-function oxidases to
form 7,8-diol-9,10-epoxide (22).

     Extensive research has been performed on the stereochemistry
of B(a)P diol-epoxide which can form two stereoisomers (Fig. 3),
named according to the position of the 7-OH group in relation to
the oxide. At left is the syn-isomer [r-7,t-8-dihydroxy-c-9,10-
oxy-7,8,9,10-tetrahydroB(a)P]. All studies to date would
implicate the anti-isomer, shown on the right of Fig. 3, as being

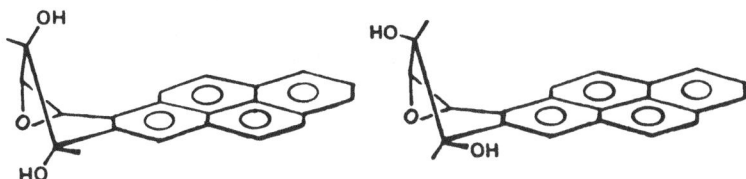

Figure 3.   Stereochemical representation of the two diastereomeric
    benzo(a)pyrene dihydrodiol-epoxides syn-isomer at left, anti-
    isomer at right.

more important in mammalian mutagenesis (5,10,28) and
carcinogenesis (24). It has been clearly demonstrated that the
anti-isomer is the major DNA-binding species with a high affinity
for the exocyclic amino group of guanine (27).

The analysis of the organic-soluble metabolites by high-
performance liquid chromatography (HPLC) shows the predominant
metabolite in cell-free systems to be 3-OH-B(a)P (Fig. 4A). The
ratio between 9-OH-B(a)P and 3-OH-B(a)P is quite variable; the 9-
OH-B(a)P is a significant metabolite in the mouse, rat, and human,
but it is relatively absent in Syrian hamster. Also, ratios of
the three metabolic dihydrodiols are species variable, with the
predominant dihydrodiol for hamster being at the 4,5-position,
whereas the 9,10-dihydrodiol is the major metabolite for the rat
(19).

Intact cell metabolism is significantly different from that
of cell-free systems (Figs. 4B). Fibroblasts utilize cytoplasmic
transferases to remove toxic phenols and quinones. In all species
studied dihydrodiols appear to be the less active substrates for
conjugation. Dihydrodiol accumulation allows a greater
probability for re-metabolism by the mixed-function oxidases to
more reactive electrophiles. The lower panel shows the
insignificant metabolic activity of human foreskin fibroblasts.

It is clear that the same profile of metabolites is formed
from B(a)P in all species that have been analyzed, and since
macromolecular binding is a presumed critical step for initiation
of malignant transformation, it becomes important to understand
the relationships between microsomal processing of carcinogens in
the cytoplasm and the readiness of the activated intermediates to
interact.

Since most human cancers are epithelial in origin and are
most probably mediated by a combination of chemical, biological
and physical components in the environment, the fate of chemical
carcinogens in cells, both susceptible and resistant to malignant
transformation, must be determined. This is especially true since
not all tissues are metabolically equivalent, and the rate of
metabolism is a function of the reactivity and hydrophilicity of
the parent chemical. The fate of the metabolic intermediates is
also a function of the selectivity of the microsomal and
cytoplasmic enzyme systems towards these electrophiles.

Figure 5 represents the range of values and the arithmetic
means for a composite of each organic solvent-soluble B(a)P
metabolite produced from the 4 tissues from each of the 8 human
donors (21). Metabolite ratios were comparable in all 4 tissues
with the tetrols, the 9,10-dihydrodiol and the quinones as major
metabolites. In all cases the 7,8-dihydrodiol was produced in

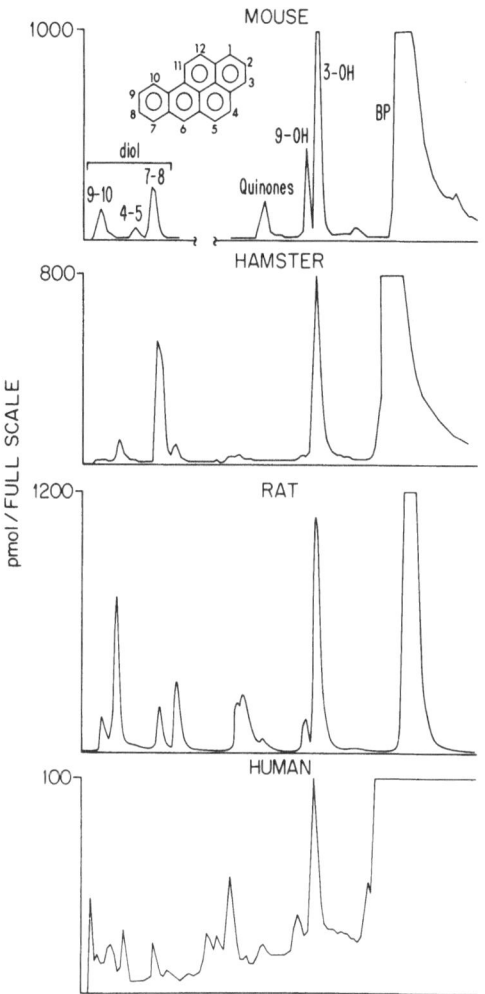

Figure 4a. Metabolism of benzo(a)pyrene by liver-microsomes from several rodent species and humans. Although the same profile of oxygenated metabolites is found among the various species, there is considerable quantitative variation, suggesting differences in metabolic rate that may be critical in determining relative susceptibility to malignant transformation between species.

these tissues, and therefore all were capable of forming the 7,8-diol-9,10-epoxide as the reactive carcinogenic intermediate of B(a)P. In general, the human tissue profiles are quite similar to the majority of non-human intact cell systems, including the formation of the 9,10-dihydrodiol as the major cellular metabolite.

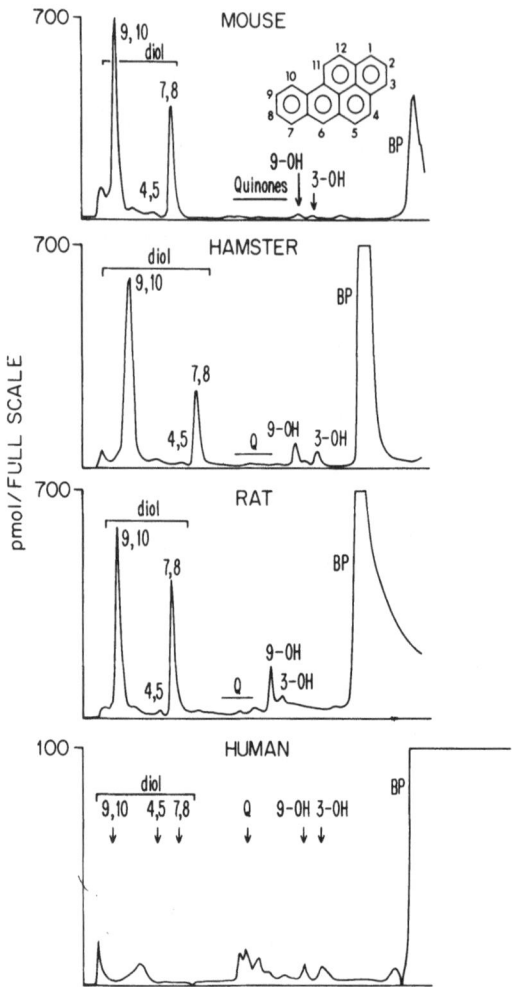

Figure 4b. Metabolism of benzo(a)pyrene by intact cells from
several rodent species and humans. Intact cells show a
considerable shift in product ratios since the cytoplasm
contains transferase activity that conjugates out cytotoxic
phenols. These profiles more closely represent physiological
conditions since all the metabolic machinery is present in an
intact cell system.

Variation in binding levels of B(a)P to DNA between the 4
tissues is seen in Fig. 6. The results clearly show a wide
variability in specific activity between donors, although there
appears to be some degree of consistency in binding in the 4
tissues from several patients. Patient no. 41 (30 years F)
exhibited high specific activity for skin, bronchus and bladder

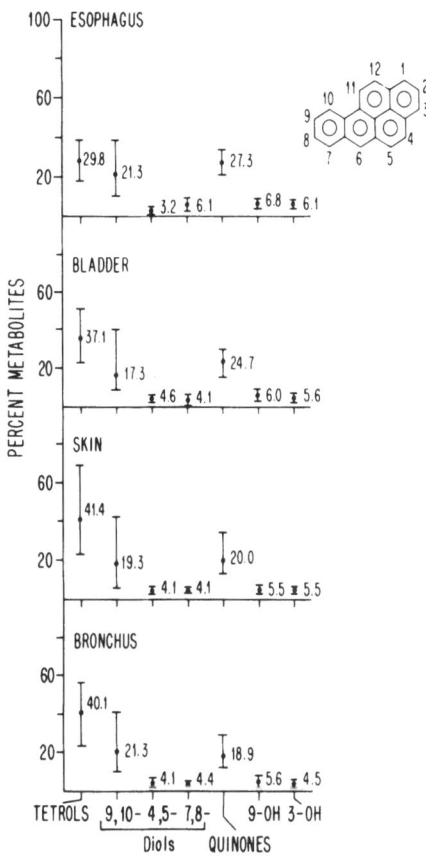

Figure 5. Human tissue explant metabolism. Composite of B(a)P
metabolites averaged from 8 donors. The numbers represent
the arithmetic mean of the respective tissue incubations from
all 8 patients. Patients Case No., age, and sex are as
follows: No. 41, 30 years F; no. 45, 63 years F; no. 52, 64
years M; no. 56, 7 years M; no. 64, 5 years M; no. 69, 10
weeks M; no. 79, 3 months M; no. 80, 57 years M; no. 62 2
months F; fibroblasts only. 4 coronary cases: 2 congenital
heart disease; 1 sudden infant death syndrome; 1 accident; 1
multiple myeloma.

(esophageal sample was not available from this patient).
Conversely, patients no. 79 (3 months M) and no. 80 (57 years M)
were consistently low in specific activity in all 4 tissues.
While a relatively high variance between patients might be
expected, due to their relative medical histories and physical
condition immediately prior to death, it is interesting to note
that the relative level of metabolic activity found in all 4
tissues from a single person may be either high or low. However,

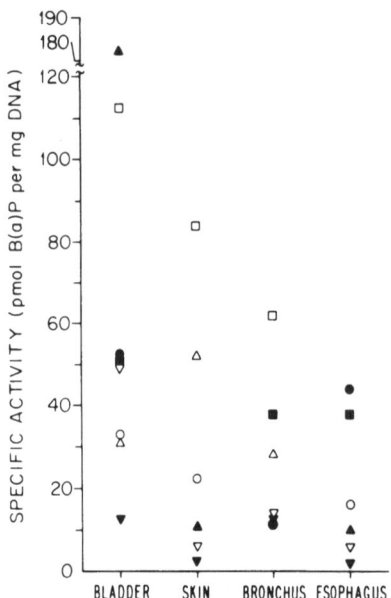

Figure 6. Comparative binding of B(a)P to DNA in human tissue
     explants. Interindividual variation of B(a)P binding to 4
     tissues from autopsy samples. In some cases insufficient
     tissue was available to perform DNA binding studies; □ , 41;
     ■ , 45; △ , 52; ▲ , 56; ○ , 64; ● , 69; ▽ , 79; ▼ ,
     80.

the relatively similar activity of tissues in culture from several
individuals in this study may reflect the genetic variance at the
Ah locus for monooxygenase induction as reported by Nebert et
al. (9).

     Nitroarenes are also prevalent polycyclic contaminants of the
environment and are becoming more widespread with the increased
use of diesel fuels. 6-Nitrobenzo(a)pyrene [6-nitroB(a)P] is an
air pollutant which can be formed in laboratory model atmospheric
chambers containing B(a)P, trace amounts of nitric acid and
nitrogen dioxide (12). It acts as a direct mutagen in the Ames
Salmonella typhimurium test although its mutagenic potential is
greatly enhanced after metabolic activation by the addition of the
liver S9 fraction (3,25), with the major metabolite identified as
3-hydroxy-6-nitroB(a)P (3).

     In order to determine if 6-nitro-B(a)P followed the same
metabolic scheme in intact cells as described with S9, we have
studied the metabolism of this compound by Syrian hamster
embryonic fibroblasts (HEF) and rat liver microsomes and its

interaction with nuclear DNA, RNA and proteins (26). Hepatic
microsomes yielded a number of UV-absorbing products, which were
separable by HPLC methods (Fig. 7). The major metabolic peak
(Peak 5) was found to be phenolic in nature due to the occurrence
of a red shift when its UV-visible spectrum was obtained under
alkaline conditions. Mass spectrometry and nuclear magnetic
resonance (NMR) studies have indicated that 3-hydroxy-6-nitroB(a)P
is the major microsomal metabolite (3).

After a 24 h incubation period of 6-nitroB(a)P with HEF,
30-35% of original radioactivity was extracted into ethyl acetate
from the medium. When analyzed by HPLC, organic solvent-soluble
products were found to be similar to those produced as a result of
microsomal metabolism, by virtue of their identical UV and
fluorescence spectra and retention times, although quantitative
differences were apparent (Fig. 8A). The main cellular products
were that of dihydrodiol (Peaks 1-4) with Peak 3 being most
prominently produced. In contrast to microsomal metabolism, only
small amounts of phenols were present.

The radioactivity remaining in the medium, after organic
solvent-soluble products were extracted into ethyl acetate,
consisted of water-soluble metabolites and comprised about half
the original radioactivity (Fig. 8B). Treatment with β-
glucuronidase released metabolites that were associated with
glucuronic acid. These products were, therefore, rendered organic
solvent-soluble and could then be extracted into ethyl acetate.

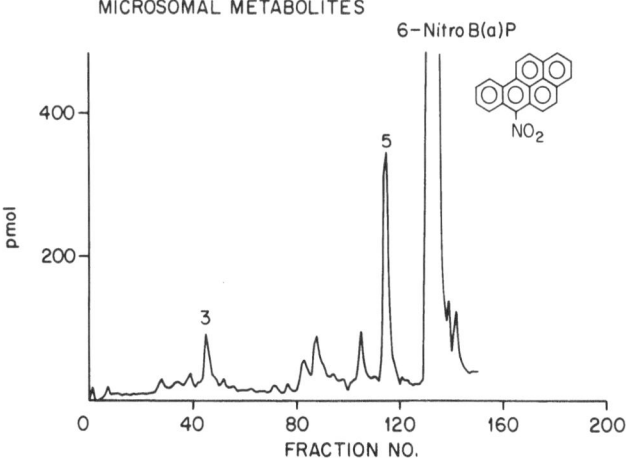

Figure 7. HPLC analysis of metabolites of 6-nitrobenzo(a)pyrene
    by incubation with hepatic microsomes from 3-
    methylcholanthrene treated rats.

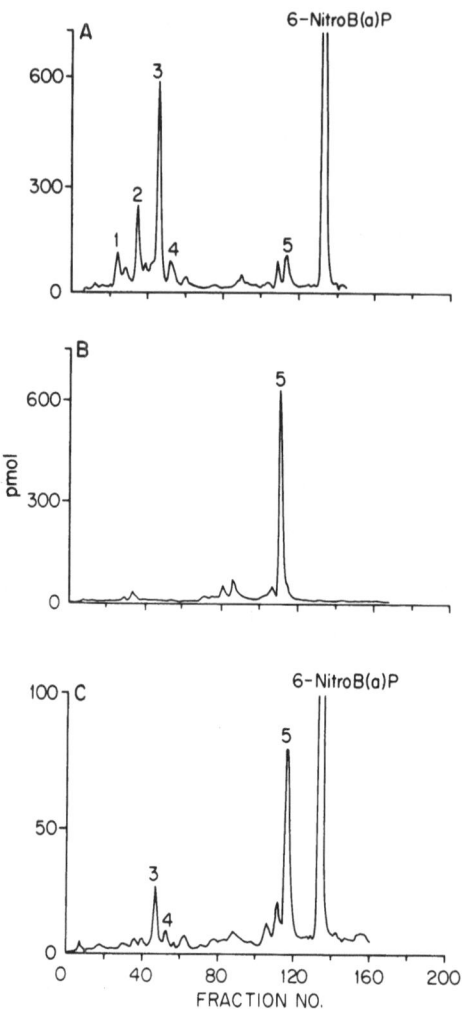

Figure 8. HPLC analysis of metabolites produced as a result of
     incubation of 6-nitrobenzo(a)pyrene with hamster embryonic
     fibroblasts. A, extracellular; B, water-soluble; C,
     intracellular.

In this case, approximately one-third of the radioactivity in the
aqueous layer was extractable into the organic solvent. HPLC
studies showed a predominant phenolic peak (Peak 5) with little
conjugation of dihydrodiols with glucuronic acid.

     Cytoplasmic metabolites consisted mainly of the phenol (Peak
5). The major dihydrodiol product (Peak 3) found in extracellular
medium is also present but in smaller quantities (Fig. 8C).

Formation of dihydrodiols as predominant products of 6-nitroB(a)P in hamster embryonic fibroblasts, but not in microsomes, may be critical in determining the mutagenic or carcinogenic properties of the compound, since it is well known that in B(a)P metabolism formation of the B(a)P-7,8-diol-B(a)P-7,8-diol-9,10-epoxide from the 7,8 diol is crucial in expressing the carcinogenic properties of the hydrocarbon.

HEF cells were then preincubated with B(a)P or 6-nitroB(a)P for 24 h, lysed, and RNA, DNA and nuclear protein separated by isopycnic separation. Determination of radioactivity from the separated fractions was used to indicate the amount of PAHs directly bound to these macromolecules. Unbound metabolites were recovered by repeated extraction with ethyl acetate prior to centrifugation.

Fig. 9 shows that both B(a)P and 6-nitroB(a)P interacted with RNA and DNA with nearly similar specific activities, although they both bind more avidly to nuclear proteins. In addition, 6-nitroB(a)P appears to have a slightly higher affinity for protein than does B(a)P. Binding to proteins in this case may adversely affect cellular mechanisms but it is also possible that such an interaction may act as a protective device in which potentially harmful metabolites become unavailable to produce cellular damage. Preliminary results from our laboratory indicated that 6-nitroB(a)P and B(a)P exhibit different selectivity toward nuclear proteins. For instance, unlike B(a)P (6), 6-nitroB(a)P has little affinity toward any core histones. Further studies are, therefore, needed to determine the biological activities of metabolites, particularly the dihydrodiols, which may be useful in predicting the potential carcinogenicity of 6-nitroB(a)P.

The mechanism that appears to be a common pathway for a structurally diverse series of chemical carcinogens is via an electrophilic intermediate. This suggests the cascade of biochemical events that lead to tumor formation is precipitated by a similar initiating event.

Fig. 10 summarizes the possible reactions of a cell when it is confronted with a foreign chemical. The process between exposure (step 1) and carcinogenesis (step 4) is long and complicated, and it is clear from this diagram that there are many intervening barriers prohibiting the cell from becoming malignantly transformed. The heavier weighted arrows indicate that most of the chemicals are transformed by microsomal enzymes in the biotransformation step into detoxification products, which are subsequently excreted as harmless metabolites. A small fraction of the chemical may survive the detoxification process and bind to cellular macromolecules. The cell, however, has a series of repair enzymes that cut out the damaged section of the

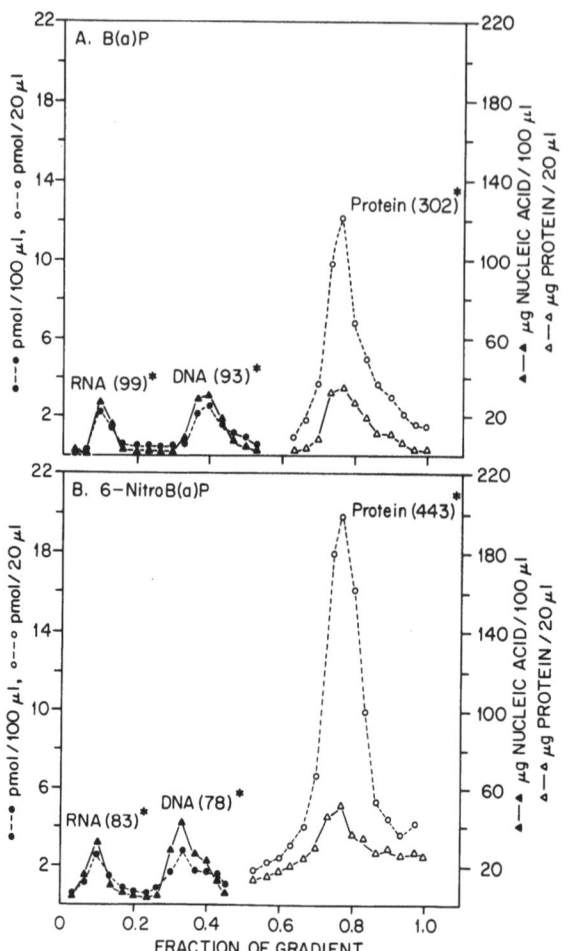

Figure 9.   Interaction of benzo(a)pyrene and 6-nitrobenzo(a)pyrene
            with nuclear macromolecules by isopycnic centrifugation.
            Specific activities measured in pmol hydrocarbon bound per mg
            nucleic acid or protein.

macromolecule and repair it by replacing the bases in exact order.
However, in some cases where the repair is incomplete or faulty, a
series of genetic lesions may ensue, such as chromosomal
aberrations, point mutation or nondisjunctions.   Any one of these
processes may lead to macromolecular binding of the carcinogen.
The question marks indicate the gaps in our knowledge of the
molecular events in cancer formation.

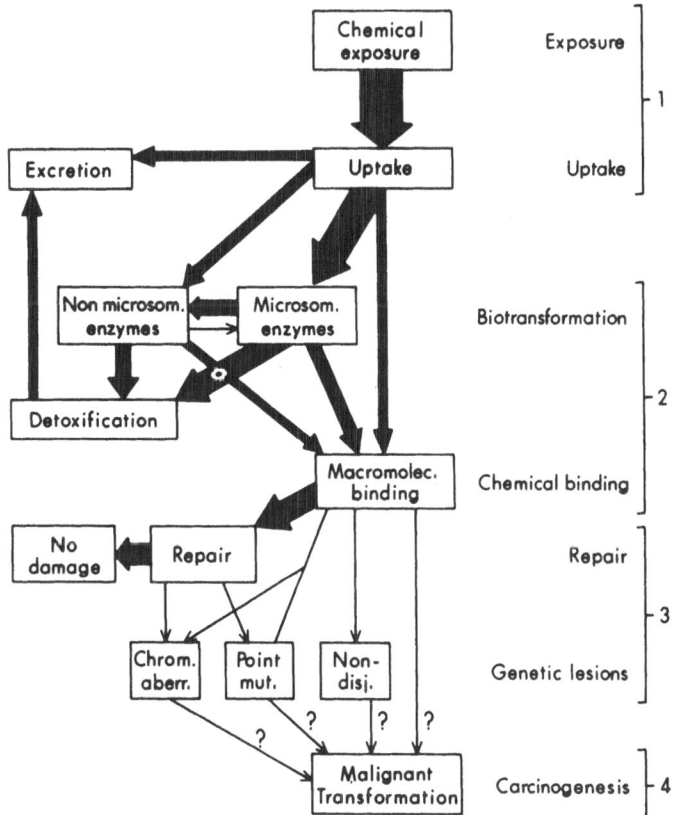

Figure 10. Metabolic processing of foreign chemicals. Heavy
    arrows indicate the major pathways taken by cells to
    deactivate and remove toxic chemicals. When macromolecular
    binding occurs, it can be repaired, leaving the cell
    unharmed. However, when repair does not occur or is faulty,
    a series of genetic lesions can occur that may lead to
    mutation and cancer.

As the data base grows and the mechanism of activation and
detoxification pathways becomes apparent, the rationale for organ,
tissue, and species specificity should become clear. Eventually
it may be possible to develop a unifying concept of the metabolic
steps required to bypass the host defense mechanism, whether
immunological or catabolic, and produce a high-risk potential for
the cell to become malignantly transformed. It is also probable
that success of a formulated cancer mechanism will require a clear
concept of the regulation of the genetic apparatus and
intracellular events taking place during cell replication.

REFERENCES

1. Badger, G.M. (1962) Mode of formation of carcinogens in human environment. National Cancer Institute Monographs 9:1-16.
2. Committee on Biologic Effects of Atmospheric Pollutants (1972) Sources of polycyclic organic matter, In Particulate Polycyclic Organic Matter, National Academy of Sciences, Washington, D.C., pp. 13-15.
3. Fu, P.P., M.W. Chou, S.K. Yang, F.A. Beland, F.F. Kadlubar, D.A. Casciano, R.H. Heflich, and F.E. Evans (1982) Metabolism of the mutagenic environmental pollutant 6-nitrobenzo(a)pyrene: Metabolic activation via ring oxidation. Biochem. Biophys. Res. Commun. 105:1037-1043.
4. Gelboin, A.V. (1980) Benzopyrene metabolism, activation, and carcinogenesis: Role and regulation of mixed-function oxidases and related enzymes. Physiol. Rev. 60:1107-1166.
5. Huberman, E., L. Sachs, S.K. Yang, and H.V. Gelboin (1976) Identification of the mutagenic metabolites of benzo(a)pyrene in mammalian cells. Proc. Natl. Acad. Sci. USA 73:607-611.
6. MacLeod, M.C., A. Kootstra, B.K. Mansfield, T.J. Slaga, and J.K. Selkirk (1980) Specificity in interaction of benzo(a)pyrene with nuclear macromolecules: Implication of derivatives of two dihydrodiols in protein binding. Proc. Natl. Acad. Sci. USA 77:6396-6400.
7. Mager, R., E. Huberman, S.K. Yang, H.V. Gelboin, L. Sachs (1977) Transformation of normal hamster cells by benzo(a)pyrene diol-epoxide. Int. J. Cancer 19:814.
8. Miller, E.C., and J.A. Miller (1976) In Chemical Carcinogens, C.E. Searle, Ed., American Chemical Society, Washington, D.C., pp. 737-762.
9. Nebert, D.W. (1978) Genetic control of carcinogen metabolism leading to individual differences in cancer risk. Biochimie 60:1019-1028.
10. Newbold, R.F., and P. Brookes (1976) Exceptional mutagenicity of a benzopyrene diol-epoxide in cultured mammalian cells. Nature 261:52-54.
11. PHS Document No. 149, Washington, D.C. U.S. Government Printing Office (1978), pp. 566-624. Survey of Compounds which have been tested for Carcinogenic Activity.
12. Pitts, J.N., Jr., K.A. Van Cauwenberghe, D. Grosjean, J.P. Schmid, D.R. Fritz, W.L. Belser, Jr., G.B. Knudson, and P.M. Hydns (1978) Atmospheric reactions of polycyclic aromatic hydrocarbons: Facile formation of mutagenic nitro-derivatives. Science 202:515-519.
13. Pullman, A., and B. Pullman (1955) Electronic structure and carcinogenic activity of aromatic molecules. Advan. Cancer Res. 3:117-169.
14. Selkirk, J.K. (1980) Chemical carcinogenesis: A brief overview of the mechanism of action of polycyclic hydrocarbons, aromatic amines, nitrosamines, and aflatoxins, In

Carcinogenesis, Vol. 5: Modifiers of Chemical Carcinogenesis, T.J. Slaga, Ed., Raven Press, New York, pp. 1-31.

15. Selkirk, J.K., G.M. Cohen, and M.C. MacLeod (1980) Glucuronic acid conjugation in the metabolism of chemical carcinogens by rodent cells, In Quantitative Aspects of Risk Assessment in Chemical Carcinogenesis, Arch. Toxicol., Suppl. 3, Springer-Verlag, pp. 171-178.

16. Selkirk, J.K., R.G. Croy, and H.V. Gelboin (1974) Benzo(a)pyrene metabolites: Efficient and rapid separation by high pressure liquid chromatography. Science 984:169-171.

17. Selkirk, J.K., Croy, R.G., Roller, P.P., and Gelboin, H.V. (1974) High pressure liquid chromatographic analysis of benzo(a)pyrene metabolism and covalent binding and the mechanism of action of 7,8-benzoflavene and 1,2-epoxy-3,3,3,-trichlorpropane. Cancer Res. 34:3474-3480.

18. Selkirk, J.K., and M.C. MacLeod (1982) Chemical carcinogenesis: Nature's metabolic mistake. BioScience 32:601-605.

19. Selkirk, J.K., M.C. MacLeod, B.K. Mansfield, P.A. Nikbakht, and K.C. Dearstone (1983) Species heterogeneity in the metabolic processing of benzo(a)pyrene, In Organ and Species Specificity in Chemical Carcinogenesis, R. Langenbach, S. Nesnow, and J.M. Rice, Eds., Plenum Publishing Corporation.

20. Selkirk, J.K., M.C. MacLeod, C.J. Moore, B.K. Mansfield, A. Nikbakht, and K. Dearstone (1982) Species variance in the metabolic activation of polycyclic hydrocarbons. Mechanisms of Chemical Carcinogenesis 331-349.

21. Selkirk, J.K., A. Nikbakht, and G.D. Stoner (1983) Comparative metabolism and macromolecular binding of benzo(a)pyrene in explant cultures of human bladder, skin, bronchus and esophagus from eight individuals. Cancer Lett. 18:11-19.

22. Sims, P., and P.L. Grover (1974) Epoxides in polycyclic aromatic hydrocarbon metabolism and carcinogenesis. Advan. Cancer Res. 20:166-274.

23. Sims, P., P.L. Grover, A. Swaisland, K. Pal, and A. Hewer (1974) Metabolic activation of benzo(a)pyrene proceeds by a diol-epoxide. Nature 252:326-328.

24. Slaga, T.J., A. Viaje, W.M. Bracken, D.L. Berry, S.M. Fischer, D.R. Miller, and S.M. LeClerc (1977) Skin-tumor-initiating ability of benzo(a)pyrene-7,8-diol-9,10-epoxide(anti) when applied topically in tetrahydrofuran. Cancer Lett. 3:23.

25. Tokiwa, H., H. Nakagawa, and Y. Onhishi (1981) Mutagenic assay of aromatic nitro compounds with Salmonella typhimurium. Mutation Res. 91:321-325.

26. Tong, S., and J.K. Selkirk (1983) Metabolism of 6-nitrobenzo(a)pyrene by hamster embryonic fibroblasts and its interaction with nuclear macromolecules. J. Toxicology and Env. Health 11:381-393.

27. Weinstein, I.B., A.M. Jeffrey, K.W. Jennette, S.H. Blobstein, R.G. Harvey, C. Harris, H. Autrup, H. Kasai, and K. Nakanishi (1976) Benzo(a)pyrene diol-epoxides as intermediates in nucleic acid binding in vitro and in vivo. Science 193:592-595.

28. Wood, A.W., P.G. Wislocki, R.L. Chang, W. Levin, A.Y.H. Lu, H. Yagi, O. Hernandez, D.M. Jerina, and A.H. Conney (1976) Mutagenicity and cytotoxicity of benzo(a)pyrene benzo-ring epoxides. Cancer Res. 36:3358-3366.

29. Yamagiwa, K., and K. Ichikawa (1915) Experimental studies on the pathogenesis of epithelial cancer. Kaiseri. Univ. Tokyo 15:295-344.

NITROSAMINE METABOLISM AND CARCINOGENESIS

R. Montesano and J. Hall[1]

Unit of Mechanisms of Carcinogenesis
International Agency for Research on Cancer
69372 Lyon Cedex 08, France

SUMMARY

The chemical class of N-nitroso compounds includes the N-nitrosamines, N-nitrosamides and N-nitrosamidines, many of which have been demonstrated to be carcinogenic, mutagenic and teratogenic in various animal species. The N-nitrosamides are unstable at physiological pH and decompose non-enzymatically to a reactive intermediate capable of reacting with cellular macromolecules. The N-nitrosamines are, however, stable at physiological pH and are metabolized by microsomal mixed function oxidases. Human, rat, monkey and trout liver possess the capacity to metabolize DNA and in most cases a lower metabolic capacity is observed in extrahepatic tissues, with a few notable exceptions. Methylbenzylnitrosamine, an esophageal carcinogen in rats, is metabolized to a greater extent in the esophagus than in the liver. Considerable evidence has accumulated that the initiation of the carcinogenic process by this group of carcinogens is linked to the metabolic competence of the target tissues or cells to convert these carcinogens into mutagenic metabolites and to the binding of these metabolites to cellular DNA. Some 12 sites have been shown to be alkylated by various methyl- and ethylating agents: $N-1$, $N-3$, $N-7$ positions of adenine, $N-3$, $N-7$ and O^6 of guanine, $N-3$, O^2 of cytosine and N^3, O^4, O^2 of thymine and the oxygens of the phosphate

[1] Recipient of an IARC-WHO Research Training Fellowship

ABBREVIATIONS: DMN, Dimethylnitrosamine; MNU, Methylnitrosourea; ENU, Ethylnitrosourea; MNNG, Methylnitrosonitroguanidine; MMS, Methylmethane-sulfonate; EMS, Ethylmethanesulfonate; MBN, Methylbenzylnitrosamine.

group. The relative extents of alkylation depend upon the
mechanism of alkylation, the agents reacting mainly by an SN2
mechanism forming more of the O-alkyl derivatives. DNA repair
processes for many of these adducts have been described and
considerable variation in the rates of removal of these adducts
exists among different animal species and cell types. Various
factors are known to influence the tissue distribution and the
metabolism of nitrosamines and the interplay of these factors is
discussed in relation to organ specific carcinogenesis.

INTRODUCTION

     The toxicity of DMN in man was recognized in 1937 by Freund,
who described the clinical and autopsy findings in two chemists
accidentally poisoned with this nitrosamine which had been used as
a solvent. Subsequently, acute and chronic liver damage has been
observed in various human beings poisoned by DMN (22,38), the
pathology being very similar to that observed in rats after acute
or chronic exposure to the same nitrosamine (7). Since these
observations, N-nitroso compounds have been shown to be
carcinogenic (18,51), mutagenic (see 56) and teratogenic (19) in
various animal species.

     The N-nitroso compounds consist of three main chemical groups
as exemplified by DMN (nitrosamines), MNU (nitrosamides) and MNNG
(nitrosoamidines) (Fig. 1). The nitrosamines are stable at
physiological pH and are metabolized mainly by microsomal mixed
function oxidases into a reactive intermediate (a methyl carbonium
ion) which reacts with various cellular macromolecules. The same
intermediate is formed non-enzymatically from MNU and in the case

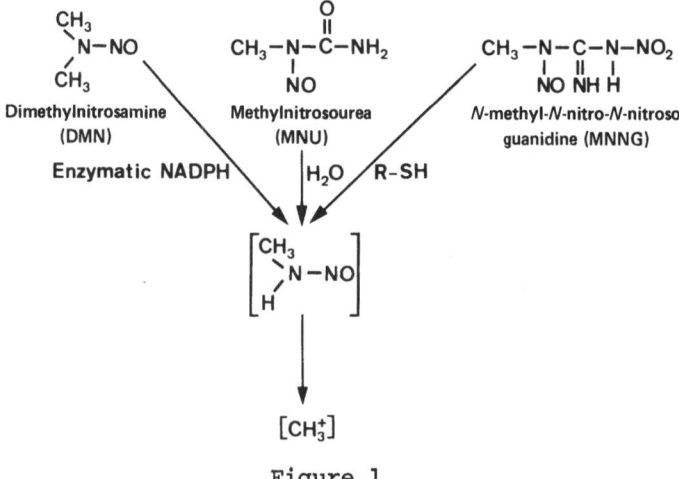

Figure 1.

of MNNG, the decomposition is catalyzed by the presence of -SH groups (see 43).

The adverse biological effects of these compounds are mediated via the formation of a reactive intermediate and its interaction with cellular macromolecules and in particular with DNA. The high degree of specificity of the N-nitroso compounds to induce tumors in a given species or organ appears to be determined by various factors: the formation of a reactive intermediate, the specific type of DNA damage induced, the level and accuracy of DNA repair processes and the rate of DNA synthesis. These various points have recently been reviewed in more detail (45,57,62,71,80,92) and they will be briefly discussed here on the assumption that initiation of the multistep process of carcinogenesis by nitrosamines is associated with damage to DNA and a mutational event.

METABOLISM AND DNA ALKYLATION

The majority of chemical carcinogens have to be metabolized into a reactive intermediate, by the microsomal mixed function oxidases, in order to have a mutagenic and a carcinogenic effect (55). This is also the case for the nitrosamines, where there is a high degree of specificity among various organs, tissues and cells in their capacity to metabolize the nitrosamine and to repair various types of DNA damage. It is unlikely that we can extrapolate either qualitatively or quantitatively this tissue-cell specificity from one species to another, although some generalizations can be made. It is therefore important to examine the metabolic capacity and DNA repair activity in various species and tissues; and the results obtained, mainly with methylating agents, are briefly discussed here.

The various forms of microsomal enzymes responsible for the metabolic activation of DMN and the nature of the metabolites formed have been described (5,54,61). DMN after administration to animals is evenly distributed within the body and it is not detectable in the blood after a few hours (85); within this time, the nitrosamine is metabolized by various organs. The end-points used to measure the metabolism are (Fig. 2): formation of formaldehyde or CO_2 and/or the initial DNA alkylation (e.g. 7-methylguanine). In the determination of DNA alkylation, it is important to distinguish the proportion of the total radioactivity which is due to alkylation, from the proportion which is the result of the incorporation of radioactive label into the purines through the C_1 pool.

Table 1 shows the results of nucleic acid alkylation obtained following in vitro incubation of liver slices from various species with [^{14}C]-DMN. The liver of Syrian golden hamster showed the

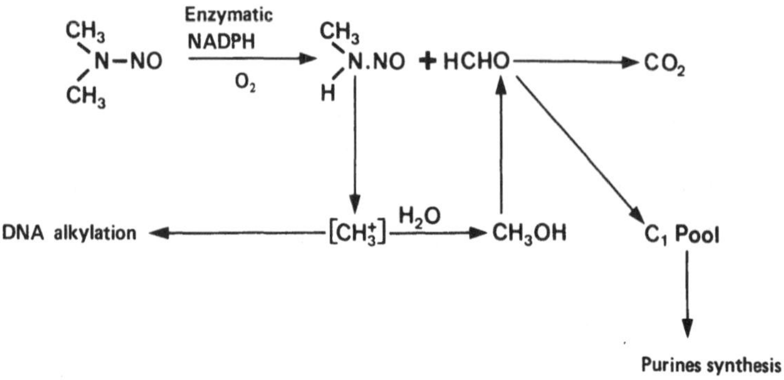

Figure 2.

highest rate of metabolism followed by rat liver. Human liver
slices have a capacity to activate DMN similar to that observed in
rat liver; lower activities were observed in the liver of trout and
monkey, these two tissues being rather resistant to the induction
of tumors by DMN (3,6). These in vitro findings parallel studies
in intact animals, where a similar ratio in the level of DNA
alkylation between rat and hamster liver has been observed (52).

The liver shows in most instances a higher capacity than
extrahepatic tissues for metabolizing nitrosamines; however, there
are some notable exceptions. Methylbenzylnitrosamine, a specific
carcinogen for rat esophagus, is metabolized by esophageal mucosa
to a methylating intermediate to a much greater extent than in rat
liver and this effect is independent of the route of administration
of the nitrosamine and of the preferential uptake into the
different organs (4,37,41). These findings strongly indicate that,
in the multistage process of carcinogenesis, the tissue-specific

Table 1. In Vitro Metabolism of Dimethylnitrosamine in Liver
Slices from Various Species Including Human Beings

Species	7-methylguanine (mol/100 mol of guanine)
Trout (Rainbow)	0.0004
Hamster (Syrian golden)	0.29
Rat (Wistar)	0.19
Monkey (Macacus cynomolgus)	0.018
Man	0.13

Data from Montesano et al., 1982.

metabolic activation is a necessary although not sufficient
requirement for tumor induction.

Following metabolism or chemical decomposition the N-nitroso
compounds react with various cellular macromolecules and in DNA
this results in the formation of adducts at various specific sites
(see Table 2). The sites and the relative percentage of alkylation
depend upon the type of alkylating agent (e.g. MNU or ENU) or the
chemical mechanism of alkylation (e.g. MMS versus MNU) and result
in a different degree of reaction with the N or O atoms of the
nucleic acid bases which is reflected in the O^6:N-7-alkylguanine
ratio (42,57,79,80). The two alkylalkane sulfonates (MMS and EMS)
react much less with the oxygens of DNA (1.1 and 14.3% of the total
alkylation) than do MNU or ENU (21 and 77%). It is also apparent
that ENU preferentially alkylates the oxygen atoms resulting in an
initial O^6:N-7-alkylguanine ratio of 0.7; the corresponding ratio
for MNU and DMN being 0.1 and that for the EMS and MMS being much
lower, indicating that the latter two compounds react
preferentially with the nitrogen atoms of DNA.

This differential reactivity with oxygen and nitrogen atoms is
dependent on the mechanism by which the different alkylating agents
react with DNA; agents that react via unimolecular nucleophilic
reaction (SN1) show a higher initial O^6:N-7-alkylguanine ratio than
agents that react via a bimolecular substitution reaction (SN2)
(42,43). The relative extent of alkylation of DNA observed in
vitro parallels remarkably well the initial levels of alkylation
observed at the various positions in cell cultures or in vivo after
exposure to alkylating agents. Small quantitative differences
arise because of differences in the secondary structure of DNA,
e.g. alkylation at the 1 position of adenine is greatly reduced in
fully hydrogen-bonded structures, and the extent of reaction at
this site may be related to the amount of single strandedness
present (53). The initial O^6:N-7-alkylguanine ratios resulting
from these four alkylating agents with native DNA in vitro
correlate well with their overall carcinogenic activity in vivo,
since MMS and EMS are very weak carcinogens as compared with ENU

Table 2. Sites of DNA Alkylation

Adenine	N.1, N.3, N.7
Guanine:	N.3, N.7, O^6
Cytosine:	N.3, O^2
Thymine:	N.3, O^4, O^2
Phosphate	

and MNU (9). However, this correlation holds only for alkylating agents that react very differently with the oxygen atoms of the DNA; for agents that produce similar initial O^6:N-7-methylguanine ratios, e.g. DMN, MNU, MNNG, all of which have a ratio of 0.11, their carcinogenic activity in vivo is determined by other additional factors, e.g. their metabolism and the repair of the adducts formed in the DNA in the target and non-target tissues. In fact, in the case of MNU and ENU, which do not require metabolic activation, no difference is observed in the initial O^6:N-7-alkylguanine ratio among target and non-target tissues.

Another important generalization that can be made on the basis of these in vitro and in vivo studies is that methylating agents have a much greater capacity to alkylate DNA than do equimolar doses of ethylating agents (79).

There is substantial evidence in prokaryotic and eukaryotic cells that the mutagenic activity of simple alkylating agents is directly related to the formation of adducts with the oxygen atoms in the DNA (17,46,80) and that the formation and persistence of O^6-methylguanine in the DNA of target tissues generally correlates well to the tumorigenic action of these agents in many animal species. O^4-methylthymine has been implicated as a promutagenic lesion and like O^6-methylguanine has been shown to induce errors during DNA synthesis in vitro (1,2) but in vivo there is limited evidence of a correlation between the formation and persistence of O-alkylpyrimidines and the induction of tumors (25,39,72,79). This is probably due to the relatively small amounts of this adduct formed and difficulties associated with its detection.

Various DNA repair processes exist to remove these and other base adducts produced by simple methyl- and ethylating agents, and the chemical, physical and biological properties of some of these enzymes have been described in detail (45,62) (Table 3). These enzymes, present in various tissues and cells, show a high degree of substrate specificity and the rates of repair (removal capacity with time) show considerable variation in different tissue and cell types.

Table 3. Repair Enzymes for DNA Methylation Adducts (44,61)

3-methyladenine glycosylase
7-methylguanine glycosylase
3-methylguanine glycosylase
O^6-methylguanine transmethylase

In liver DNA of Syrian golden hamster treated with a single dose of 25 mg/kg DNA (approx. the LD 50) 7-methylguanine has a half-life of 24 hrs and O^6-methylguanine is stable, whereas in rats after a comparable dose of DMN 7-methylguanine is not repaired to any significant extent and O^6-methylguanine is rapidly removed with a half-life of 20 hrs (52). Similar differences between these two species exist in the cases of 7-ethyl- and O^6-ethylguanine (63; Lu, Bresil, Montesano, unpublished data). Significant cellular differences in the repair of O^6-methylguanine have been observed between parenchymal and non-parenchymal cells in rat liver (86). Dose-response studies also indicate that the rate of removal depends on the extent of alkylation (64,81) and at least for O^6-methylguanine the chronic treatment with various agents affects its removal from DNA (57).

Human tissue, fibroblasts and lymphoblasts in culture appear to have a higher O^6-methylguanine repair capacity than their rodent equivalents (62,66) and in addition human cell lines in culture have been shown to fall into two classes which differ in their capabilities to repair this adduct from cellular DNA. Normal human cells have the ability to repair O^6-methylguanine (mex^+, mer^+), but some human tumors, certain tumor-derived cell lines and many lines established from cells transformed by DNA tumor viruses are deficient in this repair capacity (mex^-, mer^-) (91). Such cell lines are very sensitive to the toxic effect of alkylating agents. Repair of 7-methyl-, 3-methylguanine and 3-methyladenine are found to be similar in mex^- and mex^+ cell lines (35). It would appear that the mechanism of repair of O^6 alkyl G is similar to that observed in E. coli (45 and Lindahl this volume), i.e. the stochiometric transfer of the alkyl group to a cysteine residue within the protein molecule with the regeneration of an unsubstituted guanine in the DNA. In human lymphoid cells this activity resides in a monomeric protein with a molecular weight of 18,000 and a total of 10,000-25,000 molecules/cell are found (35).

MODULATION OF NITROSAMINE CARCINOGENESIS

The various determinants for cancer induction by nitrosamines, examined in the previous section, are important at the tissue and cellular level; however, it is necessary to consider the pharmacokinetics of the particular nitrosamine in the animal as a whole, and in particular the role of the liver (activation or detoxification) in determining the probability of extra-hepatic tissues to develop tumors. It is conceivable that a reduction in the metabolic capacity of the liver will result in a relatively higher concentration of the nitrosamine in extrahepatic tissues.

Diaz-Gomez et al. (16) and Pegg and Perry (65) have shown that after oral administration of low doses of DMN very little or none

of this nitrosamine reaches extrahepatic tissues because of the
efficient metabolism by the liver. Thus, the interaction of the
nitrosamines with extrahepatic tissues will depend upon the dose of
nitrosamine, the rate of absorption from the intestine and the
various factors which could interfere with the metabolic capacity
of the liver.

Table 4 summarizes the studies in which the effects of various
modulators on the biological effects of DMN and of some other
nitrosamines in liver and extrahepatic tissues have been examined.
Administration of a protein-restricted diet to rat results in a
decrease of liver toxicity and DNA alkylation, and as a consequence
an increase of DNA alkylation in the kidney. This change in the
pharmacokinetics of DMN is associated with an increased incidence
in a dose-related manner of kidney tumors in rats fed on a protein
restricted diet (83). Similar findings have been observed with
ethanol and disulfiram where the increased incidence of
extrahepatic tumors is paralleled by a reduction of the incidence
of liver tumors. The increased incidence of liver tumors by DMN in
rats pretreated with CCl_4 is probably the consequence of the higher
susceptibility of actively replicating cells, resulting from the
toxic effect of CCl_4 in the liver.

It is not known if the increased incidence of esophageal
cancer by MBN in rats fed with a diet deficient in zinc is due to
alteration of the pharmacokinetics of MBN or to an increased
susceptibility of the esophageal mucosa; dietary zinc deficiency is
known to induce parakeratosis and hyperkeratosis in the esophagus
of the rat (24).

In summary, the examples given in Table 4 consistently
indicate the interplay of liver and extrahepatic tissues in
determining the adverse biological effects of nitrosamines and
probably the same consideration applies to other types of
carcinogens. Recently it has been reported (28) that cirrhotic
patients who do not smoke excrete mutagens in their urine; in this
condition, the hepatic biotransformation is bypassed as a result of
the portal-systemic shunting, and the mutagenic substances might
accumulate in the body.

There is substantial evidence that human beings are exposed to
nitrosamines (47,87) and some epidemiological studies (12,13,44,89,
and these proceedings) have indicated a correlation between
increased risk of esophageal and stomach cancer and a high intake
of nitrate (a precursor of nitrosamines) or pickled vegetables and
mouldy foods containing nitrosamines (specific carcinogenic agents,
see Fig. 3). Nutritional factors, alcohol and tobacco have also
been associated as aspecific risk factors with esophageal cancer in
China and other countries (12,15,88,90,94). The relative
contribution of the three components including genetic

Table 4. Modulators of Biological Effects of Nitrosamines in Liver and Extrahepatic Tissues. (The results refer to DMN and rat except otherwise specified. However, for clarity in some cases, the animal species, the tissue and the nitrosamine were specified in full).

MODULATORS	Metabolism in whole animal	LIVER Metabolism-Nucleic Acids Alkylation in vivo	Toxicity	Mutagenicity[a]	Carcinogenicity	EXTRAHEPATIC TISSUES Metabolism-Nucleic Acids Alkylation in vivo	Carcinogenicity
Protein restricted diet	↑ 1	↑ 1	↑ 2	↑ 3,4		↑ Kidney[5]	↑ Kidney[5,6]
Ethanol	↑ 7	= 8-10,31	↑ 7	↑↓ 11, 11		↑ Kidney[8]	↑ Nasal cavities-Mice[12]
		↑ MBN[32]			↑ DEN[13]	↑ MBN-Esophagus-lung[32]	↑ DEN-Esophagus[14] NPYR-Nasal cavities, trachea-Hamster[15]
Disulfiram		↑ DMN-Mice[16] DMN[17],DEN[18]-Rat	↑ DMN-Mice rat[16] DEN-Rat[16]	↑ 19	↑ DMN,DEN[20] DBN[21]	↑ DMN,DEN[20]	↑ DMN-Nasal cavities[20] DEN-Esophagus[20] MBN-Esophagus[22] DBN-Lung[21]
Carbon tetra-chloride		↑ 23	↑ 23		↓ 24	= Kidney[23]	↑ Kidney[24]
Aminoaceto-nitrile	↑ 25	↑ 25-27	↑ 28	↑ 27	↑ 25,26,29	↑ 25	
Zinc deficiency							↑ MBN-Esophagus[30]

a/ This refers to experiments in which post-mitochondrial fractions were prepared from liver of rats treated in vivo with the various modulators and used to determine the mutagenicity in vitro of DMN in bacteria, with the exception of a study (ref. 11) carried out in vivo as a host-mediated assay.

[1]Swann & Mc Lean, 1971; [2]Mc Lean & Verschuuren, 1969; [3]Czygan et al., 1974; [4]Zeiger, 1975; [5]Swann et al., 1980; [6]Mc Lean & Magee, 1970; [7]Phillips et al., 1977; [8]Swann, 1982; [9]Schwarz et al., 1982; [10]Bolinsky et al., 1982; [11]Glatt et al., 1981; [12]Griciute et al., 1981; [13]Habs & Schmähl, 1981; [14]Gibel, 1967; [15]Mc Coy et al., 1981; [16]Schmähl et al., 1971; [17]Schmähl & Kruger, 1972; [18]Frank et al., 1980; [19]Montesano & Bartsch, 1976; [20]Schmähl et al., 1976; [21]Schweinsberg et al., 1982; [22]Schweinsberg & Bürkle, 1981; [23]Pound et al., 1973a; [24]Pound et al., 1973b; [25]Fiume et al., 1970; [26]Hadjiolov & Mundt, 1974; [27]Bartsch et al., 1975; [28]Fiume, 1962; [29]Hadjiolov, 1971; [30]Fong et al., 1978; [31]Peng et al., 1982; [32]Kourous et al., 1983

Keys-Abbreviations: =, indicates that modulators have no effect, or contradictory results were observed; DMN, dimethylnitrosamine; DEN, diethylnitrosamine; DBN, dibutylnitrosamine; MBN, methyl-buthylnitrosamine, nitrosopyrrolidine.

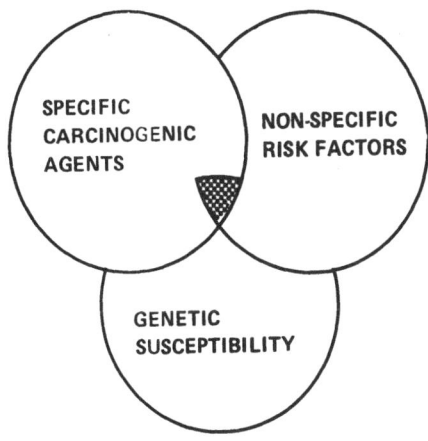

Figure 3.

susceptibility in determining the increased risk of esophageal
cancer (and of other cancers) certainly varies from one area to
another; in most cases it is difficult to assess the role of a
single component and the temporal interplay of these components in
the multifactorial origin of cancer. The recent development of
methodologies for determining individual human exposure to
carcinogenic agents (11,37,59,60) will be particularly helpful in
elucidating these problems.

REFERENCES

1. Abbott, P.J., and R. Saffhill (1977) DNA synthesis with
    methylated poly (dA-dT) templates: Possible role of 0^4-
    methylthymine as a pro-mutagenic base. Nucleic Acids Research
    4:761-768.
2. Abbott, P.J., and R. Saffhill (1979) DNA synthesis with
    methylated poly (dC-dG) templates. Evidence for a competive
    nature to miscoding by 0^6-methylguanine. Biochim. Biophys.
    Acta 562:51-61.
3. Adamson, R.H., and S.M. Sieber (1983) Chemical carcinogenesis
    studies in nonhuman primates, In Organ and Species Specificity
    in Chemical Carcinogenesis, R. Langenbach, S. Nesnow, and
    J.M. Rice, Eds., Plenum Press, New York and London, Basic Life
    Sciences, Vol. 24, pp. 129-156.
4. Archer, M.C., and G.E. Labuc (1982) On the mode of action of N-
    nitrosomethylbenzylamine, an esophageal carcinogen in the rat.
    Banbury Report 12:87-102.
5. Arcos, J.J., D.I. Davies, C.E.I. Brown, and M.F. Argus (1977)
    Repressible and inducible enzymic forms of
    dimethylnitrosamine-demethylase. Z. Krebsforsch 89:181-199.

6. Ashley, L.M., and J.E. Halver (1968) Dimethylnitrosamine-
    induced hepatic cell carcinoma in rainbow trout. J. Natl.
    Cancer Inst. 41:531-552.
7. Barnes, J.M., and P.N. Magee (1954) Some toxic properties of
    dimethylnitrosamine. Brit. J. Industr. Med. 11:167-174.
8. Bartsch, H., C. Malaveille, and R. Montesano (1975)
    Differential effect of phenobarbitone, pregnenolone-16a-
    carbonitrile and aminoacetonitrile on dialkylnitrosamine
    metabolism and mutagenicity in vitro. Chem.-Biol.
    Interactions 10:377-382.
9. Bartsch, H., B. Terracini, C. Malaveille, L. Tomatis,
    J. Wahrendorf, G. Brun, and B. Dodet (1983) Quantitative
    comparisons of carcinogenicity, mutagenicity and
    electroplilicity of ten direct-acting alkylating agents and of
    initial 0^6:7-alkylguanine ratio in DNA with carcinogenic
    potency in rodents. Mutation Res. 110:181-219.
10. Belinsky, S.A., M.A. Bedell, and J.A. Swenberg (1982) Effect of
    chronic ethanol diet on the replication, alkylation, and
    repair of DNA from hepatocytes and nonparenchymal cells
    following dimethylnitrosamine administration. Carcinogenesis
    3:1293-1297.
11. Bridges, A.B., B.E. Butterworth, and I.B. Weinstein,
    Eds. (1982) Banbury Report. Indicators of genotoxic exposure.
    Vol. 13.
12. Coordinating Groups for the Research of Esophageal Carcinoma,
    Henan Province and Chinese Academy of Medical Sciences (1975)
    Studies on relationship between epithelial dysplasia and
    carcinoma of the esophagus. Chinese Med. J. 1:110-116.
13. Cuello, C., P. Correa, W. Haenszel, G. Gordillo, C. Brown,
    M. Archer, and S. Tannenbaum (1976) Gastric cancer in
    Colombia. I. Cancer risk and suspect environmental agents.
    J. Natl. Cancer Inst. 57:1015-1035.
14. Czygan, P., H. Greim, A. Garro, F. Schaffner, and H. Popper
    (1974) The effect of dietary protein deficiency on the ability
    of isolated hepatic microsomes to alter the mutagenicity of a
    primary and a secondary carcinogen. Cancer Research
    34:119-123.
15. Day, N.E., N. Munoz, and P. Ghadirian (1982) Epidemiology of
    esophageal cancer: A review, In Epidemiology of Cancer of the
    Digestive Tract, P. Correa, and W. Haenszel, Eds., Martinus
    Nijhoff Publishers, The Hague, Boston, London, pp. 21-57.
16. Diaz-Gomez, M.I., P.F. Swann, and P.N. Magee (1977) The
    absorption and metabolism in rats of small oral doses of
    dimethylnitrosamine. Implication for the possible hazard of
    dimethylnitrosamine in human food. Biochem. J. 164:497-500.
17. Dodson, L.A., R.S. Foote, S. Mitra, and W.E. Masker (1982)
    Mutagenesis of bacteriophage T7 in vitro by incorporation of
    0^6-methylguanine during DNA synthesis. Proc. Natl. Acad. Sci.
    USA 79:7440-7444.

18. Druckrey, H., R. Preussmann, S. Ivankovic, and D. Schmähl
    (1967) Organotrope carcinogene Wirkungen bei 65 Verschiedenen
    N-nitroso verbindungen an BD-Ratten. Z. Krebsforsch.
    69:103-201.
19. Druckrey, F. (1973) Specific carcinogenic and teratogenic
    effects of "indirect" alkylating methyl and ethyl compounds,
    and their dependency on stages of ontogenic developments.
    Xenobiotica 3:271-303.
20. Fiume, L. (1962) Azione inibente dell'aminoacetonitrile
    degerazione idropica e sulla necrosi prodotte dal chloroformio
    nel regato e nel rene di ratto. Sperimentale 112:365-375.
21. Fiume, L., G. Campadelli-Fiume, P.N. Magee, and J. Holsman
    (1970) Cellular injury and carcinogenesis. Biochem. J.
    120:601-605.
22. Fleig, W.E., R.D. Fussgaenger, and H. Ditschuneit (1982)
    Pathological changes in a human subject chronically exposed to
    dimethylnitrosamine. Banbury Report 12:37-49.
23. Fong, L.Y.Y., A. Sivak, and P.M. Newberne (1978) Zinc
    deficiency and methylbenzylnitrosamine-induced esophageal
    cancer in rats. J. Natl. Cancer Inst. 61:145-150.
24. Fong, L.Y.Y. (1982) Possible relationship of nitrosamine in the
    diet to causation of cancer in Hongkong. Banbury Report
    12:473-476.
25. Fox, M., and J. Brennand (1980) Evidence for the involvements
    of lesions other than O^6-alkylguanine in mammalian cell
    mutagenesis. Carcinogenesis 1:795-799.
26. Frank, N., D. Hadjiolov, B. Bertram, and M. Wiessler (1980)
    Effect of disulfiram on the alkylation of rat liver DNA by
    nitrosodiethylamine. J. Cancer Res. Clin. Oncol. 97:209-212.
27. Freund, H.A. (1937) Clinical manifestations and studies in
    parenchymatous hepatitis. Ann. Internal Medicine
    10:1144-1155.
28. Gelbart, S.M., and S.J. Sontag (1980) Mutagenic urine in
    cirrhosis. The Lancet i:894-896.
29. Gibel, W. (1967) Experimentelle Untersuchungen zur
    Synkarzinogenese beim ösophaguskarzinem. Arch.
    Geschwulstforschung 30:181-189.
30. Glatt, H., I. de Balle, and F. Oesch (1982) Ethanol- or
    acetone-pretreatment of mice strongly enhances the bacterial
    mutagenicity of dimethylnitrosamine in assays mediated by
    liver subcellular fraction, but not in host mediated assays.
    Carcinogenesis 2:1057-1061.
31. Griciute, L., M. Castegnaro, and J.-C. Bereziat (1981)
    Influence of ethyl alcohol on carcinogenesis with N-
    nitrosodimethylamine. Cancer Letter 13:345-352.
32. Habs, M., and D. Schmähl (1981) Inhibition of the
    hepatocarcinogenic activity of diethylnitrosamine (DENA) by
    ethanol in rats. Hepato-gastroenterol. 28:242-244.

33. Hadjiolov, D. (1971) The inhibition of dimethylnitrosamine carcinogenesis in rat liver by aminoacetonitrile. $\underline{Z}$. Krebsforsch. 76:91–92.

34. Hadjiolov, D., and D. Mundt (1974) Effect of aminoacetonitrile on the metabolism of dimethylnitrosamine and methylation of RNA during liver carcinogenesis. $\underline{J}$. $\underline{Natl}$. $\underline{Cancer}$ $\underline{Inst}$. 52:753–756.

35. Harris, A.L., P. Karran, and T. Lindahl (1983) O^6-methylguanine-DNA methyltransferase of human lymphoid cells: Structural and kinetic properties and absence in repair-deficient cells. $\underline{Cancer}$ $\underline{Res}$. 43:3247–3252.

36. Hodgson, R.M., F. Schweinsberg, M. Wiessler, and P. Kleihues (1982) Mechanism of esophageal tumor induction in rats by N-nitrosomethyl-benzylamine and its ring-methylated analog N-nitrosomethyl(4-methylbenzylamine. Cancer Res). 42:2836–2840.

37. International Agency for Research on Cancer/International Programme on Chemical Safety Working Group Report (1982) Development and possible use of immunological techniques to detect individual exposure to carcinogens. $\underline{Cancer}$ $\underline{Res}$. 42:5236–5239.

38. Kimbrough, R.D. (1982) Pathological changes in human beings acutely poisoned by dimethylnitrosamine. $\underline{Banbury}$ $\underline{Report}$ 12:25–36.

39. Kleihues, P., S. Bamborschke, and G. Doeyer (1980) Persistence of alkylated DNA bases in the Mongolian gerbil ($\underline{Meriones}$ $\underline{unguiculatus}$) following a single dose of methylnitrosourea. $\underline{Carcinogenesis}$ 1:111–113.

40. Kourous, M., W. Mönch, F.J. Reiffer, and W. Dehnen (1983) The influence of various factors on the methylation of DNA by the oesophageal carcinogen N-nitroso-methylbenzylamine. I. The importance of alcohol. $\underline{Carcinogenesis}$ 4:1081–1084.

41. Labuc, G.E., and M.C. Archer (1982) Esophageal and hepatic microsomal metabolism of N-nitrosomethylbenzylamine and N-nitrosodimethylamine in the rat. $\underline{Cancer}$ $\underline{Res}$. 42:3181–3186.

42. Lawley, P.D. (1974) Some chemical aspects of dose-response relationships in alkylation mutagenesis. $\underline{Mutation}$ $\underline{Res}$. 23:283–295.

43. Lawley, P.D. (1976) Carcinogenesis by alkylating agents, In $\underline{Chemical}$ $\underline{Carcinogens}$, C.E. Searle, Ed., ACS Monograph series 173, 4:83–244.

44. Li, J.Y. (1981) The epidemiology of esophageal cancer in China. $\underline{Proc}$. $\underline{Third}$ $\underline{Pacific}$ $\underline{Rim}$ $\underline{Conference}$.

45. Lindahl, T. (1982) DNA repair enzymes. $\underline{Ann}$. $\underline{Rev}$. $\underline{Biochem}$. 51:61–87.

46. Loveless, A. (1969) Possible relevance of O-6 alkylation of deoxyguanosine to the mutagenicity and carcinogenicity of nitrosamines and nitrosamides. $\underline{Nature}$ 223:206–207.

47. Lu, S.H., A.M. Camus, C. Ji, Y.L. Wang, M.Y. Wang, and H. Bartsch (1980) Mutagenicity in $\underline{Salmonella}$ $\underline{typhimurium}$ of N-3-methylbutyl-N-1-methyl-acetonyl-nitrosamine and N-methyl-

N-benzyl-nitrosamine, N-nitrosation products isolated from corn-bread contaminated with commonly occurring moulds in Linhsien County, a high incidence area for esophageal cancer in northern China. Carcinogenesis 1:867-870.

48. McCoy, E.D., S.S. Hecht, S. Katayama, and E.I. Wynder (1981) Differential effect of chronic ethanol consumption on the carcinogenicity of N-nitrosopyrrolidine and N[1]-nitrosonornicotine in male Syrian golden hamsters. Cancer Res. 41:2849-2854.

49. McLean, A.E.M., and H.G. Verschuuren (1969) Effects of diet and microsomal enzyme induction on the toxicity of dimethyl nitrosamine. Br. J. Exp. Path. 50:22-25.

50. McLean, A.E.M., and P.N. Magee (1970) Increased renal carcinogenesis by dimethyl nitrosamine in protein deficient rats. Br. J. Exp. Path. 51:587-590.

51. Magee, P.N., R. Montesano, and R. Preussmann (1976) N-nitroso compounds and related carcinogens, In Chemical Carcinogens, C.E. Searle, Ed., ACS Monograph series, 173, 11:491-625.

52. Margison, G.P., J.M. Margison, and R. Montesano (1976) Methylated purines in the deoxyribonucleic acid of various Syrian golden hamster tissues after administration of a hepatocarcinogenic dose of dimethylnitrosamine. Biochem. J. 157:627-634.

53. Margison, G.P., and P.J. O'Connor (1979) Nucleic acid modification by N-nitroso compounds, In Chemical Carcinogens and DNA, P.L. Grover, Ed., CRC Press, Baltimore, pp. 111-159.

54. Michejda, C.J., M.B. Kroeger-Koepke, S.R. Koepke, P.N. Magee, and C. Chu (1982) Nitrogen formation during in vivo and in vitro metabolism of N-nitrosamines. Banbury Report 12:69-86.

55. Miller, E.C., and J.A. Miller (1981) Mechanisms of chemical carcinogenesis. Cancer 47:1055-1064.

56. Montesano, R., and H. Bartsch (1976) Mutagenic and carcinogenic N-nitroso compounds: Possible environmental hazards. Mutation Res. 32:179-228.

57. Montesano, R. (1982) Alkylation of DNA and tissue specificity in nitrosamine carcinogenesis, In Mechanisms of Chemical Carcinogenesis, P. Cerutti, and C. Harris, Eds., Alan R. Liss, New York, pp. 183-197.

58. Montesano, R., H. Bresil, and A.E. Pegg (1982) Metabolism of nitrosamines by human liver slices in vitro. Banbury Report 12:141-152.

59. Müller, R., J. Adamkiewicz, and M.F. Rajewsky (1982) Immunological detection and quantification of carcinogen-modified DNA components, In Host Factors in Human Carcinogenesis, H. Bartsch, and B. Armstrong, Eds., (IARC Scient. Publ.) 39:463-479.

60. Ohshima, H., B. Pignatelli, and H. Bartsch (1982) Monitoring of excreted N-nitrosomino acids as a new method to quantitate endogenous nitrosation in humans. Banbury Report 12:297-318.

61. Pegg, A.E. (1980) Metabolism of N-nitrosodimethylamine, In Molecular and Cellular Aspects of Carcinogen Screening Tests, R. Montesano, H. Bartsch, and L. Tomatis, Eds., IARC Scientific Publication 27:3-22.

62. Pegg, A.E. (1983) Reviews in biochemical toxicology. Alkylation and subsequent repair of DNA after exposure to dimethylnitrosamine and related carcinogens, In Reviews in Biochemical Toxicology, Vol. 5, E. Hodgson et al., Eds., New York, Elsevier, pp. 83-133.

63. Pegg, A.E., and B. Balog (1979) Formation and subsequent excision of O^6-ethylguanine from DNA of rat liver following administration of diethylnitrosamine. Cancer Res. 39:5003-5009.

64. Pegg, A.E., and G. Hui (1978) Removal of methylated purines from rat liver DNA after administration of dimethylnitrosamine. Cancer Res. 38:2011-2017.

65. Pegg, A.E., and W. Perry (1981) Alkylation of nucleic acids and metabolism of small doses of dimethylnitrosamine in the rat. Cancer Res. 41:3128-3132.

66. Pegg, A.E., M. Roberfroid, C. Von Bahr, H. Bresil, A. Likhachev, and R. Montesano (1982) Removal of O^6-methylguanine by human liver fractions. Proc. Natl. Acad. Sci. USA 79:5162-5165.

67. Peng, R., Y. Young Tu, and C.S. Yang (1982) The induction and competitive inhibition of a high affinity microsomal nitrosodimethylamine demethylase by ethanol. Carcinogenesis 3:1457-1461.

68. Phillips, J.C., B.G. Lake, S.D. Gangolli, P. Grasso, and A.G. Lloyd (1977) Effects of pyrazole and 3-amino-1,2,4-triazole on the metabolism and toxicity of dimethylnitrosamine in the rat. J. Natl. Cancer Inst. 58:629-633.

69. Pound, A.W., I. Horn, and T.A. Lawson (1973a) Decreased toxicity of dimethylnitrosamine in rats after treatment with carbon tetrachloride. Pathology 5:233-242.

70. Pound, A.W., T.A. Lawson, and I. Horn (1973b) Increased carcinogenic action of dimethylnitrosamine after prior administration of carbon tetrachloride. Br. J. Cancer 27:451-459.

71. Rajewsky, M.F., L.H. Augenlicht, H. Biessmann, R. Goth, D.F. Hulser, O.D. Laerum, and L.Y.A. Lomakina (1977) In, Origins of Human Cancer, H.H. Hiatt, J.D. Watson, and J.A. Winsten, Eds., Cold Spring Harbor Conf. on Cell Prolif., Cold Spring Harbor, New York, Cold Spring Harbor Laboratory, 4:709-726.

72. Scherer, E., A.P. Timmer, and P. Emmelot (1980) Formation by diethylnitrosamine and persistence of O^4-ethylthymidine in rat liver DNA in vivo. Cancer Letters 10:1-6.

73. Schmähl, Von D., F.W. Kruger, S. Ivankovic, and P. Preissler (1971) Verminderung der Toxizitat von Dimethylnitrosamin bei

Ratten und Mausen nach Behandlung mit Disulfiram,
Arzneimittel-Forschung Drug Research 21:1560-1562.

74. Schmähl, D., and F.W. Kruger (1972) Influence of disulfiram
    (tetraethylthiuramdifulfide) on the biological actions of N-
    nitrosamines, In Topics in Chemical Carcinogenesis,
    W. Nakahara, Ed., Univers. Park Press, Baltimore, pp.
    199-200.

75. Schmähl, D., F.W. Kruger, M. Habs, and B. Diehl (1976)
    Influence of disulfiram on the organotropy of the carcinogenic
    effect of dimethylnitrosamine and diethylnitrosamine in rats.
    Z. Krebsforsch. 85:271-276.

76. Schwarz, M., G. Wiesbeck, J. Hummel, and W. Kunz (1982) Effect
    of ethanol on dimethylnitrosamine activation and DNA synthesis
    in rat liver. Carcinogenesis 3:1071-1075.

77. Schweinsberg, F., and V. Burkle (1981) Wirkung von Disulfiram
    auf die Toxizitat und Carcinogenitat von N-methyl-N-
    nitrosobenzylamin bei Ratten. J. Cancer Res. Clin. Oncol.
    102:43-47.

78. Schweinsberg, F., I. Weissenberger, and V. Burkle (1982) The
    effect of disulfiram on the carcinogenicity of nitrosamines.
    IARC Scientific Publications 41:649-657.

79. Singer, B. (1982) Correlations between sites of chemical
    modification of DNA, repair, and carcinogenesis, In Primary
    and Tertiary Structure of Nucleic Acids and Cancer Research,
    M. Miwa et al., Eds., Japan Sci. Soc. Press, Tokyo,
    pp. 117-137.

80. Singer, B., and J.T. Kusmierek (1982) Chemical mutagenesis.
    Ann. Rev. Biochem. 52:655-693.

81. Stumpf, R., G.P. Margison, R. Montesano, and A.E. Pegg (1979)
    Formation and loss of alkylated purines from DNA of hamster
    liver after administration of dimethylnitrosamine. Cancer
    Res. 39:50-54.

82. Swann, P.F. (1982) Metabolism of nitrosamines: Observations on
    the effect of alcohol on nitrosamine metabolism and on human
    cancer. Banbury Report 12:53-68.

83. Swann, P.F., D.G. Kaufman, P.N. Magee, and R. Mace (1980)
    Induction of kidney tumors by a single dose of
    dimethylnitrosamine: Dose response and influence of diet and
    benzo(a)pyrene pretreatment. Br. J. Cancer 41:285-294.

84. Swann, P.F., and A.E. McLean (1971) Cellular injury and
    carcinogenesis. The effect of a protein-free high-
    carbohydrate diet on the metabolism of dimethylnitrosamine in
    the rat. Biochem. J. 124:283-288.

85. Swann, P.F., and P.N. Magee (1968) Nitrosamine-induced
    carcinogenesis. The alkylation of nucleic acids of the rat by
    N-methyl-N-nitrosourea, dimethylnitrosamine, dimethylsulphate
    and methylmethane sulphonate. Biochem. J. 110:39-47.

86. Swenberg, J.A., M.A. Bedell, K.C. Billings, D.R. Umbenhauer,
    and A.E. Pegg (1982) Cell specific differences in O^6-

alkylguanine DNA repair activity during continuous exposure to carcinogen. Proc. Natl. Acad. Sci. USA 79:5499–5502.

87. Tannenbaum, S.R. (1983) N-nitroso compounds: A perspective on human exposure. The Lancet 1:629–631.

88. Tuyns, A.J., and L.M.F. Masse (1973) Mortality from cancer of the esophagus in Brittany. Int. J. Epid. 2:242–245.

89. Yang, C.S. (1982) Nitrosamines and other etiological factors in the esophageal cancer in Northern China. Banbury Report 12:487–502.

90. Yang, C.S., J. Miao, W. Yang, M. Huang, T. Wang, H. Xue, S. You, J. Lu, and J. Wu (1982) Diet and vitamin nutrition of the high esophageal cancer risk population in Linxian, China. Nutrition and Cancer 4:154–164.

91. Yarosh, D.B., R.S. Foote, S. Mitra, and R.S. Day, III (1983) Repair of O^6-methylguanine in DNA by demethylation is lacking in mer⁻ human tumor cell strain. Carcinogenesis 4:199–205.

92. Ying, T.S., D.S.R. Sarma, and E. Farber (1981) Role of acute hepatic necrosis in the induction of early steps in liver carcinogenesis by dimethylnitrosamine. Cancer Res. 41:2096–2102.

93. Zeiger, E. (1975) Dietary modifications affecting the mutagenicity of N-nitroso compounds in the host-mediated assay. Cancer Res. 35:1813–1818.

94. Ziegler, R.G., L.E. Morris, W.J. Blot, L.M. Pottern, R. Hoover, and J.F. Fraumeni (1981) Esophageal cancer among black men in Washington, D.C. II. Role of nutrition. J. Natl. Cancer Inst. 67:1199–1206.

MODIFICATION OF CARCINOGENESIS BY DIETARY AND NUTRITIONAL FACTORS[1]

Chung S. Yang

New Jersey School, UMDNJ
Department of Biochemistry
Newark, NJ   07103 U.S.A.

SUMMARY

It is well recognized that carcinogenesis is affected to a great extent by the diet and nutritional status of the host, but the mechanisms involved, especially those related to undernutrition, are not clearly understood. In this report, the effects of overall undernutrition and specific nutrient deficiencies on carcinogenesis are reviewed. Specific examples concerning deficiencies of ascorbate, riboflavin, retinol, zinc, and protein are presented to demonstrate that these deficiencies may modify carcinogenesis by affecting carcinogen formation, carcinogen metabolism, tumor promotion, and other events. The possible effects of a cyclic or seasonal nutritional deficiency, which may be common in rural areas, on carcinogenesis are discussed. Special attention is paid to the possible modification of the cytochrome P-450 mediated monooxygenase system and carcinogen activation by dietary and nutritional factors. The effects of alcohol and butylated hydroxyanisole on the metabolism of nitrosamines and benzo(a)pyrene are discussed to illustrate the mechanisms by which these dietary factors can affect carcinogenesis. The esophageal cancer problem in China is

---

[1] This work was supported by Grants CA-16788 and CA-28298 from the U.S. National Cancer Institute. Part of the travel expenses to China was provided by Miles Laboratories, Inc.

ABBREVIATIONS: BP, benzo(a)pyrene; P-450, cytochrome P-450; NDMA, N-nitrosodimethylamine; BHA, butylated hydroxyanisole; BP-9,10-diol, 9,10-dihydroxy-9,10-dihydroBP.

discussed to relate our current understanding of carcinogenesis to possible approaches of cancer prevention.

INTRODUCTION

Diet and nutrition have been suggested as important factors in causing and affecting the genesis of cancers (13,49). The underlying mechanisms of these actions may be divided into two categories. Firstly, dietary components may contain carcinogens or their precursors. Secondly, certain dietary components or nutritional states may modify the metabolism and actions of carcinogens as well as the development and growth of tumors. This chapter deals with the second type of actions, placing special emphasis on the modification of carcinogen metabolism. Instead of reviewing the many theories and observations relating to this topic, I shall use several specific examples to highlight the important mechanisms involved and raise some questions that require further investigations.

UNDERNUTRITION MAY RETARD THE DEVELOPMENT OF CANCER

Because cancer cells require energy and nutrients for their rapid growth and division, nutrient restriction can generally inhibit the development of tumors.

Dietary and Caloric Restriction

The effects of dietary and caloric restrictions on cancer incidences were demonstrated in the pioneering work of Tannenbaum (61-63, reviewed in 11). Dietary restriction decreased the incidence of mammary tumor in DBA and C3H mice as well as spontaneous lung tumors in Swiss and ABC mice. It also inhibited benzo(a)pyrene (BP)-induced skin tumors and subcutaneous sarcomas. Similarly, caloric restriction was shown to decrease the incidence or delay the onset of mammary tumors in DBA and C3H mice as well as BP induced skin cancer. Because the restricted diet or caloric restricted diet were instituted when the mice were 9 months old or after the BP painting period (63), it appears that the inhibitory effect takes place mainly during the promotion or developmental stages of tumor formation. The inhibitory effect of restricted diet on tumor formation was also observed with rats. In a series of experiments with Sprague-Dawley rats, the total tumor risk was directly and exponentially related to caloric intake, and rats of heavier body weight had greater tumor risk than lighter rats (52). It is possible that dietary or caloric restriction may inhibit tumor development by limiting the energy supply required for the rapid proliferation of cancer cells. Other factors are likely to be involved. A recent report shows that low-calorie diet prevented the development of mammary tumors in C3H mice; this effect was accompanied by reduced circulating prolactin level, reduced murine mammary tumor virus expression, and reduced proliferation of mammary alveolar cells (53). All of these factors are closely related to carcinogenesis.

Deficiencies in Riboflavin and Other Vitamins

It is well documented that riboflavin deficiency retards the growth of a variety of spontaneous and transplanted tumors (reviewed in 51). Morris (42) demonstrated that the growth and spread of spontaneous mammary cancer in C3H mice were markedly inhibited in riboflavin deficient animals. This is apparently due to lack of flavocoenzymes required for the rapid growth of the cancer cells. It was also demonstrated that upon refeeding the deficient mice with riboflavin, the rate of the tumor growth accelerated (42). Decreased rates of growth of lymphosarcoma in mice and Walker carcinoma in rats can also be produced by certain analogs or riboflavin (51). A depression of tumor growth has also been shown with pantothenic acid deficiency (42), pyridoxine deficiency (39), and antagonists of nicotinamide (24).

In considering the practical application of these findings, it is debatable whether nutritional deprivation, of either a specific nutrient or a combination of nutrients, can be a desirable way of controlling cancer. Without specific selectivity between cancer and normal cells, the host would also suffer. The use of antifolate drugs, such as methotrexate, can also be considered as a nutritional therapy, because they deplete the cellular content of tetrahydrofolate and inhibit one-carbon metabolism. This decreases the synthesis of purine and pyrimidine bases of nucleic acids. Because cancer cells are rapidly dividing, interference in the nucleic acid metabolism has the advantage over, for example, energy metabolism for a therapeutic purpose. On the other hand, with early cancer patients that are undernourished or marginally deficient in certain vitamins, nutritional supplementation may entail the risk of enhanced tumor development.

CERTAIN DEFICIENCIES MAY ENHANCE CARCINOGENESIS

Epidemiologic studies have shown a correlation between lower dietary intake of vegetables and fruits by individuals or populations and higher incidences of cancers, especially cancers of the upper alimentary tract (49). This has been attributed to possible vitamin deficiencies in these populations. Alternatively, the correlation may be due to the presence of cancer-preventing substances in fruits and vegetables which will be discussed in a subsequent section. The association of

deficiency of specific nutrients with enhanced carcinogenesis has
been demonstrated in many epidemiological and laboratory studies.
For example, higher dietary vitamin A uptake is associated with
lower cancer incidences of the lung, bladder, and other sites
(6,25,37,38). Persons who eventually developed cancer had
significantly lower serum retinol levels at least a year before
the cancer diagnosis than the normal population (29). Natural and
synthetic retinoids have also been shown to prevent or limit
carcinogenesis of the skin and viscera in several animal models
(44,54,58). Deficiencies in riboflavin, ascorbate, lipotropes
(choline and methionine) (11), selenium (20), and zinc (15) have
also been shown to enhance carcinogenesis. Instead of reviewing
the facts individually, I shall briefly review the events known to
be involved in the genesis of cancer and then discuss how certain
nutrients or nutritional states may affect these processes.

The Multistage Nature of Carcinogenesis

Known carcinogens are mainly from dietary and environmental
sources but some are synthesized endogenously. Most carcinogens
require metabolic activation to be converted to reactive forms,
the ultimate carcinogens, which are generally electrophilic agents
(Fig. 1). They attack various nucleophilic groups in the cells
and may cause biochemical lesions on critical target molecules
such as a specific region of DNA. Some of the lesions can be
repaired, but certain specific lesions and subsequent genetic
transposition cause the "initiation" of carcinogenesis. It is
known that cell proliferation is required to "fix" the changes in
the formation of the initiated cells. A "promotion" step is

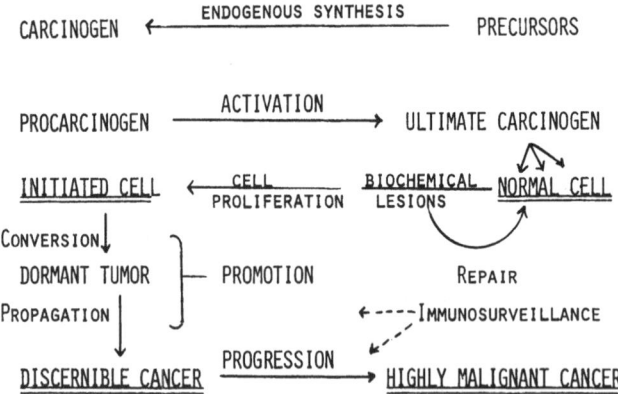

Figure 1. Multistages of carcinogenesis. The scheme depicts the
formation and activation of carcinogens, the initiation and
promotion of tumors, and the progression and possible
immunosurveillance of cancers.

generally needed to transform initiated cells into discernible
cancers (14). The promotion phase can be affected greatly by
dietary fats, acting directly or through changing hormone levels.
The possible immunosurveillance mechanisms may also be affected by
diet and nutrition. For example, it has been shown that protein
and protein-calorie restrictions decrease antibody formation, but
do not decrease or even increase cell-mediated immunities (17).

Certain Dietary Factors or Deficiencies Exert Their Effects at Specific Steps of Carcinogenesis

Understanding the multistage nature of carcinogenesis enables
us to comprehend certain complicated or seemingly contradictory
facts concerning nutrition and cancer. For example, riboflavin
deficiency is known to inhibit the growth of various cancer cells
but enhances carcinogenesis by azo dyes (27). The latter is
because riboflavin deficiency retards the metabolic elimination of
the azo dye and enhances the activation process. This point is
discussed in detail in the next section.

Vitamins C and E. The decrease in the rate of gastric cancer
in the United States during the past 40 years has been partially
attributed to the increased dietary intake of vitamin C (70).
Vitamin E has been shown to reduce the incidence of chemically-
induced tumorigenesis at several different sites (56 and papers
cited). Ascorbate may affect carcinogenesis by inhibiting the
possible in vivo nitrosation reactions that produce carcinogenic
nitroso compounds (2,41). This mode of action is shared by alpha-
tocopherol which inhibits nitrosation in a lipid environment,
whereas ascorbate is effective in an aqueous environment (36).
This is probably not the only mechanism by which these vitamins
affect carcinogenesis. Ascorbate has also been shown to inhibit
and reverse chemically-induced transformation of cells in culture
(3). Alpha-tocopherol is considered as a cellular antioxidant.
Consistent with the antioxidant mechanism is the observation that
vitamin E inhibited 7,12-dimethylbenz(a)anthracene-induced mammary
tumor in rats fed a high-polyunsaturated diet but not in those on
a low-fat diet (26).

Vitamin A. Vitamin A is required for normal cell
differentiation and a deficiency causes hyperplasia and metaplasia
of squamous cells. Vitamin A and synthetic retinoids are believed
to affect cancer incidence by inhibiting the "promotion" phase and
perhaps also the growth of tumor cells (44,54,58). At high
dosage, however, this vitamin has been shown to potentiate
chemical carcinogenesis in certain model systems (54).

In the third world, most of the vitamin A is derived from its
precursor, beta-carotene. Much of the previous work correlating
cancer rates and vitamin A intake was in fact based on beta-

carotene consumption. An inverse correlation between dietary
beta-carotene and cancer rates indeed can be established (48).
Peto et al. have suggested that beta-carotene may itself be a
cancer preventing agent (48). Although beta-carotene was shown to
inhibit carcinogenesis in animal models, it is unknown whether the
effect is solely due to its provitamin A activity or additional
mechanisms such as quenching singlet oxygen radicals. Because
beta-carotene is relatively nontoxic, it has the potential of
being used for the prevention of cancer.

Protease inhibitors. It was reported that synthetic
inhibitors of proteases reduced the incidence of chemically-
induced skin tumor in mice and a protease inhibitor-rich soybean
diet lowered breast tumor incidence in irradiated rats (reviewed
in 64). This is thought to be due to the inhibition of tumor
promotion because these protease inhibitors can also block the
superoxide radical production by polymorphonuclear leukocytes
stimulated with tumor promoters (64). Both protease action and
oxygen free-radical formation are closely linked to tumor
promotion.

MODIFICATION OF CARCINOGENESIS BY AFFECTING THE ACTIVATION OF
CARCINOGENS

Cytochrome P-450-Mediated Monooxygenase System

This enzyme system, also known as the mixed-function oxidase
system or drug metabolizing enzymes, plays key roles in the
metabolism of various xenobiotics and endogenous compounds. It
exists in liver, lung, kidney, intestine, skin and almost all
other tissues in animals. It is by far the most prominent system
in the activation of carcinogens. The system, consisting of
cytochrome P-450 (P-450) and NADPH-P-450 reductase, is embedded in
the membrane. It is found mainly in the endoplasmic reticulum
(microsomes), although lower amounts also exist in the nuclear
envelope. P-450 is a family of b-type cytochromes with molecular
weights ranging from 48,000 to 56,000. The NADPH-P-450 reductase
is a protein of 79,000 daltons containing both FAD and FMN as
coenzymes. The function of the reductase is to transfer electrons
from NADPH to P-450. The ferrous form of P-450 can then activate
molecular oxygen and allow one oxygen atom to be incorporated into
the substrate which produces or triggers a variety of reactions
such as epoxidation, hydroxylation, demethylation, etc. This is
usually referred to as the phase 1 metabolism of xenobiotics.
This enzyme system also catalyzes the reduction of various
substrates. The electrons can be donated from either the
reductase molecule or the ferrous P-450. Depending on the
substrates, the reduction can take place under either aerobic or
anaerobic conditions. Recent advances have been made in the

understanding of the induction and multiplicity of P-450 (33).
About 10 different isozymes have been purified from the livers and
lungs of rats or rabbits, with more forms to be purified and
characterized. These isozymes have different inducibility,
primary structure, substrate specificity, regiospecificity, and
responses to inhibitors. This knowledge is of great value to our
understanding of carcinogen activation.

Riboflavin Deficiency Enhances Carcinogenesis by Azo Dyes

One of the classic examples of how nutritional deficiency can
affect carcinogenesis is the observation that hepatic
carcinogenesis by azo dyes is enhanced by riboflavin deficiency
and is inhibited by riboflavin supplement (27). The observation
has been reproduced in different laboratories with different diets
and riboflavin analogs (51). A biochemical mechanism for these
results was provided by the observation that FAD played a role in
the reductive cleavage of N,N-dimethyl-4-aminoazobenzene by liver
homogenates (43) and that riboflavin deficiency decreased azo
reductase activity (71). Our current understanding of the
metabolism of this carcinogen is shown in Fig. 2. The activation
of this carcinogen is provided by the N-demethylation (40); the
reductase cleavage is a detoxification step. It appears that
riboflavin deficiency affects the detoxification pathway more
severely than the activation pathway. It is possible that both
the reduced forms of the reductase and P-450 can serve as electron
donors for the reduction. Both donors would exist at lower steady
state levels than normal in riboflavin deficiency and thus
decrease the rate of azo reduction. The azo dye appears also to
have a specific effect in depleting cellular flavin content (51).
It may induce riboflavin deficiency in animals on a normal diet.
Dietary riboflavin supplement is expected to correct these
deficiencies. In addition, cellular riboflavin or free (non-

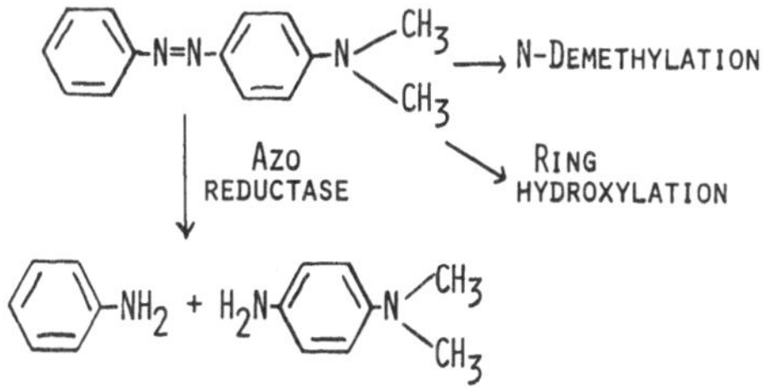

Figure 2. Metabolism of N,N-dimethyl-4-aminoazobenzene. This
    compound can be metabolized via N-demethylation, ring
    hydroxylation, and azo reduction pathways. The
    N-demethylation is believed to be the main activation
    pathway.

enzyme bound) flavocoenzymes may also serve as stimulators of azo reductase. It should be noted that the potentiation of carcinogenesis by riboflavin deficiency is only due to the specific nature of azo dye metabolism and may not be true with other carcinogens.

Riboflavin Nutrition and Skin Carcinogenesis

The effect of temporary riboflavin deficiency on tumor incidence has been studied in a two-stage skin carcinogenesis model with mice (72). Riboflavin deficiency decreased the 7,12-dimethylbenz(a)anthracene-induced tumor incidence and riboflavin supplement to a normal diet had little effect on carcinogenesis. However, when the carcinogen was given to mice during the recovery period after 4-5 weeks of riboflavin deprivation, the tumor incidence was significantly higher than in the control group maintained on a normal diet. Correlated with this is the much-enhanced skin aryl hydrocarbon hydroxylase activity in the recovery period (10). It appears that the enhanced carcinogenesis is due to the elevated carcinogen activation enzyme activity. A biochemical analysis of the enhanced hydroxylase activity in the liver has been reported (74). During the early stage of riboflavin deficiency, NADPH-P-450 reductase (assayed as cytochrome c reductase) was decreased and P-450 content increased, probably as a compensation for the decreased reductase activity; the aryl hydrocarbon hydroxylase activity (a P-450 mediated activity) was not significantly changed. When the deficiency was prolonged, the reductase activity was progressively decreased and the P-450 content was increased to reach a plateau. The hydroxylase was gradually lowered and could be enhanced by the in vitro addition of 5 to 10 μM of FMN and FAD. Upon repletion of riboflavin to the deficient mice, the reductase activity was elevated and over-shot to levels above normal after 24 hrs; whereas the P-450 content decreased. The aryl hydrocarbon hydroxylase was above normal during this recovery period and returned to normal after 2 or 3 days (74). The increase in P-450 content during the early stage of riboflavin deficiency is interesting. It is not known whether the composition of P-450 isozymes is altered. If it is, this may cause a change in substrate specificity which may enhance the activation of certain carcinogens but decrease that of others.

Dietary Protein and Carcinogen Metabolism

It has been reported that rats on a 20% casein diet had a much higher microsomal ethylmorphine N-demethylase activity than those on an isocaloric 5% casein diet (8). Humans on a high protein (44%) diet had higher rates of antipyrine and theophylline metabolism than when they were on an isocaloric low protein (10%) diet (28). The effects of dietary protein on carcinogen

metabolism and carcinogenesis are demonstrated by the observation
that a protein-free diet increased the susceptibility of rats to
the action of N-nitrosodimethylamine (NDMA) in inducing mouse
kidney tumors (60). The protein-free diet decreased the
metabolism of NDMA in the liver but decreased that of the kidney
by a smaller extent. This raised the proportion of the dose
metabolized in the kidney and increased the alkylation of DNA
which may account for the enhanced carcinogenesis. On the other
hand, low protein intake has also been shown to decrease the
incidence of hepatocarcinogenesis of aflatoxin B_1 (reviewed in 8).
This can be related to the metabolic activation of this carcinogen
because the aflatoxin-DNA adduct formation was much lower in rats
fed a 5% casein diet than that in those fed a 20% casein diet (8).
However, this may not be the only mechanism involved. Low protein
diet also had a significant effect in inhibiting the development
of hepatic gamma-glutamyl transferase foci, which is an indicator
of early hepatocarcinogenesis, even when aflatoxin treatment was
conducted when the rats were on a high protein diet (8).

Effects of Ethanol on Carcinogen Metabolism

Alcohol consumption is associated with cancers of a number of
organs and is considered as an important contributing factor to
human cancer (13). Although ethanol, wine, and spirit are not
carcinogenic in laboratory animals, ethanol has been shown to
alter the metabolism of carcinogens (47,55). Ethanol is a
competitive inhibitor of a high affinity microsomal NDMA
demethylase (47). It also inhibits the metabolism of more
lipophilic nitrosamines, although the effect is less than toward
NDMA. On the other hand, ethanol consumption (for a few days)
enhances NDMA demethylase of rat liver microsomes several fold due
to the induction of specific forms of P-450 isozymes. These
isozymes also have higher activities toward other nitrosamines
(47). The inhibitory action of ethanol may be responsible for the
altered organotropism observed when it is administered together
with nitrosamines. For example, ethanol has been shown to reduce
hepatocarcinogenesis by N-nitrosodiethylamine (21). Because the
liver is the main site of nitrosamine metabolism, inhibition of
hepatic metabolism is expected to prolong the biological half-life
of a carcinogen and increase the exposure of other organs to the
carcinogen. This analysis is consistent with the experimental
results indicating that the incidence of esophageal cancer in rats
is increased by ethanol when given together with N-
nitrosodiethylamine (16). Ethanol also modified NDMA
carcinogenesis in mice by increasing tumor induction in the nasal
cavity but decreasing hepatocarcinogenesis (19). The effect of
the induction of NDMA demethylase on carcinogenesis is not known,
although it has been shown that pretreatment of rats with ethanol
potentiates the hepatotoxicity of NDMA (35). Considering the
human populations, heavy drinkers are subjected to both the

induction effect and inhibition effect toward nitrosamine
metabolism; thus the situation is more complex.

A New Group of P-450 Inducers that Affect Carcinogen Metabolism

In past years, phenobarbital, 3-methylcholanthrene, and
polychlorinated biphenyls have been used by investigators as
classic P-450 inducers. The induction or lack of induction by
these inducers has been used as a criterion for whether P-450 is
involved in the metabolism of a certain carcinogen. This approach
has caused confusion in studying the metabolism of nitrosamines
and the controversy has been reviewed (30). In our recent
investigations on the enzymology of nitrosamines, we have
discovered a group of diet and nutrition related P-450 inducers
that enhance nitrosamine metabolism.

Rat liver microsomal NDMA demethylase activity is induced
2- to 5-fold by fasting (67), diabetes (46), acetone or
isopropanol (65), and pyrazole (66). The induction is similar but
not identical to that produced by pretreatment with ethanol (47).
Kinetic analyses indicate that this is mainly due to the induction
of a high affinity form of NDMA demethylase with K_m values of 50
to 70 μM. The induction is accompanied by the intensification of
polypeptides with molecular weights of 50,000 to 52,000. Several
lines of observation suggest that these polypeptides are P-450
isozymes and this is confirmed by the purification-reconstitution
experiments (unpublished). These isozymes are different from the
major forms induced by classic inducers. The treatments also
enhance the microsomal metabolism of N-nitrosoethylmethylamine.
The metabolism of N-nitrosomethylbenzylamine and N-
nitrosomethylaniline is enhanced by fasting, isopropanol, and
ethanol, but not by diabetes. The results suggest that, in
addition to a common form of P-450, other P-450 isozymes are also
induced by factors such as fasting, acetone, or ethanol; they may
account for the enhanced activity with nitrosamines such as N-
nitrosomethylaniline.

The mechanism of induction is still under investigation. The
fact that both fasting and diabetes cause ketogenesis and that
acetone is an inducer of NDMA demethylase suggest that ketone
bodies may be inducers for specific P-450 isozymes. These
compounds, which are produced endogenously under various
nutritional states, may be an important class of P-450 inducers
and may also affect the metabolism of other carcinogens, drugs,
and hormones. The above inducers have been shown to potentiate
the hepatotoxicity of NDMA in rats (unpublished). They are also
expected to affect carcinogenesis. Nevertheless, this possibility
remains to be investigated.

DIETARY FACTORS THAT INHIBIT CARCINOGENESIS

Studies with Animal Models

Various dietary compounds, either naturally occurring or synthetic, have been shown to inhibit chemically-induced tumorigenesis in experimental animals (68,69). These compounds, including indoles such as indole-3-carbinol and 3,3'-diindolylmethane and aromatic isothiocyanates such as benzyl isothiocyanate and phenethyl isothiocyanate, occur in edible cruciferous vegetables such as Brussels sprouts, cabbage, cauliflower, and broccoli. Lactones such as coumarines are found in a variety of vegetables and fruits. Phenol compounds such as 0-hydroxycinnamic acid, caffeic acid, and ferulic acid are plant products. Butylated hydroxyanisole (BHA) is a synthetic chemical used widely as an antioxidant food additive. Most of the anticarcinogenic experiments were performed by Wattenberg et al. using chemically-induced neoplasia in mouse lung or forestomach as the testing system (68,69). Although most of the work used BP and 7,12-dimethylbenz(a)anthracene as the carcinogens, compounds like BHA have also been shown to be protective against other polycyclic hydrocarbons, N-nitrosodiethylamine, urethane, methylazooxymethanol acetate, 4-nitroquinoline-N-oxide and uracil mustard. Some of them are direct-acting carcinogens. In most of these experiments BHA was used at a level of 0.5%, a rather high dose. However, lower dosages (0.03 or 0.1% of BHA) have recently been shown to be effective in inhibiting methylazooxymethanol acetate-induced colon and small intestinal tumors in male rats (50).

Possible Protection in Humans

Wattenberg has suggested that naturally occurring inhibitors may play a protective role in human carcinogenesis (69). Since the known inhibitors are diverse in structure and are widely occurring in the environment, it is likely that other inhibitors are also widely distributed in food products. This increases the probability for inhibitors to have an impact on human cancer. Consumption of Brussels sprouts and cabbage, which are rich in the aforementioned inhibitors, has been shown to increase the metabolism of BP and a number of drugs (45). In a case-control study, the magnitude of consumption of cabbage was shown to be inversely correlated with cancer incidence of the colon (18). It was also shown that among Singapore Chinese, the relative risk of lung cancer was less in individuals regularly consuming mustard greens and kale than in those who ate them infrequently (34). An inverse relationship between magnitude of consumption of other vegetables, including lettuce, celery, and tomatoes, and cancer or precursor lesions of the stomach has also been reported (22,23). Even though the correlation does not establish a cause-effect relationship, these results point to the importance of vegetables in cancer prevention.

Mechanisms of Anticarcinogenic Action of BHA

The actions of BHA will be used as an example to elucidate the possible mechanisms of the anticarcinogenic actions of the aforementioned compounds. BHA and butylated hydroxytoluene are antioxidants used as food additives. It has been estimated that Americans consume about 0.1 mg/kg body weight daily of these antioxidants. Butylated hydroxytoluene, which at high doses induces the liver microsomal monooxygenase system and causes enlargement of the liver, is considered toxic and banned in many countries. BHA, known to be readily excreted and nontoxic at modest levels (7) has a higher potential for practical application. The mechanisms of the anticarcinogenic action have been studied extensively. Ten years ago, we found that BHA binds to microsomal P-450 to produce a Type I binding spectrum and is an inhibitor of monooxygenase reactions with several different substrates (77,78). Recent work demonstrates that it also inhibits BP metabolism by microsomes from the lung, a target organ for BP. The inhibition is regioselective, i.e., the formation of 9,10-dihydroxy-9,10-dihydroBP (BP-9,10-diol) and 9-hydroxyBP was inhibited more severely than that of BP-4,5-diol and BP-7,8-diol, but the production of 3-hydroxyBP was not inhibited (59a).

When mice were fed a 0.5% BHA diet for several days, the microsomal monooxygenase system was modified in such a way that the profile of BP metabolites was altered (31,59). With mouse liver microsomes, the most outstanding change was the decrease in the formation of 9-hydroxyBP. The total metabolism of BP was increased but the metabolic conversion of BP-7,8-diol to BP-trans-7,8-diol-anti-9,10-oxide, the proposed ultimate carcinogen, was decreased (59). With mouse lung microsomes, the dietary BHA treatment reduced the metabolic formation of BP-9,10-diol, 9-hydroxy-BP, and BP-7,8-diol, but not the production of 3-hydroxyBP and BP-4,5-diol. The treatment decreased the total metabolism of BP as well as the conversion of BP-7,8-diol to BP-trans-7,8-diol-anti-9,10-oxide. The alteration of regioselectivity in the metabolism is probably due to the changes in the composition of P-450 isozymes in microsomes. When considered together, the above observations suggest that when mice are on a BHA diet, the liver metabolism of BP is increased mainly through enhancing the detoxification (noncarcinogenic) pathway; on the other hand, the lung metabolism of BP is decreased and the formation of the ultimate carcinogen is also decreased. When the cellular concentration of BHA is high, it inhibits the formation of the ultimate carcinogen in the lung but does not significantly inhibit the production of 3-hydroxyBP, a noncarcinogenic metabolite. Such metabolic alterations are consistent with the observation that dietary BHA decreases the formation of BP-7,8-diol-9,10-epoxide:deoxyguanosine adducts in lungs and livers of mice (1).

Dietary BHA was also shown to increase the activities of glutathione S-transferase (5- to 10-fold), UDP-glucuronyltransferase (4.6-fold), and epoxide hydrolase (11-fold) as well as the concentrations of nonprotein thiol compounds in the livers of mice (4,5,9). The increases in rats were less pronounced, and the induction of these enzyme activities was less in nonhepatic tissues in both species. These enzymes, usually referred to as phase 2 enzymes, generally catalyze the detoxification of various xenobiotics, even though some of their products can be converted to ultimate carcinogens in certain special cases (12,40). The induction of phase 2 enzymes may be a very important mechanism for the cancer prevention action of BHA, because it is effective against carcinogens with diverse structures. The alteration of P-450 composition may be an effective way of inhibiting the metabolic activation of certain compounds, but it may potentiate the phase 1 activation of other types of compounds.

Other mechanisms are probably involved in the inhibition of carcinogenesis by BHA, because it is also effective against direct acting carcinogens. This may be related to the free radical scavenging property. BHA has been reported to inhibit the tumor promoting activity of 12-o-tetradecanoylphorbol-13-acetate and benzoyl peroxide (57).

General Mechanisms of Cancer Prevention

Similarly to BHA, naturally occurring inhibitors may exert actions via any (or a combination) of the mechanisms discussed above. In addition, several plant phenols such as ellagic acid may inhibit carcinogenesis of BP by reacting with BP-7,8-diol-9,10 epoxide, the proposed ultimate carcinogen (12). Induction of phase 1 metabolism by indoles, vegetables, and other natural products has been demonstrated. Such induction is expected to have different effects on the activation of different types of carcinogens. Wattenberg has suggested that the induction of glutathione S-transferase activity may be a common feature for the inhibitors investigated in his laboratory (69). Considering the diverse structure of various inhibitors and different types of carcinogens, it is possible that different mechanisms may be involved. Nevertheless, the glutathione S-transferase hypothesis in cancer prevention is of great importance and requires further testing.

CONCLUSION

Studies with animal models have yielded a vast amount of information concerning the effects of dietary components or nutritional states on carcinogenesis. They may interfere with the endogenous synthesis of carcinogens, different steps in the metabolic activation of carcinogens, production and repair of DNA lesions, and various aspects of the tumor promotion process. They may also affect the development and growth of cancer cells by acting directly or indirectly through the actions of hormones or immunologic systems. There are still many unsolved problems in this area and there are additional complexities concerning human cancers.

Multiple Mechanisms of Actions

There are many examples showing that a certain nutrient or deficiency can affect more than one step in the carcinogenesis process. The effects may be working concertedly toward either the inhibition or enhancement of cancer formation or working in opposite directions. In animal models, the relative contributions of the individual effects can sometimes be dissociated by applying the treatment at different time periods of the experiment. In the human cancer situations, these different effects are difficult to resolve. In applying knowledge gained from animal studies, we also have to consider species differences and organ specificity in carcinogenesis. In addition to this, the carcinogens of human cancers are usually not identified. Certain human cancers are believed to have an incubation period of 20 to 30 years between the time of initiation and the appearance of detectable neoplasia which is usually preceded by a promotion period. Since the promotion is a more recent event than initiation, it may be easier to study and to intervene.

Specific Deficiency Versus General Nutritional Deficiency

In laboratory studies, animals are usually subjected to deficiency of a specific nutrient in order to isolate this factor for observation. However, in human populations deficiencies in several nutrients often occur simultaneously. These deficiencies may produce diverse or opposite effects on cellular functions and carcinogenesis. For example, zinc deficiency was found to be associated with protein deficiency; the latter enhances cell-mediated immunities while the former inhibits such functions. Riboflavin deficiency, which occurs in Linxian, Henan (76) and probably also in other areas in China, may produce the effects of caloric restriction due to the key roles of FAD and FMN in energy metabolism. This speculation remains to be examined experimentally.

Marginal, Severe, and Cyclic Deficiencies

From the previous discussions, it is seen that marginal and severe deficiencies may produce very different effects on the carcinogenesis process. This point is sometimes overlooked in considering the effects of deficiency of a certain nutrient. The work of Wynder and Chan (72) have shown the potentiation of tumor genesis when the carcinogen is applied to the mice during recovery from riboflavin deficiency. The conditions of marginal deficiency and cyclic nutrition deficiency are common, especially in rural populations, due to seasonal fluctuation in the availability of foods. In the case of riboflavin deficiency, morphological changes were also observed (73) in addition to the changes of the monooxygenase enzymes described previously (74). After the mice had been put on a riboflavin deficient diet for 3-5 weeks, atrophy of the squamous epithelium of the esophagus and stomach was noted. After 7-9 weeks, marked epithelial hyperplasia and hyperkeratosis were observed. It is not known whether these conditions would facilitate the initiation of cancer by carcinogens, or whether these changes or cell death due to severe deficiency would enhance tumor promotion. Cell division and proliferation are considered important both in the initiation and promotion of carcinogenesis. One may speculate that the possible enhanced cell growth during the recovery from a deficiency may affect carcinogenesis. These conditions may exist in rural areas when the spring crops become available after a long non-growing winter season.

Nutritional Requirement and Prevention Requirement

Even though the inverse correlation between cancer rates and the intake of certain nutrients has been observed, the quantities of these nutrients that are required for adequate protection are not known. They may be dependent on the mechanisms of action of specific compounds. For example, the effective inhibition of nitrosation by ascorbate may require the constant presence of this compound in the stomach at a quantity higher than that required to prevent scurvy. On the other hand, increased riboflavin intake above the nutritional requirement may not have any beneficial effects.

Possible Dietary and Nutritional Intervention

The esophageal cancer problem in Linxian can be used as an example to elucidate some possibilities in cancer prevention. The cancer is believed to be due to dietary and nutritional factors (32,75). Great advances in the research on dietary carcinogens have been made (32). It is also known that the dietary intake of vegetables and fruits by the population is low; deficiencies in vitamins, particularly riboflavin, have been observed (32,76). There are several possible preventive approaches. One is to remove the possible carcinogens from the diet. Certain measures, such as improving the water quality, avoiding fungal infested food and eliminating pickled vegetables, have been taken. The effectiveness of these approaches is not known, nor are the identities of the main cancer-causing substances in Linxian. A second approach is nutritional prevention. A joint China-United States intervention project has been initiated to determine whether the rate of esophageal cancer can be reduced by supplementing the local diet with ascorbate, riboflavin, vitamin A (beta-carotene), zinc and possibly other nutrients. A third possible approach is to increase the vegetable intake of the population by encouraging the cultivation of certain vegetables. If Wattenberg's theory on the relationship between glutathione S-transferase induction and cancer prevention is applicable to esophageal cancer, and this should be tested, then the massive cultivation and consumption of cruciferous vegetables might be a practical approach for cancer prevention.

REFERENCES

1. Anderson, M.W., M. Boroujerdi, and A.G.E. Wilson (1981) Inhibition in vivo of the formation of adducts between metabolites of benzo(a)pyrene and DNA by butylated hydroxyanisole. Cancer Res. 41:4309-4315.
2. Archer, M.D., S.R. Tannenbaum, T.Y. Fan, and M. Weisman (1975) Reaction of nitrite with ascorbate and its relation to nitrosamine formation. J. Natl. Cancer Inst. 54:1203-1205.
3. Benedict, W.F., W.L. Wheatly, and P.A. Jones (1980) Inhibition of chemically induced morphological transformation and reversion of the transformed phenotype by ascorbic acid in C3H/10T 1/2 cells. Cancer Res. 40:2796-2801.
4. Benson, A.M., Y.N. Cha, E. Bueding, H.S. Heine, and P. Talalay (1979) Elevation of extrahepatic glutathione-S-transferase and epoxide hydrase activities of 2(3)-tert-butyl-4-hydroxyanisole. Cancer Res. 39:2971-2977.
5. Benson, A.M., R.P. Batzinger, S.Y.L. Ou, E. Bueding, Y.N. Cha, and P. Talalay (1978) Elevation of hepatic glutathione-S-transferase activities and protection against mutagenic metabolites of benzo(a)pyrene by dietary antioxidants. Cancer Res. 38:4486-4495.
6. Bjelke, E. (1975) Dietary vitamin A and human lung cancer. Int. J. Cancer 15:561-565.
7. Branen, A.L. (1975) Toxicology and biochemistry of butylated hydroxyanisole and butylated hydroxytoluene. J. Am. Oil Chem. Soc. 52:59-63.
8. Campbell, T.C. (1982) Nutritional modulation of carcinogenesis, In Molecular Interactions of Nutrition and Cancer, M.S. Arnott, J. van Eys, and Y.-M. Wang, Eds., Raven Press, New York, pp. 359-367.

9. Cha, Y.N., F. Martz, and E. Bueding (1978) Enhancement of liver microsomal epoxide hydrase in rodents by treatment with 2(3)-tert-butyl-4-hydroxyanisole. Cancer Res. 38:4496-4498.

10. Chan, P.C., T. Okamoto, and E.L. Wynder (1972) Possible role of riboflavin deficiency in epithelial neoplasia. III. Induction of microsomal aryl hydrocarbon hydroxylase. J. Natl. Cancer Inst. 48:1341-1345.

11. Clayson, D.B. (1975) Nutrition and experimental carcinogenesis: A review. Cancer Res. 35:3292-3300.

12. Conney, A.H. (1982) Induction of microsomal enzymes by foreign chemicals and carcinogenesis by polycyclic aromatic hydrocarbons. Cancer Res. 42:4875-4917.

13. Doll, R., and R. Peto (1981) The causes of cancer. J. Natl. Cancer Inst. 66:1191-1308.

14. Farber, E. (1981) Chemical carcinogenesis. New Engl. J. Med. 305:1379-1389.

15. Fong, L.Y.Y., and P.M. Newberne (1978) Zinc deficiency and methylbenzylnitrosamines-induced esophageal cancer in rats. J. Natl. Cancer Inst. 61:145-150.

16. von Gibel, W. (1967) Experimentalle Untersuchungen zur Synkarzinogenese beim Osophagus-Karzinom. Arch. Geschwulstforsch 30:181-189.

17. Good, R.A., G. Fernandes, and N.K. Day (1982) The influence of nutrition on development of cancer immunity and resistance to mesenchymal diseases. In Molecular Interaction of Nutrition and Cancer, M.S. Arnott, J. van Eys, and Y.-M. Wang, Eds. Raven Press, New York, pp. 73-89.

18. Graham, S.H., M. Dayai, M. Swanson, A. Mittelman, and G. Wilkinson (1978) Diet in the epidemiology of cancer of the colon and rectum. J. Natl. Cancer Inst. 61:709-714.

19. Griciute, L., M. Castegnara, and J.C. Bereziat (1981) Influence of ethyl alcohol on carcinogenesis with N-nitrosodimethylamine. Cancer Lett. 13:345-352.

20. Griffin, A.C. (1979) Role of selenium in the chemoprevention of cancer. Adv. Cancer Res. 29:419-441.

21. Habs, H., and D. Schmahl (1981) Inhibition of hepatocarcinogenic activity of diethylnitrosamine by ethanol in rats. Hepato-gastroenterol. 28:242-244.

22. Haenszel, W., P. Correa, C. Cuello, N. Guzman, L. Burbano, H. Lores, and J. Muñoz (1976) Gastric cancer in Columbia: Case control epidemiological study of precursor lesions. J. Natl. Cancer Inst. 57:1021-1026.

23. Haenszel, W., M. Kuribara, M. Segi, and R.K.C. Lee (1972) Stomach cancer among Japanese in Hawaii. J. Natl. Cancer Inst. 49:969-988.

24. Herter, F.P., S.G. Weissman, and H.G. Thompson (1961) Clinical experience with 6-aminonicotinamide. Cancer Res. 21:31-37.

25. Hirayama, T. (1979) Diet and cancer. Nutr. Cancer 1(3):67-81.

26. Ip, C. (1982) Dietary vitamin E intake and mammary carcinogenesis in rats. Carcinogenesis 3:1453-1456.

27. Kansler, C.J., K. Sugiura, N.F. Young, C.R. Halter, and
    C.P. Rhoads (1941) Partial protection of rats by riboflavin
    with casein against liver cancer caused by
    dimethylaminoazobenzine. Science 93:308-310.
28. Kappas, A., K.E. Anderson, A.H. Conney, and A.P. Alvares
    (1976) Influence of dietary protein and carbohydrate on
    antipyrine and theophylline metabolism in man. Chim.
    Pharmacol. Ther. 20:643-653.
29. Kark, J.D., A.H. Smith, B.R. Switzer, and C.G. Hames (1981)
    Serum vitamin A (retinol) and cancer incidence in Evans
    County, Georgia. J. Natl. Cancer Inst. 66:7-16.
30. Lai, D.Y., and J.S. Arcos (1980) Dialkylnitrosamine
    bioactivation and carcinogenesis. Life Sci. 27:2149-2165.
31. Lam, L.K.T., and L.W. Wattenberg (1971) Effects of butylated
    hydroxyanisole on the metabolism of benzo(a)pyrene by mouse
    liver microsomes. J. Natl. Cancer Inst. 58:413-417.
32. Li, M., P. Li, and B. Li (1980) Recent progress in research on
    esophageal cancer in China. Adv. Cancer Res. 33:173-249.
33. Lu, A.Y.H., and S.B. West (1980) Multiplicity of mammalian
    microsomal cytochromes P-450. Pharmacol. Rev. 31:277-295.
34. MacLennan, R., J. DaCosta, N.E. Day, C.H. Law, Y.K. Ng, and
    K. Shanmugaratnam (1977) Risk factors for lung cancer in
    Singapore Chinese, a population with high female incidence
    rates. Int. J. Cancer 20:854-860.
35. Maling, H.M., B. Stripp, J.G. Sipes, B. Highman, W. Saul, and
    M.A. Williams (1975) Enhanced hepatoxicity of carbon
    tetrachloride, thioacetamide, and dimethylnitrosamine by
    treatment of rat with ethanol and some comparisons with
    potentiation by isopropanol. Toxicol. Appl. Pharmacol.
    33:291-308.
36. Mergens, W.J. (1982) Efficacy of vitamin E to prevent
    nitrosamine formation. Ann. N.Y. Acad. Sci. 393:61-69.
37. Mettlin, C., S. Graham, and M. Swanson (1979) Vitamin A and
    lung cancer. J. Natl. Cancer Inst. 62:1435-1438.
38. Mettlin, C., and S. Graham (1979) Dietary risk factors in
    human bladder cancer. Am. J. Epidemiol. 110:255-263.
39. Mihich, E., and C.A. Nichol (1959) The effect of pyridoxine
    deficiency on mouse sarcoma 180. Cancer Res. 19:279-284.
40. Miller, E.C. (1978) Some current perspectives on chemical
    carcinogenesis in humans and experimental animals. Cancer
    Res. 38:1479-1496.
41. Mirvish, S.S., L. Wallcave, M. Eagen, and P. Shubik (1972)
    Ascorbate-nitrite reaction: possible means of blocking the
    formation of carcinogenic N-nitroso compounds. Science
    177:65-68.
42. Morris, H.P. (1947) Effects on the genesis and growth of
    tumors associated with vitamin intake. Ann. N.Y. Acad. Sci.
    49:119-140.
43. Mueller, G.C., and J.A. Miller (1950) The reductive cleavage
    of 4-dimethylaminoazobenzene by rat liver: Reactivation of

carbon monoxide-treated homogenates by flavin adenine
dinucleotide. J. Biol. Chem. 185:145-154.

44. Nettesheim, P. (1979) Inhibition of carcinogenesis by
retinoids. Canadian Med. Assoc. J. 122:757-765.

45. Pantuck, E.J., C.B. Pantuck, W.A. Garland, B. Mins,
L.W. Wattenberg, K.E. Anderson, A. Kappas, and A.H. Connly
(1979) Effects of dietary Brussels sprouts and cabbage on
human drug metabolism. Clin. Pharmacol. Ther. 25:88-95.

46. Peng, R., P. Tennant, N.A. Lorr, and C.S. Yang (1983)
Alteration of microsomal monooxygenase system and carcinogen
metabolism by streptozotocin-induced diabetes in rats.
Carcinogenesis 4:703-708.

47. Peng, R., Y.Y. Tu, and C.S. Yang (1982) The induction and
competitive inhibition of a high affinity microsomal
nitrosodimethylamine demethylase by ethanol. Carcinogenesis
3:1457-1461.

48. Peto, R., R. Doll, J.D. Buckly, and M.B. Sporn (1981) Can
dietary β-carotene materially reduce human cancer rates?
Nature 290:201-208.

49. Reddy, B.S., L.A. Cohen, G.D. McCoy, P. Hill, J.H. Weisburger,
and E.L. Wynder (1980) Nutrition and its relationship to
cancer. Adv. Cancer Res. 32:237-345.

50. Reddy, B.S., L. Cohen, Y. Maeura, K. Fryura, and
J.H. Weisburger (1983) Effect of dietary butylated
hydroxytoluene or butylated hydroxyanisole on chemically
induced intestinal and breast tumors in rats and mice. Fed.
Proc. 42: abstract 5993.

51. Rivlin, R.S. (1973) Riboflavin and cancer: A review. Cancer
Res. 33:1977-1986.

52. Ross, M.H., and G. Bras (1965) Tumor incidence patterns and
nutrition in the rat. J. Nutrition 87:245-260.

53. Sarkar, N.H., G. Fernandes, N.T. Telang, I.A. Kourides, and
R.A. Good (1982) Low-caloric diet prevents the development of
mammary tumors in C3H mice and reduce circulating prolactin
level, murine mammary tumor virus expression, and
proliferation of mammary alveolar cells. Proc. Natl. Acad.
Sci. USA 79:7758-7762.

54. Schroder, E.W., and P.H. Black (1980) Retinoids: Tumor
preventers or tumor enhancers? J. Natl. Cancer
Inst. 65:671-674.

55. Schwarz, M., K.E. Appel, D. Schrenk, and W. Kunz (1980) Effect
of ethanol on microsomal metabolism of dimethylnitrosamine.
J. Cancer Res. Clin. Oncol. 97:233-240.

56. Shklar, G. (1982) Oral mucosal carcinogenesis in hamsters:
Inhibition by vitamin E. J. Natl. Cancer Inst. 68:791-797.

57. Slaga, T.J., A.J.P. Klein-Szanto, L.L. Triplett, and
L.P. Yotti (1981) Skin tumor-promoting activity of benzoyl
peroxide, a widely used free radical-generating compound.
Science 213:1023-1025.

58. Sporn, M.B., and D.L. Newton (1979) Chemoprevention of cancer with retinoids. Fed. Proc. 38:2528-2534.

59. Sydor, W., M.W. Chou, S.K. Yang, and C.S. Yang (1983) Regioselective inhibition of benzo(a)pyrene metabolism by butylated hydroxyanisol. Carcinogenesis 4:131-136.

59a. Sydor, W., K.F. Lewis, and C.S. Yang (1984) Effects of butylated hydroxyanisole on the metabolism of benzo(a)pyrene by mouse lung microsomes. Cancer Res. 44:134-138.

60. Swann, P.F., D.G. Kaufman, P.N. Magee, and P. Mace (1980) Induction of kidney tumors by a single dose of dimethylnitrosamine: Dose response and influence of diet and benzo(a)pyrene pretreatment. Br. J. Cancer 41:285-294.

61. Tannenbaum, A. (1940) The initiation and growth of tumors. I. Effects of underfeeding. Am. J. Cancer 38:335-350.

62. Tannenbaum, A. (1942) The genesis and growth of tumors. II. Effects of caloric restriction per se. Cancer Res. 2:460-467.

63. Tannenbaum, A. (1944) The dependence of the genesis of the induced skin tumors on the caloric intake during different stages of carcinogenesis. Cancer Res. 4:673-677.

64. Troll, W., G. Witz, B. Goldstein, D. Stone, and T. Sugimura (1982) The role of free oxygen radicals in tumor promotion and carcinogenesis, In Carcinogenesis Vol. 7, E. Hecker, N.E. Fusenig, W. Kung, F. Mark, and H.W. Thielmann, Eds., Raven Press, New York, pp. 593-597.

65. Tu, Y.Y., R. Peng, Z.-F. Cheng, and C.S. Yang (1983) Induction of a high affinity nitrosamine demethylase in rat liver microsomes by acetone and isopropanol. Chem. Biol. Interactions 44:247-260.

66. Tu, T.Y., J. Sonnenberg, K.F. Lewis, and C.S. Yang (1981) Pyrazole-induced cytochrome P-450 in rat liver microsomes: An isozyme with high affinity for dimethylnitrosamine. Biochem. Biophys. Res. Commun. 103:905-912.

67. Tu, Y.Y., and C.S. Yang (1983) A high affinity nitrosamine dealkylase system in rat liver microsomes and its induction by fasting. Cancer Res. 43:623-629.

68. Wattenberg, L.W. (1978) Inhibition of chemical carcinogenesis. Adv. Cancer Res. 26:197-226.

69. Wattenberg, L.W. (1982) Inhibition of chemical carcinogenesis by minor dietary compounds, In Molecular Interactions of Nutrition and Cancer, M.S. Arnott, J. van Eys, and Y.-M. Wang, Eds., Raven Press, New York, pp. 43-56.

70. Weisburger, J.H., E.L. Wynder, and C.L. Horn (1982) Nutritional factors and etiologic mechanisms in the causation of gastrointestinal cancers. Cancer 50:2541-2549.

71. William, J.R., P.H. Grantham, R.S. Yamamoto, and J.H. Weisburger (1970) Effect of dietary riboflavin on azo dye reductase in liver and in bacteria of cecal contents of rats. Biochem. Pharmacol. 19:2523-2525.

72. Wynder, E.L., and P.C. Chan (1970) The possible role of riboflavin deficiency in epithelial neoplasia. II. Effect on skin tumor development. Cancer 26:1221-1224.
73. Wynder, E.L., and U.E. Klein (1965) The possible role of riboflavin deficiency in epithelial neoplasia. I. Epithelial changes of mice in simple deficiency. Cancer 18:167-180.
74. Yang, C.S. (1974) Alterations of the aryl hydrocarbon hydroxylase system during riboflavin depletion and repletion. Arch. Biochem. Biophys. 160:623-630.
75. Yang, C.S. (1980) Research on esophageal cancer in China: A review. Cancer Res. 40:2633-2644.
76. Yang, C.S., J. Miao, W. Yang, M. Huang, T. Wang, H. Xue, S. You, J. Lu, and J. Wu (1982) Diet and vitamin nutrition of the high esophageal cancer risk population in Linxian, China. Nutr. Cancer 4:154-164.
77. Yang, C.S., F.S. Strickhart, and G.K. Wu (1974) Inhibition of the monooxygenase system by butylated hydroxyanisole and butylated hydroxytoluene. Life Sci. 15:1497-1505.
78. Yang, C.S., W. Sydor, M.B. Martin, and K.F. Lewis (1981) Effects of butylated hydroxyanisole on the aryl hydrocarbon hydroxylase of rats and mice. Chem.-Biol. Interactions 37:337-350.

NATURALLY OCCURRING MUTAGENS

Y. Tazima

National Institute of Genetics
Mishima, Japan

SUMMARY

There are a number of mutagens in our environment and they can be classified into two major groups: those of biological origin and those of non-biological origin. The latter group includes metals, such as mercury and cadmium, and cooked foods such as the charred surface of fish or hamburger, etc. The cooked foods are related to or produced by human activity and will not be dealt with here. Naturally occurring mutagens are those originating from microbes, plants and animals. Among them the most important and those causing the greatest concern are the products of fungi that are collectively called mycotoxins. They contaminate human foodstuffs from several sources, and some are known to have strong mutagenicity. A second group which concerns us consists of several endogenous substances produced by green plants such as pyrrolizidine alkaloids, allyl isothiocyanate, cycasin, etc. Many of these have been discovered either from a sudden outbreak of toxicosis in livestock and poultry, or from the frequent occurrence of hepatomas or cancers among inhabitants of certain districts. Surprisingly, potent mutagenicity has been discovered for flavinoids that commonly exist in the leaf constituents of green plants. Mutagens of animal origin exhibit somewhat different characteristics from mycotoxins and those occurring in plants. They are produced in the animal's body and may be mutagenic to the animal itself. A good example is dimethylnitrosamine, which is produced in an animal's stomach when it ingests foodstuffs containing nitrous acid or nitrite (ham or sausage) together with secondary amines (fish, meat, etc.).

After reviewing those naturally occurring mutagens, special attention will be focused on the epidemiological significance of mycotoxins in the human population and on the adaptation of herbivorous insects.

INTRODUCTION

There are many mutagenic substances of biological origin. These substances were first discovered to be carcinogens either from a sudden outbreak of unexpected toxicosis in livestock and poultry, or from an unusually high incidence of hepatomas or other cancers among inhabitants of certain districts. Most of these substances have later been found to be mutagenic, e.g., pyrrolizidine and some other alkaloids contained in higher plants, cycasin in cycads, bracken toxins in ferns and several kinds of mycotoxins produced by fungi. The development of short term test systems has uncovered a remarkable number of mutagenic compounds of natural origin. For instance, the flavonoids, which are contained abundantly in plant leaves have recently been found to be mutagenic.

PYRROLIZIDINE ALKALOIDS

Frequent occurrence of poisonous hepatic diseases among livestock and aboriginal inhabitants in South Africa, Central Asia and New Zealand has attracted the attention of many researchers. It was soon discovered that the alkaloids contained in some fodder plants were responsible for those hepatomas in persons and animals which had eaten the leaves of these plants. The hepatotoxic properties of these substances have stimulated pathologists and biologists to study their chemical structures, and nearly 100 compounds have been isolated and identified. The group of alkaloids consists mostly of esters of 1-hydroxymethyl-pyrrolizidine derivatives. The chemical formulae of some representative members are shown in Fig. 1. Usually the most hepatotoxic alkaloids are cyclic diesters, such as monocrotaline (III) and senecionine (IV). The double bond present in the necin moiety of the alkaloids is considered essential for their hepatotoxic action. Chemically, the alkaloids are not very reactive compounds. Their cytotoxic effects result from metabolic activation in the liver by the mixed function oxidase system.

The mutagenicity of those alkaloids was known early in 1959 in Drosophila and later in Aspergillus. Induction of chromosome breakage by these compounds has also been reported to occur in plants and mammalian cells in culture. The results obtained by Clark (1) in Drosophila are given in Table 1. He tested the mutagenicity of nine pyrrolizidine alkaloids and showed that

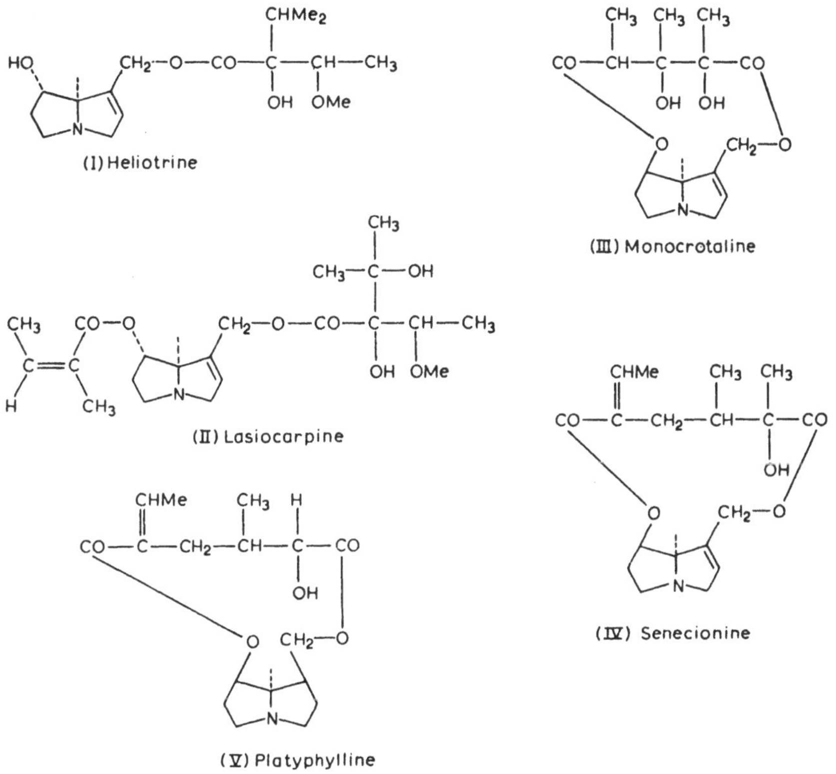

Figure 1. Pyrrolizidine alkaloids

Table 1. Sex-linked recessive lethals induced to Drosophila
after injection of 0.02 M solution
of pyrrolizidine alkaloids

Alkaloid	Brood* (%)			
	I	2	3	4
Monocrotaline	21.6	18.4		
Lasiocarpine	13.6	14.0		
Heliotrine	11.8	12.7		
Echinatine	7.9	8.4	5.0	0
Echimidine	6.7	4.9	2.7	0
Senecionine	5.0	4.3	1.2	—
Supinine	3.2	3.9	1.2	0
Jacobine	1.1	1.2	0	0
Platyphylline	0.4	2.4	0	0.6

* Each brood produced by 3-day interval.

monocrotaline, lasicarpine and heliotrine were very strongly
mutagenic in Drosophila.

With Salmonella test systems, the results of mutagenicity
testing of pyrrolizidine alkaloids are inconsistent among
researchers. According to Yamanaka et al. (11) the alkaloids
clivorine, fukinotoxin, heliotrine, lasiocarpine, ligularidine and
senkirkine, after metabolic activation, showed positive
mutagenicity with Salmonella strain TA100, but not with other
strains lacking the R-factor plasmid. Surprisingly, monocrotaline
and senecionine failed to show mutagenicity with any of the tester
strains, either with or without S9 activation.

Recently, pepasistenine, present in coltsfoot and senkirkine,
in the young flower of Tussilago farfarem, both plants belonging
to the family Senecio, have been found positive by the Ames test.
Coltsfoot is used as food in Japan (4,5). For further details see
references 2 and 3.

CYCASIN (2,3,8)

In some regions in the tropics and subtropics cycads are
utilized as foods and medicine. In those regions the frequent
occurrence of toxic symptoms and neurological disease has been
noticed, prompting intensive investigations of the constituents of
cycads.

Cycasin, methylazoxymethanol-β-D-glucoside (MAM-glucoside),
and its aglycone MAM were extracted from nuts, seeds and roots of
cycad plants. The former compound by itself was not toxic, but
became toxic to mice and guinea pigs when given enterically.
Parenteral injections did not produce toxic symptoms, nor was it
toxic in cold-blooded animals. The aglycone metabolite, MAM, was
toxic and carcinogenic independent of the route of administration.
The conversion of cycasin to the aglycone MAM in the intestinal
tract by microflora possessing a β-glucosidase has been reported.

In repeated in vitro studies cycasin was non-mutagenic,
whereas MAM was found to be mutagenic. Mutagenicity has been
confirmed both in Salmonella and Drosophila, as well as by the
induction of chromosome aberrations in onion root tip cells.
Mutagenicity of MAM is thought to be due to its N-nitroso
structure. As is well known, N-nitroso compounds (such as N-
methyl-N'-nitro-N-nitrosoguanidine) are mutagenic and their
mutagenicity is attributed to the production of an alkylating
agent which reacts on DNA or RNA.

BRACKEN TOXINS (2,3)

Bracken fronds are used as food in Japan. This plant (Pteridium aquilinum) is known to contain poisonous substance(s) which produce many harmful effects. Poisoning of cattle by feeding these bracken fronds has been reported on many occasions from parts of the world as divergent as Wales, Yugoslavia, Turkey and Brazil. It was recognized that ingestion of the plant caused the death of cattle from acute poisoning and a typical hemorrhagic syndrome. In addition, a strong carcinogenicity has been revealed in rats, quails and guinea pigs.

According to Evans (cited in 3), bracken contains a powerful mutagen to Drosophila and mice, but the final chemical agent has not been isolated yet.

FLAVONOIDS (1,7)

Recently, the mutagenicity of several members of flavonoids has been demonstrated in Salmonella. The potential importance of these compounds as environmental mutagens stems from their widespread occurrence in human foods and their use as drugs and food additives.

The flavonoids are one of the most abundant groups of natural products, comprising more than 2000 identified compounds. They include flavons and flavonols. Most of them present in plant materials are in the form of glucosides. These glucosides, particularly quercetin and kaempferol, are found in the edible portions of food plants, e.g., fruits, berries, leaf vegetables, roots, tubercles and bulbs, herbs and spices, legumes and cereal grains. The glucosides themselves are not mutagenic, but they can be hydrolyzed by suitable glycosidases to liberate the flavonols, which may show mutagenicity.

Quercetin has been shown to be mutagenic in Salmonella and Escherichia coli, and to induce gene conversion in Saccharomyces cereviceae. Mutagenicity is increased after activation with S9 mix prepared from rat liver.

The related flavonoids--morin, kaempferole, fesetin, quericitrin and rutin--are also mutagenic in microbial systems after S9 activation, while quercetin and rutin are weakly mutagenic when fed to larval or adult Drosophila melanogaster. Mutagenicity of the quercetin is also reported in cultured mammalian cells. There is, however, no evidence that flavonoids have significant carcinogenic activity in mammals.

OTHER HIGHER PLANT PRODUCTS (2,3,8)

Several groups of mutagenic compounds have been discovered so far in higher plants other than mentioned above. They are listed in Table 2.

MYCOTOXINS (8,9)

From the viewpoint of genetic hygiene the most important group of naturally occurring mutagens are mycotoxins, because these contaminants are widely present in human diet from several sources. Some are the products of fungi that grow on stored grain, while others are metabolites of contaminants in fermented food. Mycotoxins have a wide variety of chemical structures, so it is almost impossible to make a complete classification of these substances. Ueno (10) classified mycotoxins chiefly in accordance with the functional moiety but also taking into consideration the chemical structure. His grouping correlated fairly well with the biological activities. The representative compounds thus classified are illustrated in Figure 2.

Many of those compounds have been investigated for their carcinogenicity and mutagenicity. The results are summarized in Table 3. A glance at this table shows that almost all bisfurans showed positive results in Ames assay, requiring S9 activities. These results were in agreement with the known carcinogenicity of the tested compounds.

Among positive mutagenic compounds, the most potent was aflatoxin B_1, for which mutation test data have been accumulated, namely, induction of chromosome aberrations in Vicia faba and human leukocytes, induction of dominant lethal mutations in the mouse, recessive lethals in Drosophila and induction of mutations in transforming DNA, Neurospora crassa and Salmonella typhimurium. Although human data are lacking, there is no reason to suspect substantial mutagenicity of this compound in human beings.

MUTAGENICITY OF MYCOTOXINS (9,10)

Aspergillus flavus and other mycotoxin-producing fungi are found everywhere in the temperate and tropical areas. These fungi produce airborne spores and can grow on almost any types of agricultural products. Consequently, mycotoxins have been found as natural contaminants in human foodstuffs as well as in cattle fodder, e.g., in peanuts, rice, peas, corn, soybeans, wheat, etc. The fungus attacks not only growing plants but also stored grains, and produces biologically active metabolites.

Table 2. Miscellaneous mutagens contained in higher plants

Compounds	Containing organism	Use	Major biological activity	Mutagenicity	Author
Allyl isothiocyanate (Sinigrin)	Cruciferae	Foodstuffs	Insect reperent or attractants	Drosophila (m) Onion (ab)	Auerbach & Robson ('44) Swaminathan et al. ('59)
Anthraquinones large group	Rhubarb, Cassia, Aloe	Colorants in foods	Purgative agent	Salmonella (m) Onion (ab)	Review by Brown ('80)
Benzoxaziones	Corn, Wheat & Rye	-	Antifungal and insectstatic	Salmonella with S9 (m)	Hashimoto et al. ('79)
Coumarine	Graminae, Orchidaceae, Legminoceae, Compositae	Foodadditive (flavoring)	Antibiotic Spindle poisoning infertility	Onion (ab) Mouse (t)	Review by Clark ('82)
Cryptopleurine	Cryptocarya 8 families Umbeliferae	Vesicant	Inhibition of protein synthesis	Salmonella (m) Mammalian cells (ab)	de la Lande ('48)
Furanocoumarine	Rutaceae Leguminosae Moraceae	Foodstuffs(fruits) Medicine (for leukoderma)	Photodynamic inhibition of DNA synthesis	E. coli (m) Drosophila Leukocyte	Review by Scott et al. ('76)
Sesquiterpenes	Hymenoxys odorata	Fodder	Cytotoxic	Salmonella (m)	Macgreger ('77)
Safrole	Cinnamon and Sasafras	Flavoring	Hepatocarcinogenic	Salmonella with S9 (m) Human fibroblast (unscheduled DNA sy.)	McCann et al. ('75) San & Stich('75)

m: Mutation   ab : chromosome aberration   t: translocation

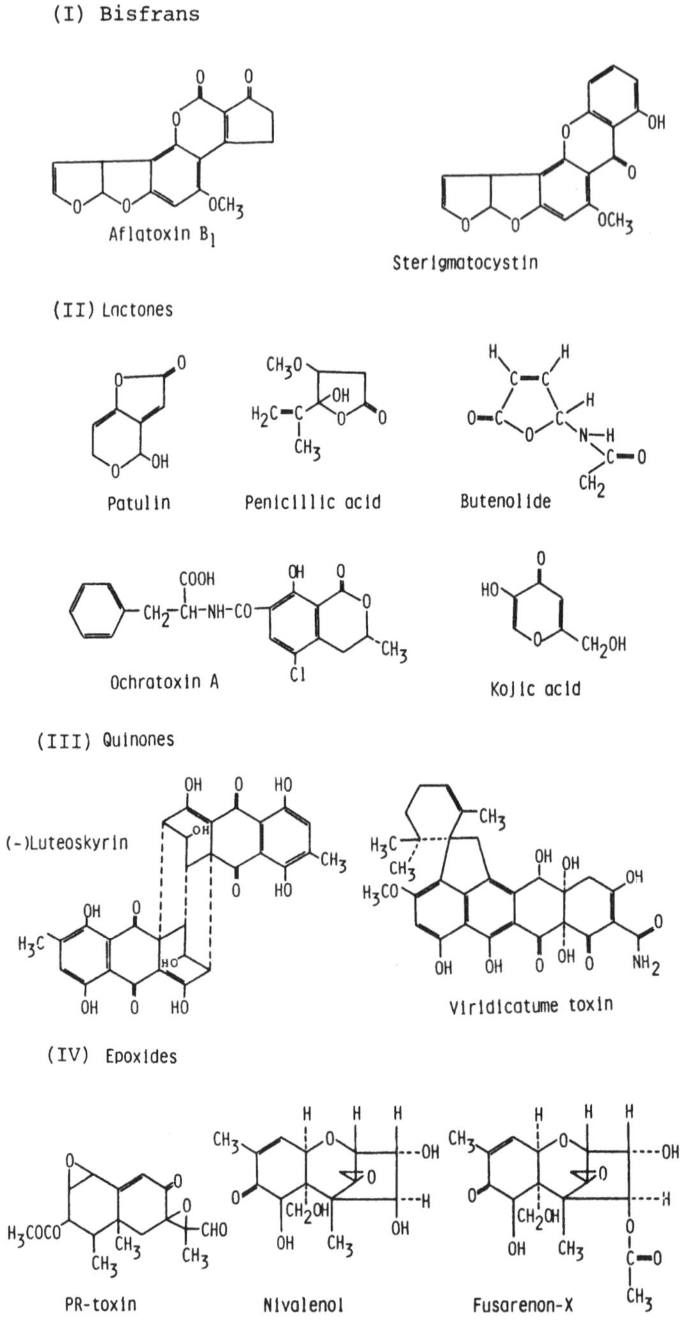

(I) Bisfrans

Aflatoxin B$_1$

Sterigmatocystin

(II) Lactones

Patulin   Penicillic acid   Butenolide

Ochratoxin A   Kojic acid

(III) Quinones

(-)Luteoskyrin

Viridicatume toxin

(IV) Epoxides

PR-toxin   Nivalenol   Fusarenon-X

Figure 2.   Chemical structure of representative mycotoxins
except the bisfuran group

(V) Halogens

Cyclochlorotine

Griseofulvin

(VI) Indoles

TR$_2$ Toxin

Fumitromorgen B

(VII) Depsidones

Mollicellin C

(VIII) Others

Zearalenone

Chaetoglobosin A

Fermented foods are an important part of the diet of people in many areas of the world. For instance, miso, a popular daily food in Japan, is a soybean paste fermented with a rice mold, A. oryzae, which is closely related to A. flavus. Several fungi are used in the production of cheese and there is a report that extracts from fungi used for the preparation of Camembert cheese produced tumors in experimental animals. Accordingly, there are innumerable sources of contamination of our foodstuffs by mycotoxins.

Epidemiological survey data gathered in Africa are summarized in Table 4 (6). There is a good correlation between the frequency of primary liver cancer in human males and the amounts of alfatoxin ingested, within the range from 1 to 35 ng per 100,000 persons annually. It is thus obvious that alfatoxins may constitute a substantial cause of the spontaneous incidence of human mutations.

Table 3.   Carcinogenicity and mutagenicity of mycotoxins

Group Mycotoxins	In Vivo Carcino-genicity	Rec-assay*	Ames Test S-9(-)	Ames Test S-9(+)	Muta-tions in Cult. Cell**
I. Bisfurans					
Aflatoxin B$_1$	+	+	-[a]	+[a,b,c]	++
Aflatoxin B$_2$	+	-	-[a]	+[a]	
Aflatoxin G$_1$	+	+	-[a]	+[a,b]	
Aflatoxin G$_2$	(-)	-	-[a]	+[a]	
Aflatoxin M$_1$	+		-[a]	+[a]	
Aflatoxin H$_1$			-[d]	+[d]	
Aflatoxin Q$_1$	(-)		-[d]	+[d]	
Aflatoxicol	+		-[d]	+[d]	
Sterigmatocystin	+	+[***]	+[a]	+[a,b,c]	+++
Dimethyl-sterigmatocystin			+[a]	+[a]	
0-Methyl-sterigmatocystin		-	-	-[c]	
0-Acetyl-sterigmatocystin		+	-	+[b]	
0-Acetyl-dihydrosterigmatocystin		+			
5-methoxy-sterigmatocystin			+[a]	+[a]	
Versicolorin A			+[a,c]	+[a,c]	
Versicolorin B			-[a]	+[a]	
6,8-Dimethylversicolorin A			-[a]	+[a]	
6,8-Dimethylversicolorin B			-[a]	+[a]	
Sterigmatin			+[a]	+[a]	
Austocystin A			-	+[c]	
Austocystin D			-	+[c]	
II. Lactones					
Patulin	+	+	-[a]	-[a,b,c]	++
Penicillic acid	+	+[***]	-	-[b,c]	+
Citrinin	+	+	-[a]	-[a,b,c]	
Oosponol	-		-[***]	-[***]	
Butenolide	-		-[***]	-[***]	-
Mycophenolic acid	-		-[a]	-[a,c]	++
Kojic acid			-	+[c]	+[e]
Austdiol			+[c]	+[c]	±[e]
III. Quinones					
(-)Luteoskyrin	+	+	-[a]	-[a,b,c]	-
(+)Rugulosin	+	+	-	-[b]	
Secalonic acid D			-	-[c]	
Viridicatum toxin			-	+[c,f]	
Auroglaucin			-	+[f]	
Emodin			-	+[f]	
IV. Epoxides					
PR-toxin		+[***]	+[a]	+[a,b]	
T-2 toxin		-	-	-	
Fusarenon-X		-	+[a]	-[a,b]	-
Crotocin (trichotecene)	+	-	-	+[b]	

(*continued*)

TABLE 3.   (Continued)

V. Halogens				
Griseofulvin	+	-	-[a]	-[a,b,c]
Ochratoxin A	+	-	-[a]	-[a,c] -
Chloropeptide	+	-	-	-[b]
Citreoviridin	-			
VI. Indoles				
Fumitremorgen B		-		-[c]
TR$_2$ toxin		-		-[c]
Cyclopiazonic acid		-		-[c]
VII. Depsidones				
Mollicellin C		+[g]		+[g]
Mollicellin E		+[g]		-[g]
Mollicellin A, B, D, F, and G		-[g]		-[g]
VIII. Others				
Zearalenone	+	-		-[b]
Cytochalasin B		-[a]		-[a]
Chaetoglobosin A	-	+[a]		-[a]
Rubratoxin B	-	-[a]		-[a]

*Ueno and Kubota (1976); Ueno (1979).

**Umeda, Tsutsui, and Saito (1977).

***Positive at pH 6.

[a]Nagao et al. (1978).

[b]Ueno, Kubota, and Nakamura (1978).

[c]Wehner et al. (1978).

[d]Wong and Hsieh (1976).

[e]Ueno (1980).

[f]Wehner, Thiel, and du Rand (1979).

[g]Stark et al. (1978).

Reprinted from Ueno (1980) by permission of the author.

Table 4.   Circumstantial evidence of a casual relationship between the incidence of primary liver cancer in humans and mean daily intake of aflatoxin

	Food Intake Aflatoxin	Primary Liver Cancer (Cases per 100,000 per Year)	
Area	(ng/kg Body Weight/Day)	Men	Women
1	<3	<1	<1
2	3-5	3	<1
3	6-8	11	3
4	10-15	13	5
5	222	35	16

Source: Neubert (1977). Reprinted by permission of the author.

Data 1-4: from studies in Kenya

Data 5: from studies in Mozambique

CONCLUSION

There is ample reason to suspect that a substantial percentage of spontaneous mutations in human beings may be caused by the naturally-occurring mutagens present in our foods. All precautionary measures should be taken to protect the human genome from these potentially mutagenic compounds.

REFERENCES

1. Brown, J.P. (1980) A review of the genetic effect of naturally occurring flavonoids, anthraquinones and related compounds. Mutation Res. 75:243-277.
2. Clark, A.M. (1976) Naturally occurring mutagens. Mutation Res. 32:361-374.
3. Clark, A.M. (1982) Endogenous mutagens in green plants, In Environmental Mutagenesis, Carcinogenesis and Plant Biology, J. Klekowski, Jr., Ed., Vol. 1, Praeger Publishers, New York, pp. 97-132.
4. Hirono, I., K. Fushimi, H. Mori, T. Miwa, and M. Haga (1973) Comparative study of carcinogenic activity of each part of bracken. J. Natl. Cancer. Inst. 50:1367-1371.
5. Hirono, I., I. Sasaoka, C. Shibuya, M. Shimizu, K. Fushimi, H. Mori, K. Kato, and M. Haga (1975) Natural carcinogenic products of plant origin. Gann Monographs on Cancer Research 17:205-217.
6. Neubert, D. (1977) Nature and levels of chemical environmental mutagens, Industrial exposure, and population at risk, In Progress in Genetic Toxicology, D. Scott, B.A. Bridges, and F.H. Sobels, Eds., Elsevier/North Holland, Amsterdam, pp. 95-115.
7. Sugimura, T., M. Nagao, T. Matsushima, T. Yahagi, Y. Seino, A. Shirai, M. Sawamura, S. Natori, K. Yoshihira, M. Fukuoka, and M. Kuroyanagi (1977) Mutagenicity of flavor derivatives. Proc. Japan Acad. 53(B-4):194-197.
8. Tazima, Y. (1974) Naturally occurring mutagens of biological origin - A review. Mutation Res. 26:225-234.
9. Tazima, Y. (1982) Mutagenic and carcinogenic mycotoxins, In Environmental Mutagenesis, Carcinogenesis and Plant Biology, E.J. Klekowski, Jr., Ed., Vol. 1, Praeger Publishers, New York, pp. 65-98.
10. Ueno, Y. (1979) Carcinogenicity of mycotoxins and Ames test. Mutagens Toxicol. 6:58-69.
11. Yamanaka, H., M. Nagao, T. Sugimura, T. Furuya, A. Shirai, and T. Matsushima (1979) Mutagenicity of pyrrolizidine alkaloids in the Salmonella/mammalian microsome test. Mutation Res. 68:211-216.

TERATOGENESIS[1]

Thomas H. Shepard[2]

Central Laboratory for Human Embryology
and Departments of Pediatrics
and Obstetrics and Gynecology
University of Washington
School of Medicine
Seattle, Washington 98195

SUMMARY

Over 50% of human conceptions are lost spontaneously. The
rate of malformations in the embryo is about 10% and in the
newborn, 2-3%. About 70% of malformations are of unknown
etiology; about 20% are inherited as Mendelian traits or are
associated with chromosomal aneuploidy; about 10% are due to
specific agents (chemicals, drugs or viruses).

There are a number of principles of teratology which help in
the management of clinical and public health problems. The
teratogenic susceptibility of experimental animals and humans is
controlled in part by the stage of development at which the
developing embryo or fetus is exposed to the agent. Examples in
humans include coumadin and androgen-masculinizing syndromes.
Species variation exists in the response to teratogens and some
controlling factors will be discussed. The dosage of the

---

[1] Supported in part by the National Institute of Child Health and
Human Development (HD00836).

[2] Professor of Pediatrics, Adjunct Professor of Obstetrics and
Gynecology and Affiliate Child Development and Mental Retardation
Center and Regional Primate Center, University of Washington.
Adjunct Professor of Environmental Health, School of Public Health
and Community Medicine, University of Washington.

teratogen is related to the damage produced; in fact, almost any
agent in high enough dose can be shown to be teratogenic in
experimental animals, even sodium chloride and sucrose.

There are problems in the interpretation of epidemiologic
studies in humans, as is illustrated by the examples of birth
control pills and caffeine.

INTRODUCTION

The biological survival of the human race is dependent upon
the success of three major processes: 1) maturation of the female
and male germ cells and their favorable union (gametogenesis and
fertilization), 2) intrauterine development of the conceptus, and
3) postnatal growth and development. Teratology is the science
dealing with the causes, mechanisms, manifestations and prevention
of intrauterine developmental defects of either structural or
functional nature. A teratogenic agent (teratogen) may be a
chemical, drug, virus or a physical or deficiency state.

Scope of Societal Problems

Incidence of reproductive loss. Although it is recognized
clinically that from 15 to 25 percent of human pregnancies end in
spontaneous abortion, the majority during the first trimester,
there are lines of evidence that over 50 percent of all
conceptuses are lost. In what is now a classic series of studies
published in the 1940's and the 1950's by Hertig, Rock and others,
a series of gravid uteri of less than four weeks' gestation was
examined following hysterectomy. Because coital and menstrual
dates were known, it was possible to study the loss of the
conceptus as related to actual gestational age. Histories could
be compared with endometrial histology, acting as a check on
menstrual dating. The criteria in these studies for immediate or
impending loss consisted of severe pathologic features such as
absence of inner cell mass or trophoblastic insufficiency for the
stage of development.

The total percentage of conceptuses in the process of loss in
these studies ranged from 34 to 43 percent. However, when Hertig
(15) reconsidered the findings, a more impressive rate of loss
emerged. It was found that in any single menstrual cycle, 15
percent of oocytes were unfertilized despite optimal conditions
for fertilization to take place. A further 10 to 15 percent were
fertilized but failed to implant. Another 30 to 35 percent loss
occurred during the first week following implantation. Therefore,
between 55 and 65 percent of oocytes were lost and only a total of
about 40 percent of the starting group succeeded in causing a
missed menstrual period.

Recent studies utilizing highly sensitive chemical tests for pregnancy have been able to detect a post-implantation conceptual loss of 43 percent among healthy women (21). In 33 percent of all the conceptions the only evidence for pregnancy was a positive human chorionic gonadotropin test on urine. Based upon the use of theoretic calculations from the rates of chromosomal defects in spontaneous abortions, Boue et al. (1) have estimated that 50 percent of all conceptuses may be lost from chromosomal errors.

In this important general area of human biology, there is an immense need for further information. The surveillance of spontaneous abortions is made very difficult by the absence (in most cases) of good control or baseline data. Unfortunately, with existing surveillance methods, the loss rate that is found is proportional to the intensity of the inquiry. More detailed reviews of this subject have been published (18,32).

Incidence of congenital defects. Approximately three percent of all human newborns have a congenital anomaly requiring medical attention, and one-third of these conditions can be regarded as life-threatening. With increasing age, over twice as many congenital defects are detected. Close to one-half of the children in hospital wards are there because of prenatally acquired malformations of one kind or another. Another three percent of our population is mentally retarded, although a certain part of this problem is caused by postnatal factors. A congenital defect, whether structural or functional, exists for an individual's entire life span. It is not a year or two, or ten years. In terms of human suffering and the financial and social dislocation of the family, this toll is tremendous. After identifying the causative mechanism, the opportunity for prevention in this field historically has proven to be outstanding. Two examples of this are the near prevention of kernicterus by preventing maternal immunization to Rh factor and of the congenital rubella syndrome by mass immunization.

Our knowledge of the cause and prevention of human malformations is extremely limited in that approximately 70 percent are of unknown causes. About fifteen percent are associated with gene mutations, five percent with chromosomal aberrations and less than ten percent are known to be due to a specific teratogenic agent. Although there are more than 600 agents known to produce congenital anomalies in experimental animals, less than 30 of these are known to cause defects in the human (29,34) (see Table 1). The reasons for this apparent wide discrepancy between the number of animal and human teratogens are discussed more fully later, but are particularly related to the immense doses which investigators are able to give experimental animals.

Table 1.   Known Teratogenic Agents in Humans

Radiation
   Therapeutic
   Atomic weapons
   Radioiodine

Drugs and environmental chemicals
   Androgenic hormones (testosterone, progestins)
   Aminopterin and methylaminopterin
   Cyclophosphamide
   Busulfan
   Thalidomide
   Mercury, organic
   Chlorobiphenyls
   Diethylstilbestrol
   Diphenylhydantoin and trimethadione
   Coumarin
   Tetracycline
   Valproic Acid

Infections
   Rubella virus
   Cytomegalovirus
   Herpes virus hominis? I and II
   Toxoplasmosis
   Syphilis
   ?Varicella virus
   Venezuelan equine encephalitis virus

Maternal metabolic imbalance
   Endemic cretinism
   Diabetes
   Phenylketonuria
   Virilizing tumors
   Alcoholism
   Hyperthermia

HISTORICAL PERSPECTIVE

There are many archaeological artifacts which give evidence that men and women of many countries and continents have been fascinated by monstrosities. Babylonian tablets give various congenital defects which were thought to be prognostic of the political and economic future of nations. A long history of how maternal impressions may cause congenital anomalies has been recorded; this idea is in part at the root of many of the guilt feelings still found in parents producing defective offspring.

A more descriptive aspect of teratology, along with the early biologic inquiry, began in the 19th century and continued through the early 20th century. Some of the scientific descriptions by Meckel, the Saint-Hilaires, Taruffi, Schwalbe and Ballantyne rival or better those of the present day. Further details of this fascinating history can be obtained from Warkany (38).

Bridging the ancient historical concept of teratology with the present-day experimental approach there existed a period of time when most malformations were considered the result of gene mutations. With the discovery by Hale (12) and by Warkany and Nelson (40) that vitamin deficiencies could cause defects in experimental animals, a new impetus to the use of experimental methods arose. In 1960 the Teratology Society in the U.S. was founded and some years later teratology societies were established in Europe and Japan.

The thalidomide tragedy of the early 1960's gave great impetus to experimental teratology, and in particular pointed out the unique teratogenic dangers of our environment. Government, and particularly the pharmaceutical industry, responded by beginning to test agents in small animals for their teratogenic potential. In the past few years with the appearance of the Ames test for mutagenicity, much pressure has been applied to produce a similar short-term tool for teratology.

PRESENT STATE OF TERATOLOGY

Principles

Teratogenic susceptibility in stage of development. The period of development when the conceptus is exposed to an agent controls, to a large extent, its sensitivity to teratogenesis (Fig. 1). Damage during the implantation and presomite periods

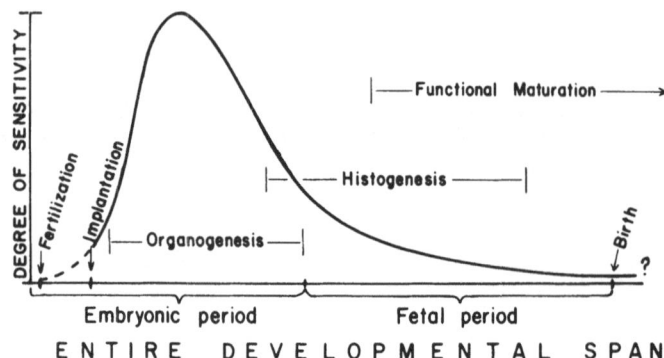

Figure 1: Curve illustrating the susceptibility to teratogenesis
     from fertilization throughout intrauterine development.
     (From 42, used by permission).

(0-17 days) generally produces little altered morphogenesis
because the ovum either dies or regenerates completely, whereas
during major organ formation (18-60 days), the embryo is highly
sensitive and exposure may produce major morphologic changes.
During the subsequent period, the fetus is less sensitive to
morphologic alterations, but changes in functional capacity, such
as intellect, reproduction or the rate and process of aging may
develop. This time specificity has been found in nearly all cases
where teratogenesis in the human has been proved and studied in
detail (see Table 2).

     Species variation in response to teratogenic agents. Some
species of animals are much more susceptible to specific
teratogenic agents than others. Aspirin, cortisone and several
vitamin deficiencies are highly teratogenic in the rodent, but
there is no solid evidence of their teratogenicity in humans. The
thalidomide epidemic would not have been prevented by testing
prenatal mice and rats, but the drug is teratogenic in rabbits,
monkeys and humans. It seems likely that there are certain
embryologic processes common to experimental animals and man.
Various pharmacologic and physiologic mechanisms that operate to
control the maternal blood concentration of a given teratogen are
shown in Figure 2. In addition, interspecies differences exist in
placental transport and embryonic metabolism. One developing
hypothesis is that the species variability is due in large part to
variations in biotransformation of the administered drug. The
hepatic microsomal monooxygenase (mixed function oxidase) system
plays a role in the metabolism of more than 70% of drugs and
chemicals. The rate of breakdown or the type of byproduct
produced by this system in different animals might help to explain

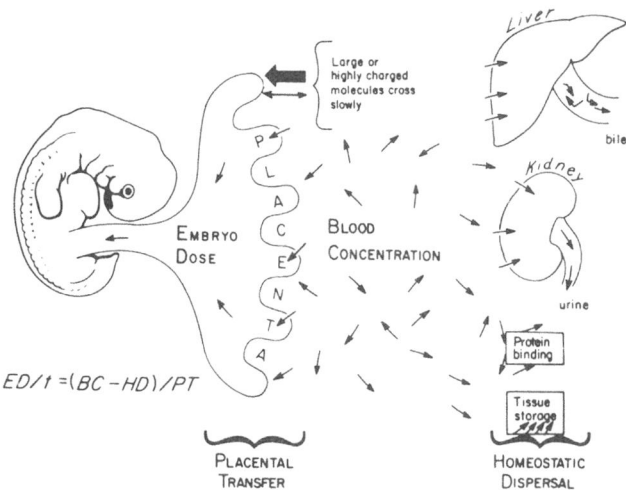

$$ED/t = (BC - HD)/PT$$

Figure 2:   Diagram of the factors that influence embryonic dose of
    a foreign chemical present in the maternal blood stream.
    Major individual and species differences in absorption and
    homeostatic dispersal control the blood concentration that is
    exposed to the placenta.   Variations in placental transfer
    exist between species but very little work has been done on
    interspecies differences in drug metabolism by the embryo.
    (From 42, used by permission).

the species specificity. The embryo itself may be relatively
inactive in this reaction but susceptible to the reactive
molecules produced by the maternal liver.

    Drug quantity and teratogenicity. Any drug given in large
enough amounts will adversely affect fetal development. This
action usually occurs through deleterious effects on maternal
health and is expressed as either embryo/fetal death, fetal growth
retardation or osseous retardation. Both sodium chloride and
sucrose given in sufficient amounts to experimental animals will
produce these embryo-fetotoxic effects. Many of the warnings in
drug inserts about potential teratogenicity are related to this
phenomenon. When extrapolating the dose response curve from
animal experiments to humans, it is important to take into
consideration the ratio (on a per kilogram basis) between the
teratogenic dose in the animal and the therapeutic or exposure
dose in the human. Figure 3 illustrates the usual relationship
between dose and the zones for teratogenicity, embryofeto
lethality and maternal lethality.

Table 2. Time Specificity of action for Human Teratogens

Teratogen	Gestational Age From Fertilization (Days)	Malformations
Rubella virus	0-60	Cataract or heart disease is more likely
	0-120+	Deafness
Thalidomide (removed from the market)	21-40	Reduction defects of extremities
Hyperthermia	18-30	Anencephaly
Male hormones (androgens, tumors, progestins)	Before 90	Clitoral hypertrophy and labial fusion
	After 90	Only clitoral hypertrophy
Coumadin anticoagulants	Before 100	Hypoplasia of nose and stippling of epiphyses
	After 100	?Mental retardation
Diethylstilbestrol	After 14	50% vaginal adenosis
	After 98	30% vaginal adenosis
	After 126	10% vaginal adenosis
Radioiodine therapy	After 65-70	Fetal thyroidectomy
Goitrogens and iodides	After 180	Fetal goiter
Tetracycline	After 120	Dental enamel staining of primary teeth
	After 250	Staining of crowns of permanent teeth

Table 3. Categories of Danger of Teratogenic Agents (FDA)

Category	Results of Animal Study	Results of Human Pregnancy Exposure	Use During Pregnancy
a	negative	negative	use if necessary
b	positive	unknown	use if necessary
	positive	negative	use if necessary
c		unknown	use if the benefit outweighs the risk
d	negative or unknown	positive	use for life-threatening situations only
x	positive	positive	Risk clearly outweighs benefit. Contraindicated for use.

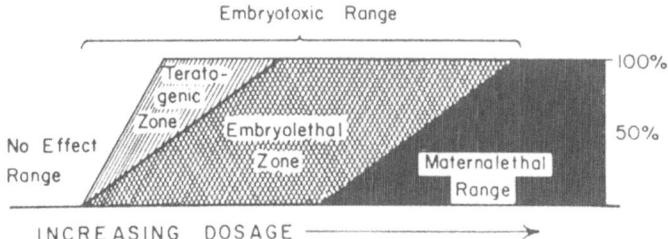

Figure 3: Diagram showing that each drug tested has a dose
spectrum ranging from no effect to one of maternal lethality.
The teratogenic zone may be broad, very small or in some
cases, non-existent (adapted from 42). Naturally occurring
rates in untreated animals have been corrected for by
subtraction from all 3 zones.

Multifactorial causes of congenital defects. According to
their etiology, congenital defects can be divided into three
general types: 1) those due to a single action of a mutated gene
(e.g., achondroplasia or phenylketonuria); 2) those due to the
single action of an environmental agent (e.g., congenital rubella
infection or aminopterin); or 3) those produced by a combination
of several or more gene defects and/or environmental agents.
These multifactorial causes probably account for the major
proportion of congenital defects. Congenital dislocation of the
hip and most forms of cleft lip and palate are commonly used
examples of multifactorially caused congenital malformations.
Fraser and his students have illustrated this mechanism in certain
inbred strains of mice and have begun to make practical human
applications (10). The A/J mouse exhibits a higher natural
incidence of cleft lip and palate than do certain other strains;
this is because of differences in the topographic relationship of
the embryonic facial processes. Other genes controlling the
development of the palate are known to predispose to clefts.
Examples such as shortening of the head, changes in mandibular
length or mechanisms by which tongue obstruction might prevent
normal palatal shelf closure have been studied using mutant genes
or inbred animal models. Most of the environmental agents (e.g.,
aspirin or cortisone) known to produce clefts are more effective
in these inbred strains. It seems imperative to accelerate the
application of these principles to man. This requires
quantitative methods for identifying the susceptible human
genotypes and the environmental agents that might contribute to
the multifactorially caused defects.

Experimental Animal Tests

Common practice is to select for testing two or three small
animal species, usually rats, mice and rabbits. The maximum dose
is usually in considerable excess such as 100 times human
therapeutic doses. A fractional dose of the maternal LD-50 (50%
of lethal dose) is used, and preferably one that causes minimal
maternal toxicity. Since nearly all agents at high dose can
produce some embryo/fetotoxicity expressed as weight loss or minor
defect, it has been recommended that this level be determined and
taken into consideration in dosing.

Three general time periods for dosing have been recommended
(17). The first is a general test for reproductive effects where
both the male and female are started on the drug before pregnancy.
Midway during pregnancy, one-half of the females are sacrificed
and the number of corpora lutea of pregnancy (ovulated eggs) and
the number and state of implantation sites are examined. A
decrease in ovulation, fertilization or implantation, as well as
an increase in early embryonic death can be determined. The
remainder of the pregnant females are allowed to go to term and to
litter. The offspring are examined and reared, and subsequently
rebred to measure any intergenerational change in reproduction.

A second type of experiment calls for administration of the
compound during varying periods of major embryonic organ
development and then sacrifice of the mother just before
parturition to measure and examine the fetus for visceral and bone
defects. A third approach is to administer the drug during the
latter part of pregnancy and to observe the offspring in the
perinatal and postnatal period. A more complete discussion of
these tests is available (26,43).

Subhuman Primate Testing

Embryonic growth and placental function are more like the
human in subhuman primates than in the rodent and other small
animals. Although it would seem that these species would best
approximate the human, their cost, the long duration of pregnancy
and the availability of sufficient numbers of pregnant animals
mitigate against their use as a standard animal. Testing in these
species may be indicated for agents which must be used during a
human pregnancy (antihypertensives, hypoglycemics,
anticonvulsants) or agents which are likely to be inadvertantly
taken during pregnancy and for testing agents about which a
question of safety arises after regulatory approval for widespread
exposure.

In Vitro Tests

Study of ovigenesis and spermatogenesis. Following the identification of the very exacting nutritional requirements for mammalian ova by Brinster (3), the possibility was realized for in vitro study of the effects of specific environmental contaminants. For instance, Brinster and Cross (2) using mouse ova exposed to copper were able to establish a dose response and a lethal concentration. This type of analysis applied to other chemicals, and coupled with the knowledge of intrauterine concentrations found in exposed workers, could be a great help in establishing safe exposure levels. An approximation of uterine levels in exposed women might be obtained by analysis of curettage tissue removed for unrelated health reasons.

The in vitro tests of mammalian eggs have been applied to the study of abnormal chromosome division (6). There is evidence that accumulated radiation exposure and delayed ovulation may be associated with increased chromosome imbalances in the human ova (for discussion and references see 34).

Sperm counts, along with the study of their morphology and motility, have been used in monitoring men at risk. Unfortunately, these parameters are not well correlated with fertility. A new test which appears to be better associated with the state of fertility utilizes the frequency with which human sperm penetrate hamster eggs in vitro (28). Infertility associated with hypospermia has been linked to workplace exposure to the pesticide 1,2-dibromo-3-chloropropane (DBCP) (41). At apparent airborne concentrations between 0.4 to 1.0 ppm of DBCP, chronically exposed workers had reduced sperm concentrations and were infertile.

Cell culture. Individual cells are grown in much the same way as bacteria. All the organization of the original tissue in such cultures is lost, and cell multiplication and growth in uniform populations are the dominant interests. The technique is very useful in predicting alteration in cell replication and has been used by some laboratories to screen new chemicals and drugs. The discovery of the important biologic action of cytochalasins by S.B. Carter (4) of Imperial Chemical Industries, Ltd. resulted from such a routine test and this led to the finding that these chemicals prevented cellular, but not nuclear division. More specific questions about the growth of cells could be answered by using special cell lines (teratocarcinoma, neuronal and endocrine) or cultures of different tissues, i.e., fetal lung or myocardium.

Organ culture. Small pieces of differentiating organs or limb buds removed from experimental animals or from human abortion material can be maintained for a number of days in culture.

During this time the growth and differentiation can be assessed during exposure to environmental toxicants. The technique has been used mainly for cartilaginous or endocrine tissues, but also for renal and cardiac rudiments. Explanted thyroid tissue from human abortuses has been used to determine the period of development when iodide can be concentrated. This is an important question since therapeutic amounts of ^{131}I, if administered to a thyrotoxic woman, can be concentrated by the fetal thyroid and cause its destruction. This test was done by adding radioactive iodide to the growth medium and measuring its incorporation into the explant (31). Concentration commenced in thyroid tissue from fetuses of 70 to 75 days of age.

Whole embryo culture. The use of whole culture of embryos from mammals has been feasible since D.A.T. New (24) demonstrated in 1967 normal growth of pre and somite stage rat embryos maintained in a serum medium. A major attraction of this method is that a dose-response curve can be established and the no effect level may prove valuable to regulatory agencies faced with the problem of safety standards. A number of agents have been studied including hyperthermia, ethanol, trypan blue, B vitamins, anesthetics, alkylating and autonomic stimulating drugs. A suggestion has been made that the serum of at-risk workers could be tested with this system (5).

Many drugs and chemicals require bioactivation by liver enzymes (the P450 monooxygenase system). Fantel et al. (8) have shown that this fraction of liver can be added directly to the embryo culture system and will activate teratogenic activity. In Figure 4, this bioactivation system used with cyclophosphamide, an antitumor agent, is illustrated. Besides expanding the number of chemicals which show in vitro activity against embryos, hopefully species variation (see second principal above) may be better understood by using maternal P450 systems from teratogenically sensitive and insensitive species.

Other short term tests. Many new test schemes are being described and evaluated. Since most of these are in preliminary stages, the reader is referred to a series of papers from a symposium published in Vol. 2, part 3 of Teratogenesis, Carcinogenesis and Mutagenesis. Several reviews on the use of various in vitro tests in teratology have appeared (7,31,44). Although the main use of in vitro techniques has been to learn about mechanisms of action, there has been the expectation that a simple, inexpensive prescreen for predicting teratogenic activity in humans could be developed. Unfortunately, this is unlikely since the basic initiating causes of teratogenic activity involve many biologic processes which include cell replication and cleavage, cell differentiation, muscle innervation, as well as

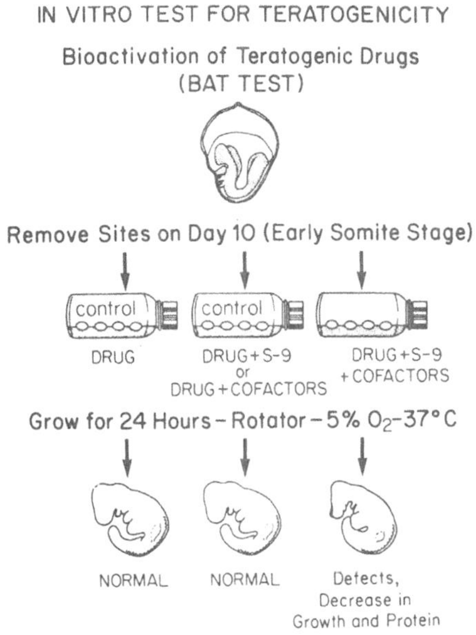

IN VITRO TEST FOR TERATOGENICITY

Bioactivation of Teratogenic Drugs
(BAT TEST)

Remove Sites on Day 10 (Early Somite Stage)

control   control

DRUG   DRUG + S-9   DRUG + S-9
                     or   + COFACTORS
             DRUG + COFACTORS

Grow for 24 Hours – Rotator – 5% O_2 – 37° C

NORMAL   NORMAL   Defects,
                                  Decrease in
                                  Growth and Protein

Figure 4:   Diagram giving an example of an in vitro test of a
      teratogen exposed in vitro to a growing rat embryo. The drug
      used was bioactivated by addition to the medium of a P450
      monooxygenase liver fraction (S-9). In the bottle containing
      drug alone and in the bottles containing S-9 alone or
      cofactors alone, the embryos grew in a normal fashion over a
      24 hour period. In the bottle containing the complete system
      (drug, S-9 and cofactors), there was a dose response relation
      with decreased protein, growth and an increase in
      malformations. This example was drawn from experiments using
      cyclophosphamide (8).

cell adhesion, motility and interaction (22). The complexity of
these causes of altered development is given in Figure 5.

     Even though these tests may eventually be a help in
predicting teratogenicity in humans, most teratologists believe
that testing of pregnant animals will continue to be necessary
because there is little doubt, as has been sadly learned, that the
ultimate test subject for a new agent may be the human.

Present Defenses Against Teratogenic Agents

     It is possible to envision these existing defenses against
teratogenicity as walls or hurdles (Fig. 6). The standard
teratogenicity testing in pregnant laboratory animals would

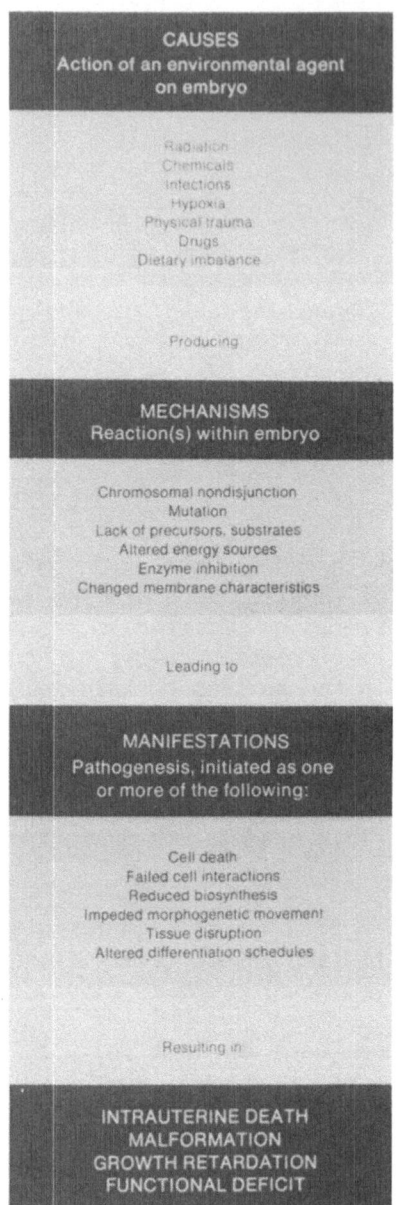

Figure 5: Diagram showing the relationship among cause, mechanism and manifestation of a teratogenic agent (adapted from 42).

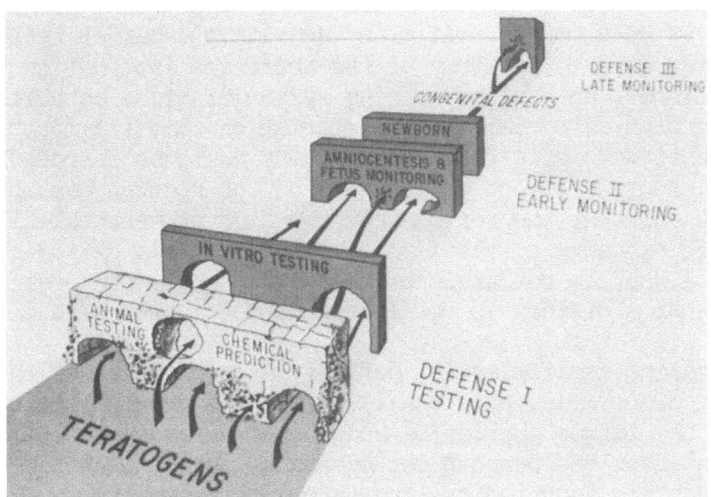

Figure 6:   Perspective of our major defenses against teratogenic
    agents although the in vitro test wall is currently used
    after animal tests, it should precede animal testing in the
    future.   (From 30, used by permission.)

represent part of the first barrier. Created only after the
devastating effects of thalidomide in the early 1960's this
defense is held by a few to be effective, but is recognized by the
majority of scientists to have severe limitations. Animal tests
of new drugs may have prevented the introduction of new human
teratogens. Various aspects of these tests have been described
above. The use of chemical structure and pharmacologic activity
(also a part of this first defense) as a predictor of
teratogenicity is theoretically of great promise, but to date has
seen little practical application except, perhaps, in revealing
the teratogenic effect of chemotherapeutic agents. Since extreme
variations in drug clearance exist between individuals but not
between identical twins, it is very probable that an individual's
pharmacologic defenses are under strong genetic control. This
knowledge could lead to identification and protection of women who
might be more teratogenically vulnerable than others to common
environmental exposures. Pharmacologic variations in the
metabolism of diphenylhydantoin might account for the small
proportion of women in which the drug is teratogenic. The second
defense, early monitoring of the fetus and newborn, is necessary
because the first defenses are still inadequate. Although fetal
monitoring and amniocentesis are able to reduce the number of
severely crippling genetic and morphogenetic defects, the
disadvantage of this form of monitoring is that it is after
conception, and many people believe that therapeutic abortion is

not a morally acceptable tool. One would expect that prenatal
losses might be a more dramatic and sensitive index of response to
teratogenic agents. By study of the abortuses lost during the
first trimester, an earlier warning system might be established
that would provide epidemiologic information some six months
before the effects of a teratogenic agent would be detected in
newborns. Another advantage to the study of spontaneous abortuses
is that it shortens the period from the time of maternal
teratogenic exposure to the time of inquiry. This period may be
less than a week as compared to the 7-8 month time interval
involved when a history is obtained from the mother of a newborn.

Monitoring facilities for defects in neonates exist in many
countries. Generally, only easily recognized physical defects are
recorded; the larger portion of congenital diseases (60%) is
identified after the newborn period and so is not included. Minor
changes in brain function or long term carcinogenesis would be
missed. Although this monitoring system produces variable results
because of artifacts associated with data collection, a continuous
recording registration of time and place of congenital defects
should provide an important warning of teratogenic action by a new
chemical, physical or infectious agent. In the United States a
computerized system for measuring gross congenital defects from
1500 hospitals (with an average of 1 million births per year) is
maintained by the Center for Disease Control and Commission of
Professional Hospital Activities.

Late monitoring (defense 3) is illustrated best by the
discovery of vaginal carcinomas in young women exposed in utero to
diethylstilbestrol. The carcinomas are thought to develop because
of misplacement of genital epithelium during embryogenesis. This
defense, late monitoring, is manned generally by the alert medical
practitioner and in Appendix 1 comments and suggestions for
improving this system will be made.

More than 2,000 new chemicals are synthesized or otherwise
produced each year, and as many as 200 of these may find their way
into the human environment in measurable amounts. These cannot
all be tested. In Appendix 2, I outline the selection of agents
to be tested by government agencies and industry, and the way in
which the dangers posed by teratogens are categorized.

Interpretation of Epidemiologic Findings

There is evidence that about one-half of all human
conceptions terminate in spontaneous abortion, and the incidence
of reproductive waste increases in proportion to the
sophistication and care of the interviewers or methods used in the
study. This means that any form of publicity will increase the
recorded losses and give the false impression of an epidemic.

Factors which increase false positive findings. There are
many preliminary published reports that associate drugs or other
agents with congenital malformations. The conclusions drawn from
these reports are either of borderline significance or are
produced by artifacts of collection, questionable statistical
analysis or both. This incomplete information, released through
news reports or drug package inserts, may lead to panic in exposed
pregnant women and their doctors. Subsequent, detailed studies
which often do not support the original report are either omitted
or given only brief mention. Through biased and inaccurate
reporting several lay publications have created concern and even
panic in the minds of women using a common antinauseant
(Bendectin) during pregnancy while considerable experience with
this drug has supported the opinion by most experts that the drug
is not teratogenic. Positive rather than the negative
associations are naturally more likely to be written up by the
investigator, and also more likely to be accepted by scientific
journal editors.

Regardless of how hard one tries to be impartial, there is a
distinct tendency to be over enthusiastic about positive
associations and at the same time to be less rigorous and critical
when searching for malformations in the control groups.
Collecting positive and interesting associations can easily create
a snowball effect in that the initial observation, after being
broadcast, tends to attract other case associations; in this
manner the size of the exposed population remains unknown and
consequently the true risk is not assessable. Double-blind
studies are nearly impossible to carry out since the person who is
being interviewed will generally mention the presence of his
health problems.

Other false positive biases that occur in the collection of
data include improved recall of events by parents of malformed
children; increased defect rates found by multiple observers
(i.e., the hierarchy of examiners in the university hospitals
where at-risk pregnancies are likely to be referred); and an
increased rate found when a full autopsy is performed. These last
two factors are especially important when considering rare
conditions such as diabetes which tends to be seen more often in
large medical centers. As an example, the newborn of the diabetic
mother is more likely to die and complete autopsy examinations can
inflate the number of malformations in this group.

By use of computers, statistical analyses of many hundreds of
associations between agents and particular malformations can be
performed easily. The great majority of these associations are
made without a prior hypothesis. By the laws of chance alone a
small proportion, perhaps 5 out of each 100, will fall into what
is considered to be a statistically significant group. These

results are reported frequently in the literature and cause another increase in falsely positive associations.

Another serious drawback in epidemiologic studies is the extremely large number of cases needed to prove teratogenicity at a certain probability level. This is especially true where only limited numbers of women are susceptible to the teratogenic action of a drug. Given the extreme genetic heterogeneity of the human population, there is always the possibility that a single individual or a small number of individuals will have a heightened susceptibility that could lead to excessive levels of a relatively safe drug or agent or of a toxic metabolite and, consequently, to embryo-feto toxicity.

Several improvements in the collection and analysis of data have recently been made. In larger studies it is often possible to match the affected mother to a cohort with similar parity, age and social class. In the analysis of data from the Collaborative Perinatal Project[1], a mathematical model was developed to adjust for confounding variables; variables that are related both to rates of drug usage and to malformations (14). Another technique that controls memory bias is the use of a control group of parents who have offspring with types of congenital defects that are known to be unrelated to the drug under study. For example, a study of an agent may include parents of infants with Down syndrome as controls. Another more expensive but not foolproof way is illustrated by the work of Milkovich and Van den Berg (20) which linked computer-stored prescription data on mothers to the findings in their newborns. Some interesting positive associations for meprobamate and chlordiazepoxide were found, but another large prospective study based upon history taking failed to confirm them (13). It is of interest that in the computerized study of Milkovich and Van den Berg (20) all of the significant prescriptions were written by physicians treating women for conditions other than pregnancy, while none were given by the obstetrician. This highlights the fact that much of the teratogenic period of pregnancy occurs before definite diagnosis of pregnancy is made.

FUTURE NEEDS FOR DEVELOPMENT OR EXPANSION

Nearly all the recommendations in this section are dependent upon our ability to validate and transmit information. This includes the entire spectrum of our society from the school age child through pregnant women, administrators and scientists (Appendix 1).

Short Term Tests

Small animal testing for effects on reproduction by new and other untested agents must be continued even though they are not completely predictive for humans. Although a number of biologic systems for short term tests in vitro are mentioned above, none has yet met all the criteria needed for a prescreen. These criteria are: 1) simplicity, 2) lack of expense, 3) rapidity, 4) basing upon multiple mechanisms and 5) most importantly, accurate prediction of known and future human teratogens. The culture of mammalian embryos in association with bioactivation of compounds by maternal enzymes seems to come closest at present to filling these criteria. The growth period of these cultures needs to be extended beyond embryogenesis in order to study late developmental aspects (palate and skeletal formation) and the effect on the chorioallantoic placenta.

The development of short term tests for teratogenicity is hampered by the existence of multiple mechanisms which contribute to the molecular and pathogenic expression of abnormal development (see Figure 4). Unlike mutagenesis, and probably carcinogenesis, the cell targets include other cell organelles than those in the nucleus.

Linkage or Association Between Environmental Agents and Human Disease or Disabilities

In the past 50 years, disease states with short incubations have been fairly easily linked with causative factors. There is now a challenge to associate early exposures to long-term health events. For instance, women who smoke heavily appear to enter menopause at an earlier age than non-smokers. Is this also true for women who have their only exposure to smoking during prenatal life? Are there any prenatal determinants to the onset and course of atherosclerosis and hypertension in old age? What percent of childhood malignancies are causally related to prenatal exposures? Besides diethylstilbestrol, and possibly ionizing radiation, there is some evidence appearing which suggests that diphenylhydantoin may be a transplacental carcinogen.

Triangulation From Existing Data Bases

Triangulation is a navigational technique which allows a traveler to plot his location and subsequent course by determination of his position in respect to certain known points such as stars or coastal markers. Similarly, scientists are finding that the cause and prevention of certain disease states can be determined by linkage of three fixed but expanding data bases (Figure 7). These data bases can be labeled as universal

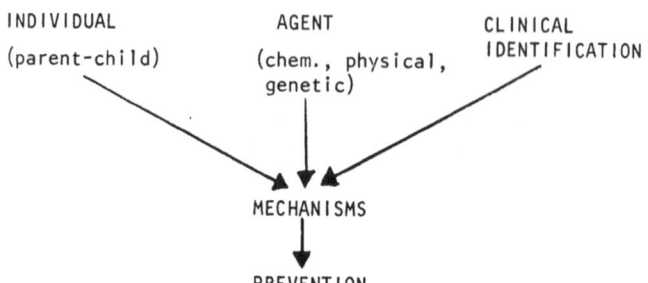

Figure 7:   Diagram of a scheme whereby three data bases
            (individual patient, agents and clinical syndromes) might be
            used to determine the mechanisms which cause congenital
            defects and their prevention.

identifiers since they are world-wide. They consist of 1)
patients, 2) agents, and 3) disease syndromes. At birth an
individual should receive a unique health identification number to
be used throughout life. This identification number could be used
for all health-related records. At the same time, such a system
should incorporate proper safeguards for protection of privacy.
An example of the use of this system could be the identification
and recording of the health numbers of offspring of pregnant women
exposed to workplace agents of unknown long-term toxicity. There
is a reasonable chance that some existing workplace toxicants will
be identified later as agents which can cross the placenta and
initiate changes which over a long term will produce cancer or
other pathology in the offspring. Such a multigenerational
identification system could foster public health measures to warn
these individuals and through early detection to protect their
health.

The second universal identifier system, causative agents, is
partly in place since unique numbers are being assigned to
chemicals (CAS, Chemical Abstract Service numbers). There is,
however, a delay in the assignment of CAS numbers to many physical
and infectious agents.

The area of congenital syndrome identification (syndromology)
is providing a great deal of useful and important data. An
example of the utility of specific syndrome identification is the
fetal alcohol syndrome which appears to account for a major part
of our population in which a specific cause of mental retardation
can be assigned. McKusick's Catalog of Mendelian Inheritance in
Man (19) represents a good updated annotated system for labeling
human mutations and their associated syndromes.

Some disease states have been controlled without knowledge of their cause and this is because of the discovery of effective treatment. An example of this is the surgical correction of congenital pyloric stensis. Congenital rubella is partly controlled by immunizing young women, but we do not know the exact mechanism by which the fetus is in some cases protected by natural defenses. Since we lack the intimate knowledge of mechanisms of pathogenesis, these two congenital syndromes are still not completely preventable.

The concept of linkage of these three data bases was proposed and more fully discussed in a public document produced for federal legislators, Human Health and the Environment, Some Research Needs (23).

Identification and Investigation of Teratogenic Outbreaks

The life span of an epidemic. The time course of an epidemic can be plotted in sequence from the introduction of the causative agent through hypothesis testing to removal of the agent (Figure 8). The time course of the thalidomide epidemic in different countries varied from 2 years to over 4 years. The process was lengthened by the then slow international exchange and assimilation of scientific reports and by delay in the mechanisms leading to removal of the drug from the market. Much improvement has since occurred in these two areas. An example of the slow association between a clinical syndrome and agent is that of the oral coagulants (39). These dicoumarin anticoagulants were associated with two case reports of infants born with very small noses in 1968. The rare linkage between the use of anticoagulants during pregnancy and the unusual reduction in the size of the nose should have signaled an association or at least a concerted search for further supportive evidence. Instead, it was not confirmed until 1973 when additional associations were finally brought to light at a meeting dealing with malformation syndromes. Parenthetically, it is of interest that an animal model for this human teratogen does not exist. The association between alcoholism and the fetal alcohol syndrome was delayed literally for centuries. These examples emphasize the continued need for sound monitoring of our population in order to make epidemics visible against the background of malformations and also for exchange of information between experimental teratologists, clinical syndromologists and regulatory agencies.

A proposal for triage, investigation and management of toxicologic outbreaks. Since such a system should combine all toxicologic dangers (organ specific, carcinogenesis, mutagenesis and teratogenesis), the phrase toxicologic is used here. An executive secretary and small permanent staff should be located geographically and administratively near to the National

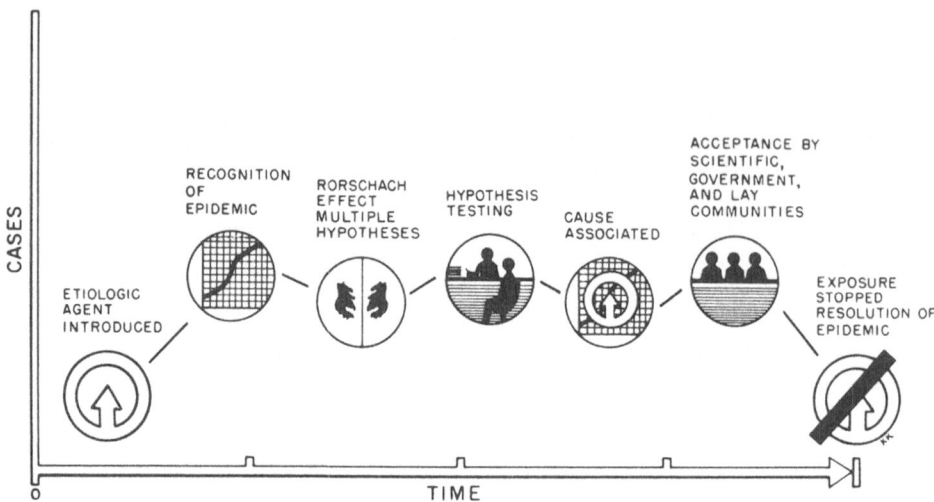

Figure 8: Events between the introduction of a new agent causing
           epidemic disease and the ending of the epidemic, the life
           span of an epidemic. The vertical plot indicates the number
           of cases. The Rorschach stage is named after the psychiatric
           ink blot test where seemingly unstructured ink blots take on
           meaningful shape when studied. The puzzling circumstances
           associated with an early outbreak are equally without meaning
           until hypotheses are generated and studied. (Modified from
           6).

Toxicology Program, Center for Disease Control or other federal
body. These resource people should be consulted by telephone for
purposes of triaging new reports of hazards and for appointment of
ad hoc committees to validate, investigate and take action on
toxicants. Active members of relevant research societies should
be called on to help in these processes.

ACKNOWLEDGEMENTS

     The author thanks his coworkers and several teratologist
friends for their original ideas incorporated into this chapter.
Drs. Alan Fantel, Philip Mirkes and Trent Stephens offered
valuable criticisms.

APPENDIX 1

Sources of Information on Reproductive Hazards

Publications. An excellent general introduction
discussion of teratology is available in Wilson's Environment and
Birth Defects (42). The annotated Catalog of Teratogenic Agents
(34) includes about 1300 hundred agents that have been studied for
teratogenesis in animals and man. Schardein's book on teratogenic
drugs (29) is another useful source. Handbook of Teratology
edited by J.G. Wilson and F.C. Fraser, a comprehensive set of 4
volumes, summarizes many general and specific topics as well as
techniques important in teratology.

Heinonen et al. (14) reported on the outcome of more than
50,000 pregnancies in which drug exposures were known and lists
their risk rates. Hunt (16) has addressed problems of pregnant
women in the work place. Congenital Malformations: Notes and
Comments (38), Recognizable Patterns of Human Malformations (35)
and Syndromes of the Head and Neck (11) help to define and extend
the description of human defect syndromes. McKusick's Mendelian
Inheritance in Man (19) includes more than 3,000 congenital
syndromes produced by gene mutations.

Computerized information retrieval. The National Library of
Medicine (NLM) maintains a number of data bases including Toxline
which lists references dealing with toxic properties. These
listings are on line to most regional medical libraries. The
subjects are stored by CAS numbers (Chemical Abstract Service
number), and in one or two days, it is possible to have a listing
of articles dealing with the reproductive toxicologic aspects of
certain chemicals. The Environmental Teratogen Information Center
(ETIC), maintained by the National Institute of Environmental
Health Sciences, contributes to this service (Toxline) and also
maintains its own answering service at Research Triangle Park,
North Carolina.

Exchange of scientific information. A large and inaccessible
area of scientific data is that of negative findings.
Pharmaceutical companies often fail to release results of safety
tests of their new products. In addition, scientific journals do
not publish many manuscripts on agents which are not toxic in some
way. In Japan, industries and others are encouraged by the
government to publish their safety tests and at least one journal,
Oyo Yakuri, is available for publication of these results.

Scientific exchanges with other countries are recognized as
important. As a result of this, the computerized retrieval of
literature references on toxicologic substances (Toxline) of the
National Library of Medicine is available during slack hours

through a number of outlets in Western Europe, Central and South
America and Africa. The library tapes are also accessible in
Japan and Australia. These countries contribute in a significant
way to collection and validation of these data bases, particularly
with scientific articles originating in their own countries.
Printing of texts by computer and other means of reducing the
publication time and cost of scientific information should be
fostered.

Training. Most medical students have only modest exposure to
toxicology and over one-half of the medical schools in the United
States have no specific courses or trained faculty for teaching
teratology. It is interesting that those schools with very strong
programs in developmental biology seem to be the weakest in
teratology, which in itself is a form of applied developmental
biology. Human embryology should be reinstated as a medical
school requisite.

APPENDIX 2

Selection of Agents to be Tested existing federal regulatory agencies have their own investigative capabilities. The larger of these agencies include the Center for Disease Control (CDC), Environmental Protection Agency (EPA), Food and Drug Administration (FDA), National Cancer Institute (NCI), the National Institute of Child Health and Human Development (NICHD), National Institute of Environmental Health Sciences (NIEHS) and the National Institute of Occupational Health (NIOSH). Other departments of government, such as Transportation, Labor, Energy, Defense and Agriculture also foster some teratology work. The National Center for Toxicologic Research (NCTR) in Arkansas, funded through the FDA and EPA, has a mandate to develop new tests for toxic substances and, in particular, to do long term, low dose experiments in large numbers of experimental animals (referred to by some as mega mouse studies). All of these agencies choose agents which they believe to be important to their regulatory or mandated programs. A recent development in toxicology testing has been the formation of a National Toxicology Program (NTP) for selecting and testing chemicals and developing new test methods. A chemical selection unit located at NCTR prepares summaries of data on agents which may be nominated by government, industry or private individuals. The prioritizing and testing of these chemicals is performed under the NTP.

The Toxic Substances Control Act (TOSCA) charges the EPA to identify and test a specific number of chemicals and some of these tests will be of teratogenic nature. Training and money for toxicologists was included in this legislation.

Industry and academia. Many pharmaceutical and larger chemical producers have elaborate facilities for teratology testing of their products. Although many of these results are published, some degree of secrecy enters due to the competitive nature of the enterprise. The faculty of the universities and other research units contribute many of the new observations and techniques to teratology as well as acting as consultants to both industry and government.

Categorization of the danger of teratogenic agents. Some methods of regulating exposure to known or suspected teratogens have been developed. The FDA has published five categories of danger for drugs that may be taken by pregnant women (9). The classification is based upon animal and human exposure and takes into account the absence of toxicologic information (Table 1). The FDA recommends that where data are available they be summarized and referenced in the accompanying package insert for the drug.

The International Agency for Research in Cancer (IARC) of the World Health Organization has taken up the practice of reviewing reproductive and teratologic data and assessment of hazards of selected chemicals. This organization produces many useful publications based upon ad hoc committees drawn from experts in the field of carcinogenesis, mutagenesis and teratogenesis.

Since the Delaney Clause was passed by the U.S. Congress forbidding the use of known animal carcinogens in food substances, a similar exclusion of teratogens was considered. Because nearly every agent in sufficiently large doses can produce embryofeto toxicity, it seemed unwise to have a similar regulation for teratogens. In one of the very few political activities taken by the Teratology Society, the membership was surveyed on this question and the overwhelming opposition was recorded in several scientific journals (36).

REFERENCES

1. Boue, J., A. Boue, and P. Lazar (1976) Retrospective and
    prospective epidemiological studies of 1,500 karyotyped
    spontaneous human abortions. Teratology 12:11-26.
2. Brinster, R.L., and P.C. Cross (1972) Effect of copper on the
    preimplantation mouse and embryo. Nature 238:398-399.
3. Brinster, R.L. (1975) Teratogen testing using preimplantation
    mammalian embryos, In Methods for Detection of Environmental
    Agents that Produce Congenital Defects, T.H. Shepard,
    J.R. Miller, and M. Marois, Eds., North Holland-American
    Elsevier, Amsterdam.
4. Carter, S.B. (1967) Effects of cytochalasins on mammalian
    cells. Nature 213:261-264.
5. Chatot, C.L., N.W. Klein, J. Piatek, and L.J. Pierro (1980)
    Successful culture of rat embryos on human serum: Use in the
    detection of teratogens. Science 207:1471-1473.
6. Donahue, R.P. (1975) Chromosomal anomalies and the meiotic
    divisions of the oocyte, In Methods for Detection of
    Environmental Agents that Produce Congenital Defects,
    T.H. Shepard, J.R. Miller, and M. Marois, Eds., North
    Holland-American Elsevier, Amsterdam.
7. Ebert, J., and M. Marois, Eds. (1976) Tests of teratogenicity
    in vitro, In Proceedings of the Woods Hole Conference 1975,
    North-Holland-American Elsevier, Amsterdam.
8. Fantel, A.G., J.C. Greenaway, M.R. Juchau, and T.H. Shepard
    (1979) Teratogenic bioactivation of cyclophosphamide in
    vitro. Life Sciences 25:67-72.
9. Federal Register (1979) Vol. 44, No. 124, Tuesday, June 26,
    37462-37466.
10. Fraser, F.C. (1969) Gene-environment interactions in the
    production of cleft palate, In Methods for Teratologic
    Studies in Experimental Animals and Man, H. Nishimura,
    J.R. Miller, and M. Yasuda, Eds., Igaku Shoin Ltd., New York.
11. Gorlin, R.J., J.J. Pindborg, and M.M. Cohen, Jr. (1976)
    Syndromes of the Head and Neck, 2nd edition, McGraw-Hill Book
    Co., New York.
12. Hale, F. (1937) Relation of maternal vitamin A deficiency to
    microphthalmia in pigs. Texas St. J. Med. 33:228-232.
13. Hartz, S.C., O.P. Heinonen, S. Shapiro, V. Siskind, and
    D. Slone (1975) Antenatal exposure to meprobamate and
    chlordiazepoxide in relation to malformations, mental
    development and childhood mortality. New Engl. J. Med.
    292:726-728.
14. Heinonen, O.P., D. Slone, and S. Shapiro (1977) Birth Defects
    and Drugs in Pregnancy, Publishing Sciences Group, Littleton,
    Massachusetts.
15. Hertig, A.T. (1967) The overall problem in man, In Comparative
    Aspects of Reproductive Failure, K. Benirschke, Ed.,
    Springer-Verlag, New York.

16. Hunt, V.R. (1979) Work and the Health of Women, CRC Press, Inc., Boca Raton, Florida.

17. Kelsey, F.O. (1974) Present guidelines for teratogenic studies in experimental animals, In Congenital Defects: New Directions in Research, D.T. Janerich, R.G. Skalko, and I.H. Porter, Eds., Academic Press, New York.

18. Leridon, H. (1977) Human Fertility, the Basic Concepts, Chicago Press, Chicago, Illinois.

19. McKusick, V.A. (1983) Mendelian Inheritance in Man, 6th Edition, The Johns Hopkins University Press, Baltimore, Maryland.

20. Milkovich, L., and B.J. Van den Berg (1974) Effects of prenatal meprobamate and chlordiazeproxide hydrochloride on human embryonic and fetal development. New Engl. J. Med. 291:1268-1271.

21. Miller, J.F., E. Williamson, J. Glue, Y.B. Girdon, J.G., Grudzinskas, and A. Sykes (1980) Fetal loss after implantation. Lancet 2:554-556.

22. Moscona, A.A. (1975) Invited discussion: Embryonic cell and tissue cultures as test systems for teratogenic agents, In Methods for Detection of Environmental Agents that Produce Congenital Defects, T.H. Shepard, J.R. Miller, and M. Marois, Eds., North Holland-American Elsevier, Amsterdam.

23. Nelson, N., and J.L. Whittenberger (1977) Human health and the environment - Some research needs. Report of the Second Task Force for Research Planning in Environmental Health Science, DHEW Publication No. NIH 77-1277.

24. New, D.A.T. (1978) Whole embryo explants and transplants, In Handbook of Teratology, Vol. 4, J.G. Wilson, and F.C. Fraser, Eds., Plenum Press, New York.

25. Oakley, G.P., Jr. (1976) Birth defect surveillance in the search for and evaluation of possible human teratogens. Birth Defects: Original Article Series, Vol. XII, No. 5.

26. Palmer, A.K. (1978) The design of subprimate animal studies, In Handbook of Teratology, Vol. 4, J.G. Wilson, and F.C. Fraser, Eds., Plenum Press, New York.

27. Rodier, P.M. (1978) Behavioral teratology, In Handbook of Teratology, Vol. 4, J.G. Wilson, and F.C. Fraser, Eds., Plenum Press, New York.

28. Rogers, B.J., H. Van Campen, M. Ueno, H. Lambert, R. Bronson, and R. Hale (1979) Analysis of human spermatozoal fertilizing ability using zona-free ova. Fertility and Sterility 32:664-670.

29. Schardein, J.L. (1976) Drugs as Teratogens, CRC Press, Inc., Cleveland, Ohio.

30. Shepard, T.H. (1974) Teratogenicity from drugs--an increasing problem. Disease-a-Month June, 1-32.

31. Shepard, T.H., and D. Pious (1978) Cell, tissue and organ culture as teratologic tools, In Handbook of Teratology,

Vol. 4, J.G. Wilson, and F.C. Fraser, Eds., Plenum Press, New
York.

32. Shepard, T.H., and A.G. Fantel (1979) Embryonic and early
fetal loss, In Clinics in Perinatology, Symposium on Fetal
Disease, J. Warshaw, Ed. 6:219-243.

33. Shepard, T.H. (1979) Teratogenicity of therapeutic agents, In
Current Problems in Pediatrics, L. Gluck, Ed., Year Book
Publishers, Inc., Chicago, Ill.

34. Shepard, T.H. (1983) A Catalog of Teratogenic Agents, 4th ed.,
The Johns Hopkins University Press, Baltimore, Maryland.

35. Smith, D.W. (1976) Recognizable Patterns of Human
Malformation, 2nd ed., W.B. Saunders Co., Philadelphia,
Pennsylvania.

36. Staples, R.L. (1974) Teratogens and the Delaney Clause.
Science 185:813.

37. Teratology Society (1980) Teratology 21:A1-A77.

38. Warkany, J. (1971) Congenital Malformations: Notes and
Comments, Year Book Medical Publishers, Chicago, Illinois.

39. Warkany, J. (1976) Warfarin embryopathy. Teratology
14:205-209.

40. Warkany, J., and R.C. Nelson (1940) Appearance of skeletal
abnormalities in the offspring of rats reared on a deficient
diet. Science 92:383-384.

41. Whorton, D., R.M. Krauss, S. Marshall, and T.H. Milby (1977)
Infertility in male pesticide workers. Lancet 2:1259-1261.

42. Wilson, J.G. (1973) Environment and Birth Defects, Academic
Press, New York.

43. Wilson, J.G. (1975) Critique of current methods for
teratogenicity testing in animals and suggestions for their
improvement, In Methods for Detection of Environmental Agents
that Produce Congenital Defects, T.H. Shepard, J.R. Miller,
and M. Marois, Eds., North Holland-American Elsevier,
Amsterdam.

44. Wilson, J.G. (1978) Feasibility and design of subhuman primate
studies, In Handbook of Teratology, Vol. 4, J.G. Wilson, and
F.C. Fraser, Eds., Plenum Press, New York.

A POSSIBLE MECHANISTIC LINK BETWEEN TERATOGENESIS AND CARCINOGENESIS: INHIBITED INTERCELLULAR COMMUNICATION[1]

James E. Trosko and Chia-cheng Chang

Department of Pediatrics and Human Development
College of Human Medicine Michigan State University
East Lansing, MI 48824-1317

SUMMARY

The integrity of the DNA of each cell is necessary for normal functioning of the cell and the organism. DNA damage, if not repaired, might influence the expression of genes in that cell, as well as act as cytotoxic and mutagenic substrates. Genetic, developmental, physiological, nutritional and environmental factors are known to influence both the amount and kind of DNA damage, as well as its repair. Not all gene or chromosomal mutations are the result of agents which damage DNA. Factors which could alter the fidelity of DNA replication of normal DNA templates, as well as the segregation of chromosomes, could lead to both gene and chromosomal mutations. Since carcinogenesis is a multi-step process, involving the transformation of a normal cell to a premalignant cell, with subsequent clonal expansion and phenotypic evolution of that premalignant cell to a malignant cell, DNA damage and faulty DNA repair probably play a significant role in carcinogenesis through the production of mutations and cell killing. Cell killing due to inadequate DNA repair of DNA damage (or due to any other means) could act as a mitogenic stimulus by inducing regenerative hyperplasia. This hyperplasia stimulus, by amplifying initiated cells and creating opportunities for other unrepaired DNA lesions to be substrates for new mutations during cell division, might be responsible for converting the pre-malignant cell to the malignant cell, thus completing the carcinogenic process. Inhibition of intercellular

[1]Research was supported by granst to J.E.T. from the National Cancer Institute (CA 21104, CA 26803) and the EPA (R808587010).

529

communication by a wide variety of physical, chemical and biological factors has been postulated to disrupt the regulation of proliferation and differentiation in stem cells. Agents which interrupt intercellular communication during early organogenesis have the potential to be teratogens, while if they are present in the developed, initiated organism they have the potential to be tumor promoters. Consequently, the observed linkage between teratogens and carcinogens might be mediated by their ability, via several mechanisms, to interfere with intercellular communication.

INTRODUCTION

Several observations have been made which suggest a possible connection between carcinogenesis and teratogenesis. Of course, these observations might only reflect a coincidental, but not causal, relationship between these two biological processes.

One of the major pieces of evidence linking carcinogenesis and teratogenesis is the fact that many "carcinogens" are teratogens (19,47). Several known teratogens have not been shown to be "carcinogens" (e.g., thalidomide). Because of the complexities of testing chemicals as carcinogens and teratogens (26) and because of several potential mechanisms leading to both carcinogenesis (70) and teratogenesis (78), experimental evidence supporting or contradicting this potential connection is generally weak.

Another observation which appears to suggest a connection between these two processes is the frequent appearance of congenital malformations with individuals predisposed to cancer [e.g., Fanconi's anemia, xeroderma pigmentosum, Down syndrome, retinoblastoma, Bloom's syndrome, etc.] (4,47,76).

Lastly, there are reports that there is an increased incidence of congenital malformations in children with cancer (32,36,55). By themselves, these observations can be considered potentially weak for a variety of reasons. However, taken together, there seems to be some basis for believing that these two disease processes might share some common mechanism(s).

In the following, a speculative analysis of some current hypotheses of the mechanisms of carcinogenesis and teratogenesis will be made. Specifically, the potential roles of mutations, cell death and inhibited intercellular communication, and the various means leading to these cellular consequences, will be analyzed in early organogenesis and during the multistep carcinogenic process.

INITIATION AND PROMOTION MODEL OF CARCINOGENESIS

Empirical observations on the "natural history" of cancer seem to indicate, in both experimental animals and in "naturally" occurring human cancers, that a tumor is the end result of a complex evolution of a normal cell which has "gone wrong" (14).

Other empirical studies indicate that there may be a clonal original or tumors (12), in spite of the obvious heterogeneity of phenotypes and genotypes within a tumor (51). Carcinogenesis seems to describe the process by which a normal stem cell, which has the phenotype of contact inhibition and appropriate regulatory control of its ability to proliferate and/or differentiate, loses these abilities, and acquires the phenotypes of tissue invasiveness and metastasis. When a normal cell is transformed to a premalignant cells, it appears to have started on its way to losing its ability to "orchestrate the available repertoire of gene capabilities in a manner appropriate to the whole organism at any given time" (53). One can infer that this initial change is insufficient to acquire all the phenotypic changes associated with an invasive and metastatic tumor cell. In addition, the observation of the latency of tumors suggests that the initial abnormal premalignant phenotype can, in some fashion, be suppressed or compensated (see later discussion).

Experimental observations on various rodent models indicated that this complex carcinogenic process could be dissected into at least two distinct phases, namely the "initiation" and "promotion" phases (3) [a "progression" phase could also be included (52)]. Operationally, initiation refers to the biological conversion of a single normal cell to a "premalignant cell" by means of a rather "irreversible" or stable mechanism, whereas promotion (and progression) refer to the clonal amplification of this single premalignant cell to the invasive, metastatic state by agents or conditions which by themselves are not "carcinogenic" (see 66).

The mechanism(s) of initiation and promotion is(are) obviously not known, nor is it axiomatic that all cancers must conform to this initiation/promotion model of their genesis. Initiators must be agents or conditions which could induce the rather stable and irreversible genomic change. Although mutations (gene or chromosomal), by definition, produce stable and "irreversible" changes in the genome, and although most mutagens are either carcinogenic initiators or "complete" carcinogens (see 70 for further discussion on these points), clearly one can conceive of stable, non-mutagenic changes in the genome, which could be induced by non-mutagens. Promoters, by helping to "release" a premalignant cell from its latent state and by clonally amplifying the number of initiated cells, are acting, at least, as "selective mitogens", not as mutagens (see 68 for

exceptions to this generalization). They must have other
properties related to their promoting ability (75).

The somatic mutation theory of cancer in its many forms (65),
as well as nonmutational theories (5), has been offered to explain
carcinogenesis. In either opposing theory, the multi-stage,
initiation/promotion observations of carcinogenesis must be
explained. Simple single gene/chromosomal mutations or gene
modulations seem insufficient to explain this complex evolution of
phenotypes occurring during carcinogenesis. Multi-genetic
mutation models (2,29), as well as attempts to integrate both
mutational and non-mutational events into a model (64), have been
offered.

In summary, the initiation and promotion model of
carcinogenesis seems to be consistent with observations of its
multi-staged nature. The molecular mechanisms are not yet known
for either of these processes. However, understanding the
molecular bases for mutagenesis and gene modulations would seem to
be appropriate for understanding how genes could be altered in
either information content or expression.

GENE, ONCOGENES AND CARCINOGENESIS

Starting with the assumption that the multistage nature of
carcinogenesis can be explained by initiation and promotion
models, several additional observations and concepts must now be
examined.

Clearly, it has been known for a long time that there are
many genes and chromosomes which can influence animal and human
carcinogenesis. Many human syndromes, genetically predisposing
individuals to cancer, have been noted (40). Some, such as Down
syndrome, have a chromosomal imbalance. Others, such as xeroderma
pigmentosum or retinoblastoma, are due to recessive mutations.
The fact that the cancer-prone genetic syndromes of xeroderma
pigmentosum (43) and Bloom's (73) exhibit higher induced mutation
frequency and spontaneous mutation rate, respectively, than cells
from normal individuals, also implicates mutations with
carcinogenesis. The theoretical fit of a "two mutational hit"
model for hereditary retinoblastoma adds more evidence for the
role of genes (29). Recent observations of specific chromosomal
mutations with many kinds of cancer cells (58) implicate specific
genes with the carcinogenic process. The well established
observations that most mutagens which have been tested properly
can be either complete carcinogens or carcinogenic initiators (45)
also implicate mutated genes and chromosomes in carcinogenesis.

However, the most recent observation that "oncogenes" of human tumor cells differ from the normal oncogene by a single base change, directly implicate mutations with some aspect of the carcinogenic process (56,62). Of course, these observations do not imply that all cancers must have a mutational basis, nor that mutations are responsible for the whole carcinogenic process.

However, because so many phenotypes, all of which are under genetic control, are needed to affect the transition of a normal stem cell, which can appropriately regulate its specific proliferative and differentiation functions, to an invasive and metastatic cell, one can reasonably postulate that there are several "oncogenes". For each stem cell of a given tissue, there are probably tissue specific "oncogenes", as well as several "oncogenes" regulating a basic phenotypic function (see below) needed to regulate proliferation and differentiation.

GENETICS OF INITIATION AND PROMOTION

Building on the assumptions of the initiation and promotion model of carcinogenesis and of the role of genetics and mutations in carcinogenesis, the manner in which genetics might influence carcinogenesis must be examined. Of course, as mentioned earlier, the detailed molecular mechanisms of initiation and promotion are not yet understood. However, agents which influence mutagenesis by damaging DNA, inhibiting error-free DNA repair or by rendering DNA replication error-prone do seem to be carcinogenic initiators (65). It should be stressed that not all mutations will be due to agents which damage DNA. A very misleading idea is that "genotoxins" are agents which interact with or damage DNA to cause mutations. Clearly, there are physical conditions or chemical agents which can influence gene and chromosomal mutations without interacting with, or damaging, DNA [e.g., aphidicolin, a specific inhibitor of DNA polymerase α, induces endoreduplication of chromosomes (23); mutations can alter the fidelity of DNA polymerase (33)].

Promoters seem to be agents or conditions which act to stimulate the multiplication of initiated cells to bring about a "critical mass" of dysfunctional cells (68). Among the many phenotypic responses to various promoters (8), their membrane-perturbing activity seems most relevant (75). From our vantage point, the observations that tumor promoters inhibit intercellular communication seems very relevant to their ability to induce hyperplasia and/or selective proliferation of the initiated cells (81) and to alter patterns of differentiation (68).

To reiterate, the process of mutagenesis and the process of intercellular communication would seem to be very important in the

initiation and promotion phases of carcinogenesis. Since both
processes are dependent on various enzymes and protein structures
(i.e., DNA repair and replication enzymes; gap junction proteins;
gap junction regulatory enzymes), there must be many genes
affecting these two fundamental processes. Since each gene is
potentially mutable (either through germ line or somatic cell line
mutations), one would predict that mutations could occur affecting
mutation production and cell-cell communication.

Some of these mutations, which occur in the germ line, would
be expected to predispose the individual genetically to either or
both initiation and promotion phases of carcinogenesis (71). For
example, using this conceptual perspective, one can now ask how
xeroderma pigmentosum, albinism, Bloom's, Down's, or
neurofibromatosis syndromes might affect the carcinogenic process.
Is it because they contain germ line mutations which affect
initiation (i.e., the mutation rate and frequency in their cells),
or do they aid promotion by affecting the regulation of
proliferation or differentiation of cells?

If a person inherits a mutated gene which: (a) makes the DNA
more prone to damage (e.g., albinos with a lack of melanin in the
skin would incur more UV-induced DNA damage/unit incident exposure
than normal pigmented cells); (b) affects DNA repair [e.g.,
xeroderma pigmentosum (43)]; (c) affects the fidelity of DNA
replication [e.g., possibly Bloom's syndrome (73)]; and (d)
sensitizes the cells to mutagens [e.g., Fanconi's (27)], then
these could legitimately be conceived of as "initiator-prone"
syndromes. As will be seen later, "initiator-prone" syndromes, in
a manner similar to physical or chemical mutagens, can be
characterized as acting as "initiators" and "complete"
carcinogens, depending on whether significant cytotoxicity is
associated with the mutagenic events.

On the other hand, individuals may inherit genes which: (a)
cause tissue specific hyperplasia because of an over production of
a mitogen/growth factor [e.g., possibly neurofibromatosis (71)];
(b) have a metabolic blockage which produces a metabolite of
promoter-capacity; (c) cause an underproduction of growth
suppression; or (d) have a defective ability to regulate
proliferation or differentiation. Because hormones have been
implicated as possible promoters (79), and because hormones do
affect proliferation and differentiation in specific tissues,
genetic imbalances of hormones are associated with predisposition
to certain hormone-target tissue cancers (48).

Down's syndrome individuals do not seem to be defective in
their ability to repair certain types of DNA damage, nor do they
seem to be hypermutable (82). On the other hand, their
predisposition to leukemia might be related to some imbalance in

interferon function (44), which has been shown to have promoter-like action in vitro (37). The altered patterns of differentiation and local hyperplasia in neurofibromatosis seem to indicate the endogenous local production of promoter-like materials, since promoters do affect control of proliferation and differentiation.

To reiterate, this concept that various genes can influence either or both the initiation and promotion phases of carcinogenesis may be very helpful in identifying the molecular disturbances in these cancer-prone individuals (e.g., are they initiator- or promoter-sensitive?).

GENETICS OF DNA REPLICATION/REPAIR AND MUTAGENESIS

At this point, based on the assumptions that there are many genes which affect the initiation and promotion phases of carcinogenesis, and that these genes are subject to mutations, it is important to understand the mechanisms of mutagenesis in human cells. In addition, ultimately it will be important to identify the gene(s), which when mutated, transform(s) the normal phenotype of a cell to the pre-malignant and malignant phenotypes ("oncogenes").

Mutations are dependent on cell replication. DNA replication is necessary to "fix" mutations in the new daughter cells. Gene and some chromosomal mutations are the result of unrepaired DNA lesions acting as substrates for error-prone repair or replication (67). Gene and chromosomal mutations can also be due to faulty replication of non-damaged DNA templates, either by imbalances in nucleotide pools (343) or by infidelity of replicating enzymes (77). Our ignorance of the details of normal error-free DNA repair and DNA replication in human cells precludes our outlining those genetic and environmental factors needed to ensure error-free repair and replication of genes and chromosomes. There are probably several, as yet unidentified, mechanisms influencing mutagenesis in both genes and chromosomes.

Work on the excision-repair defective, skin-cancer-prone xeroderma pigmentosum individuals has implicated unrepaired base damage and excision repair in mutagenesis (43). Because of the large number of genetic complementation groups of xeroderma pigmentosum (28), there are obviously several DNA sequences influencing mutagenesis via error-prone DNA repair. "Liquid holding" type experiments in mammalian and human cells have suggested that the normal excision repair process is relatively error-free (49,59). Our knowledge of the "error-process" of other repair mechanisms (e.g., single strand break repair) is more limited.

DNA polymerase α, and its gene, plays a major role in semiconservative replication (60), although it seems to be able to
contribute under certain conditions to DNA excision repair (46).
DNA polymerase β, and its gene, seems to be the major DNA repair
replication enzyme needed for excision repair (74), although it,
also, can contribute to semi-conservative replication (46).
Interestingly, at least in Chinese hamster V79 cells, the
relatively error-free DNA polymerase β can be made "error-prone",
by conditions which inhibit the DNA polymerase α (23). Studies
have shown that, in the absence of functional DNA polymerase α,
when a cell has to divide, the DNA polymerase β is able to
replicate chromosomes. Unfortunately, however, the DNA polymerase
β does not seem to be able to "shut off" after it replicates the
complete set of chromosomes. As a result, endoreduplication of
the chromosomes occurs.

It might be of interest to note that many carcinogens seem to
be able to induce endoreduplication (23). Many tumors have
endoreduplicated chromosomes in their cells (22). Several cancer-
prone syndromes have endoreduplicated cells in their cells (30).
It may well be that carcinogens which damage DNA, force the DNA
polymerase α to stop replication due to the unrepaired lesions.
Under severe circumstances, the DNA polymerase β, which probably
evolved to function around DNA lesions, could now replicate the
chromosome. An "error-free" DNA polymerase β, on the gene level,
can be made to be error-prone on the chromosome level. From an
evolutionary perspective, this probably had great survival value,
since if the environmental conditions were harsh enough to cause
substantial DNA damage, one way to survive is to produce mutant
offspring that might be adaptable to the new environment. When
this occurs in somatic cells, however, the mutant offspring might
be a cell that has selective growth advantage (e.g., a cancer
cell).

Bloom's syndrome is also a cancer-prone syndrome (16) and it,
also, seems to be a "hypermutable" syndrome (73). However, in the
case of Bloom's syndrome the mechanistic basis of its hypermutable
spontaneous mutation rate is unknown. Most studies have not
implicated any detectable defect in DNA repair (1). Some have
suggested defective DNA replication (18) or possibly the exogenous
production of mutagens (11).

Only future studies on other cancer-prone, "chromosome-
fragile", or mutagen-sensitive syndromes will bring us more
information on the role of various genes and enzymes affecting
error-free DNA replication and repair.

When mutations are discussed in relation to carcinogenesis,
one significant point is usually forgotten, namely, cytotoxicity.
If lesions are induced in DNA, some of them are repaired. Others

which are not repaired either do not affect the functioning of the DNA, or they are substrates for mutations or cell killing. Some of the mutations are also lethal. There are studies now indicating that some DNA damage may be a major cause of mutagen-induced cell killing and some may induce the mutations (31). The importance of mutagen-induced cell killing has been discussed in terms of its relationship to tumor promotion. In other words, if tumor promotion is dependent on inhibited intercellular communication and sustained hyperplasia, then when a mutagen kills cells, the mutagen is acting not only as an initiator, but an "indirect promoter" [i.e., it is a "complete" carcinogen (70)].

To illustrate this point, it has been argued that xeroderma pigmentosum syndrome provides strong evidence against the role of DNA repair and mutations in cancer (5). The argument is that if the lack of DNA repair and mutations are the cause of cancer, then why are there so few internal cancers in xeroderma pigmentosum? The reasoning is that environmental mutagens should induce DNA damage internally. Although there does seem to be large numbers of internal tumors in xeroderma pigmentosum individuals (33), there seems to be a very reasonable explanation for the predominance of the tumors in the skin (63). Since the greatest flux of environmental mutagens seems to contact the skin (i.e., ultraviolet light), not only will the DNA of skin cells have more lesions than internal cells, but also these lesions will probably kill many more cells in the skin than chemical-induced lesions internally. The compensatory hyperplasia induced in the skin by sunlight will probably act to promote previously initiated cells, as well as create more opportunities for new mutations which are dependent on cell division (68).

In summary, we do know that errors in both DNA replication and repair can produce mutations. The number of genes and the ways environmental factors could modulate the mutation production in human cells are clearly not known. The molecular mechanisms leading to the various kinds of gene and chromosomal mutations are only beginning to be understood.

GENETICS OF INTERCELLULAR COMMUNICATION

If we accept the hypothesis that mutations do affect carcinogenesis, the genes affecting both the "initiated" phenotype and the invasive and metastatic phenotypes seem to be some of the important ones on which to focus one's attention.

Attempts are now being made to isolate "pure clones" of initiated cells (17). However, too little is known at present to speculate on what these genes do in the cell. Tumor promoters are known to affect membranes to cause hyperplasia, among other things

(8). However, as was pointed out previously, most, if not all, malignant cells do not appear to communicate properly (7). The fact that some cells of a tumor can communicate might be due to the fact that not all cells of a tumor are malignant (68). It seems very plausible to speculate that tumor promoters cause normal and premalignant cells to "mimic" the malignant phenotype. As long as the promotion is present, the normal and premalignant cells can "escape" normal regulatory mitotic controls (82). If, during the promotion of the pre-malignant cell, genomic changes occur which "stabilize" the malignant phenotype (i.e., inability to communicate), then tumor promoters will not be needed.

There seem to be several genes controlling gap-junction structure and function. Although the gap-junction is a highly conserved structure in all multicellular organisms, we can only surmise that the gene(s) and protein(s) are highly conserved. We and others have isolated mutants deficient in gap-junction mediated metabolic cooperation (41). However, it is not yet known if any of these mutants have defective gap junctions or whether the genes and enzymes which regulate their function have been mutated. It has been shown that, at least for one mutant which was unable to perform intercellular communication, modulating c-AMP levels of the cell restored normal gap-junction mediated intercellular communication (13). This implies that if the gap-junctional mediated intercellular communication phenotype is necessary for normal growth and differentiation, mutation of any of the several genes which could alter that phenotype (i.e., cyclic AMP levels) could contribute to the carcinogenic process.

There is inferential evidence that membrane-triggered signals can influence gene expression (20) and enzyme activity (21,57). Since tumor promoters are known to affect membranes (75) and to affect differentiation and proliferation in various cell types (83), these affects on inhibited intercellular communication, a membrane-dependent process, might be related to the activation of gene expression and enzyme activity. Therefore, chemicals which interfere with intercellular communication could be considered "gene modulators". It could also be surmised that these chemicals would be species, tissue and cell type specific [i.e., affects either the epithelial or fibroblast types] (35).

It is this last property of many tumor promoters (i.e., the ability to modulate gene expression and enzyme activity via membrane-trigger reactions, including inhibition of intercellular communication) that might be an important determinant linking carcinogenesis to teratogenesis.

MUTAGENESIS, CYTOTOXICITY AND INHIBITED INTERCELLULAR
COMMUNICATION IN TERATOGENESIS

In order for a multicellular organism to develop normally from a single fertilized egg , and to maintain a complex of specialized functions,
a delicate orchestration and integration of a variety of intercellular
communication mechanisms [i.e., systemic, portal, local, long and short-
half-life types (54] must be maintained to ensure the required proliferation and differentiation of cells (39). In theory, disruption of these
cell-cell communication processes, including cell-cell adhesion and
cell-substrate adhesion (69), could lead to embryonic or fetal toxicity,
or to congential defects (69).

Considering the three major kinds of cellular responses to radiations,
chemicals or viruses (i.e., cell death, mutations or gene "modulation",
one can reason how they could, in theory, affect normal development (Fig.1)

Mutations of specific "developmental" genes in the stem cell
of given organs during early development, if dominant and viable
but not too severe, could lead to an abnormal organ development.
Mutagens, depending on concentration and the type of mutation
formed, are potentially cytotoxic. The removal of critical cells
or a critical number of cells during organogenesis could also, in
theory, lead to teratogenesis (69).

Inhibition of intercellular communication, either by non-
cytotoxic membrane perturbing agents or by cytotoxic agents
(mutagens or non-mutagenic cytotoxins), could also lead to
abnormal proliferation or differentiation of cells during
organogenesis (25). It has been reported that many inhibitors of

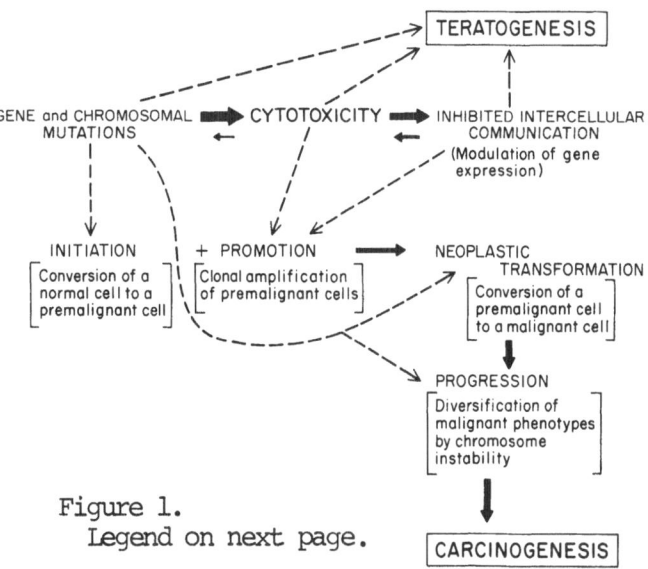

Figure 1.
Legend on next page.

Figure 1. A heuristic scheme to classify chemicals on the basis
     of three biological endpoints, mutagenicity, cytotoxicity and
     inhibition of intercellular communication. The arrows
     between the biological endpoints are designated to mean that
     chemicals can have multiple biological consequences in the
     direction indicated (i.e., mutagens can kill cells and the
     death of cells can cause the modulation of gene expression in
     some surviving cells). In addition, some chemicals which can
     inhibit intercellular communication at noncytotoxic levels
     could, at much higher concentrations, kill cells, possibly
     also inducing some chromosomal mutations at these high doses.
     It further depicts how mutagenesis, cell death and disrupted
     intercellular communication can play a role in teratogenesis
     and carcinogenesis. An early conceptus or fetus, exposed to
     a mutagen, at a low dose, could conceivably have a critical
     gene mutated in a stem-cell leading to a congenital defect.
     At high, cytotoxic doses of a mutagen, fetal toxicity would
     be expected. Exposure to nongenotoxic, but cytotoxic
     chemicals could also, conceivably, lead to congenital defects
     or fetal toxicity, depending on the type and amount of cells
     killed by the exposure. Chemicals which inhibit
     intercellular communication could disrupt regulation of
     proliferation and differentiation of tissues if given at a
     critical period of development. The consequence would lead
     either to congenital defects in specific organs or embryo to
     fetal toxicity, depending on the time, amount, distribution,
     metabolism or receptors for the chemical. These same
     chemicals and actions of chemicals can influence
     carcinogenesis as either complete or "incomplete"
     carcinogens. A low dose of a mutagen could initiate or
     mutate a stem cell of a given tissue. If that initiated
     cell, which is held "in check" by surrounding and
     communicating normal cells, is allowed to escape the
     "metabolic cooperation" of the communication normal cell,
     then it can conceivably clonally expand itself. During that
     clonal expansion, additional genetic changes could occur
     which "stabilize" the neoplastic phenotype which resists the
     suppressive influence of normal tissue. Cytotoxic doses of
     the mutagen itself could act as an "indirect" promoter by
     allowing surviving, but initiated, cells to repopulate the
     dead tissue (i.e., compensatory hyperplasia) and therefore it
     would be considered a "complete" carcinogen. Alternatively,
     removal of cells by non-genotoxic chemicals, surgery, or
     normal cell death could act as a "promoter" to the surviving
     initiated stem cell. Lastly, exposure of initiated tissue to
     non-cytotoxic, but intercellular communication, inhibitors
     could cause promotion of the initiated cell to lead to
     cancer.

intercellular communication are both teratogens and tumor
promoters (69). Several interesting chemicals, 2,4,5,2',4',5'-
hexabromobiphenyl and 3,4,5,2',4',5'-hexabromobiphenyl, are tumor
promoters for rat liver but by different mechanisms (24,25). The
2,4,5,2',4',5'-HBB is relatively noncytotoxic, whereas the
3,4,5,3',4',5'-HBB is very cytotoxic (72). The 2,4,5,2',4',5'-
HBB, which inhibits metabolic cooperation without being
noncytotoxic, is a powerful tumor promoter at doses which do not
cause liver damage (72). The 3,4,5,3',4',5'-HBB does not act as a
tumor promoter at non-cytotoxic doses, where it also does not
directly inhibit intercellular communication (72). However, at
cytotoxic concentrations in the rat liver, it is a weak-to-
moderate tumor promoter, presumably because it acts as an inducer
of compensatory hyperplasia (24).

If gap-junction structure/function is disrupted by tumor
promoters (80), then the observation that certain anti-tumor
promoters, such as retinoic acid, can increase the number of gap
junctions (10) in certain cells would support the interconnection
between carcinogenesis and teratogenesis. Retinoic acid has also
been shown to be teratogenic (42). A possible explanation is that
increases in gap junctions are as teratogenic as decreases in gap
junctions.

Viruses can, of course, kill cells. They are known
teratogens and potential "carcinogens". The src gene can induce
the transformed phenotype in certain cells. Src transformed cells
have disrupted intercellular communication (61). The src gene
product appears to have been demonstrated to have protein kinase
activity (6). The tumor promoter, 12-tetradecanoyl phorbol-13-
acetate, has a membrane receptor which appears to be a protein
kinase (50). All of these observations again would be consistent
with the hypothesis that viruses could be both teratogens and
tumor promoters by cytotoxic or non-cytotoxic inhibition of
intercellular communication (68).

REFERENCES

1. Amed, F.E., and R.B. Setlow (1978) Excision repair in ataxia
   telangiectasia, Fanconi's anemia, Cockayne syndrome and
   Bloom's syndrome after treatment with ultraviolet radiation
   and N-acetoxy-2-acetylamino-fluorene. Biochim. Biophys. Acta
   521:805-817.
2. Ashley, D.J. (1969) The two "hit" and multiple "hit" theories
   of carcinogenesis. Br. J. Cancer 23:313-328.
3. Berenblum, I., and P. Shubik (1947) A new, quantitative
   approach to the study of the stages of chemical
   carcinogenesis in the mouse's skin. Br. J. Cancer 1:383-391.

4. Bonaiti-Pellie, C., M.L., Briard-Guillemot, J. Feingold, and
    J. Frezal (1975) Associated congenital malformations in
    retinoblastoma. Clinical Genetics 7:37-39.
5. Cairns, J. (1981) The origin of human cancers. Nature
    289:353-357.
6. Collet, M.S., and R.L. Erikson (1978) Protein kinase activity
    associated with the avian sarcoma virus Src gene product.
    Proc. Natl. Acad. Sci. USA 75:2021-2024.
7. Corsaro, C.M., and B.R. Migeon (1977) Comparison of contact-
    mediated communication in normal and transformed human cells
    in culture. Proc. Natl. Acad. Sci. USA 74:4476-4480.
8. Diamond, L., T.G. O'Brien, and G. Rovera (1978) Tumor
    promoters: Effects on proliferation and differentiation of
    cells in culture. Life Science 23:1979-1988.
9. Edelman, G.M. (1983) Cell adhesion molecules. Science
    219:450-457.
10. Elias, P.M., S. Grayson, I.M. Caldwell, and N.S. McNutt (1980)
    Gap junction proliferation in retinoic acid-treated human
    basal cell carcinoma. Lab. Invest. 42:469-474.
11. Emerit, I., and P. Cerutti (1981) Clastogenic activity from
    Bloom's syndrome fibroblast cultures. Proc. Natl. Acad. Sci.
    USA 78:1868-1872.
12. Fialkow, P.J. (1976) Clonal origin of human tumors. Biochim.
    Biophys. Acta 458:384-421.
13. Flagg-Newton, J.L., G. Dahl, and W.R. Loewenstein (1981) Cell
    junction and cyclic AMP: 1. Upregulation of junctional and
    membrane permeability and junctional membrane particles by
    administration of cyclic nucleotide or phosphodiesterase
    inhibitor. J. Memb. Biol. 63:105-121.
14. Foulds, L. (1954) The experimental study of tumor progression:
    A review. Cancer Res. 14:327-339.
15. Freese, E. (1982) Use of cultured cells in the identification
    of potential teratogens. Teratogenesis, Carcinogenesis and
    Mutagenesis 2:355-360.
16. German, J., D. Bloom, and E. Passarge (1977) Bloom's syndrome,
    V. Surveillance for cancer in affected families. Clin.
    Genet. 12:162-168.
17. Hanigan, H.M., and H.C. Pitot (1982) Isolation of α-glutamyl
    transpeptidase positive hepatocytes during the early stages
    of hepatocarcinogenesis in the rat. Carcinogenesis
    3:1349-1354.
18. Hand, R., and J. German (1975) A retarded rate of DNA chain
    growth in Bloom's syndrome. Proc. Natl. Acad. Sci. USA
    72:758-762.
19. Harbison, R.D. (1978) Chemical-biological reactions common to
    teratogenesis and mutagenesis. Env. Health Persp. 24:87-100.
20. Hirata, F., and J. Axelrod (1980) Phospholipid methylation and
    biological signal transmission. Science 209:1082-1090.
21. Hiwasa, T., S. Fujimara, and S. Sakiyama (1982) Tumor
    promoters increase the synthesis of a 32,000 dalton protein

in BALB/c 3T3 cells. Proc. Natl. Acad. Sci. USA
79:1800–1804.

22. Houston, E.W., W.C. Levin, and S.E. Ritzmann (1964)
Endoreduplication in untreated early leukemia. Lancet
ii:496–497.

23. Huang, Y., C.C. Chang, and J.E. Trosko (1983) Aphidicolin
induces endoreduplication in Chinese hamster cells. Cancer
Res. 43:1361–1364.

24. Jensen, R.K., S.D. Sleight, S.D. Aust, J.I. Goodman, and
J.E. Trosko (in press) Hepatic tumor promotion by
3,3',4,4',5,5'-hexabromobiphenyl: The interrelationship
between toxicity, induction of microsomal drug metabolizing
enzymes and tumor promotion. J. Appl. Pharm. Toxicol.

25. Jensen, R.K., S.D. Sleight, J.I. Goodman, S.D. Aust, and
J.E. Trosko (1982) Polybrominated biphenyls as promoters in
experimental hepatocarcinogenesis in rats. Carcinogenesis
3:1183–1186.

26. Johnson, E.M. (1980) Screening for teratogenic potential: Are
we asking the proper question? Teratology 21:259.

27. Kano, Y., and Y. Fujiwara (1982) Higher induction of twin and
single sister chromatid exchanges by cross-linking agents in
Fanconi's anemia cells. Human Genet. 60:233–238.

28. Keijzer, W., N.G.J. Jaspers, P.J. Abrahams, A.M.R. Taylor,
C.F. Arlett, B. Zelle, H. Takebe, P.D.S. Kinmont, and
D. Bootsma (1979) A seventh complementation group in
excision-deficient xeroderma pigmentosum. Mutat. Res.
62:183–190.

29. Knudson, A.G. (1971) Mutation and cancer: Statistical study of
retinoblastoma. Proc. Natl. Acad. Sci. USA 68:820–823.

30. Knudson, A.G., Jr. (1977) Genetic predisposition to cancer, In
Origins of Human Cancer, H.H. Hiatt, J.D. Watson, and
J.A. Winsten, Eds., Cold Spring Harbor Laboratory, New York,
pp. 45–52.

31. Konze-Thomas, B., R.M. Hazard, V.M. Maher, and J.J. McCormick
(1982) Extent of excision repair before DNA synthesis
determines the mutagenic but not the lethal effect of UV-
radiation. Mut. Res. 94:421–434.

32. Koyabyashi, N., T. Furukawa, and T. Takatsu (1968) Congenital
anomalies in children with malignancy. Pediatr. Univ. Tokyo
16:31–37.

33. Kraemer, K.H. (1980) Oculo-cutaneous and internal neoplasms in
xeroderma pigmentosum: Implications for theories of
carcinogenesis, In Carcinogenesis: Fundamental Mechanisms and
Environmental Effects, B. Pullman, P.O.P. T'so, and
H. Gelboin, Eds., D. Reidel, Amsterdam, pp. 503–507.

34. Kunz, B.A. (1982) Genetic effects of deoxyribonucleotide pool
imbalances. Environ. Mut. 4:695–725.

35. Lechner, J.F., and H.E. Kaighn (1980) EGF growth promoting
activity is neutralized by phorbol esters. Cell Biology
Intern. Reports 4:23–28.

36. Li, F.P. (1978) Host factors in childhood cancers. Semin. Oncol. 5:17-23.

37. Little, J.B. (1977) Radiation carcinogenesis in vitro: Implications for mechanisms, In Origins of Human Cancer, H.H. Hiatt, J.D. Watson, J.A. Winsten, Eds., Cold Spring Harbor, New York, pp. 923-939.

38. Liu, P.K., C.C. Chang, J.E. Trosko, D.K. Dube, G.M. Martin, and L.A. Loeb (1983) Mammalian mutator mutant with an aphidicolin-resistant DNA polymerase α. Proc. Natl. Acad. Sci. USA 80:797-801.

39. Loewenstein, W.R. (1979) Junctional intercellular communication and the control of growth. Biochim. Biophys. Acta 560:1-65.

40. Lynch, H.T., H. Guirgis, P. Lynch, J. Lynch, and R. Harris (1977) Familial cancer syndromes: A survey. Cancer 39:1867-1868.

41. MacDonald, C. (1982) Genetic complementation in hybrid cells derived from two metabolic cooperation defective mammalian cell lines. Exp. Cell Res. 738:303-310.

42. Maden, M. (1982) Vitamin A and pattern formation in the regenerating limb. Nature 295:672-675.

43. Maher, V.M., and J.J. McCormick (1976) Effect of DNA repair on the cytotoxicity and mutagenicity of UV-irradiation of chemical carcinogens in normal and xeroderma pigmentosum cells, In Biology of Radiation Carcinogenesis, J.M. Yahus, R.W. Tennent, and J.D. Regan, Eds., Raven Press, New York, pp. 129-145.

44. Maroun, L.E. (1980) Interferon action and chromosome 21 trisomy. J. Theoret. Biol. 86:603-606.

45. Miller, J.A., and E.C. Miller (1977) Ultimate chemical carcinogens as reactive mutagenic electrophiles, In Origins of Human Cancer, Book B, H.H. Hiatt, J.D. Watson, and J.A. Winsten, Eds., Cold Spring Harbor Laboratory, New York, pp. 605-627.

46. Miller, M.R., and D.N. Chinault (1982) Evidence that DNA polymerases α and β participate differentially in DNA repair synthesis induced by different agents. J. Biol. Chem. 257:46-49.

47. Miller, R.W. (1977) Relationship between human teratogens and carcinogens. J. Nat. Cancer Inst. 58:471-474.

48. Mulvihill, J.J. (1975) Congenital and genetic diseases, In Persons at High Risk of Cancer, J.F. Fraumeni, Ed., Academic Press, New York, pp. 3-37.

49. Nakamo, S., H. Yamagami, and R. Takaki (1979) Enhancement of excision-repair efficiency by conditioned medium from density-inhibited cultures in V79 Chinese hamster cells. Mut. Res. 62:369-381.

50. Niedel, J.E., L.J. Kuhn, and G.R. Vanderbark (1983) Phorbol diester receptor copurifies with protein kinase. Proc. Natl. Acad. Sci. USA 80:36-40.

51. Owens, A.H., D.S. Coffey, and S.B. Baylin, Eds. (1982) Tumor Cell Heterogeneity, Academic Press, New York.
52. Pitot, H.C., T. Goldsworthy, and S. Moran (1981) The natural history of carcinogenesis: Implications of experimental carcinogenesis in the genesis of human cancer. J. Supramolecular Struct. Cellul. Biochem. 17:133-146.
53. Potter, V.R. (1978) Hormonal induction of enzyme functions, cyclic AMP levels and AIB transport in Morris hepatomas and in normal liver systems, In Morris Hepatomas: Mechanisms of Regulation, H.P. Morris, and W.E. Criss, Eds., Plenum Press, New York, pp. 59-87.
54. Potter, V.R. (1980) Initiation and promotion in cancer formation: The importance of studies on intercellular communication. Yale J. Biol. Med. 53:367-384.
55. Purtilo, D.T., L. Paquin, and T. Gindhart (1978) Genetics of neoplasia-impact of ecogenetics on oncogenesis. Amer. J. Pathol. 91:609-681.
56. Reddy, E.P., R.K. Reynolds, E. Santos, and M. Barbacid (1982) A point mutation is responsible for the acquisition of transforming properties by the T24 human bladder carcinoma oncogene. Nature 300:149-152.
57. Roth, R.A., and D.J. Cassell (1983) Insulin receptor: Evidence that it is a protein kinase. Science 219:249-301.
58. Rowley, J.D. (1983) Human oncogene locations and chromosome aberrations. Nature 301:290-291.
59. Simons, J.W.I.M. (1979) Development of a liquid holding technique for the study of DNA repair in human diploid fibroblasts. Mut. Res. 59:273-283.
60. Spadari, S., and A. Weissbach (1974) The inter-relations between DNA synthesis and various DNA polymerase activities in synchronized HeLa cells. J. Mol. Biol. 86:11-20.
61. Steinberg, M., and V. Defendi (1981) Patterns of cell communication and differentiation in SV40 transformed human keratinocytes. J. Cell. Physiol. 109:153-159.
62. Tabin, C.J., S.M. Bradley, C.I. Bargmann, R.A. Weinberg, A.G. Papageorge, E.M. Scolnick, R. Dhar, D.R. Lowy, and E.H. Chang (1982) Mechanism of activation of a human oncogene. Nature 300:143-149.
63. Trosko, J.E. (1981) Cancer causation. Nature 290:356.
64. Trosko, J.E., and C.C. Chang (1978) Environmental carcinogenesis: An integrative model. Quart. Rev. Biol. 53:115-141.
65. Trosko, J.E., and C.C. Chang (1979) Chemical carcinogenesis as a consequence of alterations in the structure and function of DNA, In Chemical Carcinogens and DNA, Vol. II, CRC Press, Boca Raton, Florida, pp. 181-260.
66. Trosko, J.E., and C.C. Chang (1981) An integrative hypothesis linking cancer, diabetes and atherosclerosis: The role of mutations and epigenetic changes. Med. Hypotheses 6:455-468.

67. Trosko, J.E., and C.C. Chang (1981) The role of radiation and chemicals in the induction of mutations and epigenetic changes during carcinogenesis. Advances in Radiation Biology, Vol. 9, Academic Press, New York, pp. 1-36.
68. Trosko, J.E., C.C. Chang, and A. Medcalf (in press) Mechanisms of tumor promotion: Potential role of intercellular communication. Cancer Investigation.
69. Trosko, J.E., C.C. Chang, and M. Netzloff (1982) The role of inhibited cell-cell communication in teratogenesis. Teratogenesis, Carcinogenesis and Mutagenesis 2:31-45.
70. Trosko, J.E., C. Jone, and C.C. Chang (1983) The role of tumor promoters on phenotypic alterations affecting intercellular communication and tumorigenesis. New York Acad. Sci. 407:316-327.
71. Trosko, J.E., V.M. Riccardi, C.C. Chang, S.T. Warren, and M.H. Wade (in press) Genetic predispositions to initiation on promotion phases in human carcinogenesis, In Genetics, Biomarkers and Cancer, H.T. Lynch, and H.A. Guirgis, Eds., Van Nostrand, Reinhold, New York.
72. Tsushimoto, G., J.E. Trosko, C.C. Chang, and S.D. Aust (1982) Inhibition of metabolic cooperation in Chinese hamster V79 cells in culture by various polybrominated biphenyl (PBB) congeners. Carcinogenesis 3:181-186.
73. Warren, S.T., R.A. Schultz, C.C. Chang, M.H. Wade, and J.E. Trosko (1981) Elevated spontaneous mutation rate in Bloom's syndrome fibroblasts. Proc. Natl. Acad. Sci. USA 78:3133-3137.
74. Wawra, E., and I. Dolejs (1979) Evidences for the function of DNA polymerase β in unscheduled DNA synthesis. Nucleic Acids Res. 7:1675-1686.
75. Weinstein, I.B., L.S. Lee, P.B. Fisher, A. Mufson, and H. Yamasaki (1979) Action of phorbol esters in cell culture: Mimicry of transformation altered differentiation and effects on cell membranes. J. Supramolecular Struct. 2:194-208.
76. Welshimer, K., and M. Swift (1982) Congenital malformations and developmental disabilities in ataxia telangiectasia, Fanconi's anemia, and xeroderma pigmentosum families. Am. J. Hum. Genet. 34:781-793.
77. Weymouth, L.A., and L.A. Loeb (1978) Mutagenesis during in vitro DNA synthesis. Proc. Natl. Acad. Sci. USA 75:1924-1928.
78. Wilson, J.G. (1977) Current status of teratology, In Handbook of Teratology, J.G. Wilson and F.C. Fruser, Eds., Plenum Press, New York, pp. 47-74.
79. Yager, J.D., and R. Yager (1980) Oral contraceptive steroids as promoters of hepatocarcinogenesis in female Sprague-Dawley rats. Cancer Res. 40:3680-3685.
80. Yancey, S.B., J.E. Edens, C.C. Chang, and J.P. Revel (1982) Decreased incidence of gap junctions between Chinese hamster

V79 cells upon exposure to the promoter, TPA. _Expt._ _Cell_
Res. 139:329-340.

81. Yotti, L.P., C.C. Chang, and J.E. Trosko (1979) Elimination of
    metabolic cooperation in Chinese hamster cells by a tumor
    promoter. _Science_ 206:1089-1091.

82. Yotti, L.P., T.W. Glover, J.E. Trosko, and D.J. Segal (1980)
    Comparative study of X-ray and UV-induced cytotoxicity, DNA
    repair and mutagenesis in Down's syndrome and normal
    fibroblasts. _Pediat._ _Res._ 14:88-92.

83. Yuspa, S.H., T. Ben, H. Hennings, and U. Lichti (1982)
    Divergent responses in epidermal basal cells exposed to the
    tumor promoter 12-0-tetradecanoylphorbol-13-acetate. _Cancer_
    Res. 42:2344-2349.

GERM CELL TOXICITY: SIGNIFICANCE IN GENETIC AND

FERTILITY EFFECTS OF RADIATION AND CHEMICALS[1]

E. F. Oakberg

Biology Division
Oak Ridge National Laboratory
Oak Ridge, TN 37830

SUMMARY

The primordial germ cells originate in the region of the
caudal end of the primitive streak, root of the allantois, and
yolk sac splanchnopleure, and migrate to the gonadal ridges where
they divide to form the oogonia of the female and gonocytes of the
male. In the female, the transition to oocytes occurs in utero,
and the female mammal is born with a finite number of oocytes that
cannot be replaced. By contrast, the gonocytes of the male
initiate divisions soon after birth to form the spermatogonial
stem cells, which persist throughout reproductive life of the male
and are capable of regenerating the seminiferous epithelium after
injury. As a result of these basic differences in gametogenesis,
the response of the male and female to radiation and chemicals is
different. Any loss of oocytes in the female cannot be replaced,
and if severe enough, will result in a shortening of the
reproductive span. In the male, a temporary sterile period may be
induced owing to destruction of the differentiating spermatogonia,
but the stem cells are the most resistant spermatogonial type, are
capable of repopulating the seminiferous epithelium, and fertility
usually returns. The response of both the male and female changes

[1] Research sponsored by the Office of Health and Environmental
Research, U.S. Department of Energy under contract W-7405-eng-26
with the Union Carbide Corporation.

549

with development of the embryonic to the adult gonad, and with
differentiation and maturation in the adult. The primordial germ
cells, early oocytes, and differentiating spermatogonia of the
adult male are unusually sensitive to the cytotoxic action of
noxious agents, but each agent elicits a specific response owing
to the intricate biochemical and physiological changes associated
with development and maturation of the gametes. The relationship
of germ cell killing to fertility is direct, and long-term
fertility effects can be predicted from histological analysis of
the gonads. The relationship to genetic effects, on the other
hand, is indirect, and acts primarily by limiting the cell stages
available for testing, by affecting the distribution of
mitotically active stem cells among the different stages of the
mitotic cycle, and thereby changing both the type and frequency of
genetic effects observed.

INTRODUCTION

     The level of our understanding of the effect of radiation,
chemicals, pollutants, and noxious agents in general on the
gonads, and the relationship of these effects to fertility and to
transmission of genetic damage is dependent upon our understanding
of the normal process of germ cell development. Some of the
problems to be discussed here were phrased over a century ago, but
it is only within the last thirty years that the direct lineage
between primordial germ cells and the definitive gametes has been
firmly established, and significant progress in describing
spermatogonial stem cell renewal has come only within the last
twelve years. On the basis of these advances, much progress has
been made in characterizing the response of the different stages
of gametogenesis, and in applying these data to an understanding
of the hazards due to fertility and possible genetic damage due to
radiation and chemical exposure. Many questions concerning normal
germ cell development remain, however, and we also are woefully
ignorant of the basic mechanisms involved in mutagenic and
cytotoxic action of radiation and chemicals.

Normal Gametogenesis: Male

     The primordial germ cells originate in the region of the
caudal end of the primitive streak, root of the allantoic
mesoderm, and yolk sac splanchnopleure. From there, they migrate
to the germinal ridges by route of the dorsal mesentery
(12,19,60,99). Mitotic division occurs both during migration of
the germ cells and after they reach the germinal ridges. The
initial stages are the same in both sexes, but diverge at about
12-13 days in the mouse, when testis and ovary can be
distinguished morphologically (60). Mitotic activity of the
gonocytes decreases at 14-18 days in the male and only rare cells

are in division at birth (3,60). DNA synthesis begins in the first few hours after birth (75), the definitive stem cell population and differentiating spermatogonia appear, and the entire process of spermatogenesis is initiated (75,93). As spermatogonia differentiate into spermatocytes, their numbers are replenished by division and differentiation of the stem cells through the process termed stem cell renewal.

The problem of identification of the stem cell and description of stem cell renewal has had a long history, for it was recognized by Benda over a century ago that the continued production of spermatozoa over the reproductive life of a male mammal required continuous renewal of the spermatogonial population (4). The spermatogonia were first described by LaValette (51), and Regaud (84) was the first to recognize that the A spermatogonia were the most primitive cell type of the

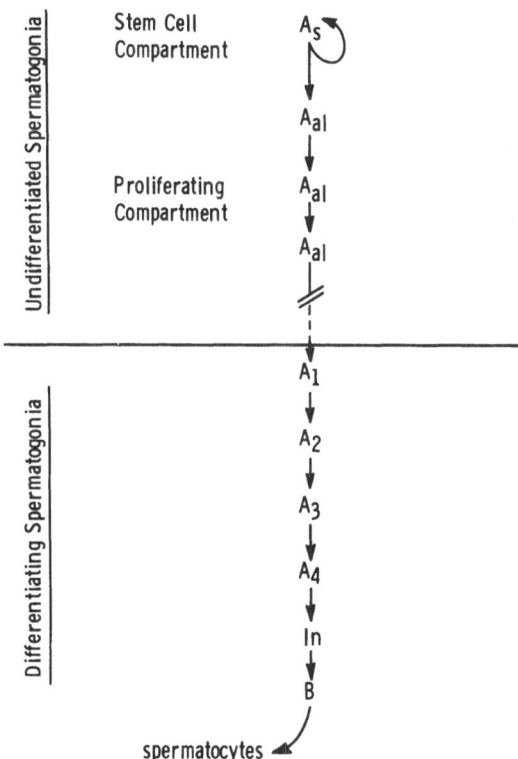

Fig. 1. Schematic representation of spermatogonial stem cell renewal (79).

seminiferous epithelium, a conclusion which was supported by subsequent experiments with radiation (85). For excellent reviews of the literature prior to 1965, see Hannah-Alava (40), Clermont (13) and Roosen-Runge (87). Classification of stages of the cycle of the seminiferous epithelium by the technique of Leblond and Clermont (52) led to the identification of 4 classes of type A spermatogonia by Monesi (61), and a 5th class was identified by the study of segments of seminiferous tubules mounted in toto (14). A meticulous study of tubule whole mounts, in the rat (42), a similar study in the mouse (86) and long-term ^{3}H-thymidine labeling in the mouse (71) demonstrated that this 5th class of spermatogonia, given the term A_o by Clermont and Bustos-Obregon (14), actually are the active stem cells of the testis (42,46,71,86), and maintain their own numbers by the process of division and also replenish the generations of differentiating spermatogonia. These cells have been given the term A_s (A stem, A single) to describe both their stem cell role and their occurrence as single, isolated cells in the tubule (79).

In whole-mounts, the interphase nucleus of the stem cells is small (Figs. 2,4), ovoid or spherical in shape, with heterogeneous finely granular chromatin and an inconspicuous nucleolus (42). They sometimes have an irregular, lobed nucleus suggestive of cells with ameboid properties. In sections they have an oval, darkly-staining nucleus with granular chromatin and an indistinct nucleolus (71) (Figs. 5,6). A study of their cell-cycle properties revealed at least two populations of A_s spermatogonia in the rat, one with a cell-cycle time of about 50 hrs, and a second, long-cycling population that could have a cell-cycle time as long as 13 days (43). Labeling with ^{3}H-thymidine demonstrated that these cells eventually divide and form labeled daughter cells. A comparable analysis has not been made in the mouse, but heavily labeled A_s (Fig. 8) spermatogonia are frequently observed 207 hrs (one cycle of the seminiferous epithelium) after ^{3}H-TdR labeling, and these cells also divide (Fig. 9) to form both labeled A_s and A_{pr} (Fig. 10) spermatogonia (71,73,79). The above observations, both in the rat and mouse, are positive proof of both the long-cycling property of A_s spermatogonia and their active stem-cell role. The initial step in differentiation is not known, but the formation of a pair of cells connected by a cytoplasmic bridge appears to be an irreversible step toward the production of spermatozoa (45). However, since both slow and fast-cycling A_s, A_{pr}, and A_{al} spermatogonia all have the same nuclear morphology, or positive proof that only the long-cycling A_s spermatogonia have a stem-cell role has not yet been possible.

On the basis of the above studies on the stem cells, Huckins (42,44) and Oakberg and Huckins (79) have proposed the model of stem cell renewal shown in Figure 1. The A_s (Figs. 2,5) spermatogonia can divide to form more A_s cells, or cytokinesis can be incomplete, forming a pair of cells (A_{pr}) (Figs. 3,4,10) connected by a cytoplasmic bridge. With subsequent division, groups of connected A_{al} cells 2^n in number are formed, still with the nuclear morphology of the A_{pr} and A_s. Collectively, these all are referred to as "undifferentiated" type A spermatogonia because of their similar nuclear morphology (44), but formation of the pair probably is the initial step in an irreversible developmental sequence. The A_{al} in stages 5, 6, and 1 transform into A_1 (Figs. 11,16,17) spermatogonia which divide at the end of stage 1 to form A_2 (Figs. 12,18), with subsequent divisions giving rise to A_3 (Figs. 13,19), A_4 (Figs. 14,20) In (Figs. 15,21) and B (Figs. 16,22) spermatogonia. The B spermatogonia divide to form preleptotene spermatocytes (Fig. 11) which enter DNA synthesis in stage 1 and begin the long meiotic prophase which culminates in the meiotic divisions in stage 3 with formation of spermatids. Although the above description is specific for the rat and mouse, the basic principles apply to all mammals, and probably most members of the animal kingdom (40), with species specific morphology of the type A spermatogonia and late spermatids. B spermatogonia, primary spermatocytes, the meiotic divisions, and early spermatid stages have similar morphology in all mammals.

The development of the different cell types is closely synchronized, and gives rise to the characteristic cellular associations represented by regions of the tubules which are in different stages of development. The complete series of tubule stages constitutes one cycle of the seminiferous epithelium. On the basis of the acrosome development in spermatids, Leblond and Clermont (52) described 14 stages in the cycle of the seminiferous epithelium of the rat, and Oakberg (66) described 12 in the mouse. Arrangement of the stages on the basis of acrosome development automatically reveals the developmental sequence of differentiating spermatogonia and spermatocytes. Not only the cell type, but the number of cells in any given stage can be predicted, and comparison of counts in control and treated animals is used in assessing the effects of cytotoxic agents (65,66,73,78,79). However, only 6 stages of the cycle, identifiable on the basis of the 6 generations of differentiating spermatogonia, can be recognized in whole mounts. For this reason, Oakberg and Huckins (79) proposed the classification given in Table 1. The spermatogonia characteristic of these stages are represented in Figures 11-16 for whole-mounts, and in Figures 17-22 for sections.

The concept of a spermatogenic wave progressing along the tubule was proposed early in the study of the testis (4,18,84), but irregularities such as reversal of the progression of stages make it more practical to think of the succession of stages in the dimension of time, i.e., a given spot in the tubule will pass through all stages in the cycle of the seminiferous epithelium

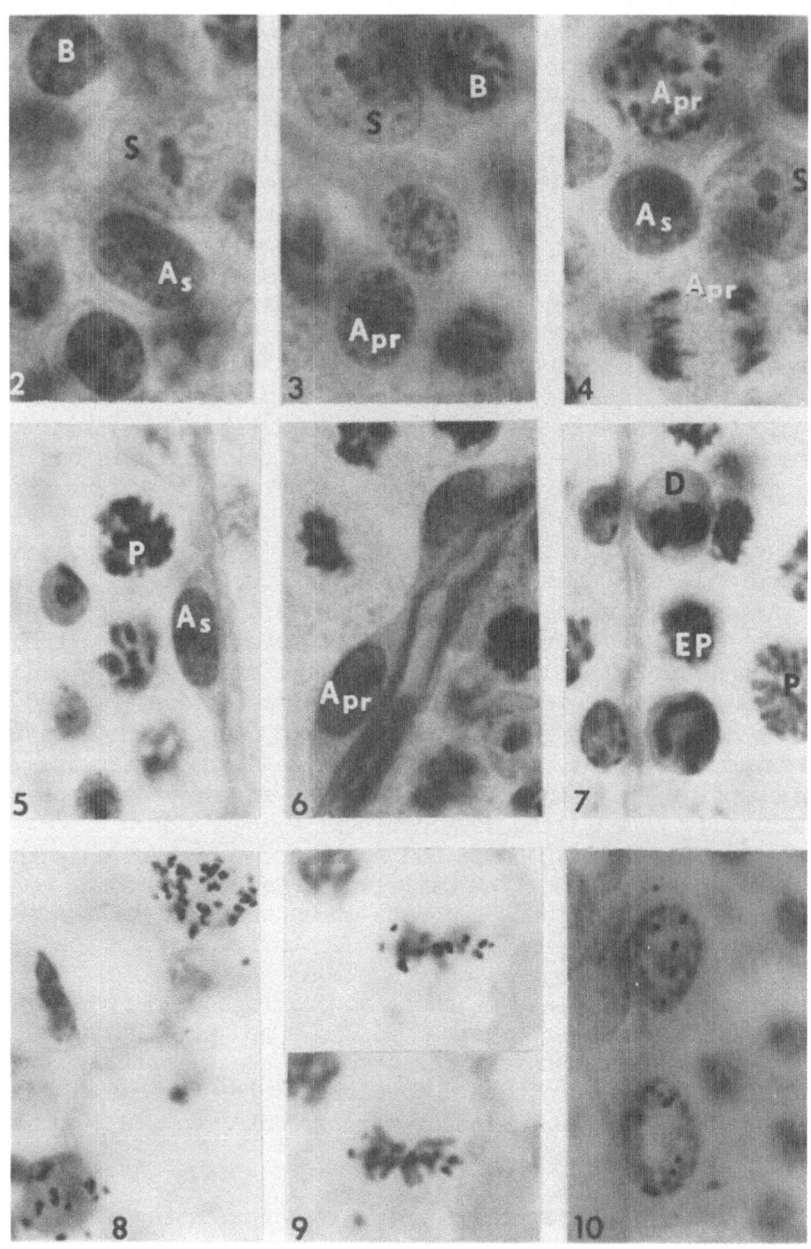

during a time interval which is a constant for the species. This
is E207 hours in the mouse, 13 days in the rat (depending upon
strain) and 16 days in man. Four complete cycles of the
seminiferous epithelium are required for development of the A_1
spermatogonia of mouse and rat into mature spermatids in the
testis (52,66). At least one additional cycle is required for
development of A_s into A_1 spermatogonia, so the duration of
spermatogenesis is E43 days in the mouse, 65 days in the rat, and
E90 days in man. Study of the cellular associations after
irradiation (67) progression of ^{3}H-thymidine labeled spermatocytes
in irradiated testes (20), and after 6-mercaptopurine injection
(76), and endocrinological studies (15) all indicate that the rate
of spermatogenesis is a species constant that is not affected by
experimental procedures. There is evidence, however, that
spermatogonial development proceeds more rapidly in the juvenile
animal (41).

Once the cellular associations were described and the
duration of the cycle of the seminiferous epithelium determined,
it was obvious that information on the duration of each stage of
the cycle would allow observation of effects on selected cell
types at any desired stage of subsequent development. For this
purpose, the 207 hr duration of the cycle was apportioned among
the 12 stages on the basis of the frequency distribution of a
sample of 200 randomly selected tubule cross sections from each of
12 control mice (67), and in a later study, of 16 control mice for
the frequency distribution of the six stages described by Oakberg
and Huckins (79) (Table 1). From this, the time required for each
cell stage to develop into mature spermatids and be released from
the tubule was calculated (Table 2). The seven days required for
transit from testis to ejaculate (78) was added to these estimates
to give the time at which specific stages are available for
fertilization. It is clear that all stages except the stem cell

----

Fig. 2-10. Stem and A_{pr} spermatogonia of the mouse, and labeled
    spermatogonia 207 hrs after ^{3}H-TdR injection. Figs. 2-4 from
    whole mount, 5-10 from sections. Fig. 2, A_s spermatogonium
    in stage 7 among B gonia; Fig. 3, A_{pr} spermatogonium in stage
    6 with B gonia; Fig. 4, one A_s interphase and two A_{pr}
    spermatogonia in division; Fig. 5, A_s spermatogonium in
    tubule section; Fig. 6, A_{pr} spermatogonia; Fig. 7, 2
    degenerating spermatogonia; Fig. 8, two labeled
    spermatogonia, 1 heavy; Fig. 9, labeled metaphase; Fig. 10,
    labeled A_{pr} spermatogonia. Key to symbols for Figs. 2-22.
    A_s, A_s-spermatogonia; A_1-A_4, A_1-A_4-spermatogonia; In, In-
    spermatogonia; B, B-spermatogonia; PL-preleptotene
    spermatocyte; EP-early pachytene spermatocyte, P, pachytene
    spermatocyte; S-II, secondary spermatocyte; S, Sertoli cell;
    Sptd-spermatid.

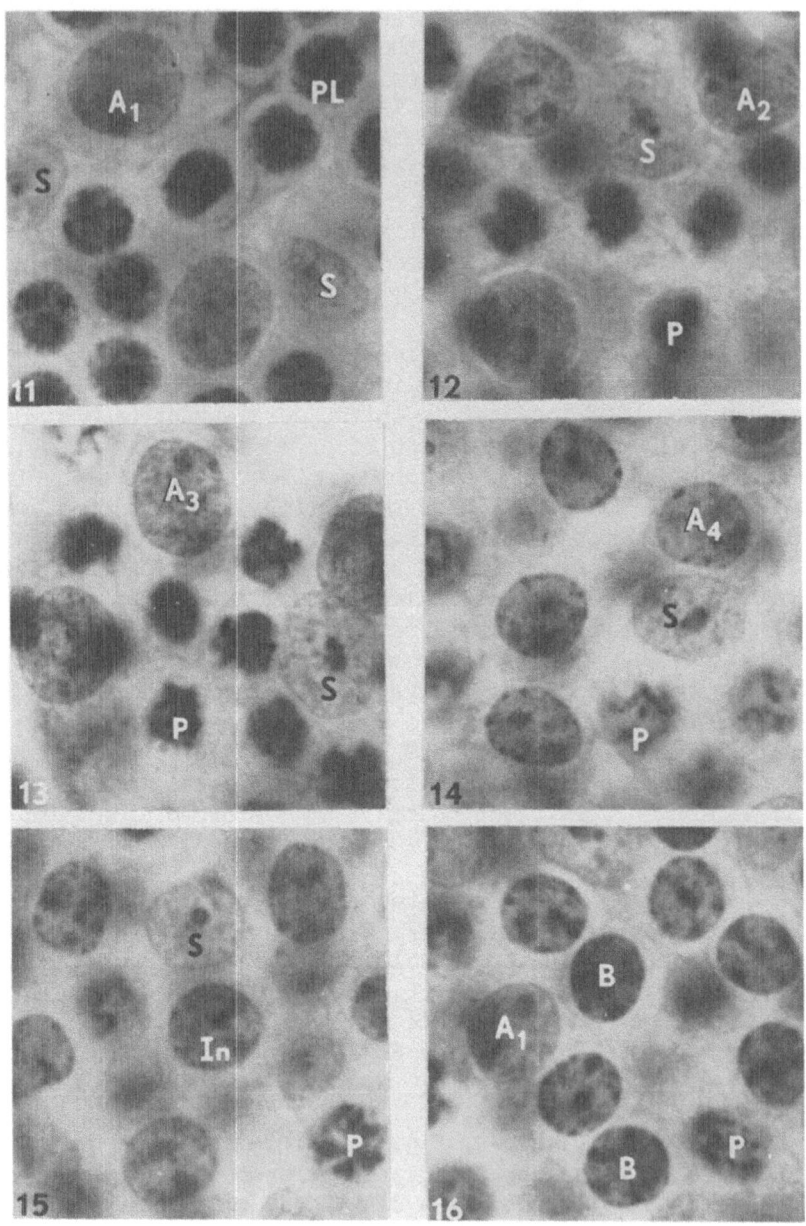

have short lives in comparison to the total reproductive span, and
the stem cell therefore is the single most important cell type in
both reproductive and genetic effects of radiation and chemicals
in males.

Radiation Response

Prenatal and neonatal. The immature rat testis has three
periods of high sensitivity to radiation-induced sterility. The
first, in the 13.5-17.5 day embryo, is associated with high
mitotic activity of the primordial germ cells, and is common to
both sexes (2,55). The second and most sensitive period in the
male occurs during the period of low mitotic activity of the germ
cells just before birth until 2 days postpartum in the rat (2).
The induced sterility can be traced to killing of germ cells and
resulting deficiency of spermatogonia in the adult. Resistance
increases as the first type A spermatogonia appear, but a third
period of sensitivity occurs at 17 days. A similar response
occurs in the mouse, but at slightly younger ages. In contrast to
the rat, the newborn mouse is not sterilized by doses of several
hundred rads (94). Comparable information is not available for
other species, including man, but similar patterns of response are
likely to occur and particular care should be taken to avoid
radiation exposure of the fetal, neonatal, and juvenile testis.

Adult. Males are initially fertile after acute radiation
exposure owing to continued development of cells irradiated as
spermatozoa, spermatids, and spermatocytes (72,85). A period of
temporary infertility then ensues with doses of 20-300 R depending
upon genetic strain and species, and fertility usually returns as
surviving stem cells repopulate the seminiferous epithelium. The
dose of 20 R is for man, and is not based on demonstrated lack of
fertility, but is a result anticipated from the known effect of
small acute radiation exposure on sperm count in men (88). The
sterile period is of longer duration after high exposures owing to
killing of spermatocytes, extensive gonial killing, induction of
very high levels of dominant lethals in spermatids, and slow
repopulation of the seminiferous epithelium when the number of
surviving spermatogonia is low (72).

Development of spermatids is not affected by doses as high as
1500 R in the mouse (78), 32,000 R or more is required to affect
the motility of human sperm, and the fertilizing capacity of

Fig. 11-16. Differentiating spermatogonia of the mouse as seen in
tubule whole mounts. A_1, A_2, A_3, A_4, In and B mark
respective classes of differentiating spermatogonia; PL,
preleptotene spermatocytes; P, pachytene spermatocytes; S,
Sertoli cells.

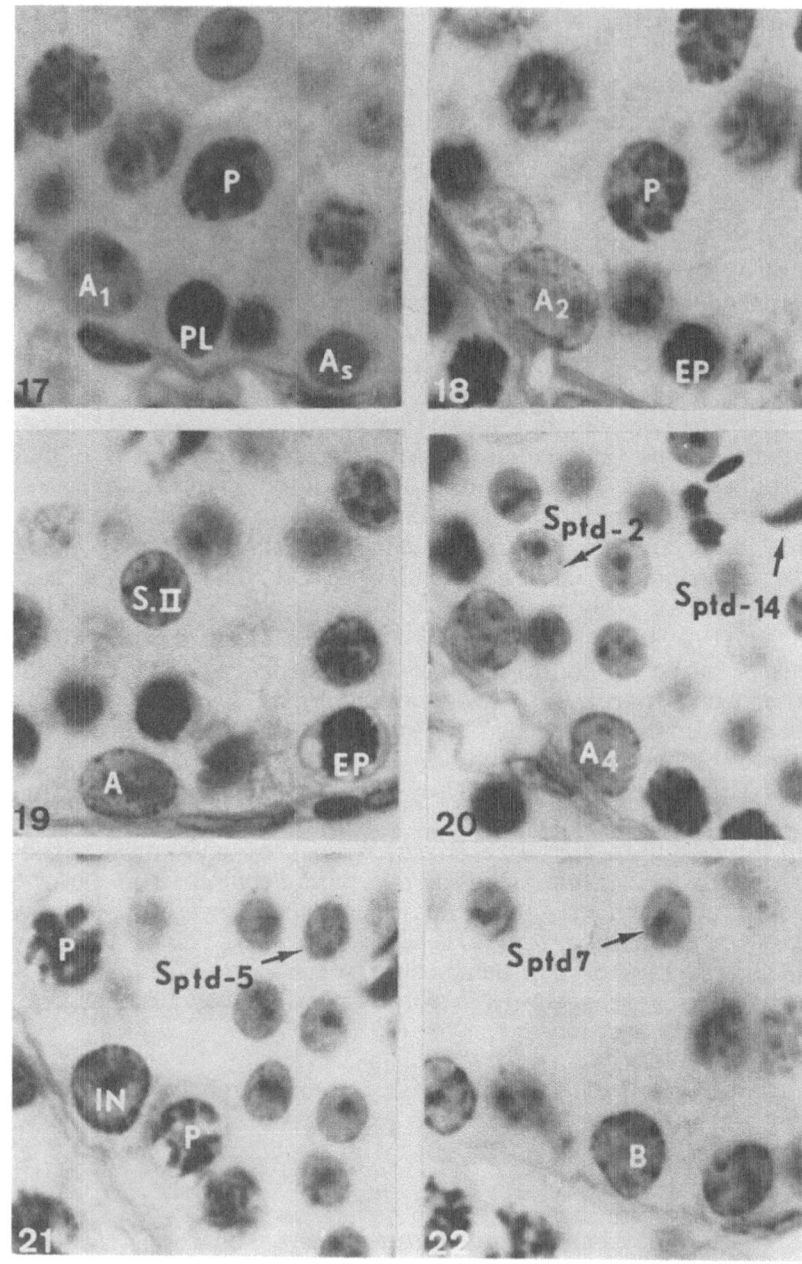

rabbit sperm is normal after an exposure of 65,000 R (11). Thus
spermatids and spermatozoa are extremely resistant to direct
effects of radiation on their development and function. They are
sensitive, however, to the induction by radiation and certain
chemical mutagens of presumed point mutations, and to chromosome
breakage with resultant high frequency of dominant lethality and
chromosome rearrangements (7,31).

Chromosome breakage in primary spermatocytes of the mouse
shows the same changes in sensitivity with meiotic stage that have
been observed in other species, including plants (77,98).
Chromosome breakage is highest in diakinesis-metaphase I, and
lowest in preleptotene and leptotene. Sensitivity to cell-
killing, as measured by production of spermatids, shows the
reverse pattern, with diakinesis-metaphase I being the most
resistant and preleptotene the most sensitive. As a result of
induced chromosome breakage, spermatids derived from spermatocytes
exposed to 100 R or more show an abnormal size distribution
ranging from micronuclei to obviously polyploid nuclei owing to
various levels of aneuploidy and heteroploidy (78,85).

The induction of a sterile period and subsequent recovery of
fertility both arise from the spermatogonial response. This was
described by Regaud and Lacassagne (85), but analysis of the
response of the different spermatogonial classes became possible
only after description of the cycle of the seminiferous epithelium
and timing of germ cell development were available. It has now
been demonstrated that the primary radiation response of the
rapidly cycling A_s, A_{pr}, A_{al} and all differentiating spermatogonia
is cell death (61,62,65,72,73). Observation of early intervals
revealed a high incidence of necrotic spermatogonia at 12-18 hrs
(62,65). Cell division is rare at this time, and study of early
stages of degeneration indicated late interphase or early prophase
as the stages where degeneration occurs (62,65). This varies
somewhat for the different generations of spermatogonia (62).
With early spermatogonia A_1, death is delayed until the cells
approach their first post-irradiation division several days later,
and the long-cycling stem cells continue to degenerate up to 7 or
8 days after irradiation. In the meantime, division of surviving
cells has begun, and the concurrence of continued degeneration and
repopulation makes determination of the minimum numbers of

Figs. 17-22. Differentiating spermatogonia of the mouse as seen
    in tubule cross sections. Sptd-2, 5, 6, and 14, steps 2, 5,
    6, and 14 of spermiogenesis. A_1, A_2, A_3, A_4, In and B mark
    respective classes of differentiating spermatogonia; A_s,
    spermatogonial stem cell; EP, early pachytene, P, pachytene;
    Sptd-2, Sptd-5, Sptd-7, Sptd-14, different steps of
    spermiogenesis.

Table 1. Cellular associations at each stage of the cycle of the seminiferous epithelium, and duration of each stage in hours.

Stage of Cycle		1	2	3	4	5	6
spermatogonia	stem cells (A_s)	A_s	A_s	A_s	A_s	A_s	A_s
	proliferating	A_{pr},A_{al}	A_{pr},A_{al}	A_{pr},A_{al}	A_{pr},A_{al}	A_{pr},A_{al}	A_{pr},A_{al}
	differentiating	A_1	A_2	A_3	A_4	In	B
spermatocytes	1st layer	PL-L	L-P	P	P	P	P
	2nd layer	P	P	Dip,Dia,MI,SII,M-II			P
spermatids	1st layer	7,8,9	10,11	11,12,1	1,2	3,4	5,6
	2nd layer	16	16	13	13,14	14,15	15
frequency of stage (percent)*		31	16	14	15	11	13
duration in hrs		64	33	29	31	23	27

PL   pre-leptotene
P   pachytene
Dip   diplotene
Dia   diakinesis   } spermatocytes
M-I   First meiotic division
M-II   Second meiotic division
SII   Secondary spermatocyte

1-16   Spermatid stages (from Oakberg, 1956a)

*Based on a sample of 3,200 tubules; 200 from each of 16 mice.

Table 2. Duration of each cell type in hours, time required to develop into mature spermatids, and time required to reach the ejaculate in the mouse.

Cell Type	Life-span (days)	Interval to release of mature spermatids from testis (days)	Interval to appearance of mature spermatozoa in the ejaculate (days)
Spermatogonia			
A_s	total reproductive span	~42*	~49*
A_{pr}, A_{al}	8.6*	35-42*	42-49*
A_1	2.7	33-35	40-42
A_2	1.4	32-33	39-40
A_3	1.2	31-32	38-39
A_4	1.3	30-31	37-38
In	1.0	29-28	36-37
B	1.0	28-29	35-36
Spermatocytes**			
preleptotene	1.3	26-28	33-35
leptotene	1.3	25-26	32-33
pachytene	~9.0	16-25	23-32
diplotene	0.9	15-16	22-23
meiotic divisions	0.8	14-15	21-22
spermatids	14	0-14	7-21
Spermatozoa			0-7

*The exact times required for transformation of A_s into spermatogonia A_{pr} and A_{al}, and for the A_{pr} to form A_1 spermatogonia, is not known, but it is assumed that at least one cycle of the seminiferous epithelium is required.

**Zygotene has been omitted from Tables 1 and 2 because recent data indicates that the synaptonemal complex is present at the time of preleptotene DNA synthesis (37,38). Since contraction of the chromosomes is gradual, the division between leptotene and pachytene is arbitrary.

surviving cells impossible. From the above, it follows that the
stem cell is the most important cell type in radiation response.
Other spermatogonia have short life spans, and owing to their high
sensitivity to cytotoxic agents also are likely to be eliminated
by cell death.

The stem cells surviving radiation doses of 150 R or more are
almost exclusively from the long-cycling compartment (46,79).
Furthermore, cells in DNA synthesis are sensitive (73), and
selective killing restricts the survivors to a progressively
narrowing segment of the cell cycle with increasing dose. As a
result, the mitotic index is initially reduced, and initiation of
recovery is delayed at high doses. The pattern of cell death in
stem cells is similar for that of A_1 spermatogonia, and as a
result, minimum cell count is not reached until 7 or 8 days after
treatment (Table 3). This is further indication that the cell
cycle time of the stem cell approximates one cycle of the
seminiferous epithelium (207 hrs). It is of note that the effect
of agents that induce slight but still definite reduction in stem
cell numbers is detectable only at 207 hrs after treatment.

The survival curve for undifferentiated spermatogonia is
smooth throughout the 100-1000 R range (73,90) (Fig. 23). The
representation of different stages of the mitotic cycle among the
survivors, however, is dependent upon both dose and dose-
fractionation. Labeling with ^{3}H-TdR 24 hrs prior to irradiation
suggests comparable percentage labeling in controls and
spermatogonia present 207 hrs after doses of 100-600 R. The
number of labeled cells is significantly reduced below control at
1000 R, and enhanced if the dose is given in two 500 R fractions

Table 3. Survival of A_s spermatogonia 120 and 207 hrs
after exposure to 100-1000 R x-rays.

Dose - R	Percentage Survival	
	120 h	207 h
0 (c)	-	-
100		70.7
300	22.9	47.6
500	11.1	21.2
600	13.6	13.2
1000	2.6	1.6
500 + 500	5.0	2.1

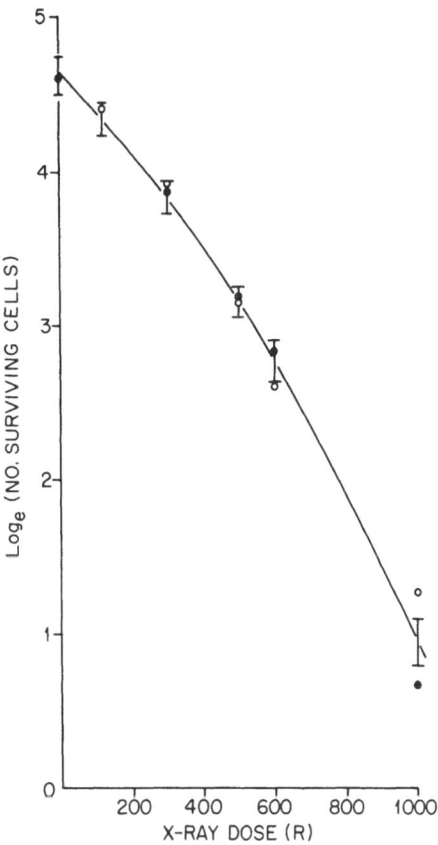

Fig. 23. Survival of A$_s$ spermatogonia 207 hrs after irradiation. ●, Expt. 1; ○, Expt. 2 [from Oakberg (72)].

24 hrs apart. Thus at the highest doses used, distribution of survivors among stages of the mitotic cycle is changed, and may be a factor in the reduced number of mutations/roentgen/locus observed with a single exposure of 1000 R, and the enhanced rate observed with the 500 + 500 R dose (73). Such relationships must be accepted with caution, however, for the number of mutations is so low that it is impossible to relate them to a specific stage of the stem cell cycle. The results of labeling prior to irradiation also have significant bearing on models of stem cell renewal and repopulation after radiation-induced depletion of spermatogonia (79).

The fact that the cells surviving radiation are capable of repopulating the seminiferous epithelium is clear demonstration of their stem cell role; the fact that many of them are labeled if

^{3}H-thymidine is given before irradiation is irrefutable evidence
that the stem cell of the testis is in continuous cycle (73,79).
Furthermore, the data yield information on mitotic activity of the
stem cells during repopulation. Various authors have suggested
that stem cells enter a period of rapid division in order to
replace the depleted spermatogonial populations. First of all, if
this were so, one would not expect to see labeled cells 207
(Figs. 8-10) and 414 hrs after labeling (73), for label would be
diluted beyond recognition by rapid divisions. Also, there is no
effect of dose on percent labeled cells over the 100-600 R range,
yet few stem cells are killed by 100, and many are killed by 600
R. Finally, the relative positions of all doses except control
and 500 + 500 R is maintained from 207 to 414 hrs (73). This
suggests that stem cell cycle properties have not been affected in
the 0-17 day interval, yet repopulation already has begun. All of
the above considerations suggest that radiation exposure does not
affect the cell cycle properties of A_s spermatogonia.

Mitotic activity of stem cells. The above conclusion is in
agreement with recent data demonstrating the lack of effect of
radiation on the circadian rhythm of cells in the corneal
epithelium (89) and bone marrow (64). The effect of fractionation
interval on frequency of chromosome breakage, length of sterile
period, and frequency of specific locus mutations is more likely
the result of partial synchronization of the stem cells by killing
of sensitive stages of the mitotic cycle. This is supported by
the fact that 75% of the cells are labeled by a single injection
of ^{3}H-TdR 3 days after 150 R, and 5 days after 300 R. This
suggests that mitotic activity has an oscillating pattern with
time after exposure, and the effect of fractionation interval is a
reflection of the response of the preponderant stage at any given
interval. Eventually these oscillations are damped as the
synchrony is lost and the population returns to a normal
distribution among the mitotic stages. That the synchrony is lost
is proof that cell cycle times are variable, and the division of
the A_s spermatogonia into only slow- and fast-cycling classes
probably is an oversimplification. Differences observed with
short fractionation intervals also may be influenced by either
damage of repair mechanisms of the cell, or by failure of repair
before the second dose is given.

A possible complication in the study of fractionation effects
could arise if mitotic activity of the stem cells had a circadian
rhythm. Recently, circadian rhythms have been shown to be a
significant factor in mitotic activity after irradiation of the
corneal epithelium and bone marrow (64,89). Absence of a
circadian rhythm in spermatogonia was reported by Bullough (9),
and later we confirmed his observation (unpublished data). The
procedures used, however, did not identify the stem cell.
Recently we initiated such a study in the mouse using segments of

tubules mounted in toto. The highest mitotic rate of A_s spermatogonia of E2% was observed at 10 and 11:00 a.m., but owing to high variability among mice, this was not significantly higher than the overall mean of 1.4%. Similar results were obtained for DNA synthesis where mice were given ^{3}H-TdR, irradiated with 300 R X-rays 24 hrs later, and killed 207 hrs after labeling. The labeling percentages varied from 14 to 21%, but did not differ from the mean of 18%. Since the slides for exposure to 300 R were available, sections were scored for effect of time of day on survival of A_s spermatogonia 183 hrs after irradiation. Again there was no indication of a circadian rhythm, with number of stem cells per 200 tubule cross-section ranging from 33 to 41 with a mean of 36.7. On the basis of these results, it appears that the seminiferous epithelium is different from other cell-renewal tissues of the body in that mitotic activity is not influenced by a circadian rhythm. Therefore, time of day can be ignored in fractionation experiments and in the time that tissues are taken for observation. One must be aware, however, that the data on mitotic index in controls includes both slow- and fast-cycling A_s spermatogonia, and there is a possibility that a circadian rhythm in the slow-cycling cells is masked by the larger, more rapidly dividing compartment.

Stage of cycle of the seminiferous epithelium, however, does have an effect both on mitotic index, where the 1.7% observed in stage 5 is significantly higher than for other stages, and in percentage of labeled cells observed 207 hrs after the combined ^{3}H-TdR labeling and radiation treatment, where frequency of labeled cells is low in stages 1 and 2 of the cycle. Therefore, the common practice of selecting only certain stages of the cycle of the seminiferous epithelium for study could lead to erroneous conclusions.

Response to chemicals. Information on the normal process of gametogenesis and the basic techniques for study of the testis are the principal carryovers from radiation work to the investigation of the effects of chemicals. Because of the large number of chemicals tested, and the wide differences in their response, a complete coverage of the literature will not be attempted. Instead, a few selected examples will be used to illustrate general principles important in evaluating the effects of chemicals on the testis.

That certain chemicals are cytotoxic to the germ cells and can produce sterility in the male has been known since the work of Jackson and colleagues (47,48,81) in the 50's. The action was termed radiomimetic because of the induction of temporary sterile periods with subsequent recovery, but more refined analysis showed that response to radiation differed from response to chemicals and, furthermore, that the response to a specific compound was unique (47,48,81). In spite of the numerous papers published in the following 30 years, and the number of chemicals tested, this statement is still valid, and we still do not understand the bases for the differences in response to even closely related compounds. A comparison of the effects of some of the more thoroughly tested chemicals is presented in Table 4. It is clear that all compounds inducing mutations in stem cells are cytotoxic, but that not all cytotoxic agents induce mutations in stem cells. Comparison with dominant lethality and heritable translocations shows the same lack of correspondence with cytotoxic action on the undifferentiated spermatogonia. Therefore, killing of testicular cells cannot be used to predict genetic effects, but it can demonstrate the presence of cytotoxic substances even in the absence of demonstrable dominant lethality, translocations, or gene mutations.

We have studied only two compounds in detail, 6-mercaptopurine because it was suggested that it may alter the rate of gametogenesis (34), and ethyl nitrosourea because of its high mutagenicity for spermatogonial stem cells (92). The results with both of these chemicals reveal how the response shifts with germ cell stage, undoubtedly in relation to changes in organization of the DNA and associated biochemical and physiological processes occurring during differentiation.

One-hundred fifty mg/kg 6-mercaptopurine had no effect on undifferentiated spermatogonia (A_s, A_{pr}, A_{al}). A_1 cells were slightly reduced in numbers, but A_3 spermatogonia (counted as In and B cells 72 hrs later), were reduced to 52% of control. A_4-In spermatogonia, scored as preleptotene spermatocytes at 72 hrs, showed no decrease from control, yet these cells show chromatid and iso-chromatid breaks in diakinesis at 14-15 days (34). The response obviously is changing with spermatogonial differentiation, but what makes the A_3 gonia sensitive to immediate cell death when the precursor A_1 and A_2 gonia and the A_4, which are derived from the division of A_3, show no early cell death? Also, what is unique about the late A_4 and In cells that results in a delayed effect expressed as chromosome breakage at days 14 and 15? Type B spermatogonia, which are derived from the In, do not show this effect (76).

It has been demonstrated repeatedly that radiation and chemicals have no effect on the rate of gametogenesis or on minimum sperm transport time (20,67,76). This is demonstrated in Figure 24, where appearance of labeled sperm in the ejaculate is compared for males given 150 mg/kg 6-mercaptopurine and controls. Labeled sperm reached the ejaculate at the same time in both groups,, confirming the conclusion of earlier workers. Labeled sperm persisted in the ejaculate for a longer time in the 6-mercaptopurine treated males, however, indicating an increase in

Table 4. Genetic and fertility effects of chemicals on the male.

Chemical	Dominant Lethals	Translocations	Mutations in stem cells	Fertility effects	Spermatogonia killing	Reference
Triethylenemelamine (TEM)	Yes	Yes	Yes	Transient sterility	Yes	21, 33, 47, 97
Nitrogen mustard	Inconclusive	Yes		Semi-sterile		24
Cyclophosphamide	Yes	Yes	No	Short sterile period	Yes	21, 54, 96
Ethyl methanesulfonate	Yes	Yes	No	Transient sterility	Yes	5, 10, 21, 35, 36
Methyl methanesulfonate	Yes	Yes	No	Transient sterility		8, 53
Mitomycin C	Yes	Yes	Yes	Transient sterility	Yes	21, 26, 53
Ethylnitrosourea (ENU)	Yes	Yes	Yes	Long period of sterility	Yes	26, 92
n-Propylmethanesulfonate	Inconclusive		No			21, 22
Isopropylmethanesulfonate	Yes	Only at negligible frequencies	No	Transient sterility	Yes	21, 22, 31
6-Mercaptopurine	Yes	No	No		Yes	28, 76
Tris(1-aziridinyl)phosphine sulfide (Thio-TEPA)	Yes	Yes	No	Transient sterility	Yes	28, 53
Tris(1-aziridinyl)phosphine oxide (TEPA)	Yes	Yes	No	Transient sterility	Yes	27, 30, 53
Natulan (Procarbazine)	Yes	Yes		Transient sterility	Yes	21, 23, 27, 95

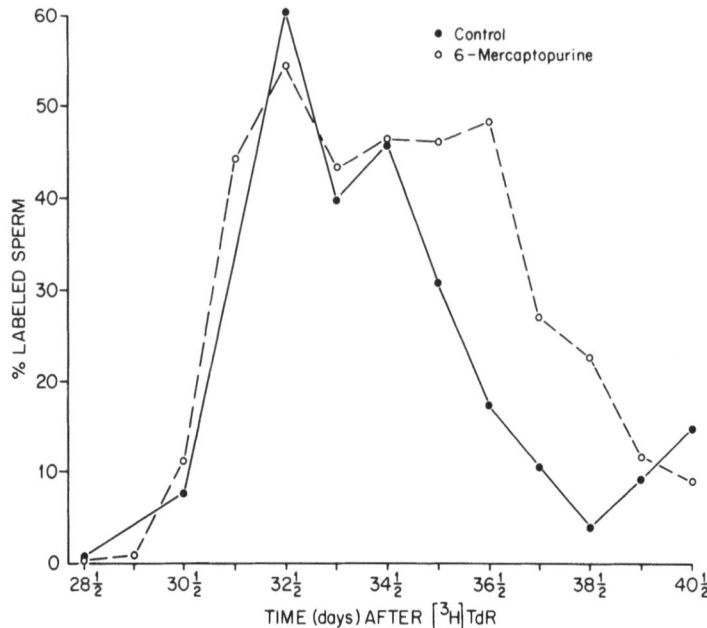

Fig. 24. Frequency of labeled spermatozoa in the ejaculate 28.5 – 40.5 days after ^{3}H-thymidine injection in controls and in experimental mice given 150 mg/kg 6-mercaptopurine intraperitoneally 1 hr after ^{3}H-TdR [from Oakberg et al. (75)].

maximum sperm transport time, most likely as a result of oligospermia (76). That passage through the epididymis and vas deferens is longer when sperm numbers are low had not previously been demonstrated, but could have been inferred from radiation data, where induction of a sterile period requires higher doses than would be predicted on the basis of spermatogonial killing.

Since spermatogenesis and minimum sperm transport time are unaffected by 6-mercaptopurine, preleptotene spermatocytes, which show no detectable chromosome breakage at diakinesis-metaphase I, are nevertheless responsible for the dominant lethality observed 32.5 – 35.5 days after 150 mg/kg 6-mercaptopurine. Conversely, the chromatid and isochromatid breaks observed on day 14 apparently do not result in dominant lethality (76). Our previous experience with radiation response of spermatogonia was used to resolve this question. An exposure of 150 R of X-rays, known to be an LD_{100} for A_4-In and B spermatogonia was given prior to 150 mg/kg 6-mercaptopurine, thus limiting dominant lethality to a spermatocyte response. The response was the same as observed with 6-mercaptopurine alone, demonstrating that it clearly was a preleptotene response (76). Dominant lethals were higher at 36.5 - 41.5 days in the combined treatment, most likely as a result of the severe oligospermia induced.

The results with 6-mercaptopurine demonstrate that the response, in terms of cell lethality, chromosome breakage, and dominant lethality changes dramatically with progression through development of even closely related cell types. Furthermore, the minimum times estimated for cells in the testis to complete development and to reach the ejaculate are valid; sperm transport time in the epididymis and vas deferens can be increased in treated males, however, leading to a mixture of cell stages different from controls at certain intervals after treatment. Finally, the relationship between detectable chromosomal damage at diakinesis metaphase I does not always conform to expectations developed from radiation or results from other chemicals such as TEM.

One of the first observations we made with ENU was how long it took for the stem cell count to reach a minimum. In our first dose-curve experiment mice were killed 3 days after injection, but the dose response was flat for 100, 150, 200, and 250 mg/kg. This was unexpected, for the length of the sterile period was known to increase with dose. Our results could be explained, however, if cell death in the long-cycling stem cell compartment was delayed. A second experiment showed this to be true, with the lowest value observed 8 days after both 50 and 100 mg/kg (Table 5). This is different from radiation, where time at which the lowest number of cells is observed is dose-dependent (Table 3). Such a long delay in reaching a minimum makes estimation of the number of surviving cells difficult, for recovery and continued cell loss overlap (77). That this was the case was shown by ^{3}H-TdR labeling where recovery of DNA synthesis began at 48 hrs, and had reached control levels by 72 hrs after injection of 100 mg/kg ENU, yet cell numbers are declining rapidly at this time. Labeling was above control (41-56%) from 6-9 days, but not as high as the 75% observed 3 days after 150 R and 5 days after 300 R.

In contrast to results with radiation, there was no difference in the number of labeled cells observed at 207 hrs in mice given ^{3}H-TdR 24 or 1 hr before ENU with respective values of 2.5 and 2.3% labeled spermatogonia compared to 18% for controls. This is quite different from the radiation results, where labelling of cells 207 hrs after 100-600 R given 24 hrs after ^{3}H-TdR is at the control level (21%), but reduced to only 6% if irradiation occurs while the cells still are in S (73). This result suggests that synchronization of stem cells by killing sensitive stages of the cell cycle should be less marked after ENU than after irradiation.

Table 5. Survival of A_s spermatogonia 3–16 days after
50 and 100 mg/kg ethylnitrosourea

Dose mg/kg	Time after injection (days)	Percentage of control
0	–	–
50	3	43.5
50	5	36.0
50	8	27.8
50	12	65.6
50	16	102.4
100	3	32.8
100	5	14.5
100	8	8.3
100	12	49.3
100	16	77.8

An effect previously not observed with radiation or chemicals was observed for spermatocytes exposed to ENU in leptotene (77). No change was seen until 4 days later, when degeneration occurred in early pachytene. The sensitive stage was not long, but reduced the number of pachytene spermatocytes from 35 per tubule to zero in some cases. Spermatocytes in preleptotene and early pachytene were unaffected, methylnitrosourea (MNU) did not show this effect, but it was induced by 455 mg/kg hydroxyethylnitrosourea (HENU). Delayed effects of this nature have been observed previously for chemicals, for example, the delay in occurrence of chromosome breakage after 6-mercaptopurine and TEM (33,34) and the expression of cell lethality in spermatogonia after ENU (77). However, this phenomenon is not limited to chemicals, for irradiated spermatogonia often do not degenerate until they reach late interphase or early prophase of their first post-irradiation division (62,65), and primary spermatocytes show no detectable damage until they reach diakinesis-metaphase I, when many cells degenerate (78). For irradiation, it appears as if expression of damage is delayed until critical stages in development are reached, but for chemicals, one cannot distinguish between delay in expression of initial damage and binding to sensitive sites with subsequent induction of lethality.

Finally, comparison of 75 mg/kg MNU, 57-455 mg/kg HENU, and 50-250 mg/kg ENU revealed similar cytotoxic effects on the spermatogonia of the mouse (77). On the basis of the testis response, one would never predict that only ENU would be highly mutagenic in stem cells. This is another demonstration of the observation that all of the agents producing mutations in stem cells also are cytotoxic, but not all agents that kill stem cells are mutagens (Table 4).

The exquisite sensitivity of certain spermatogonial stages to cytotoxic agents provides a sensitive test for the presence of chemicals in the testis even in the absence of genetic and fertility effects. This can have significant impact on interpretation of the lack of effect of a given chemical, for though not of importance in a laboratory test on a specific strain of experimental animals, the magnitude of the response and the significance of a small effect can vary with both species and genotype.

Normal Oogenesis

The initial stages of gametogenesis, including site of origin of the primordial germ cells, their migration to the genital ridges, and mitotic division are the same as in the male (2,3,12,19,55,60). At about 13 days in the mouse, however, the female germ cells begin to enter meiotic prophase (19,60), and by birth, all oocytes are in late pachytene, diplotene, and the arrested state of diffuse diplotene (dictyate). All oocytes enter the diffuse diplotene stage within a few days after birth. The same pattern of gametogenesis occurs in all species, but on different time schedules associated with different lengths of gestation. The rabbit is unique in that oogonial divisions continue until about 2 weeks after birth, but in most mammals, the female is born with her total supply of oocytes. The stage of arrest is in diplotene of meiosis, but the degree of chromosome condensation varies from the condensed nucleus of the guinea pig, the more typical diplotene of the human, to the highly diffuse dictyate stage of the mouse and some other rodents (74). Response of the female to cytotoxic and mutagenic agents differs greatly from the male, for except in specific embryonic and fetal stages, the germ cells are almost exclusively in diplotene of meiosis, and oogonia do not persist after birth or the early post-natal period. Therefore, any loss of oocytes cannot be remedied, and may have irreparable effects on both physiology and fertility of the adult.

At the time of arrest in development, the oocyte is surrounded by a few follicle cells with small, oval nuclei and a flattened cytoplasm that completely envelops the oocyte (6,60,80,82). The stimuli that initiate follicular growth are not known, but at intervals a few follicles begin to increase in size through division of the follicle cells, the oocyte begins to enlarge, formation of the zona pellucida occurs (74), the metabolic activity of the oocyte, as indicated by RNA synthesis, increases (70), and after several weeks, mature Graafian follicles are formed. Most follicles and their contained oocytes are destined to degenerate; only a few complete development and are ovulated. Since the oocyte pool is fixed at birth, this normal attrition results in a continued decline in oocyte number with age (69,83), and in experiments where number of oocytes are counted, it is essential that the controls are matched for both strain and age. The rate of loss is the same in pregnant and virgin females (69), but can be accelerated by exposure to radiation and chemicals (Fig. 25). The increased rate of loss can continue long after a single radiation exposure, and could result either from effects on the oocyte, the follicle, the relationship of the follicle cells to the oocyte, or the ovarian milieu in which the follicle develops.

Different terminologies have been applied to the stages of follicular and oocyte growth, and as a result it is often difficult to compare data presented in the literature. The most precise system of classification is that proposed by Pederson and Peters (82), a slightly modified version of which is presented in Table 6 and illustrated in Figures 26-36.

Radiation Response

Response of the female, both in terms of genetic and fertility effects, is different from the male owing to basic differences in gametogenesis (2,3,55). The two sexes have a common response only during the early divisions of the primordial germ cells (2,3); once meiotic prophase is initiated in the ovary, the paths of gametogenesis diverge, and so likewise the response of the gonads to radiation and chemicals.

The ovary is sensitive during mitosis of the primordial germ cells, then becomes more resistant as the oocytes enter meiotic prophase (3). In all species so far investigated, oocytes pass through a very sensitive stage in early diplotene (3,17,68,83), and in the case of the mouse, hamster, and rat, oocyte arrest occurs in this stage. In other species, the sensitive stage appears to be of short duration, and asynchrony of development protects the prenatal female from the sterilizing effect of acute radiation exposure (17). Continuous or fractionated exposure in utero can induced permanent sterility, however, and some species are even more sensitive than the mouse when exposure occurs in utero. In the mouse, oocytes that have just reached the diffuse diplotene (dictyate) stage are more sensitive than the arrested oocyte of the adult. An LD_{50} of only 8.4 R has been estimated for stage 1 oocytes of the 10-day-old mouse (Fig. 37). By 14-21 days,

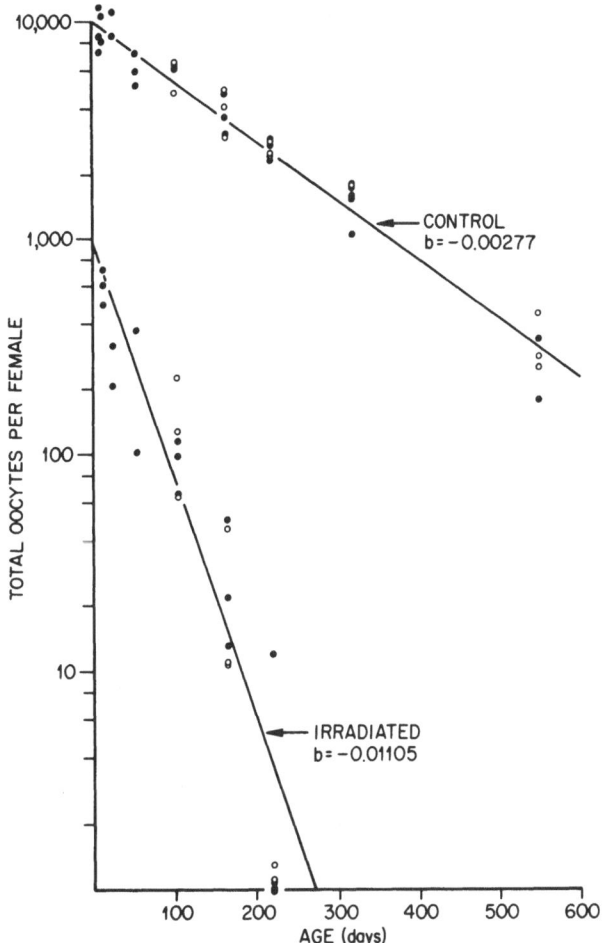

Fig. 25. The effect of 25 R X-rays at 10 days of age on decrease in oocyte number with age in 101 x C3H female mice [from Oakberg (68)].

The arrested oocyte of the adult mouse is extremely sensitive to radiation-induced cell death (63,69), whereas comparable stages in other species are quite resistant. As the follicle and its contained oocyte grows in the mouse, resistance to radiation increases (63) and once the growing oocytes have attained a follicle with a single layer of cuboidal cells, they have both a common nuclear morphology and a similar radiation response in all mammals so far investigated (1). In the mouse, at least one litter is obtained after an exposure of 400 R, indicating high resistance of follicles that will be ovulated at the next estrus. On the other hand, a dose of 50 R will induce permanent sterility

Table 6. Classification of follicles in the adult mouse ovary

Follicle Stage	Characteristics	Illustration
1	0 to 3 or 4 follicle cells visible	Fig. 26
2	1 complete layer of flattened follicle cells	Fig. 27
3a	single layer of cuboidal cells, no zona	Fig. 28
3b	single layer of cuboidal or low columnar cells, zona formation beginning	Fig. 29
4a	1 to 1 1/4 layers of follicle cells, zona complete but thin	Fig. 30
4b	1 1/4 to 2 1/4 layers of cells, zona completely surrounds oocyte but still not fully developed	Fig. 31
5a	2 1/4 to 3 1/4 follicle cell layers	Fig. 32
5b	>3 1/4 follicle cell layers, oocyte at mature size, zona fully developed	Fig. 33
6	many layers of follicle cells, antrum formation initiated	Fig. 34
7	large follicle, 1 or 2 small antra	Fig. 35
8	mature Graafian follicle with large antrum and thin wall (normal stage 8 follicles occur only in proestrus)	Fig. 36

after about 4 litters owing to destruction of arrested oocytes in the smallest follicles. Response of mature oocytes is similar in the human female, where one or two menstrual cycles occur after acute radiation exposures of 300 - 400 R (49,100). A period of amenorrhea then sets in, but in contrast to the mouse, ovulations begin again several months to a year later owing to development of the radiation-resistant "arrested" oocytes (49,100). Because of these differences, the female mouse often is considered a poor model for the human (1). This is true as far as fertility is concerned but, as will be shown in the following paragraphs, has little relationship to genetic effects.

After an exposure of 50 R of X-rays, all the mutations obtained in the mouse are from conceptions occurring in the first 7 weeks after exposure (91); the observed mutation rate in litters conceived more than 7 weeks after irradiation is actually (but not significantly) below the control rate. In evaluating the relevance of genetic data in the mouse to other species, it therefore becomes important to identify the stage in oocyte

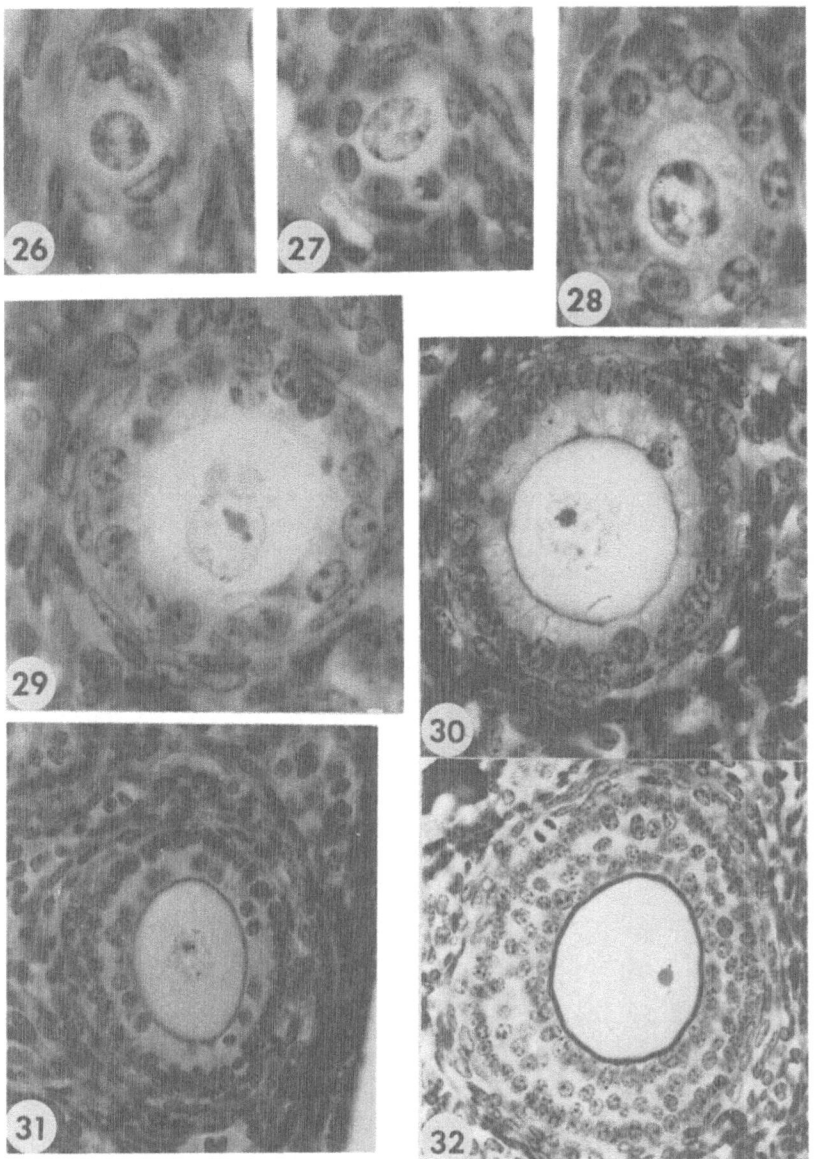

Figs. 26-32. Stages 1-5a in oocyte development of the mouse
(Table 6). 26, stage 1; 27, stage 2; 28, stage 3a; 29, stage
3b; 30, stage 4a; 31, stage 4b; 32, stage 5a. All follicles
normal. Figs. 26-29, X970; Fig. 30, X700; Figs. 31-32,
X400.

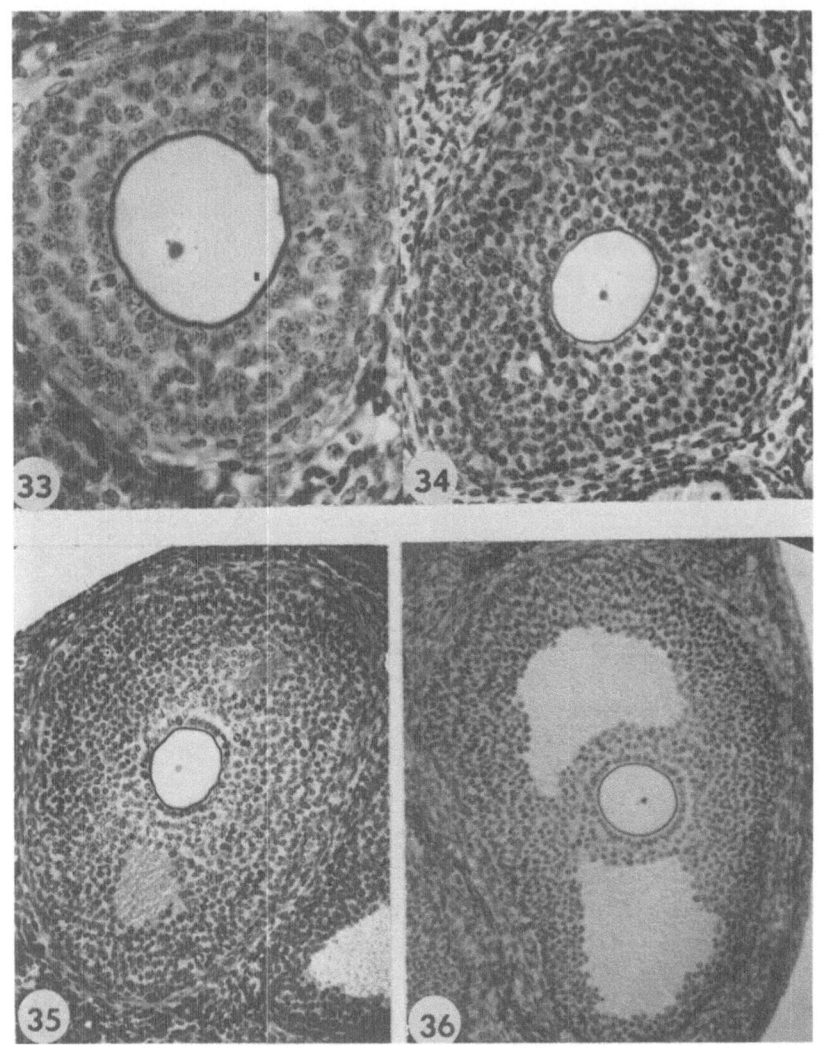

Figs. 33–36. Stages 5b–8 in oocyte development of the mouse
(Table 6). Fig. 33, stage 5b; 34, stage 6; 35, stage 7; 36,
stage 8. All follicles normal. Fig. 33, X400; Fig. 34,
X300; Figs. 35–36, X200.

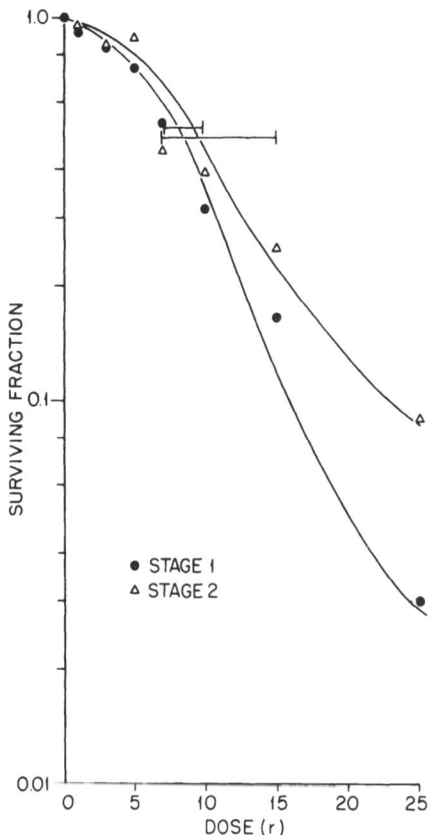

Fig. 37. Gamma-ray sensitivity of stage 1 and 2 oocytes in 10-
day-old 101 x C3H female mice [from Oakberg (68)].

development at which this shift in mutational response occurs.
Timing of oocyte growth by labeling the zona pellucida with N-
[^{3}H]acetyl-D-glucosamine, D-[1-^{3}H] glucosamine, and N-[1-^{3}H]
fucose indicated that the change occurs after the oocytes have
initiated growth, and in a stage that is similar both in nuclear
morphology and in radiation response in all mammals (74). The
early oocyte that is sensitive to killing in the mouse therefore
has no bearing on species comparisons for genetic effects, and the
mouse may be a better model for mammals than is commonly thought.
It appears logical that a cell with as long a life span as the
oocyte (months in mice, decades in the human) would evolve
efficient repair mechanisms, and the only comparison that has been
made suggests that this indeed may be true, for guinea pigs have
arrested oocytes that are resistant to radiation-induced killing,
yet they also show a low frequency of dominant lethals just as do
early oocytes in mouse and hamster (16).

Early oocytes of the mouse make a sensitive test system for measuring the cytotoxicity of radiation and chemicals (Fig. 37) and for demonstrating the presence of such agents even in the absence of genetic and fertility effects. Quantitation of follicle stages 1-3b is straightforward, for degeneration in controls is rare among these stages. In treated mice, degeneration is rapid and oocytes can be counted 72 hrs after treatment without difficulty. As the follicle acquires multiple layers of cells, however, the frequency of degenerating follicles in control mice increases, and in some stages, especially 5 and 6, the majority of follicles are atretic. It has long been known that degeneration is the common fate of most follicles and their contained oocytes, with only a few reaching ovulation. This causes no problem in the normal course of reproduction, and there is some evidence that degenerating follicles serve a hormone secreting role. However, this high level of atresia does cause serious problems in evaluating the effect of noxious agents on the ovary, especially since no exact criteria exist for classifying follicles as normal or atretic. A complete continuum from the unquestionably normal to the obviously degenerate with corresponding variations in frequencies of necrotic cells and apparently normal mitoses occurs in all follicle stages of more than two layers. Quantitation of the growing follicles therefore is highly subjective.

The effect of oocyte loss is expressed at the end of the reproductive span (29,69). An experiment where 25 R of X-rays was given to 10-day-old females is used as an example (Fig. 38). Oocyte numbers were reduced to only 3.9% of control at 56 days, when the females were mated, yet reproductive performance was reduced to only 35% of control (69). Reproduction in the irradiated group ceased by 38 weeks, whereas matched controls continued to breed for 75 weeks (Fig. 38). Oocyte counts given in Fig. 25 reveal two interesting phenomena: (1) oocyte loss occurs at a more rapid rate in the irradiated than in the control females, and (2) sterility occurs in the irradiated females because the oocyte supply is exhausted whereas it occurs from physiological reasons in controls, for several hundred oocytes remain in the ovaries of control females at the end of their reproductive cycle. Such effects may be difficult to detect in other species, such as the human female where number of births is low, and usually limited to the early part of the reproductive span. Physiological consequences of an earlier than normal onset of menopause because of depletion of the oocyte pool, however, deserve serious consideration.

Chemicals. Investigation of the effects of chemicals on the ovary is much more limited than for the testis, but the data are adequate to demonstrate effects that are a hazard to reproductive performance. In addition to direct effects on the ovary, the

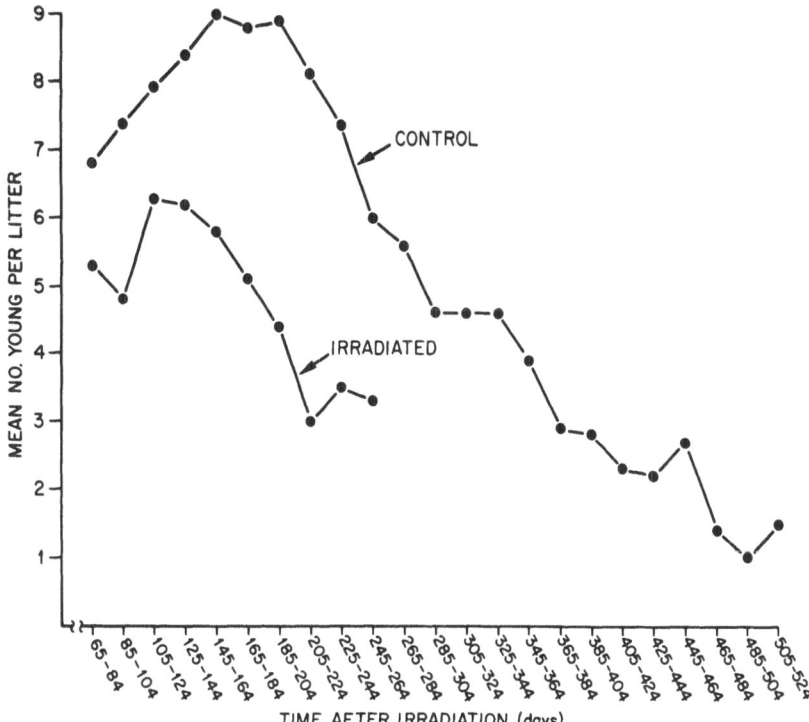

Fig. 38.   Effect of 25 R X-rays at 10 days of age on litter size
           and reproductive span of 101 x C3H hybrid mice [from Oakberg
           (68)].

response of females is complicated by possible physiological
effects on ovulation, implantation, and maintenance of the embryo
and fetus.

   Embryo and fetus.   Only a few studies have been made on the
embryo and fetus, but they are adequate to demonstrate that
dividing primordial germ cells and oogonia are more sensitive to
the cytotoxic action of chemicals than the oocytes of either the
late fetus or adult (59).   For example, isopropyl methane
sulfonate (IMS) induces sterility in the female offspring if given
to the pregnant rat on day 13 or 14, is less effective on days 15
and 16, and has no effect on fertility if given on day 20.
Azathioprine and its active metabolite, 6-mercaptopurine, also are
toxic to oocytes of the developing ovary, but not in the adult
ovary (56,59).   Those compounds showing oocyte toxicity in the
adult (cyclophosphamide, benzo[a]pyrene, IMS), have proven to be
cytotoxins in the developing ovary as well (59).   Owing to higher
sensitivity to both radiation and chemicals, exposure of the fetal
and neonatal ovary can have serious effects on subsequent

Table 7. Genetic and fertility effects of chemicals in the female.

Chemical	Presumed Dominant lethal effect	Mutations	Fertility effects	Oocyte killing	Reference
Benzo[a]pyrene	No		Reduced fertility	Yes	28, 57, 58
Cyclophosphamide			Sterility after prenatal or post-natal exposure	Yes	56
Azathioprine			Sterility to female if exposed in utero	No	56
6-Mercaptopurine			Sterility to female if exposed in utero	No	56
Ethyl methanesulfonate	Yes, for some strains	Yes	Large reduction of litter size; later onset	No	26, 29, 32
Methyl methanesulfonate	Yes, for some strains	Yes	Slight litter size reduction	No	26, 29, 32
Triethylenemelamine	Yes	Yes	Permanent sterilizing effect	Yes	29, 32
Dimethylbenzanthracene (DMBA)				Yes	50
3-methylcholanthrene		Yes	Sterility	Yes	39, 57
Isopropyl methanesulfonate	Yes	Yes	Later onset sterility	Yes	29, 30, 32
Myleran (1,4 di-methanesulfonoxybutane)	Yes	Yes	Permanent sterility	Yes	29, 32

reproductive performance. Yet this is a neglected area of
research. Furthermore, a compound like 6-mercaptopurine which has
no effect on fertility of the adult mouse, but can sterilize both
males and females in utero (59), demonstrates that risk of
chemical exposure cannot be fully evaluated by tests on the adult.

Adult. Compared to the male, relatively few studies have
been made on the effect of chemicals on female reproduction, but
they reveal the same complexity of response observed in the male,
with the added variable of genotype (25,29,32,35,58). Female mice
of different strains vary not only in their sensitivity to the
direct action of cytotoxic agents, but also in their ability to
repair genetic damage induced in the male after the sperm has
entered the egg. Also, analysis of fertility effects in females
can be affected by possible physiological effects on ovulation,
implantation, and maintenance of the embryo.

With the exception of IMS, which kills follicles with an
antrum (32), chemicals that affect female fertility do not
interrupt continued development and ovulation of oocytes that have
initiated growth, but once ovulated, they are not replaced owing
to destruction of the "arrested" oocytes in small follicles. This
response is similar to that observed for radiation exposure, and
myleran, 9, 10 - Dimethyl - 1, 2 -Benzanthracene,
Triethylenemelamine (TEM), cyclophosphamide, benzo[a]pyrene, and
ethylnitrosourea are examples of this class of chemicals (Table
7). Reduction in fertility is expressed at the end of the
breeding span, and it is important that total reproductive
performance be assessed in order to detect an effect (29,69).
This is especially true for compounds such as 6-mercaptopurine,
methylmethanesulfonate (MMS), chloro-9-[3-(ethyl-2-
chloroethyl)amino propylamino] acridine dihydrochloride (ICR-170),
ethyl methanesulfonate (EMS), and N-methyl-N'-nitro-N-
nitrosoguanidine (MNNG) that have only a slight effect (29).

In cases where neither fertility nor genetic effects are
observed, reduction in number of oocytes often can be detected by
counts of serial ovarian sections. The procedure is the same as
described for X-rays, and has been used extensively by Mattison
(56,57,58,59) and others in an analysis of the effect of
polycyclic aromatic hydrocarbons on the ovary. As pointed out for
X-rays, these effects may not affect reproduction in laboratory
animals, but they may have serious physiological consequences in
other species, for example, an earlier onset of the menopause in
women. Also, we have to recognize that there not only are species
differences in response to chemicals but that genotype also is a
significant variable (35,58).

MECHANISM OF CELL DEATH

In spite of decades of work, the mechanisms involved in cell death remain obscure. It is known that chromosome breakage leading to aneuploidy or heteroploidy of daughter cells can lead to cell death, but this cannot be a major factor in spermatogonial or oocyte killing, for most spermatogonia degenerate in late interphase or prophase before they divide (62,65), and the oocytes most sensitive to cell-killing are arrested in diplotene of meiotic prophase. The response can be immediate with detectable changes in the cells as early as 3 hrs after irradiation, or it can be delayed, as demonstrated by the response of stem cells and spermatocytes to both radiation and chemicals. For example, a single exposure of 25 R Co^{60} γ rays results in accelerated oocyte loss up to 200 days later (Fig. 25). Furthermore, one must account for the changes in response with differentiation and development, the differences in response between even closely related spermatogonial classes, the unique response of different meiotic prophase stages, and the changing response during spermiogenesis. Also, what occurs during growth of the mouse oocyte to change it from a cell refractive to mutation induction and sensitive to cell killing to a cell sensitive to mutation induction and resistant to cell killing by radiation? It is easy to associate response to a noxious agent with morphologically recognizable cell stages, but this may be possible only because of correlated and unknown biochemical and physiological differences involved in the intricate biological process of differentiation.

The organization of the chromosomes of the arrested oocyte has received considerable attention in reference to cell killing (1,70,74) but this hypothesis is difficult to reconcile with the fact that the oocyte is in diplotene of meiotic prophase, with a 4 N complement of DNA. It is difficult to see how doses of only a few rads could induce sufficient chromosome damage to kill the cell, with degenerative changes evident even minutes after irradiation. The high sensitivity of the mouse oocyte suggests that the sensitive target either is the entire cell, or that there are a large number of small targets, inactivation of any one of which will lead to cell death. The work of Parsons (80) with radiation, and of Gulyas and Mattison with chemicals (39) indicate that the mitochondria are the first cellular organelles to be affected. This would also agree with the destruction of growing oocytes with ^{3}H-uridine (70), for the beta radiation would be concentrated in the RNA-rich cytoplasmic organelles.

In the male, the existence of A_{pr}, A_{al} and all differentiating spermatogonia in a syncytial arrangement provides an amplification mechanism, for these cells either survive or degenerate in groups (45). The observed progression of degenerative changes in groups of cells with cytoplasmic connections strongly suggests that inactivation of one cell is sufficient to kill the entire group. Thus the effective target size is much larger than the single cell, and as a result, the effect of doses as low as 3 rads of γ rays can be detected on the A_4 and In spermatogonia of the mouse.

Even though we do not yet understand the mechanism of cell killing, the exquisite sensitivity of the primordial germ cells, oogonia, gonocytes of the embryo and fetus, and the differentiating spermatogonia and arrested oocytes of the mouse to cytotoxins make these excellent test systems for detecting the presence of noxious agents. The relationship of this cytotoxicity to fertility is direct. Killing of germ cells in the embryo and fetus affects lifetime reproductive performance in both males and females. Even though temporary sterility may be induced in the adult male, fertility eventually returns if a sufficient number of stem cells survive. In the female, however, oocytes cannot be replaced, and reduction in oocyte numbers can result in premature onset of sterility.

The relationship of cell killing to genetic effects is more complex and less direct, and acts primarily by altering the frequency distribution of stem cells among stages of the mitotic cycle, and by affecting the cell stages available for testing, and thereby the types and frequencies of genetic effects observed in the offspring.

ACKNOWLEDGEMENTS

The author is grateful to Miss C. D. Crosthwait for her help in compiling the literature on chemical effects, and to W. M. Generoso and G. A. Sega for reviewing the manuscript.

REFERENCES

1.   Baker, T.G. (1973) The effects of ionizing radiation on the mammalian ovary with particular reference to oogenesis, In Handbook of Physiology, Section 7: Endocrinology, Volume II, Female Reproductive System, Part 1, R.O. Greep, B.A. Astwood, S.R. Geiger, Eds., American Physiologcal Society, Washington, D.C., pp. 349-361.
2.   Beaumont, H.M. (1960) Changes in the radiosensitivity of the testis during foetal development. Int. J. Rad. Biol. 2:247-256.
3.   Beaumont, H.M. (1961) Radiosensitivity of oogonia and oocytes in the foetal rat. Int. J. Rad. Biol. 3:59-72.
4.   Benda, C. (1887) Untersuchungen über den Bau des funktionerierenden Samenkanälchens einiger Säugetiere und

Folgerungen für die Spermatogenese dieser Wirbeltierklasse. Arch. Anat. Microsc. Morphol. Exp. 30:49-110.

5.   Bhattacharjee, D., T.K. Shetty, and K. Sundaram (1978) Cytotoxicity of ethyl methanesulfonate in mice spermatogonia. Experientia 35(5):630-631.

6.   Brambell, F.W.R. (1927) The development and morphology of the gonads of the mouse. I. The morphogenesis of the indifferent gonad and of the ovary. Proc. Roy. Soc. Lond. B 101:391-408.

7.   Brenneke, H. (1937) Strahlenschädigung von Mäuse - und Rattensperma, beobachtet an der Frühent wicklung der Eier. Strahlentherapie 60:214-238.

8.   Brewen, J.G., H.S. Payne, K.P. Jones, and R.J. Preston (1975) Studies on chemically induced dominant lethality. I. The cytogenetic basis of MMS and induced dominant lethality in post-meiotic male germ cells. Mut. Res. 33:239-250.

9.   Bullough, W.S. (1948) Mitotic activity in the adult male mouse, Mus musculus L. The diurnal cycles and their relation to waking and sleeping. Proc. Roy. Soc. B 135:212-233.

10. Cattanach, B.M., C.E. Pollard, J.H. Isaacon (1968) Ethyl methanesulfonate-induced chromosome breakage in the mouse. Mut. Res. 6:297-307.

11. Chang, M.C., D.M. Hung, and E.B. Romanoff (1957) Effects of radio-cobalt irradiation of rabbit spermatozoa in vitro on fertilization and early development. Anat. Rec. 129:211-229.

12. Chiquoine, A.D. (1954) The identification, origin, and migration of the primordial germ cells in the mouse embryo. Anat. Rec. 118: 135-146.

13. Clermont, Y. (1968) Differenciation et evolution des cellules sexuelles 1. La lignee male. Cinetique de la spermatogenese chez les Mammiferes, In La Physiologie de la Reproduction les Mammiferes, A. Jost, Ed., Editions du centre National de la Recherche Scientifique, Paris, pp. 7-60.

14. Clermont, Y., and E. Bustos-Obregon (1968) Re-examination of spermatogonial renewal in the rat by means of seminiferous tubules mounted "in toto". Am. J. Anat. 122:237-248.

15. Clermont, Y., and S.C. Harvey (1965) Duration of the cycle of the seminiferous epithelium of normal, hypophysectomized and hypophysectomized - hormone treated albino rats. Endocrinol. 76:80-89.

16. Cox, B.D., and M.F. Lyon (1975) X-ray induced dominant lethal mutations in mature and immature oocytes of guinea-pigs and golden hamsters. Mut. Res. 28:421-436.

17. Dobson, R.L., and J.S. Felton (1983) Female germ-cell loss from radiation and chemical exposures. Symposium on "Reproductive Toxicology", D.R. Mattison, Ed., Am. J. Indust. Med. 4:nos. 1 and 2.

18. Ebner, V. von (1871) Untersuchungen über den Bau der Samenkanälchen und die Entwicklung der Spermatozoiden bei den Säugethieren und beim Menschen. Leipzig.

19. Eddy, E.M., S.M. Clark, D. Gong, and B.A. Fenderson (1981) Review article: Origin and migration of primordial germ cells in mammals. Gamete Res. 4:333-362.

20. Edwards, R.G., and J.L. Sirlin (1958) The effect of 200r of X-rays on the rate of spermatogenesis and spermiogenesis in the mouse. Exp. Cell Res. 15:522-528.

21. Ehling, U.H. (1978) Specific-locus mutations in mice, In Chemical Mutagens, Vol. 5, A. Hollaender, and F.J. de Serres, Eds., Plenum Publishing Co., New York, pp. 233-256.

22. Ehling, U.H., D.G. Doherty, and H.V. Malling (1972) Differential spermatogenic response of mice to the induction of dominant lethal mutations by n-propyl methanesulfonate and isopropyl methanesulfonate. Mut. Res. 15:175-184.

23. Ehling, U.H., and A. Neuhauser (1979) Procarbazine-induced specific-locus mutations in male mice. Mut. Res. 59:245-256.

24. Falconer, D.S., B.M. Slizynski, and C. Auerbach (1951) Genetical effects of nitrogen mustard in the house mouse. Jour. of Genetics 10:81-88.

25. Generoso, M. (1969) Chemical induction of dominant lethals in female mice. Genetics 61:461-470.

26. Generoso, W.M., personal communication.

27. Generoso, W.M., J.B. Bishop, D.G. Gosslee, G.W. Newell, C.J. Sheu, and E. von Halle (1980) Heritable translocation test in mice. Mut. Res. 76:191-215.

28. Generoso, W.M., K.T. Cain, and S.W. Huff (1978) Chromosomal aberration effects of benzo[a]pyrene in male and female germ cells of mice. Mut. Res. 53:126.

29. Generoso, W.M., S.K. Stout, and S.W. Huff (1971) Effects of alkylating chemicals on reproductive capacity of adult female mice. Mut. Res. 13:171-184.

30. Generoso, W.M., K.T. Cain, and S.W. Huff (1978) Inducibility by chemical mutagens of heritable translocations, In Advances in Modern Toxicology, W.G. Flamm and M.A. Mehlman, Eds., Hemisphere, Washington, D.C., pp. 109-129.

31. Generoso, W.M., S.W. Huff, and K.T. Cain (1979) Relative rates at which dominant-lethal mutations and heritable translocations are induced by alkylating chemicals in postmeiotic male germ cells of mice. Genetics 93:163-171.

32. Generoso, W.M., S.W. Huff, and S.K. Stout (1971) Chemically induced dominant-lethal mutations and cell killing in mouse oocytes in the advanced stages of follicular development. Mut. Res. 11:411-420.

33. Generoso, W.M., M. Krishna, R.E. Sotomayor, and N.L.A. Cacheiro (1977) Delayed formation of chromosome aberrations in mouse pachytene spermatocytes treated with triethylenemelamine (TEM). Genetics 85:65-72.

34. Generoso, W.M., R.J. Preston, and J.G. Brewen (1975) 6-mercaptopurine, an inducer of cytogenetic and dominant-lethal effects in premeiotic and early meiotic germ cells of male mice. Mut. Res. 28:437-447.

35. Generoso, W.M., and W.L. Russell (1969) Strain and sex variations in the sensitivity of mice to dominant-lethal induction with ethyl methanesulfonate. Mut. Res. 8:589-598.

36. Generoso, W.M., W.L. Russell, S.W. Huff, S.K. Stout, and D.G. Gosslee (1974) Effects of dose on the induction of dominant-lethal mutations and heritable translocations with ethyl methanesulfonate in male mice. Genetics 77:741-752.

37. Grell, R.F., and E.E. Generoso (1982) A temporal study at the ultrastructural level of the developing pro-oocyte of Drosophila melanogaster. Chromosoma 87:49-75.

38. Grell, R.F., E.F. Oakberg, and E.E. Generoso (1980) Synaptonemal complexes at premeiotic interphase in the mouse. Proc. Natl. Acad. Sci. USA 77:6720-6723.

39. Gulyas, B.J., and D.R. Mattison (1979) Degeneration of mouse oocytes in response to polycyclic aromatic hydrocarbons. Anat. Rec. 193:863-880.

40. Hannah-Alava, A. (1965) The premeiotic stages of spermatogenesis, In Advances in Genetics, E.W. Caspari, Ed., Vol. 13, Academic Press, New York, pp. 157-224.

41. Huckins, C. (1965) The initiation of spermatogenesis in the testis of the Wistar albino rat. Ph.D. dissertation, McGill University, Montreal, Canada.

42. Huckins, C. (1971a) The spermatogonial stem cell population in adult rats. I. Their morphology, proliferation and maturation. Anat. Rec. 169:533-558.

43. Huckins, C. (1971b) The spermatogonial stem cell population in adult rats. III. Evidence for a long-cycling population. Cell Tissue Kinet. 4:335-349.

44. Huckins, C. (1972) Spermatogonial stem cell behavior in rodents, In Biology of Reproduction, Basic and Clinical Studies, Vol. III, J.T. Velardo and B.A. Kasprow, Eds., Pan American Congress of Anatomy, New Orleans, Louisiana, pp. 395-421.

45. Huckins, C. (1978) Spermatogonial intercellular bridges in whole-mounted seminiferous tubules from normal and irradiated rodent testes. Am. J. Anat. 153:97-122.

46. Huckins, C., and E.F. Oakberg (1978) Morphological and quantitative analysis of spermatogonia in mouse testes using whole mounted seminiferous tubules II. The irradiated testis. Anat. Rec. 192:529-542.

47. Jackson, H. (1958) The effects of radiomimetric chemicals on the fertility of male rats. The Journal of the Faculty of Radiologists IX(4):216-220.

48. Jackson, H., B.W. Fox, and A.W. Craig (1961) Antifertility substances and their assessment in the male rodent. J. Reprod. Fertil. 2:447-465.

49. Jacox, H.W. (1939) Recovery following human ovarian irradiation. Radiology 32:538-545.

50. Krarup, T. (1969) Oocyte destruction and ovarian tumorigenesis after direct application of a chemical

carcinogen (9:10–dimethyl–1:2–Benzanthrene) to the mouse ovary. Int. J. Cancer 4:61–75.

51. LaValette, St. George von (1876) Uber die Genese der Samenkorper 1. Arch. Anat. Microsc. Morphol. Exp. 1:403–414.

52. Leblond, C.P., and Y. Clermont (1952) Definition of the stages of the cycle of the seminiferous epithelium in the rat. Ann. N.Y. Acad. Sci. 55:548–573.

53. Leonard, A. (1976) Heritable chromosome aberrations in mammals after exposure to chemicals. Rad. and Environm. Biophys. 13:1–8.

54. Lu, C.C., and M. Meistrich (1979) Cytotoxic effects of chemotherapeutic drugs on mouse testis cells. Cancer Res. 39:3575–3582.

55. Mandl, A.M. (1964) The radiosensitivity of germ cells. Biol. Rev. 39:288–371.

56. Mattison, D.R., L. Chang, S.S. Thorgeirsson and K. Shiromizu (1981) The effects of cyclophosphamide, azathioprene, and 6-mercaptopurine on oocyte and follicle number in C57BL/6N mice. Res. Commun. Chem. Path. Pharmacol. 31:155–161.

57. Mattison, D.R., and M.S. Nightingale (1980) The biochemical and genetic characteristics of murine ovarian aryl hydrocarbon (benzo[a]pyrene) hydroxylase activity and its relationship to primordial oocyte destruction by polycyclic aromatic hydrocarbons. Tox. & Appl. Pharm. 56:399–408.

58. Mattison, D.R., and M.S. Nightingale (1981) Murine ovarian benzo[a]pyrene metabolism and primordial oocyte destruction: A survey of nine inbred strains, In Dynamics of Ovarian Function, N.B. Schwartz, and M. Hunzicker-Dunn, Eds., Raven Press, New York, pp. 89–94.

59. Mattison, D.R., and M.S. Nightingale (1982) Prepubertal ovarian toxicity. Banbury Report 11:395–409.

60. Mintz, B. (1960) Embryological phases of mammalian gametogenesis. J. Cell Comp. Physiol. 56; Supp. 1:31–48.

61. Monesi, V. (1962) Autoradiographic study of DNA synthesis and the cell cycle in spermatogonia and spermatocytes of mouse testis using tritiated thymidine. J. Cell. Biol. 14:1–18.

62. Monesi, V. (1962) Relation between X ray sensitivity and stages of the cell cycle in spermatogonia of the mouse. Rad. Res. 17:809–838.

63. Murray, J.M. (1931) A study of the histological structure of the mouse ovaries following exposure to roentgen irradiation. Am. J. Roentgenol. Radium Ther. Nucl. Med. 25:1–45.

64. Nečas, E. (1982) Stem cell (CFU-s) proliferation in sublethally irradiated mice. Cell Tissue Kinet. 15:667–672.

65. Oakberg, E.F. (1955) Degeneration of spermatogonia of the mouse following exposure to X-rays, and stages in the mitotic cycle at which cell death occurs. J. Morph. 97:39–54.

66. Oakberg, E.F. (1956a) A description of spermiogenesis in the mouse and its use in analysis of the cycle of the

seminiferous epithelium and germ cell renewal. Am. J. Anat. 99:391-413.

67. Oakberg, E.F. (1956b) Duration of spermatogenesis in the mouse and timing of stages of the cycle of the seminiferous epithelium. Am. J. Anat. 99:507-516.

68. Oakberg, E.F. (1962) Gamma-ray sensitivity of oocytes of immature mice. Proc. Soc. Expl. Biol. Med. 109:763-767.

69. Oakberg, E.F. (1966) Effect of 25R of X-rays at 10 days of age on oocyte numbers and fertility of female mice, In Radiation and Aging, P.J. Lindop, and G.A. Sacher, Eds., Taylor & Frances LTD, London, England, pp. 293-306.

70. Oakberg, E.F. (1968) Relationship between stages of follicular development and RNA synthesis in the mouse oocyte. Mut. Res. 6:155-165.

71. Oakberg, E.F. (1971) Spermatogonial stem-cell renewal in the mouse. Anat. Rec. 169:515-532.

72. Oakberg, E.F. (1975) Effects of radiation on the testis, In Handbook of Physiology, Section 7: Endocrinology, Volume V, Male Reproductive System, R.O. Greep, E.B. Astwood, D.W. Hamilton, and S.R. Geiger, Eds., American Physiological Society, Washington, D.C., pp. 233-243.

73. Oakberg, E.F. (1978) Differential spermatogonial stem-cell survival and mutation frequency. Mut. Res. 50:327-340.

74. Oakberg, E.F. (1979) Timing of oocyte maturation in the mouse and its relevance to radiation-induced cell killing and mutational sensitivity. Mut. Res. 59:39-48.

75. Oakberg, E.F. (1981) The age at which the long-cycling spermatogonial stem-cell population is established in the mouse, In Development and Function of the Reproductive Organs, A.G. Byskov, and H. Peters, Eds., Excerpta Medica, Amsterdam, pp. 149-152.

76. Oakberg, E.F., Crosthwait, C.D., and Raymer, G.D. (1982) Spermatogenic stage sensitivity to 6-mercaptopurine in the mouse. Mut. Res. 94:165-178.

77. Oakberg, E.F., and C.D. Crosthwait (1983) The effect of ethyl-methyl- and hydroxyethyl-nitrosourea on the mouse testis. Mut. Res. 108:337-344.

78. Oakberg, E.F., and R.L. DiMinno (1960) X-ray sensitivity of primary spermatocytes of the mouse. Intern. J. Radiation Biol. 2:196-209.

79. Oakberg, E.F., and C. Huckins (1976) Spermatogonial stem cell renewal in the mouse as revealed by ^{3}H-thymidine labeling and irradiation, In Stem Cells of Renewing Cell Populations, A.B. Cairnie, P.K. Lala, and D.G. Osmond, Eds., Academic Press, New York, pp. 287-302.

80. Parsons, D.F. (1962) An electron microscope study of radiation damage in the mouse oocytes. J. Cell. Biol. 14:31-48.

81. Partington, M., B.W. Fox, and H. Jackson (1964) Comparative action of some urethane sulphonic esters on the cell population of the rat testis. Exptl. Cell Res. 33:78-88.

82. Pedersen, T., and H. Peters (1968) Proposal for a classification of oocytes and follicles in the mouse ovary. J. Reprod. Fertil. 17:555-557.

83. Peters, H. (1969) The effect of radiation in early life on the morphology and reproductive function of the mouse ovary, In Advances in Reproductive Physiology, A. Mclaren, Ed., Academic Press, New York, pp. 149-185.

84. Regaud, C. (1901) Etudes sur la structure des tubes seminferes et sur la spermatogenese chez les Mammiferes. Arch. Anat. Micr. 4:101-156.

85. Regaud, C., et A. Lacassagne (1927) Effects histophysiologique des Rayons de Roentgen et de Becquerel-Curie sur les tissues adultes normaux des animaux superieurs. Arch. de l'Institute due Radium, Vol. 1.

86. Rooij, D.G. de (1973) Spermatogonial stem cell renewal in the mouse 1. Normal situation. Cell Tissue Kinet. 6:281-287.

87. Roosen-Runge, E.C. (1962) The process of spermatogenesis in mammals. Biol. Rev. 37:343-377.

88. Rowley, M.J., D.R. Leach, G.A. Warner, and C.G. Heller (1974) Effect of graded doses of ionizing radiation on the human testis. Rad. Res. 59:665-678.

89. Rubin, N.H. (1982) Influence of the circadian rhythm in cell division on radiation-induced mitotic delay in vivo. Rad. Res. 89:65-76.

90. Ruiter-Bootsma, A.L. de, M.F. Kramer, D.G. de Rooij, and J.A.G. Davids (1976) Response of stem cells in the mouse testis to fission neutrons of 1 MeV energy and 300 k V X-rays, methodology, dose-response studies, relative biological effectiveness. Rad. Res. 67:56-68.

91. Russell, W.L. (1965) Effect of the interval between irradiation and conception on mutation frequency in female mice. Proc. Natl. Acad. Sci. USA 54:1552-1557.

92. Russell, W.L., E.M. Kelly, P.R. Hunsicker, J.W. Bangham, S.C. Maddux, and E.L. Phipps (1979) Specific-locus test shows ethylnitrosourea to be the most potent mutagen in the mouse. Proc. Natl. Acad. Sci. USA 76(11):5818-5819.

93. Sapsford, C.S. (1962) Changes in the cells of the sex cords and seminiferous tubules during the development of the testis of the rat and mouse. Aust. J. Zool. 10:178-195.

94. Selby, P.B. (1973) X-ray-induced specific-locus mutation rate in newborn male mice. Mut. Res. 18:63-75.

95. Sharma, R.K., G.T. Roberts, F.M. Johnson, and H.V. Malling (1979) Translocation and sperm abnormality assays in mouse spermatogonia treated with procarbazine. Mut. Res. 67:385-388.

96. Sotomayor, R.E., and R.B. Cumming (1975) Induction of translocations by cyclophosphamide in different germ cell

stages of male mice: Cytological characterization and transmission. Mut. Res. 27:375–388.

97. Steinberger, E. (1962) A quantitative study of the effect of an alkylating agent (triethylenemelamine) on the seminiferous epithelium of rats. J. Reprod. Fertil. 3:250–259.

98. Walker, H.C. (1977) Comparative sensitivities of meiotic prophase stages in male mice to chromosome damage by acute X- and chronic gamma-irradiation. Mut. Res. 44:427–432.

99. Witschi, E. (1948) Migration of the germ cells of human embryos from the yolk sac to the primitive gonadal folds. Contrib. Embryol. Carnegie Inst. 32:67–80.

100. Zimmer, K. (1953) Rontgenbestrahlung der Ovarien und nachfolgenden Konzeption. Strahlen Therapie 92:117–122.

THE SCIENCE OF TOXICOLOGY - SCOPE, GOALS AND FOUR CASE STUDIES

B. E. Matter

Preclinical Research, Toxicology, SANDOZ, Ltd
Basel, Switzerland

SUMMARY

Toxicology is a multidisciplinary science that deals with the study of the harmful actions of chemical substances on biological material. The scope of toxicology is very wide, and contains three principal categories: environmental (pollution, residues, industrial hygiene); economic (medicines, food, food additives, pesticides, dyestuffs, chemicals); and forensic (intoxication, diagnosis, therapy).

The goal of toxicology is to contribute to the general knowledge of the harmful actions of chemical substances, to study their mechanisms of action, and to estimate their possible risks to humans on the basis of experimental work on biological test systems. An overall assessment of the toxicological profile of natural or man-made chemical substances consists of acute, subacute and chronic toxicity studies, mutagenicity, carcinogenicity and teratogenicity studies, and a series of specially designed experiments. In this paper, the relevance of these toxicological studies as well as the place of mutagenicity, carcinogenicity, and teratogenicity studies within the frame of toxicological evaluation are discussed.

As the day-by-day increase in scientific knowledge directly or indirectly linked to toxicology is so great, it is impossible for any single individual to be an expert in more than a limited area of the entire toxicological field. As this author's main responsibility is the toxicological evaluation of drugs, he will concentrate on the toxicological profile of four drugs developed by our company, in particular their mutagenic, carcinogenic and

teratogenic effects. These are Endralazine, an antihypertensive;
ergotamine, a vasoactive drug for the treatment of migraine;
bromocriptine, a prolactin inhibitor; and cyclosporin A, an
immunosuppressive agent.

I. INTRODUCTION

Toxicology is a science that deals with the study of the
harmful actions of chemical substances on biological material. It
can also be defined as the sum of what is known regarding poisons:
the scientific study of poisons, their actions, their detections,
and the treatment of conditions produced by them.

In ancient times, toxic substances were usually of natural
origin, such as animal venoms, and plant, animal, and microbial
poisons. As human and industrial activities grew and a great
number of chemical substances were produced, the field of
toxicology expanded rapidly in our modern society.

Toxicology is a multidisciplinary science. It borrows from
other sciences such as medicine, pharmacology, pathology, general
and molecular biology, physiology, genetics, biochemistry,
chemistry and statistics. It had to await the development of some
of these disciplines in order to become a modern, quantitative
science. Approaches and activities in toxicology may differ
depending on whether we are dealing with natural substances, air
and water pollution, residues, workplace exposure to industrial
chemicals, food-additives, pesticides, dyestuffs, or drugs.

The wide scope of toxicology, the vastness of the subjects
mentioned above and the rapid day-by-day increase in knowledge
that is directly or indirectly related to toxicology, precludes
the possibility that a single individual can absorb, or retain,
more than a small fraction of the total knowledge presently
available. The goal of toxicology is to estimate the possible
health risks to humans of chemical substances, to contribute to
the general knowledge of the harmful actions of chemical
substances, and to study their mechanisms of action. This is
essential in order to establish preventive measures since one can
only prevent risks which are already known and which have already
been identified, and the objective can be achieved by the use of
experimental biological model systems, or else by epidemiological
surveys in humans or animals.

Innumerable books on special topics in toxicology have been
published, and toxicologists can choose among many journals on
toxicology in order to read the latest information or to publish
their results. A number of useful handbooks exist that provide a
general overview on this complex field (5,6,7,14,18,25,27,31).

The aim of this paper is to give, firstly, a short synopsis on
various principles and aspects of toxicology. In the second part,
four case studies are presented on the toxicological evaluation of
drugs, with special emphasis on questions of mutagenicity,
carcinogenicity and teratogenicity.

II. ALL CHEMICAL SUBSTANCES ARE TOXIC

All chemical substances are toxic if they are given at a
sufficiently high dose level (19). Too much or too little oxygen
or water, or too many or too few vitamins may be harmful. We
should, therefore, not ask if a particular chemical compound is
toxic, because toxicity may be an intrinsic property of the
substance. The important question is whether there is a risk
attached to the use of a particular chemical substance and this
largely depends upon the qualitative and quantitative aspects of
its toxicity and its application, use and exposure to man.

This implies that no chemical substance, whether natural or
man-made, is entirely safe, and likewise, that no chemical agent
should be considered entirely harmful. Such a concept is based on
the premise that virtually any chemical can be permitted to come
into contact with a biological material or process without
producing a harmful effect on it, provided the concentration of
the chemical is below an effective level. If, however,
sufficiently greater concentrations of the chemical are allowed to
come into contact with biological materials or processes, it is
reasonable to expect undesirable effects. The single most
important factor, therefore, that determines the potential
harmfulness (or safeness) of a compound is the relationship
between the concentration of the compound and the effect it
produces.

III. EXPERIMENTAL MODEL SYSTEMS TO MEASURE TOXICITY

Experimental toxicology is concerned with the design and
execution of tests which determine the potential of a substance to
cause a certain type of injury, and with the development of data
to warrant conclusions as to the potential risks of the substance
for man. Since the goal of toxicology is to ensure that a
substance presents the least possible harm to humans, experimental
model systems should ideally be designed to detect any and all
toxic effects. In reality, this is not possible, because the
problem of designing toxicity experiments involves one major
uncertainty: the uncertainty of whether the test subjects chosen
are indeed appropriate models from which to extrapolate the
results to humans.

In order to detect a wide spectrum of toxic effects, it is at present an accepted procedure to use a battery of experimental model systems. Animals of choice are mice, rats, rabbits, dogs and monkeys. In specific areas such as mutagenicity testing other organisms, e.g., bacteria, fungi, insects etc. may be used. Or else, in environmental toxicology, studies on fish and/or birds are common practice. Today, there is a trend towards making more use of in vitro systems involving cell, organ, or embryo cultures in order to reduce the number of higher mammals in toxicological experiments.

In every toxicity study, different groups of test subjects are treated with different dose-levels of the test substance, the doses ranging from non-toxic to toxic levels. As a general rule three or more treatment groups (i.e., 3 dose-levels) are used and the effects compared to a non-treated group of test-subjects acting as a control. It is common practice nowadays in all leading toxicology institutes dealing with testing of chemical substances to perform the experiments according to the guidelines for "Good Laboratory Practice". These guidelines have been issued by health authorities of various industrialized countries. The guidelines are meant to be followed in order to assure the quality of the personnel, the testing facilities, the test system care facilities, equipment, handling of test subjects, planning/ performance of experiments etc. as well as data storage.

In the following sections, a short survey is given of the major types of toxicity experiments currently needed in order to evaluate the overall toxicological profile of a single test compound.

A. Acute Toxicity

One of the classic model systems in the field of toxicology is the acute toxicology study. In this study, laboratory animals are administered relatively high single doses of a test substance. Mortality is then recorded within a period of 7-14 days (5). Acute toxicity experiments provide important background information on what might happen should this compound be used in extremely high quantities or be misused for a suicide attempt.

In these studies the LD_{50} or LD_5 values representing the hypothetical single dose that would kill 50% or 5% of the animals respectively can be calculated from experimental data by statistical means. So far it has been a generally accepted procedure to classify chemical substances on the basis of the LD_{50} values. However, the LD_{50} value cannot be regarded as a biological constant; it is influenced by many factors, such as animal species and strain, age and sex, diet, temperature, caging regime, season, circadian rhythms, experimental protocol, etc. In

order (a) to gain more relevant scientific information and (b) to
reduce the number of test animals, an improved acute toxicity test
has recently been proposed (32). The main purpose of this newly
developed test is to obtain much more scientific information on
the basis of fewer animals.

B. Long-Term Toxicity Studies

The purpose of long-term toxicity studies is to investigate
the toxic effects of a substance, other than just mortality, in
relation to dose and time. In these experiments, the three or
more dose levels are administered daily, to groups of anywhere
from 5-50 animals, depending on the experiment, over periods of
weeks, months, or years.

There is certainly no consistent relationship between the
LD_{50} values obtained in acute toxicity studies and the severity of
effects found in long-term toxicity studies, since the cause of
toxicity after a single high dose may be entirely different from
that encountered after repeated administration of the compound at
sublethal doses. Furthermore, harmful actions are, of course, not
limited to death alone, and disturbances of various biological
mechanisms can lead to a wide variety of effects which do not
necessarily end in mortality. A number of effects are adaptive in
nature which means the effects seen shortly after the beginning of
treatment disappear during prolonged treatment.

Long-term toxicity studies are valuable in detecting direct
toxic effects on organs and tissues. These are often first
recognized during the experiment by changes seen with organ
function tests, e.g. clinical chemical parameters, hematological
disturbances, electrocardiogram, etc. Proof of specific organ
toxicity, however, can only be made at the end of a study when the
animals are killed and their organs examined histopathologically.
If a specific organ toxicity is found, its occurrence and/or
degree may manifest itself dose-dependently.

In long-term toxicity studies, a number of animals are
usually allowed to survive for a certain period of time without
further compound administration so that the reversibility of any
toxic effects can be studied. It should be stressed here that
many of the toxic effects observed during prolonged treatment with
chemical substances may, indeed, be reversible.

The reversibility of some of the effects caused by test
substances indicates that not all effects are necessarily toxic in
nature. In the case of drugs, for example, some effects are
manifestations of their known pharmacological activity. A case
has been reported where a health authority nearly banned a
proposed drug on the ground that it caused "leucopenia" in rats

(21). The drug, a useful and effective antibiotic, had benefited
rats in the test by reducing the incidence and severity of various
infections which they carried, and in doing so reduced the
pathologically high white cell counts seen in untreated controls
to more normal levels.

Some controversy exists as to the optimal duration of a long-
term study. Experimental toxicologists claim that 6-12 month
studies - as opposed to 1-3 month studies - do not reveal, in
general, any additional information on direct organ toxicity.

C. Mutagenicity, Carcinogenicity, and Teratogenicity Studies

Most papers in the present Symposium Proceedings address
principles of mutagenicity, carcinogenicity and teratogenicity.
Therefore, it is not deemed necessary to provide much further
detail here. There are some distinct features - as compared to
acute and long-term toxicity - that make mutagenesis,
carcinogenesis, and teratogenesis specialized fields. Carcinogens
differ from poisons in that their manifestation of biological
action is delayed for many years. Mutagens interact with the
process of heredity, thereby possibly affecting future
generations. Teratogens interact with the processes of
reproduction and organogenesis which means they may affect only
the first generation offspring. For an overall assessment of the
toxicological profile of a test compound it is, therefore,
indispensable to take these three fields into account.

D. Other Relevant Areas of Research

In addition to the experimental systems described above, it
is often necessary to carry out other types of experiments (5).
In reproduction toxicology it is not only feasible to study
teratogenic effects, but also to evaluate embryotoxicity in
general, adverse effects on fertility, birth, and on newborn
animals during the weaning phase.

Other areas involve investigation of local tolerance, for
example on skin, in order to evaluate whether a chemical compound
may produce irritations and allergic reactions. Relatively new
research areas in toxicology include neurotoxicity, behavioral
toxicity, and immunotoxicity. These areas are still in their
infancy and various laboratories all over the world are presently
engaged in developing, characterizing and validating new
methodologies.

E. Factors That Influence Toxicity - The Need to Carry Out
Specially Designed Studies

As briefly mentioned under (A) above, there are various factors (chemical, biological, experimental) that may influence the toxic response of chemical substances. The more that is known about these factors, the more specific the questions that can be asked, and the more specific the answers that can be obtained by means of special toxicity studies.

When an animal is exposed to a chemical compound, the chemical has to overcome several barriers (membranes) which to a certain extent block free transfer. First of all, it must be absorbed. Secondly it may be evenly or unevenly distributed in the body. The distribution depends to a great extent on the polarity of the substance, its degree of ionization in solution and its solubility in lipid material. Chemicals may be deposited or stored in some tissues, they may bind to receptors or other macromolecules, they may accumulate or may be strongly or rapidly excreted.

Chemicals may also be altered by specific or non-specific enzymatic systems present in various organs. Enzymes involved in biotransformation of chemicals exist according to the genetic template characteristics of each strain and species of organisms. Reduced enzyme activity may result in insufficient metabolism of the chemical which in turn may accumulate in the body and cause severe toxic actions. Increased enzyme activity may lead to quicker metabolic activation or detoxification of the chemical in question. It is possible that there are many as yet undiscovered genetically-induced enzyme-deviants which would account for rare individual responses to chemical agents (i.e., idiosyncrasies, see below). Experimental factors which influence toxicological events are, for example, route, site, rate and time of compound administration.

One of the most important fields that must be addressed in toxicological research is pharmacokinetics. Pharmacokinetic measurements (absorption, distribution, metabolism, excretion) may not only bridge a gap between in vitro and in vivo experiments; they are necessary for designing optimal experimental protocols and interpreting results derived from in vivo studies. For example, it is rather difficult to interpret results derived from extensive animal experiments without knowing whether and to what extent the chemical substance is absorbed, activated, detoxified or excreted in the animal body. Another difficulty of interpretation arises when nothing is known about whether qualitative or quantitative metabolic differences exist between different species and/or man.

In principle, all the factors mentioned above must be taken into account in order to characterize and understand toxic effects better and in order to extrapolate experimental findings to man.

It is obvious that the information on the mechanism of action can
only be gained by specially designed studies which go beyond the
"standard" battery of toxicological tests mentioned in this
chapter.

IV. PREDICTABILITY OF EXPERIMENTAL MODEL SYSTEMS TO MAN

The fundamental problem in toxicology lies in the question of
the applicability to man of knowledge gained with experimental
model systems. Conclusions based on results from animal
experiments cannot necessarily be extrapolated to man, mainly for
the following reasons:

(i) The massive dosage employed in toxicological
experiments often evokes alarming effects which
should not necessarily be taken at full face value
since the maximum human exposure to a substance may
be orders of magnitude lower than that encountered
in animals.

(ii) No adequate animal model systems as yet exist for
several important iatrogenic diseases, or for
relatively harmless but nonetheless critical
untoward effects (i.e., headache, nausea,
perspiration, flushes, disturbance of visual
accommodation, etc.).

(iii) It is uncertain whether idiosyncratic events can be
detected with the number of experimental subjects
that are feasible and practical in laboratory
investigations.[1]

---

[1]Direct toxic effects and idiosyncrasies: In acute or chronic
toxicity tests, specific toxic effects usually occur dose-
dependently. This means that above a certain dose-level an
increasing number of animals or test subjects will show a
particular effect, or different degree of effect. Although
biological variability of animal populations exists the
predictability of what may happen to humans should they take the
substance at similarly high dose-levels under similar conditions
is relatively good. Most of the toxicologically-caused human
disasters were indeed due to specific toxic actions, and
experimental toxicological methods have subsequently been
developed in order to prevent possible future disasters.

The biological variation of human populations, however, is much
larger than that of laboratory animals. A wide variety of
untoward effects of chemical substances in humans are only seen -
by means of epidemiological studies - in a very small fraction
(i.e., 0.01-1%) of the population at risk. Such effects include,
for example, liver necrosis, sensitization reactions, allergies,
etc. Such a rare effect is called an idiosyncrasy which is, by
definition, an abnormal susceptibility to some chemical substances
that is peculiar to an individual.

It is obvious that all standard toxicological examinations have
severe limitations, since the number of animals and test subjects
used in our experiments is generally far too small to detect
idiosyncratic events. To give an example: an animal test model
for the detection of a particular idiosyncrasy, if it existed,
would involve many more than 100 animals per dose group to be able
to find at least one animal showing this toxic response, assuming
an identical incidence of toxic effects in animals and man (30).
For technical and ethical reasons, however, it is not feasible to
use such high numbers of animals for toxicological experiments.
The only possible solution to this problem may be to develop in
vitro experiments on specially designed cell lines or cell
cultures, provided adequate metabolic systems are applied.

> (iv) There is lack of feedback from human
>   epidemiological studies. Without studies in man,
>   the correctness of our techniques of measuring
>   biological effects such as toxicity in experimental
>   model systems can never be verified.

These general considerations indicate that there is still
plenty of room left for improving research in toxicology.
However, it is also noteworthy that the predictability of
experimental model systems to man is not as bad as it may sound,
particularly if one takes into account single chemical entities or
certain substance-classes for which a tremendous amount of
information is already available (see Section VIII below).
Indeed, the degree of predictability can often be enhanced by
asking the right question, selecting the appropriate experimental
models, and by interpreting the results as carefully as possible.

V. THE ESTIMATION OF RISKS[2]

---

[2]"Risk" is the probability that a chemical substance will produce
harm under specified conditions (5).

Few headlines in scientific journals or newspapers are so
alarming, perplexing, complex and often personal in their
implications as those concerning safety issues. Lowrance (15) has
defined the term safety as judgement of the acceptability of
risks: "A thing is safe if its risks are judged to be
acceptable." This situation implies that toxicologists cannot
measure "safety" or "risk" in strictly scientific terms in the
same way as they measure, for example, acute toxicity. They can
only assess probabilities and possible consequences of events but
not their importance for humans. As discussed below, other
methodologies are required to evaluate risks. In addition, since
all personal and social activities involve risks, it is clear that
there can be no hope of reducing risks to zero.

Based on these considerations it is evident that even if one
carried out all the toxicological or epidemiological trials one
could dream of, one could not guarantee that a certain natural or
man-made chemical substance to which man is exposed is absolutely
safe. The presence of chemical substances in our environment will
therefore always entail an element of risk.

It goes beyond the task of this paper to provide an extensive
outline of how risks can be assessed. Readers are therefore
advised to consult the published literature (15,16). There are,
essentially, four lines of investigation which lead to an overall
risk assessment: (i) to define the conditions of exposure; (ii)
to identify the adverse effects (by means of toxicological
experiments and, if possible, epidemiological studies in man), and
the conditions under which these effects are observed; (iii) to
relate exposure to effect; and (iv) to weigh one's conclusions
against other situations and risks.

Another dimension in this process must be added considering
the benefits to society of certain chemical substances: it is
evident that the conclusions made for two substances entailing
similar toxic effects may be quite different depending on their
differing benefits.

VI. THE ROLE OF SCIENTISTS, CONSUMERS, AND SOCIETY

Human toxicological disasters (e.g., the thalidomide
catastrophy) or relevant findings in experimental test systems
have surely two main consequences. First, a considerable research
effort is made in universities, health agencies and industry.
This retrospective research is designed to study the injuries and
mechanisms of action involved by means of experiments in
laboratory animals or other test subjects and to reproduce them.
These findings then constitute a basis for prospective testing of

other chemical substances for potentially harmful effects and for assessing and reducing risks as efficiently as possible.

Second, there is the impact of legislation. In response to public opinion and scientific expertise the administrative institutions of many countries, acting through their public health officials, have exerted an increasing influence on the development and use of man-made chemical substances, not only by stricter supervision, but also by issuing detailed guidelines and specifications for toxicological experiments. The purpose of these guidelines and testing specifications is to prevent damage being caused by these substances and to reduce the overall risks imposed by them.

Guidelines and standard tests are not without merits. They ensure that a certain comprehensive set of data will be gathered for chemical substances before humans are exposed to them and they provide a base set of data for comparing toxic effects of different substances. If, however, routine standardized studies are regarded as being all that is needed to characterize the toxicity of chemical substances, if regulatory agencies of various countries devise their own favored guidelines and requirements, then research into the mechanisms of action of toxicity and into better risk assessments will certainly stagnate. As guidelines become more variable and thus more demanding, the scientist's time and energy will be squandered by ever more tedious routine work, and the quality of this research will deteriorate. As a consequence, scientists of imagination, needed to solve new problems of new chemical substances, will be discouraged from entering this field. Let us remember that the number, type and specifications of standard testing protocols in toxicology do not necessarily guarantee improved safety for man.

Guidelines should be flexible. Only a minimal base set of test systems should be required by health authorities and there should be international harmonization. This should permit sufficient time and capacity for the scientists to select those methods which seem appropriate to solve his or her particular problems. Depending on the results obtained with the base set of test systems, the next step should, if possible, be designed to elucidate mechanisms of action.

Scientists have indeed special responsibilities to the rest of society with respect to safety aspects of chemical substances. They alone are in a position to adapt the planned procedures to the particular objective of the research and to the current state of scientific knowledge. It is precisely this sense of responsibility that spurs the scientist to improve the toxicological evaluation of chemical substances.

There are two kinds of risks which ought to be on the
conscience of the scientific community. Firstly, risks that can
be significantly reduced by applying new methodologies or by
improving existing ones. Secondly, possible consequences of toxic
effect appear so grave that prudence dictates great caution, even
before the risks have been clearly identified.

The consumer or exposed individual shares a part of the
responsibility. It has been known for several hundred years that
all chemical substances may, under certain circumstances, be
harmful (19). He should know that the chronic inhalation of
tobacco smoke, or that consuming huge amounts of alcoholic
beverages or rotten food may indeed do him more harm,
toxicologically, than artificial sweeteners, food preservatives,
or minute amounts of other chemicals in air or drinking water.
The patient, when he takes drugs, must be aware that he should
take only as much as is indicated in the directions. To take 4
tablets instead of one may well involve a greater risk instead of
a greater benefit.

Finally, the society represented by the health officials
shares a part of the responsibility. Those chemical substances
which bear a proven risk or which are highly suspect of doing so
must be identified and their use restricted. But the regulations
on which decisions have to be based should be well balanced in
order to reduce risks without, at the same time, being prohibitive
to new developments in basic research.

It is obvious that balanced views can only be expected when
scientific experts from universities, health agencies and
industries as well as the consumer share an understanding of the
basic problems that are inherent in the toxicological evaluation
of chemical substances.

VII.   TOXICOLOGICAL EVALUATION OF NEW DRUGS

Toxicology has developed into three principal divisions,
depending on the interests of those organizations active in this
field (14). These divisions are the environmental (pollution;
residues in food, air, water; workplace exposure), economic
(drugs, food additives, pesticides), and forensic (intoxications,
diagnosis and therapy). Each has its academic qualifications, its
specific research aims and approaches, and its types of
toxicologists. As this author's main responsibility is the
toxicological evaluation of drugs, he will concentrate, in the
following, on some aspects relating to this field.

The specific task of industrial scientists working in
pharmaceutical research is the discovery of new drugs to be used

in the treatment and prevention of disease. Since the pharmaceutical industry has a vital interest in developing effective drugs which cause the least possible harm to humans, it is obvious that the toxicological evaluation of a candidate drug is a very important feature of its characterization. Ideally, experimental toxicological tests should be designed by industry itself, according to the specificities of each new drug, particularly as it is the company which has developed a drug which has also gathered the greatest information concerning it. However, as has been discussed before, certain standard toxicological procedures issued by governmental regulatory agencies in various countries (1) must be performed as directed. Since pharmaceutical industries like to export effective drugs, it is evident that the guideline which satisfies the most demanding health authority is the one which then is adopted (31).

The following discussion about toxicological evaluation of new drugs deals with the generally accepted approach. It should be noted, however, that this approach may not necessarily be similar to all new chemical entities or to that used by other pharmaceutical firms.

The stages in the development of a new drug are: chemical synthesis or molecular modifications, pharmacological research, toxicological studies, clinical trials, pharmaceutical development, production and marketing (2,10).

Out of hundreds of new chemical entities synthesized every year by our firm, only relative few - approximately 30/year - are found to be sufficiently interesting to justify development. Out of these, only about one or two substances are finally introduced on the market.

First, in the preclinical phase, pilot studies are carried out in order to provide feedback to the pharmacologist who must decide which compound he should select for further development. In this phase relatively little is known about the biological activity of the compound in question, and since one cannot always make any deductions concerning its chemical structure and its possible harmful effects, the pilot studies are standardized experiments. These consist of an acute toxicity test in the mouse or rat, a four-week toxicity study in a few dogs using several dosages, involving examinations during the "In Life" phase and morphological and histological post-mortem examinations on all organs. The Salmonella Ames mutagenicity test is usually also performed. It is evident that highly toxic or mutagenic compounds or compounds showing specific organ toxicity have little chance of being developed further.

In the second phase of preclinical toxicology, when the decision to begin clinical trials has been made, a number of potentially useful drug-candidates are thoroughly investigated per annum, before clinical trials start, in a battery of experimental test systems using a larger number of animals. This battery involves acute toxicity tests in three species, long-term (4 - 26 weeks) toxicity and teratology tests in two species each, and, possibly, more mutagenicity studies and further experiments depending on the outcome of the standard tests. Acute and local tolerance tests, as well as gestation and fertility tests, may also be performed during this phase. The amount of toxicological information deemed necessary to permit clinical trials depends to a great extent on the length of time the drug will be administered to the patient. 2, 4, 13 or 26 week toxicity studies in laboratory animals are usually made when the administration to humans is to be of 1 day, 1 week, 4 or 13 weeks' duration, respectively (31).

The third phase of toxicological experiments is performed when the drug-candidate reveals efficacy and good tolerance in man. The studies include long-term chronic toxicity and carcinogenicity usually lasting for 78 weeks (mouse), 104 weeks (rat), or 52 weeks (dogs), thorough fertility tests, and, if necessary, other specially designed studies to evaluate and characterize particular findings obtained with standard tests. The results of these studies enable both the industry and the health authorities to make their final decisions as to whether or not a drug may be introduced on the market.

Most large laboratories which undertake large-scale animal toxicology are already equipped with, or presently in the process of becoming equipped with, on-line computer facilities for dealing with a variety of clinical, hematological, and histopathological data-processing and all relevant information is usually evaluated by existing or specially designed statistical methods.

The whole package of toxicological evaluation of drugs mentioned above provides the reader with an example of what is currently required before a new single chemical entity can be tested in man and subsequently registered and marketed as a pharmaceutical speciality. Since a drug is developed in a stepwise procedure, the toxicological evaluation may extend over a period of 7-10 years, and the present day prices may run approximately $1 million to $2 million U.S.

When a new drug is marketed, the safety-evaluation procedures will continue, taking the form of post-marketing surveillance. The Drug Monitoring Centers of pharmaceutical firms are responsible for documenting and investigating all side effects which are reported to them. Also these centers perform - if

necessary - world-wide epidemiological studies to answer questions
specific to any special circumstances pertinent to the drug's use.

Two particularly positive aspects of toxicological
experiments of medicines should be emphasized. Firstly,
toxicological evaluation of drugs has a distinct advantage
compared to other environmental substances. The usefulness and
relevance of some of the test systems involved can be evaluated by
direct observation in man, thus providing both impetus and a
scientific background for improvements. The situation concerning
the correlation of adverse effects in humans and the results of
animal tests are not, however, completely satisfactory. There are
two principal difficulties. On the one hand, a preparation which
has been proven to be toxic in animal tests is not normally given
to man, and so it is hardly possible to obtain information
concerning its effects in man. On the other hand, when severe
adverse effects (e.g. idiosyncrasies) occur in man after the
administration of new drugs, they have obviously not all been
detected by standard toxicity studies.

Secondly, since pharmacological studies in animals are
usually acute or subacute in nature, the chronic toxicity studies
often provide the only opportunity to evaluate the possible
beneficial aspects of long term drug administration in animals.
An example is provided in Case IV below.

It has been estimated that predictions from animals to man
concerning the presence or absence of a particular toxic sign are,
approximately, 75% accurate (13). This figure which is in
agreement with our experience on limited number of substances,
seems to be fairly satisfactory until one considers either those
effects occasionally seen in animals which cannot easily or
quickly be recognized in man (e.g., mutagenicity, carcinogenicity,
teratogenicity), or those signs and adverse effects which occur in
man but cannot be detected in animals. Most of the latter
represent sensitization/allergy and idiosyncrasy phenomena such as
aplastic anemia, thrombocytopenic urticaria, etc. or less severe
but undesired signs such as nausea, dry mouth, flushes, dizziness,
etc. The question, then, is how to improve the predictability of
experimental test systems for man.

In the past ten years, full-scale standard toxicity studies
have been a part of the required investigation of a new drug and,
with constant improvements, they have provided an ever-increasing
amount of useful information. One of the main goals of these
classic standard studies is to demonstrate the absence of harmful
effects below a certain toxic dose level. The establishment of a
"no-toxic-effect level" is a practical means of estimating the
margin between toxic and desired effects. The larger this margin,

the higher the chance that the clinician does not detect adverse side effects in clinical trials.[3]

In recent years there has been a positive trend towards a more rational toxicological evaluation and regulation of drugs (8). The aim of this new approach is to learn as much as possible about the characteristics of a new drug by means of specially designed experiments, and to relate all biological effects to dose and blood plasma levels or levels in various organs and tissues. Furthermore, the aim is to determine whether the observed biological effect could be of clinical relevance, and to define circumstances which might lead to adverse effects in man. Recent research has proven that it pays to ask questions, to design an experiment in such a way that reasonable answers can be expected and that a greater number of correct predictions can be made. This trend may be illustrated by the following four case studies with special emphasis on mutagenic, carcinogenic, and teratogenic effects.

VIII. FOUR CASE STUDIES ON TOXICOLOGICAL EVALUATION

1: Cyclosporin A

Cyclosporin A (CS-A, CAS 59865-13-3), a cyclic undecapeptide of fungal origin, represents a novel and potent immunosuppressive drug (26). The available mechanism of action points towards an effect on an early phase of the immune response, namely a preferential inhibition of the activation of T-lymphocytes. CS-A is successfully used in clinical organ-transplantation trials for prevention of graft rejection. Since CS-A belongs to a new class of immunosuppressants, an extensive toxicological evaluation was deemed necessary. An overall review of the toxicological profile of CS-A has recently been published (22).

To summarize, the major drug-related findings were seen in rats and rhesus monkeys treated orally for 13 or 104 weeks with high doses of CS-A. They consisted of nephrotoxicity in rats, and - to a lesser extent - in rhesus monkeys. In addition, in rats, there was evidence of hepatotoxicity. As to the nephrotoxicity, both functional and histopathological investigations indicated tubular damage in the absence of glomerular involvement. The

---

[3] For those readers who are not actively engaged in development of drugs it must be emphasized that what is not seen in experimental toxicity studies is as important as what is, since the absence of major toxicity problems below a certain dose-level is a prerequisite for the clinical testing, development, registration, and marketing of a new drug.

hepatotoxicity was reflected in functional tests as well as histopathologically by the presence of centrolobular degenerative liver changes. Although similar lesions were not seen in dogs treated for 52 weeks, liver and kidney appeared to be the main target organs for toxicity. These findings are indeed relevant for the clinical use of CS-A, since impairment of renal function and minor adverse hepatic reactions have occasionally been reported in transplant patients (26). At present, a series of additional animal experiments have been initiated in order to evaluate the mechanism(s) that may be involved in the development of nephrotoxicity. The question whether (S-A stimulates the renin-angiotensin-aldosterone (RAAS) directly and thereby induces adverse renal reactions, or whether the RAAS is stimulated by Na[+] loss due to a renal tubulotoxic effect of the compound has not yet been answered.

Other interesting drug-related findings were inflammatory changes observed at relatively high doses of CS-A, the gums and skin being the most affected. An atrophic gingivitis was found in rats whereas a hypertrophic gingivitis and periodontitis developed in dogs after prolonged treatment. Hyperplastic changes of the gums have also been reported in clinical trials. The etiology of these changes has not yet been elucidated. Furthermore, a generalized papillomatosis with chronic inflammatory cells in the cutis was found in several dogs. In no case was a malignant transformation observed; in fact, these lesions regressed during the later phases of drug treatment as well as in the recovery animals.

The most interesting outcome of the toxicity studies, however, was not the above findings, but the total lack of certain toxic responses frequently found with some classical immunosuppressive drugs. Firstly, adverse hematological reactions due to myelotoxicity, which represent a serious complication of conventional immunosuppressive therapy with cytostatic drugs, were observed neither in experimental animals nor in clinical trials with CS-A alone. Moreover, it was shown that CS-A not only lacks myelotoxicity, but actually allows immunological recovery in recipients of bone marrow grafts.

Secondly, there was a total lack of irreversible toxic effects such as mutagenicity, carcinogenicity, and teratogenicity in a series of experimental systems (16,22). In particular, no increase of lymphoreticular tumors occurred, as commonly found in immunodeficient rodents. In early clinical studies, where CS-A was combined with other immunosuppressants, a relatively high incidence of lymphomas associated with Epstein-Barr virus (EBV) was reported but subsequent clinical experience now shows that the incidence of lymphomas is lower in CS-A treated patients than in those given conventional immunosuppressive therapy (29).

The results of the toxicity studies with CS-A indicate that the predictability of findings from animals to man is surprisingly good. It is obvious that the possible risk of nephro- and hepatotoxicity as well as of malignancy associated with EBV and other oncogenic viruses in immunosuppressed patients must be carefully monitored.

2: Endralazine

Endralazine (CAS 65322-72-7), a hydrazine derivative, is a potent antihypertensive drug with distinct peripheral vasodilator activity. The toxicological characteristics of endralazine have been carefully evaluated during the past seven years. In acute, subacute and chronic oral toxicity studies using mice, rats and dogs endralazine - despite its good absorption - was found to be of relatively low toxicity without causing any organ toxicity. Furthermore, in reproduction studies, endralazine did not reveal embryotoxic or teratogenic effects in rats and rabbits, and was without any adverse effects on fertility, weaning and lactation phase (unpublished data).

On the basis of reports indicating that some hydrazine derivatives were found to be carcinogenic in rodents (28), extensive mutagenicity[4] and carcinogenicity studies were carried out with endralazine. Detailed results will shortly be published (17). In summary, endralazine revealed mutagenic effects in Salmonella typhimurium strains TA 1535, TA 1538 and TA 100 above concentration levels of 25 μMol/plate, both in the presence and absence of a rat liver homogenates (S9). Similar findings were reported by another research group (4). On the basis of these findings, a series of additional genotoxicity and mutagenicity studies was carried out in order to verify these results. Similar concentration levels of endralazine were used to score for 6-thioguanine resistant clones in V79 Chinese hamster fibroblasts treated both in the presence and absence of an S9 rat liver homogenate. No mutagenic activity was found. Several in vivo animal experiments were performed in which endralazine was given intraperitoneally to mice and hamsters in order to enhance genotoxic action. These studies consisted of: (i) the mouse coat color spot test in order to measure mutagenicity in embryonic pigment precursor cells; (ii) sister chromatid exchange tests in bone marrow cells of mice; (iv) chromosomal analysis in bone marrow and spermatogonia of treated mice; and finally (v) an assay to measure unscheduled DNA synthesis in sperm and spermatids of mice as an indicator of DNA damage. Although in all of these experiments the doses employed approached sub-lethal levels, neither mutagenic nor genotoxic effects of endralazine were

_____

[4]For the methods mentioned in this paper see ref. 11.

recorded. Another group of researchers found endralazine slightly
genotoxic in the mouse DNA-elution assay and the mouse bone marrow
sister chromatid exchange test after intraperitoneal application
of endralazine - at doses that killed 50% of the animals (4). In
these experiments, however, it is not clear whether these effects
were due to a direct action of endralazine or the consequence of
the severe general intoxication caused by extremely high (lethal)
dose levels of endralazine. In the carcinogenicity studies, oral
daily doses of 3-150 mg endralazine per kg body weight (i.e.,
15-700 times the maximum therapeutic dose) were given to mice and
rats for 78 and 104 weeks, respectively. There was no indication
of a drug-related increase in the incidence of benign or malignant
tumors (17). In particular, there was no evidence of increased
incidences of the tumor types reported with other hydrazine
derivatives (24).

Despite the mutagenic activity of endralazine per se in the
Salmonella assay in concentrations above 25 µMol/plate, the
cumulative weight of evidence suggests that endralazine has
neither genotoxic nor carcinogenic potential in other experimental
in vitro/in vivo systems.

The fact is that when therapeutic, up to sublethal dose-
levels of endralazine are applied to mammals, the peak plasma and
tissue concentrations of endralazine and its main metabolite
(hydrazone) measured are by a factor of 10^3 - 10^5 lower than
mutagenic concentrations applied to Salmonella typhimurium. In
other words, the results fit well together if one considers that
concentration levels up to 25 µMol/plate of endralazine failed to
produce mutations in Salmonella typhimurium.

It can be concluded, therefore, that any genotoxic/
carcinogenic risk to humans receiving therapeutic doses of
endralazine must be considered as very low indeed.

3: Ergotamine

Ergotamine (CAS 379-79-3) is a vasoactive drug used for the
treatment of migraines and vascular headaches and, in obstetrics,
as a uterotonic vasoconstricting and hemostatic agent. Ergotamine
is a synthetic derivative of the ergot alkaloid family (23). It
is worth mentioning a few details of the history of ergot since
ergotamine provides a perfect example of the rule that the dose
differentiates a poison and a remedy (19). Ergot is the name for
the compacted resting stage, the sklerotium, of the fungus
Claviceps purpurea, which is a parasite of rye and other cereals.
In medieval times, the consumption of fungus-infected cereals led
to severe intoxications mainly in the form of gangrene (ergotism)
which even took the form of epidemics. These intoxications
diminished to zero as their cause became known, as agricultural

standards improved, and when the potato became one of the major
foodstuffs in Europe (3).

Medical doctors and pharmacists of the nineteenth and
twentieth centuries became interested in studying the biological/
medical actions of ergot in more detail, basing their research on
the observation made way back in the seventeenth century that
ergot might be medically useful in childbirth. Ergot alkaloids
were isolated in order to discover their active principles. This
work led to the era of modern ergot research - chemical,
pharmacological, toxicological, and clinical. Derivatives were
made by means of semi- or fully synthetic procedures. Today, a
number of important drugs based on ergot are widely used in
medicine and obstetrics, and the mode of action of these drugs is
relatively well understood (3).

When ergotamine is given to animals at doses which far exceed
therapeutic doses, toxic effects which are mainly the result of
its pharmacodynamic property can be seen. For example, in
subacute and chronic toxicity studies the occurrence of gangrene
in various animal species correlated closely with the
vasoconstrictor potential (9). In this case, human intoxication
is indeed well predictable from animal experiments.

In standard reproduction toxicological studies in pregnant
mice, rats and rabbits it was found that high oral doses given
during the organogenetic phase affected the maternal weight gain
and caused an increase in prenatal mortality as well as evidence
of fetal retardation and anomalies in rats (9). This result was
observed at about the time of the thalidomide tragedy and caused
considerable concern, because the majority of patients using
ergotamine preparations are women.

Two series of experiments were carried out in order to test
whether this effect was due to a direct toxic effect on the embryo
(as with thalidomide) or to an exaggerated pharmacodynamic action
(9). In the first, ergotamine was given to pregnant rats on
single days only between days 4 and 19 of gestation. A phase
specificity of embryotoxicity which reached a maximum on day 14,
as well as characteristic anomalies after day 14 such as cleft
palates, were found. In the second experiment, temporary clamping
of the main uterine vessels in the rat on day 14 of gestation
produced similar effects in the embryo. Thus, the interruption of
uterine blood flow, leading to hypoxia (a teratogenic factor) and
subsequent impairment of nutrition, is obviously the cause of
embryotoxic and teratogenic effects, and it is irrelevant whether
this effect is achieved mechanically or pharmacologically. The
hypothesis that embryotoxicity in rats given ergotamine may also
be caused by α-adrenoceptor stimulation producing hypoxia was
strengthened by further special experiments.

The hypothesis that the embryotoxicity was due to a direct
specific effect of ergotamine on the embryo could be rejected on
the basis of the results of experiments in which ergotamine was
applied intra-amniotically. Indeed, it was found that the intra-
amniotic dose of ergotamine needed to produce a similar
embryolethal effect on day 14 of gestation was 50 times higher
than the amount of radioactive-labelled substance reaching the
fetus in the placenta following oral administration of an
embryolethal dose tot he dams. Furthermore, it could be
demonstrated that ergotamine reduced the transplacental passage of
^{3}H-L-leucine, as a marker for uteroplacental blood supply, and
that this effect could be antagonized by phenoxybenzamine.

In summary, these and many other experiments have
demonstrated that the embryotoxic effects of ergotamine are
related to its pharmacodynamic properties, i.e., vasoconstriction
of uterine vessels and ∝-adrenoceptor stimulation, and not to a
direct specific embryotoxic action (9). Due to the relatively low
therapeutic dose, compared to the doses used experimentally, and
to the precautions taken in ergotamine therapy, epidemiological
surveys have indeed failed to indicate embryotoxic effects in
humans.

4: Bromocriptine

Bromocriptine (CAS 22260-51-1) inhibits the secretion of the
anterior pituitary hormone, prolactin, without affecting other
pituitary hormones (growth hormone, gonadotropins, thyrotropin)
except in patients with acromegaly where it lowers raised blood
levels of growth hormone; this effect may be due to a dopaminergic
mechanism. Bromocriptine is used in medicine for the treatment of
galactorrhea, paramenia, subfertility and infertility caused by
prolactin disturbances, acromegaly, parkinsonism and pituitary
tumors (23).

Although bromocriptine belongs to the ergot alkaloid family,
it lacks the prominent uterotonic, pressor, and vasoactive
characteristics of ergotamine and other derivatives of this class
of compounds. It was, therefore, expected that the toxicological
profile of bromocriptine would be different from other ergot-
derivatives. Extensive experimental and clinical studies have
been undertaken in the past ten years, and the pharmacological,
toxicological and clinical characteristics of bromocriptine have
been reviewed (20).

Bromocriptine was found to be of relatively low toxicity in
acute, subacute and chronic toxicity studies. It produced several
signs such as emesis and excessive salivation in dogs, which
reflect the nausea and vomiting that are occasionally seen in man.
Results of mutagenicity, reproduction and fertility studies were

inconspicuous, and there was no specific organ toxicity except the
following in the rat.

In a 52-week oral toxicity study in rats at doses of 5-82 mg/
kg bromocriptine per day the post mortem evaluation revealed a
decrease in pituitary weight and an increase in the number of
cystic follicles with decreased luteal tissue in the ovaries, at
all dose levels. These findings were associated with some
squamous metaplasia of the endometrium, indicating a drug-effect
on the pituitary-gonadal axis, resulting in a picture of estrogen-
dominance. Examinations of the "recovery" animals after a 5 week
drug-free period revealed almost complete reversal of these
effects. Similar effects were not seen in mice or dogs treated
with bromocriptine for 78 or 52 weeks, respectively.

In the rat carcinogenicity study, where bromocriptine was
given daily for 100 weeks at doses of 1.8, 9.9, and 44.5 mg/kg
(25-600 times the therapeutic dose inhibiting prolactin) the only
severe untoward findings - not entirely unexpected - were in
uteri, where the endometrial changes seen after one year's
administration of bromocriptine had progressed to uterine
neoplasia. It is noteworthy that the total incidence of all tumor
types in rats decreased with increasing dose-levels. Similar
uterine effects were not seen in the mouse carcinogenicity study.

A series of additional experiments showed that as female rats
age, the cyclicity of reproductive events deteriorates and
prolonged periods of pseudo-pregnancy alternate with periods of
persistent estrus. Bromocriptine, owing to its prolactin-
inhibiting action, prevented the occurrence of pseudo-pregnancy
and, correspondingly, increased the incidence and duration of
persistent estrus. The progesterone-estradiol ratio was lower,
and estrogen dominance prevailed in bromocriptine-treated female
rats showing endometrial metaplasia and uterine neoplasia. The
special hormonal changes in the rat are therefore considered
responsible for the development of uterine neoplasia.

The fundamental difference between women and female rats as
regards the aging process of their reproductive functions makes it
clear that the uterine effects produced by bromocriptine in rats
cannot occur in women as they either have cyclic menstrual
activity or, after menopause, involution of both ovaries and
uterus. In hyperprolactinaemic women with disturbed menstrual
cycles bromocriptine restores cyclicity rather than prevents it.

Post-marketing surveillance in the form of gynecological
assessments at regular intervals has been carried out for eight
years, and there is indeed no indication of endometrial effects of
bromocriptine in women. Thus, it was clear to us as well as to
the Drug Regulatory Authorities and their scientific consultants

that the uterine findings in rats were species-specific and, therefore, not relevant for the human situation.

Another aspect of the extensive toxicological evaluation should perhaps be mentioned. As has previously been discussed, chronic toxicity experiments often provide the only opportunity to evaluate the possible beneficial aspects of long term drug administration in animals. Thus, in these studies bromocriptine prevented significantly the development of chronic renal disease, polyarteritis nodosa, adrenal cortical adenomas and mammary tumors in rats. Each of these observations has initiated clinical and other research activities which may well lead to the identification of new indications for bromocriptine in the future.

IX. CONCLUSIONS

Standard toxicity studies such as those mentioned in this paper which, for the most part, are required for drugs by health authorities, provide a useful base-set of toxicological data. The relevance of toxic effects to man can sometimes be evaluated quickly, and decisions as to the further development and clinical testing of a drug candidate can be made. In cases where the relevance of certain early findings is questionable or unknown, two directions can be taken.

Firstly, prudence may dictate that further development be stopped. However, anxiety may sometimes be taken for prudence, and because of this a company may miss a good chance of developing a useful drug. Remember that a very anxious company could have found reasons enough to stop further development of the drugs mentioned in cases 1-4.

Secondly, scientists may question the relevance to man of their findings in other mammalian species and design a series of special experiments such as, for instance, have been described in the four case-studies above. They do this by characterizing initial findings, extending the experiments to other organisms and species and by evaluating a drug's possible mechanisms of toxic action.

It is exactly this second direction which helps both to enrich the field of toxicological research and to prevent apparently wrong decisions from being made on the basis of insufficient or misleading information. This second way is pure basic research and cannot, therefore, be subject to guidelines and regulations (12). Nonetheless, since even the best toxicity experiments alone can never guarantee the absolute safety of a drug, post-marketing surveillance of patients receiving drugs will be of increasing importance in the years to come.

One may argue now that the present state of the art of
toxicological evaluation of new products that are either given
deliberately to man (drugs) or to which for some reason human
exposure cannot be reduced or stopped, is relatively well advanced
compared to the situation with old products or all those other
natural and man-made chemical compounds that fall into the
division of "environmental toxicology". Indeed, there are
hundreds and thousands of natural and man-made chemical substances
present in our environment - to which man is or is not exposed -
that we have never had the chance to evaluate thoroughly in this
regard. It is an illusion, however, to believe that such a task
is possible, due to the limited worldwide scientific and
laboratory capacity. Apart from this consideration, it may not
even be worthwhile to perform such a task because only a minority
of all chemical substances in the universe may be important enough
to mankind to be extensively studied. Attempts, therefore, have
been and are being made on national and international levels, in
order to identify those relatively few natural and man-made
substances which are believed to be the real troublemakers in
affecting human health. At least for these substances a broad
toxicological evaluation and risk estimation should and could be
made.

Furthermore, as we are all aware, toxicology is still a young
science, the presently favored and required experimental systems
are in many ways not optimal and our knowledge on toxicological
issues is scanty. Important and difficult problems can often not
be solved by the application of existing standard experiments or
rapid and simplistic screening techniques. The study of possible
mechanisms of toxic actions of a single chemical compound can only
be carried out on the basis of specially designed methods by a
team of scientists dealing with the same problem for several
months or years.

A considerable effort has to be made in the coming years in
order to identify the most important problems and to understand
toxic effects and their underlying mechanisms better, and to find
the means to analyze and estimate possible risks for man.
Progress in this field can best be expected when common sense is
used and when scientists from universities, health agencies and
industry share an understanding of the important basic principles
and problems inherent in the toxicological evaluation of chemical
substances. This International Workshop on the Principles of
Environmental Mutagenesis, Carcinogenesis and Teratogenesis is
certainly an important contribution towards that goal.

REFERENCES

1. Alder, S., C. Janton, and G. Zbinden (1981) Preclinical Safety Requirements in 1980, Brochure, Institute of Toxicology, Schwerzenbach/Zürich, Switzerland.
2. Berde, B. (1974) Industrial research in the quest for new medicines. Clin. Exp. Pharmacol. Physiol. 1:183-195.
3. Berde, B., and H.P. Schild, Eds. (1978) Ergot Alkaloids and Related Compounds, Handbook of Experimental Pharmacology, Vol. 49, Springer-Verlag, Berlin, Heidelberg, New York.
4. De Flora, S., P. Zanacci, C. Becmicelli, A. Camoirano, M. Cavaima, L. Sciaba, E. Cagelli, P. Faggin, and G. Brambilla (1982) In vivo and in vitro genotoxicity of three antihypertensive hydrazine derivatives. Environmental Mutagenesis 4:605-619.
5. Doull, J., C.D. Klaassen, and M.O. Amdur (1980), Casarett and Doull's Toxicology, The Basic Science of Poisons, Macmillan Publishing Company, Inc., New York.
6. Galli, C.L., S.D. Murphy, and R. Paoletti, Eds. (1980) The Principles and Methods in Modern Toxicology, Elsevier/North Holland, Biomedical Press, Amsterdam, New York, Oxford.
7. Gorrod, J.W., Ed. (1981) Testing for Toxicity, Taylor and Francis, Ltd., London.
8. Graham-Smith, D.G. (1982) Preclinical toxicological testing and safeguards in clinical trials. Eur. J. Clin. Pharmacol. 22:1-6.
9. Griffith, R.W., J. Hodel Ch. Grauwiler, K.H. Leist, and B.E. Matter (1978) Toxicological considerations, In Ergot Alkaloids and Related Compounds. Handbook of Experimental Pharmacology, Vol. 49, B. Berde, and H.P. Schild, Eds., Springer-Verlag, Berlin, Heidelberg, New York, pp. 805-851.
10. Gross, F. (1976) The present dilemma of drug research. Clin. Pharmacol. Therap. 19:1-10.
11. Hollaender, A., Ed. (1971-1980) Chemical Mutagens, Principles and Methods for Their Detection, Vols. 1-6, Plenum Press, New York, London.
12. Laurence, D.R., A.E.M. McLean, and M. Weatherall, Eds. Safety and Testing of New Drugs, Prediction and Performance, Academic Press, London, (in press).
13. Litchfield, J.T. (1962) Evaluation of the safety of drugs by means of tests in animals. Clin. Pharmacol. and Therap. 3:665.
14. Loomis, T.A., Ed. (1978) Essentials of Toxicology, 3rd edition, Lea & Febiger, Philadelphia.
15. Lowrance, W.W. (1976) Of Acceptable Risk. Science and the Determination of Safety, William Kaufmann, Inc., Los Altos, California.
16. Matter, B.E., P. Donatsch, R.R. Racine, B. Schmid, and W. Suter (1982) Genotoxicity evaluation of cyclosporin A, a new immunosuppressive agent. Mutation Research 105:257-264.

17. Matter, B.E., W. Suter, R.R. Racine, P. Donatsch, B. Schmid, and B.P. Richardson. Mutagenicity and carcinogenicity evaluation of endralazine, a new antihypertensive drug. Mutation Research, (in press).

18. McLean, A.E.M. (1979) Hazards from chemicals: Scientific questions and conflicts of interest. Proc. R. Soc. Lond. B 205:179-197.

19. Paracelsus (1493-1541): "All substances are poisons, there is none which is not a poison. The right dose differentiates a poison and a remedy."

20. Richardson, B.P., I. Turkalj, and E. Flückiger (1983) Bromocriptine, In Safety Testing of New Drugs, Prediction and Performance, D.R. Laurence, A.E.M. McLean, and M. Weatherall, Eds., Academic Press, London, (in press).

21. Roe, F.J.C. (1981) Testing in vivo for general chronic toxicity and carcinogenicity, In Testing for Toxicity, J.W. Gorrod, Ed., Taylor and Francis Ltd., London, pp. 29-43.

22. Ryffel, B., P. Donatsch, M. Madörin, B.E. Matter, G. Rüttimann, H. Schön, R. Stoll, J. Wilson (1983) Toxicological evaluation of cyclosporin A. Arch. Toxicol. 53: 107-141.

23. Thorner, N.O., E. Flückiger, and D.B. Calne, Eds. (1980) Bromocriptine, Raven Press, New York.

24. Toth, B. (1975) Synthetic and naturally occurring hydrazines as possible cancer causative agents. Cancer Res. 35:3693-3697.

25. Tu, A.T., Ed. (1980) Survey of Contemporary Toxicology, John Wills & Sons, New York.

26. White, D.J.G., Ed. (1982) Cyclosporin A, Elsevier Biomedical Press, Amsterdam, New York, Oxford.

27. WHO, Environmental Health Criteria 6 (1978) Principles and Methods for Evaluating the Toxicity of Chemicals, World Health Organization, Geneva.

28. WHO/IARC (1974) Monographs on the evaluation of carcinogenic risk of chemicals to man. Vol. 4, pp. 127-281.

29. Wood, A.J. (1982) Cyclosporin A as sole immunosuppressive agent in recipients of kidney allografts from cadaver donors. The Lancet 8289:57-60.

30. Zbinden, G. (1964) The problem of the toxicological evaluation of drugs in animals and their safety in man. (Editorial), Clinical Pharmacol. and Therap. 5:537-545.

31. Zbinden, G. (1973, 1976) Progress in Toxicology, Vols. I, II, Springer-Verlag, New York, Heidelberg, Berlin.

32. Zbinden, G., and M. Flury-Roversi (1981) Significance of the LD_{50}-test for the toxicological evaluation of chemical substances. Arch. Toxicol. 47:77-99.

MUTAGENICITY OF PESTICIDES

Yasuhiko Shirasu, Masaaki Moriya,
Hideo Tezuka, Shoji Teramoto, Toshihiro
Ohta, and Tatsuo Inoue

Institute of Environmental Toxicology
Kodaira, Tokyo 187, Japan

SUMMARY

Microbial mutagenicity screening studies employing
Ames tests were done on 228 pesticides. As a result,
50 compounds showed mutagenicity and five of them
required metabolic activation for their activities.
Among the various compound groups, organic phosphates,
halogenated alkanes and dithiocarbamates contained
mutagens at a higher ratio. Mutagenicity of these
pesticides was less potent than that of other mutagens.
Cytogenetic studies on the pesticides which were
positive in microbial assays revealed a good
correlation between the ability to induce sister
chromatid exchanges or chromosomal aberrations and the
mutagenic potency in bacteria. Dominant lethal studies
on 1,2-dibromo-3-chloropropane (DBCP) and ethylene
dibromide (EDB) disclosed that DBCP gave positive
results in rats but not in mice, although EDB was
negative in both species. DBCP was also positive in a
sex-linked recessive lethal test using Drosophila
melanogaster. Occupational exposure to pesticides,
especially gaseous ones like DBCP, in industry or in
agriculture seems hazardous from the genetic
toxicological point of view.

Microbial mutagenicity screening studies were done
on 228 pesticides and related chemicals, including 88
insecticides, 60 fungicides, 62 herbicides, 12 plant-
growth regulators, 3 metabolites, and 3 other related
compounds. The results of our earlier screening

studies were previously published (4,5). The reverse
mutation tests using Salmonella typhimurium and
Escherichia coli WP2 hcr with and without metabolic
activation were employed.

Out of 228 compounds (Table 1), 50 chemicals (22%)
showed mutagenicity. They were 25 insecticides (28%),
20 fungicides (33%), 3 herbicides (5%), 1 plant-growth
regulator and 1 other compound. Organic phosphates,
halogenated alkanes and dithiocarbamates contained
mutagens at higher ratios. Forty-five of the mutagenic
pesticides were direct-acting mutagens and five were
indirect ones which require S-9 mix for their activity
(3) (Table 2).

The mutagenicity of some positive pesticides was
investigated by an in vitro metabolic activation system
employing S-9 mix (1). In this study, the mutagenicity
of captan, captafol, folpet and 2.4-dinitrophenyl
thiocyanate (NBT) disappeared, whereas mutagenic
activity of 5-nitro-1-naphthonitrile (NNN) increased.
S-9 fraction, cysteine and rat blood also effected the
disappearance of mutagenicity (Table 3).

Cytogenetic and dominant lethal studies were
conducted on captan which showed the most potent
activity in the microbial assays (10). The cytogenetic
analysis of human diploid fibroblast cells treated with
captan revealed no chromosomal aberrations. Captan
also could not induce chromosomal aberrations in the
bone marrow cells of male rats. Dominant lethal
studies on captan failed to produce a positive
response.

We conducted a series of mutagenicity tests on
ethylenethiourea (ETU), a decomposition product of
fungicidal ethylene-bis-dithiocarbamates (6). ETU was
weakly mutagenic for TA1535 without metabolic
activation, inducing about a 4-fold increase in the
number of revertants with the higher doses. Neither
the in vitro cytogenetic studies using a Chinese
hamster cell line (Don) nor the in vivo chromosomal
analysis of the rat bone marrow cells could reveal any
abnormalities attributable to ETU treatment. Dominant
lethal studies in mice also gave negative results.

Interactive mutagenicities of ETU plus sodium
nitrite were investigated (5). The reaction mixture of
ETU and sodium nitrite at acidic conditions was

Table 1. Pesticides tested for microbial mutagenicity

INSECTICIDE (88)

Acephate, Chlorfenvinphos, Chlorpyrifos, Chlorpyrifosmethyl, Cyanofenphos, Diazinon, Dichlofenthion, Dichlorvos, Dimethoate, Dimethylvinphos, Disulfoton, EPBP, EPN, ESP, Ethion, Etrimfos, Fenitrothion, Formothion, Isofenphos, Isoxathion, Malathion, Mecarbam, Monocrotophos, Naled, Phenthoate, Phosmet, Proxim, Pirimiphosmethyl, Propaphos, Prothiophos, Pyridaphenthion, Salithion, Tetrachlorvinphos, Thiometon, Trichlorfon, Vamidothion, Chlorfenethol, Chlorobenzilate, Chloropropylate, p,p'-DDT, Dicofol, Phenisobromolate, Tetradifon, Chloropicrin, DBCP, D-D, EDB, EDC, Methyl bromide, Aldrin, α-BHC, Dieldrin, Dienochlor, Endosulfan, Endrin, Heptachlor, BPMC, Butoxycarboxim, Carbaryl, Carbofuran, Isoprocarb, Methomyl, MPMC, MTMC, Oxamyl, Pirimicarb, Propoxur, XMC, Metamammonium, Allethrin, Nicotine sulfate*, Petroleum oil*, Piperonyl butoxide*, Permethrin, Polynactins, Pyrethrins, Rotenone*, Diflubenzuron, Amitraz, Benzomate, Binapacryl, Chlorfenson, Cyhexatin, DCIP, Methylisothiocyanate, Propargite, Quinomethionate, Vendex

FUNGICIDE (60)

Edifenphos, IBP, Phosdifen, Pyrazophos, Amobam, DDC, EMSC, Ferbam, Mancozeb, Maneb, Milneb, Polycarbamate, Thiram, TTCA, Zineb, Ziram, Benomyl, Blasticidin S, Griseofulvin, Kasugamycin, Oxytetracycline, Polyoxin B, Polyoxin D-Zn, Validamycin A, Anilazine, Calcium polysulfide*, Captafol, Captan, Carboxin, Chloroneb*, Chlorothalonil, Copper sulfate*, DAPA, Dazomet, Dichlofluanid, Dimethirimol, Dinocap, Dithianon, Echlomezol, Fentin hydroxide, Folpet, Fthalide, Guazatine, Hymexazol, Isoprothiolane, MAF, NBT, NNN, Oxine-copper, Oxycarboxin*, Phenazine oxide, o-Phenylphenol, Probenazole, Quintozene, Sulfur*, Thiabendazole, Thiophanatemethyl, Tricyclazole, Triforine, Zinc sulfate*

HERBICIDE (62)

Amiprofos-Methyl*, Bensulide, Glyphosate, Krenite*, Piperophos, Asulam, Chloropropham, Cycloate, EPTC, Orthobencarb, Phenmedipham, Swep, Bifenox, Chlomethoxynil, Chlorthal-dimethyl, CNP, 2,4-D, Dicamba*, Dinoseb acetate, MCPA, MCPB, Mecoprop, Nitrofen, PCP, Phenothicl, Bromacil, Dimuron, Diuron, Lenacil, Linuron, Methabenz-thiazuron, Methyldimuron, Siduron, Terbacil, Thiochlormethyl, Cyanazine, Dimethametryn, Metribuzin, Alachlor, Benefin, Butachlor, Chlorthiamid*, Diphenamid, Napropamide, Trifluralin, Propanil, Propyzamide, ACN, Amitrole, AMS*, Bentazon, Dalapon*, Diquat dibromide, Ioxynil octanoate, Methoxyphenone, Molinate, Oxadiazon, Paraquat dichloride, Sodium chlorate, Sodium fluoride*, TCA*, Triclopyr*

GROWTH REGULATOR (12)

Ethephon, Gibberellin A$_3$, Gibberellin A$_4$, 4-CPA, 4-CHMPA, Aminozide, HEH, Maleic hydrazide, Benzyladenine, Indolebutyric acid, NAA, OED,

OTHER (3)   METABOLITE (3)

Ferric oxide*, Sodium polyacrylate, ETU   p,p'-DDD, p,p'-DDE, Heptachlor epoxide

*: Pesticide was tested with only TA98 and TA100.

Table 2. Mutagenic potency of pesticides

No.	Pesticide	Strain	Revertants/nmole
Direct Mutagen (- S-9 mix)			
1	Captan	TA100	93.67
2	NBT	TA100	58.28
3	Folpet	TA100	15.00
4	Dienochlor	TA100	12.04
5	NNN	TA100	9.39
6	DAPA	TA98	5.22
7	Captafol	WP2 hcr	3.11
8	Nitrofen	TA100	2.63
9	EMSC	TA100	2.00
10	Ferbam	TA100	1.33
11	Polycarbamate	TA100	1.33
12	Ziram	TA100	1.24
13	Dichlofluanid	WP2 hcr	1.10
14	Thiram	TA100	0.84
15	Chlomethoxynil	TA100	0.79
16	DDC	TA100	0.79
17	TTCA	TA100	0.58
18	Echlomezol	TA100	0.35
19	Phosmet	TA100	0.34
20	Salithion	TA100	0.14
21	EDB	TA100	0.092
22	DBCP	TA100	0.085
23	Fenitrothion	TA100	0.074
24	HEH	WP2 hcr	0.044
25	Chlorfenvinphos	TA100	0.038
26	Dichlorvos	TA100	0.027
27	MAF	TA98	0.020
28	Polyoxin D·Zn	TA100	0.018
29	Thiometon	TA1535	0.013
30	D-D	TA100	0.0087
31	Dinocap	TA98	0.0073
32	Trichlorfon	TA100	0.0067
33	Monocrotophos	TA100	0.0064
34	ESP	TA100	0.0060
35	Vamidothion	TA100	0.0046
36	Dimethoate	TA100	0.0045
37	Pirimiphosmethyl	TA100	0.0040
38	Binapacryl	TA100	0.0036
39	Allethrin	TA100	0.0031
40	Formothion	TA100	0.0029
41	Carbofuran	TA98	0.0023
42	Disulfoton	TA1535	0.0014
43	Acephate	TA100	0.0013
44	ETU	TA1535	0.00065
Indirect Mutagen (+ S-9 mix)			
1	Oxine-copper	TA100	1.30
2	Chloropicrin	TA100	0.37
3	Phenazine oxide	TA1537	0.13
4	Butachlor	TA100	0.046
5	EDC	TA1535	0.0023

Table 3.   Effects of rat-liver metabolizing factors cysteine, and blood on the mutagenicity of pesticides

Pesticide	μmole/ plate	Tester strain	Revertants per plate				
			Incubation with				
			H₂0	S-9 mix	S-9 fraction	Cysteine	Blood
Captan	0.15	WP2 hcr	3200	30	111	19	32
		TA1535	268	6	31	4	6
Captafol	0.15	WP2 hcr	158	31	24	31	21
Folpet	0.15	WP2 hcr	1320	50	60	21	19
		TA1535	219	8	35	6	14
NBT	0.15	TA1535	2580	64	70	85	39

mutagenic in S. typhimurium strains TA1535, TA100 and Escherichia coli strain WP2 hcr. The host-mediated assay with S. typhimurium strain G46 also revealed interactive mutagenicity of ETU and sodium nitrite simultaneously administered into the mouse stomach (Table 4). Chromosomal aberrations were induced in the bone marrow cells of mice given simultaneous treatment with ETU and sodium nitrite. A single or 5-day oral administration of these two chemicals induced dominant lethality, increasing pre-implantation losses in weeks 5 and 6 (8). N-nitroso-ETU, a possible derivative from ETU and sodium nitrite, caused similar effects. Furthermore, N-nitroso-ETU was revealed to be a carcinogen (2).

Table 4.   Interactive mutagenicity of ETU plus NaNO₂ in the host-mediated assay using S. typhimurium G46

Chemical	No. of mice	Revertants per 10⁸ survivors (mean ± S.D.)
Gum arabic (2%)	6	0.59 ± 0.21
NaNO₂ (50 mg/kg)	5	0.38 ± 0.12
ETU (150 mg/kg)	6	0.50 ± 0.35
ETU + NaNO₂ (150 + 50 mg/kg)	5	15.37 ± 4.19[a]
Nitroso-ETU (125 mg/kg)	6	82.64 ± 17.71[b]

[a]; Significantly different from ETU and NaNO₂ alone (p < 0.01).

[b]; Significantly different from gum arabic (p < 0.001).

Cytogenetic studies were performed on 10
pesticides and N-nitroso-ETU which were positive in the
microbial reversion assays (9). A good correlation was
observed between the ability to induce sister chromatid
exchanges or chromosomal aberrations in the Chinese
hamster V79 cells and the mutagenic potency in bacteria
(Fig. 1).

Dominant lethal studies on 1,2-dibromo-3-
chloropropane (DBCP) and ethylene dibromide (EDB)
revealed that DBCP gave positive results in rats but
not in mice, although EDB was negative in both species
(7). DBCP induced post-implantation losses in weeks 4
and 5, exerting its adverse effect on early spermatids
(Fig. 2).

Our latest studies revealed that DBCP induced sex-
linked recessive lethal mutations in Drosophila
melanogaster (Fig. 3). Occupational exposure to
pesticides, especially gaseous ones like DBCP, in
industry or in agriculture seems hazardous from the
genetic toxicological point of view.

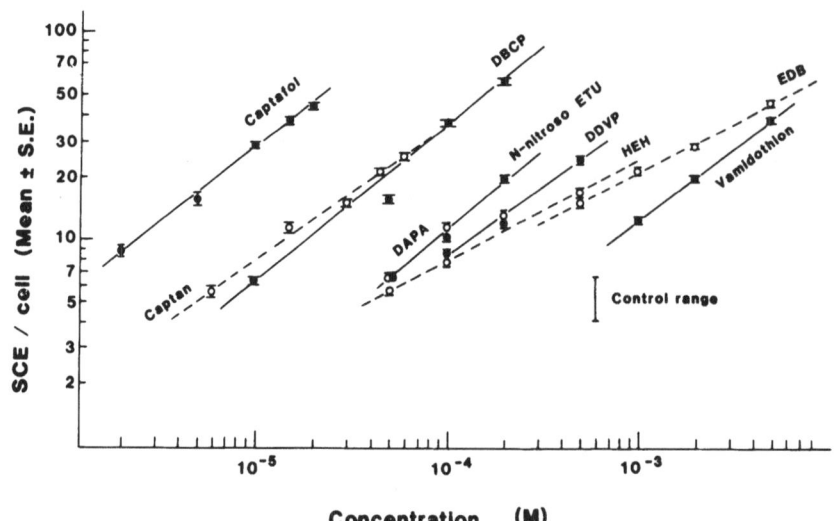

Figure 1. Dose-response relationships between the
frequency of sister chromatid exchanges in
cultured Chinese hamster V79 cells and the initial
concentration of each chemical. The lines in the
figure were drawn by eye. Each point represents
the mean ± standard error.

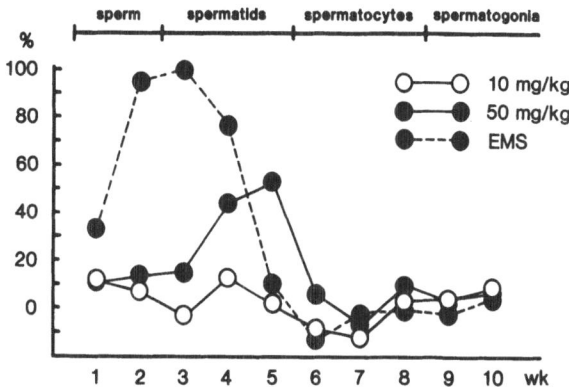

Figure 2. Frequency of dominant lethal mutations induced by DBCP in rats.

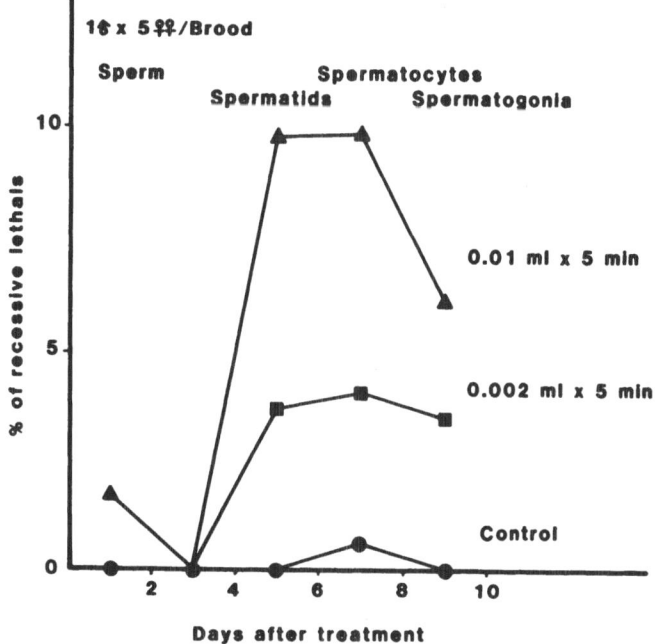

Figure 3. Induction of sex-linked recessive lethals in Drosophila males by 5-minute treatment with gaseous DBCP.

REFERENCES

1. Moriya, M., K. Kato, and Y. Shirasu (1978) Effects
    of cysteine and a liver metabolic activation
    system on the activities of mutagenic pesticides.
    Mutation Res. 57:259-263.
2. Moriya, M., K. Mitsumori, K. Kato, T. Miyazawa, and
    Y. Shirasu (1979) Carcinogenicity of N-nitroso-
    ethylene-thiourea in female mice. Cancer Let.
    7:339-342.
3. Moriya, M., T. Ohta, K. Watanabe, T. Miyazawa,
    K. Kato, and Y. Shirasu (1983) Further
    mutagenicity studies on pesticides in bacterial
    reversion assay systems. Mutation Res.
    116:185-216.
4. Shirasu, Y., M. Moriya, K. Kato, A. Futuhashi, and
    T. Kada (1976) Mutagenicity screening of
    pesticides in the microbial system. Mutation
    Res. 40:19-30.
5. Shirasu, Y., M. Moriya, K. Kato, F. Lienard,
    H. Tezuka, S. Teramoto, and T. Kada (1977)
    Mutagenicity screening on pesticides and
    modification products: A basis of carcinogenicity
    evaluation, In Origins of Human Cancer, Cold
    Spring Harbor Conferences on Cell Proliferation,
    H.H. Hiatt, J.D. Watson, and J.A. Winsen, Eds.,
    Cold Spring Harbor Laboratory, Vol. 4,
    pp. 267-285.
6. Teramoto, S., M. Moriya, K. Kato, H. Tezuka,
    S. Nakamura, A. Shingu, and Y. Shirasu (1977)
    Mutagenicity testing on ethylenethiourea.
    Mutation Res. 56:121-129.
7. Teramoto, S., R. Saito, H. Aoyama, and Y. Shirasu
    (1980) Dominant lethal mutation inducted in male
    rats by 1,2-dibromo-3-chloropropane (DBCP).
    Mutation Res. 77:71-78.
8. Teramoto, S., A. Shingu, and Y. Shirasu (1978)
    Induction of dominant-lethal mutations after
    administration of ethylenethiourea in combination
    with nitrite or of N-nitroso-ethylenethiourea in
    mice. Mutation Res. 56:335-340.
9. Tezuka, H., N. Ando, R. Suzuki, M. Terahata,
    M. Moriya, and Y. Shirasu (1980) Sister chromatid
    exchanges and chromosome aberrations in cultured
    Chinese hamster cells treated with pesticides
    positive in microbial reversion assays. Mutation
    Res. 78:177-191.
10. Tezuka, H., S. Teramoto, M. Kaneda, R. Henmi,
    N. Murakami, and Y. Shirasu (1978) Cytogenetic and
    dominant lethal studies on captan. Mutation Res. 57:201-207.

GENETIC TOXICOLOGY OF 14 AGENTS CAUSALLY

ASSOCIATED WITH CANCER IN HUMANS

Michael D. Waters[a], Neil E. Garrett[b],
Christine M. Covone-de Serres[c], Barry E. Howard[a], and
H. Frank Stack[b]

[a]Genetic Toxicology Division
 Health Effects Research Laboratory
 U.S. Environmental Protection Agency
 Research Triangle Park, NC 27711

[b]Northrop Services, Inc. - Environmental Sciences
 Research Triangle Park, NC 27709

[c]Genetics Curriculum
 University of North Carolina
 Chapel Hill, NC 27514

SUMMARY

     The purpose of this report is to summarize the currently
available qualitative and quantitative information, obtained from
genetic bioassay systems, for 14 agents or groups of agents that
have been classified as human carcinogens by the International
Agency for Research on Cancer. These compounds are of particular
interest in relating evidence of carcinogenic effects in man and in
experimental animals to qualitative and quantitative data obtained
from short-term genetic bioassays. The results of this study may be
used to aid in the selection of appropriate bioassays for suspected
human mutagens or carcinogens and as a comparative reference for the
evaluation of other agents with potential genetic activity in man.
The data base for the 14 known carcinogens consists of test
responses for approximately 100 test systems. Test systems are
organized according to classes of genetic or genetically related
activity and subdivided according to the phylogenetic level of the
indicator organisms. Quantitative values for responses in each of
these test systems are represented for each compound in graphic form
by computer. An evaluation of the genetic spectra was achieved by

applying simple pattern recognition techniques and statistical
methods.

1. INTRODUCTION

Qualitative and quantitative results from genetic bioassay
systems for 14 agents or groups of agents that have been classified
by the International Agency for Research on Cancer (IARC) in IARC
Supplement I (1979) as causally associated with cancer in
humans[97,112] are summarized in this report. Recently, IARC has
published an updated list[98] of known human carcinogens including
the 14 agents reviewed here and four others: azathioprene,
chlorambucil, 1,4 butanediol dimethanesulfonate (Myleran), and
treosulfonate. IARC also lists several groups or mixtures of agents
as known human carcinogens, but these have not been considered in
the present evaluation.

The IARC evidence for human carcinogenicity is derived from
case or clinical reports of individual patients exposed to the
agents in question, descriptive epidemiological studies where cancer
incidence has varied with exposure to the agent, and case-control
and cohort epidemiological studies. Thus adequate data of
carcinogenicity in humans are available for the 14 agents examined
in this report, and the IARC classifications were not based solely
on evidence of animal carcinogenicity or activity in short-term
tests. These compounds are of particular interest for studying the
relationships between carcinogenicity in man and the qualitative and
quantitative data obtained from short-term genetic bioassays.
Currently more than 100 test systems or bioassays are used for
detection of agents that produce genetic or genetically related
effects. Analysis of the genetic effects on the 14 human
carcinogens provides a reference or standard and may aid in the
selection of appropriate bioassays for the evaluation of other
agents with potential to cause human cancer or mutation.

The bioassay information was organized according to classes of
genetic or related activity and, where applicable, subdivided
according to the phylogenetic level of organization of the indicator
organism. The set of tests and responses for a given chemical were
then plotted so as to define a profile or spectrum of genetic
activity. The quantitative values for responses in each of these
test systems were represented for each compound in graphic form by
computer.

A quantitative assessment of the test data is essential if we
are to select properly those chemicals that should be subjected to
further evaluation from among the many chemicals active in
short-term genetic bioassays. Ultimately, quantitative evaluations
of genetically mediated effects must be coupled with accurate

estimates of cellular or tissue dose or dose to the DNA, because
this combination of information forms the basis of all current
models for quantitative risk assessment.

The present report considers the qualitative and
semiquantitative aspects of the assessment problem as it relates to
short-term tests. A computerized data base was assembled for the
14 agents based upon information compiled and evaluated by the EPA
GENE-TOX Program[80,247] and a substantial amount of additional data
obtained from the recent literature and evaluated by the authors.
These data were applied in combination to generate an overall
summary of the genetic bioassay data that are available for each
agent.

The report presents an abbreviated discussion of background
material and clinical, industrial and epidemiological findings for
each of the 14 agents. The agents are grouped according to target
organ specificity and the chemical structures of the agents are
discussed. The short-term test results are presented in tabular as
well as graphic form. A detailed analysis of the graphic spectra of
test results obtained for these agents has been presented
previously[73]. The thrust of the present effort has been the
compilation and preliminary analysis of a relatively large body of
short-term data, and an integrated quantitative assessment of the
information presented remains for the future.

2. CHEMICAL SELECTION

The 14 agents or groups of agents were selected for evaluation
by reference to the IARC Monographs on the Evaluation of
Carcinogenic Risk of Chemicals to Humans[97,112]. Chemicals, groups
of agents, or industrial processes are discussed in these
Supplements. The IARC Working Group concluded that the first 23 of
the chemicals or groups of chemicals are carcinogenic for humans.
This category was used only when sufficient evidence was found to
support a causal association between exposure to the agent and
cancer. The 14 carcinogenic agents for which bioassay data are
presented here and their Chemical Abstracts Service (CAS) registry
numbers are as follows:

    4-aminobiphenyl (92-67-1)
    arsenic and certain arsenic compounds (7440-38-2)
    asbestos (1332-21-4)
    benzene (71-43-2)
    benzidine (92-87-5)
    N,N-bis(2-chloroethyl)-2-naphthylamine (chlornaphazine)
      (494-03-1)
    bis(chloromethyl) ether (542-88-1) and technical grade
      chloromethyl methyl ether (107-30-2)

chromium and certain chromium compounds (7440-47-3)
cyclophosphamide (50-18-0)
diethylstilbestrol (56-53-1)
melphalan (148-82-3)
mustard gas (505-60-2)
2-naphthylamine (91-59-8)
vinyl chloride (75-01-4)

3. LITERATURE SEARCH AND DATA EVALUATION

A search of the major computer-based bibliographic systems was
performed by the EPA Library at Research Triangle Park, NC. Toxline
was used as the primary source of literature citations for the
following reasons: (1) comprehensiveness (includes Biological
Abstracts, Chemical Abstracts, Index Medicus, Pesticide Abstracts,
the Environmental Mutagen Information Center, the Environmental
Teratology Information Center, Research Projects in
Toxicology--Smithsonian Science Information Exchange); (2) inclusion
of abstracts; (3) use of CAS registry numbers for reliable retrieval
of specific chemical substances; and (4) the cost is significantly
less than that of accessing individual sources directly.

The search process also included Cancerlit, the National
Technical Information Service file (government reports), and
Excerpta Medica (a European biomedical file that is particularly
strong in chemical and drug toxicity information).

From each report the reviewers abstracted information for any
of the 14 agents evaluated in this study. Reports in abstract form
without quantitative data were excluded. The overall test result
was recorded together with an appropriate assay code and publication
identification number. The use of an exogenous metabolic activation
system was indicated. Each report was evaluated a second time by a
different reviewer for assurance of completeness and accuracy. The
information was then put into 100 separate files, each identified by
a distinct three-letter code given to the short-term bioassays. The
files were established on floppy disks with a Tektronix 4050 series
computer and Tektronix 4907 file manager. The data in the test
reports were then re-examined by adding dose-related information
from the reports to the computer files.

These data have also been coded on data entry forms,
keypunched, and verified before entry into a data base maintained at
the EPA Univac 1110 Data Center at Research Triangle Park, NC. The
data were processed through the use of a System 2000 (Intel Corp.,
Austin, TX) data management software package. Results are retrieved
through a data selection program written in COBOL that allows
reports to be generated in matrix format on the basis of chemical
name, CAS number, and/or assay code.

For purposes of standardization with similar literature evaluations, the three-character EPA GENE-TOX bioassay codes were used when applicable; other three-character codes were assigned as needed. The sequence of the bioassays, the bioassy codes and the definition of each bioassay used in this report are shown in Table I. Codes not specified by GENE-TOX appear in parentheses.

With the bioassays arranged according to classes of genetic or related activity, phylogenetic level, and type of test, a discrete set of tests and responses defines a profile or spectrum of genetic activity. The genotoxic data for each chemical may thus be presented in graphical form. Because of the range of doses encountered in 100 bioasssays, the logarithm of the dose was used in calculations and graphical representations. The graphical representation of the data was formed by using an x-axis of 100 unit values corresponding to the different test systems and y-axis values corresponding to the negative logarithmic value of the observed effective or ineffective doses[73]. Detailed aspects of dose-response relationships were not recorded because test results were often reported in the published papers as a single dose. Thus, either the lowest effective dose or the maximum dose producing no effect was recorded for each test agent and bioassay system. This allowed comparison of results generated at a single dose with other results in the data base. For negative results, the highest dose reported yielding the negative response was utilized. Positive results were recorded for the least effective dose (LED) and 10^{-5} times the LED value was used in the logarithmic transformation.

Dose information from the reports was often given in various units and these were converted to μg/ml. Gaseous compounds such as vinyl chloride reported on a volume-to-volume basis were converted to mass per volume according to the ideal gas laws. A volume of 2 ml was used for the top agar in the microbiological plate incorporation assays. A volume of 1 ml was used for DNA repair-deficient bacteria spot tests that lacked uniform dispersion of the test compound. Doses for in vivo bioassays were expressed in units of mg/kg body weight of the exposed animal. Under the logarithmic transformation, a dose range is defined as 0 to -5 units for ineffective dose (1-100,000 ppm) and 0 to +5 units for effective doses (100,000-1 ppm). Because of differing results for a given bioassay of a chemical, the subset of data corresponding to the majority of the test results (+ or -) was determined, and the mean value of the effective or ineffective doses was plotted.

If available, the dose for positive effects was extracted from the reports where results were significant at the P < 0.05 level. If the probability value was not available, the reviewers made a

TABLE I.   GENETIC BIOASSAYS USED TO SCORE
FOR GENOTOXICITY

NUMBER	CODE	GENETIC ENDPOINT

POINT/GENE MUTATION ASSAYS

Prokaryotic Assays

Salmonella typhimurium

1	SA5	TA1535
2	SA7	TA1537
3	SA8	TA1538
4	SA9	TA98
5	SA0	TA100

Escherichia coli

| 6 | WP2 | WP2, reverse mutation |
| 7 | WPU | WP2 uvrA, reverse mutation |

Lower Eukaryotic and Body Fluid Assays

Saccharomyces cerevisiae

| 8 | YEF | forward mutation |
| 9 | YER | reversion test |

Schizosaccharomyces pombe

| 10 | YEY | forward mutation |
| 11 | YEZ | reversion test |

Neurospora crassa

12	NEF	forward mutation
13	NER	reversion test
14	BFU	Body fluids, urine

Higher Eukaryotic In Vitro Assays

Chinese hamster

15	(CHL)	lung
16	CHO	ovary
17	V7H	lung (V-79) HGPRT locus
18	V70	lung (V-79) ATPase locus
19	L5T	Mouse lymphoma (L5178Y) TK locus

Higher Eukaryotic In Vivo Assays

20	ARM	Arabidopsis mutation
21	TRM	Tradescantia mutation
22	SRL	Drosophila melanogaster sex-linked recessive lethal test
23	MST	Mouse spot test
24	SLP	Mouse specific locus test, postspermatogonial stages
25	SLT	Mouse specific locus test, all stages
26	HMA	Host-mediated assay

PRIMARY DNA DAMAGE ASSAYS

Prokaryotic Assays

Escherichia coli Pol A (W3110-P3478)

27	REP	spot test
28	RER	spot test, well
29	RET	liquid suspension test
30	REW	Bacillus subtilis rec (H17-M45), spot test
31	REC	DNA repair-deficient bacteria

Escherichia coli

| 32 | (WPR) | WP100 uvrA⁻ rec⁻ or other rec⁻ strains |
| 33 | (ECO) | differential toxicity, miscellaneous strains |

NUMBER	CODE	GENETIC ENDPOINT

Lower Eukaryotic Assays

                 Saccharomyces cerevisiae

| 34 | YEC | gene conversion |
| 35 | YEH | homozygosis (through recombination or gene conversion) |

Higher Eukaryotic Assays

                 Unscheduled DNA synthesis in mammals

36	(UDB)	human bone marrow
37	UDH	human diploid fibroblasts
38	(UDL)	HeLa cells
39	UDS	all cell types, in vitro
40	(UDX)	Xeroderma pigmentosum cells
41	UDP	rat primary hepatocytes
42	(UDM)	mouse, in vivo

                 Inhibition of DNA synthesis

43	(IDL)	HeLa cells
44	(IDP)	rat primary hepatocytes
45	(IDR)	rodent
46	(DBH)	DNA strand break, human

CHROMOSOMAL EFFECTS ASSAYS

Sister Chromatid Exchange

47	SCC	Chinese hamster ovary (CHO) cells, transformed
48	SCF	Human fibroblasts, normal
49	SCH	HeLa cells, transformed
50	SCM	Human lymphoblastoid cells, transformed
51	SCP	Chinese hamster fibroblasts, transformed
52	SCV	Chinese hamster lung fibroblasts (V-79cells), transformed
53	SCE	In vitro and in vivo
54	SCL	Human lymphocytes in vitro
55	SC2	In vitro, all animals except human
56	SC3	In vivo, all animals except human
57	SC4	In vivo, human cells

Aneuploidy

| 58 | NEN | Neurospora crassa |

                 Drosophila melanogaster

| 59 | DAC | whole sex chromosome loss |
| 60 | DAP | partial sex chromosome loss |

Chromosomal Aberrations In Vitro

                 Mammalian cytogenetics

61	CYU	Chinese hamster ovary (CHO) cells
62	CYV	Syrian golden hamster
63	CYY	mouse
64	CYZ	human
65	CYC	all cell types
66	CYH	lymphocytes, human
67	(CYX)	Xeroderma pigmentosum cells

Chromosomal Aberrations In Vivo

                 Mammalian cytogenetics

68	CYB	bone marrow studies, all animals
69	CYL	lymphocyte or leukocyte studies, all animals
70	CYG	spermatogonial stem cells treated, spermatocytes observed
71	CYO	oocyte or early embryo studies

(continuted)

TABLE I.   (CONTINUED)

NUMBER	CODE	GENETIC ENDPOINT
72	CYS	spermatogonia treated, spermatogonia observed
73	CYT	differentiating spermatogonia or spermatocytes treated, differentiating spermatocytes observed
74	HOC	Hordeum cytogenetics
75	TRC	Tradescantia cytogenetics

Micronuclei
76	(MNC)	In vitro
		In vivo
77	(MNH)	hamster
78	(MNM)	mouse
79	(MNR)	rat

Chromosomal Damage In Vivo
		Dominant lethal test
80	(DLD)	Drosophila melanogaster
81	DLM	mouse
82	DLR	rat
		Heritable (reciprocal) translocation
83	DHT	Drosophila melanogaster
84	(MHT)	mouse

CELLULAR TRANSFORMATION ASSAYS

		Syrian hamster embryo
85	CTC	clonal assay
86	CTF	focus assay
87	CTS	transformation strains
88	CTB	BALB/c3T3 cells
89	CTH	C3H10T1/2 cells
90	CTL	Established cell lines
91	CTK	AKR/ME cells
92	CTR	RLV/Fischer rat embryo cells
93	CTV	Virus enhancement
94	CT7	SA-7/SHE cells

SPERM MORPHOLOGY ASSAYS

95	SPA	rat
96	SPH	human
97	SPI	mouse
98	SPR	rabbit
99	SPS	sheep
100	SPF	mouse F_1 assay

conservative judgment (response > 2 x control value) of the dose required for the effect.

4. HUMAN EXPOSURE, CARCINOGENICITY, AND MUTAGENICITY

Table II represents the total data base in its condensed form as it is discussed in subsequent sections. In Table II, where systems or strains àre combined under a specific column, the letter(s) in the three-character code appropriate to a specific entry in the body of the table appear immediately above the test result(s). The detailed qualitative data base and listing of supporting references for 24 known or suspected human carcinogens has been published previously[248].

To facilitate an integrated discussion of the condensed data base, the agents under investigation are grouped by target organ specificity. Chemical structures are provided for each group. Because the genetic bioassay results have been condensed, not all citations pertinent to each statement regarding a qualitative response are provided. Instead, an appropriate reference that supports the statement is given.

4.1 Human Respiratory Tract Carcinogens

Seven compounds examined in this study are either volatile or undergo condensation on the surfaces of respirable particles and enter the human respiratory tract. Three of these compounds are highly volatile alkylating agents: bis(chloromethyl) ether, chloromethyl methyl ether, and mustard gas. One is the highly reactive substituted alkene, vinyl chloride. (The results of the literature review for vinyl chloride are presented in Section 4.4 Other Carcinogens.) Two metals (arsenic and chromium) and the group of particulate minerals considered collectively as "asbestos" complete this group of agents that appear to induce cancer of the human respiratory tract.

BIS(CHLOROMETHYL)ETHER   CHLOROMETHYL METHYL ETHER   VINYL CHLORIDE   MUSTARD GAS

$$\begin{array}{cccc} \text{CH}_2\text{-Cl} & \text{CH}_3 & \text{H} \quad \text{Cl} & \text{CH}_2\cdot\text{CH}_2\text{Cl} \\ \text{O} & \text{O} & \text{C=C} & \text{S} \\ \text{CH}_2\text{-Cl} & \text{CH}_2\text{-Cl} & \text{H} \quad \text{H} & \text{CH}_2\cdot\text{CH}_2\text{Cl} \end{array}$$

Sufficient data are not available to adequately analyze the genetic activity spectra of the highly volatile alkylating agents bis(chloromethyl) ether, chloromethyl methyl ether, and

TABLE II. CONDENSED BIOASSAY DATA BASE[a]

Compound	Point/Gene Mutation Assays													Primary DNA Damage Assays			
System:	SA_	WP_	YE_	NE_	(B_)	CH_	V7_	L5T	_RM	SRL	MST	SL_	HMA	RE_	(WPR)	(ECO)	YE_
Test/Strain:	57890	2U	FRYZ	FR	SWY DMMR	(L)O	HO		AT			PT		PRTWC			OH
4-aminobiphenyl	-/+	-/-			S R ±/+		H 0/-						-/0	T 0/+			H +/+
arsenic compounds	-/-	±/-	2U +/0					+/+		-/0				W +/0	+/0	-/0	H +/+
asbestiform minerals	-/-	-/-				(L) +/+											
benzene	-/-	-/-							T +/0	-/0				C -/-			H -/-
benzidine	-/+				S R -/+					+/0					-/+		H -/-
bis(chloromethyl) ether	0/+																
chlornaphazine	+/+									+/0							
chloromethyl methyl ether																	
chromium compounds	±/±	+/0	Y +/0		SY HMR +/0						+/0			PW ±/0		+/0	H +/+
cyclophosphamide	-/+	±/0	F +/0	F +/0			0 +/0	-/+	A -/0	+/0	+/0	P +/0	+/0	PT 0/+			CH ±/+
diethylstilbestrol	-/-	-/-			S H -/0		HO -/-	+/+								-/0	
melphalan	+/+				S H -/0			+/+		+/0				PR -/+			
mustard gas		U +/0		R +/0						+/0			+/0			+/0	
2-naphthylamine	-/+	F +/0	F +/0	FR -/-	S R -/+		H 0/-			±/0			+/0	TC -/-	-/+	-/0	CH ±/±
vinyl chloride	±/+	2U -/-	RY -/+	FR -/-	S HR -/-	O -/+	HO -/+		T +/0	+/0	-/0		+/0	PTC -/+			CH -/+

[a]System and test/strain codes are defined in Table I. Test results are shown without metabolic activation/with metabolic activation and are expressed as follows: + = positive; - = negative; ± = different results in two or more tests; 0 = no result entered. In results from whole-animal systems, the 0 entry for "with metabolic activation" indicates that exogenous metabolic activation is not required.

| | Primary DNA Damage Assays | | | | | Chromosomal Effects Assays | | | | | | | | | | | | | |
Compound	UD_ (B)H(L)S(X)	UDP	(UDM)	(ID_) LPR	(DBH)	SC_ CFPV	SCL	SC_ 234	NEN	DA_ CP	CY_ UVYZC	CYH	(CYX)	CY_ BL	CYG	CY_ OST	_C HO TR	(MNC)	(MN_) HMR
4-aminobiphenyl	H −/0																		R −/0
arsenic compounds	H −/−	H +/0					+/0				Z +/0	+/0		BL +/0				−/0	
asbestiform minerals										CP ±/0	UVC +/0	+/0		B −/0					M −/0
benzene				R +/0				3 +/0						BL ±/0					M +/0
benzidine				L +/0			±/+					±/0				S −/0			MR ±/0
bis(chloromethyl) ether																			
chloronaphazine																			
chloromethyl methyl ether																			
chromium compounds	HS +/0				+/0	CF ±/0					UVYZ +/0	+/0						±/0	M +/0
cyclophosphamide	(L) 0/+		+/0	L −/+	−/+	CFP +/+	−/0	234 +/0		CP −/0	UZC ±/+	−/+		BL +/0	+/0	OS +/0			HMR +/0
diethylstilbestrol	(L) 0/+					FP ±/0		3 −/0	−/0		C −/0	−/0		B ±/0					M ±/0
melphalan	(B) +/0				+/0			24 +/0			C +/0								
mustard gas										CP +/0	C +/0								
2-naphthylamine	H 0/+																		MR ±/0
vinyl chloride							−/+	34 ±/0		CP −/0				BL ±/0					M +/0

(continued)

TABLE II. (CONTINUED)

Compound	Chromosomal Effects Assays					Cellular Transformation Assays				Sperm Assays	
System:	(DLD)	DLM	DLR	DHT	(MHT)	CT_	CT_	CT_	CT7	SP_	SPF
Test/Strain:						CFS	BHL	KR		AHIRS	
4-aminobiphenyl						C +/0		KR +/0			
arsenic compounds		-/0				C +/0			+/0		
asbestiform minerals							B +/0				
benzene									-/0	I +/0	
benzidine						C +/0		KR +/0	+/0	I -/0	
bis(chloromethyl) ether											
chlornaphazine									+/0		
chloromethyl methyl ether									+/0		
chromium compounds						C +/0	B +/0	R +/0	+/0		
cyclophosphamide		-/0	+/0		±/0	C +/0	BHL -/+	R +/0		AHIS +/0	-/0
diethylstilbestrol						C -/0	B -/0	R +/0		HI ±/0	
melphalan							H +/0				
mustard gas	+/0			+/0							
2-naphthylamine						C +/0		KR +/0	+/0	I -/0	
vinyl chloride	-/0	-/0	-/0	-/0							

bis(β-chloroethyl) sulfide (mustard gas). Considerably more
bioassay data are available for vinyl chloride.

The inorganic compounds As and Cr are positive in one or more
mutational tests at doses $\leq$ 1μg/ml. Arsenic is positive at these
levels for sister chromatid exchange (SCE) in human lymphocytes, for
chromosomal aberration in human cells including lymphocytes in vitro
and for virus-enhanced cell transformation. The chromium compounds
are active at concentrations $\leq$ 1 μg/ml. These include: unscheduled
DNA synthesis (UDS) in human diploid fibroblasts and other cell
systems; sister chromatid exchange in CHO cells and normal human
fibroblasts ($\leq$ 0.1 μg/ml); chromosomal aberrations in human cells,
including lymphocytes in culture, and hamster and mouse cells
$\leq$ 0.1 μg/ml). Chromium is also positive at doses $\leq$ 0.1 μg/ml in the
BALB/c3T3 mouse cell transformation system, and $\leq$ 1.0 μg/ml in the
SHE clonal and SA7 virus-enhanced transformation systems.

    4.1.1 Bis(chloromethyl) Ether and Chloromethyl Methyl Ether.
Dichloromethyl ether, or bis(chloromethyl) ether (BCME), is a highly
volatile compound sold as a laboratory chemical and used as a
chloromethylating reaction mixture in the preparation of ion
exchange resins[172]. Thiess et al. reported in 1973 on 6 cases of
lung cancer in 18 testing laboratory workers who used BCME[103,226].
Five of these six men were moderate smokers. However, of the total
cancers reported, the majority were oat-cell carcinomas, a tumor
type not usually seen in smokers.

Chloromethyl methyl ether (CMME) also is highly volatile and
frequently is contaminated with BCME. CMME is also used in the
preparation of ion exchange resins. One study reported an increased
risk of lung cancer for workers exposed to CMME[67]. Fourteen men
developed lung cancer and 12 of these had oat-cell carcinomas.

BCME gives positive results in Salmonella strain TA100 in the
presence of S-9 activation[9]. No other point/gene mutation,
primary DNA damage, or chromosomal effects data are reported. Cell
transformation assays have been performed with CMME. The compound
enhances viral transformation of Syrian hamster embryo cells in the
absence of exogenous activation[33].

BCME and CMME have not been studied extensively for genetic
activity in short-term tests. This may be because of case reports
indicating that these compounds are exceedingly dangerous, with a
high percentage of oat-cell carcinomas developing in exposed
workers. The paucity of data precludes consideration of the
relevance of specific genetic bioassay systems for these compounds.

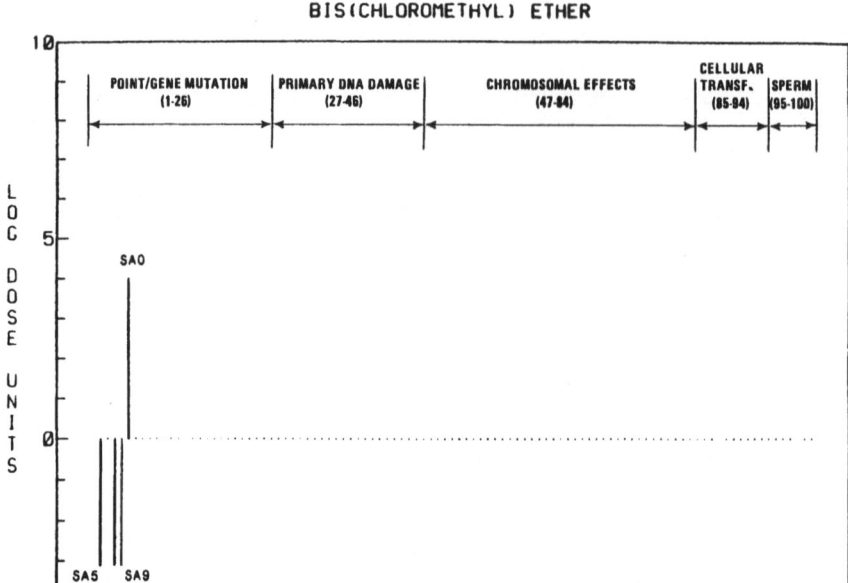

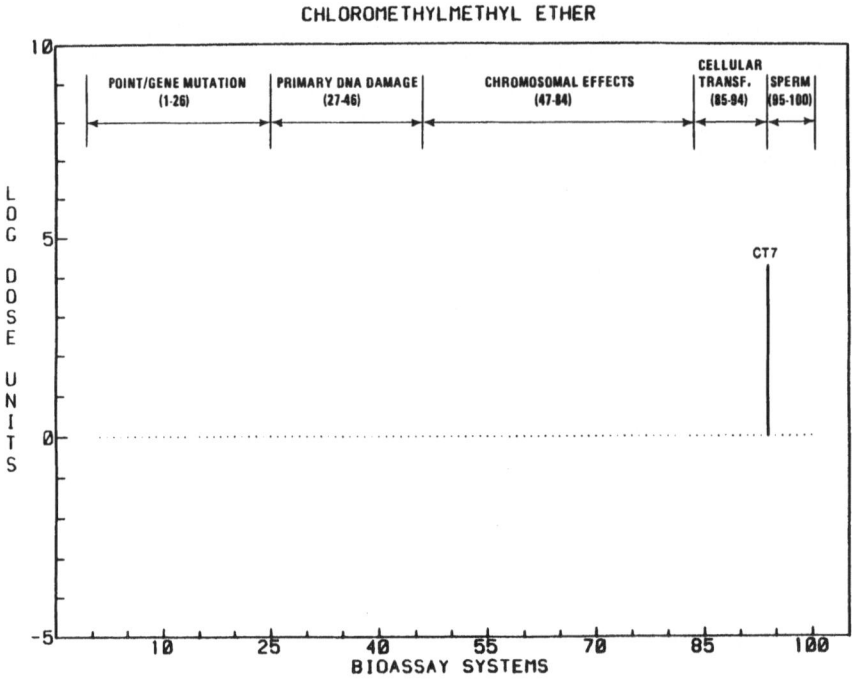

4.1.2. Mustard Gas. Mustard gas or sulfur mustard is a deadly vesicant that has been used as a war gas. The compound causes conjunctivitis and blindness in man. It is an important model compound in biological studies requiring the use of direct-acting alkylating agents.

Wada et al. reported that factory workers engaged in the manufacture of mustard gas had an elevated incidence of respiratory neoplasia[244]. In this study, 33 deaths resulted from neoplasia of the respiratory tract. Thirty of these were histologically confirmed. Only 0.9 deaths were expected. A study of World War I veterans indicated that the risk of death from lung cancer for men gassed with the compound was significantly increased above that of the controls[166]. A similar study by Case and Lea indicated that persons exposed to mustard gas also suffered from chronic bronchitis[32]. These authors indicated that an increased risk of lung and pleural cancer appeared to be associated with the chronic bronchitis.

Mustard gas gives positive results in E. coli WP2 uvrA and in a host-mediated assay using L5178Y cells in mice in the absence of S-9[30]. The gas also produces reverse mutations in Neurospora[220] and sex-linked recessive lethality in Drosophila[11]. In assays for primary DNA damage, mustard gas is

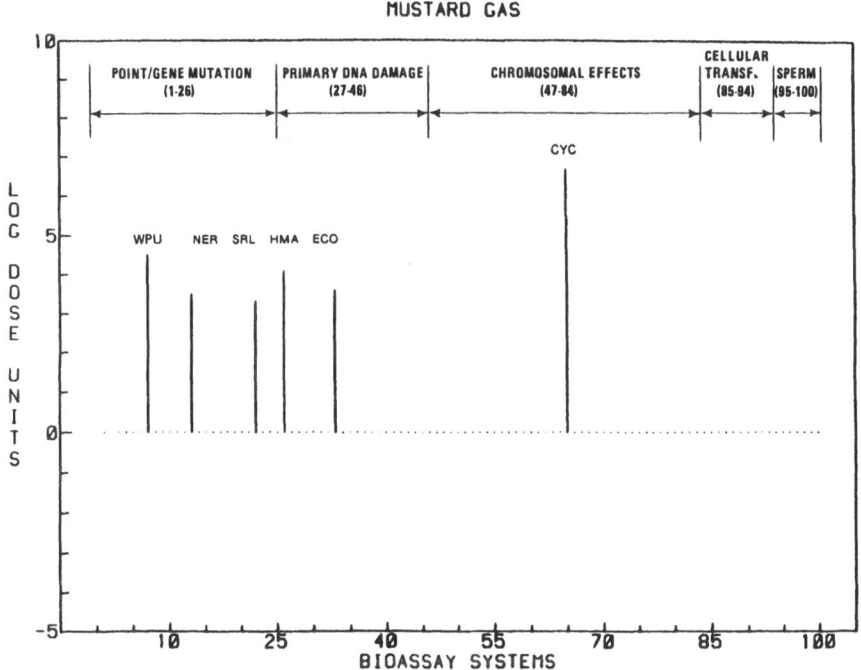

more toxic to repair-deficient E. coli strains than to the
corresponding competent strains[132]. In assays for chromosomal
effects, mustard gas produces positive responses in mammalian cells
in vitro[199] and in the Drosophila aneuploidy[170], dominant
lethal[11], and heritable translocation[170] tests. The genetic
toxicology data base indicates that exogenous metabolic activation
is not required to produce point mutations, primary DNA damage, or
chromosomal effects.

    4.1.3 Arsenic Compounds. Arsenic is used in metal alloys, in
semiconductor devices, and in the manufacture of certain types of
glass. The compound has been used medically in the treatment of
infections and various types of skin disorders. Arsenic is also
used in pesticides for agricultural processes.

    Hutchinson (1887-8) first called attention to the relationship
between certain arsenic-containing drugs and skin cancer[96].
Subsequently, a variety of reports appeared concerning
arsenic-containing drugs and skin cancer. For example in 1913
Pye-Smith reported 31 such cases, 24 apparently of medical
origin[182]. Neubauer reviewed the reports of cancer resulting from
arsenic[162]. Three major categories are of interest: (1) cancer
resulting from arsenic-containing drugs; (2) cancer caused by
arsenic in drinking water; and (3) cancer caused by occupational
exposure to arsenic.

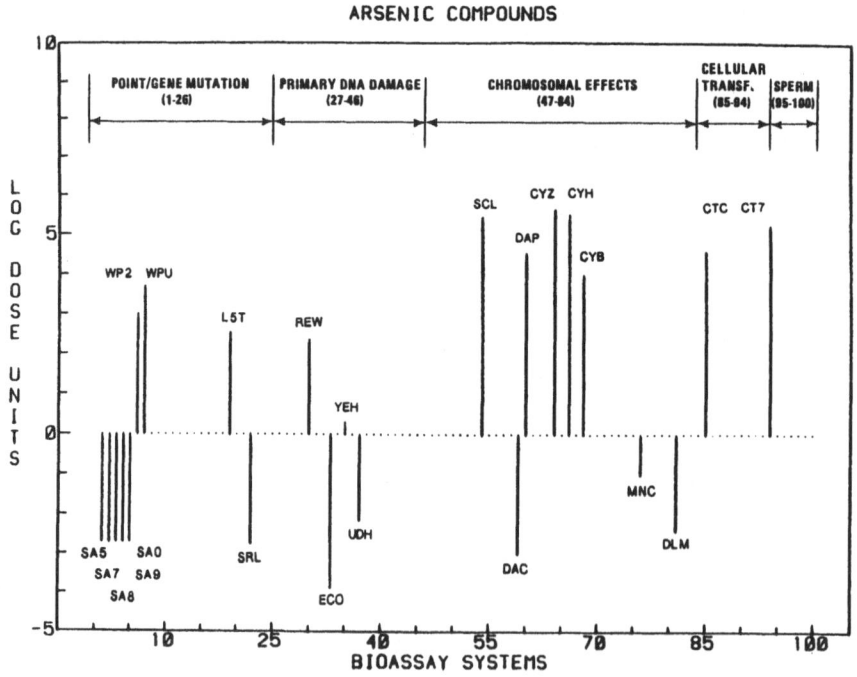

A study of occupational lung cancer presumably due to arsenic was reported by Kuratsune and coworkers[127]. The only significant difference between a studied group of cancer patients and the control group was that 11 of 19 who died from lung cancer had been employed as smelters in a local copper refinery where they had received heavy exposure to arsenic trioxide. Recently arsenic was found to be concentrated seven-fold in the lung tissue of smelter workers[250]. Other studies of occupational exposure in mines and smelters have been catalogued[111].

Arsenic released into the environment by various industrial processes may contribute to the total cancer burden[133]. One area of concern is air pollution. Mortality rates from lung cancer are significantly increased in counties in which copper, lead, or zinc smelting or refining industries contribute to atmospheric arsenic[24]. Residents of an area near a pesticide plant that has produced arsenic-containing materials since the early 1900s were reported to have a significantly increased incidence of lung cancer[142].

The trivalent arsenic compounds as a group give uniformly negative results in the standard Salmonella tester strains in the absence of S-9 metabolic activation[180]. Other reports of tests in Salmonella in the presence of metabolic activation are also negative. Trivalent arsenic compounds induce reverse mutations in E. coli WP2 and WP2 uvrA⁻ in the absence of exogenous metabolic activation[164]. Trivalent arsenic induces forward mutations at the thymidine kinase (TK) locus in the mouse lymphoma system [180]. Neither the trivalent compound nor the pentavalent compound causes sex-linked recessive lethality in Drosophila[58,237].

With regard to primary DNA damage, trivalent and pentavalent arsenic give positive results in relative toxicity assays in B. subtilis rec⁻ strains[164]. Pentavalent arsenic induces enhanced mitotic recombination or gene conversion in S. cerevisiae with or without metabolic activation[210]. The compound does not induce unscheduled DNA synthesis (UDS) in human embryonic lung fibroblasts with or without activation[210]. Trivalent arsenic induces sister chromatid exchanges (SCEs) in lymphocytes of patients treated with the compound[28]. Arsenic also produces chromosome aberrations in bone marrow of mice[218], in lymphocytes of exposed patients,[175] and in cultured human lymphocytes and fibroblasts[160]. Micronuclei are not induced in vitro[49].

A cell transformation clonal assay in SHE cells is reported to be positive[52]. In addition, trivalent arsenic, among other metals, enhances viral transformation of cells in culture[35].

In summary, the most convincing positive results indicate
primary DNA damage and chromosomal effects in mammalian cells
in vitro and in vivo. Results of cell transformation assays also
are positive.

4.1.4 Chromium Compounds. Chromium compounds are used in the
dyeing and tanning industries to improve the stability and affinity
of dyes to textiles and various polymers. Chromium compounds are
also used in coloring and hardening marble, in polishing metals, in
coloring glass, and in printing fabrics. Several of the compounds,
including chromic sulfate and chromium trioxide, are used in chrome
plating. Chromium is used in the manufacture of chrome steel,
chrome nickel steel alloys, and stainless steel.

Chromium compounds are strong nasal and pulmonary irritants and
have been reported to cause bronchiogenic carcinoma. They may also
cause gastrointestinal irritations and renal injury. Chromium is
the only one of the essential trace metals that is retained in
insoluble form in the lung[197]. Cancer of the lung in German
chromate workers was first reported in 1911 and 1912[12,13]. The
duration of exposure was often quite long, sometimes 40 years, and
the time between the beginning of the exposure and onset of cancer
was often 30 to 40 years. In 1948 Machle and Gregorius reported
that almost 22% of all deaths in the United States chromate industry
were due to cancer of the respiratory system[139]. This was

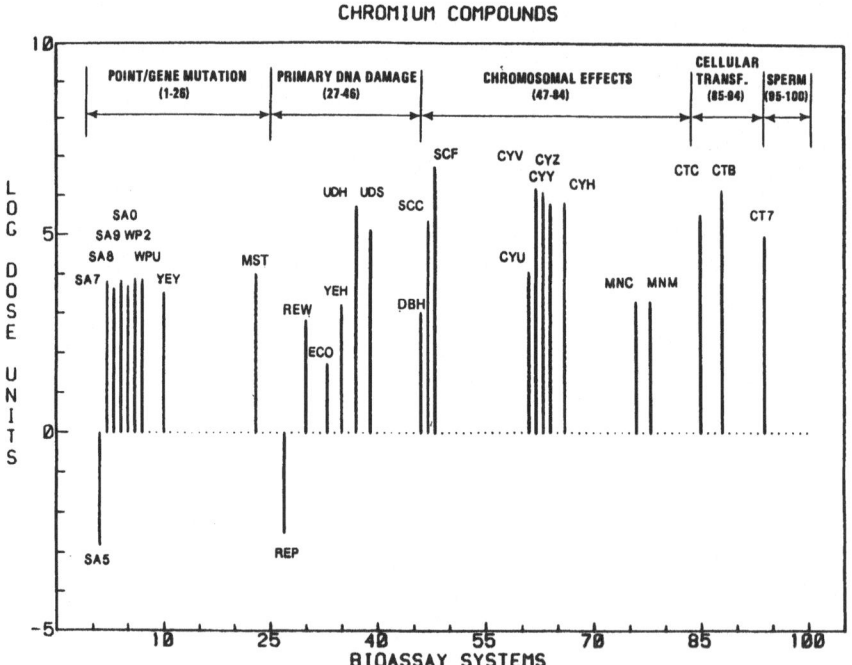

16 times the expected rate. The death rate for cancer of the lung in various chromate plants was increased over normal values by as much as a factor of 50. A report on the bichromate production industry in Great Britain corroborated the findings pertaining to the health hazards involved in that industry in Germany and the U.S.[21]. More recently, a follow-up study of 2715 workers in chromate production factories showed an increased risk of lung cancer[5]. Lung cancer in Japanese chromate workers also has been reported[168].

Hexavalent chromium compounds have tested mainly positive in all five standard Salmonella tester strains both in the presence and absence of S-9[161,176]. Potassium dichromate induces forward mutation and gene conversion in S. pombe[25]. Hexavalent chromium and stainless steel welding fume particles containing chromium compounds give positive results in the mouse coat color spot test, a somatic tissue assay[123].

Results of primary DNA damage assays in repair-deficient strains of E. coli Pol A and B. subtilis rec are mixed[161,164]. Chromium enhances homozygosis in S. cerevisiae with and without metabolic activation[161]. Hexavalent chromium compounds induce UDS in cloned mouse cells[185] and in human diploid fibroblasts[251] in culture in the absence of exogenous activation.

SCEs are induced in CHO cells and in human fibroblasts after exposure to low levels of hexavalent chromium in the absence of exogenous metabolic activation[140]. Trivalent chromium does not induce SCEs in CHO cells[140]. In vitro cytogenetic assays in CHO cells, mouse cells, Syrian hamster cells, and human fibroblasts show positive results without exogenous metabolic activation[134,140,236]. Hexavalent chromium produces aberrations in metaphases of human cells at a concentration of 2 μM[140]. Mixed results are obtained in micronucleus assays in cells in culture[62]. Hexavalent chromium induces micronuclei in the mouse in vivo[252].

Chromium gives positive results without exogenous metabolic activation in cell transformation assays using primary SHE cells[52] and established cell lines[185]. Positive results are also reported for the SA-7 viral enhancement assay[33].

The genetic toxicology data base consists of mostly positive results for hexavalent chromium compounds and mixed results for trivalent chromium compounds. With hexavalent compounds, independent and concurring positive responses are reported for induction of reverse mutations in the Salmonella system, for SCEs, and for chromosome abnormalities in CHO and mouse cells. The hexavalent compounds are positive while the trivalent compounds are negative in E. coli reverse mutation and B. subtilis differential toxicity assays, in SCE assays in CHO cells in vitro, and in chromosomal aberration assays in mouse and Chinese hamster cells in vitro.

4.1.5. Asbestiform Minerals. Although asbestos occurs naturally, mining operations introduce asbestos as a pollutant in both water and air. Also, asbestos fibers have been used to make filters and are thus introduced inadvertently in the manufacture of processed food, beverages, and drugs.

An association between respiratory exposure to asbestos and lung cancer was reported as early as 1935[77,138]. Doll presented evidence from an epidemiological study showing a ten-fold excess risk of lung cancer for asbestos textile workers[54]. In 1960, Wagner and coworkers reported the occurrence of mesothelioma in asbestos mine workers and in non-mining populations close to the mines[245]. Selikoff et al. reported other studies of the association between asbestos exposure and neoplasia[201]. In a study population of 632 insulation workers, 45 died of cancer of the lung pleura when only 6.6 such deaths were expected. Twenty-nine men died of cancer of the stomach, colon, and rectum when only 9.4 such deaths were expected. Hammond and coworkers, reporting on cancer among insulation workers, noted a significantly increased incidence of intraabdominal neoplasias[84]. The occurrence of gastrointestinal cancer was also reported by Kleinfeld et al.[122].

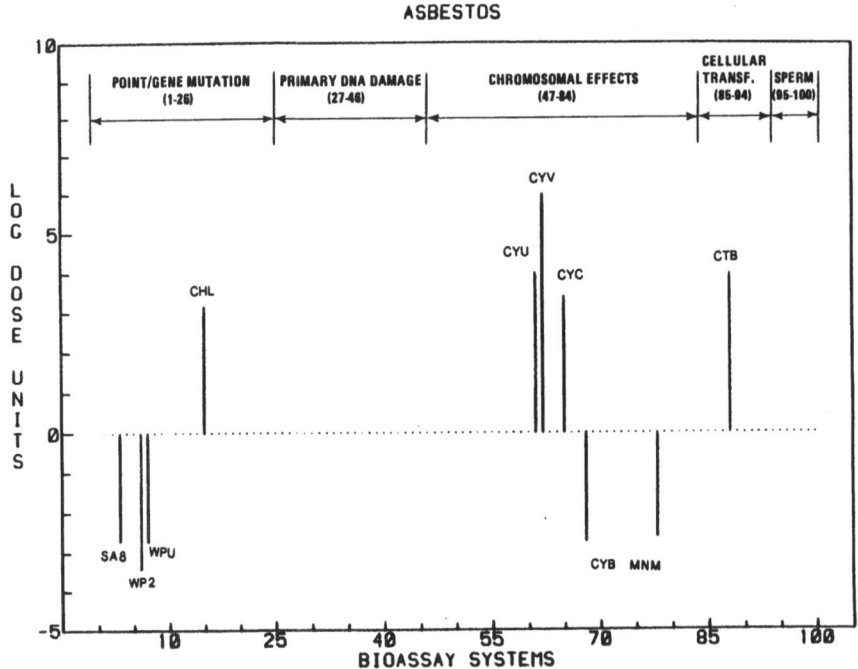

Gerber reported on asbestos exposure in neoplastic disorders of the hematopoietic system[75] and many additional epidemiological studies on asbestos and lung cancer have been summarized[108]. Studies of workers in asbestos manufacture, insulation, and shipyards have provided the most concrete evidence for the association of asbestos with lung cancer. Several authors have reported on the synergistic effects of smoking and occupational exposure to asbestos in the induction of neoplasia. Selikoff et al. concluded that asbestos workers who smoke have 8 times the risk of developing lung cancer when compared with all other workers, and 92 times the risk when compared with nonsmokers who do not work with asbestos[202].

For the purpose of assessing qualitative genetic and related bioassay data, the results reported for amosite, crysotile, and crocidolite were grouped under the category "Asbestiform Minerals." Particulate samples of these minerals do not enter bacteria; hence, results in Salmonella strains TA1535 and TA1538 are negative both without and with S-9 activation[36]. Results in E. coli strain WP2 and WP2 uvrA⁻ are also negative both without and with S-9 activation[36]. The CHL strain of Chinese hamster lung cells responds to this class of agents to exhibit forward mutation at the HGPRT locus[92]; however, the observed mutation frequency is low and may result via a clastogenic mechanism.

The results of in vitro cytogenetic assays without S-9 metabolic activation in SHE cells[130], Chinese hamster cells[93,213], and human lymphocytes[238] are positive. Oral or intraperitoneal administration of chrysotile asbestos does not induce micronuclei in the mouse[130]. Foci of multilayered growth indicative of cell transformation are observed in the BALB/c3T3 assay in the absence of exogenous metabolic activation[212].

In summary, nearly half of the bioassay results are negative. Positive responses have been confirmed only in independent investigations of cytogenetic abnormalities in mammalian cells in culture. Possibly because of their particulate nature, these materials are largely refractory to analysis by in vitro genetic bioassay systems.

4.2. Human Hematolymphopoietic System Carcinogens

Two of the agents examined in this report produce cancer of the bone marrow or of the lymphopoietic system. Melphalan is an alkylating agent and benzene is believed to be metabolized to reactive intermediates that can alkylate. That hematolymphopoietic tissue is the common target of these compounds is interesting in

view of the different routes of human exposure. The chemical
structures of melphalan and benzene are shown below.

MELPHALAN   BENZENE

HOOCCHCH₂———⟨   ⟩———N(CH₂CH₂Cl)₂
      |
      NH₂

In the genetic activity spectra benzene shows more negative
than positive responses whereas largely positive results are shown
for melphalan. The potency of melphalan is reflected in the number
of spectral lines where the corresponding least effective dose was
less than 1 μg/ml: in the mouse lymphoma forward mutation system, a
sister chromatid exchange system, a mammalian cytogenetics system,
and a cell transformation system.

4.2.1. Benzene. The major source of benzene in the United
States is petroleum. Gasolines contain small quantities of benzene,
usually less than 5%, although special motor fuels may contain up to
30%. Benzene has been heavily utilized in the manufacture of
rubber, paint, and dry-cleaning fluids. The compound has also been
used as the basis for fast-drying inks and in the manufacture of
plastics. Currently about 87% of benzene is used in the chemical
industry for synthesis of compounds such as styrene, phenol, and
cyclohexane[104]. Environmental exposures to benzene occur from
coke-oven emissions, automotive emissions, and at gas stations[83].

Benzene is a bone marrow poison that has been used to treat
leukemia, polycythemia vera, and malignant lymphoma[153]. The
compound resembles chloramphenicol in that it causes aplastic anemia
and, in some cases, acute leukemia[115]. As early as 1928, Delore
and Borgomano described a case of benzene leukemia[50]. Since then,
approximately 150 cases have been identified[69].

In 1976 Vigliani and Forni reviewed the relationship between
benzene and leukemia[241] and the relationship is strengthened by
the major epidemiological study of Ishimaru et al.[117]. Benzene
exposure and leukemia also are discussed in a recent case-control
study of leukemia in the United States rubber industry[254].

The development of leukemia from benzene exposure generally
results from chronic benzene poisoning. Numerous outbreaks of
benzene poisoning have occurred and these have led to the
prohibition of the use of benzene as a solvent in certain industries
and in some materials such as inks and glues. In one study of
shoe-industry workers in which the average exposure period was
9.7 years, 26 of approximately 28,500 employees developed
pre-leukemia or leukemia[4]. A follow-up study of this population

showed that 6 of 44 pancytopenic patients developed leukemia after chronic exposure to benzene[3].

There is considerable interest in monitoring for chromosomal changes in workers exposed to benzene[68]. Dean, in a review of the genetic toxicology of benzene and related compounds, reported that the interpretation of chromosome monitoring studies is complicated by a lack of quantitative data on the amount of benzene exposure[47]. This is less of a problem when the clinical syndromes are present. The metabolism of benzene in vivo and in vitro has been the subject of extensive research and has been reviewed by Snyder and Kocsis[215].

Benzene gives negative results for reverse mutation in the five standard Salmonella tester strains both in the absence and in the presence of S-9 activation[43]. The compound causes forward mutation in the Tradescantia stamen hair assay, a somatic tissue assay[196]. However, when fed to Drosophila larvae, benzene does not cause sex-linked recessive lethality[167].

In primary DNA damage assays in bacteria, benzene gives negative results in the absence and in the presence of S-9[225]. The compound does not cause enhanced mitotic recombination in yeast strain D3 in either the absence or presence of S-9[43].

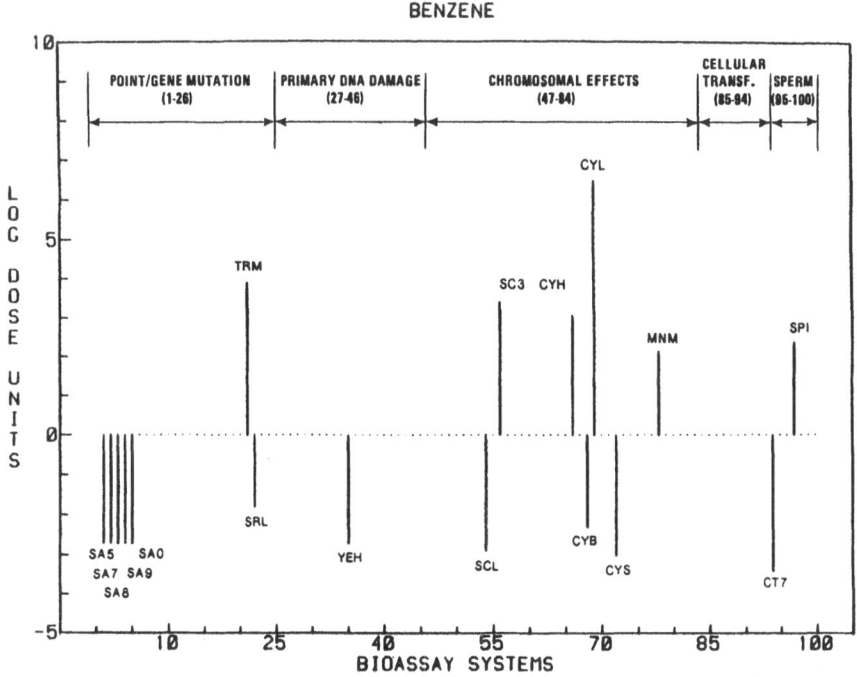

Both positive and negative results are reported in assays for
SCEs after in vitro and in vivo exposure; however, there are more
reports of negative results. According to one report, exogenous
metabolic activation is not required for the induction of SCEs in
human peripheral lymphocytes in vitro[51]. Both positive[124] and
negative[76] results are obtained in cytogenetic studies following
exposure of human lymphocytes to benzene in vitro; millimolar
concentrations are necessary for positive effects. Positive[154]
and negative[229] results are obtained in in vivo studies on bone
marrow although the majority of these tests are negative. Results
from in vivo cytogenetic studies of lymphocytes are mostly
positive[177]. Benzene induces micronuclei in the mouse[90] and
alters the morphology of mouse sperm[232]; however, it fails to
enhance the viral transformation of SHE cells[34].

In summary, the results in the present data base are mixed.
Although there is considerable interest in monitoring human subjects
for benzene poisoning, the results of studies of chromosomal effects
are discordant. Both positive and negative results are reported in
SCE studies and in classic cytogenetic evaluations. The data base
does contain five positive reports of induction of micronuclei in
the mouse. Thus, this test may be a valuable indicator of benzene
damage to the hematopoietic system.

4.2.2. Melphalan. Melphalan, an alkylating nitrogen mustard,
is phenylalanine-substituted mechlorethamine. The compound is used
therapeutically in the treatment of multiple myeloma.

Several reports indicate that melphalan is a
leukemogen[48,60,120]. These investigations detail five case
studies and summarize a larger number of cases in which patients
developed leukemia during therapy for multiple myeloma. These case
studies indicate a causal relationship between the development of
leukemia and therapeutic treatment with the alkylating agent.
However, a number of questions have been raised concerning this
relationship, such as whether the leukemia developed independently
or was merely an unrecognized aspect of multiple myeloma[131].
Patients with a variety of disorders who were treated with
alkylating agents including melphalan have developed acute
leukemia[128], but the incidence of leukemia after melphalan
treatment is low. Case studies of cancer induction after melphalan
treatment have been reviewed[107].

Melphalan gives positive results in Salmonella strain TA1535
with and without S-9[17]. Urine from exposed humans is not
mutagenic in Salmonella[155]. The compound induces forward
mutations in L5178Y cells without exogenous activation[143], and is
also active in the Drosophila sex-linked recessive lethal test[64].

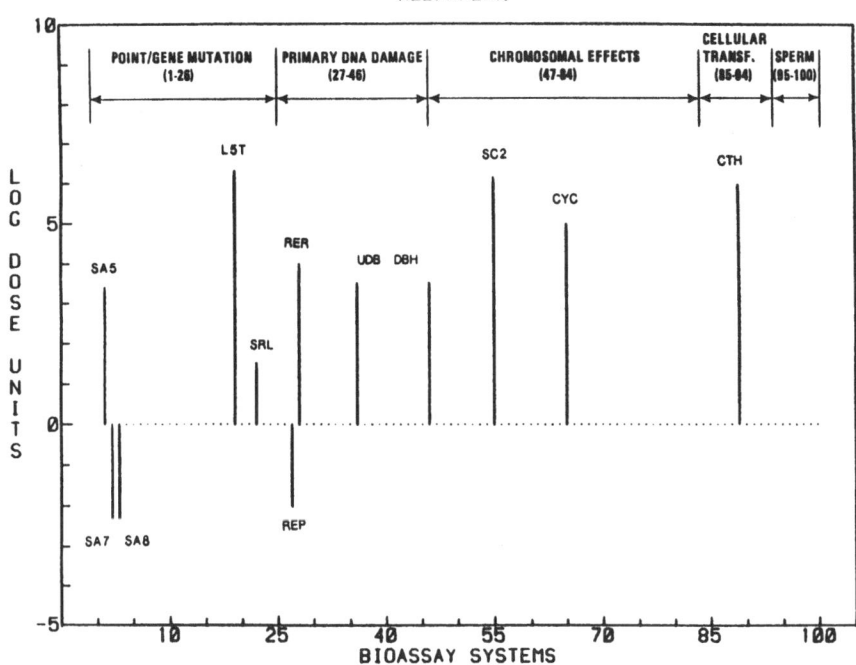

MELPHALAN

S-9 metabolic activation is necessary to elicit DNA damage in DNA polymerase mutants of E. coli[136]. UDS is observed in human bone marrow cells without activation[135].

Human lymphocytes from an exposed patient are reported to show a weak positive SCE response[129]; a strong positive dose-response is obtained in A(T$_1$)C1-3 cloned hamster cells[14]. Fibroblasts cultured in vitro exhibit gross chromosomal aberrations[18] and DNA strand breaks[165] when exposed to melphalan in the absence of exogenous activation. Melphalan transforms C3H10T1/2 cells exposed in the absence of an exogenous activation system[18].

In summary, most of the results are positive. Independent and corroborative results exist in only some of the cases. The data are unlike those for benzene in that melphalan causes point/gene mutation in in vitro and in vivo systems.

4.3.   Human Bladder Carcinogens

Of the human bladder carcinogens considered, three are aromatic amines (4-aminobiphenyl, benzidine, and 2-naphthylamine) and two are nitrogen mustards (chlornaphazine and cyclophosphamide). The latter two compounds are used in cancer chemotherapy. Chlornaphazine is a derivative of 2-naphthylamine. In contrast to several types of the respiratory tract and hematolymphopoietic system carcinogens, four of the five bladder carcinogens are multi-ring structures[248].

BENZIDINE

H_2N —◯—◯— NH_2

CHLORNAPHAZINE

$N(CH_2CH_2Cl)_2$

4-AMINOBIPHENYL

—◯—◯— NH_2

2-NAPHTHYLAMINE

NH_2

CYCLOPHOSPHAMIDE

$N(CH_2CH_2Cl)_2$

     The chemicals 4-aminobiphenyl, benzidine, and 2-naphthylamine have been evaluated in 15, 18, and 22 different bioassays, respectively, where adequate dosimetry information is available. The structure-activity relationships for these primary aromatic amines have been reviewed[183]. Thirteen mutation tests show agreement between one or more pairs of the primary amines. One of these tests is negative for effects on mouse sperm, SPI. Positive results are obtained in the prokaryotic gene mutation assays SA8, SA9, SAO, the urine test BFU, and the higher eukaryotic gene mutation assay SRL. Primary DNA damage is positive for the prokaryotic system RET, WPR, and for unscheduled DNA synthesis in higher eukaryotes (UDP). The primary amines are also positive in the mammalian cell transformation systems CTC, CTK, CTR, and CT7. Benzidine and 2-naphthylamine are positive in the Syrian hamster embryo (CTC) and AKR/ME (CTK) cells at concentrations equal to or less than 1 μg/ml. In addition benzidine is positive at this level for unscheduled DNA synthesis in HeLa cells.

     Cyclophosphamide has been thoroughly evaluated in bioassays (more than 190 test results) and the results are mainly positive. Chlornaphazine has not been adequately tested although the results to date match those of cyclophosphamide. Chlornaphazine has been reported positive in the virus-enhanced cell transformation assay (CT7) at 0.1 μg/ml, although the results should be corroborated.

     4.3.1. 4-Aminobiphenyl. 4-Aminobiphenyl is structurally similar to benzidine; recent information suggests that it is no longer commercially produced[99]. One of the first occupational studies of human bladder cancer concerned 4-aminobiphenyl[150]. This study of 171 male workers revealed 19 cases (11.1%) of bladder tumors. A more recent study[151] expanded the study population to 315 male workers. Of these, 53 (16.8%) were found to have bladder tumors.

     4-Aminobiphenyl has not been well tested in genetic bioassay systems. For those systems in which it has been examined, the

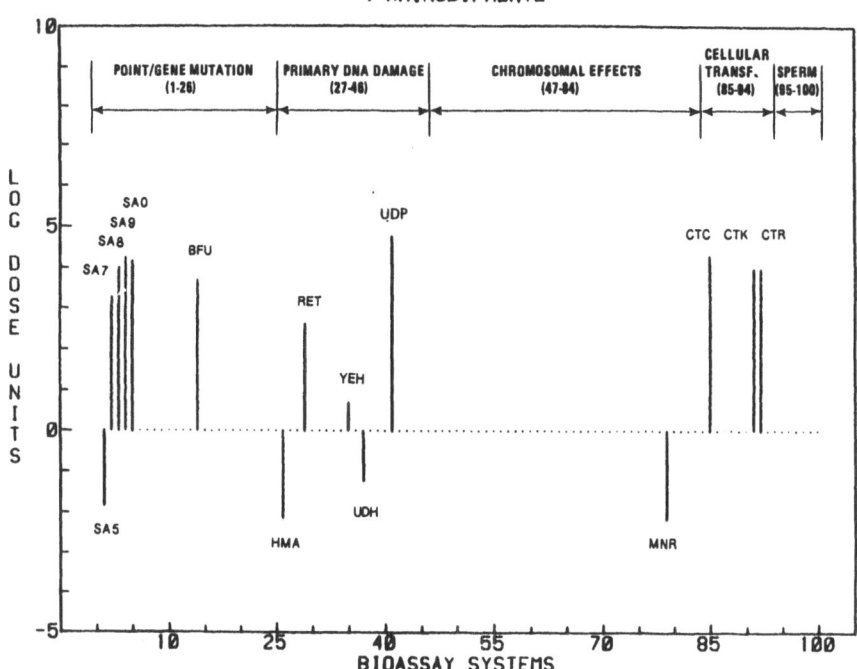

4-AMINOBIPHENYL

profile of biological activity resembles that of the structurally similar compound benzidine.

In the presence of S-9 metabolic activation, 4-aminobiphenyl gives largely positive results in the standard Salmonella tester strains[207]. Without metabolic activation, results are negative[184]. The compound is apparently not mutagenic at the HGPRT locus in V-79 Chinese hamster lung cells in the presence of activation[126]. A body fluid analysis using Salmonella to detect mutagens in urine of rats exposed to 4-aminobiphenyl shows positive results with S-9 activation and negative results without activation[26]. In another study, activation was not required for positive results[184]. One host-mediated assay shows negative results[211].

In primary DNA damage assays using repair-proficient and -deficient strains of E. coli, 4-aminobiphenyl gives positive results in the presence of S-9 metabolic activation[191]. It enhances mitotic recombination in strain D3 of S. cerevisiae in the absence and in the presence of S-9 metabolic activation[208]. Exogenous metabolic activation is not required for induction of UDS in primary rat hepatocytes[253]. Thus it appears that 4-aminobiphenyl causes genetic damage in prokaryotic and eukaryotic in vitro systems.

No chromosomal aberration assays are reported. However, micronucleus tests in the rat give negative results[235]. Several cell transformation assays show positive results in the absence of exogenous metabolic activation. These include SHE cells in a clonal assay[178] and in mouse[188] and rat[71] embryo cultures. The mouse and rat embryo cells were infected with the AKR leukemia and Rauscher leukemia viruses, respectively.

In summary, the results of the genetic tests reveal a response pattern similar to that for benzidine. Most of the test results are positive. 4-aminobiphenyl has been extensively tested in the Salmonella test and clearly causes point mutations in the presence of metabolic activation. In addition, several of the cell transformation tests exhibit positive results.

4.3.2. Benzidine. Unlike 4-aminobiphenyl, benzidine is currently an important industrial chemical. One and one-half million pounds were produced in the United States in 1972[37], and more than 250 dyes are derived the compound[100].

Benzidine produces kidney damage and can cause bladder tumors. The relationship between dyes and the induction of bladder tumors has been known since 1895[187]. In 1954 Case et al. reported on the relative frequency of bladder tumors among workers in the British dye industry[31]. Workers exposed to benzidine were 19 times as likely to develop bladder tumors as were control workers not exposed. Haley reviewed the literature concerning the specific problems associated with benzidine use[82]. Scott and Williams described the prevalence of bladder tumors in industrial settings involving benzidine exposure[200]. They found the latency period for development of bladder cancer in workers to be approximately 16 years. Several outbreaks of bladder cancer in industrial settings have been reported. A typical report was provided by Zavon et al.[257]. A series of cases was detected in 1958 in one manufacturing firm, where one case of bladder cancer was noted after more than 25 years of benzidine manufacture. After the first case was detected, further study revealed that 25 of 35 exposed men eventually developed tumors. The route of exposure was thought to be the respiratory system. Presumably exposure of the bladder occurs via the urine; free benzidine was detected following acid hydrolysis of urine from workers who weighed benzidine-derived dyes[149]. Hueper reported other incidences of benzidine-induced bladder cancer in several countries[95].

In the standard Salmonella tester strains, benzidine generally gives positive results in the presence of S-9 metabolic activation and negative results without S-9[43,72,224]. However, the majority of the test results are negative for the TA1535 and TA1537 assays[6,42]. Urine from rats exposed to the compound gives

BENZIDINE

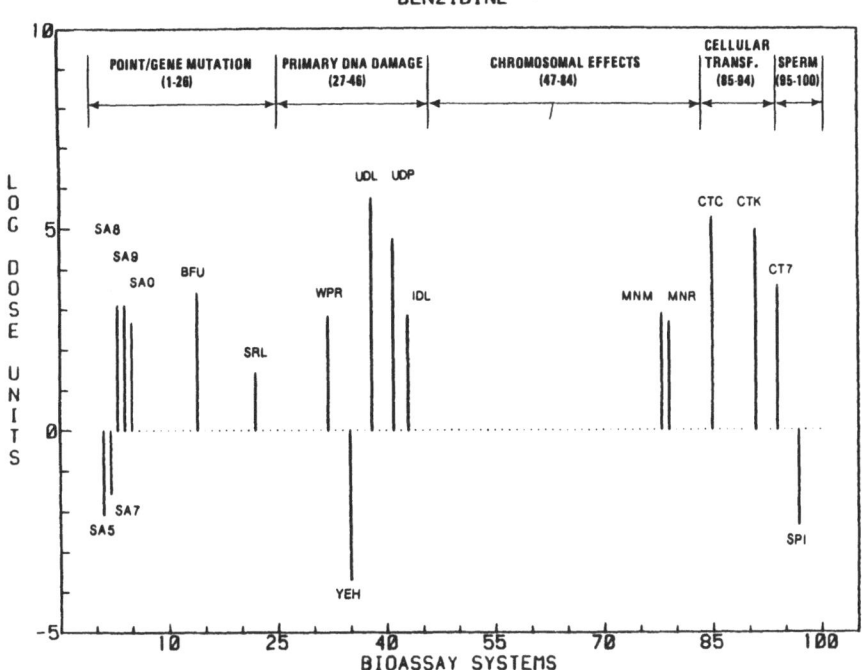

positive results in <u>Salmonella</u> in the presence of S-9[26]. The results of the sex-linked recessive lethal test in <u>Drosophila</u> are positive[63].

Benzidine does not induce enhanced mitotic recombination in S. cerevisiae D3 with or without S-9 activation[209]. The compound induces UDS in rat primary hepatocytes[253] in the absence of exogenous metabolic activation and in HeLa cells[141] with activation. Benzidine also inhibits DNA synthesis in HeLa cell cultures in the absence of an added metabolic activation system[188].

Conflicting positive and negative results are obtained in the micronucleus test[39,235], although the majority are positive. Results from several kinds of cell transformation assays are reported. The SHE cell clonal assay shows positive results without exogenous activation[179], as do the AKR/ME mouse embryo cell system[188] and the Fisher RLV rat embryo cell system[233]. Benzidine enhances the viral transformation of SHE cells in culture[34], but fails to alter the morphology of mouse sperm[232].

In summary, more than half of the findings for benzidine in the genetic toxicology data base are derived from the <u>Salmonella</u> test system, in which benzidine gives mainly positive results when an exogenous source of metabolic activation is supplied. Consistently

positive responses are obtained for effects relating directly to DNA
synthesis and cell transformation. Although many of these results
have not been confirmed in independent studies, the data suggest
that short-term in vitro genetic bioassays are useful for evaluating
benzidine-like compounds, including the closely related analogue
4-aminobiphenyl.

    4.3.3. 2-Naphthylamine. 2-Naphthylamine is an intermediate in
the manufacture of dyes and antioxidants; it is also present in coal
tar[234]. The compound is recognized as a cause of malignant tumors
of the bladder[152]; there are several case reports on the
association of 2-naphthylamine with bladder cancer in
workmen[94,101]. Goldwater et al. presented the startling finding
that 12 out of 48 workmen exposed to 2-naphthylamine developed
bladder cancer[79]. Dye industry workers exposed to 2-naphthylamine
were reported to be 61 times as likely as control workers to develop
bladder tumors[31].

    In Salmonella strains TA1535, TA1538, TA98, and TA100,
2-naphthylamine gives negative results in the absence of S-9 but
positive results in the presence of S-9[207,221]. Similar results
are obtained in E. coli rec-[113]. Salmonella mutagenicity tests on
urine of rats exposed to the compound give positive results with
exogenous metabolic activation[26]. 2-Naphthylamine induces forward

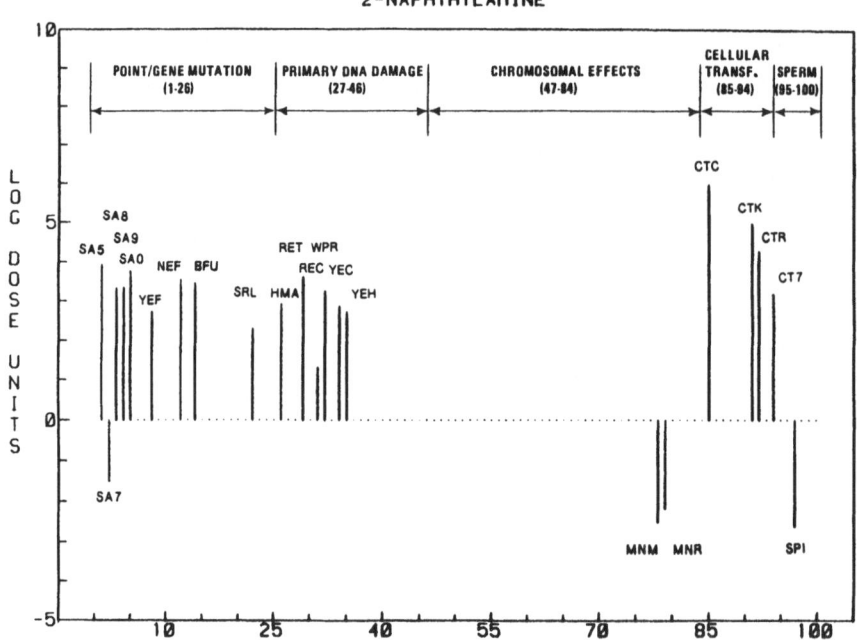

mutations in S. cerevisiae in the presence of an enzyme-free hydroxylation system[145]. Resting conidia of Neurospora crassa require microsomal activation to show a positive response, while growing conidia apparently have the ability to metabolize the compound endogenously[169]. V-79 Chinese hamster lung cells do not respond to 2-naphthylamine in the presence of S-9 metabolic activation[126]. A host-mediated assay using Salmonella gives positive results without S-9 activation[211].

Repair-deficient bacteria require S-9 metabolic activation to exhibit a differential toxic response as compared to the wild-type strains[191,225]. Primary DNA damage as evidenced by enhanced mitotic recombination[146] or gene conversion[29] can be observed in yeast, although the reported results are varied[208]. A qualitative result is available indicating UDS in human diploid fibroblasts[156]. No gross chromosomal aberration assays are reported. Micronucleus tests in the mouse and rat show largely negative results[194,235], although one positive result is reported for the mouse system[121].

Cell transformation is observed in SHE cells[179], in AKR/ME mouse embryo cells[188], and in RLV/1706 Fischer rat embryo cells[71] in the absence of exogenous activation. Enhancement of viral transformation is observed in SHE cells infected with SA-7 virus in the absence of exogenous metabolic activation[33]. 2-Naphthylamine does not alter the morphology of mouse sperm[27].

In summary, the present genetic bioassay data base for 2-naphthylamine consists of approximately 60 reports. Nearly half of the findings are derived from the Salmonella test system, in which the compound produces base-pair substitution and frameshift reactions. Eleven test results are presented for a single Salmonella tester strain (TA100). The data are not as substantial for the other test systems, and many of the results must be confirmed. Overall, the pattern of test responses for 2-naphthylamine is similar to that for benzidine and 4-aminobiphenyl. All three compounds are associated with increased incidence of carcinoma of the bladder.

4.3.4. Chlornaphazine. Chlornaphazine or dichloroethyl-β-naphthylamine has been used as a chemotherapeutic agent in the treatment of leukemia[240]. The cytostatic activity of the compound resides in the bis-(2-chloroethyl) reactive group, and it was the first cytostatic drug shown to be carcinogenic in man[195]. It has been used therapeutically for control of polycythemia vera because it inhibits hematopoiesis[227]. Presently, however, the compound does not have wide therapeutic usage.

Chlornaphazine is a derivative of β-naphthylamine, a well-known bladder carcinogen. Several reports have indicated that patients treated with chlornaphazine can develop the type of bladder cancer seen in workers exposed to β-naphthylamine. In 26 patients treated with high doses of chlornaphazine for polycythemia vera, there were 2 cases of bladder carcinoma and 1 of renal carcinoma[38]. In another study, 10 cases of bladder tumors were observed among 61 patients treated with chlornaphazine[228]. Other case studies of the association between chlornaphazine use and bladder cancer have also been catalogued[102].

The present data base consists of a positive response in Salmonella strain TA100[17] and positive responses for sex-linked recessive lethality in Drosophila[65] and enhancement of viral transformation of SHE cells in culture[33].

4.3.5. Cyclophosphamide. Cyclophosphamide is an alkylating nitrogen mustard and a derivative of mechlorethamine. Cyclophosphamide's reactive moiety is the bis-(2-chloroethyl) group also found in chlornaphazine. The compound is widely used clinically for treatment of Hodgkin's disease, and is also used as an antineoplastic agent in the treatment of a variety of diseases including malignant lymphomas and leukemias, neuroblastoma, and carcinoma of the breast. It is used as an immunosuppressant in a variety of nonmalignant diseases, including rheumatoid arthritis, nephrotic syndrome in children, and chronic hepatitis.

Wall and Clausen[246] described five patients who received large doses of cyclophosphamide over a relatively long period of time and subsequently developed carcinomas of the urinary bladder. The tumors were fatal in four of the five patients. The occurrence of secondary cancers in certain patients from the pre-chemotherapy era makes it difficult to attribute secondary tumors directly to drug therapy. However, other cases have been reported in which cyclophosphamide was administered for nonmalignant diseases and malignant tumors developed[106]. There have been at least 10 cases of primary malignant tumors arising from cyclophosphamide treatment and 16 reports of secondary tumors. Several authors have expressed concern over the widespread use of cyclophosphamide in nonneoplastic diseases[219,246].

Cyclophosphamide gives positive results in Salmonella strains TA1535[143], TA1538[9], TA98[27], and TA100[40] in the presence of S-9 metabolic activation but not in its absence, although the majority of the results for strains TA1538 and TA98 are negative[40,155]. It is direct-acting in forward mutation assays in S. cerevisiae [147]. Urine from mice, rats, and humans exposed to cyclophosphamide gives positive results in Salmonella or yeast test systems without exogenous activation[155,198,205,206,222]. There

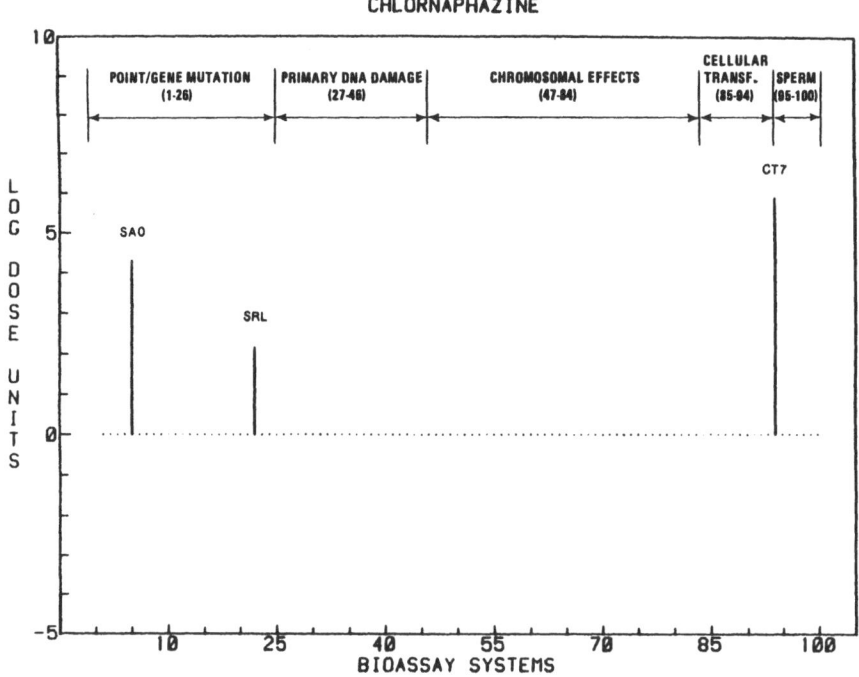

CHLORNAPHAZINE

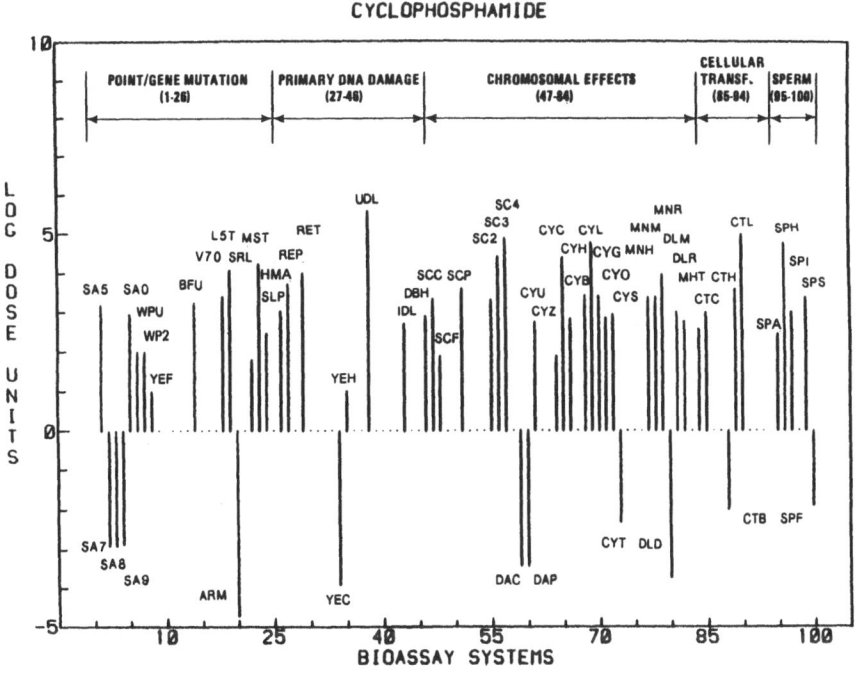

CYCLOPHOSPHAMIDE

are additional reports of positive results in host-mediated systems[66,214]. Positive results are obtained in the L5178Y mammalian cell mutagenesis assay in the presence of S-9[41], and also obtained without S-9 in Chinese hamster lung cells at the ouabain locus[214]. Cyclophosphamide gives negative results in the in vivo Arabidopsis assay[159] and positive results in the sex-linked recessive lethal test in Drosophila[243]. Results of the mouse spot test[123] and the mouse specific locus test[45] also are positive.

The results of primary DNA damage assays in repair-deficient bacteria are positive[192], and results for UDS in HeLa cells are positive when S-9 is added to the test system[141]; UDS is also observed in male germ cells of mice after exposure to cyclophosphamide[216]. Positive results are obtained for mitotic recombination[147] and gene conversion[29] in yeast, although more negative results have been reported for the gene conversion assay[66,205].

Cyclophosphamide induces SCEs with and without S-9 in cells in culture[19] in various animal cell types after exposure in vivo[125]. Cytogenetic changes are observed in vitro in mammalian and human cells[91,114] and in vivo in bone marrow[218], human lymphocytes[53], oocytes[85], and spermatogonia and spermatocytes[78]. The compound induces micronuclei in the mouse[27], rat[235], and Chinese hamster[190]. Positive results are also obtained in dominant lethal tests in the mouse[218] and rat[223]. Positive[46] and negative[74] results are reported for the heritable translocation test in the mouse, and negative results are obtained in the Drosophila dominant lethal test[242].

Cell transformation is observed in the C3H10T1/2 cell line after treatment in the presence of S-9[19]. S-9 is not required for transformation of SHE cells[89] or RLV/1706 Fischer rat embryo cells[59]. Altered sperm morphology is observed in the mouse[27], rat[2,223], sheep[116], and man[173]; however, an assay for altered sperm morphology in the F_1 generation in mice is reported to give negative results[230].

In summary, the present data base consists of 193 test results, 31 of which are from the Salmonella system. Results are positive in more than 82% of the reported tests, and positive responses are reported for each of the major categories of genetic damage. If prokaryotic gene mutation is not considered, 88% of the results are positive. Many of these responses have been confirmed by independent investigations.

4.4 Other Human Carcinogens

Two agents, diethylstilbestrol (DES) and vinyl chloride, were
not assigned according to one of the above target organ
specificities. The chemical structures for DES and vinyl chloride
are shown below.

DIETHYLSTILBESTROL   VINYL CHLORIDE

The genetic activity spectrum for DES is characterized by a
considerable number of positive and negative responses. The
compound is active at concentrations less than 1 μg/ml for
unscheduled DNA synthesis in HeLa cells (UDL) and in the Syrian
hamster embryo clonal assay (CTC). Positive results are obtained at
a dose less than 0.1 μg/ml for sister chromatid exchange in normal
human fibroblasts (SCF). Vinyl chloride also exhibits many positive
and negative test results; it requires metabolic activation in many
of the test systems and is also carcinogenic in the liver, which
catalyzes metabolic activation reactions.

4.4.1 Diethylstilbestrol. Diethylstilbestrol (DES) is used in
human medicine to produce the physiological effects of the natural
estrogens. The compound is also used as a growth promoter for
cattle and sheep, although under restricted conditions[256]. Herbst
and colleagues originally reported the consequences of intrauterine
exposure to DES[88]. Intrauterine exposure causes clear cell
adenocarcinoma--an exceedingly rare tumor of the vagina. However,
between 1966 and 1969, seven young women from 15 to 22 years of age
were diagnosed with adenocarcinoma of the vagina. The most
significant aspect of the study was that the patients' mothers had
been treated with DES starting during the first trimester of
pregnancy. Presently over 300 cases have been recorded in the
registry of clear cell adenocarcinoma of the vaginal tract of young
females[87,189]. Clear cell adenocarcinoma can be detected
cytologically, and nonneoplastic changes of the vaginal tract can
also be shown to be associated with prenatal exposure to DES[189].
Transitional changes between the nonneoplastic and fully developed
cancers have not been carefully investigated. Herbst et al.
estimated the risk to exposed female subjects at about 1 in 1,000 to
1 in 10,000[87]. They indicated that the development of tumors is
more rare than previously assumed and is extremely rare even among
DES-exposed individuals.

In summary, the available information suggests that cancer may occur in the vaginal tracts of human females exposed in utero to DES but that such occurrences are very rare[110].

DES gives negative results in the five standard Salmonella tester strains[9], in E. coli WP2 and WP2 uvr⁻[40], and in V-79 Chinese hamster lung cells in vitro[56]. The compound is mutagenic, however, at the TK locus in L5178Y mouse lymphoma cells with and without metabolic activation[41]. DES induces UDS in HeLa cells in the presence of activation[141].

In assays for SCE, positive results are obtained in human fibroblasts[193] and negative results are obtained in Chinese hamster cells exposed in vitro[1]. Negative results are reported for SCE in the mouse in vivo[118]. The results of an assay for induction and aneuploidy in Neurospora are negative[81]. Conflicting results are obtained in in vitro and in vivo cytogenetic assays[22,23] and in the mouse micronucleus assay[157,194].

With regard to cell transformation, results are positive in the RLV/1706 Fischer rat embryo cell system[59] in the absence of exogenous metabolic activation. Results are negative for the SHE cell clonal assay and for BALB/c3T3 in the absence of activation[59]. Both positive and negative effects on sperm morphology are observed in the mouse[231,255] and in man[10,20].

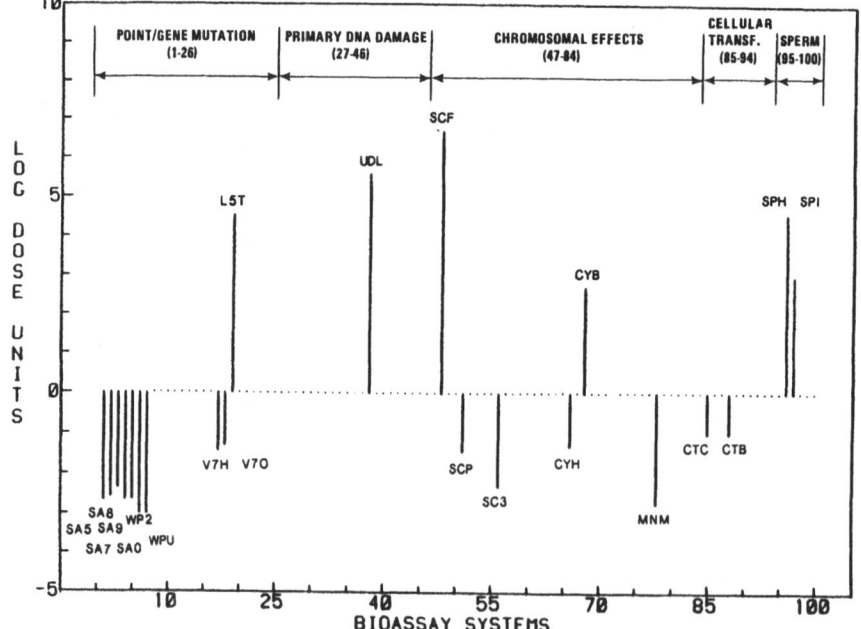

In contrast with many of the other agents examined in this report, DES elicits predominantly negative responses. All results from tests in Salmonella are negative. Most of the other test results have not been confirmed by independent studies. This variety of test results is consistent with the fact that DES is only rarely causally associated with tumor incidence in the human female.

4.4.2. Vinyl Chloride. Over 90% of vinyl chloride monomer is used in the production of vinyl chloride homopolymer and copolymer resins used in the building and construction industries. Workers exposed to vinyl chloride monomer have developed a variety of medical disorders[105,109]. Creech and Johnson first reported four fatal cases of angiosarcoma of the liver among men who worked in the manufacture of polyvinyl chloride and copolymers[44]. Spirtas and Kaminski reviewed 64 cases of angiosarcoma associated with vinyl chloride[217]; Monson et al. analyzed 161 deaths among workers at plants using vinyl chlorides and found an increase of 50% in cancer deaths[158]. These included cancers of the liver, biliary tract, lung, and brain. However, other investigators reported that the increased incidence of vinyl chloride-induced cancer is much less[163]. Waxweiler et al. studied 1,294 individuals to determine the neoplastic risk to workers exposed to vinyl chloride [249]. For all malignant cancers the standard mortality ratio was 184. An excess of cancer was found in four organ systems: the central nervous system; the respiratory system; the hepatic system; and the hematolymphopoietic system. In 14 cases involving biliary and liver cancer, the cancer was angiosarcomic in form. Extensive epidemiological studies of workers exposed to vinyl chloride monomer have been undertaken[70,186]. An increased fetal mortality was observed for the wives of workers who had been exposed to the compound[109], and other studies have indicated an increased risk of birth defects in children of parents residing in the vicinity of vinyl chloride production and polymerization plants[105,109].

Vinyl chloride has been examined repeatedly in the Salmonella reverse mutation test. The gaseous agent returns mixed results in Salmonella strains TA1535 and TA100 in the absence of S-9 metabolic activation but generally positive responses in the same strains in the presence of S-9[148]. The results of reverse mutation assays in S. cerevisiae are negative[203]; however, the results of forward mutation assays in S. pombe are positive in the presence of S-9 and negative in its absence[137]. Negative results are obtained in Neurospora mutation tests[57]. Urine from exposed rats and humans gives negative results in Salmonella[144]. Positive results are obtained with S-9 in CHO cells at the HGPRT locus[15] and in V-79 cells at the HGPRT and ATPase loci[55]. The results of in vivo gene mutation assays are positive in Tradescantia[196] and in the Drosophila sex-linked recessive lethal test[239]. However, vinyl chloride apparently does not induce somatic gene mutation in

VINYL CHLORIDE

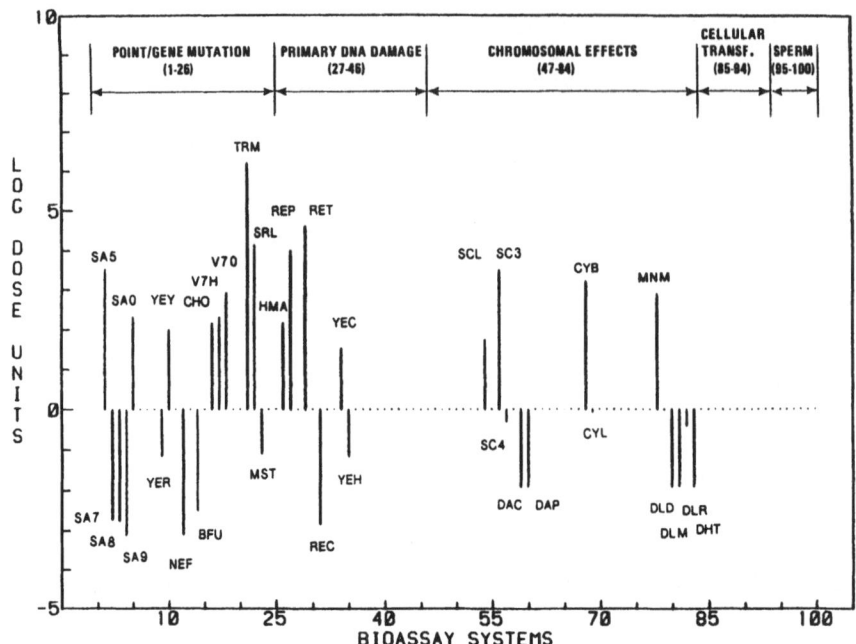

BIOASSAY SYSTEMS

mammals: negative results are reported for a coat color spot test in mice[174]. Results are positive in host-mediated activation systems using yeast as the indicator[137].

Negative results are obtained in bacterial primary DNA damage assays in the absence of S-9[61]; in the presence of S-9, results are positive[192]. Tests for enhanced mitotic recombination[203] or gene conversion[137] in yeast show negative results without S-9 and some positive results with S-9.

Vinyl chloride fails to induce SCEs in cultured human lymphocytes from exposed workers[8,86], but in vitro studies using the same cell type show positive results in the presence of S-9[8]. SCEs are induced in the bone marrow cells of Chinese hamsters after in vivo exposure at high concentrations (1.25% v/v)[16]. Negative responses are reported for a Drosophila aneuploidy test[239]. Cytogenetic abnormalities are produced in bone marrow of hamsters[16]; mixed results are obtained in human lymphocytes[86,181]. Micronuclei are produced in the mouse in vivo[119]. Results are negative in dominant lethal tests in the rat[204] and mouse[7] and in Drosophila[239]; results of the heritable translocation test in Drosophila are also negative[239].

In summary, vinyl chloride clearly produces gene mutations and primary DNA damage when metabolic activation is employed. The compound induces SCEs, gross chromosomal aberrations, and micronuclei in vivo. Overall, approximately half of the test results in the data base are positive. Unfortunately many of the reports present conflicting results.

5. CONCLUSIONS

This study has examined the genetic and related bioassay data available for 14 agents or groups of agents known to have human carcinogenic potential. To facilitate an examination of the available information, agents were divided into major categories based on the target organ or system affected, and the examination of bioassay data followed this classification. Two other compounds, diethylstilbestrol and vinyl chloride, could not be assigned to any one of these target organs.

Most of the data available for the 14 agents have been developed in relatively few screening tests, and most of those involved the use of microorganisms. The data base for known or suspected human carcinogens is not unlike the genetic toxicological data base as a whole. The preponderance of information concerns a limited number of compounds that have been evaluated in relatively few test sytems.

To maximize the utility of genetic bioassay systems and of the overall genetic toxicological data base, it is important that information be developed on reference compounds. It is toward that end that the present data base on known human carcinogens was established. One important measure of the validity of a genetic bioassay is the strength of the correlation between the test results and the evidence for carcinogenicity of the compound or similar compounds in man. While the number of compounds for such human carcinogenicity data is small, these agents represent a diversity of chemical classes and hence a valuable resource for comparative mutagenesis assessment and test system validation.

It is widely recognized that chemicals produce an array of potentially damaging effects on genetic material. As the amount of bioassay data has increased it is possible to test the hypothesis that similar chemical compounds or those possessing specific reactive groups will exhibit similar spectra of genetic activity. As the data base concerning human carcinogens is enlarged, an analysis of the spectra of genetic activity may provide information on the interrelationships among chemicals and their effects.

6. ACKNOWLEDGMENTS

The authors gratefully acknowledge the help of Michael R. Gross. They also acknowledge the outstanding cooperation of John Wassom and the staff of the Environmental Mutagen Information

Center, Oak Ridge National Laboratory, Oak Ridge, TN, in identifying
and obtaining references contained in the GENE-TOX data base.

REFERENCES

1. Abe, S. and M. Sasaki (1977) Chromosome aberrations and sister
    chromatid exchanges in Chinese hamster cells exposed to
    various chemicals. JNCI 58:1635-1641.

2. Adams, P. M., J. D. Fabricant, and M. S. Legator (1981)
    Cyclophosphamide-induced spermatogenic effects detected in the
    F_1 generation by behavioral testing. Science 211:80-82.

3. Aksoy, M. and S. Erdem (1978) Follow-up study on the mortality
    and the development of leukemia in 44 pancytopenic patients
    with chronic exposure to benzene. Blood 52:285-292.

4. Aksoy, M., S. Erdem, and G. Din Col (1974) Leukemia in
    shoe-workers exposed chronically to benzene. Blood
    44:837-841.

5. Alderson, M.R., N. S. Rattan, and L. Bidstrup (1981) Health of
    workmen in the chromate-producing industry in Britain. Br. J.
    Ind. Med. 38:117-124.

6. Ames, B. N., W. E. Durston, E. Yamasaki, and F. D. Lee (1973)
    Carcinogens are mutagens: A simple test system, combining
    liver homogenates for activation and bacteria for detection.
    Proc. Natl. Acad. Sci. U.S.A. 70:2281-2285.

7. Anderson, D., M. C. E. Hodge, and I. F. H. Purchase (1976)
    Vinyl chloride: Dominant lethal studies in male CD-1 mice.
    Mutation Res. 40:359-370.

8. Anderson, D., C. R. Richardson, I. F. H. Purchase, H. J.
    Evans, and M. L. O'Riordan (1981) Chromosomal analysis in
    vinyl chloride exposed workers: Comparison of the standard
    technique with the sister-chromatid exchange technique.
    Mutation Res. 83:137-144.

9. Anderson, D. and J. A. Styles (1978) An evaluation of 6
    short-term tests for detecting organic chemical carcinogens,
    Appendix II. The bacterial mutation test. Br. J. Cancer
    37:924-930.

10. Andonian, R. W. and R. Kessler (1979) Transplacental exposure
    to diethylstibestrol in men. Urology 13:276-279.

11. Auerbach, C. and J. M. Robson (1947) The production of
    mutations by chemical substances. Proc. Royal Soc. Edinburgh
    Sect. B 62:271-283.

12. Baetjer, A. M. (1950) Pulmonary carcinoma in chromate
     workers. I. A review of literature and report of cases.
     Arch. Ind. Hyg. 2:487-504.

13. Baetjer, A. M. (1950) Pulmomary carcinoma in chromate workers.
     II. Incidence on basis of hospital records. Arch. Ind. Hyg.
     2:505-516.

14. Banerjee, A. and W. F. Benedict (1979) Production of sister
     chromatid exchanges by various cancer chemotherapeutic agents.
     Cancer Res. 39:797-799.

15. Barsky, F. C., J. D. Irr, and D. F. Krahn (1979) Mutagenicity
     of gases in the Chinese hamster ovary cell assay. Environ.
     Mutagen. 1:167.

16. Basler, A. and H. Rohrborn (1980) Vinyl-chloride: An example
     for evaluating mutagenic effects in mammals in vivo after
     exposure to inhalation. Arch. Toxicol. 45:1-7.

17. Benedict, W. F., M. S. Baker, L. Haroun, E. Choi, and
     B. N. Ames (1977) Mutagenicity of cancer chemotherapeutic
     agents in the Salmonella/microsome test. Cancer Res.
     37:2209-2213.

18. Benedict, W. F., A. Banerjee, A. Gardner, and P. A. Jones
     (1977) Induction of morphological transformation in mouse
     C3H10T1/2 clone 8 cells and chromosomal damage in hamster
     A(T₁)Cl-3 cells by cancer chemotherapeutic agents. Cancer
     Res. 37:2202-2208.

19. Benedict, W. F., A. Banerjee, and N. Venkatesan (1978)
     Cyclophosphamide-induced oncogenic transformation, chromosomal
     breakage, and sister chromatid exchange following microsomal
     activation. Cancer Res. 38:2922-2924.

20. Bibbo, M., W. B. Gill, F. Azizi, R. Blough, V. S. Fang,
     R. L. Rosenfeld, G. F. B. Schumacher, K. Sleeper, M. G. Sonek,
     and G. L. Wied (1977) Follow-up study of male and female
     offspring of DES-exposed mothers. Obstet. Gynecol. 49:1-8.

21. Bidstrup, P. L. and R. A. M. Case (1956) Carcinoma of the lung
     in workmen in the bichromates-producing industry in Great
     Britain. Br. J. Ind. Med. 13:260-264.

22. Bishun, N., S. Forster, N. Valera, and D. C. Williams (1980)
     The clastogenic effects of diethylstilboestrol on ascitic
     tumor cells in vivo. Microbiol. Lett. 13:27-31.

23. Bishun, N. P., N. Smith, H. Eddie, and D. C. Williams (1977) Cytogenetic studies and diethylstilboestrol. Mutation Res. 46:211-212.

24. Blot, W. J. and J. F. Fraumeni, Jr. (1975) Arsenical air pollution and lung cancer. Lancet 2:142-144.

25. Bonatti, S., M. Meini, and A. Abbondandolo (1976) Genetic effects of potassium dichromate in Schizosaccharomyces pombe. Mutation Res. 38:147-150.

26. Bos, R. P., R. M. E. Brouns, R. Van Doorn, J. L. G. Theuws, and P. T. Henderson (1980) The appearance of mutagens in urine of rats after the administration of benzidine and some other aromatic amines. Toxicology 16:113-122.

27. Bruce, W. R. and J. A. Heddle (1979) The mutagenic activity of 61 agents as determined by the micronucleus, Salmonella, and sperm abnormality assays. Can. J. Genet. Cytol. 21:319-334.

28. Burgdorf, W., K. Kurvink, and J. Cervenka (1977) Elevated sister-chromatid exchange rate in lymphocytes of subjects treated with arsenic. Hum. Genet. 36:69-72.

29. Callen, D. F. and R. M. Philpot (1977) Cytochrome P-450 and the activation of promutagens in Saccharomyces cerevisiae. Mutation Res. 45:309-324.

30. Capizzi, R. L., B. Papirmeister, J. M. Mullins, and E. Cheng (1974) The detection of chemical mutagens using the L5178Y/Asn⁻ murine leukemia in vitro and in a host-mediated assay. Cancer Res. 34:3073-3082.

31. Case, R. A. M., M. E. Hosker, D. B. McDonald, and J. T. Pearson (1954) Tumors of the urinary bladder in workmen engaged in the manufacture and use of certain dyestuff intermediates in the British chemical industry, Part I. The role of aniline, benzidine, α-naphthylamine and β-naphthylamine. Br. J. Ind. Med. 11:75-104.

32. Case, R. A. and A. J. Lea (1955) Mustard gas poisoning, chronic bronchitis and lung cancer: Investigation into the possibility that poisoning by mustard gas in 1914-18 war might be a factor in production of neoplasia. Br. J. Prev. Soc. Med. 9:62-72.

33. Casto, B. C. (1981) In: Advances in Modern Environmental Toxicology, Vol. 1, Mammalian Cell Transformation by Chemical Carcinogens (N. Mishra, V. Dunkel, and M. Mehlman, Eds.), Princeton Junction, NJ, pp. 241-271.

34. Casto, B. C. and G. G. Hatch (unpublished) In Vitro Study of the Nature of the Interaction Between Chemical and Viral Carcinogens.

35. Casto, B. C., J. Meyers, and J. A. DiPaolo (1979) Enhancement of viral transformation for evaluation of the carcinogenic or mutagenic potential of inorganic metal salts. Cancer Res. 39:193-198.

36. Chamberlain, M. and E. M. Tarmy (1977) Asbestos and glass fibers in bacterial mutation tests. Mutation Res. 43:159-164.

37. Chem. Eng. News Feb. 11, 1974, 12.

38. Chievitz, E. and T. Thiede (1962) Complications and causes of death in polycythaemia vera. Acta Med. Scand. 172:513.

39. Cihak, R. (1979) Evaluation of benzidine by the micronucleus test. Mutation Res. 67:383-384.

40. Cline, J. C. and R. E. McMahon (1977) Detection of chemical mutagens: Use of concentration gradient plates in a high capacity screen. Res. Commun. Chem. Pathol. Pharmacol. 16:523-533.

41. Clive, D., K. O. Johnson, J. F. S. Spector, A. G. Batson, and M. M. M. Brown (1979) Validation and characterization of the L5178Y/TK$^{+/-}$ mouse lymphoma mutagen assay system. Mutation Res. 59:61-108.

42. Commoner, B. (1976) Reliability of Bacterial Mutagenesis Techniques to Distinguish Carcinogenic and Noncarcinogenic Chemicals. EPA-600/1-76-022, U.S. Environmental Protection Agency.

43. Cotruvo, J. A., V. F. Simmon, and R. J. Spanggord (1977) Investigation of mutagenic effects of products of ozonation reactions in water. Ann. N.Y. Acad. Sci. 298:124-140.

44. Creech, J. L., Jr. and M. N. Johnson (1974) Angiosarcoma of liver in the manufacture of polyvinyl chloride. J. Occup. Med. 16:150-151.

45. Cumming, R. B. and M. F. Walton (1971) Genetic effects of cyclophosphamide in the germ cells of male mice. Genetics 68:S14.

46. Datta, P. K., H. Frigger, and E. Schleiermacher (1970) In: Chemical Mutagenesis in Mammals and Man (F. Vogel and G. Rohrborn, Eds.), pp. 194-213, Springer-Verlag, Berlin and New York.

47. Dean, B. J. (1978) Genetic toxicology of benzene, toluene, xylenes and phenols. Mutation Res. 47:75-97.

48. DeBock, R. F. K. and M. E. Peetermans (1977) Leukemia after prolonged use of melphalan for non-malignant disease. Lancet 1:1208-1209.

49. De Brabander, M., R. Van deVeire, F. Aerts, S. Geuens, and J. Hoebeke (1976) A new culture model facilitating rapid quantitative testing of mitotic spindle inhibition in mammalian cells. JNCI 56:357-363.

50. Delore, P. and C. Borgomano (1928) Leucemie aigue au cours de l'intoxication benzenique: Sur l'origine toxiquede certains leucemies aigues et leur relations avec les anemies graves. J. Med. Lyon 9:227-233.

51. Diaz, M., N. Fijtman, V. Carricarte, L. Braier, and J. Diez (1979) Effect of benzene and its metabolites on SCE in human lymphocyte cultures. In Vitro 15:172.

52. DiPaolo, J. A. and B. C. Casto (1979) Quantitative studies of in vitro morphological transformation of Syrian hamster cells by inorganic metal salts. Cancer Res. 39:1008-1013.

53. Dobos, M., D. Schuler, and G. Fekete (1974) Cyclophosphamide-induced chromosomal aberrations in nontumorous patients. Humangenetik 22:221-227.

54. Doll, R. (1955) Mortality from lung cancer in asbestos workers. Br. J. Ind. Med. 12:81.

55. Drevon, C. and T. Kuroki (1979) Mutagenicity of vinyl chloride, vinylidene chloride and chloroprene in V79 Chinese hamster cells. Mutation Res. 67:173-182.

56. Drevon, C., C. Piccoli, and R. Montesano (1981) Mutagenicity assays of estrogenic hormones in mammalian cells. Mutation Res. 89:83-90.

57. Drozdowica, B. Z. and P. C. Huang (1977) Lack of mutagenicity
     of vinyl chloride in two strains of Neurospora crassa.
     Mutation Res. 48:43-50.

58. Dugatova, G., S. Podstavkova, and M. Trebaticka (1978)
     Influence of arsenic on Drosophila melanogaster, II. Test on
     recessive lethal and other mutations affecting vitality and
     located in X chromosome and on the occurrence of chromosome
     aberations. Acta F.R.N. Univ. Comend. Genetica 9:79-87.

59. Dunkel, V. C., R. J. Pienta, A. Sivak, and K. A. Traul (1981)
     Comparative neoplastic transformation responses of Balb/c3T3
     cells, Syrian hamster embryo cells, and Rauscher murine
     leukemia virus-infected Fischer 344 rat embryo cells to
     chemical carcinogens. JNCI 67:1303-1315.

60. Einhorn, N. (1978) Acute leukemia after chemotherapy
     (melphalan). Cancer 41:444-447.

61. Elmore, J. D., J. L. Wong, A. D. Laumbach, and U. N. Streips
     (1976) Vinyl chloride mutagenicity via the metabolites
     chlorooxirane and chloroacetaldehyde monomer hydrate.
     Biochem. Biophys. Acta 442:405-419.

62. Fabry, L. (1980) Relationship between the induction of
     micronuclei in marrow cells by chromium salts and their
     carcinogenic properties. C. R. Soc. Biol. 174:889-892.

63. Fahmy, M. J. and O. G. Fahmy (1977) Mutagenicity of hair dye
     components relative to the carcinogen benzidine in Drosophila
     melanogaster. Mutation Res. 56:31-38.

64. Fahmy, O. G. and M. J. Fahmy (1976) Cytogenetic analysis of
     the action of carcinogens and tumor inhibitors in Drosophila
     melanogaster, V. Differential genetic response to the
     alkylating mutagens and X-radiation. J. Genet. 54:146-164.

65. Fahmy, O. G. and M. J. Fahmy (1970) Gene elimination in
     carcinogenesis: Reinterpretation of the somatic mutation
     theory. Cancer Res. 30:195-205.

66. Fahrig, R. (1973) Metabolic activation of mutagens in mammals
     host-mediated assay utilizing the induction of mitotic gene
     conversion in Saccharomyces cerevisiae. Agents Actions
     3:99-110.

67. Figueroa, W. G., R. Raszkowski, and W. Weiss (1973) Lung
     cancer in chloromethyl methyl ether workers. New England J.
     Med. 288:1096-1097.

68. Forni, A., E. Pacifico, and A. Limonta (1971) Chromosome
    studies in workers exposed to benzene or toluene or both.
    Arch. Environ. Health 22:373-378.

69. Forni, A. and E. C. Vigliani (1974) Chemical leukemogenesis in
    man. Ser. Haemat. 7:211-223.

70. Fox, A. J. and P. F. Collier (1977) Mortality experience of
    workers exposed to vinyl chloride monomer in the manufacture
    of polyvinyl chloride in Great Britain. Br. J. Ind. Med.
    34:1-10.

71. Freeman, A. F., E. K. Weisburger, J. H. Weisburger,
    R. G. Wolford, J. M. Maryak, and R. J. Huebner (1974)
    Transformation of cell cultures as an indication of the
    carcinogenic potential of chemicals. JNCI 51:799-808.

72. Garner, R. C., W. L. Walpole, and F. L. Rose (1975) Testing of
    some benzidine analogues for microsomal activation to
    bacterial mutagens. Cancer Lett. 1:39-42.

73. Garrett, N. E., H. F. Stack, M. R. Grose, and M. D. Waters
    (submitted) Analysis of the spectrum of genetic activity
    produced by 24 known or suspected human carcinogens.

74. Generoso, W. M., K. T. Cain, S. W. Huff, and D. G. Gosslee
    (1978) In: Advances in Modern Toxicology, Vol. 5,
    pp. 109-129, Hemisphere Publishing Corporation, Washington, DC
    and London.

75. Gerber, M. A. (1970) Asbestosis and neoplastic disorders of
    the hematopoietic system. Am. J. Clin. Pathol. 53:204-208.

76. Gerner-Smidt P. and U. Friedrich (1978) The mutagenic effect
    of benzene, toluene and xylene studied by the SCE technique.
    Mutation Res. 58:313-316.

77. Gloyne, S. R. (1935) Two cases of squamous carcinoma of the
    lung occurring in asbestosis. Tubercle 17:5.

78. Goetz, P., A. M. Malashenko, and N. I. Surkova (1980)
    Chromosome aberrations induced by cyclophosphamide in meiotic
    cells of male mice. Tsitologiya i Genetika 14:29-35.

79. Goldwater, L. J., A. J. Rosso, and M. Kleinfeld (1965) Bladder
    tumors in a coal-tar dye plant. Arch. Environ. Health 11:814.

80. Green, S. and A. Auletta (1980) Editorial introduction to the reports of "The Gene-Tox Program." An evaluation of bioassays in genetic toxicology. Mutation Res. 76:65-168.

81. Griffiths, A. J. F. (1981) In: Short-Term Tests for Chemical Carcinogens (H. F. Stich and R. H. C. Sans, Eds.), pp. 187-199, Springer-Verlag, New York/Berlin.

82. Haley, T. J. (1975) Benzidine revisited: A review of the literature and problems associated with the use of benzidine and its congeners. Clin. Toxicol. 8:13-42.

83. Haley, T. J. (1977) Evaluation of the health effects of benzene inhalation. Clin. Toxicol. 11:531-548.

84. Hammond, E. C., I. J. Selikoff, and J. Churg (1965) Neoplasia among insulation workers in the United States with special reference to intra-abdominal neoplasia. Ann. N.Y. Acad. Sci. 132:519-525.

85. Hansmann, I. (1974) Chromosome aberrations in metaphase II--Oocytes stage sensitivity in the mouse oogenesis to amethopterin and cyclophosphamide. Mutation Res. 22:175-191.

86. Hansteen, I., L. Hillestad, E. Thiis-Evensen, and S. S. Heldaas (1978) Effects of vinyl chloride in man: A cytogenetic follow-up study. Mutation Res. 51:271-278.

87. Herbst, A. L., P. Cole, T. Colton, S. J. Robboy, and R. E. Scully (1977) Age-incidence and risk of diethylstilbestrol-related clear cell adenocarcinoma of the vagina and cervix. Am. J. Obstet. Gynecol. 128:43-50.

88. Herbst, A. L., H. Ulfelder, and D. C. Poskanzer (1971) Adenocarcinoma of the vagina. Association of maternal stilbestrol therapy with tumor appearance in young women. New England J. Med. 284:878-881.

89. Hirakawa, T., M. Tanaka, and S. Takayama (1979) Morphological transformation of hamster embryo cells by cancer chemotherapeutic agents. Toxicol. Lett. 3: 55-60.

90. Hite, M., M. Pecharo, I. Smith, and S. Thornton (1980) The effect of benzene in the micronucleus test. Mutation Res 77:149-155.

91. Huang, C. C., K. McKernan, J. R. Pantano, and S. R. Sirianni
     (1980) An in vitro metabolic activation assay using liver
     microsomes in diffusion chambers: Induction of sister
     chromatid exchanges and chromosome aberrations by
     cyclophosphamide or ifosfamide in cultured human and Chinese
     hamster cells. Carcinogenesis 1:37-40.

92. Huang, S. L. (1979) Amosite, chrysotile, and crocidolite
     asbestos are mutagenic in Chinese hamster lung cells.
     Mutation Res. 68:265-274.

93. Huang, S. L., D. Saggioro, H. Michelmann, and H. V. Malling,
     (1978) Genetic effects of crocidolite asbestos in Chinese
     hamster lung cells. Mutation Res. 57:225-232.

94. Hueper, W. C. (1942) Occupational Tumors and Allied Diseases.
     Thomas, Springfield, IL.

95. Hueper, W. C. (1969) Occupational and Environmental Cancers of
     the Urinary System. Yale University Press, New Haven, CT.

96. Hutchinson, J. (1887) Br. Med. J 11:1280; J. Hutchinson (1888)
     Trans. Path. Soc. London. 39:352.

97. IARC (1979) Chemicals and Industrial Processes Associated with
     Cancer in Humans. Supplement I to Vols. 1-20 of IARC
     Monographs, IARC, Lyon, France.

98. IARC (1982) Chemicals and Industrial Processes Associated with
     Cancer in Humans. Supplement 4 to Vols. 1-29 of IARC
     Monographs, IARC, Lyon, France.

99. IARC Monographs, Vol. 1, IARC, Lyon, France (1972) pp. 74-79.

100. IARC Monographs, Vol. 1, IARC, Lyon, France (1972) pp. 80-86.

101. IARC Monographs, Vol, 4, IARC, Lyon, France (1974) pp. 97-111.

102. IARC Monographs, Vol. 4, IARC, Lyon, France (1974)
     pp. 119-124.

103. IARC Monographs, Vol. 4, IARC, Lyon, France (1974) pp.
     231-245.

104. IARC Monographs, Vol. 7, IARC, Lyon, France (1974)
     pp. 203-221.

105. IARC Monographs, Vol. 7, IARC, Lyon, France (1974)
     pp. 291-318.

106. IARC Monographs, Vol. 9, IARC, Lyon, France (1975)
      pp. 135-156.

107. IARC Monographs, Vol. 9, IARC, Lyon, France (1975)
      pp. 167-180.

108. IARC Monographs, Vol. 14, IARC, Lyon, France (1977)
      pp. 11-106.

109. IARC Monographs, Vol. 19, IARC, Lyon, France (1979)
      pp. 377-438.

110. IARC Monographs, Vol. 21, IARC, Lyon, France (1979)
      pp. 173-231.

111. IARC Monographs, Vol. 23, IARC, Lyon, France (1980)
      pp. 39-141.

112. IARC Working Group (1980) An evaluation of chemicals and
      industrial processes associated with cancer in humans based on
      human and animal data. IARC Monographs 1 to 20, Cancer Res.
      40:1-12.

113. Ichinotsubo, D., H. F. Mower, J. Setliff, and M. Mandel (1977)
      The use of rec bacteria for testing of carcinogenic
      substances. Mutation Res. 46: 53-62.

114. Ikeuchi, T., K. Sugimura, and M. Sasaki (1979) Evaluation of
      mutagen-metabolizing capacity of cultured mammalian cells, as
      revealed by the induction of chromosome aberrations and sister
      chromatid exchanges. Jpn. J. Hum. Genet. 24:186-187.

115. Infante, P. F., R. A. Rinski, J. K. Wagoner, and R. J. Young
      (1977) Leukemia in benzene workers. Lancet 2:76-78.

116. Inskeep, E. K., J. C. Herrington, and I. L. Lindahl (1971)
      Effects of cyclophosphamide in rams. J. Animal Sci.
      33:1022-1025.

117. Ishimaru, T., H. Okada, T. Tomiyasu, T. Tsuchimoto,
      T. Hoshino, and T. Ichimaru (1971) Occupational factors in the
      epidemiology of leukemia in Hiroshima and Nagasaki. Am. J.
      Epidemiol. 93:157-165.

118. Ivett, J. L. and R. R. Tice (1981) Cytogenetic effects of
      diethystilbestrol-diphosphate (DES-dp) in murine bone marrow.
      Environ. Mutagen. 1:184.

119. Jenssen, D. and C. Ramel (1980) The micronucleus test as part of a short-term mutagenicity test program for the prediction of carcinogenicity evaluated by 143 agents tested. Mutation Res. 75:191-202.

120. Karchmer, R. I., M. Amare, W. E. Larsem, A. G. Mallouk, and G. G. Caldwell (1974) Alkylating agents as leukemogens in multiple myeloma. Cancer 33:1103-1107.

121. Kirkhart, B. (1981) In: Evaluation of Short-Term Tests for Carcinogenesis: Report of the Internation Collaborative Program, Progress in Mutation Research, Vol. 1 (F. J. de Serres and J. Ashby, Eds.), pp. 698-704, Elsevier/North-Holland Biomedical Press, New York.

122. Kleinfeld, M., J. Messite, and C. Kooymann (1967) Mortality experience in a group of asbestos workers. Arch. Environ. Health 15:177-180.

123. Knudsen, I. (1980) The mammalian spot test and its use for the testing of potential carcinogenicity of welding fume particles and hexavalent chromium. Acta Pharmacol. Toxicol. 47:66-70.

124. Koizumi, A., Y. Dobashi, Y. Tachibana, K. Tsuda, H. Katsunuma (1974) Cytokinetic and cytogenetic changes in cultured human leukocytes and HeLa cells inducted by benzene. Ind. Health 12:23-27.

125. Korte, K. (1980) Comparative analysis of chromosomal aberrations and sister-chromatid exchanges in bone marrow cells of Chinese hamsters after treatment with aflatoxin B_1, patulin and cyclophosphamide. Mutation Res. 74: 164.

126. Krahn, D. F. (1977) Rat liver homogenate-mediated toxicity and induction of 6-thioguanine-resistance in V79 Chinese hamster cells by chemical carcinogens. Diss. Abstr. Ins. B37:3726.

127. Kuratsune, M., S. Tokudome, T. Shirakusa, M. Yoshida, T. Tokumitsu, T. Hayano, and M. Seita (1974) Occupational lung cancer among copper smelters. Int. J. Cancer 13:552-558.

128. Kyle, R. A., R. V. Pierre, and E. D. Bayrd (1975) Multiple myeloma and acute leukemia associated with alkylating agents. Arch. Int. Med. 135:185-192.

129. Lambert, B., U. Ringborg, A. Lindblad, E. Harper,
      M. Nordenskjold, and B. Werelius (1979) Sister-chromatid
      exchanges in smoking and non-smoking control subjects,
      patients receiving cancer chemotherapy and laboratory workers
      exposed to organic solvents. Mutation Res. 64:138.

130. Lavappa, K. S., M. M. Fu, and S. S. Epstein (1975) Cytogenetic
      studies on chrysotile asbestos. Environ. Res. 10:165-173.

131. Law, I. P. and J. Blom (1977) Second malignancies in patients
      with multiple myeloma. Oncology 34:20-24.

132. Lawley, P.D. and P. Brookes (1968) Cytotoxicity of alkylating
      agents towards sensitive and resistant strains of Escherichia
      coli in relation to extent and mode of alkylation of cellular
      macromolecules and repair of alkylation lesions in
      deoxyribonucleic acids. Biochem. J. 109:433-447.

133. Leonard, A. and R. R. Lauwerys (1980) Carcinogenicity,
      teratogenicity and mutagenicity of arsenic. Mutation Res.
      75:49-62.

134. Levis, A. G. and F. Majone (1979) Cytotoxic and clastogenic
      effects of soluble chromium compounds on mammalian cell
      cultures. Br. J. Cancer 40:523-533.

135. Lewensohn, R. and U. Ringborg (1979) Induction of unscheduled
      DNA synthesis in human bone marrow cells by bifunctional
      alkylating agents. Blood 54:1320-1329.

136. Longnecker, D. S., T. J. Curphey, S. T. James, D. S. Daniel,
      and N. J. Jacobs (1974) Trial of a bacterial screening system
      for rapid detection of mutagens and carcinogens. Cancer Res.
      34:1658-1663.

137. Loprieno, N., R. Barale, S. Baroncelli, C. Bauer,
      G. Bronzetti, A. Cammellini, G. Cercignani, C. Corsi,
      G. Gervasi, C. Leporini, R. Nieri, A. M. Rossi, G. Stretti,
      and G. Turchi (1976) Evaluation of the genetic effects induced
      by vinylchloride monomer (VCM) under mammalian metabolic
      activation: Studies in vitro and in vivo. Mutation Res.
      40:85-96.

138. Lynch, K. M. and W. A. Smith (1935) Pulmonary asbestosis:
      Carcinoma of the lung in asbestos-silicosis. Am. J. Cancer
      24:56.

139. Machle, W. and F. Gregorius (1948) Cancer of the respiratory
      system in the United States chromate-producing industry. Pub.
      Health Rep. (Washington) 63:1114-1127.

140. Macrae, W. D., R. F. Whiting, and H. F. Stich (1979) Sister
      chromatid exchanges induced in cultured mammalian cells by
      chromate. Chem.-Biol. Interact. 26:281-286.

141. Martin, C. N., A. C. McDermid, and R. C. Garner (1978) Testing
      of known carcinogens and noncarcinogens for their ability to
      induce unscheduled DNA synthesis in HeLa cells. Cancer Res.
      38:2621-2627.

142. Matanoski, G. M., E. Landau, and J. Seifter (in press) Cancer
      Mortality in an Industrial Area of Baltimore. U. S.
      Environmental Protection Agency.

143. Matheson, D., D. Brusick, and R. Carrano (1978) Comparison of
      the relative mutagenic activity for eight antineoplastic drugs
      in the Ames Salmonella/microsome and TK +/- mouse lymphoma
      assays. Drug Chem. Toxicol. 1:277-304.

144. Mattern, I. E. and W. B. Van der Zwaan (1977) Mutagenicity
      testing of urine from vinylchloride (VCM) treated rats using
      the Salmonella test system. Mutation Res. 46:230-231.

145. Mayer, V. W. (1972) Mutagenic effects induced in Saccharomyces
      cerevisiae by breakdown products of 1-naphthylamine and
      2-naphthylamine formed in an enzyme-free hydroxylation system.
      Mutation Res. 15:147-153.

146. Mayer, V. W. (1973) Induction of mitotic crossing over in
      Saccharomyces cerevisiae by breakdown products of
      dimethylnitrosamine, diethylnitrosamine, 1-naphthylamine and
      2-naphthylamine formed by an in-vitro hydroxylation system.
      Genetics 74:433-442.

147. Mayer, V. W., C. J. Hybner, and D. J. Brusick (1976) Genetic
      effects induced in Saccharomyces cerevisiae by
      cyclophosphamide in vitro without liver enzyme preparations.
      Mutation Res. 37:201-212.

148. McCann, J., V. Simmon, D. Streitwieser, and B. N. Ames (1975)
      Mutagenicity of chloroacetaldehyde, a possible metabolic
      product of 1,2-dichloroethane (ethylene dichloride),
      chloroethanol (ethylene chlorohydrin), vinyl chloride, and
      cyclophosphamide. Proc. Natl. Acad. Sci. U.S.A. 72:3190-3193.

149. Meal, P. F., J. Cocker, H. D. Wilson, and J. M. Gilmour (1981)
      Search for benzidine and its metabolites in urine of workers
      weighing benzidine-derived dyes. Br. J. Ind. Med. 38:191-193.

150. Melick, W. F., H. M. Escue, J. J. Naryka, R. A. Mezera, and
      E. R. Wheeler (1955) The first reported case of human bladder
      tumors due to a new carcinogen--Xenylamine. J. Urol.
      (Baltimore) 74:760.

151. Melick, W. F., J. J. Naryka, and R. E. Kelly (1971) Bladder
      cancer due to exposure to para-aminobiphenyl: A 17-year
      follow up. J. Urol. (Baltimore) 106:220.

152. The Merck Index, 8th ed., Merck and Co., Rahway, NJ (1968)
      p. 717.

153. The Merck Index, 8th ed., Merck and Co., Rahway, NJ (1968)
      p. 128.

154. Meyne, J. and M. S. Legator (1980) Sex-related differences in
      cytogenetic effects of benzene in the bone marrow of Swiss
      mice. Environ. Mutagen. 2:43-50.

155. Minnich, V., M. E. Smith, D. Thompson, and S. Kornfeld (1976)
      Detection of mutagenic activity in human urine using mutant
      strains of Salmonella typhimurium. Cancer 38:1253-1258.

156. Mitchell, A. D. (1976) Potential Prescreens for Chemical
      Carcinogens: Unscheduled DNA Synthesis. Task 2 (Final Report
      under Contract N01/CP-33394), Stanford Research Institute,
      Stanford, CA.

157. Molina, L., S. Rinkus, and M. S. Legator (1978) Evaluation of
      the micronucleus procedure over a 2-yr period. Mutation Res.
      53:125.

158. Monson, R. R., J. M. Peters, and M. N. Johnson (1974)
      Proportional mortality among vinyl-chloride workers. Lancet
      2:397-398.

159. Muller, A. J. (1965) A survey on agents tested with regard to
      their ability to induce recessive lethals in Arabidopsis.
      Arabidopsis Information Service 2:22-24.

160. Nakamuro, K. and Y. Sayato (1981) Comparative studies of
      chromosomal aberration induced by trivalent and pentavalent
      arsenic. Mutation Res. 88:73-80.

161. Nestmann, E. R., T. I. Matula, G. R. Douglas, K. C. Bora, and
      Kowbel, D. J. (1979) Detection of the mutagenic activity of
      lead chromate using a battery of microbial tests. Mutation
      Res. 66:357-365.

162. Neubauer, O. (1947) Arsenical cancer: A review. Br. J.
      Cancer 1:192-251.

163. Nicholson, W. J., E. C. Hammond, H. Seidman, and
      I. J. Selikoff (1975) Mortality experience of a cohort of
      vinyl chloride-polyvinyl chloride workers. Ann. N. Y. Acad.
      Sci. 246:225-230.

164. Nishioka, H. (1975) Mutagenic activities of metal compounds in
      bacteria. Mutation Res. 31:185-189.

165. Nordenskjold, M., S. Soderhall, and P. Moldeus (1979) Studies
      of DNA-strand breaks induced in human fibroblasts by chemical
      mutagens/carcinogens. Mutation Res. 63: 393-400.

166. Norman, J. E. (1975) Lung cancer mortality in World War I
      veterans with mustard-gas injury: 1919-1965. JNCI
      54:311-317.

167. Nylander, P., H. Olofsson, B. Rasmuson, and H. Svahlin (1978)
      Mutagenic effects of petrol in Drosophila melanogaster,
      I. Effects of benzene and 1,2-dichloroethane. Mutation Res.
      57:163-167.

168. Ohsaki, Y., S. Abe, K. Kimura, Y. Tsuneta, H. Mikami, and
      M. Murao (1978) Lung cancer in Japanese chromate workers.
      Thorax 33:372-374.

169. Ong, T. and F. J. de Serres (1972) Mutagenicity of chemical
      carcinogens in Neurospora crassa. Cancer Res. 32:1890-1893.

170. Oster, I. I. (1958) Interactions between ionizing radiation
      and chemical mutagens, Z. Indukt. Abstamm. Vererbungsl 89:1-6.

171. Painter, R. B. (1978) DNA synthesis inhibition in HeLa cells
      as a simple test for agents that damage human DNA. J.
      Environ. Pathol. Toxicol. 2:65-78.

172. Pasternack, B. S., R. E. Shore, and R. E. Albert (1977)
      Occupational exposure to chloromethyl ethers. J. Occup. Med.
      19:741-746.

173. Pennisi, A. J., C. M. Grushkin, and E. Lieberman (1975) Gonadal function in children with nephrosis treated with cyclophosphamide. Am. J. Dis. Child 129:315-318.

174. Peter, S. and G. Ungvary (1980) Lack of mutagenic effect of vinyl chloride monomer in the mammalian spot test. Mutation Res. 77:193-196.

175. Petres, J., D. Baron, and M. Hagedorn (1977) Effects of arsenic on cell metabolism and cell proliferation: Cytogenetic and biochemical studies. Environ. Health Perspect. 19:223-227.

176. Petrilli, F. L. and S. De Flora (1977) Toxicity and mutagenicity of hexavalent chromium on Salmonella typhimurium. Appl. Environ. Microbiol. 33:805-809.

177. Picciano, D. (1979) Cytogenetic study of workers exposed to benzene. Environ. Res. 19:33-38.

178. Pienta, R. J. (1979) In: Carcinogens: Identification and Mechanisms of Action (A. C. Griffin and C. R. Shaw, Eds.), pp. 121-141, Raven Press, New York.

179. Pienta, R. J., J. A. Poiley, and W. B. Lebherz (1977) III. Morphological transformation of early passage golden Syrian hamster embryo cells derived from cryopreserved primary cultures as a reliable in vitro bioassay for identifying diverse carcinogens. Int. J. Cancer 19:642-655.

180. Piper, C. E., N. E. McCarroll, and T. J. Oberly (1978) Mutagenic activity of an organic arsenical compound detected with L5178Y mouse lymphoma cells. Environ. Mutagen. 1:165.

181. Purchase, I. F. H., C. R. Richardson, and D. Anderson (1975) Chromosomal and dominant lethal effects of vinyl chloride. Lancet 2:410-411.

182. Pye-Smith, R. J. (1913) Proc. Royal Soc. Med. Clin. Sec. 6:229.

183. Radomski, J. L., W. L. Hearn, and T. Radomski (1979) Structure-activity relationship amongst the primary aromatic amines in the induction of bladder cancer, In: Toxicology and Occupational Medicine (W. B. Deichmann, Ed.), pp. 201-208, Elsevier/North Holland.

184. Radomski, J. L., W. L. Hearn, T. Radomski, H. Moreno, and
      W. E. Scott (1977) Isolation of the glucuronic acid conjugate
      of N-hydroxy-4-aminobiphenyl from dog urine and its mutagenic
      activity. Cancer Res. 37:1757-1762.

185. Raffetto, G., S. Parodi, C. Parodi, M. De Ferrari, R. Troiano,
      and G. Brambilla (1977) Direct interaction with cellular
      targets as the mechanism for chromium carcinogenesis. Tumori
      63:503-512.

186. von Reinl, W., H. Weber, and E. Greiser (1977) Epidemiological
      study on mortality of VC-exposed workers in Federal Republic
      of Germany. Medichem. (Germany) September 2-8.

187. Rehn, L. (1895) Blasengeschwulste bei fuchsin-arbeitern.
      Arch. Klin. Chirugie 50:588.

188. Rhim, J. A., D. K. Park, E. K. Weisburger, and
      J. H. Weisburger (1974) Evaluation of an in vitro assay system
      for carcinogens based on prior infection of rodent cells with
      nontransforming RNA tumor virus. JNCI 52:1167-1173.

189. Robboy, S. J., R. E. Scully, W. R. Welch, and A. L. Herbst
      (1977) Intrauterine diethylstilbestrol exposure and its
      consequences. Arch. Pathol. Lab. Med. 101:1-5.

190. Rohrborn, G. and A. Basler (1977) Cytogenetic investigations
      of mammals. Comparison of the genetic activity of cytostatics
      in mammals. Arch. Toxicol. 38:35-43.

191. Rosenkranz, H. and L. A. Poirier (1979) Evaluation of the
      mutagenicity and DNA-modifying activity of carcinogens and
      noncarcinogens in microbial systems. J. Natl. Cancer Inst.
      62:873-892.

192. Rosenkranz, H. S. and Z. Leifer (1980) In: Chemical Mutagens,
      Principles and Methods for Their Detection Vol. 6 (F. J. de
      Serres and A. Hollaender, Eds.), Plenum Press, New York,
      pp. 109-147.

193. Rudiger, H. W., F. Haenisch, M. Metzler, F. Oesch, and
      H. R. Glatt (1979) Metabolites of diethylstilboestrol induce
      sister chromatid exchange in human cultured fibroblasts.
      Nature 281:392-394.

194. Salamone, M. F., J. A. Heddle, and M. Katz (1981) In:
     Evaluation of Short-Term Tests for Carcinogenesis: Report of
     the International Collaborative Program, Progress in Mutation
     Research, Vol. 1 (F. J. de Serres and J. Ashby, Eds.),
     pp. 686-697, Elsevier/North-Holland Biomedical Press,
     New York.

195. Sax, N. I. (1981) Cancer Causing Chemicals, Van Nostrand
     Reinhold, New York.

196. Schairer, L. A., J. Van't Hof, C. G. Hayes, R. M. Burton, and
     F. J. de Serres (1978) In: Application of Short-Term
     Bioassays in the Fractionation and Analysis of Complex
     Environmental Mixtures (M. D. Waters, S. Nesnow, J. L.
     Huisingh, S. S. Sandhu, and L. Claxton, Eds.), pp. 419-440,
     Plenum Press, New York.

197. Schroeder, H. A. (1970) A sensible look at air pollution by
     metals. Arch. Environ. Health 21:798-806.

198. Schubert, A. (1979) Host-mediated assay and urinary assay with
     the same mice for the detection of chemical mutagens in
     Saccharomyces cerevisiae. Biol. Zbl. 94:451-454.

199. Scott, D., M. Fox, and B. W. Fox (1974) The relationship
     between chromosomal aberrations, survivals and DNA repair in
     tumour cell lines of differential sensitivity to X-rays and
     sulphur mustard. Mutation Res. 22:207-221.

200. Scott, T. S. and M. H. C. Williams (1957) The control of
     industrial bladder tumors. Br. J. Ind. Med. 14:150-163.

201. Selikoff, I. J., J. Churg, and E. C., Hammond (1964) Asbestos
     exposure and neoplasia. JAMA 188:22-26.

202. Selikoff, I. J., E. C. Hammond, and J. Churg (1968) Asbestosis
     exposure, smoking and neoplasia. JAMA 204:106.

203. Shahin, M. M. (1976) The non-mutagenicity and
     -recombinogenicity of vinyl chloride in the absence of
     metabolic activation. Mutation Res. 40:269-272.

204. Short, R. D., J. L. Minor, J. M. Winston, and C. Lee (1977) A
     dominant lethal study in male rats after repeated exposures to
     vinyl chloride or vinylidene chloride. J. Toxicol. Environ.
     Health 3:965-968.

205. Siebert, D. (1973) A new method for testing genetically active
      metabolites: Urinary assay with cyclophosphamide (Endoxan,
      Cytoxan) and Saccharomyces cerevisiae. Mutation Res.,
      17:307-314.

206. Siebert, D. and U. Simon (1973) Genetic activity of
      metabolites in the ascitic fluid and in the urine of a human
      patient treated with cyclophosphamide. Mutation Res.
      21:257-262.

207. Simmon, V. F. (1971) In vitro mutagenicity assays of chemical
      carcinogens and related compounds with Salmonella typhimurium.
      JNCI 62:893-899.

208. Simmon, V. F. (1979) In vitro assays for recombinogenic
      activity of chemical carcinogens and related compounds with
      Saccharomyces cerevisiae D3. JNCI 62:901-909.

209. Simmon, V. F., S. L. Eckford, and A. F. Griffin (1978) In:
      Proceedings of a Conference: Ozone/Chlorine Dioxide Oxidation
      Products of Organic Materials, pp. 126-133.

210. Simmon, V. F., A. D. Mitchell, and T. A. Jorgenson (1977)
      Evaluation of Selected Pesticides as Chemical Mutagens:
      In Vitro and In Vivo Studies. EPA-600/1-77-028, U.S.
      Environmental Protection Agency.

211. Simmon, V. F., H. S. Rosenkranz, E. Zeiger, and L. A. Poirier
      (1979) Mutagenic activity of chemical carcinogens and related
      compounds in the intra-peritoneal host-mediated assay. JNCI
      62:911-918.

212. Sincock, A. M. (1977) In: Origins of Human Cancer, Book B,
      pp. 941-954, Cold Spring Harbor, NY.

213. Sincock, A. and M. Seabright (1975) Induction of chromosome
      changes in Chinese hamster cells by exposure to asbestos
      fibers. Nature 257:56-58.

214. Sirianni, S. R., M. Furukawa, and C. C. Huang (1979) Induction
      of 8-azaguanine-and ouabain-resistant mutants by
      cyclophosphamide and 1-(pyridyl-B)-3,3-dimethyltriazene in
      Chinese hamstser cells cultured in diffusion chambers in mice.
      Mutation Res. 64:259-267.

215. Snyder, R. and J. J. Kocsis (1975) Current concepts of chronic
      benzene toxicity. CRC Crit. Rev. Toxicol. 3:265-288.

216. Sotomayor, R. E., G. A. Sega, and R. B. Cumming (1976)
      Unscheduled DNA synthesis in the germ cells of male mice
      treated in vivo with chemical mutagens requiring metabolic
      activation. Mutation Res. 38:395.

217. Spirtas, R. and R. Kaminski (1978) Angiosarcoma of the liver
      in vinyl chloride/polyvinyl chloride workers, Update of NIOSH
      Register. J. Occup. Med. 20:427-429.

218. Sram, R. J. (1976) Relationship between acute and chronic
      exposures in mutagenicity studies in mice. Mutation Res.
      41:25-42.

219. Steinberg, A. D., P. H. Plotz, S. M. Wolff, V. G. Wong,
      S. G. Agus, and J. L. Decker (1972) Cytotoxic drugs in
      treatment of nonmalignant diseases. Ann. Int. Med.
      76:619-642.

220. Stevens, C. M. and A. Mylroie (1950) Biological action of
      mustard gas compounds. Mutagenic activity of beta-chloroalkyl
      amines and sulphides. Nature 166:1019.

221. Sugimura, T., S. Sato, M. Nagao, T. Hahagi, T. Matsushima,
      Y. Seino, M. Takeuchi, and T. Kawachi (1976) In: Fundamentals
      in Cancer Prevention (P. N. Magee et al., Eds.) pp. 191-215,
      University of Tokyo Press/University Park Press,
      Tokyo/Baltimore.

222. Suling, W. J., R. F. Struck, C. W. Woolley, and W. Shannon
      (1978) Comparative disposition of phosphoramide mustard and
      other cyclophosphamide metabolites in the mouse using the
      Salmonella/mutagenesis assay. Cancer Treat. Rep.
      62:1321-1328.

223. Sykora, I., K. Rezabek, D. Pokorna, and D. Gandalovicov (1979)
      In: Evaluation of Embryotoxic, Mutagenic, and Carcinogenic
      Risks of New Drugs, Proceedings of a 1976 Symposium,
      pp. 263-266.

224. Tanaka, K., S. Marui, and T. Mii (1980) Mutagenicity of
      extracts of urine from rats treated with aromatic amines.
      Mutation Res. 79:173-176.

225. Tanooka, H. (1977) Development and applications of Bacillus
      subtilis test systems for mutagens, involving DNA repair,
      deficiency and suppressible auxotrophic mutations. Mutation
      Res. 42:19-32.

226. Thiess, A. M., W. Hey, and H. Zeller (1973) Zur toxikologie von dichlordimethylather--Verdacht auf kanzerogene wirkung auch beim menschen. Zbl. Arbeitsmed. 23:97.

227. Thiede, T., E. Chievitz, and B. C. Christensen (1964) Chlornaphazine as a bladder carcinogen. Acta Med. Scand. 175:721.

228. Thiede, T. and B. C. Christensen (1969) Bladder tumors induced by chlornaphazine. Acta Med. Scand. 185:133-137.

229. Tice, R. R., D. L. Costa, and R. T. Drew (1980) Cytogenetic effects of inhaled benzene in murine bone marrow: Induction of sister chromatid exchanges, chromosomal aberrations, and cellular proliferation in DBA/2 mice. Proc. Natl. Acad. Sci. U.S.A. 77:2148-2152.

230. Topham, J. C. (1980) Chemically-induced transmissible abnormalities in sperm-head shape. Mutation Res. 70:109-114.

231. Topham, J.C. (1980) The detection of carcinogen-induced sperm head abnormalities in mice. Mutation Res. 69:149-155.

232. Topham, J. C. (1980) Do induced sperm-head abnormalities in mice specifically identify mammalian mutagens rather than carcinogens. Mutation Res. 74:379-387.

233. Traul, K. A. and J. S. Wolff (1979) Unpublished report, John L. Smith Memorial on Cancer (Pfizer, Inc.), Contract No. N01-CP-55703.

234. Treibl, H. G. (1967) In: Encyclopedia of Chemical Toxicology, 2nd ed. (R. E. Kirk and D. F. Othmer, Eds.), Vol. 13, John Wiley and Sons, New York, p. 708.

235. Trzos, R. J., G. L. Petzold, M. N. Brunden, and J. A. Swenberg (1978) The evaluation of sixteen carcinogens in the rat using the micronucleus test. Mutation Res. 58:79-86.

236. Umeda, M. and M. Nishimura (1979) Inducibility of chromosomal aberrations by metal compounds in cultured mammalian cells. Mutation Res. 67:221-229.

237. Valencia, R. (1977) Mutagenesis Screening of Pesticides Using Drosophila, WARF Institute final report prepared for U.S. EPA under contract 68-01-2474.

238. Valerio, F., M. De Ferrari, L. Ottaggio, E. Repetto, and L. Santi (1980) Cytogenetic effects of Rhodesian chrysotile on human lymphocytes in vitro. IARC Sci. Publ. 30:485-489.

239. Verburgt, F. G. and E. Vogel (1977) Vinyl chloride mutagenesis in Drosophila melanogaster. Mutation Res. 48:327-336.

240. Videbaek, A. (1964) Chlornaphazin (Erysan®) may induce cancer of the urinary bladder. Acta Med. Scand. 176:45.

241. Vigliani, E. C. and A. Forni (1976) Benzene and leukemia. Environ. Res. 11:122-127.

242. Vogel, E. (1975) Mutagenic activity of cyclophosphamide, trofosfamide, and ifosfamide in Drosophila melanogaster, Specific induction of recessive lethals in the absence of detectable chromosome breakage. Mutation Res. 33:221-228.

243. Vogel, E. (1975) Mutagenic activity of cyclophosphamide, trofosfamide, and ifosfamide in Drosophila melanogster, Specific induction of recessive lethals in the absence of detectable chromosome breakage. Mutation Res. 33:221-228.

244. Wada, S., M. Miyanishi, Y. Nishimoto, S. Kambe, and R. W. Miller (1968) Mustard gas as a cause of respiratory neoplasia in man. Lancet 1:1161-1163.

245. Wagner, J. C., C. A. Sleggs, and P. Marchand (1960) Diffuse plural mesothelioma and asbestos exposure in the North-Western Cape Province. Br. J. Ind. Med. 17:260.

246. Wall, R. L. and K. P. Clausen (1975) Carcinoma of the urinary bladder in patients receiving cyclophosphamide. New England J. Med. 293:271-273.

247. Waters, M. D. and A. Auletta (1981) The GENE-TOX Program:
     Genetic activity evaluation. J. Chem. Inf. Comput. Sci.
     21:35-38.

248. Waters, M. D., N. E. Garrett, C. M. Covonne-de Serres,
     B. E. Howard, and H. F. Stack (1983) Genetic toxicology of
     some known or suspected human carcinogens, In: Chemical
     Mutagens. Principles and Methods for Their Detection Vol. 8
     (F. J. de Serres and A. M. Hollaender, Eds.), pp. 261-341,
     Plenum Press, New York.

249. Waxweiler, R. J., W. Stringer, J. K. Wagoner, J. Jones,
     H. Falk, and C. Carter (1975) Neoplastic risk among workers
     exposed to vinyl chloride. Ann. N. Y. Acad. Sci. 271:40-48.

250. Wester, P. O., D. Brune, and G. Nordberg (1981) Arsenic and
     selenium in lung, liver, and kidney tissue from dead smelter
     workers. Br. J. Ind. Med. 38:179-184.

251. Whiting, R. F., H. F. Stich, and D. J. Koropatnick (1979) DNA
     damage and DNA repair in cultured human cells exposed to
     chromate. Chem.-Biol. Interact. 26:267-280.

252. Wild, D. (1978) Cytogenetic effects in the mouse of 17
     chemical mutagens and carcinogens evaluated by the
     micronucleus test. Mutation Res. 56:319-327.

253. Williams, G. M. (1978) Further improvements in the hepatocyte
     primary culture DNA repair test for carcinogens: Detection of
     carcinogenic biphenyl derivatives. Cancer Lett. 4:69-75.

254. Wolf, P. H., D. Andjelkovich, A. Smith, and H. Tyroler (1981)
     A case-control study of leukemia in the U.S. rubber industry.
     J. Occup. Med. 23:103-108.

255. Wyrobek, A., L. Gordon, and G. Watchmaker (1981) In:
      Evaluation of Short-Term Tests for Carcinogenesis: Report of
      the International Collaborative Program, Progress in Mutation
      Research, Vol. 1 (F. J. de Serres and J. Ashby, Eds.),
      pp. 712-717, Elsevier/North Elsevier North-Holland Biomedical
      Press, New York.

256. Young, C. L. (1978) Cancer Control Monograph,
      Diethylstilbestrol, Project 4418, SRI International,
      Stanford, CA.

257. Zavon, M. R., U. Hoegg, and U. Bingham (1973) Benzidine
      exposure as a cause of bladder tumors. Arch. Environ. Health
      27:1-7.

SENSITIVITY, SPECIFICITY AND ACCURACY OF THE <u>ARABIDOPSIS</u>

ASSAY IN THE IDENTIFICATION OF CARCINOGENS

G. P. Rédei[*], Gregoria N. Acedo[*], and S. S. Sandhu[†]

[*]University of Missouri
117 Curtis Hall
Columbia, MO 65211 and
[†]United States Environmental Protection Agency
Health Effects Research Laboratory, MD-68
Research Triangle Park, NC 27711, USA

SUMMARY

In 61 laboratories using a variety of prokaryotic and eukaryotic short-term assays a maximum of 42 chemicals were tested. The purpose of the study was to identify the best battery of assays for screening carcinogens. None of the assay systems correctly identified all the proven carcinogens or "noncarcinogens". The individual assays were characterized by sensitivity (% of correct identification of carcinogens), specificity (% of correct identification of "noncarcinogens"), predictivity (% of carcinogens correctly identified among all the compounds tested) and hypersensitivity (% of compounds classified as carcinogens among all the compounds tested.

INTRODUCTION

Practically all human adults are conscious of the dangers of chemical hazards in the environment. Only a very small fraction of the population is cognizant, however, of the reliability of classification of the various agents as harmful or harmless to human health.

Direct assays of chemicals regarding carcinogenicity to human beings are impossible, and the theoretically best tests using animals are also impractical for large scale surveys. The direct assays with animals are too expensive, and because of their long duration they cannot keep up with the deluge of new compounds continually being identified or synthesized. According to some estimates, about one new chemical becomes known every two minutes (10), and there are too many among the over 6,000,000 compounds (11) accumulated in the recent past that have not been adequately examined regarding their biological properties.

Cancer has been in existence for millenia. Tumors were observed on 5,000-year-old Egyptian mummies and on dinosaurs which lived 18 million years ago. Fossil records indicate the presence of tumors on plants in the carboniferous strata dating back 50 million years (20). The ancient Greek philosopher-physicians wrote extensively about human cancer, and they entertained the view that it was caused by environmental effects. Theodor Boveri in 1902 suggested "that malignant tumors might be the result of a certain abnormal condition of the chromosomes". Also, he found it "conceivable that in a species with a very small number of chromosomes, malignant tumors cannot appear at all" (ref. 4, p. 46).

Auerbach and Robson (1944) have shown that the carcinogenic mustards are mutagenic to the low chromosome number Drosophila (3), and subsequently the mutagenic effects of carcinogens were demonstrated for a number of eukaryotes and prokaryotes (19). Nevertheless the relation between mutation and carcinogenicity did not receive universal recognition then (5) and some doubts still remain today (24; see also Mutation Res. 99:73-9).

At present a wide variety of prokaryotic and eukaryotic short-term mutagenicity assays are available (9,21, see also Gene-Tox Reports in Mutation Res. vols. 87 and 99). None of the short-term assays known today is capable of correctly identifying all the carcinogens (15). The problems with the proper identification of "noncarcinogens" are even greater because there are no valid criteria for defining what a noncarcinogen is. Very weak carcinogens are extremely difficult to distinguish from noncarcinogens because "spontaneous" incidence of cancer blurs the differences.

The majority of the short-term assays are designed to detect alterations in the genetic material (genotoxic effects) that are directly or indirectly related to the expression of neoplasia or malignant growth. It may be assumed that most of these short-term assays miss carcinogenic agents which act through metabolic regulatory means (epigenetic effects).

The short-term assays differ also in their ability to activate procarcinogens to become proximate or ultimate carcinogens. Most of the prokaryotic assays have a rather limited activating power and rely on the activating enzymes provided

through the microsomal fraction of rodent liver homogenates. Even
several eukaryotic assays (e.g. yeast or animal cell cultures)
benefit by the use of the exogenous microsomal enzymes. Although
these supplements to the in vitro tests enhance the efficiency of
identification of carcinogenicity, the exogenous material may not
entirely duplicate the metabolic processes taking place in intact
animals. Carcinogenicity depends on a variety of metabolic
pathways (activations and detoxifications), and it is expected to
involve structural and developmental processes which may be organ-
specific (26). Therefore, theoretically the most "native" systems
of detection are expected to be the most reliable. In practice,
unfortunately, this expectation has not been realized.

   An alternative to the biological tests would be the chemical
definition of the molecular fragments which contribute to the
assembly of carcinogenic compounds. Such an inquiry should be
rewarding in computer-assisted classification of chemicals and
their active groups but at this time neither the statistical
approach nor the considerations of the organic chemistry can
provide reliable predictions (8).

PERFORMANCE OF MUTAGENICITY ASSAY SYSTEMS IN TESTS WITH
CARCINOGENS AND NONCARCINOGENS

   If none of the short-term assays are capable of correctly
identifying all the chemicals regarding carcinogenic properties,
one would like to have a battery of tests which through
complementary performance would identify as large a fraction as
possible of all carcinogens. Similarly, one may develop a
reasonably safe system for granting clearance to most of the
noncarcinogens. The latter goal is actually more difficult to
achieve because defining noncarcinogenicity is similar to proving
that there are no witches. The information provided by a battery
of three hypothetical assays can be represented as shown in Table
1. Although none of the individual assays needs to be perfect,
the 2/3 majority of the correct answers may be acceptable for
reliable classification of the various carcinogenic agents. A
larger battery, comprising several types of tests, is expected to
be more reliable than the one shown (Table 1); the larger number
involves, however, more time and cost.

   Within an international cooperative program (7) a maximum of
42 chemicals were tested by several modifications of the best
established (15) bacterial mutation assays and a large number of
other microbial and eukaryotic tests, including intact animals and
in vitro studies, in order to find the best complements to the
Salmonella/S9 histidine reversion systems (7).

Table 1. A hypothetical battery of assays that is considered
        satisfactory although each member of the battery is
        unsatisfactory alone. Filled and opened circles indicate
        correct and incorrect results, respectively. Correct results
        are positive responses for carcinogens and negative responses
        for noncarcinogens.

		ASSAYS		
		1	2	3
AGENTS				
	A	●	●	o
Carcinogens	B	●	o	●
	C	o	●	●
	D	●	●	o
Noncarcinogens	E	●	o	●
	F	o	●	●

1. Chemicals Studied

The chemicals were selected to include representatives of
some of the major classes of organic compounds. When possible, a
structurally closely related but presumably noncarcinogenic analog
was also chosen for each proven carcinogen. Among the compounds
used were some widely-used but presumably harmless ones such as
sucrose, ascorbic acid and methionine. Special consideration was
given to the selection of batches of the compounds with the
highest available purity.

Of the 42 chemicals tested, 25 carcinogens were classified as
such on the basis of information obtained from reliable direct
carcinogen assays with animals. This group included 4-
nitroquinoline-N-oxide, benzidine, 4-dimethylaminobenzene (butter
yellow), benzo(a)pyrene, β-propiolactone, 9,10-dimethylanthracene
(carcinogenicity information had not been originally entirely
convincing), chloroform, 2-acetylaminofluorene, dimethylcarbamoyl
chloride, 2-naphthylamine, N-nitrosomorpholine, urethane,
methylazoxymethanol acetate, DL-ethionine, hydrazine sulphate,
hexamethylphosphoramide (HMPA), ethylenethiourea,
diethylstilbestrol, safrole, cyclophosphamide, epichlorhydrine, 3-
aminotriazole, 4,4'-methylenebis(2-chloroaniline) [MOCA], o-
toluidine hydrochloride and auramine (technical grade).

The noncarcinogens were classed in groups I, II and III,
indicating in this order the strength of evidence regarding their
lack of carcinogenic activities. Group I contained only
isopropyl-N-(3-chlorophenyl)carbamate. Group II included 4-
dimethylaminoazobenzene-4-sulphonic acid Na salt, 1,1,1-
trichloroethane, dimethylformamide, 1-naphthylamine, and
diphenylnitrosamine. In Group III we find 3-methyl-4-

nitroquinoline–N–oxide, 3,3',5,5'-tetramethylbenzidine, γ-butyrolactone, dinitrosopentamethylene tetramine, azoxybenzene and sucrose.

The third major group consisted of compounds which remained "unclassified" because of the lack of sufficient evidence regarding carcinogenic or noncarcinogenic properties: pyrene, anthracene, 4–acetylaminofluorene, L–methionine and ascorbic acid (sodium ascorbate).

2. Mutagenicity Assay Systems

The 61 assays included in the program are identified in Table 2. The description of the Arabidopsis embryo assays can be found in an accompanying paper in this volume. The other assays are identified in more detail in reference 7. The Salmonella plate assays were basically similar or identical to the Ames test, albeit some of the laboratories used their own preferred protocols. One of the Salmonella assays (No. 14) used also the supplement norharman (abbreviated NH, 9H–pyrido [3,4–b]indole), a compound which enhanced the mutagenicity of some chemicals and reduced that of others. The fluctuation assays were similar to the plate assays but used liquid media.

The prokaryotic repair assays utilized either Rec$^-$ bacteria or polA$^-$ Escherichia coli and scored the survival. Assay no. 26 was the so called "zoro" test. Lysogenic E. coli contained the γ prophage so constructed that the bacterial gal$^-$ gene was fused to the cro repressor of the phage. The system included also the phage repressor cI. When the cI repressor is present the cro gene is shut off; when the cro gene is expressed cI is shut off. In the former case, on galactose medium in the presence of an indicator stain, white colonies are formed; in the latter case the colonies appear red. These two phenotypes of the colonies may result either because of a mutation in the cI gene (genotoxic effect) or by a proteolytic cleavage of the cI–coded repressor protein (epigenetic effect). Assay no. 27 measured γ phage induction in E. coli.

The eukaryotic in vitro assays included both forward and reverse mutation, mitotic recombination, gene conversion, increase in aneuploidy, and genetic repair in yeast. Three assays used the criteria of unscheduled DNA synthesis (an indication of repair) in cultured human fibroblasts or HeLa cancer cells, respectively. Three assays measured the frequency of sister chromatid exchange in cultured Chinese hamster ovary cells. Chromosome aberrations were scored in a cultured rat liver epithelial–like cell (RL) system, and in Chinese hamster ovary cells in assays nos. 46 and 47, respectively. Forward mutation at the thymidine kinase locus was scored in mouse lymphoma cell line L5178Y (assay 48), in

Table 2. Summary of the results of 61 assays with 42 chemicals. Filled and open circles stand for correct and incorrect results, respectively. Question marks indicate ambiguity and dots represent lack of data. The screen distinguished the data obtained with "non carcinogens". The <u>Arabidopsis</u> data are original, the rest is based on reference no. 7.

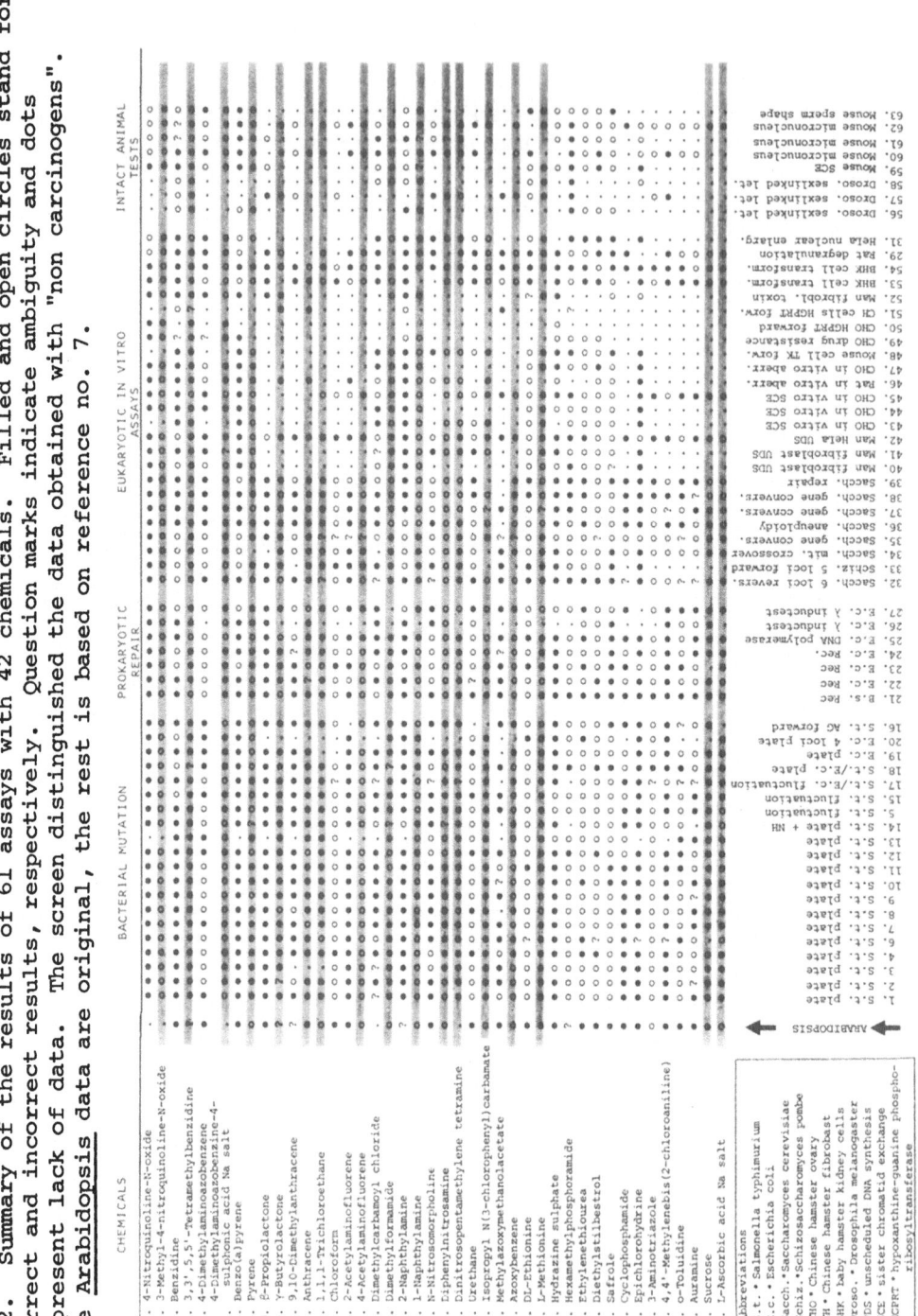

Chinese hamster ovary cells at the loci controlling resistance to
azaadenine, thioguanine, ouabain, and fluorodeoxyuridine (assay
no. 49), for thioguanine resistance in Chinese hamster ovary cell
line CHO-K$_1$-BH$_4$ (assay no. 50), for hypoxanthine-
guaninephosphoribosyl transferase deficiency in Chinese hamster
cell line V79 (assay no. 51), for diphtheria toxin resistance (HF
Dip^r) in human fibroblast culture (assay no. 52). Baby hamster
kidney cell transformation was scored in BHK21 C13/HRCl (assay
no. 53) and in BHK-21 cell lines (assay no. 54), respectively.

A nongenetic criterion, displacement of polysomes from the
endoplasmic reticulum of rat cells, was used by assay no. 29, and
the nuclear enlargement of HeLa cell and human fibroblast was
scored in assay no. 31.

The intact animal assays included three sex-linked recessive
lethal tests with Drosophila using the FM6 stock (heterozygous for
multiple inversions and carrying the w, y, and B markers, the
Berlin K wild type carrying the y, mei-9^a, and mei-41^{D5} (thus
deficient in excision and postreplication repair), and the
standard Basc female system mated to Berlin K males, respectively
(assay nos. 56, 57, 58). Assay no. 59 was involved in the
determination of sister chromatid exchange in the bone marrow (and
liver cells) of CBA/J male mice. Micronucleus tests were carried
out on bone marrow cells of B6C3F$_1$ mice (assay no. 60), on bone
marrow or erythrocytes of male ICR mice (assay no. 61), and on the
bone marrow of CD-1 mice (assay no. 62), respectively. Mouse
sperm abnormality was assayed on B6C3F$_1$/CRL animals.

The Arabidopsis assays were conducted in our laboratory in
Columbia, Missouri, and the results have not been reported yet in
detail; the outcome of the other assays is summarized on the basis
of the publication of the final report (7) where the laboratories
and investigators are identified.

3. TEST PERFORMANCE

Not all the laboratories involved in this project tested all
the 42 chemicals and not all the tests permitted a definite
conclusion regarding the activity of the individual compounds. In
this summary, some assays were not included because of the
incompleteness of the data. In the following, we will compare the
performance of the single plant assay (Arabidopsis) with bacterial
mutation, prokaryotic repair, a diverse group of in vitro
eukaryotic assays and 8 intact animal assays (Table 2).

In Table 2 the results of each assay are shown by a filled
circle if the classification agreed with the expectation based on
direct carcinogenicity information. The open circles indicate

that the results were in disagreement with the expectation, the
dots signal lack of data for a particular test and the question
marks indicate that the information provided by the test was
ambiguous. The screened lines identify the compounds for which
there was no evidence for carcinogenicity at the stage of the
design of the program, and therefore we will consider these
chemicals as "noncarcinogens". Thus if a particular assay
classifies them as inactive in inducing mutation or meeting other
criteria of activity, the result is considered correct and is
represented by a filled circle.

Ideally, Table 2 should have provided 42 x 61 = 2562 bits of
useful information. Some of the chemicals were not tested by all
assays and some of the results were ambiguous. Therefore in the
evaluation only the reasonably certain pieces of information were
considered. For the analysis of the data the classification of
Cooper, Saracci and Cole (6) was adopted. Sensitivity of an assay
was determined by multiplying by 100 the fraction of the number of
carcinogens correctly identified/total number of carcinogens
unambiguously tested. This index defined the power of
identification of the true carcinogens (true positives).
Specificity was calculated by multiplying by 100 the fraction of
the number of correctly identified "noncarcinogens"/total number
of "noncarcinogens" unambiguously classified. This index defined
the power of identification of the "noncarcinogens" (true
negatives). Accuracy represents the fraction of the correct
answers/all unambiguous tests, multiplied by 100. This estimate
combines in a single figure the correct positive and correct
negative data.

The sensitivity of a test thus defines the frequency of
successful identification of a particular group of carcinogens by
an assay or by a group of assays. Obviously this parameter is not
directly applicable to another group of chemicals, but from the
view-point of health hazards it is still the most useful criterion
of a good assay system.

The specificity index has more meaning from the view-point of
the economist than from that of the biologist. It can be used to
grant clearance to an industrially important compound but it will
never prove that a chemical is absolutely innocuous to a mammalian
system. Obviously, it is desirable to be aware of this index
because modern societies are dependent on a large number of
chemicals and some reasonable compromises are unavoidable.

The accuracy estimates may indicate the overall versatility
and utility of an assay, although this also may be quite
misleading. Two assays may have the same accuracy index, yet one
which is highly sensitive to carcinogens may classify as such a
large number of "noncarcinogens", while the other may acquire a

high accuracy figure through failing to identify many carcinogens and giving negative scores (i.e. "true" noncarcinogen rating) to most of the presumable "noncarcinogens".

a. Sensitivity

    Table 3 summarizes the sensitivity (true positive indexes of the five major groups of assays to 25 carcinogens. The overall sensitivity of the 61 assays is disappointingly low; only 58% of the unambiguous answers were correct. Thus many dangerous compounds could have been misclassified on the basis of these results. Surprisingly, the intact animal tests involving mutational and cytogenetic criteria missed about two thirds of the compounds. This can be considered particularly bad because most of these assays are relatively slow and expensive. The best assay systems of bacterial mutation performed about as well as the prokaryotic repair assays or the in vitro eukaryotic assays which are generally considered inferior to the Salmonella/S9 and similar systems. Arabidopsis, the only higher plant for which data are shown, performed extremely well, properly identifying 19/20 of the carcinogenic chemicals.

    Table 3 shows great variation in the facilities by which the different chemicals are identified. Some compounds, such as 4-nitroquinoline-N-oxide (1), benzo(a)pyrene (7), 2-acetylaminofluorene (15), methylazoxymethanolacetate (26) and epichlorhydrine (36), were detectable by an overall efficiency of about 80%. On the other hand chloroform (14), ethionine (28), diethylstilbestrol (33) and 3-aminotriazole (37) were all identified by only 25% efficiency or less.

    The data (Table 3) provide very little evidence that any type of assay would be particularly effective in identifying carcinogens in cases where the other types of tests are ineffective or show poor performance. Remarkably, the plant assay (Arabidopsis) seems to be an exception because it identified correctly 3 or the 4 carcinogens which were missed by the great majority of the other assays.

    On the brighter side of the picture, it is gratifying to see that 5 of the carcinogens were identified correctly by 100% of the bacterial mutation assays.

b. Specificity

    On the 17 "noncarcinogens" 63% of all the assays gave the expected classification based on the original information (Table 4). The sensitivity of the intact animal assays was the highest (85), and that of Arabidopsis the lowest (25). The three other types of assays were again close to each other and in between the

Table 3. Sensitivities of the various types of assays to proven carcinogens. In the Arabidopsis assay +, - and ? stand for correct, incorrect and ambiguous results respectively. Non-Arabidopsis data were derived from reference no 7.

Chemicals	Arabidopsis Embryo Assay	Bacterial Mutation	Prokaryotic Repair	Eukaryotic Systems	Intact Animals	Total
1. 4-Nitroquinoline-N-oxide		100	100	91	20	89
3. Benzidine	+	85	57	70	33	70
5. 4-Dimethylaminoazobenzene	+	45	0	65	40	47
7. Benzo(a)pyrene	+	100	71	70	83	82
9. β-Propiolactone	+	100	100	92	33	67
11. 9,10-Dimethylanthracene	?	88	60	70	0	71
14. Chloroform	+	6	17	27	0	14
15. 2-Acetylaminofluorene	+	100	67	77	100	87
17. Dimethylcarbamoyl chloride		81	71	63	40	68
19. 2-Naphthylamine	?	95	71	75	29	75
21. N-Nitrosomorpholine	+	72	43	50	50	58
24. Urethane	+	26	33	41	100	36
26. Methylazoxymethanolacetate	+	77	100	90	100	86
28. DL-Ethionine	+	11	29	33	40	25
30. Hydrazine sulphate	+	80	83	73	0	71
31. Hexamethylphosphoramide	?	11	17	48	88	38
32. Ethylenethiourea	+	20	57	38	0	30
33. Diethylstilbestrol	+	16	13	24	33	23
34. Safrole	+	20	86	38	17	35
35. Cyclophosphamide	+	84	57	78	100	78
36. Epichlorohydrine	+	100	80	86	20	84
37. 3-Aminotriazole	-	5	17	46	0	18
38. 4,4'-Methylenebis(2-chloroaniline)	+	83	100	55	67	77
39. o'Toluidine	+	29	43	60	0	39
40. Auramine	+	46	57	50	0	47
Total	95	59	58	60	37	58

Table 4. Specificities of the various types of assays to presumably noncarcinogenic chemicals. For symbols see Table 3. Non-Arabidopsis data were derived from reference no. 7.

Chemicals	Arabidopsis Embryo Assay	Bacterial Mutation	Prokaryotic Repair	Eukaryotic Systems	Intact Animal Tests	Total
2. 3-Methyl-4-nitroquinoline-N-oxide		0	0	5	60	8
4. 3,3',5,5'-Tetramethylbenzidine	?	100	71	92	88	91
6. 4-Dimethylaminoazobenzine-4-sulfonic acid Na salt		45	71	71	100	64
8. Pyrene	-	58	71	53	50	63
10. γ-Butyrolactone	?	94	67	71	100	83
12. Anthracene	+	89	67	92	100	87
13. 1,1,1-Trichloroethane	-	94	83	73	100	86
16. 4-Acetylaminofluorene	-	0	33	50	100	26
18. Dimethylformamide	-	89	100	76	100	85
20. 1-Naphthylamine	-	30	50	36	100	42
22. Diphenylnitrosamine	-	72	0	6	100	44
23. Dinitrosopentamethylene tetramine		94	66	56	75	74
25. Isopropyl N(3-chlorophenyl)carbamate	-	100	33	43	100	70
27. Azoxybenzene	-	37	17	42	67	38
29. L-Methionine	+	90	67	79	100	84
41. Sucrose	+	100	100	75	100	92
42. L-Ascorbic acid Na salt	-	84	67	58	100	74
Total	25	69	56	57	85	63

two extremes represented by the intact animal and the Arabidopsis
test, respectively. The noncarcinogens, 3,3',5,5'-
tetramethylbenzidine (4), γ-butyrolactone (10), anthracene (12),
1,1,1-trichloroethane (13), dimethylformamide (18), L-methionine
(29) and sucrose (41) all scored better than 80. Five presumed
noncarcinogens, namely 3-methyl-4-nitroquinoline-N-oxide (2), 4-
acetylaminofluorene (16), 1-naphthylamine (20),
diphenylnitrosamine (22) and azoxybenzene (27) obtained
specificity ratings below 50 and therefore they seem to be
mutagenic (and possibly carcinogenic) in spite of the fact that
the original information indicated otherwise. We may recall that
no. 2 and no. 27 had the class III (the least certain)
noncarcinogen classification. Nos. 20 and 22 were originally
believed to belong to class II of the noncarcinogens, whereas
no. 16 was considered as a compound for which the carcinogenicity
information was inadequate. In addition, compound no. 6 scored
only 45 by the bacterial mutation tests combined and it also
appears to be a mutagen (and possibly a carcinogen).

The Arabidopsis embryo assay identified four of these six
compounds as mutagenic and two of them were not tested. Obviously
a reclassification, by taking into consideration the overall
results of these 61 assays, would have improved substantially the
specificity rating of the Arabidopsis assay and would have lowered
the scores of the intact animal test systems.

c. Accuracy

    The accuracy of the five major types of assays is illustrated
in Table 5. In spite of its low specificity, the Arabidopsis
assay displayed the highest accuracy. Among the other groups of
assays, the accuracy of the bacterial mutation tests was the best.
This confirms the generally recognized usefulness of these assays
(15). The intact animal assays, as a group, appeared to be the
least accurate.

    If we rank the accuracy of individual types of assays, rather
than groups of assays, a completely different order emerges (Table
6). The Arabidopsis test is surpassed in accuracy by four other
assays, and surprisingly the bacterial mutation assays (the 12
Salmonella plate assays and the 3 fluctuation tests) occupy the
9th and the 10th positions with equal ratings. The human
fibroblast assays using diphtheria toxin resistance rise to the
top. We must note, however, that this apparently superior
performance may be due to the few compounds tested, and that fact
that most of the carcinogens which are very hard or impossible to
identify with the other assays were not tested. Such a criticism
does apply to the four other assays rated among the top five for
accuracy. The relatively low rating of the bacterial mutation
assays is due to the fact that several of the carcinogens included

Table 5. Accuracy of the five types of assays for the
    identification of carcinogenicity of chemicals. The body of
    the table shows the number of observations in accordance with
    expectation/total number of tests with conclusive results.
    Non-Arabidopsis data were derived from reference no. 7.

| | The groups of assays with their numbers | | | | | | | |
Chemicals	1 Arabidopsis	20 Bacterial Mutation	7 Prokaryotic Repair	25 Eukaryotic Systems	8 Intact Animals	61 Total	Carcinogenicity	Success of Identification of Compounds
1. 4-Nitroquinoline-N-oxide	0/0	0/19	0/7	1/22	3/5	4/53	yes	8
2. 3-Methyl-4-nitroquinoline-N-oxide	0/0	19/19	7/7	21/22	2/5	49/53	no	93
3. Benzidine	1/1	17/20	4/7	16/23	2/6	40/57	yes	70
4. 3,3',5,5'-Tetramethylbenzidine	?	19/19	5/7	22/24	7/8	53/58	no	91
5. 4-Dimethylaminoazobenzene	1/1	9/20	0/6	13/20	2/5	25/52	yes	48
6. 4-Dimethylaminoazobenzene-4-sulphonic acid Na salt	0/0	9/20	5/7	15/21	4/4	33/52	no	64
7. Benzo(a)pyrene	1/1	19/19	5/7	16/23	5/6	46/56	yes	82
8. Pyrene	0/1	11/19	5/7	10/19	6/6	32/52	no	62
9. β-Propiolactone	1/1	18/18	7/7	11/12	1/3	38/41	yes	93
10. γ-Butyrolactone	?	17/18	4/6	10/14	3/3	34/41	no	83
11. 9,10-Dimethylanthracene	?	15/17	3/5	7/10	0/3	25/35	yes	71
12. Anthracene	1/1	16/18	4/6	12/13	1/1	34/39	no	87
13. 1,1,1-Trichloroethane	0/1	16/17	5/6	8/11	2/2	31/37	no	84
14. Chloroform	1/1	1/17	1/6	3/11	0/2	6/37	yes	16
15. 2-Acetylaminofluorene	1/1	18/18	4/6	10/13	2/2	35/40	yes	88
16. 4-Acetylaminofluorene	0/1	0/18	2/6	6/12	2/2	10/39	no	26
17. Dimethylcarbamoyl chloride	0/0	13/16	5/7	12/19	2/5	32/47	yes	68
18. Dimethylformamide	0/1	16/18	4/4	16/21	4/4	40/48	no	83
19. 2-Naphthylamine	?	18/19	5/7	18/24	2/7	43/57	yes	75
20. 1 Naphthylamine	0/1	6/20	3/6	8/22	5/5	22/54	no	41
21. N-Nitrosomorpholine	1/1	13/18	3/7	7/14	2/4	26/44	yes	59
22. Diphenylnitrosamine	0/1	13/18	0/6	1/17	3/3	17/45	no	38
23. Dinitrosopentamethylene tetramine	0/0	17/18	4/6	10/18	3/4	34/46	no	74
24. Urethane	1/1	5/19	2/6	7/17	2/2	17/45	yes	38
25. Isopropyl N(3-chlorophenyl)carbamate	0/1	19/19	2/6	6/14	1/1	28/41	no	68
26. Methylazoxymethanolacetate	1/1	10/13	5/5	9/10	1/1	26/30	yes	87
27. Azoxybenzene	0/1	7/19	1/6	5/12	2/3	15/41	no	37
28. DL-Ethionine	1/1	2/19	2/7	7/21	2/5	14/53	yes	26
29. Methionine	1/1	18/20	4/6	15/19	5/5	43/51	no	84
30. Hydrazine sulphate	1/1	16/20	5/6	16/22	0/4	38/53	yes	72
31. Hexamethylphosphoramide	?	2/18	1/6	10/21	7/8	20/53	yes	38
32. Ethylenethiourea	1/1	4/20	4/7	8/21	0/6	17/55	yes	31
33. Diethylstilbestrol	1/1	3/19	2/7	5/21	2/6	13/54	yes	24
34. Safrole	1/1	4/20	6/7	8/21	1/6	20/55	yes	36
35. Cyclophosphamide	1/1	16/19	4/7	7/9	1/1	29/37	yes	78
36. Epichlorohydrin	1/1	18/18	4/5	18/21	1/5	42/50	yes	84
37. 3-Aminotriazole	0/1	1/19	1/6	5/11	0/3	7/40	yes	18
38. 4,4'-Methylenebis(2-chloroaniline)	1/1	15/18	7/7	6/11	2/3	31/40	yes	78
39. o-Toluidine	1/1	5/17	3/7	6/10	0/2	15/37	yes	41
40. Auramine	1/1	7/15	4/7	5/10	0/2	17/35	yes	49
41. Sucrose	1/1	18/18	6/6	9/12	1/1	35/38	no	92
42. Ascorbic acid	0/1	16/19	4/6	7/12	1/1	28/39	no	72
Accuracy of Tests	68.8	63.0	57.1	58.9	55.0	60.3		

in the project were selected on the basis of some known
difficulties of identification with the best established assays
(2). These 42 chemicals do not represent an entirely random
sample of the compounds common in the human environment.

Table 6. Ranking the assays on the basis of accuracy judged by
the results of this study involving 15 or more observations.
Non-Arabidopsis data were derived from reference no. 7.

Rank	Assays	No. of Labs	Correct Answers Number of Tests	Accuracy
1	Human cells,diphteria toxin resistance	1	12/16	75.0
2	Saccharomyces, reversion 6 loci	1	26/36	72.2
3	Bacillus subtilis repair	1	30/42	71.4
4	Escherichia coli plate assay	1	27/39	69.2
5	Arabidopsis, forward mutation, many loci	1	22/32	68.8
6	BHK cell transformation	2	52/79	65.8
7	Schizosaccharomyces, forward, 5 loci	1	21/32	65.6
8	CHO chromosome aberrations	1	13/20	65.0
9	Rat liver cell degranulation	1	22/34	64.7
10	HeLa unscheduled DNA synthesis	1	27/42	64.3
11	Salmonella/S9, histidine reversion	12	294/471	62.4
12	Bacterial fluctuation tests	3	73/117	62.4
13	Mouse, TK forward mutation	1	13/21	61.9
14	E. coli DNA-polymerase	1	25/42	59.5
15	Mouse micronucleus, intact animal	3	51/86	59.3
16	HeLa nuclear enlargement	1	14/24	58.3
17	E. coli - λ inductests	2	35/60	58.3
18	Saccharomyces aneuploidy	1	23/41	56.1
19	Saccharomyces gene conversion	3	63/113	55.8
20	Drosophila, sex-linked lethal	3	20/36	55.6
21	Saccharomyces repair	1	22/40	55.0
22	Mouse, SCE, intact animal	1	11/20	55.0
23	Rat, in vitro chromosomal aberrations	1	13/24	54.2
24	Mouse sperm shape	1	9/17	52.9
25	E. coli repair	3	63/120	52.5
26	Saccharomyces, mitotic crossingover	1	19/38	50.0
27	CHO, sister chromatid exchange	3	37/75	49.3
28	Human fibroblast UDS	2	20/42	47.6

d. Complementarity of the Assays

        For the purpose of selecting the best battery of assays, we
must know which chemical group of carcinogens cannot be safely
identified with the bacterial mutation assays that are the best
established. Eleven carcinogenic compounds included in this
project were generally or very frequently missed by the bacterial
mutation assays as well as by the prokaryotic repair tests (Table
7). The overall sensitivities to these 11 compounds of the
bacterial mutation assays, prokaryotic repair, eukaryotic systems

Table 7. Experimental results with 11 carcinogens which are hard to identify by bacterial mutation assasy as shown by these data. Figures in the body of the table represent sensitivity, and the signs +, -, 0 and ? indicate correct, incorrect identification or lack of information or ambiguity, respectively. Non-Arabidopsis data were derived from reference no 7.

Carcinogens which are Hard to Identify by the Bacterial Mutation Assays	Groups of Assays				Top Six Assays Based on Overall Accuracy						Other Types of Assays Frequently Considered				
	Bacterial Mutation	Prokaryotic Repair	Eukaryotic Systems	Intact Animal Assays	Man, Fibroblast, Diphtheria Toxin Resistance (52)	Saccharomyces Reversion at 6 Loci (32)	Bacillus subtilis Rec Reapir (21)	Escherichia coli, plate assay (19)	Arabidopsis Forward Mutation	Baby Hamster Kidney Cell Transformation (54)	Chinese Hamster Ovary Cell Chromosomal Aberrations (47)	Saccharomyces Repair (39)	Drosophila Sex-Linked Recessive Lethals (57)	Chinese Hamster Ovary Cell, Sister Chromatid Exchange (45)	Schizosaccharomyces, Forward Mutation (33)
5. 4-Dimethylaminoazobenzene	45	0	65	67	+	+	-	+	+	+	-	+	o	+	+
14. Chloroform	14	17	27	0	o	-	-	o	+	-	o	-	-	-	+
24. Urethane	26	33	41	100	o	+	+	+	+	+	o	+	o	-	-
28. DL-Ethionine	11	29	33	40	?	+	-	-	+	+	-	+	-	-	+
31. Hexamethylphosphoramide	11	17	48	88	o	+	+	-	?	+	-	+	+	+	-
32. Ethylenethiourea	20	57	38	0	o	+	+	-	+	+	-	+	o	-	-
33. Diethylstilbestrol	15	29	24	33	o	+	+	-	+	-	+	-	o	-	-
34. Safrole	20	67	38	17	-	+	+	-	+	+	-	+	o	-	+
37. 3-Aminotriazole	6	17	46	0	o	-	+	-	-	+	o	+	-	+	o
39. o-Toluidine	29	43	60	0	o	?	+	-	+	+	o	+	o	+	o
40. Auramine	58	57	50	0	o	?	+	-	+	-	o	+	o	+	o
Overall Sensitivity	18	37	42	36	.	78	72	.	90	72	.	81	.	45	50

and intact animal assays were 18, 37, 42 and 32, respectively. It is interesting to compare how the top six assays, so rated on the basis of overall accuracy, detected the activity of these compounds. The assay which appeared no. 1 in Table 6 correctly identified only a single one out of the 9 which were unambiguously classified. Obviously such an assay cannot be employed successfully in a battery of tests involving prokaryotic assay systems. Also, the bacterial plate assay (19) which was rated fourth, identified correctly only two of this group of chemicals. The Arabidopsis embryo test, the Saccharomyces reversion assay (32), the Bacillus subtilis repair test (21) and one of the two baby hamster kidney cell transformation assays (54) showed 90, 78,

72 and 72 sensitivity, respectively. Also the Saccharomyces repair test (39) which had rather low overall rank in Table 6, correctly identified 9 of these 11 carcinogens and thus displayed a sensitivity of 81 to these compounds. The performance of the sister chromatid exchange assay using Chinese hamster ovary cells (45) was much better in sensitivity than the prokaryotic assays yet it identified correctly only 5 out of the 11 compounds. The Drosophila sex-linked recessive lethal assays tested only 4 of the 11 compounds and missed 3 out of the 4, therefore they cannot be adequately classified but do not appear to show much promise regarding complementarity to the prokaryotic tests.

e. Metabolic Activation

   All the prokaryotic assays and most of the in vitro assays, involving lower eukaryotes or cell cultures of higher animals, require the addition of activating systems for the proper identification of promutagens (procarcinogens). The intact animal assays rely on the metabolic system of the animal used. In the Arabidopsis assays, the chemicals were administered either in water or in dilute dimethylsulphoxide solution and no exogenous activating enzymes were used.

   On the basis of the information provided by this project, it seems certain that the carcinogens 4-dimethylaminoazobenzene, benzidine, 2-acetylaminofluorene, benzo(a)pyrene, 4,4'-methylenebis (2-chloroaniline), 9,10-dimethylanthracene and auramine require activation for genetic effects. All of these compounds were found mutagenic, however, for Arabidopsis. Similarly, the "noncarcinogens" 4-acetylaminofluorene and 1-naphthylamine generally required activation in prokaryotes to become mutagenic; in yeasts their effects were unclear but in Arabidopsis both compounds appeared mutagenic. It has been known for some time that higher plants can activate promutagens (12,13,25).

CONCLUSIONS

   From a biological viewpoint, sensitivity is the most important criterion of any assay system and this parameter is independent of the number of carcinogens included in the test. The sensitivity is, however, dependent on the structure of the chemicals involved. Thus, the sensitivity estimates obtained for one group of chemicals cannot be safely extrapolated for another group even when exactly the same biological assay is used.

   The specificity estimate is very important from the industrial or agricultural viewpoints but frequently it is an indication of the insensitivity of an assay system in identifying mutagens (carcinogens) of low effectiveness. The data in Table 2 indicate that 3-methyl-4-nitroquinoline-N-oxide (2), 4-dimethylaminoazobenzine-4-sulphonic acid Na salt (6), 4-acetylaminofluorene (16), 1-naphthylamine (20), diphenylnitrosamine (22) and azoxybenzene (27) are mutagenic in the majority of tests. The specificity of the tests toward these six compounds was below 50 (Table 4). If these six compounds are reclassified as carcinogens, the sensitivity ratings of the assay systems may substantially change as shown in parentheses in Table 8. Several other "noncarcinogens" such as pyrene and ascorbic acid may also be suspected of being low efficiency mutagens.

The accuracy may be considered as a useful estimate of the overall efficiency of the various assay systems but this parameter must not be extrapolated for any test performance beyond this group of chemicals.

It was somewhat surprising that the highest overall sensitivity was displayed by a plant assay (Arabidopsis), yet the data were in agreement with a survey based on the information reported by several laboratories on 56 known or suspected carcinogens and neoplastic agents of which 49 (87.5%) were mutagenic (17). This good performance may be due to the very large number (thousands) of loci scanned for forward mutation (16).

It is disappointing that the overall sensitivities of the best established assays (15) failed to identify about 40% of the carcinogens. This low success rate may be due to the nature of the chemical sample: some of the compounds were selected on the basis of difficulties of identification by the prokaryotic assays (2).

Table 8. Synopsis of the parameters concerned with utility of the groups of assay systems. The numbers in parenthesis show sensitivities when some of the "noncarcinogens" were reclassified as carcinogens according to considerations described in the text. Non-arabidopsis data were derived from reference no. 7.

Parameters	Arabidopsis Embryo Assay	Bacterial Mutation	Prokaryotic Repair	Eukaryotic Systems	Intact Animal Tests
Sensitivity	95	59	58	60	37
Specificity	25 (58)	69 (80)	56 (71)	57 (68)	85 (57)
Accuracy	69	63	57	59	55
Predictivity	67	73	67	67	82
Hypersensitivity	88	47	53	53	28

One may also look at the data from a more optimistic viewpoint because some of the best individual assays have performed far better than the group averages. The short-term assays have, however, some inherent problems. The metabolic systems (26) even among mammals are somewhat different and even larger differences exist between taxonomically distant organisms. Purchase (14) noted that 2-naphthylamine is a potent carcinogen for man and three laboratory species of mammals but it failed to induce cancer in rats and rabbits. Also, among 250 compounds tested with mouse and rats 8% were carcinogenic only to mice and 7% only to rats, whereas 44% were carcinogenic and 38% were noncarcinogenic to both (14). Tomatis et al. (23) indicated that according to epidemiological evidence hematite dust causes lung cancer in man but mouse, hamster, guinea pig and rat experiments all appeared negative. Similar observations were made with arsenic. A recent and very extensive study in England found that 4-chloromethylbiphenyl, a well-proven potent bacterial mutagen, turned out to be a relatively weak mutagen in mammalian in vitro systems and inactive in intact animal assays including direct carcinogenicity assays (1).

Although during the last 20 years very substantial progress has been made in the rapid screening of carcinogens and mutagens, much basic research is needed on the mechanisms of carcinogenesis and mutagenesis before we can expect further advances. Some approaches for the future have recently been outlined (22) but a much deeper understanding of the structural and functional organization of the genome is necessary (18) before a major breakthrough can be achieved.

ACKNOWLEDGEMENT

Contribution from the Missouri Agric. Exp. Sta. Journal Series No. 9369.

REFERENCES

1. Ashby, J., P.A. Lefevre, B.M. Elliott, and J.A. Styles (1982) An overview of the chemical and biological reactivity of 4CMB and structurally related compounds: Possible relevance to the overall findings of the UKEMS 1981 study. Mutation Res. 100:417-433.
2. Ashby, J., and D. Paton (1981) Selection, preparation, and purity of the test chemicals. pp. 8-15, in ref. no. 7.
3. Auerbach, C., and J.M. Robson (1944) Production of mutation by allyl isothiocyanate. Nature 154:81-82.
4. Boveri, T. (1929) The Origin of Malignant Tumors, Williams and Wilkins, Baltimore, Maryland.

5. Burdette, W.J. (1955) The significance of mutation in relation
    to the origin of tumors: A review. Cancer Res. 15:201–226.
6. Cooper, J.A., II, R. Saracci, and P. Cole (1979) Describing
    the validity of carcinogen screening tests. Br. J. Cancer
    39:87–89.
7. de Serres, F.J., and J. Ashby, Eds. (1981) Evaluation of
    Short-Term Tests for Carcinogens. Report of the
    International Collaborative Program. Elsevier/North Holland,
    New York.
8. Hodes, L. (1981) Computer-aided selection of compounds for
    antitumor screening: Validation of a statistical-heuristic
    method. J. Chem. Inf. Comput. Sci. 21:128–132.
9. Hollaender, A., Ed. (1971 on) Chemical Mutagens. Principles
    and Methods for Their Detection. Plenum, New York.
10. Maugh, T.H. (1983) How many chemicals are there? Science
    199:162.
11. Maugh, T.H. (1983) How many chemicals are there? Science
    220:293.
12. Plewa, M.J. (1978) Activation of chemicals into mutagens by
    green plants: A preliminary discussion. Env. Health
    Persp. 27:45–50.
13. Plewa, M.J., D.L. Weaver, L.C. Blair, and J.M. Gentile (1983)
    Activation of 2-aminofluorene by cultured plant cells.
    Science 219:1427–1429.
14. Purchase, I.F.H. (1980) Inter-species comparison of
    carcinogenicity. Br. J. Cancer 41:454–468.
15. Purchase, I.F.H. (1982) An appraisal of predictive tests for
    carcinogenicity. Mutation Res. 99:53–71.
16. Redei, G.P. (1982) Gene number estimates based on mutation
    frequencies in Arabidopsis. Genetics 100, Suppl. 1, Pt. 2,
    s56–s57.
17. Redei, G.P., M.M. Redei, W.R. Lower, and S. Sandhu (1980)
    Identification of carcinogens by mutagenicity for
    Arabidopsis. Mutation Res. 74:469–475.
18. Robertson, M. (1983) What happens when cellular oncogenes
    collide with immunoglobulin genes. Nature 302:474–475.
19. Scherr, G.H., M. Fishman, and R.H. Weaver (1954) The
    mutagenicity of some carcinogenic compounds for Escherichia
    coli. Genetics 39:141–149.
20. Selkirk, J.K., and M.C. MacLeod (1982) Chemical
    carcinogenesis: Nature's metabolic mistake. BioScience
    32:601–605.
21. Stich, H.F.A., and R.H.C. San, Eds. (1981) Short-Term Tests
    for Chemical Carcinogens, Springer-Verlag, New York.
22. Streisinger, G. (1983) Extrapolations from species to species
    and from various cell types in assessing risks from chemical
    mutagens. Mutation Res. 114:93–105.
23. Tomatis, L., C. Agthe, H. Bartsch, J. Huff, R. Montesano,
    R. Saracci, E. Walker, and J. Wilbourn (1978) Evaluation of
    carcinogenicity of chemicals: A review of the monograph

program of the International Agency for Research on Cancer (1971–1977). Cancer Res. 38:877–885.

24. Weisburger, J.H., and G.M. Williams (1980) Chemical carcinogens, in Casarett and Doull's Toxicology, J. Doull, C.D. Klaassen, and M.O. Amdur, Eds., Macmillan, New York, pp. 84–138.

25. Wildeman, A.G., and R.N. Nazar (1982) Significance of plant metabolism in the mutagenicity and toxicity of pesticides. Can. J. Genet. Cytol. 24:437–449.

26. Wright, A.S. (1980) The role of metabolism in chemical mutagenesis and chemical carcinogenesis. Mutation Res. 75:215–241.

ENVIRONMENTAL STUDIES IN SWEDEN

Claes Ramel

Wallenberg Laboratory
University of Stockholm
S-106 91 STOCKHOLM, Sweden

SUMMARY

In Sweden public attention was focused on environment pollution problems by observations in the 1950s of a pronounced decrease in the populations of several bird species traceable to methyl mercury poisoning. This was the starting signal for a broad interdisciplinary investigation of mercury pollution. These studies revealed the methylation of mercury by aquatic microorganisms, the accumulation of methyl mercury in the food chain, and genetic effects of mercury compounds, notably deviating chromosome numbers from inactivation of the spindle fiber mechanism as demonstrated both in experimental organisms and in man. Chemical analyses of wild life samples were subsequently used as a general monitoring method to discover persistent chemical pollutants; in that way the wide contamination by polychlorinated biphenyls, PCB, was brought to light for the first time. In the 1970s environmental studies have more directly involved human toxicological problems, particularly in the work environment. Cancer induction and other genotoxic effects have been in the center of attention in Sweden as in other countries, especially emphasized by epidemiological observations of occupational cancer induction, such as by vinyl chloride and asbestos. The health effects of air pollution have also received much attention and several interdisciplinary research projects have been launched, concerning car exhausts and the emission from coal-fired as compared to oil-fired plants. The last-mentioned issue involves the problem of acidification of water and soil from sulphur-containing oil, which has been a problem of great concern in Sweden.

INTRODUCTION

Environmental studies may comprise a wide array of activities connected with nearly all aspects of life. This presentation will, however, be restricted to the problems related to chemicals and chemical pollution, with special emphasis on long term effects - mutagenic, carcinogenic and teratogenic.

In most countries public attention has been focused on environmental pollution problems through specific issues, usually brought up by the mass media. In Sweden the issue which had the most immediate impact in this respect was the contamination of the environment with mercury, which was brought to light in the 1950s (25). This mercury problem is of general interest as it serves as an illustrative example of the behavior of a persistent chemical pollutant and the magnified effect along the food chain when the pollutant has the ability to accumulate in organisms.

The Mercury Problem

The mercury story in Sweden started with the discovery by ornithologists in the 1950s that some seed-eating birds, notably the yellow bunting, Emberiza citrinella, were drastically diminishing in number. This was attributed to the treatment of seed for sowing with anti-fungicides containing methyl mercury, which was introduced in Sweden in the 1940s. It was also verified by K. Borg in 1956 (4) that seed-eating birds had high levels of mercury. It was, however, also shown that predatory birds, such as hawks and falcons, had extremely high levels of mercury, evidently from secondary poisoning from seed-eating prey. There obviously was an accumulation of mercury at higher trophic levels and the question arose whether humans were also at risk in that connection. Alarmingly high levels of mercury were in fact found in some food items, such as eggs and pheasants, as a consequence of the use of methyl mercury in agriculture. In 1967 the use of methyl mercury in agriculture was therefore forbidden.

Further research on mercury contamination in Sweden strongly substantiated the suspicion of a human risk, but from another angle. It was revealed by Johnels et al. (15) that fresh water fishes, in particular predatory species such as the pike, Esox lucius, contained astonishingly high concentrations of mercury, but this accumulation of mercury did not show any connection with the use of methyl mercury in agriculture, but rather with industrial release of metallic mercury, inorganic mercury and phenyl mercury. When Westöö (39) developed a chromatographic method for the analysis of methyl mercury, the surprising result was obtained that essentially all mercury in the fish occurred as methyl mercury, in spite of the fact that the industries had only released other forms of mercury. This problem was resolved by the

discovery that microorganisms in the water environment methylate
mercury (13). The fact that mercury in fish, used for human
consumption, occurred as methyl mercury was alarming, especially
in view of the epidemic intoxication of humans through methyl-
mercury-containing seafood in Minamata in Japan (the Minamata
disease). It was in fact reported that people with a high
consumption of mercury-containing fish in Sweden had an elevated
frequency of chromosomal aberrations in their lymphocytes (34).

The Swedish health authorities were forced to act in order to
prevent any intoxication from fish consumption. Therefore, waters
with pike containing more than 1 mg/kg were blacklisted for
fishing in 1967. An expert committee was appointed in 1968 to
evaluate the toxicological risks from methyl mercury in fish and
to recommend appropriate actions. The report was published in
1971 (2). The Minamata data had shown that the lowest exposure of
methyl mercury giving rise to manifest intoxication was about 0.3
mg Hg/day, which corresponds to 0.2 μg/g in total blood. On the
basis of this information, the pattern of fish consumption in
Sweden and other data of relevance, five models were constructed,
as shown in Fig. 1. It was decided that the model c in Fig. 1,
with fish containing 1 mg Hg/kg being prohibited for consumption,
was to be used as the basis of administrative action. In this way
the preliminary blacklisting of waters with pike containing more
than 1 mg Hg/kg was established for the future and this regulation
was furthermore supplemented with a recommendation of a maximum of
one meal a week of fish from any mercury-contaminated waters.

These risk calculations, performed in 1968-1971, were based
on cases of acute intoxications. However, the situation has
recently changed, since it has been found both in Japan (9) and in
Iraq (17) that methyl mercury exerts a subacute effect at
considerably lower dosages after exposure in utero, leading to
mental retardation. In fact, according to recent calculations (3)
it cannot be ruled out that methyl mercury presently may be
responsible for some cases of mental retardation in Sweden.

The biological effects of methyl mercury have been studied
intensely, particularly as a consequence of the intoxication
catastrophies in Japan, Iraq and elsewhere. Although the well
known neurological effects of methyl mercury have been in the
center of attention, its effects on the genetic material must also
be taken into consideration. Organic mercury compounds in general
are extremely potent inhibitors of the spindle fiber mechanism,
causing aneuploidy. Methyl mercury produces such effects in plant
material in far lower concentrations than colchicine and induces
nondisjunction of the sex chromosomes in Drosophila (24). Also,
methyl mercury interacts with DNA and functions as an efficient
enzyme inhibitor. What the consequence of this is to humans is
not known, but recently there are reports indicating a

carcinogenic (18) and cocarcinogenic effect (19) of methyl
mercury.   Also a comutagenic effect has been reported, possibly
through an inactivation of DNA repair enzymes (20).

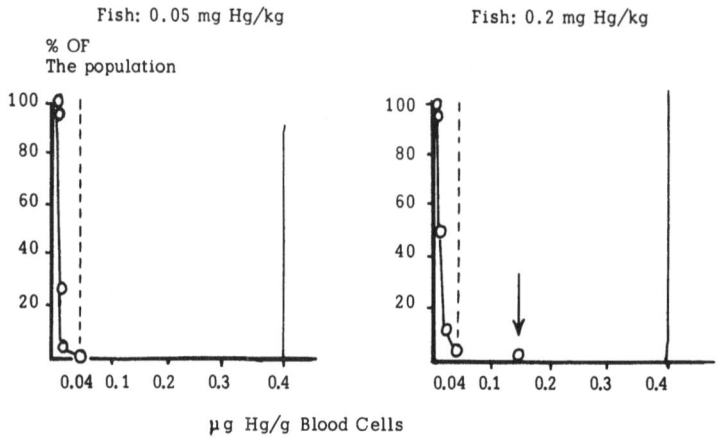

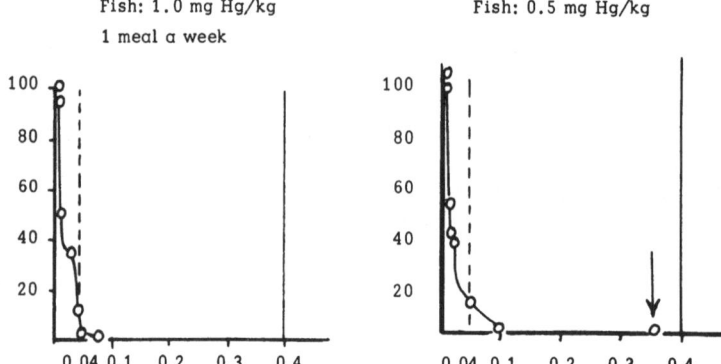

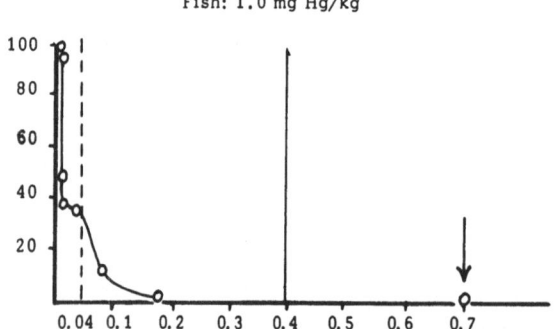

Persistent Contaminants in the Aquatic Environment

   The mercury contamination in the late 1950s focused attention
in Sweden on persistent chemical pollutants, and this initiated
considerable research activity even at so early a date.   The
experience of mercury contamination led to the search for other
chemical pollutants in the aquatic environment.   Rachel Carson's
book "Silent Spring", published in the United States in 1962,
served as another support for this line of research, in particular
concerning DDT.   Monitoring of biological samples revealed a high
contamination of animals by DDT, particularly in the Baltic Sea.
However, the investigations by a chemical analytical group under
Dr. S. Jensen at University of Stockholm identified another
chlorinated pollutant, which later has turned out to be an even
more serious threat to the higher trophic levels of the aquatic
food chain; PCB or polychlorinated biphenyls (11,12).

   The bioaccumulation, biochemical behavior and physiological
effects of PCB have given a strong indication that this highly
persistent contaminant in the water environment is primarily
responsible for the rapid decline of the populations of some birds
and mammalian species in Sweden, especially the seals in the
Baltic Sea.   Although humans are also subjected to contamination
by PCB and DDT, negative consequences have been suspected but not
established, at least not at the dose levels found in Sweden.   It
may, however, be pointed out that both PCB and DDT have been shown
as animal carcinogens although the mechanism is obscure, as these
compounds have little or no mutational effects.

   The strong emphasis on problems concerning the aquatic
environment in environmental research in Sweden during the 1960s
was influential both in the establishment of the Swedish
Environmental Protection Agency in 1973 and in directing its early

Figure 1.   Estimated distribution of Hg levels in blood in the
       Swedish population when Hg-contaminated fish contains the
       following levels of methyl Hg: (a) 0.05 mg Hg/kg.   Free
       consumption. (b) 0.2 mg/Hg/kg.   Free concumption. (c) 1.0 mg
       Hg/kg.   At the most one meal a week of Hg contaminated
       fish. (d) 0.5 mg Hg/kg.   Free consumption. (e) 1.0 mg Hg/kg.
       Free consumption.   the lowest acute toxic dose, 0.4 µg Hg/g
       blood cells, and the highest acceptable dose, 0.04 µg Hg/g
       blood cells are indicated by vertical lines.   The arrow
       indicates the level of extremely high fish consumption (after
       ref. 2).

research policy. The elucidation of problems concerning the aquatic environment has also initiated an extensive investment in the cleaning of Swedish waters from pollutants by means of sewage treatment plants. However, some polluting industries inevitably have caused and still cause serious problems. Among those the paper and pulp industries may be mentioned, as they are of great importance in the Swedish economy. The bleaching of paper and pulp by chlorine results in the release in the aquatic environment of mutagenic components, the exact identity of which, so far, has not been established (32).

Air Pollution

Perhaps the most serious pollution problem in Sweden from an ecological point of view is the acidification of waters from the release of sulphur into the ambient air through the use of fossil fuels. This problem is an international one in the sense that about 75 percent of the sulphur in the air over Sweden emanates from outside the country. The acidification has caused changes of the fauna and flora in many inland waters and has also resulted in negative ecological effects on woods. The Swedish Environmental Protection Agency has supported research and development in order to counteract the negative effects of this acidification process. One way of counteracting the acidification of waters has been a large-scale dumping of lime in threatened lakes.

So far this paper has dealt with pollution problems of a more general nature, for which the direct health hazard to humans has not been the only, or in some cases not even the most, important issue. Increasing evidence of the long-term health hazards of chemicals, such as hereditary effects, cancer and teratogenic effects, has focused the primary attention on this aspect of the problem during the 1970s. Contamination of the aquatic environment with methyl mercury and cadmium led to manifest human intoxications in Japan and therefore served as a powerful demonstration of chemical health hazards, and other issues had a similar influence in other parts of the world. In Sweden, as well as in some other European countries, the teratogenic catastrophy of the sedative drug thalidomide during the 1960s was of great importance in turning attention toward potential health hazards to humans.

In the 1970s health hazards from air pollution have been in the forefront of attention and several fairly large-scale research projects have been launched by Swedish governmental agencies. The exhaust from cars is a particularly important source of hazardous air pollution, and a governmental Committee on Automotive Air Pollution was formed in 1977 in order to organize a research project to deal with it. Exhausts from different types of cars, fuels and driving cycles have been analyzed with respect to chemicals of particular importance from a health point of view, such as carbon monoxide, nitrogen oxides, polyaromatic hydrocarbons, nitrites and aldehydes (6). Exhaust samples were also used for extensive studies of the mutagenicity in the Salmonella/microsomal assay of Ames (27). The mutagenicity tests comprised comparisons of diesel, gasoline and less conventional fuels such as methanol and propane (LPG-Liquified Petroleum Gas). The mutagenicity of the different fuels could be divided into one high mutagenicity group, consisting of diesel; a medium mutagenicity group, consisting of leaded and lead-free gasoline as well as alcohol/gasoline; and a low mutagenicity group, consisting of methanol and propane. The difference in mutagenicity, expressed as revertant Salmonella colonies, between the high and medium mutagenicity groups and between the medium and low mutagenicity groups was roughly one order of magnitude. If the car was equipped with a three-way catalyst the mutagenicity of gasoline diminished to about a tenth. The data thus indicate that there are two ways of preventing an excessive release of mutagenic substances through the exhausts, that is, by using alcohol fuels or by equipping gasoline driven cars with a three-way catalyst.

The mutagenicity of diesel exhaust has been reported as being considerably higher than that of gasoline. However, most investigations of this kind have dealt with particulate samples. For this reason the Swedish project paid special attention to the mutagenicity property of the gas phase. The contribution of the gaseous phase to the mutagenicity is fairly low for diesel, approximately ten percent, but usually higher for gasoline - sometimes up to 50 percent. It should, however, be emphasized that the bioavailability of the particulate phase in vivo may be lower as suggested by several data (22) in which case the gas phase will contribute proportionally even more to the genotoxic effect.

The oil crisis has forced Sweden as well as other countries to plan for future alternative energy sources. Even though nuclear energy is responsible for a significant part of the power production today, the Swedish people decided in a national referendum in 1978 gradually to discard that source of energy. Coal constitutes an obvious alternative, but the environmental problems connected with a large escalation of the use of coal are difficult to predict and require further investigations. An extensive interdisciplinary governmental project - the Coal Health Environmental Project - was therefore started in 1977 and is now complete and the final report has recently appeared (37). The use of coal for power plants involves a series of possible environmental disturbances, beside the emission problems from the actual coal burning, for instance leachates from ash, sludge, and other waste disposals, and the discharge of coal dust in the air. The long-term effects of emissions from both coal and oil fired

plants on the genetic material have been investigated by means of
short-term tests with bacteria (1), mammalian cell cultures and
Drosophila (14). Although there was a correlation in mutagenicity
between bacteria and cell cultures, the latter system tended to be
a more sensitive indicator of the mutagenicity, probably dependent
on the fact that several samples were highly toxic to bacteria but
less toxic to mammalian cells. An evaluation of the genotoxic
hazards from coal- as compared to oil-fired plants has been made
on the basis of such mutagenicity tests, supplemented with
chemical analyses of the content of known mutagenic and
carcinogenic compounds in emission samples. The results indicate
that the release of mutagenic substances varies considerably
between plants, depending on their size and operation. The
emission of mutagenic substances is low and probably insignificant
from large and well-functioning plants, while emissions from ill-
functioning plants exhibit high mutagenicity and a high release of
certain mutagenic compounds, such as polyaromatic hydrocarbons.
Emission samples from coal- and oil-fired plants have also been
used in cancer tests on hamsters by means of intratracheal
installation (23). No indication of an increased tumor frequency
from either oil or coal emission samples has been noticed.

The general conclusion from the Coal Health Environmental
Project has been that a replacement of oil with coal as a fuel
will not increase health hazards from air pollution, provided
well-functioning plants are used. Although it cannot be excluded
that leaching from waste disposals may result in the release of
some heavy metals, calculations of the amounts have not given any
ground for concern, with the possible exception of mercury.

The Work Environment

The rapid development of the chemical industries and the
increased use of synthetic chemicals has directed a large part of
environmental research facilities towards problems connected with
occupational exposure to chemicals. Long-term effects and
particularly cancer induction have caused particular concern.
Specific events have also led to intensified research activity.
One such key event in several countries, including Sweden, was the
detection of an increased incidence of angiosarcoma of the liver
and other tumors among workers in vinyl chloride factories. The
subsequent discovery of the mutagenic effects of vinyl chloride
(28) and its effect on the chromosomes of exposed people (7) was
followed by extensive analyses of the genetic and biological
effects of vinyl chloride and its metabolites at the Wallenberg
Laboratory, University of Stockholm (8,21,29). These studies
ramified to other related chlorinated compounds, particularly 1.2-
dichloroethane (30,31).

The increasing concern about occupational health hazards has had several important legislative and administrative consequences. Of importance in the present context was the formation of the Swedish Work Environment Fund in 1972, with an annual budget of about $100 million U.S. The purpose of this fund is to support research and education for occupational safety and health and it is financed by taxes levied on Swedish industries. A large number of grants have been given to projects of varying magnitude concerning long-term health effects. One example of a project dealing with long-term effects of chemicals within one specific type of industry is a Swedish-Finnish project on the rubber industries (26). This project has involved (a) epidemiological investigations of cancer frequency, and (b) teratogenic effects in rubber workers, monitoring the exposed workers for chromosomal effects and urine mutagenicity. Furthermore, mutagenicity testing has been performed for a large number of chemicals used in the rubber industry as well as vulcanization samples collected in the factories or produced experimentally. The purpose of the investigations has been to trace compounds and processes which may imply a risk for cancer and mutations, and may be of relevance for the elevated cancer frequency reported from rubber industries in many countries. The monitoring of workers both with respect to urine mutagenicity (35) and effects on sister chromatid exchanges in lymphocytes (36) identified the weighing and mixing processes as well as vulcanization as hazardous from a genotoxic point of view. The genotoxic effects of vulcanization gases indicated by the monitoring of workers, was verified experimentally both by experimental samples and inducstrial samples (10). The experimental mutagenicity tests on bacteria revealed a large number of mutagenic components among practically all groups of rubber additives, such as vulcanization chemicals, accelerators, antioxidants and retardants (10). The thiurams were subjected to a special investigation because of their extensive use and their peculiar mutagenic action. Tetraethylthiuram disulfide (TETD) and the corresponding methyl compound (TMTD) do not seem to act directly on DNA but in an indirect way as inhibitors of the radical protection enzymes, superoxide dismutase and glutathione peroxidase. This interpretation was supported by the fact that the mutagenicity of TMTD was strongly enhanced by an increased oxygen pressure and diminished by a decreased oxygen pressure. Also TMTD increased the mutagenicity of a radical quinone, vitamin K (33).

The results from this research project have been put to direct use in the rubber industries in Finland and Sweden. It has been possible by technical improvements of processes involving the release of mutagenic substance, to diminish the exposure of the workers to these substances.

It can be concluded that the emphasis of adverse effects by
environmental chemicals has shifted since the 1950s, when the
problems were revealed by the biological effects of pesticides.
In the 1960s pollution of waters became a central issue, but long
term health hazards of drugs were also displayed by the
thalidomide disaster.   In the early 1970s the discovery of vinal
chloride carcinogenicity among exposed workers played an
equivalent role to direct the attention to health hazards in the
working environment.   Although all of these issues no doubt still
remain, one can forsee that the importance of food items for the
development of cancer, as indicated both by epidemiological and
experimental data, will make this area a major subject for
research in the 1980s.

REFERENCES

1. Alfheim, I. and M. Möller (1983) Mutagenicity in emissions
    from coal and oil-fired boilers,as detected by the
    Salmonella/microsome assay. KHM Technical Report 60. The
    Swedish State Power Board, Stockholm.
2. Berglund, F., M. Berlin, G. Birke, U. von Euler, L. Friberg,
    B. Holmsteadt, E. Jonsson, C. Ramel, S. Skerfving,
    Å. Swensson and S.Tejning (1971). Methyl mercury in fish. A
    toxicologic-epidemiologic evaluation of risks. Report from
    an expert group. Nordisk Hygienisk Tidskrift, Suppl. 4.
3. Berlin, M. (1983). Health effects of mercury emission from
    coal combustion (Swedish with English summary). Coal Health
    Environment, Technical Report 79.
4. Borg, K., H. Wanntorp, K. Erne and E. Hanko (1966). Mercury
    poissoning in Swedish wildlife. J. Appl. Acol. Suppl.
    3:171-172.
5. Donner, M., K. Husgafvel-Pursiainen, D. Jenssen and A. Rannug
    (1983). Mutagenicity of rubber additives and curing fumes.
    Scand. J. Work Environm. Health 9:27-37.
6. Egebäck, K.E., and B.M. Bertilsson (Eds.) (1983). Chemical
    and biological characterization of exhaust emissions from
    vehicles fueled with gasoline, alcohol, LPG and diesel.
    National Swedish Environment Protection Board Report SNV PM
    No 1635.
7. Funes-Cravioto, F., B. Lambert, J. Lindsten, L. Ehrenberg,
    A.T. Natarajan and S. Osterman-Golkar (1975). Chromosome
    aberrations in workers exposed to vinyl chloride. Lancet
    i:459.
8. Göthe, R., C.J. Calleman, L. Ehrenberg, and C.A. Wachtmeister
    (1974). Trapping with 3,4-dichlorobenzenethiol of reactive
    metabolites formed in vitro from the carcinogen vinyl
    chloride. Ambio 3:233-236.
9. Harada, M., T. Fujino, and K. Kabashina (1977). A study on
    methylmercury concentration in the umbilical cords of the

inhabitants born in the Minamata area. Brain Develop. 9:79-84.

10. Hedenstedt, A., C. Ramel, and C.A. Wachtmeister (1981). Mutagenicity of rubber vulcanization gases in Salmonella typhimurium. J. Toxicol. Environ. Health 8:805-814.

11. Jensen, S. (1966). Chlorinated biphenyls in nature. Nordisk Biocid-Information 7.

12. Jensen, S. (1972). The PCB story. Ambio 1:45-53.

13. Jensen, S., and A. Jernelöv (1969). Biological methylation of mercury in aquatic organisms. Nature 223:753-754.

14. Jenssen, D., and J. Magnusson (1983). Mutagenicity of flue gas emission from coal- and oil-combustion installations - V79 hamster cells and Drosophila (Swedish with English summary) KHM Technical Report 61. The Swedish State Power Board, Stockholm.

15. Johnels, A.G., T. Westermark, W. Berg, P.I. Persson, and B. Sjöstrand (1967). Pike (Esox lucius L.) and some other aquatic organisms in Sweden as indicators of mercury contamination in the environment. Oikos 18:323-333.

16. Lindbohm, M-L., K. Hemminki, P. Kyyrönen, I. Kilpikari, and H. Vainio (1983). Spontaneous abortions among rubber workers and congenital malformation in their offspring. Scand. J. Work Environ. Health 9:85-90.

17. Marsh, D.O., G.J. Myers, T.W. Clarkson, L. Amin-Zaki. and S. Tikriti (1980). Fetal methylmercury poisoning: Clinical and toxicological data on 29 cases. Ann. Neurol. 7:348-353.

18. Mitsumori, IK., K. Maita, T. Saito, S. Tsuda, and Y. Shirasu (1981). Carcinogenicity of methylmercury chloride in ICR mice: Preliminary note on renal carcinogenesis. Cancer Lett. 12:305-310.

19. Nixon, J.E., L.D. Coller, and J.H. Exon (1979). Effect of methyl mercury chloride on transplacental tumors induced by sodium nitrate and ethyl urea in rats. J. Natl. Cancer Inst. 63:1057-1063.

20. Önfelt, A., and D. Janssen (1982). Enhanced mutagenic response of MNU by post-treatment with methylmercury, caffeine or thymidine in V79 Chinese hamster cells. Mutation Res. 106:297-303.

21. Osterman-Golkar, S., D. Hultmark, D. Segerbäck, C.J. Calleman, R. Göthe, L., Ehrenberg and C.A. Wachtmeister (1977). Alkylation of DNA and protein in mice exposed to vinyl chloride. Biochem. Biophys. Res. Commun. 76:259-266.

22. Pepelko, W.E., R.M. Danner and N.A. Clark (Eds.) (1980). Proc. of an Int. Symp. on Health Effects of Diesel Engine Emissions. EPA-600/9-80-075, Washington, D.C.

23. Persson, S.A., M. Ahlberg, L. Berghem, E. Könberg, F. Bergman and G.F. Nordberg (1983). Long-term carcinogenic studies on Syrian golden hamsters. A comparison between particulate emissions from a coal-fired power statation and an oil-fired district heating power station. An evaluation of results

over 75 weeks (Swedish with English summary). KHM Technical
Report 65. The Swedish State Power Board, Stockholm.

24. Ramel, C. (1972). Genetic effects, in "Mercury in the
Environment", L. Friberg and J. Vostal, eds., CRC Press, Boca
Raton, Florida, pp. 169-181.

25. Ramel, C. (1973). The mercury problem - a trigger to
environmental pollution control. Mutation Res. 26:341-348.

26. Ramel, C., and H. Vainio (1983). The Finnish-Swedish project
on genotoxic hazards in the rubber industry. Conclusions and
recommendations. Scand. J. Work Environ. Health 9:91-93.

27. Rannug, U. (1983). Data from short-term tests on motor
vehicle exhausts. Environ. Health Perspect. 47:161-169.

28. Rannug, U., A. Johansson, C. Ramel, and C.A. Wachtmeister
(1974). The mutagenicity of vinyl chloride after metabolic
activation. Ambio 3:194-197.

29. Rannug, U., R. Göthe, and C.A. Wachtmeister (1976). The
mutagenicity of chloroethylene oxide, chloracetaldehyde, 2-
chloroethanol and chloroacetic acid, conceivable metabolites
of vinyl chloride. Chem.-Biol. Interactions 12: 251-263.

30. Rannug, U., and C. Ramel (1977). Mutagenicity of waste
products from vinyl chloride industries. J. Toxicol.
Environ. Health 2:1019-1029.

31. Rannug, U., A. Sundvall, and C. Ramel (1978). The mutagenic
effect of 1,2-dichloroethane on Salmonealla typhimurium. I.
Activation through conjugation with glutathione in vitro.
Chem.-Biol. Interactions 20:1-16.

32. Rannug, A., U. Rannug, and C. Ramel, K.E. Eriksson, and
K. Kringstad (1981). Mutagenic effects of effluents from
chlorine bleaching of pulp. J. Toxicol. Envron.Health
7:41-55.

33. Rannug, A., U. Rannug, and C. Ramel. Genotoxic effects of
additives in synthetic elastomers with special reference to
the mechanism of action of thiurames and dithiocarbamates.
Proc. Int. Symp. on Occupational Hazards Related to Plastics
and Synthetic Elastomers, J. Järvisalo, P. Pfäffli and
H. Vainio, eds, Alan R. Liss, New York, (in press).

34. Skerfving, S., A. Hansson, and J. Lindsten (1970).
Chromosomal breakage in human subjects exposed to methyl
mercury through fish consumption. Arch. Environ. Health
21:133-139.

35. Sorsa, M., K. Falck, and H. Vainio (1982). Detection of
worker exposure to mutagens in the rubber industry by use of
the urinary mutagenicity assay, in "Environmental Mutagens
and Carcinogens", T. Sugimura, S. Kondo, and H. Takebe,
eds., Univ. of Tokyo Press, Alan R. Liss, New York,
pp. 323-329.

36. Sorsa, M., J. Mäki-Pakkanen, and H. Vaino (1982).
Identification of mutagen exposures in the rubber industry by
the sister chromatid exchange method. Cytogen. Cell
Genet. 33:68-73.

37. Swedish Coal-Health-Environment Project (1983). Final
     Report, April 1983. The Swedish State Power Board,
     Stockholm.
38. Victorin, K., U.G. Ahlborg, M. Stahlberg and S. Honkasalo
     (1983). Mutagenic dust emissions from a pulerized coal-fired
     installation--long-term variations (Swedish with English
     summary). KHM Technical Report 63. The Swedish State Power
     Board, Stockholm.
39. Westöö, G. (1966) Determination of methyl mercury compounds
     in foodstuffs. I. Methyl mercury compounds in fish,
     identification and determination. Acta Chem. Scand.
     20:2131-2137.

RESEARCH PROGRESS ON ENVIRONMENTAL MUTAGENESIS,

CARCINOGENESIS AND TERATOGENESIS IN CHINA

C. C. Tan and J. L. Hsueh

Institute of Genetics
Fudan University
Shanghai, People's Republic of China

SUMMARY

China is a developing country, having a population of over one billion people. Of course, she has her own special conditions of climate, environment, flora and fauna. Along with the progress of modernization, the country is also facing the problem of environmental pollution through the release of industrial wastes into the air and waterways. This, together with population pressure and agricultural problems, has already received the wide attention of the Chinese public. China is prepared to exercise quality control and environmental monitoring, and the application of basic scientific methodology to the solution of practical problems has been found essential to the development of genetic toxicology. Some 30 to 40 independent qualified laboratories for identifying mutagens are now operating in China. They are distributed over more than 15 provinces. Most of these assays are used to help identify mutagens, some can also identify carcinogens. Of these laboratories, two-thirds are working mutagenetic assays and the other one-third is doing either carcinogenetic or teratogenetic assays. The in vitro assays use a variety of cell types ranging from bacterial to human, from somatic cells to germ cells; other tests can be done directly on insects, rodents and plants. With a regulatory framework gradually taking shape, mutagenicity testing is becoming a big business. Nowadays, all new drugs, pesticides, food additives, contraceptives, and even certain suspect traditional Chinese medicinal herbs are subjected to the Ames test, micronucleus test, chromosome aberrations and SCE analysis of mouse bone-marrow cells in routine screening procedures. At present, our research studies in the field of environmental mutagenesis, carcinogenesis and teratogenesis place more emphasis on the practical and applied aspects. In the long run, however, it is conceivable that attention will also be directed to the quantitative approaches and mechanism studies in order to contribute substantially to closing the gap between the basic and applied works.

INTRODUCTION

The human environment is being subjected to radiation exposures and an ever-increasing number and variety of chemicals. These agents have played an important role in industry, agriculture and public health. However a significant proportion of them have been found to be mutagenic. Studies carried out during the past two decades have documented the existence of mutagenic substances among such things as food additives, pesticides, drugs, cosmetics and industrial compounds. At the same time, studies have also shown a striking correlation between mutagenic and carcinogenic potential of most of these compounds. All these findings have underscored the need to reduce these genetic and carcinogenic hazard in the environment and offer protection to the human population. With the rapid development of modern industry during the past decades, China is now also facing the problem of environmental pollution. The public is more and more concerned about the potential dangers thus involved. The growing awareness of these facts led to the establishment of a new branch of science--genetic toxicology, which is now gaining importance rapidly.

In China, genetic toxicology was initiated in the early 1970s. Since its start, tests for mutagenicity of various environmental chemicals have been conducted in a number of institutions. As an example, take the traditionally used oral male contraceptive gossypol. At one time reports were contradictory as to whether this was mutagenic or not. In the Institute of Genetics of Fudan University, a series of experimental studies with gossypol was performed using in vitro biological systems (Table 1). All the results showed that gossypol acetate at a clinical dosage is indeed a safe male contraceptive (4-6, 15, 16).

In the assessment and evaluation of the effects of chemical mutagens various test systems are now available in China. Fast bacterial assay systems like those developed for Salmonella or E. coli are used routinely for the primary identification of mutagenic activity, whether it be an environmental mutagen, food, drug or pesticide. If the agent under study is found to damage DNA, it is desirable to continue testing by employing various

Table 1. The genetic effects of gossypol in various
          short term tests.

Assays	Results
Ames test	– (0.1-10µg/ml, w or w/o S9)
SCE of human lymphocyte in vitro	– (0.1-10µg/ml, w or w/o S9)
SCE of human lymphocyte in vivo	– (20mg/day for 85 days)
SCE of spermatogonia in mice	– (1-50mg/kg.b.w. for 19 days)
Chromosome aberration of spermatogonia in mice	– (1-50mg/kg.b.w. for 19 days)
SCE of bone-marrow in mice	– (4mg/kg.b.w. for 4 days)
Chromosome aberration of spermato-gonia and spermatocytes in rats	– (5mg/kg.b.w. for 9 days)
Micronucleus of lymphocytes in rats	– (20mg/kg.b.w. for 9 days)
Dominant lethal mutation in rats	– (20mg/kg.b.w./5 days a week for 30 days
Teratogenesis in rats	– (2-12mg/kg.b.w./day for 10 days after the sixth day of pregnancy)

eukaryotic test systems, from simple sub-mammalian assays (such as
Drosophila and yeast) to mammalian cells in culture, in order to
detect chromosomal aberrations, sister chromatid exchanges or
mutations at specific loci. Recently, a rapid and simple
procedure for the measurement of unscheduled DNA synthesis (UDS)
in cultured human cell lines was developed in the Zhejiang Medical
University at Hangchou (13, 14). In this procedure, cell
synchronization and inhibition of semiconservative DNA synthesis
of the ^{14}C-TdR prelabelled human amnion (FL) cells were achieved
by the combination of arginine-starvation and hydroxyurea (HU)
treatment. Cells thus prepared were exposed to the test-
chemicals, with or without rat liver microsomal S-9 activation, in
the presence of ^{3}H-TdR for 5 hours. The HU-resistant
incorporation of the latter into repaired DNA was measured in the
acid insoluble fraction of the target cells by liquid
scintillation counting. Six known mutagens/carcinogens were
selected to be tested by this improved technique for validation
studies—three direct mutagens MNNG (N-methyl-N'-Nitro-N-
Nitrosoguanidine), MMS (methyl methanesulfonate), and MMC
(mitomycin C); three indirect mutagens BP (benzo(a)pyrene), AFB_1
(Aflatoxin B_1) and CPP (Cyclophosphamide) (Fig. 1). Among 14
chemicals (7 pesticides, 4 food additives, 2 medicines and 1
extract of molded rice) so far tested, three were found to be able
to induce UDS in FL cells. These are bis(0,0-diethyl-
phosphinothioyl)-disulfide, an organic phosphorous germicide and
acaricide, which also gave a positive result in the Ames test; 4-

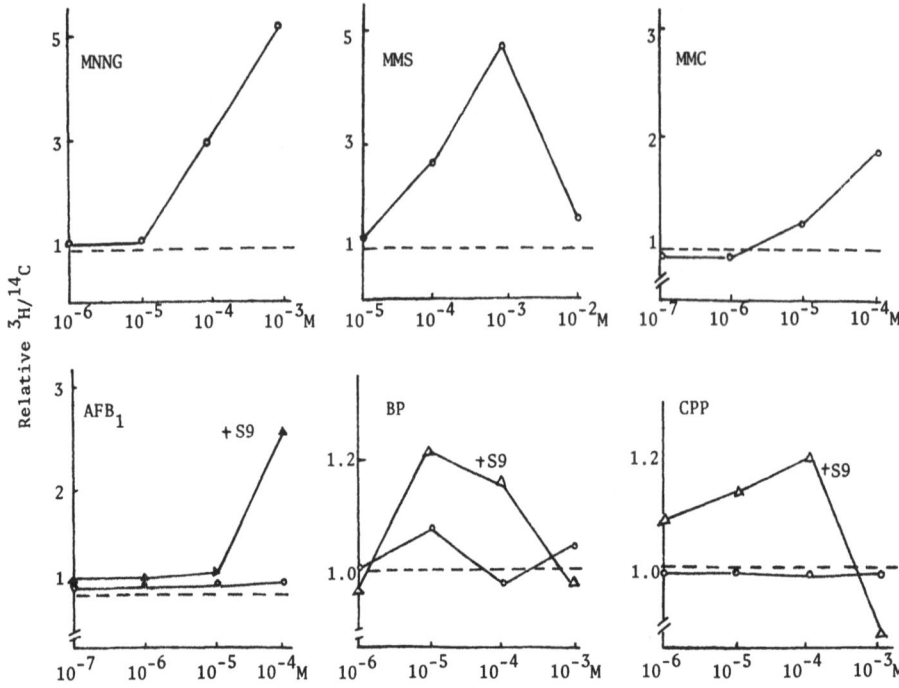

Figure 1. The dose response curves of UDS induction elicited
           by 6 known mutagens/carcinogens: MNNG,
           MMS, MMC, AFB$_1$, BP, and CPP.

chlorophenoxyacetic acid, a plant growth regulating factor
negative caramelized malt sugar, a brown coloring material widely
used as a food additive in soybean sauce which gives a positive
result in the Ames test (Fig. 2) (3). The improved method could
be used for screening the DNA-damaging effects of environmental
chemicals.

Radiation protection continues to be a great concern among
geneticists. Our general guideline is that all unnecessary
exposures should be avoided and that all necessary exposures
should be kept as low as possible. Although medical radiation
remains the largest man-made source of gonadal exposures, more
recently similar concern has been expressed over radiation
exposures that may result from nuclear fuels and from the
discharge of radioactive wastes from nuclear reactors. With a
view to accurately assessing the possible genetic hazard involved,
a survey was carried out in North China in persons exposed to low
levels of radiation in the uranium miners, in the reactor and
reprocessing plant (1-2). Detailed analysis can be seen in Tables
2-5. These results indicate that the exposed groups show a

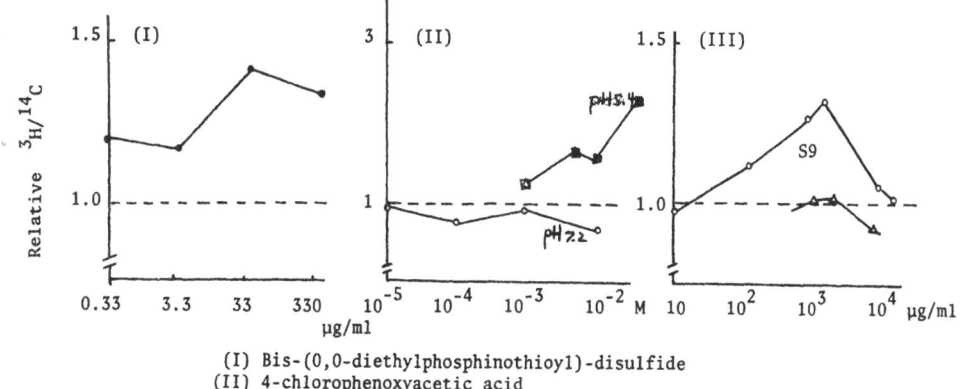

(I) Bis-(0,0-diethylphosphinothioyl)-disulfide
(II) 4-chlorophenoxyacetic acid
(III) Caramelized malt sugar

Figure 2. The dose-response curves of UDS induction
elicited by 3 above chemicals of unknown
mutagenicity/carcinogenicity.

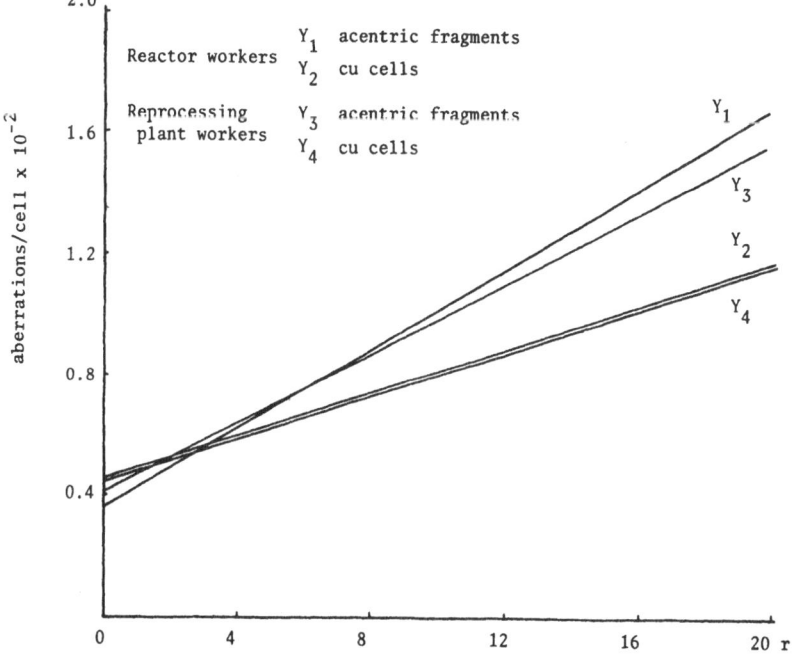

Figure 3. Relationship between dosage and aberration frequencies
for acentric fragments (Y_1), cu cells (Y_2) in reactor
workers; (Y_3) acentric fragments (Y_4) cu cells in
reprocessing plant workers.

Table 2. Dose of radiation exposure in different regions.

	Site	μr/hr
Uranium mine	open area	400-500
	tunnel	200-1500
Residential area (as control)		40-60

Table 3. Chromosomal aberrations in uranium miners compared to control group.

Group	No. of Subject	Number of cells examined	Chromosomal aberration						
			Fragments	Deletion	Translocation	Dicentric	Ring	Total cells	%
Control	37	6609	24					23	0.34
Uranium miners	39	7293	56	3	3	2	1	62	0.86

Table 4. Yields of chromosome aberrations in control, nuclear industry workers and radiologists.

Groups	Persons	cells examined	No. of Chr.-type aberrations					Cells with chr.-type aberrations	
			fragments	minutes	translocation	dicentrics	rings	No.	%
Control	86	15806	49/47c	2	1			50	0.31
Uranium miners	71	13421	103/96c	7	8	8	3	122	0.91
Reactor and reprocessing plant workers	97	19607	110/92c	11/10c	14	6	1	123	0.61
Radiologists	167	32172	225/205c	47/38c	8	19/18c	4/3c	272	0.83

Table 5. Frequencies of micronucleus in different
groups of workers of uranium mine.

Groups	Persons	Cells examined	Lymphocyte micronucleus (‰) mean ± S.E.
Uranium miners	29	58,000	0.55 ± 0.10
Ground workers	18	36,000	0.80 ± 0.15
Electricians	6	12,000	0.25 ± 0.14

significant increase, both in acentric fragment and dicentrics, as
compared to the control group. In other words, chromosomal
aberrations were found to be positively correlated with the level
of radiation exposure. They can be described by a linear dose
response model, $Y = a + bX$. These data will be useful and
important for environmental radiation monitoring (Fig. 3).

A preliminary epidemiological study by the Shanghai Institute
of Cell Biology in collaboration with several other institutions
in Shanghai on cancer incidences (1961-1970) among 146,292 workers
of 89 factories in Shanghai with over 420,000 Shanghai inhabitants
as controls, revealed excessive cancer risks to workers in certain
workshops of rubber tire factories (7-12). Three two-year in situ
animal exposures staged in the First Workshop of Rubber Tire
Factory A with 195 experimental and 62 control rats showed that
compounding and banbury mills for mastication and mixing were the
origin of carcinogenic pollutants (Table 6). Results of this
study prompted reconstruction of the workshop and suspension of

Table 6. In-situ animal exposures in Factory A.

Place of exposure	No. of rats	Carcinoma		Sarcoma	
		No.	%	No.	%
Banbury mill for mastication	68	24 ($P<0.01$)	34.3	7	10.3
Banbury mill for mixing	65	21 ($P<0.01$)	32.3	3	4.6
Compounding room	62	17 ($P<0.05$)	27.4	4	6.5
Natural control	62	2	3.2	4	6.5

Duration of exposure: 1.5yrs (compounding room 1 yr.) experiment: 2 yrs.

the use of antioxidant D in Shanghai. All 6 chronic experiments
carried out in different institutes by feeding, injection and
inhalation of technical and pure phenyl-2-naphthylamine (PBNA)
indicated the carcinogenicity, especially to the lungs, of PBNA to
rats and mice (Table 7). Repeated subcutaneous injections of both
technical and chemically pure PBNA dissolved in DMSO resulted in
male ICR mice 45% malignant tumors in 10 months. Determinations
of atmospheric concentrations of PBNA at different sites in two
rubber tire factories suggested a causal relationship between PBNA
in the air and lung cancers of the workers. The L-isomer of PBNA,
N-phenyl-L-naphthylamine (PANA) is the main constituent of
antioxidant A, also a widely used rubber additive, the
carcinogenicity of which had never been suspected before. Its
biological effect was compared with PBNA in chronic injection
experiments with ICR and TAI mouse strains. Both tests showed
similar carcinogenic potency. PANA had a tendency to induce
haemangiosarcomas. Previous unilateral nephrectomy enhanced
induction of renal haemangiosarcomas by both chemicals. The
approximately similar carcinogenic potency of PANA and PBNA
suggests other routes of metabolic activation than dephenylation
of PANA and PBNA in mice.

Table 7. Essential results of animal experiments on
Carcinogenicity of PBNA.

PBNA	Animal used ($\male$)	Method of experiment	Result		
			Group	Malignant tumor per all animals	Main tumor
Technical pure	Albino rats	Gastric intub. 18mo exp. 24mo	Expt.	27/57	lung and others
			oil control	6/43	sarcoma
Partial purified	Wistar rats	ditto large dose 12mo exp. 18mo	Expt.	29/57	lung and kidney
			oil control	5/25	sarcoma
			natural control	1/26	sarcoma
Technical pure	C57BL mice	inhalation 12mo exp. 22mo	Expt.	23/51	lung
			natural control	7/47	sarcoma
Technical pure	C57BL mice	feed in diet 18mo exp. 24mo	Expt.	48/67	lung and liver
			natural control	4/19	sarcoma
Technical pure	ICR mice	Subcut. injection of DMSO for 1 wk exp. 10mo	Expt.	9/20	lung
			DMSO control	0/20	-
Chemical pure	ICR mice	ditto	Expt.	9/20	lung
			DMSO control	0/20	-

The detection of environmental teratogens has also become the subject of wide-spread concern in China. As most families now favor the idea of having only one child, the desire for a healthy fetus seems to be particularly strong. Many surveys have been carried out in different regions in China to estimate the congenital malformation rate of the newborn. The malformation rate varies between 0.9 - 2.8%, as reported by different sources. The most likely interpretation for the lower incidence might be due to different diagnostic criteria used or an inadequate reporting system, while the highest prevalence might be attributed to some other factors in which environment might play its part. Anyway, these figures seem to be much the same as those reported by most countries. With a view to early detection of congenital malformation and other genetic defects, much importance has now been attached to prenatal diagnosis so as to reduce the malformation rate and protect pregnant women from giving birth to defective infants. Prenatal diagnosis previously only available in limited numbers of laboratories has now passed into routine clinical use in many hospitals. For the most part, it relies upon cytogenetic and biochemical analysis of cultured amniotic fluid constituents and amniotic cells. As these techniques are familiar to most readers, they will not be further elaborated. In China, the most common indication for prenatal diagnosis is a fetus at risk from a chromosomal abnormality. As the increased incidence of trisomic offspring, especially trisomy 21, is definitely linked to advanced maternal age, these pregnant women are usually the candidates for chromosomal analysis. If an aneuploid fetus is found, the pregnancy will be terminated. The second group of patients tested in prenatal diagnosis are women who gave birth to a previous child with neural tube disorders. Along with amniotic fluid studies, non-invasive techniques like ultrasonography and measurement of maternal serum AFP are now also attempted. We hope that with the use of these diagnostic tools the incidence of congenital malformation, which brings so much suffering to the families affected, will be further decreased.

Environmental agents that cause birth defects have been extensively reviewed by Dr. Thomas Shepard in this Workshop; so I will not go into further detail. What I would like to say at this moment is that although a number of agents such as mercury, aminopterin, methotrexate, warfarin, busulfan etc. are known to be teratogenic or fetopathic in nature, yet well-matched, comprehensive clinical trials with an aim to link these environmental agents to increased malformation of infants have not so far been tried. One of the difficulties we have encountered is the absence of an on-going monitoring system for the surveillance of possible environmental teratogenic factors and it is quite obvious that watching the changes in frequency of malformation alone would not provide a clue to the association with environmental factors unless they are measured simultaneously.

In the study of teratogenesis, it is reported by Shanghai
First Medical College and several other institutions that
substances which are teratogenic to animals are not necessarily so
to humans. Here we have an example to illustrate this.
Methylene-Bis Amino-Thiodiazole, MATDA, (Thiodaqual), a pesticide,
useful in killing Xanthomonus oryzae for the prevention of
"bacterial leaf blight" disease in rice, has been found to be a
potent teratogenic agent in both experimental mice and rats. It
has been reported that doses as low as 0.1 - 1 mg/kg B.W. could
induce teratogenesis in these animals. These findings prompted an
extensive investigation in the regions where Thiodaqual had been
used. Two counties where Thiodaqual was employed were chosen as a
study group whereas another two counties free of Thiodaqual
contamination served as control. A total of more than two million
rural inhabitants were investigated. Analysis of the data showed
that there was no evidence of increased induced teratogenesis in
the study group in terms of frequency of fetal loss, congenital
malformed infants and infant death, despite the fact that 15-18
tons of Thiodaqual were consumed by each county for 7 consecutive
years (Table 8). Failing a better explanation, it could be
suggested that the susceptibility to Thiodaqual of experimental
animals might be much greater than that of humans. The amount of
Thiodaqual that may possibly enter into the human body is only one
thousandth of the dose for the animal experiment. Nicotinic acid
amides may also antagonize the teratogenic effects of Thiodaqual.

Table 8. Comparison of pregnancy and birth defect among
observed and control groups (1979-1981).

Population group	No. of pregnancy	No. of spont. abortion	Birth and premature birth				Birth defect	
			living birth	still birth	dead fetus	Total	No.	‰
Observe	74,437	647 (8.69%)	72,693	775	322	73,790	713	9.66
Control	36,452	354 (9.71%)	35,612	330	156	36,098	520	14.41
Total	110,889	1,001 (9.03%)	108,305	1,105	478	109,888	1,233	11.22

REFERENCES

1. Cao, Z.Y. (1979) The observations of lymphocyte chromosome
   aberrations in uranium workers (I). Radiation Protection
   1:21-24.

2. Cao, Z.Y. (1979) The observations of lymphocyte chromosome aberrations in uranium workers (II). Radiation Protection 1:25-28.

3. Ding, C., Y.N. Yu, and X.R. Chen (1983) DNA-damaging effect of 4-chlorophenoxyacetic acid in cultured mammalian cells. J. Zhejiang Med. Univ. (in press).

4. Hsueh, J.L., and W. Xiang (1983) A concise report on environmental mutagenesis research in our laboratory. International Workshop on Environmental Mutagenesis, Carcinogenesis, and Teratogenesis, Shanghai, People's Republic of China, May 1983.

5. Li, C.B., F. Ding, Z.R. Ma, and S.Y. Zhao (1981) The determination of the male contraceptive gossypol as a non-mutagenic agent by the Ames test and SCE method. Fudan J. (Natural Science) 20(4):361-365.

6. Tsui, Y.C., M.R. Creasy, and M.A. Hulten (1983) The effect of gossypol on human lymphocytes in vitro: Cell kinetics, traditional chromosome breakage, micronuclei and sister chromatid exchange. J. Med. Genet. 20:81-85.

7. Wang, H.W., R.W. Dzeng, and D. Wang (1982) The carcinogenicity of N-phenyl-2-naphthy-L-amine on ICR mice. Acta Biol. Exper. Sinica 15:199-207.

8. Wang, H.W., R.J. Shen, and R.W. Dzeng (1981) The carcinogenic activity of antioxidant D on C57/BL mice. Acta Biol. Exper. Sinica 14:129-135.

9. Wang, D., and H.W. Wang (1981) Experimental studies on carcinogenicity of refined antiox D on Wistar rats. Tumor 1:1-2.

10. Wang, H.W., and D. Wang (1981) The synergistic action of antioxidant D and carbon tetrachloride on rat liver. Acta Biol. Exper. Sinica 14:371-377.

11. Wang, H.W., D. Wang, P.X. Tang, R.J. Shen, and R.W. Zeng (1981) The carcinogenic activity of antioxidant D on albino rats. Acta Biol. Exper. Sinica 14:77-85.

12. Wang, D., R.W. Zeng, and H.W. Wang (1982) Comparison of carcinogenicity of antiox D to TA-1 and ICR strains of mice. Shanghai Commun. Animal Husb. Vet. Med. 2:9-11.

13. Yu, Y.N., C. Ding, X.R. Chen, and Q.G. Li. A rapid and simple procedure for measurement of unscheduled DNA synthesis in cultured human cell line suitable for screening the DNA-damaging effects of chemicals. Scientia Sinica (submitted for publication).

14. Yu, Y.N., C. Ding, X.R. Chen, Q.G. Li, X.M. Yao, and S.Z. Zhang. Caramel, a DNA-damaging agent. Chinese Med. J. (submitted for publication).

15. Zhang, Z.S., X.X. Pan, M.M. Wang, and Y.I. Yao (1981) Genetic studies of the gossypol I. The comparative studies of cytogenetic effects of gossypol acetate on male germ cells and lymphocytes in blood of the micronucleus test in rats. Reproduction and Contraception 1(1):33-36.

16. Zhou, J.M., Q. Lu, and S.H. Jiang (1982) Effects of gossypol
    on germ cell chromosomes in mice. Reproduction and
    Contraception 2:49-52.

A STRATEGY OF APPROACH TO CANCER CONTROL IN CHINA

You Hui Zhang

Cancer Institute
Chinese Academy of Medical Sciences
Beijing, People's Republic of China

SUMMARY

In China cancer has become the second leading cause of death in men and the third leading cause of death in women. Each year approximately 700,000 people die of cancer. Mass screenings have picked up many early cases for radical treatment with good results; however, the early detection program, though effective in certain types of cancer, needs a tremendous amount of man-power and money, making it impractical in a large developing country like China. In the past decade or so, we have been pursuing an approach to cancer control which focuses primarily on prevention, especially in cancer-prone areas. Clues provided by pioneering epidemiologic surveys have indicated that both environmental and host factors contribute to the causation of cancer. A number of carcinogens have been demonstrated in the environment and some are synthesized de novo from precursor substances in the body. Although they are shown to be carcinogenic in experimental animals, it is meaningful only if they are so also in man. People exposed to carcinogens are not equally vulnerable; susceptibility as revealed by family studies can be explained, in part, by defects in DNA repair and/or HLA-associated Ir gene defects. Nutritional imbalance and local precursor lesions are among the acquired host factors modifying human resistance to carcinogens. These two factors are perhaps causally related. This offers ample opportunity to carry out nutritional intervention, particularly in high-risk subjects, for the prevention of human cancer.

According to the recent census, China has a population of over 1 billion. Although the cancer mortality rate in China, adjusted to the world population, is not as high as that of the

European countries, the huge population makes the absolute number
of cancer deaths a great burden to the health care in this
country. Data collected in the period from 1975 to 1978 indicated
that the number of all cancer deaths was 700,000 per year: 160,000
due to stomach cancer, 157,000 due to esophageal cancer and
100,000 due to liver cancer (7). Clearly, China is faced with a
gigantic task of cancer control.

The majority of cancer patients, once diagnosed, are advanced
cases. Their prognosis is usually poor. This is particularly
true in patients with cancer of the esophagus and liver. The 5-
year survival rate of esophageal cancer patients in highly
selected cases was 30% treated with radical surgery, whereas in
those whose tumor could not be resected and was treated with
radiotherapy, the rate was only 8%. The survival rate of liver
cancer patients was even worse. Efforts have been made in the
past decade to diagnose and treat patients in the early stages of
cancer development. This so-called secondary prevention has
proven to be quite encouraging. Early, asymptomatic or
subclinical cancers of the esophagus, liver and nasopharynx were
successfully detected. When adequately treated in time, a
considerable number of these cases have been cured. However, to
find early cancer patients it is necessary to carry out regular
screening among the general population with simple tests specific,
or relatively specific, for a particular type of cancer. Such
screening must be done once or twice a year lest new cases escape
detection. This, of course, needs tremendous amounts of manpower
and money. A rough estimate indicates that detecting one early
esophageal cancer patients costs about ⌐000-1500. Further,
patients so detected may refuse treatment because of apparently
good health. They may not be properly treated because therapeutic
facilities are limited, especially in the remote rural districts.
Therefore, when one evaluates the usefulness of the secondary
prevention of cancer on the basis of cost versus benefit, it can
hardly be considered an effective and practical way to approach
cancer control in a developing country like China.

Most of us agree that prevention is better than cure. In the
case of cancer this is perhaps even more generally accepted.
Here, by prevention I mean primary etiologic prevention. To
achieve this goal it is of paramount importance to search for the
carcinogen(s) responsible for the cancer in question. This
necessitates, first of all, epidemiologic studies to provide
guidelines on which the etiologic studies are to be based. A
nation-wide retrospective survey of cancer deaths in China (7) has
revealed a peculiar geographic distribution of cancers, especially
cancer of the esophagus, liver and nasopharynx. In some
districts, the cancer death rate is extraordinarily high whereas
in others it is quite low. The difference may be several- to
hundred-fold. This provides a unique situation to do comparative

epidemiologic studies between the high- and low-rate areas to elucidate the possible cause(s) of cancer. The geographic clustering of cancer suggests the possible existence of carcinogen(s) in the environment. Both the macro- and the micro-environment have to be considered, but we have focused our attention more on the latter, on the life style in relation to diet, dietary habit and food hygiene, since 60% of all cancer deaths are due to cancer of the upper alimentary tract.

To search for carcinogen(s) in the environment is not an easy job. Thanks are due to those who developed various short-term in vitro systems for the biological assay of mutagens. These assays are usually used as the first step in screening. Once mutagenicity has been demonstrated, the mutagen(s) is isolated and its chemical structure identified. The substance(s) involved can then be synthesized in greater amounts for subsequent in vivo carcinogenesis study in experimental animals. This is the commonly-adopted sequence of approach. For instance, pickled vegetables, a favorite food of the inhabitants in the high-incidence area of esophageal cancer, were at the outset of the epidemiologic study incriminated as the main cause, because there was positive correlation between the amount and duration of consumption of pickles and the mortality rate from esophageal cancer. Pickle extracts were tested for mutagenicity and were found to induce (a) an increase in mutants in S. typhimurium TA 98 and TA 100 strains; (b) an increase in sister chromatid exchange (SCE) in Syrian hamster embryo cells in vitro; and (c) an increase in the frequency of 6-thioguanine-resistant mutants in cultured Chinese hamster V79 cells (2,17). Chemical analysis of the pickle extracts revealed the presence of nitrate, nitrite and secondary amines. Nitrosamines (NAs) were identified on thermal energy analyzer (TEA), and mass spectrometry (MS) confirmed the presence of dimethyl nitrosamine (DMNA), diethyl nitrosamine (DENA) and dimethylthiotetranitrosodiiron (Roussin red methyl ester, RRME). The first two nitrosamines are known carcinogens whereas RRME is a dye synthesized in the last century by Roussin but never before identified as a natural product (28). RRME, with four NO-groups in the molecule, may behave as NA precursor but itself is not mutagenic or only weakly so (1,17). It was found capable of exerting a significant promoting effect in vitro on C3H/10T1/2 cells initiated by 3-methylcholanthrene (1). In mice, RRME per se was non-tumorigenic, but greatly promoted the induction of papilloma of the fore-stomach by MBNA (12).

A slightly different sequence of approach was occasionally adopted. In this case, substances were first identified and characterized and then tested for mutagenicity and carcinogenicity. Here is an example. Staple foods, such as corn flour, collected from different households in high-incidence areas of esophageal cancer, were examined for NAs. Although the presence of a number of NAs was demonstrated (6), the quantity was usually too low to be significant. Observing that people in the region very often ate steamed corn bread which had become visibly moldy during storage, investigators inoculated the fungus Fusarium moniliforme isolated from the moldy food into corn bread. After incubation for 8 days, upon addition of a minute quantity of sodium nitrite, appreciable amounts of NAs were formed (9,10). They were shown on MS to be DMNA, MBNA and a hitherto unidentified NA, N-1-methylacetonyl-N-3-methylbutyl-nitrosamine (MAMBNA) (20). DENA was found in corn bread inoculated with Aspergillus flavus. This newly discovered NA was tested for mutagenicity and was found (a) to be positive in the Ames test, (b) to induce 8-azaguanine-resistant mutants in V79 cells and (c) to induce in vitro transformation in golden hamster lung fibroblasts (18). In animal experiments, papillomas of the fore-stomach were induced in mice treated with MAMBNA for 5 months. Papillomas and early carcinomas of the fore-stomach were also induced in rats fed with $\underline{F}$. moniliforme-inoculated corn bread and $NaNO_2$ (11).

The above-mentioned investigations strongly suggest that NAs in food may be the cause of esophageal cancer prevalent in certain parts of China. However, the mere demonstration of the occurrence of carcinogens in the environment is far from enough to pinpoint the cause of a human malignant disease. It is absolutely necessary to explore further their carcinogenic effect on human beings since the results obtained from experimental animals cannot be extrapolated directly to man. Moreover, environmental NAs are not the sole source of NAs. They can also be synthesized in vivo from precursors. The in vivo synthesis of NAs was demonstrated in the stomach of rats and pigs fed with secondary amines and nitrites before slaughtering (8). When proline was given orally to man, N-nitrosoproline could be demonstrated in the urine, substantiating in vivo nitrosation in man. It is, therefore, necessary to ascertain whether NAs and/or their metabolites can be detected in the human body. Recently, in collaboration with the International Agency for Research on Cancer, N-nitroso-sarcosine (NS) was identified by TEA in the 24-hour urine samples collected from high-incidence area of esophageal cancer in significantly higher quantities than in the urine samples from low-incidence areas (16). NS is a nitrosamine which selectively induces cancer of the esophagus in rats. NS has not as yet been demonstrated in the environment and is apparently synthesized within the body. Whether DMNA, DENA, MBNA and MAMBNA can be found in the human body remains to be determined. If their level can be monitored we will be able to get a clear idea of the level of exposure to these carcinogens on an individual basis. At this time, in fact, the concentration of urinary aflatoxin is being monitored by radioimmunoassay in areas where liver cancer is commonly seen (31). A prospective study has just started to correlate the level of exposure to aflatoxin and the incidence of liver cancer. This is a reliable way to assess the significance of mycotoxin exposure in the etiology of human liver cancer.

Most carcinogens are precarcinogens in the sense that they must be metabolically activated before they can effect carcinogenesis. It is a routine practice that in the in vitro biological assays of mutagens, liver microsome preparation of mice or rats is added to the assay system to provide the cytochrome P450 activity. The difference in cytochrome P450 activity among various species and organs may explain, at least in part, the species variation in the resistance or susceptibility to a certain carcinogen and its organ specificity. Most NAs so far discovered are precarcinogens and require metabolic activation. It is, therefore, important to know if the cytochrome P450 of human origin is capable of metabolically activating NAs. Even more important is to demonstrate the enzyme activity in the normal counterpart of malignant cells and its capability to activate NAs in situ. Preliminary studies have demonstrated P450 activity in human fetal liver but its level is low, as compared to that of the rat (14). As cytochrome P450 is an induction enzyme, it is not impossible that the low activity seen in human fetal liver may be due to lack of induction. When 'normal' human esophageal epithelium obtained from surgical specimens was incubated for 8 hours with ^{3}H-MAMBNA or ^{14}C-MBNA, the DNA extracted from the epithelium was found to have been incorporated with the radioisotopes (19). Although it remains to be determined on which position (O^6 or N^7) of the guanine molecule alkylation occurred, the formation of NA-DNA adduct is circumstantial evidence indicating the ability of human esophageal epithelium to metabolically activate NAs. Thus, NAs are potentially carcinogenic to man. More direct support may be obtained from in vitro malignant transformation study of cells of human origin. A fibroblast cell line derived from 'normal' human esophageal tissue at the vicinity of cancer has undergone malignant transformation induced by MNNG (29). MAMBNA was found to be able to induce malignant transformation of baby hamster lung fibroblasts (30). Human cells, especially epithelial in origin, have not yet been successfully transformed in vitro by MAMBNA or other NAs.

Carcinogenesis can be considered as a series of complicated interactions between carcinogens and the host. A variety of host factors can alter the individual response to environmental carcinogens so as to render the host more or less susceptible. These host factors may be inherited or acquired, systemic or local. Esophageal cancer has been noted for its tendency to family clustering. This phenomenon per se may be explained on the basis of hereditary predisposition and/or common exposure to environmental carcinogens. Recent family studies (5) showed that the relative risk (RR) to esophageal cancer varied with the consanguinity in relation to the patients, being highest among

first degree blood-relatives. Non-blood relatives, such as wives
of cancer patients, showed only a slightly increased RR though
having lived for decades in the high-risk families. The
difference in RR between these two extreme conditions was several-
fold. This suggests that carcinogens in the environment more
easily lead to malignancy when acting on genetically susceptible
hosts.

What is the underlying molecular mechanism of the inherited
susceptibility? Preliminary investigation (4) revealed that the
ability to repair DNA after UV damage was very much impaired in
esophageal cancer patients from high-risk families. Even normal
individuals from high-risk families showed impaired DNA syntheses
but less severe in degree. Of course, defects in DNA repair may
not be the sole explanation. It may also be related to an HLA-
associated Ir gene defect. HLA typing has provided some clue.
The risk of esophageal cancer was found to be correlated with the
high frequency of HLA-A2 and B40 which are in linkage
disequilibrium (38).

Among the acquired host factors that may modify human
susceptibility to environmental carcinogens, particularly worth
mentioning are nutritional imbalance and local pathologic
processes.

Many years ago, people began to study the relationship
between trace elements and cancer. In high-incidence areas of
esophageal cancer, contents of molybdenum (Mo), magnesium (Mg) and
zinc (Zn) in the environment as well as in the human body were
found to be decreased and the magnitude of the decrease correlated
with the cancer mortality rate (24). Mo has received special
attention because it is a co-factor of the enzyme nitrate
reductase, the activity of which affects nitrite contents in
vegetables. Mo, in the form of ammonium molybdate, was shown to
inhibit the in vitro mutagenic effect and in vivo carcinogenic
effect of nitroso-sarcosine ethyl ester (NSEE) (23). Mo
deficiency produced by feeding an antagonist, on the contrary,
facilitated tumor induction by NSEE (22). In rats orally treated
with ^{14}C-DENA, Mo in a dose of 1 mg/day for 30 days markedly
inhibited DNA alkylation in the liver whereas high doses (5 mg/
day) produced the opposite effect (21). Consistent with these
results were the findings that low doses of Mo increased, whereas
high doses decreased, demethylation of MAMBNA and MBNA, an
important step in the metabolism of NAs in the body (15). The
effect of both doses of Mo on cytochrome P450 activity was not
significant. It is thus conceivable that Mo deficiency may be an
important contributing factor in human esophageal cancer and Mo
supplementation may be of prophylactic value, but the dosage seems
crucial.

Selenium is another trace element which attracts cancer etiologists' attention. Sero-epidemiologic studies have shown a negative correlation between serum selenium levels and mortality rates from human liver cancer (37), colon cancer (34) and lung cancer (39). Experiments in mice showed that sodium selenite increased the cAMP level in hepatoma cells but not in normal hepatocytes of the tumor-bearing host owing to a selective inhibitory effect on the phosphodiesterase activity of the tumor cells (33).

Vitamin deficiency is another noteworthy host factor that may have an important bearing on carcinogenesis. Owing to an insufficient intake of animal protein and fat, and lack of fresh vegetables and fruits, especially during the winter, the general level of vitamin A, B_2 and C is low among the inhabitants of high-incidence areas of esophageal cancer although overt symptoms and signs of vitamin deficiency are not common (21). Retinol and riboflavin deficiency may produce changes in the epithelium which render it more vulnerable to environmental carcinogens. Ascorbic acid deficiency may, among other things, facilitate NA synthesis in vivo.

In the past, chronic irritation with resultant inflammation was considered to be the cause of cancer. Nowadays, it is still looked on as a pre-malignant lesion of many cancers. For instance, chronic inflammatory lesions of the uterine cervix, if left untreated, may eventually develop into cancer. In high-incidence areas of esophageal cancer, esophagitis was found in a high proportion of normal individuals (25). Its cause has not been well elucidated. It may be due to physical trauma caused by hot food and/or coarse food. Silica fragments in the millet bran were found capable of penetrating into the mucosa (26). Vitamin deficiency may be another cause which makes the mucosal epithelium less resistant to invasion by micro-organisms. Fungi (mostly Candida albicans) have been demonstrated in the esophageal epithelium of patients with severe dysplasia and early cancer (32). Even less clear is the mechanism by which chronic inflammation promotes carcinogenesis. Stimulation of anchorage-dependent cell growth and activation of alveolar macrophages were taken as two possible mechanisms by which asbestos promotes lung cancer (27). It is likely that silica fibers have a similar effect on esophageal cancer. Chronic inflammation, regardless of its cause, is almost always accompanied by mononuclear cell infiltration. Silica fiber as foreign body, if present, would also preferentially stimulate local accumulation of macrophages. A pilot study disclosed that LPS-activated macrophages of mice were competent to metabolically activate benzo(a)pyrene and MBNA in an in vitro SCE assay in V79 cells (3). Further study has to be done before any definitive conclusion can be reached. To study the causal relationship between inflammation and carcinogenesis is undoubtedly an important field of investigation which will help explain the organ selectivity of carcinogens and why chronically inflamed tissue is apt to become malignant.

In summary, in the past ten years or more, we have devoted ourselves to exploring the possible causes of human cancer in order to effectively conduct etiologic prevention. We have since developed a strategy of approach, as formulated above, the most important links of which are: (a) to identify the putative carcinogens and promoters in the external environment as well as in the body; (b) to provide indirect and/or direct evidence that these carcinogens can indeed initiate carcinogenesis in human cells; and (c) to demonstrate high-risk individuals with an increased susceptibility, be it inherited or acquired. When essential information from these studies is available, the time will be ripe for cancer prevention. It can be done on a population basis but major attention should be focused on high-risk individuals. Measures should preferably be directed to carcinogens but complete elimination of carcinogens is not always feasible. Those measures directed to contributing factors may also be beneficial since the etiology of cancer is usually multi-factorial. In the latter case, nutritional intervention seems most promising. Sodium selenite has been employed in an intervention trial for liver cancer among ducks in a high-incidence area of human liver cancer (35). Attempts have been made to increase the content of organic selenium in the crops so as to provide more selenium in the diet in low-selenium areas. Preliminary trials gave good results: one gram of sodium selenite per mu (1/15 of a hectare) of land increased the selenium content in the maize 5-fold (35). A program of nutritional intervention with vitamins and zinc for esophageal cancer has also been started (36). Obviously, intervention trials must be based on solid results of research on cancer etiology and the outcome of the trials will, in turn, test and verify etiologic hypotheses of cancer.

REFERENCES

1. Cheng, S.J., M. Sala, M.H. Li, I. Courtois, and I. Chouroulinkov (1981) Promoting effect of Roussin's red identified in pickled vegetables from Linhsien China. Carcinogenesis 2:313-319.
2. Cheng, S.J., H. Sala, M.H. Li, M.Y. Wang, J. Pot-Deprun, and I. Chouroulinkov (1980) Mutagenic, transforming and promoting effect of pickled vegetables from Linhsien County, China. Carcinogenesis 1:685-692.
3. Cheng, S.J., D.M. Sun, and Y.H. Zhang (1983) (unpublished data).
4. Ding, J.H., and M. Wu (1983) (to be published).

5. Ding, J.H., M. Wu, X.Q. Wang, X. Luo, and J.M. Wu (1983) Hereditary susceptibility to esophageal cancer in Linxian. Nat. J. Med. China 63:213-215.
6. Ji, C., and M.H. Li (1983) Naturally occurring nitrosamines in the pickled vegetables and corn flour in Linxian. Chinese J. Oncology 5:73.
7. Li, B., and J.Y. Li (1980) National survey of cancer mortality in China. Chinese J. Oncology 2:1-10.
8. Li, M.H., S.H. Lu, C. Ji, S.J. Cheng, M.Y. Wang, C.L. Jin, and Y.L. Wang (1978) In vitro and in vivo synthesis of nitrosamines, an experimental study. Collection of Papers for the 20th Anniversary of the Cancer Institute and Ritan Hospital. pp. 75-79.
9. Li, M.X., S.X. Lu, C. Ji, M.Y. Wang, S.J. Cheng, and C.L. Jin (1979) Formation of carcinogenic N-nitroso compounds in corn bread inoculated with fungi. Scientia Sinica 22:471-477.
10. Li, M.H., S.H. Lu, C. Ji, Y. Wang, S.J. Cheng, and G. Tian (1980) Experimental Studies on the Carcinogenicity of Fungus-Contaminated Food from Linxian County. Japan Science Society Press, Tokyo, pp. 139-148.
11. Li, M.H., G.Z. Tian, S.X. Lu, S.P. Guo, C.L. Jin, and Y.L. Wang (1982) Forestomach carcinoma induced in rats by cornbread inoculated with Fusarium moniliforme. Chinese J. Oncology 4:241-244.
12. Lu, S.H. (1982) (unpublished data).
13. Lu, S.H. (1982) (personal communication).
14. Lu, S.H. (1983) (personal communication).
15. Lu, S.H. (1983) (personal communication).
16. Lu, S.H. (1983) (to be published).
17. Lu, S.H., A.M. Camus, L. Tomatis, and H. Bartsch (1981) Mutagenicity of extracts of pickled vegetables collected in Linhsien County, a high-incidence area for esophageal cancer in northern China. J. Nat. Cancer Inst. 66:33-36.
18. Lu, S.H., A.M. Camus, Y.L. Wang, M.Y. Wang, and H. Bartsch (1980) Mutagenicity in Salmonella typhimurium of N-3-methyl-butyl-N-1-methylacetonyl nitrosamine and N-methyl-N-benzyl-nitrosamine, N-nitrosation products isolated from corn bread contaminated with commonly occurring moulds in Linhsien County, a high-incidence area for esophageal cancer in northern China. Carcinogenesis 1:867-870.
19. Lu, S.H., M.H. Li, C. Ji, and S.J. Cheng (1983) Experimental study of the formation of nitrosamines in mouldy food and their carcinogenicity, an overview. Chinese J. Oncology 5:76-78.
20. Lu, S.X., L.X. Li, C. Ji, M.T. Wang, Y.L. Wang, and L. Huang (1979) A new N-nitroso compound, N-3-methylbutyl-N-1-methyl-acetonyl nitrosamine, in corn bread inoculated with fungi. Scientia Sinica 22:601-607.
21. Lu, S.H., and P. Lin (1982) Recent research on the etiology of esophageal cancer in China. Z. Gastroenterol. 20;361-367.

22. Luo, X.M. (1983) (personal communication).
23. Luo, X.M. (1983) (personal communication).
24. Luo, X.W. (1978) Preliminary analysis of hair molybdenum
    content in high-and low-incidence area for esophageal cancer
    in Henan Province. Collection of Papers for the 20th
    Anniversary of the Cancer Institute and Ritan Hospital. pp.
    120-125.
25. Munoz, N., A. Grassi, Q. Shan, M. Crespi, G.Q. Wang, and
    Z.C. Li (1982) Precursor lesions of esophageal cancer.
    Lancet April 17, pp. 876-879.
26. O'Neill, C.H., Q.Q. Pan, G. Clark, F.S. Liu, G. Hodges, M. Ge,
    P. Jordan, Y.M. Chang, R. Newman, and E. Toulson (1982)
    Silica fragments from millet bran in mucosa surrounding
    esophageal tumor in patients in northern China. Lancet May
    29, pp. 1202-1206.
27. Stoker, M., C. O'Neill, S. Berryman, and V. Waxman (1968)
    Anchorage and growth regulation in normal and virus-
    transformed cells. Int. J. Cancer 3:683-693.
28. Wang, G.H., W.X. Zhang, and W.G. Chai (1980) The
    identification of natural Roussin red methyl ester. Acta
    Chim. Sinica 38:95-102.
29. Wu, M., X.Q. Wang, and S.D. Li (1980) Chemically-induced in
    vitro malignant transformation of human esophagus
    fibroblasts. Scientia Sinica 23:658-662.
30. Wu, M., D.F. Wu, X.Q. Wang, S.D. Li, X. Luo, S.H. Lu,
    Y.L. Wang, and C. Ji (1982) Malignant transformation of baby
    hamster lung fibroblasts induced in vitro by a new
    nitrosamine compound. Scientia Sinica 7:738-743.
31. Wu, S.M., J.Z. Yang, and Z.T. Sun (1983) Studies on
    immunoconcentration and immuno-assay of aflatoxins. Chinese
    J. Oncology 5:81-84.
32. Xia. Q.J. (1978) Fungal invasion in the tissue of esophagus
    and its possible relation to esophageal cancer. Nat. J. Med.
    China 38:392-395.
33. Yu, S.Y. (submitted for publication) Acta Acad. Med. Sinica.
34. Yu, S.Y. (1982) (unpublished data).
35. Yu, S.Y. (1983) (personal communication).
36. Yu, S.Y. (1983) (personal communication).
37. Yu, S.Y., L.M. Wang, and W.J. Qian (1982) Modulating effect of
    sodium selenite on cAMP content in hepatoma cells in mice.
    Chinese J. Oncology 4:50.
38. Zhu, X.K., Z.X. Li, Q.Q. He, X.Z. Hu, and M.H. Cao (1982) A
    preliminary report on the possible association of esophageal
    cancer and HLA. Chinese J. Oncology 4:257-261.
39. Zhu, Y.J., Y.Q. Liu, C. Hou and S.Y. Yu (1982) Blood selenium
    level in different population groups in high-incidence area
    for lung cancer. Chinese J. Oncology 4:158.

TOXICOLOGICAL RESEARCH IN PUBLIC HEALTH:

THE CHINESE EXPERIENCE

S. Z. Hsueh

Toxicological Laboratory
Institute of Preventive Medicine
Shanghai First Medical College
Shanghai, People's Republic of China

SUMMARY

As the public health service and the general health status of the population in China have been significantly improved during the past three decades, toxicology has gained importance and is making advances to meet the challenges produced by economic and social development.

The first toxicological laboratory was established in 1954 in Beijing in the Institute of Hygiene, Chinese Academy of Medical Sciences. During the early 1960s and late 1970s the number of such facilities increased to more than 50, and many courses are being offered in medical colleges and institutes.

Beginning with the traditional method of assessing LD_{50} in rodents, modern sophisticated cellular and molecular investigations on toxicokinetics and mechanisms, predictive prescreening tests done in a couple of days (e.g., Ames test, micronuclei and SCE scoring), life-span carcinogenicity bioassay in intact animals etc. are regularly applied. Meanwhile, the scope of study has been broadened from simple toxicity estimation to comprehensive safety evaluation, research into etiological factors in carcinogenesis (e.g., lung cancer in tin miners, malignancy in rubber workers), teratological surveys in vast populations (e.g., to evaluate the effect of thiodiazol), psycho-behavioral performance tests etc.

Besides drugs and the extracts of some herbal medicines, pesticides, food additives, atmospheric and aquatic metal pollutants, organic solvents and monomers of synthetic resins are the major objects of the toxicologists' attention. Great benefits have accrued in: 1) reducing the dangers from prophylaxes and curing poisonings (e.g. parathion); 2) turning hazards into profits (e.g. gossypol); 3) evaluating the safety of widely-used chemicals (pesticides and additives); 4) setting guidelines and standards for controlling exposure; 5) discovering the etiological factors of endemic diseases (fluorosis, highly concentrated "endemic" malignancies).

Research activities in China in the field of toxicology are characterized by: 1) much attention being focused, at the human level, on the practical problems related to promoting health, e.g., the endemic acute cardio-myopathy has been largely prevented by the supplementation of trace selenite; 2) intimate collaboration between laboratory work and field survey and between animal experiments and epidemiological investigations.

Thanks to the development and progress made in bio-medical sciences, tremendous advances have been made in the practice and theory of public health. I would like to mention some aspects of the field of public health in China.

To begin with, it would be proper to sketch the health status of the population as a whole, showing the developments of the last 30 years. Certain salient facts and representative figures can speak for themselves. The vital statistics of Shanghai County are an example (6) (Tables 1-4).

The characteristic features of the changes are the prolongation of life expectancy, the decline of the death rate, a shift in the major causes of death, control of the birth rate and the rate of population increase. All these transitions occurred in a relatively short period of time. The demographic indices approximate those of developed countries.

As the health status of the population and the public health service in China have been improved significantly in the past three decades, toxicology--the science of helping people to recognize risks and to-protect themselves against hazards--is gaining importance and making advances to meet the challenges produced by economic and social development.

Table 1. Birth and death rates of residents in Shanghai County, per 1,000.

Year	Birth Rate	Death Rate	Infantile Mortality	Neonatal Mortality
1953	40.57	10.3	–	–
1962	25.08	8.7	34.33	16.39
1970	18.53	4.6	11.2	4.7
1980	14.79	6.2	15.8	10.6

Table 2. Life expectancy at birth of the reisdents in Shanghai County.

Year	Male	Female
1951	42.7 yr.	46.7 yr.
1961	61.8	67.3
1970	70.0	75.2
1980	69.7	75.2

Table 3. Leading causes of death 1960–1962 and 1978–1980.

Rank	Cause of Death	Annual Rate 10^5
	1960–1962	
1	Infectious diseases	78.46
2	Accidents	66.45
3	Respiratory disease	48.84
4	Digestive disease	40.26
5	Neonatal deaths	28.78
6	Malignant tumors	23.46
7	Cerebrovascular disease	18.71
8	Heart disease	13.54
9	Endocrine disease	6.73
10	Urinary disease	6.31
	1978–1980	
1	Malignant tumors	142.30
2	Cerebrovascular disease	105.49
3	Heart disease	103.06
4	Respiratory disease	68.45
5	Accidents	49.38
6	Infectious diseases	34.68
7	Digestive disease	28.00
8	Mental/Neurologic dis.	13.00
9	Urinary disease	10.32
10	Neonatal deaths	7.96

Table 4. Comparison of selected demographic indices.

Demographic Index	Shanghai County	China	World[*]	Developing[*] Countries	Developed[*] Countries
Birth Rate 10^{-3}	15	18	28	32	16
Death Rate 10^{-3}	6	6	11	12	9
Increase Rate 10^{-3}	9	12	17	20	7
Infantile Mortality 10^{-3}	16	56	97	109	20
Life Expectancy at Birth (yrs)	75.2	68	62	58	72

1980 Data

[*] 1981 World Population Data Sheet

The first toxicological laboratory in China was established in 1954 in Beijing in the Institute of Hygiene, Chinese Academy of Medical Sciences. During the late 1960s and early 1970s the number of such facilities increased to more than fifty (4). Most of the activities in this field dealt with industrial chemicals and pesticides, owing to the support and encouragement by the government toward improving the working conditions in factories, especially during the beginning of Reconstruction. Starting with the conventional method of assessing LD_{50} in rodents, toxicities of many substances were tested, e.g., heavy metals such as lead and mercury, organic solvents such as benzene and carbon disulfide, aromatic amines, plastic monomers or intermediates, etc. (4). The results obtained were compared with each other and used as the bases of decision-making in the production of chemicals with less risk and of selective pesticides. Dipterex, dimethoate, Dichlorvos and later acephate were chosen as substitutes for highly toxic Parathion and Demeton. As a consequence, the occupational intoxication cases due to organic phosphate insecticides were quickly reduced (13).

At the end of the 1950s, in an attempt to eradicate Schistosomiasis, furanpromide, a derivative of nitrofuran, was synthesized by the Institute of Parasitology. After passing through toxicity tests, it was put into clinical use particularly for the treatment of acute infection of schistosoma. Hundreds of thousands of patients received furanpromide in the 1960s, and apparently no serious adverse effects were noticed. The results of toxicity tests seemed quite reliable at that time, and no one had looked more deeply into other aspects, such as the mutagenic potential of this compound. In the middle 1970s, it was revealed that most derivatives of nitrofurans, including furanpromide, are strong mutagens (14). This is just one of the many examples which demonstrate the importance of toxicology. Necessity has inspired scientists in many institutions in a wide variety of areas such as

biology, genetics, medicine and hygiene, in environmental
protection and birth control. Fortunately, praziquantal, a drug
without mutagenic and teratogenic potential, has been developed as
a substitute for many mutagenic anti-schistosomial drugs such as
niridazole, hycanthone and furanpromide (11) (Table 5).

In addition to mutagenesis, carcinogenesis became a matter of
much concern, as the cancer mortality rate rose quickly. The
problem of teratogenesis has been raised in the evaluation of a
pesticide, amino-thiadiazole. Research activities in the areas of
mutagenesis, carcinogenesis and teratogenesis highlighted the
contribution of toxicology, and led to the new branch—genetic
toxicology. Hundreds of chemicals were tested using such assays
as the Ames test, in vitro and in vivo sister chromatid exchanges,
micronucleus and chromosomal aberration analysis, unscheduled DNA
synthesis, dominant lethal test in mice, and sex-linked recessive
lethal test in Drosophila melanogaster. Modern sophisticated
cellular and molecular investigations on toxicokinetics and
mechanism explorations, such as rapid short-term prescreening
tests, life span carcinogenicity bioassay in intact animals (8,16)
(Tables 6 and 7), environmental monitoring on biological
materials, etc. are generally applied. From 1978 to 1981, a
nation-wide collaborative toxicity assessment program, organized
by the Institute of Hygiene and including about seventy
pesticides, was implemented. There are 10 papers dealing with
pesticide toxicity testing submitted to the present meeting.
Meanwhile, the scope of the toxicological studies has been
broadened from simple toxicity estimation to comprehensive safety
evaluation, research into etiological factors in carcinogenesis
(e.g., lung cancer in tin miners and farmers in Xian-Wei country,
Yuen-Nan Province; malignancies in workers of rubber factories,
etc.), and epidemiological surveys covering vast areas and
population sizes, e.g., investigation on the teratogenic effect of
amino-thiadiazole. The benefits derived from this toxicological
research may be summarized as follows:

Table 5. Mutagenicity and Teratogenicity Potentials
of some antischistosomal drugs.

	Mutagenicity	Teratogenicity
Niridazole	+	+
Hycanthone	+	+
Furapromide	+	nt
Praziquantol	−	−

nt: not tested

Table 6. Result of long-term feeding carcinogenicity
assay of CDM in mice.

Group and Treatment	No. of Animals necropsied	Neoplasm-bearing Animals no.	%	Malignancy-bearing Animals no.	%	Benign Tumor-bearing Animals no.	%
A.negative Control	50	0	–	0	–	0	–
B.CDM 20ppm	50	8	16	0	–	8	16
C.CDM 100ppm	50	22	44	5	10	17	34
D.CDM 300ppm	50	36	72	15	30	21	42
E.positive control p-^{CT*}300ppm	50	31	62	13	26	18	36

* p-^{CT}: para-Chloro-o Toluidine, also major metabolite of CDM in animal
CDM. Abbreviation of Chlordimeform

Table 7. Types and nature of neoplasma observed

Group	Hemangio-endothedioma Sarcoma	Angioma	Lung A-C	A	Liver C	Spleen Lympho-Sarcoma	Kidney A-C	A	Gastric A-C	Intestinal C	Others
B	0	6	0	0	0	0	0	0	1	1	0
C	0	15	3	3	1	0	1	1	0	0	1
D	2	24	6	4	2	1	1	0	0	4	2
E	10	19	4	0	1	0	4	1	0	0	0

Abbreviation: A-C: adenocarcoma C: carcinoma A: adenoma

1.   Reducing the dangers from prophylaxes and finding the
cure for poisonings. Occupational intoxication of organic
phosphate insecticides that prevailed in the early 1960s was
prevented by the knowledge of the route of entry, through
toxicological research, introduction of routine screening of the
blood cholinesterase activity, demonstration of the effectiveness
of measures in preventing percutaneous permeation, development of
low mammalian toxic selective pesticides, and a special antidote--
the oxime cholinesterase reactivators (13).

2.   Turning hazards into profits. Gossypol is another
interesting story. It was responsible for the endemic heat stroke
and hypokalemic paralysis in the early 1960s in some provinces
where cotton planting was being introduced (2). Now, having been
extracted from the cotton seed, gossypol has become a well known
and effective contraceptive. On the other hand, the removal of
gossypol also makes the detoxified cotton seed meal a nutritive

food resource with a high protein content. According to
nutritionists, it is better than proteins from peanuts.

3. Reducing exposure to mutagens, carcinogens and
teratogens. Efforts have been made to reduce the human exposures
to such substances as aflatoxins, home-made salted cabbages, and
carcinogenic industrial chemicals. We have not found
epidemiological evidence that correlates chlorinated hydrocarbon
insecticides to primary liver cancer (9), but the production
yield, the additive residue and concentration of DDT in human body
fat are gradually decreasing.

4. Evaluating the degree of safety of the widely adopted
chemicals, especially pesticides and additives. The story of
methylene-bis-(2 amino-1,3,4,-thradiazole) (MATDA) is an
extraordinary example both for a laboratory routine test performed
on rodents and for a field survey observing the direct effect on
human beings (5) (Tables 8 and 9). Although there appears to have
been no warning signals for pregnancy and no birth defect has been
observed at present among inhabitants living in areas where such a
pesticide was applied. MATDA must be kept under a close watch.
This may also be an appropriate case for Risk/Benefit Analysis, a
methodology which we are now learning.

5. Setting standards and guidelines for introducing
chemicals into the community. Research on the toxicity and
poisoning of pyrolysates derived from heated Teflon and its
analogs is a successful case in point (10). At present, a program
of consolidation and revision of Hygienic Standards for Toxicants
and Pollutants is progressing well. There are three special
subcommittees of occupational, environmental and food toxicants,
respectively, in the Ministry of Public Health.

6. Discovering etiological factors of endemic diseases. The
etiology of such diseases as the Keshan disease (3), the Kasching-
Beck disease (12), endemic fluorosis (1), and in special

Table 8. End of pregnancies and birth defects among
observed population exposed to MATDA and
control populations 1979-1981.

Population Group	No. of Pregnancy	No. of Spont. Abortion	Birth and Premature Birth				Birth Defect	
			Living Birth	Still Birth	Dead Fetus	Total	No.	‰
Observe	74,437	647(8.6%)	72,693	775	322	73,790	713	9,66
Control	36,452	354(9.71%)	35,612	330	156	36.098	520	14.41
Total	110,889	1001(9.01%)	108,305	1,105	478	019,888	1,233	11.22

Table 9.   Comparison of the site of birth defect recorded.

Site or System Defect Occur	Observe		Control	
	number	‰	number	‰
Nervous System	118	1.60	109	3.02
Face and Neck	33	0.45	38	1.05
Trunk	42	0.57	32	0.89
Digestive System	188	2.55	80	2.22
Heart	92	1.25	78	2.16
Uro-Genital System	20	0.27	17	0.47
Extremities	36	0.49	29	0.80
Finger and Toe	120	1.63	56	1.55
Others	64	0.87	81	2.24
Total	713	9.66	520	14.41

instances, some highly concentrated "endemic" malignancies (9) has been investigated extensively.   These toxicological researches have provided a solid basis for early detection, cure and prevention of such diseases.

In short, the toxicological research activities in China in recent years are characterized by:

(1) much attention being focused, at the human level, on the practical problems related to promoting health, e.g. successful prevention of the Keshan disease by supplementation of trace amounts of selenium in foods (7) (Table 10).

(2) a close collaboration between laboratory work and field survey, and between animal experiments and epidemiological investigations (5,7,9).

However, toxicological research in China is at the beginning stage.   More basic, theoretical and exploratory work and improvement in coordination among programs are needed.   As anticipated, the International Workshop on Environmental Mutagenesis, Carcinogenesis and Teratogenesis held in Shanghai will further promote such efforts.   It should be recorded as one of the major steps taken in this direction.

Table 10. Incidence rate and prognosis of Keshan
disease after Se supplementation

Group	Year	Subjects	Cases	Incidence	Death
Control	1974	3,985	54	13.55	27
	1975	5,445	52	9.55	26
	1976	212	1	4.72	0
	Total	9,642	107	11.10	53
Treated	1974	4,510	10	2.22	0
	1975	6,767	7	1.03	1
	1976	12,579	4	0.32	2
	1977	12,749	0	–	0
	Total	36,605	21	0.57	3

$X^2 = 63.7$ $P < 0.001$

Data from CMJ 92(7):472, 1979

Improvement of Clinical Feature of Endemic
Flurorosis after Change Drinking Water

Fluoride in	before 1975	1.5-7.5mg/l
Drinking water	after 1975	0.8-1.2mg/l

Estimated reduction   4.5-23.25mg/capita,day
of fluoride intake

Improvement of clinical   78.04%
    feature   From Chinese J. Endemiol.
                                      2(1):50-53, 1983

REFERENCES

1. Dai, G.J., J.Y. Chang, D.F. Liu, F.L. Chen, D.Z. Yue,
   H.X. Hong, J.Y. Wei, and G.M. Su (1883) Observations on the
   clinical and X-ray changes of the endemic fluorosis cases
   with low fluoride drinking water after 5 years. Chinese
   J. Endemiol. 2:50–53.

2. Dong, E.T. (1981) Investigation on the etiology and
   mechanism on the endemic hypokalemic paralysis. Chinese
   Med. J. 61:542.

3. Gu, B.Q. (1983) Pathology of Keshan Disease –– A
   comprehensive review. Chinese Med. J. 96:251-261.

4. Gu, X.Q. (Editor-in-Chief) Industrial Toxicology Vol I
   (1976). Ibid Vol II (1976). Shanghai: Shanghai Sci. and
   Tech. Publ.

5. Gu, X.Q., C. Qien, Q.M. Lu, and Z.Y. Hong (1983)
   Epidemiological survey on the teratogenic effect of
   Thiodiazol. Paper submitted to IWEMCT May 25 – June 1, 1983,
   Shanghai.

6.   Gu, X.Y. and M.L. Chen (1982) Vital statistics. Am. J.
     Public Health 72(Suppl.):19–23.
7.   Keshan Disease Research Group of the Chinese Academy of
     Medical Science. (1979) Observations on the effect of sod
     selenite in prevention against Keshan Disease. Chinese
     Med. J. 92:471–476.
8.   Li, F., S.Z. Xue, Y.L. Wang (1983) Study of the
     carcinogenicity of chlordimeform. III. Long-term feeding
     test in mice. Paper submitted to IWEMCT May 25 – June 1,
     1983, Shanghai.
9.   Li, W.G., Y.Y. Liu, Y.M. Hu, B.F. Ye, J.W. Chen, and C.B. Liu
     (1983) Analysis of the epidemiological factors of liver
     cancer in Jiangsu, Zhejiang and Shanghai. Chinese
     J. Oncology 5:48–51 (in Chinese).
10. Liang, Y.X., P.K. Lu, X.H. Jiang, Z.Q. Chen, J.B. Zhu, and
     L.D. Fang. (1980) A preliminary study for the recommendation
     of MAC of pyrolysis product of Teflon in the air of working
     premises. Acta Acad. Med. Primae Shanghai 7:276–281.
11. Ni, Y.C., B.R. Shao, C.Q. Zhan, Y.Q. Xu, S.H. Ha, and
     P.Y. Jiao (1982) Mutagenic and teratogenic effects of anti-
     schistosomal Praziquantol. Chinese Med. J. 95:494–498.
12. Sci. Res. Group of Kaschin-Beck Disease (1982) Effect and
     mechanism of Selenium in the prevention and cure of Kaschin-
     Beck patients. Chinese J. Endemiol. 1:145–149.
13. Shih, J.H., X.Q. Gu, Z.Q. Wu, Y.L. Wang, Y.H. Chang and
     S.Z. Xue (1983) Field investigation and prevention of
     poisoning of Parathion and Demeton in rural areas around
     Shanghai. Chinese J. Ind. Hyg. and Occup. Dis 1:17–23.
14. Ong, T.M., D. F. Callen, S.L. Huang, R.P. Batzinger, and E.
     Bueding (1977) Mutation Induction by the anti-schistosomal
     drug F30066 in various test systems. Mutation Res. 43:37–42.
15. Wong, H.W., X.J. You, Y.H. Qu, W.F. Wang, D. Wang, Y.M. Long,
     J.A. Ni, and R.W. Dzeng (1983) Search for carcinogenic
     agents in Shanghai rubber industry. Paper submitted to
     IWEMCT, May 25 – June 1, 1983, Shanghai.
16. Pan, Z.M. (1983) Induction of lung cancer by the dust of
     tin ore containing arsenic on Wistar rat. Paper submitted to
     IWEMCT, May 25 – June 1, 1983, Shanghai.

ENVIRONMENTAL MUTAGENESIS RESEARCH AT FUDAN UNIVERSITY

J. L. Hsueh and W. Xiang

Institute of Genetics
Fudan University
Shanghai, People's Republic of China

SUMMARY

In the late 1950s we began to work on the radiosensitivity
and radiocytogenetics of the germ cells of the male Rhesus monkey,
Macaca mulatta. Since monkeys are closely related to man, the
data obtained from the monkeys may be of value in predicting what
might occur in man. In the late 1970s, rapid strides in
industrialization occurred in China after the "cultural
revolution". Studies on environmental mutagens are very important
for devising ways to protect humans from hazardous effects of
these agents that may produce genetic diseases, birth defects,
cancer and premature aging. Today we have very good short term
test systems developed in our laboratory for the detection of
mutagens. These include the Ames test, cytogenetic assays of the
mammalian cells and human lymphocyte system, micronucleus test
using bone-marrow cells and human lymphocytes. Detection of SCE
has often been used in biological assaying of environmental
pollution factors. So far, in our laboratory such tests have been
done with the somatic cells in vitro (mammalian cells and human
lymphocytes) and in vivo (bone-marrow) and even with spermatogonia
in vivo. Since the genetic hazard to the filial generation is
directly related to germ cells, people are more concerned about
the effects of various harmful environmental chemicals on the germ
cells than on somatic cells. We adopted an improved simple method
for the detection of SCE of mouse bone-marrow cells and
spermatogonia in vivo. Recently, we have developed a technique of
SCE of peripheral lymphocytes of Bufo bufo gargarizans, an animal
widely distributed throughout China, for environmental monitoring.
We propose that this technique could be used as a sensitive and

reliable in vivo screening system in environmental monitoring studies.

In recent years people have become exposed to chemicals either environmentally, occupationally, therapeutically, or intentionally due to life style. Many of these chemicals are suspected of being mutagenic. It is very difficult to study chromosomal aberrations or point mutations in germ cells of human beings. In the late 1950s, our Institute began to work on the radiosensitivity and radiocytogenetics of the germ cells of the male Rhesus monkey, Macaca mulatta (14). Since monkeys are closely related to man, the data obtained from the monkeys may be of value in predicting what might occur in man. Using different doses of gamma-radiation, ranging from 5 to 400 roentgen units, the chromosome aberrations induced in the spermatogenesis of the monkey, exhibit an exponential dose-effect relationship (Table 1 and Fig. 1). We then showed (2,15) the quantitative and morphological changes of the various types of cells in the testes of the monkey after exposure to 200r of X-irradiation (Fig. 2). For the purpose of comparison, the testes were taken out separately at intervals: 5, 10, 20, 30, 40, 50 and 60 days after irradiation. The degree of radiosensitivity appears to be in inverse relation to the progress of cellular maturation, i.e., it is the highest in the spermatogonial stage and becomes lowered gradually in the order of primary spermatocytes, secondary spermatocytes and spermatids down to spermatozoa, the last of which proves to be the least sensitive. The third piece of work (4) showed the result of single and fractionated dosage of exposures to X-irradiation. The materials consisted of 12 testes, which were obtained from 12 different monkeys and were set into three different groups: (a) the control, (b) a single exposure to 200r X-irradiation, and (c) fractionated dosage of exposures with 20r each to a total of 200r for 10 consecutive days. The results showed clearly that there exists a marked difference in the frequencies of chromosome aberrations between the control and the irradiated animals, and the frequency is significantly higher in the single exposure group than in the fractionated dosage group.

In the late 1970s, rapid strides in industrialization occurred in China after the "cultural revolution". With increasing use of nuclear power, as well as the increase in environmental chemical pollutants, the question of this influence on the mutation rate in the human population is of growing concern. There is a need to evaluate their potential toxicity for individuals residing within polluted areas. Studies on environmental mutagens are very important for devising ways to protect human beings from hazardous effects that may produce genetic diseases, birth defects, cancer and premature aging. Without data carefully collected from a variety of test systems, regulatory actions may be either inadequate or excessive. Today

Table 1.   A comparison of the chromosome aberrations obtained after exposure of different dosages of γ-radiation and theor controls.

Monkey No.		8	7	5	2	3	9	10	11	12
Radiation dosage (r)		5	10	25	50	100	150	200	300	400
Treated series										
Total number of figures observed		8 704	9 100	8 570	24 983	4 798	8 672	8 480	4 260	4 243
Number of aberrant cells		8	16	6	16	11	24	35	48	59
Number of bridges	1	6	11	6	9	10	17	29	40	41
	2	2	4	—	4	—	7	4	5	12
	3	—	—	—	—	1	—	1	—	—
	4	—	1	—	—	—	—	—	—	1
	Total	10	23	6	17	13	31	40	50	69
Number of fragments		—	—	—	3	—	—	1	3	5
% of aberrant cells		0.091912	0.175824	0.070012	0.064043	0.229262	0.276753	0.412736	1.126761	1.390526
% of bridges		0.114890	0.252747	0.070012	0.068046	0.270946	0.357472	0.471698	1.173710	1.626208
Control series										
Total number of figures observed		8 446	8 966	9 873	10 430	8 475	8 554	9 129	3 634	4 181
Number of aberrant cells		9	4	3	7	6	8	6	1	—
Number of bridges	1	8	4	3	7	5	4	4	—	—
	2	1	—	—	—	1	4	2	1	—
	Total	10	4	3	7	7	12	8	2	—
Number of fragments		—	—	—	—	—	—	—	—	—
% of aberrant cells		0.106559	0.044613	0.030386	0.067114	0.070796	0.093523	0.065725	0.027518	0.000000
% of bridges		0.118399	0.044613	0.030386	0.067114	0.082596	0.140285	0.087633	0.055036	0.000000
% differences of aberrant cells		−0.014647	+0.131211	+0.039626	−0.003071	+0.158466	+0.183230	+0.347011	+1.099243	+1.390526
% differences of bridges		−0.003509	+0.208134	+0.039626	+0.000932	+0.188350	+0.217187	+0.348065	+1.118674	+1.626208

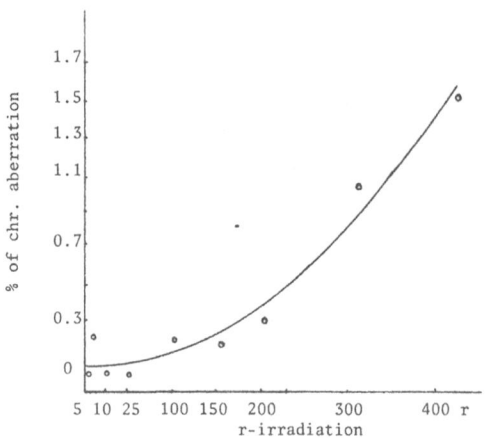

Figure 1.   Increase of chromosome aberrations by γ-irradiation.

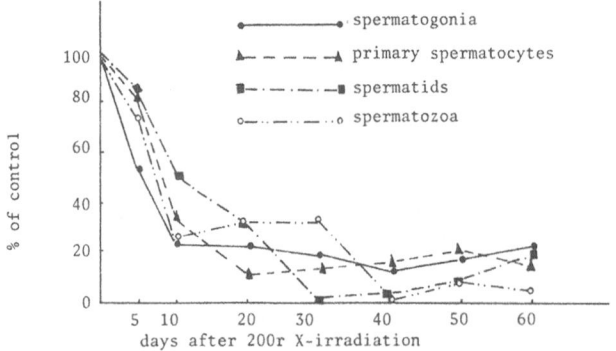

Figure 2.   The effect of 200r X-irradiation on
spermatogenesis in the Rhesus monkey (Macaca niulator)

we have very good short term test systems developed in our
laboratory for the detection of mutagens.

In the Ames test seventy different chemicals including food
additives, oral contraceptives, anti-cancer drugs, pesticides and
other chemical reagents have been tested for their mutagenicity in
the Salmonella–microsome mutagenicity test.   The percentage of
mutagens identified in the tested chemicals is 39% (Table 2) (17).

In cytogenetic assays the cytological and histological
principles involved in detecting chromosome damage microscopically
have developed from our earlier studies on the effects of ionizing

Table 2. The use of Ames test in screening
mutagenicity for 70 chemicals

No. of Samples		Ames test	
		Positive	Negative
Food additives	11	3	8
Oral contraceptives	13	1	9
Anti-cancer drugs	8	6	2
Pesticides	7	2	5
Flame-retardant additives	3	1	2
Chemical Reagents	20	10	10
Others	8	3	5
Total	70	26	44

radiations on chromosomes. Cytogeneticists have employed a
variety of test materials and test methods to measure the effects
of mutagens. In addition to the classic chromosome aberrations,
other endpoints (such as sister chromatid exchanges (SCE) and
micronucleus test) have been developed for assays. A cytogenetic
technique currently in wide use is the analysis of SCEs. SCE
analysis appears to be a very good screening tool (5).
Cytogenetic evaluations cannot be used alone to define mutagens,
since not all mutagens produce DNA lesions which lead to the
formation of chromosome aberrations or SCEs. A comparison of SCE
assay and the Ames test, with or without S9 activation, is
presented for 11 agents including the male oral contraceptive,
Gossypol (9.18). The result would have been substantially the
same had the SCE analysis and <u>Salmonella</u> assays been used. Both
assays gave negative results to Gossypol (Tables 3 and 4). For

Table 3. A comparison of sister chromatid exchange
(SCE) and the Ames test:Gossypol acetic acid.

Treatment	Ames test				SCE	
	TA 98$_{S9}$	TA 100$_{S9}$	TA 98	TA100	S9	-S9
0			25.7 ± 2.0	167.3 ± 8.6		6.97 ± 0.65
S9	37.3 ± 2.9	153.3 ± 5.3			6.86 ± 0.25	
0.1	29.7 ± 2.8	108.7 ± 6.9	17.7 ± 1.2	149.0 ±15.2	6.60 ± 0.43	7.00 ± 0.40
1.0	30.5 ± 2.5	141.3 ± 6.9	17.7 ± 1.2	141.7 ±14.6	6.43 ± 0.40	7.53 ± 0.51
10.0	27.7 ± 2.4	164.0 ± 5.9	14.3 ± 3.2	159.7 ±28.7	7.33 ± 0.50	7.41 ± 0.58
100.0	–	–	–	–		

The nagative results in TA 1535, TA 1537, TA 1538 were also studied.

Table 4. Sister chromatid exchange (SCE) and micronucleus (MU) test in gossypol formicum users

Group	No. of subjects	No. of cells observed	SCE/cell ± S·E·M·	No. of subjects	No. of cells observed	No. of cells with micronuclei	% of MU
Control	12	316	4.26 ± 0.13	12	12480	3	0.24
Subject	11	275	3.56 ± 0.14	13	13722	5	0.36

Gossypol formicum oral-taken subjects were 20mg/day for 3 months and then
200-300mg/month maintence for another 23 months.

assays of chromosome aberrations and SCE established mammalian
cell lines and human lymphocytes in culture were generally used.
Cells from the Chinese hamster cell line V79 were pulse-treated
with Alfatoxin (AF) B_1, B_2, G_1 and G_2 at various doses, with or
without the metabolic activation system S9-mix. Only with
metabolic activation, all these toxins caused significant
increases in frequency of SCEs in a dose-dependent manner. Based
on the rate of SCE induction, the relative order of potency was
established as: $AFB_1 > G_1 > G_2 > B_2$. Induction of chromosome
aberrations in V79 cells after pulse treatment with AFB_1 and G_1
plus S9 mix was also studied. A dose-and time-dependent increase
of aberrations was observed for both compounds. AFB_1 again is a
more potent aberration inducer than AFG_1 (1). Induction of SCE
and chromosome aberrations in V79 cells after treatment with a
radiation frequency field (RF field) (13) and organophosphorus
pesticides were also studied (3). Using SCE as an index, the
cytogenetic effect of ozone of low concentration on human
lymphocytes was examined (20). Ozone (over 0.15 ppm)
significantly enhanced the incidence of SCE. The effect increased
with the concentration of ozone and the duration of the treatment.
Beyond a definite range, the effect could decrease, while the
inhibition of cell growth appeared. According to the result of
the SCE detection, we reevaluated the critical concentration of
ozone toxicity; 0.1 ppm was suggested (Fig. 3).

Retinoids are known to have the ability to prevent or delay
the induction of tumors in animals and in man and have been
referred to as chemopreventative agents. Treatment of V79 cells
with the 2 retinoids, Retinol (Rol) or Retinyl acetate (Race), at
various doses for 1 hr or 26 hrs caused no obvious increase of SCE
frequency (8,67). There was no difference whether the metabolic
activation system of S9 mix was added or not. Cells treated with
various doses of cyclophosphamide (CPP, 0.75 – 6 μg/ml) plus S9
for 1 hr, were all significantly inhibited by concurrent addition

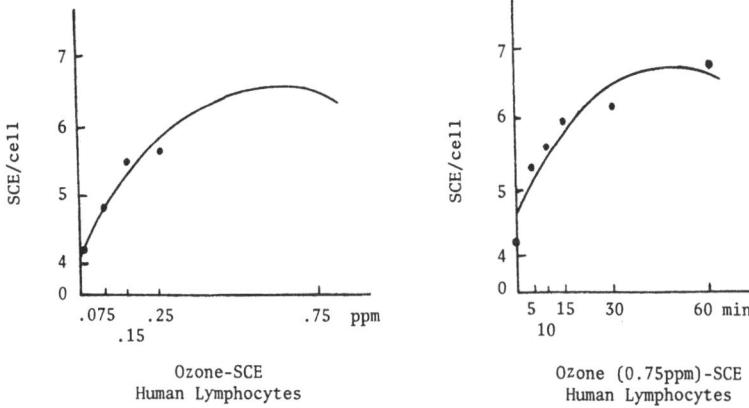

Figure 3. Relationship between SCE frequencies and
different treatments of Ozone

of a fixed dose of Rol (16 μg/ml) (Table 5). There was no
difference in SCE frequency when harvested at three different time
intervals (Table 6). In addition, a dose-dependent inhibition was
evident when cells were treated with a fixed dose of CPP (3 μg/ml)
plus S9 and various doses of Rol (1-32 μg/ml) for 1 hr (Table 7).
For mutations to diphtheria toxin resistance (DT^r), cultures of
V79 cells were treated with diffusion chamber containing CPP
(10-80 μg/ml) plus S9 or CPP plus S9 and a fixed dose of Rol or

Table 5. The effect of Retinol on SCE frequencies
in Chinese hamsters V79 cells treated with
cyclophosphamide (CPP)

Treatment	Mean No. SCE/cell (± SEM)	
	Exptl. I	Exptl. II
S9 + CPP_6	71.5(3.2)	60.4(1.9)
S9 + CPP_6+ R_{16}	44.9(2.5)	37.2(1.4)
S9 + CPP_3	53.0(2.6)	44.5(1.3)
S9 + CPP_3+ R_{16}	37.9(1.6)	23.6(0.8)
S9 + $CPP_{1.5}$	28.1(1.3)	25.7(0.7)
S9 + $CPP_{1.5}$+ R_{16}	22.1(0.8)	15.7(0.6)
S9 + $CPP_{0.75}$	20.5(0.8)	17.1(0.7)
S9 + $CPP_{0.75}$+ R_{16}	12.8(0.6)	11.0(0.5)

Table 6. Inhibition of SCE by R(μg/ml) in V79 cells treated with cyclophosphamide (CPP, 3βg/ml) or Aflatoxin B$_1$ (AFB, 0.1μg/ml) and harvested at three different time invervals

Treatment	SCE/cell ± SEM harvested after indicated hours of treatment		
	26	28	30
0	7.4 ± 0.4	8.1 ± 0.3	7.3 ± 0.4
S-9	7.5 ± 0.4	8.0 ± 0.4	7.2 ± 0.3
S-9 + R$_{16}$	7.7 ± 0.4	8.4 ± 0.4	7.7 ± 0.4
S-9 + CPP	52.8 ± 1.6	NA	46.9 ± 1.6
S-9 + CPP + R$_{16}$	27.2 ± 1.1	29.8 ± 1.0	23.9 ± 0.8
0	7.6 ± 0.3	7.8 ± 0.3	7.6 ± 0.3
S-9	7.7 ± 0.3	8.4 ± 0.3	7.7 ± 0.3
S-9 + R$_{32}$	8.0 ± 0.3	8.5 ± 0.4	8.4 ± 0.3
S-9 + AFB	29.2 ± 1.1	28.9 ± 1.0	29.8 ± 1.6
S-9 + AFB + R$_{32}$	20.8 ± 1.0	19.5 ± 0.7	21.2 ± 0.9

160.7-Race (8 μg/ml) for 24 hrs. The results showed that treatment of V79 cells with Rol or Race alone, CPP at 10 or 20 μg/ml, with or without Rol or Race, all had higher frequencies of DT^r mutation as compared to the controls. There was no significant or consistent difference between cultures treated with CPP alone and Rol or Race

Table 7. Inhibition of sister chromatid exchanges by different dosage of Retinol in V79 cells treated with cyclophosphamide (CPP, 3μg/ml)

Treatment	Mean No. SCE/cell (± SEM)	
	Experiment I	Experiment II
0	7.1 (0.3)	6.5 (0.4)
S-9	-	5.7 (0.4)
R$_{16}$	6.3 (0.3)	-
S-9 + R$_{16}$	7.0 (0.1)	-
S-9 + CPP$_3$	25.3 (1.1)	31.1 (1.0)
S-9 + CPP$_3$+ R$_1$	-	30.3 (1.1)
S-9 + CPP$_3$+ R$_2$	26.5 (1.2)	29.7 (0.9)
S-9 + CPP$_3$+ R$_4$	24.7 (1.1)	27.5 (0.9)
S-9 + CPP$_3$+ R$_8$	20.7 (1.0)	26.8 (0.9)
S-9 + CPP$_3$+ R$_{16}$	19.6 (1.0)	22.3 (0.9)
S-9 + CPP$_3$+ R$_{32}$	-	22.4 (0.9)

plus CPP. However, a comparison between cultures treated with Rol
or Race plus CPP at higher doses, i.e. 40 and 80 μg/ml, showed
that the DT^r mutation frequencies were much lower than those
treated with CPP alone. These results indicated that at certain
conditions, Rol and/or Race have the ability to suppress SCE or
DT^r mutation frequencies induced by CPP. This suggests that the
action of retinoids is not limited to the widely accepted role of
preventing or reversing promotion step but also provides the
initiation step of carcinogenesis.

The most important technique is the lymphocyte culture
system, which involves a short term in vitro culture of in vivo
exposed cells; one advantage of in vitro studies is the
possibility of using higher concentrations of agents than may be
tolerated in vivo. The major disadvantage, of course, is that
many of the cell types used in vitro may be unable to metabolize a
potential carcinogen, or a chromosome damaging agent, to its more
active metabolite or ultimate carcinogen (Table 8). In our
laboratory we use SCE as an index to detect the mutagenicity of
anti-tumor drugs and it may be possible to use it as a method for
screening new drugs (12). The effects of Camptothecin,
Harringtonine, Fluorouracil and Mytomycin C on SCEs in human
gastric carcinoma cells and normal human lymphocytes were studied.
It was observed that all of them could increase SCE frequency in
both kinds of cells. But they induced SCEs in the gastric
carcinoma cells more than in the normal lymphocytes and the
differences were found to be statistically significant. At a
lower concentration (10^{-7} μg/ml), Camptothecin and Fluorouracil

Table 8. The frequencies of sister chromatid exchanges (SCE)
induced by varies chemicals in
human lymphocytes

	SCE frequency	
	$-$ S9	$+$ S9
Methotrexatum Natricum	$-$	
Daunomycin	$+$ (1 x 10^{-5} μg/ml)	
Furapronidum	$-$	$+$ (1 x 10^{-5} μg/ml)
M-170	$+$ (1 x 10^{-3} μg/ml)	$+$ (1 x 10^{-4} μg/ml)
Pyronaridinum	$-$	$-$
Cyclophophamide	$-$	$+$ (1 x 10^{-5} μg/ml)
Mitomycin C	$+$ (2 x 10^{-2} μg/ml)	
Harringtonine	$-$	
Camptothecin	$-$	
Flurouracil	$-$	
Gossypol	$-$	

could remarkably increase SCEs in the tumor cells but not in the normal human lymphocytes. This result is in good agreement with the fact that they are specifically used for treating stomach cancer. Therefore, we consider that the method we described may be useful for screening new anti-stomach cancer drugs (Fig. 4).

In individuals exposed to ionizing radiations and chemical mutagens, blood samples taken after exposure provide populations of cells that are normally in a resting phase, but which can be made to divide in short-term culture and reveal any chromosome damage sustained as a consequence of exposure. In the same way, blood samples taken from the body have been exposed to radiation or treated with chemical mutagens in vitro and the responses obtained from whole body exposed in vitro and in vivo are shown to be qualitatively and quantitatively identical. In China, the analysis of somatic chromosome damage in lymphocytes has been used routinely to estimate the possibly hazardous chemical exposures occurring in various operations and job categories. This technique provides a reliable monitoring system for workers.

Another screening technique developed to assess the induction of chromosome damage is the test for production of micronuclei. The theoretical basis for the micronucleus test is the hypothesis that broken chromosomes or chromatid fragments may lag behind intact chromosomes during the anaphase of mitosis. At telophase, daughter nuclei are formed. If the broken and lagging chromatin is not included in the main nucleus during telophase, micronuclei are formed in the cytoplasm. Thus, it is believed that the

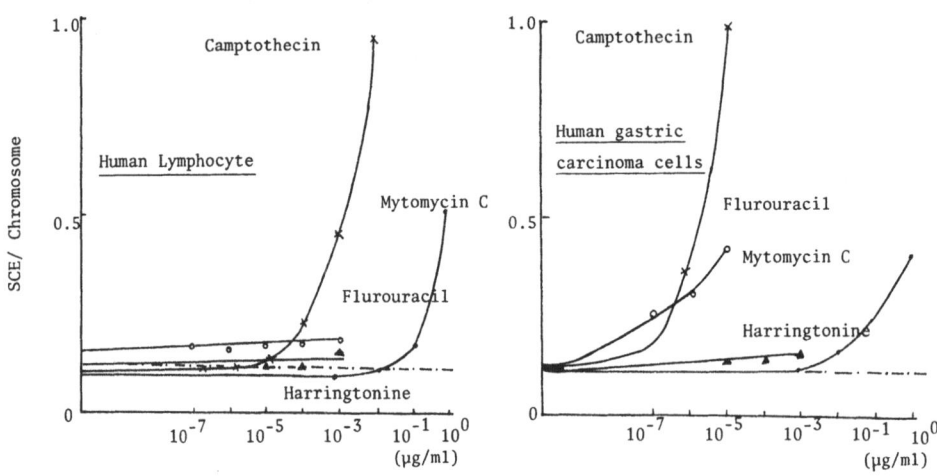

Figure 4.   The effect of 4 drugs on SCE frequency in
human gastric carcinoma cells and
normal human lymphocytes

frequency of cells containing micronuclei following chemical or
radiation treatment is an indicator of clastogenic activity
(Fig. 5).

Detection of SCE has often been used in the biological
assaying of environmental pollution factors. So far, most of such
tests have been done with the somatic cells in vitro, little has
been done with bone-marrow cells in vivo and even less with germ
cells in vivo. Since the genetic hazard to the filial generation
is directly related to germ cells, people are more concerned about
the effects of various harmful environmental chemicals on the germ
cells than on somatic cells. We adopted an improved simplified
method for the detection of SCE of mouse bone-marrow and
spermatogonia in vivo (Table 9) (10). It consists of: (a) before
hypotonic treatment, preparing cell suspensions pretreated with
collagenase solution in order to make germ cells more readily
dissociate from testis tissue; (b) using the method of a single
subcutaneous implantation of the tablet in place of the continuous
intraperitoneal injection for 14 times at 1 hr intervals; (c)
using 5-iodo-2'-deoxyuridine (IudR) instead of Brdu. Two tissues
may be subject to direct examination, the bone-marrow and the
testicular tissue with the spermatogonial mitoses. One advantage
of using bone-marrow is that we are able to study mitoses that
have developed in vivo, where the tissue is flushed with blood so
that substances reaching the blood will expose the bone marrow
cells with no further barrier. The disadvantages are that
sampling is more complicated than drawing blood, that the number

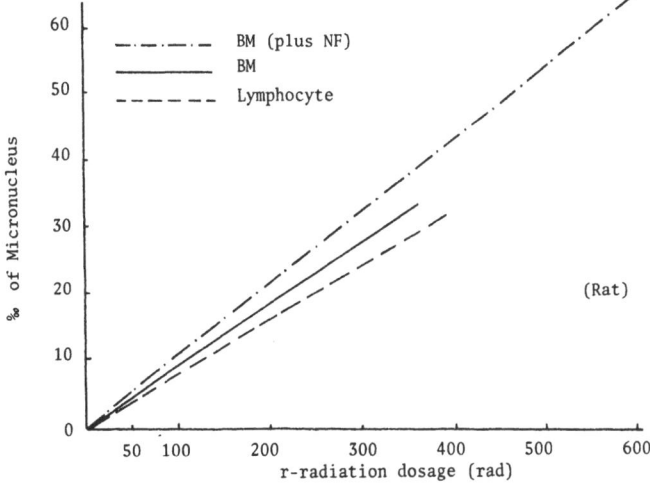

Figure 5. The frequency of micronuclei induced
by γ-irradiation (NF = Nuclear fragment).

Table 9.   The frequencies of sister chromatid exchanges
(SCE) induced by methyl methanesulfonate (MMS)
in mouse spermatogonia

MMS dose (mg/kg.b.w.)	No. of Spermatogonia observed	SCE/cell±SEM
0	60	2.83±0.18
5	60	3.88±0.22
10	60	4.30±0.23

of suitable metaphases is often low, and that the chromosomes are
short and not easily scored.

Levels of SCE induced by Gossypol were investigated in
spermatogonia, spermatocyte and bone-marrow cells in a short term
in vivo test in mice (19). Adult male ICR mice were given
Gossypol acetic acid (GAA) at the respective doses of 0, 1, 2.5,
5, 7.5, 10, 20, 30, 40 and 50 mg/ml per kg body weight per day for
19 consecutive days. With the lower dosages of GAA (1–10 mg), no
significant differences were found between the treated mice and
the controls (P > 0.05), both in the frequencies of SCE in
spermatogonia and in the percentage of chromosomal aberrations
either in the spermatogonia or in the spermatocyte. But with
larger doses of GAA, namely 20–50 mg per kg body weight per day
(about 50–125 times the clinical dose), the frequency of SCE in
the spermatogonia was found significantly higher than that in the
controls, with P values less than 0.005. GAA at the dosage of 4
mg/kg b.w./day (10 times higher than the clinical dose) for 4
consecutive days was treated for its ability to induce SCE in
bone-marrow cells of ICR mice. Though the mean yield of SCE in
the treated mice (3.09 ± 0.27/cell) is a little higher than in the
control (2.96 ± 0.29/cell), the statistical analysis is not
significant (P > 0.05). The results suggest that a small dose of
GAA does not produce any observable effect in germ cells and bone–

Table 10. The frequencies of sister chromatid exchanges
(SCE) in lymphocytes of <u>Bufo</u> <u>bufo</u> gargarizans
from different areas

Samples	Total cells	SCE/cell± SEM
Lushan	100	7.70 ± 0.23
Huongshan	30	7.83 ± 0.48
Fudan campus	30	8.30 ± 0.55 ($P > 0.05$)
Certain factory: Residential area	100	7.34 ± 0.24 ($P > 0.05$)
Oxidation pond	100	13.91 ± 0.33 ($P < 0.001$)
Waste water treatment plant	100	11.78 ± 0.27 ($P < 0.001$)

marrow cells, and it may be inferred that after the withdrawal of
the drug, there should be no irreversible cytogenetic effect on
the second generation or on the patient himself. However, it
should not be overlooked that treatment with large doses of GAA
could increase the frequency of SCE in the spermatogonia.

Recently, we developed the technique of SCE of peripheral
lymphocytes of the toad, <u>Bufo</u> <u>bufo</u> gargarizans, for environmental
monitoring (11). Two known mutagens, Cyclophosphamide and methyl
methane sulphonate, were administered intraperitoneally as
positive controls. Several animal samples taken from different
localities were analyzed (Table 10). The results showed that the
SCE frequencies of the oxidation pond of a certain factory were
markedly higher than in other areas. Generally, the females were
more sensitive than males. We propose that the SCE of lymphocytes
of <u>B</u>. <u>b</u>. gargarizans, an animal widely distributed throughout
China, could be used as a sensitive and reliable in vivo screening
system in environmental monitoring studies.

REFERENCES

1. Batt, T.R., J.L. Hsueh, H.H. Chen, and C.C. Huang (1980)
   Sister chromatid exchanges and chromosome aberrations in V79
   cells induced by Aflatoxin B_1, B_2, G_1, G_2 with or without
   metabolic activation. Carcinogenesis 1:759-763.

2. Chang, C.S., J.J. Hsueh, M.C. Liu, and L.H. Liu (1978) The
   effect of single and fractionated dosage of X-irradiation in
   the spermatogenesis of the Rhesus monkey, Macaca mulatta.
   Fudan J. (Natural Science) 1:20-24.

3. Chen, H.H., J.L. Hsueh, S.R. Sirianni, and C.C. Huang (1981)
   Induction of sister chromatid exchange and cell cycle delay
   in cultured mammalian cells treated with eight
   organophosphorus pesticides. Mutation Res. 88:307-316.

4. Hsueh, J.L., C.S. Chang, L.H. Lue, and M.C. Liu (1978)
   Observations on the histological changes in the testes of the
   monkey, Macaca mulatta after single and fractionated dosage
   of exposure to X-irradiation. Acta Biol. Exper. Sinica
   11:81-86.

5. Hsueh, J.L., H.H. Chen, and C.C. Huang (1981) Effect of tumor
   promoter 12-o-tetradecanoyl-phorbol 13-acetate on induction
   of sister chromatid exchange in Chinese hamster V79 cells
   treated with mutagens. Mutation Res. 81:387-394.

6. Hsueh, J.L., and C.C. Huang (1981) Inhibitory effect of
   Retinol and Retinol acetate on cyclophosphamide induced
   sister chromatid exchange and diphtheria toxin resistant
   mutants (DT^r) in V79 cells. J. Cell Biol. 91(2),pt. 2:380.

7. Hsueh, J.L., and C.C. Huang (1981) Induction of diphtheria
   toxin resistant mutants (DT^r) in CHO cells cultured in
   diffusion chambers (DC) in mice injected with
   cyclophosphamide (CPP) and in CHO cells treated with CPP plus
   S9 mix in DC. Annual Report of Institute of Genetics, Fudan
   University, Shanghai, China.

8. Huang, C.C., J.L. Hsueh, H.H. Chen, and T.R. Batt (1982)
   Retinol (vitamin A) inhibits sister chromatid exchange and
   cell cycle delay induced by cyclophosphamide and Aflatoxin B_1
   in Chinese hamster V79 cells. Carcinogenesis 3:1-5.

9. Li, C.B., F. Ding, Z.R. Ma, and S.Y. Zhao (1981) The
   determination of the male contraceptive Gossypol as a non-
   mutagenic agent by the Ames test and SCE method. Fudan
   J. (Natural Science) 20:361-365.

10. Lu, Q., S.H. Jiang, X.R. Jiang, W. Xiang, and Z.D. Liu (1982)
    An improved method of sister chromatid differentiation
    staining in vivo. Kexue Tongbao 27:81-82.

11. Ma, Z.R., Y.L. Liu, C.B. Li, and S.Y. Zhao. A comparative
    study on induction of sister chromatid exchanges in human
    gastric carcinoma cells and lymphocytes exposed to anti-tumor
    drugs. (Submitted for publication).

12. Qin, S.Z., X.F. Qiu, J.M. Zhou, R.H. Jian, X.J. Hsueh, and
    X.M. Fu (1983) Increased sensitivity of cell strains from

Chinese hamster cells to sister chromatid exchange induced by RF fields. Rad. Protect. (in press).

13. Tan, C.C., T.T. Liu, C.S. Chang, X.F. Qiu, S.T. Lai, and J.L. Hsueh (1964) Radiation genetics of the Rhesus monkey, Macaca mulatta. I. The effect of different dosages of r-radiation on the chromosome aberrations in the spermatogenesis of Macaca mulatta. Scientia Sinica XIII, 4:611-619.

14. Tan, C.C., S.Y. Chao, C.S. Chang, and T.T. Liu (1964) Radiation genetics of the Rhesus monkey, Macaca mulatta II. The effects of X-irradiation on the spermatogenesis of the monkey, Macaca mulatta. Scientia Sinica XIII, 8:1253-1264.

15. Wen, S.E., Q. Lu, S.E., and W. Xiang (1983) The application of the sister chromatid exchange of lymphocyte of Bufo bufo gargarizans for environmental mutagenicity assay. Acta Genet. Sinica 10:291-294.

16. Zhang, Z.S., Q. Lu, Y.B. Tan, X.R. Jiang, H.H. Zheng, R.L. Wang, M.M. Wang, and Y.L. Yiao (1983) Sister chromatid exchanges and micronucleus test in Gossypol Formicum users. Acta Genet. Sinica 10(2):157-160.

17. Zhao, S.Y., C.B. Li, and C.C. Tan (1981) The use of Salmonella microsome and sister chromatid exchanges in screening mutagenicity and carcinogenicity for seventy different chemicals. Third International Conference on Environmental Mutagens, Sugimura, T., Kondo, S., and Takebe, H. (eds.), Tokyo, University of Tokyo Press and New York, Alan R. Liss, Inc.

18. Zhao, S.Y., X.F. Qiu, C.B. Li, S.Z. Qin, and X.M. Fu (1981) Detection of mutagenicity of Daunomycin and Methotrexatum matricum by Ames test and sister chromatid exchanges. Hereditas (Beijing) 3(1):12-13.

19. Zhou, J.M., Q. Lu, and S.H. Jiang (1982) Effects of Gossypol on germ cell chromosomes in mice. Reproduct. Contracept. 2:49-52.

20. Zhou, J.M., X.F. Qiu, Z.J. Wang, and R.H. Jian (1981) The effect of ozone on sister chromatid exchange of human lymphocytes. Acta Biol. Exper. Sinica 14:313-316.

THE ROLE OF RISK ASSESSMENT IN REGULATORY DECISIONS

IN THE UNITED STATES

Gordon W. Newell

Electric Power Research Institute
Palo Alto, California 94303

SUMMARY

Risk assessment is an art that attempts to produce estimates
of the probability of adverse effects for a particular exposure.
Thus, a risk assessment may develop a series of probabilities for
illness or death which could be associated with varying degrees of
exposure to a substance such as an air pollutant, coffee, or the
use of an automobile. The regulation of environmental toxicants
is a difficult and complex procedure, burdened by scientific
uncertainties that permit wide differences in the interpretation
of scientific evidence. Two major components of a federal
regulatory strategy for carcinogens involve the assessment of
health risks from suspected carcinogens, and the regulatory
processes that convert such estimates of risk into appropriate
regulatory action. There are at least ten Public Laws in the
United States that provide for the use of risk assessment. In
some cases, risk assessment may be used to set Federal Standards;
in others, state- and situation-specific regulations prevail. The
probability of adverse health effects in relation to the extent of
exposure is considered here, involving the analysis and ,
combination of epidemiological and other human data with animal
effects data. Specific problems considered include the choice of
data, combination of data, translation of data from experimental
to actual settings, conversion of the results from animal
experiments to human response, and the extrapolation of
experimental results from high doses to estimated responses at
low-ambient concentrations. The various methods used to consider
such data in developing a risk assessment are considered.
Frequently, there is disagreement about the best method to use for
particular circumstances. Thus, there exists a great deal of

uncertainty about risk assessments. Some knowledge of the extent
of the uncertainty is important in interpreting risk assessment
results.

INTRODUCTION

When making regulatory decisions about materials that may
impose biological burdens on human beings, the process of
developing an acceptable risk assessment may be fraught with
controversy and emotional challenge. Nevertheless, in the United
States there is an increasing use of this procedure. It serves as
an important tool in bridging the gap between information
developed with subhuman species, and extrapolation of such
information to human beings. For the most part, risk assessment
practices use cancer as the end-point (18). Although other
physiological or biochemical measurements can be used, cancer is
the process through which biological risk assessments are
developed today. The following discussion is based on this
premise.

The regulation of environmental carcinogens is a difficult
and complicated procedure. It deals with agents that, in many
cases, are of major economic and social importance. Hence, there
may be vested interests that are strongly opposed to regulatory
action. This in turn breeds a multi-sided adversarial situation
between industry and environmentalist groups, with government
regulatory agencies the principal actors; Congress, the Executive
branch of the Federal Government, and the judiciary also have
major roles. The climate of public opinion in these matters plays
a critical role, particularly with regulatory agencies and
Congress. Extreme positions tend to be taken by industry and the
environmentalists; their positions, particularly on scientific
policy issues, seem to be adopted as much from strategy
considerations as from scientific convictions (1).

Government agencies naturally tend to push for regulatory
action unless there are substantial forces in opposition.
Environmentalist groups serve the important function of prodding
the regulatory agencies to take regulatory action by threats of
legal action. Sometimes, industry will fight a regulation with
every scientific, legal and political weapon it can muster (4).
The scientific community is far from having a unified voice, and
it has its vociferous and polarized splinter groups. It is not
wonder that it is difficult to develop a moderate and sensible
approach to regulating carcinogens.

The two major components of federal regulatory strategy for
carcinogens involve the assessment of health risks from suspected

carcinogens and the regulatory processes which translate these
estimates of risk into appropriate regulatory action (17).

THE EVOLUTION OF CARCINOGEN REGULATION IN THE UNITED STATES

In the 1970s there was a major thrust towards the development
of a regulatory program for chemical carcinogens. Although there
are more than ten public laws that are concerned with risk
assessment, only four regulatory agencies are principally involved
on a regular basis with chemical carcinogens (32); three of these
began functioning only within the past 7 years; they are (i) The
Environmental Protection Agency (EPA); (ii) Occupational Safety
and Safety Health Administration (OSHA); and (iii) The Consumer
Protection and Safety Commission (CPSC). The fourth regulatory
agency has been in existence since the early 1900s. This is the
Food and Drug Administration (FDA).

Three factors have stimulated the regulation of carcinogens.
The first is the failure to develop an effective cure for cancer
in spite of major efforts by the National Program for the Conquest
of Cancer. Since there seems to be little possibility in the
immediate future of finding a cure for cancer, the focus has
shifted to taking effective preventive measures against agents
known to produce cancer (14,15).

The second factor is the now generally accepted evidence that
cancer for the most part is caused by environmental factors.
Indeed, epidemiologists estimate that 70 to 90% of all human
cancers may derive from exposure to chemical carcinogens present
in the environment (10). The air we breathe, the food we eat, the
water we drink, the work we do and whether or not we smoke
cigarettes, all have a bearing on our potential risk of developing
cancer (32). Epidemiology studies show that cancer patterns
differ markedly in various parts of the world, where such
differences cannot be explained by genetic factors. For example,
the incidence of liver cancer is very high in parts of Africa, but
is relatively low amongst blacks in the U.S. (19). Furthermore,
the cancer patterns in migrant populations show marked changes
toward imitating those of the new homeland. Such examples suggest
that a large percentage of cancers are caused by environmental
agents, and it provides the basis for hope that these
environmental agents can be identified and eventually controlled
(40).

A third factor which provides impetus for the regulation of
environmental chemicals is a demonstration that large numbers of
carcinogenic chemicals are present in the environment. Much of
this evidence was developed through the bioassay program initially
undertaken by the National Cancer Institute (36). Several hundred

chemicals have been tested by this animal procedure and about half
of them have been shown to be carcinogenic. The increasing use of
short-term in vitro tests, such as the Ames mutagenesis assay,
allows the rapid development of information about suspect
carcinogens.

The history of the regulation of chemical carcinogens is
quite recent. The only precedent has been that of ionizing
radiation. Some years ago the approach to setting standards for
permissible exposure to ionizing radiation involved the use of
safety factors applied to observed responses. It was during the
controversy over the health effects of fallout from atomic weapons
testing in the 1950s that the concept of linear nonthreshold dose-
response relationships was developed. An analysis of the leukemia
incidence in survivors of the atomic bombing in Japan showed a
dose-response relationship that was consistent with the linear
nonthreshold pattern. The significance of this type of a dose-
response suggests that there is no such thing as a safe dose.
There is also a close correlation between carcinogenesis and
mutagenesis. Both actions appear to cause irreversible damage to
DNA. An important feature of the linear nonthreshold dose-
response relationship is that it predicts a significant increase
in cancer when large numbers of people are exposed even at
exceedingly low dose levels--as is the case with fallout from
atomic tests or exposure to many environmental chemical
carcinogens (27).

The Delaney Clause of the Food and Drug Act passed in 1958
(12) represents a key translation of the nonthreshold concept from
the area of ionizing radiation to that of chemical carcinogens.
It bans any food additives that shows carcinogenic effects in
humans and animals. It also embodies the concept that there is no
such thing as a safe dose of a carcinogen. On the other hand,
many carcinogens have not been banned because of their social and
economical importance.

In the early 1970s a series of new laws were introduced that
dealt with the control of environmental pollutants, particularly
carcinogens. For instance, within the Environmental Protection
Agency, public laws deal with carcinogens in six areas: (i) Air
Pollution (Clean Air Act); (ii) Pesticides (Federal Insecticide
Fungicide and Rodenticide Act); (iii) Pollution of Water Bodies
(Federal Water Pollution Control Act); (iv) Drinking Water
(Federal Drinking Water Act); (v) Toxic Substances (Toxic
Substances Control Act, TSCA); (vi) Toxic Waste (Resource Recovery
and Reclamation Act). The Occupational Safety and Health
Administration (OSHA) and the Consumer Product Safety Commission
(CPSC) were both formed in the 1970s and have their own laws
dealing with the control of carcinogens. The Department of
Agriculture also is involved with carcinogens in food; for

example, nitrates in preserved meat. Every component of the
environment in the U.S. is covered by one or more of these
existing laws, with the noticeable exceptions of cigarettes and
liquor.

TERMINOLOGY AND DEFINITIONS

Although risk assessment has become a subject about which
there have been numerous publications in recent years, no standard
definitions have evolved. Frequently, one can find the same
concepts described under different names. A committee of the
National Research Council/National Academy of Sciences published a
document on the use and practice of risk assessment in the federal
government (29). Since this report is the most recent with regard
to U.S. government practices in risk assessment, much of the
following commentary has been taken from that document. Also
included were definitions of various terms appropriate to this
discussion. Several of them are included here for reference.

Risk Assessment is the use of a factual base to identify
potential adverse health effects resulting from the exposure of
individuals to hazardous materials. It includes several elements:
the description of potential adverse health effects based on the
evaluation of results from epidemiologic, clinical, toxicologic
and environmental research; extrapolation from those results to
predict the type and to estimate the extent of health effects in
humans under given conditions of exposure; judgment as to the
number and characteristics of persons exposed to various
intensities and durations; and summary judgments on the existence
and overall magnitude of the public health problem.

Risk Management is the process of integrating the results of
risk assessment with engineering data, and with social, economic
and political concerns. It is a process of evaluating alternative
regulatory actions and selecting the most appropriate from among
them. The selection process necessarily requires the use of value
judgments on such issues as the acceptability of risk and
reasonableness of the cost of control.

The Steps in Risk Assessment are: (i) Hazard identification:
the process of determining whether exposure to an agent can cause
an increase in incidence of an adverse health effect (e.g.,
cancer, birth defect). Positive responses to such questions
typically are taken as evidence that an agent may pose a cancer
risk for any exposed humans. (ii) Dose-response assessment: the
process of characterizing the relation between the dose of an
agent and the incidence of an adverse health effect; such an
assessment usually requires extrapolation from high to low dose
and extrapolation from animals to humans. (iii) Exposure

assessment: the process of measuring or estimating the intensity, frequency, and duration of human exposures to an agent present in the environment. It describes the magnitude, duration, schedule and route of exposures and the size, nature and classes of human populations exposed. (iv) Risk characterization: performed by combining the exposure and dose-response assessments. It also provides an estimate of the magnitude of a public health problem.

EVOLUTION OF RISK ASSESSMENT IN THE REGULATORY PROCESS

Through the 1960s the regulatory programs addressed exposure to toxic chemicals, but were directed mainly at the risk of poisoning and other acute effects. Such an approach gave little recognition to problems that might have been associated with smaller and repeated exposures. Cancer, birth defects and other conditions were seldom seen as preventable by government intervention. The often-cited estimate that a large proportion of all cancers may be attributed to human exposure to toxic agents originated fairly recently, and it was not until the 1970s that regulatory agencies focused their attention on cancer and other chronic health risks (3). Increasingly, epidemiologic investigations have either confirmed the findings of animal experiments or provided evidence that would link exposure to particular chemicals with particular chronic health effects (17). Also, the increase in newly suspected chemicals was accompanied by the development of instruments and procedures that permitted the detection of chemicals at extremely low concentrations (2). However, public policies are not as immediately adaptable to rapid changes as are social priorities and scientific advances. Many of the fundamental difficulties of regulatory risk assessment have resulted from attempts to bend old laws and policies to fit newly perceived risks. Government agencies are frequently beset by the inherent limitations of uncertainty along with practical constraints which limit decisive actions because of external pressures (11).

Because our knowledge is limited, conclusive direct evidence of chemical threats to human health is rare. Fewer than 30 agents are definitely linked with cancer in humans (38); in contrast, some 1500 substances are reportedly carcinogenic in animal tests. We know even less about most chemicals; only about 7000 of the over 5 million known substances have ever been tested for carcinogenicity (21).

Ethical considerations prevent deliberate human experimentation with potentially dangerous chemicals. The length of the latency period of cancer and certain other effects also greatly complicate epidemiologic studies of uncontrolled human exposures. Therefore, animal models must be used to investigate

whether exposure to a chemical is related to the incidence of
health effects, and the results must be extrapolated to humans.
When the risk involves birth defects, or a serious disease such as
cancer, feelings are likely to run high, particularly if the
groups exposed to a chemical are mobilized to express themselves
in agency deliberations. Such groups insist that regulatory
action need not await conclusive evidence of cause and effect.

Much of the recent controversy in the U.S. about perceived
adverse health effects due to populations being exposed to man-
made chemicals is generated by public opinion; the concerns
reflect a conflict in values between different groups in society,
particularly with regard to the relative importance of economic
factors on the one hand, and health protection in the formation of
regulatory decisions on the other. Different groups will
inevitably disagree about the degree of risk that is defined as
acceptable in a particular case. Some critics question whether
current practices adequately safeguard the quality of the
scientific interpretations needed for risk assessment. Since the
scientific base is still evolving, and there are great external
pressures and large uncertainties to be addressed in each
decision, some see a danger that scientific interpretation of risk
assessments will be distorted by policy considerations, and they
seek new institutional safeguards against such distortions.

VARIABILITY OF RISK ASSESSMENT PRACTICES AMONGST GOVERNMENT REGULATORY AGENCIES

The different organizational structures and agencies
responsible for regulating toxic substances have produced a
diversity in approaches to risk assessment, but some common
patterns can be discerned. First, most agencies have not
attempted to separate employees engaged in assessing risk from
those responsible for identifying and evaluating regulatory
responses. The two functions often are housed in one
organizational unit; i.e., where there are groups responsible for
preparing integrated assessments of risk, as well as defining
recommended regulatory responses (29).

Second, almost every regulatory agency in the government has
developed its own assessment methodology for substances that are
candidates for regulation. This operational independence is a
product of several factors, including a lack of formal mechanisms
for inter-agency collaboration, the desire of agency policy-makers
to retain the authority for policy discretion when reaching
conclusions based on risk assessment, and the belief that the
diversity of types of exposure for which an agency is responsible
makes collaborative risk assessment impractical.

Third, although the regulatory agencies view themselves as ultimately responsible for any risk assessment they conduct, it is often necessary for them to augment their own staff with consultants and contractors when performing a risk assessment. In addition, outside scientists often are called upon to review assessments produced by agency staff. Such consultation may take place informally, but often occurs through special advisory panels and committees. These advisory groups can exercise considerable influence and, although legally they are only advisory, share to some extent the agency's authority to reach conclusions about risk (4).

The diverse approaches to risk assessment among the various regulatory agencies have led to a dissatisfaction among the regulated. Such concerns have stimulated the introduction of legislation in Congress which directs the development of consistent guidelines and practices for risk assessment throughout the Government (28). The following discussion highlights practices used by various regulatory agencies for risk assessment, and describes their strengths and weaknesses (29).

Until 1976 OSHA had only a few personnel in the health sciences; however, the Directorate of Health Standards Programs has since become staffed primarily by health professionals, including industrial hygienists, responsible for performing risk assessments and for preparing standards. For most of its history, OSHA has not had formal guidelines for carcinogenic risk assessment. Instead, agency staff conducted their assessment by choosing options for the components of risk assessment on a case-by-case assessment. However, the generic guidelines for identification and classification of carcinogens proposed in 1980 were intended to replace criteria used in individual cases with generic guidelines that would then be applied consistently to all risk assessments of potential carcinogens (30). Although for many years OSHA did not perform quantitative risk estimates for use in setting standards for carcinogens, it now intends to do so whenever appropriate. Historically, OSHA has done a less thorough job than any other agency in developing relevant scientific information and independent peer-review of its information before issuing a notice of proposed rule-making.

The Bureau of Foods of the Food and Drug Administration (FDA) assesses risk associated with thousands of new and existing products every year. In 1981, the Bureau evaluated 65 food additives, 2 color additives and approximately 45 animal-drug petitions. Although no effort is made to evaluate the benefits that an additive might provide, the Bureau must be satisfied that the additive achieves its intended effects. In carrying out this evaluation, the Division of Toxicology first determines a "no-observed effect" concentration for an additive. Then, with the application of a so-called safety factor, the staff determines a permissible "extent of use" in human food, or what is referred to as an "acceptable daily intake". Methods for evaluating potential carcinogens have undergone substantial change since the early 1970s. Several years ago, the Bureau formed a Cancer Assessment Committee to evaluate the carcinogenicity of substances being considered for approval or regulation, and to perform risk assessments. The Committee does not follow comprehensive written guidelines, although it does follow general guidelines that have been used in previous decisions (29).

The Carcinogen Assessment Group (CAG) of the Environmental Protection Agency (EPA) was created in 1976 by the EPA administrator to implement generic and uniform agency guidelines for carcinogenic risk assessment. The CAG staff are separate from, and independent of, the risk management function, a unique example of intra-agency separation. This body also serves as an example of an internally centralized assessment body in that it performs assessments for several different regulatory programs within EPA. Currently, all CAG assessments are done by in-house staff, although in the past some were done by consultants (3,29).

The Occupational Safety and Health Act of 1970 created two organizations: OSHA and the National Institute for Occupational Safety and Health (NIOSH). NIOSH's primary functions include the conduct of research and the development of criteria for recommendations to OSHA for occupational health standards. Although Congress intended a close coupling between NIOSH's recommendations and OSHA's standards, relatively few NIOSH criteria documents have led to OSHAs standards. OSHA conducts its own assessments of risk as well as setting standards, and NIOSH does risk assessment and recommends standards, but the relationship between NIOSH and OSHA that has existed since 1976 represents, in some sense, duplication, rather than true extra-agency separation.

AN ATTEMPT AT CONSISTENT CARCINOGENIC RISK GUIDELINES FOR REGULATORY AGENCIES

In view of the types problems and inconsistencies just described, the President's Interagency Regulatory Liaison Group, in 1977, established a work group to develop a document that would represent an interagency consensus on the scientific aspects of carcinogenic risk assessment. In February of 1979, the IRLG published a report entitled, "Scientific Bases for Identification of Potential Carcinogens and Estimation of Risk" (17).

The report was the most comprehensive set of guidelines that had been developed up to that time for interagency use. It

addressed the components of hazard identification and dose-
response assessment, with a discussion of exposure assessment and
risk characterization. The report, however, had no official legal
status. Despite its lack of regulatory authority, the report was
the first evidence that all Federal regulatory agencies could
agree on options applicable to the identification of carcinogenic
hazards and the measurements of risk. Shortly after the report's
publication in the Federal Register, the President's Regulatory
Council adopted it as the scientific basis of the Counsel's
government-wide statement on the regulation of chemical
carcinogens. The Council viewed IRLG guidelines as a major step
in decreasing inconsistencies, eliminating duplication of effort,
and reducing the lack of coordination among agencies on
carcinogenic risk assessment.

Although the document received a great deal of public
comment, a published review of these comments has never appeared.
Industry in the United States criticized the document because it
was published in final form and adopted before the comments could
be presented and revisions considered. The present administration
disbanded the IRLG effort in 1981. Thus, it is likely that a
formal review of the public comments will never appear.

A similar task force was established in 1982, but through the
White House Office of Science and Technology Policy. A draft
document on the scientific basis of risk assessment has been
prepared by government scientists and has been circulated for
comment to scientists within and outside the government. A final
version is expected by late 1983. A second part of the policy
document will include a set of principles for conducting risk
assessment. It is anticipated that this document will serve as a
reference for the identification of carcinogens and the later
agencies (31).

CURRENT RESEARCH IN RISK ESTIMATION METHODOLOGY

We must not delude ourselves into thinking that current
techniques of quantitative risk assessment have unqualified
application in the making of regulatory decisions. The
quantification of human risk, on the basis of the results of
laboratory studies in animals, should be approached with great
caution. We must not lose sight of the fact that animal studies
serve primarily as a qualitative surrogate for humans, and that
any attempts to quantify responses beyond the realm of biological
certainty are open to question (20). Thus, it is important that
all who are concerned with implementing and using risk assessment
information (such as regulators and industrialists, as well as the
public) understand the magnitude of uncertainty that surrounds

tnis process. To think otherwise would create a false impression
of the realities of safety evaluation, as it is now practiced.

As we look to the future, it is critically important that we
attain a deeper understanding of the molecular events in
toxicology. With a deeper understanding of the biological
process, we will be able to put greater reliance on quantitative
risk estimation through the development of more realistic and
reliable methods for extrapolation. Examples of research being
conducted to improve risk assessment methodology are included
below.

The large uncertainties associated with human risk estimates
have prompted suggestions that estimations of low risk be replaced
by ranking of chemicals with respect to their carcinogenic
potency. Peto et al. (33) have proposed the TD_{50} as a numerical
description of carcinogenic potency. It is a dose rate in
milligrams/kilogram of body weight/day which cause tumors in 50%
of an animal species which theoretically has no incidence of
spontaneous tumors. The TD_{50} may be estimated by interpolation
between observed data points. There are, however, certain hurdles
which must be overcome to rank human chemical risks accurately
when using animal potency data. One important aspect is an
increased incorporation of species-specific data on chemical
uptake, distribution, metabolism and elimination. To date, such
kinds of data have played too small a role in risk assessment. We
need standard, acceptable protocols for experiments to generate
such data (39).

To determine how much chemical potencies can tell us about
human risk, we need to test the metabolic ratio hypothesis among
both males and females of several mammalian species (41). Crump
(8) has recommended the development of a human data base from
available epidemiologic cancer data. Such an evaluation would
estimate the potency of chemicals from which suitable data are
available, by comparing human information with corresponding
estimates made from animal data. With this information, one then
could empirically determine the best procedures for animal-to-
human extrapolation. Then the various dose measures of: mg/kg of
body weight/day; or ppm in the diet; or mg/m^2 body surface/day; or
mg/kg of body weight/lifetime could be studied to determine how
well any one of them might predict human potencies from animal
data. Both the National Academy of Sciences (25) and Crouch and
Wilson (6) have done preliminary studies using these types of dose
measurements, but there has been no definitive investigation.

To estimate relative safety for humans in the absence of
human data, toxicity information obtained from animal studies is
frequently used to extrapolate information from animal to man,
even though large quantitative and qualitative interspecies

differences may exist (6,8,37). An improved understanding of such interspecies differences or similarities is needed to estimate in a more meaningful manner a chemical's toxicity for man. Greater use of pharmacokinetic techniques could aid in identifying different or comparable metabolic mechanisms among animals (within and among species) at different dose levels, and could provide guidance information at the low dose region below dose extrapolation methodology. Greater use of pharmacokinetic information should improve the reliability of risk assessment models, particularly in the enhancement of structure activity relationship information (11,13,23).

A new area of considerable activity is reflected in the work of McCann and co-workers (22) in which they are developing procedures for estimating mutagenic potency from dose-response data developed in bacterial short-term tests. The potential for short-term tests to provide quantitative information about carcinogenesis is much less certain than is the potential to provide qualitative information. Attempts have been made, however, to show that some quantitative correlation in potency between short-term bacterial tests and animal cancer tests may exist. If potency from short-term tests could be used to make rough predictions of potency in animal cancer tests, further risk evaluation could be more easily focused on chemicals with the greatest potential hazard (16).

CONCLUDING REMARKS

The process of risk assessment, as performed by and for U.S. federal regulatory agencies, has been developing rapidly in recent years. However, rapid change is bound to lead to misunderstandings about the use of risk assessment in regulatory policy-making, particularly if risk assessment is mistakenly construed as a strictly scientific undertaking. Much of the criticism of risk assessment stems from dissatisfaction with regulatory outcomes, and many proposals for change are based largely on the unwarranted assumption that altering the administrative arrangement for risk assessment would lead to regulatory outcomes that critics would find less disagreeable. Because risk assessment is only one aspect of risk management decision-making, however, even greatly improved assessments would not eliminate the frequent dissatisfactions with risk management decisions.

Perhaps the basic problem with risk assessment is not its administrative setting, but rather the paucity and uncertainty of scientific knowledge of the health hazards addressed. We conclude by summarizing the recent recommendations made by a committee of the National Academy of Sciences concerning ways in which risk

assessment might be enhanced within government regulatory bodies:
(i) implementation of changes to assure that risk assessments take
full advantage of available scientific knowledge, while
maintaining the diverse organizational approaches necessary to
accommodate the varied requirements of various federal regulatory
programs; (ii) standardization of analytical procedures among
federal programs, through the development and use of uniform
guidelines; and (iii) creation of a mechanism that will insure
orderly and continuing review of risk assessment procedures as
scientific understanding of hazards improves. This might best be
accomplished through the establishment of a continuing Board on
Risk Assessment Methods within the organizational structure of the
National Academy of Sciences--National Research Research Council.

REFERENCES

1. Albert, R.A. (1980) Toward a more uniform federal strategy for
    the assessment and regulation of carcinogens. Report
    prepared for the Office of Technology Assessment, Washington,
    D.C.
2. Ames, B.N. (1979) Identifying environmental chemicals causing
    mutations and cancer. Science 204:587-593.
3. Anderson, E.L. (1982) Quantitative methods in use in the
    United States to assess cancer risk. Paper presented at the
    Workshop on Quantitative Estimation of Risk to Human Health
    from Chemicals. Rome, Italy, July 12.
4. Barr, J.T., D.H. Hughes, and R.C. Barnard (1981) The use of
    risk assessment in regulatory decision-making: Time for a
    review. Regul. Toxicol. Pharmacol. 1:264-276.
5. Cornfield, J. (1977) Carcinogenic risk assessment. Science
    198:693-699.
6. Crouch, E., and R. Wilson (1979) Interspecies comparison of
    carcinogenic potency. J. Toxicol. Environ. Health
    5:1095-1118, and appendices.
7. Crouch, E., and R. Wilson (1981) Regulation of carcinogens.
    Risk Anal. 1:47-57, 59-66.
8. Krump, K.S. (1982) Methods for estimating human cancer risks
    from non-human data. Presented at EPRI Workshop on Risk
    Assessment for Toxic Substances, December 1-2.
9. Krump, K.S. (1983) An improved procedure for low-dose
    extrapolation. In C.R. Richmond, P.J. Walsh and
    E.D. Copenhaver, Eds. Health Risk Analysis, Proceedings of
    the Third Life Sciences Symposium. The Franklin Institute
    Press, Philadelphia, PA. pp. 381-392.
10. Doll, R., and R. Peto (1981) The causes of cancer:
    Quantitative estimates of avoidable risks of cancer in the
    United States today. J. Nat. Cancer Inst. 66:1191-1308.
11. Electric Power Research Institute (1981) Conference
    Proceedings: Environmental Risk Assessment. How new

regulations will affect the utility industry. EPRI EA-2064,
Palo Alto, California.

12. Federal Food, Drug, and Cosmetics Act, The Delaney Clause,
Section 409 (c)(3)(A).

13. Gaylor, D.W. (1979) The ED_{01} study. Summary and conclusion.
J. Environ. Pathol. Toxicol. 3:179-183.

14. Gori, G.B. (1980) The regulation of carcinogenic hazards.
Science 208:256-261.

15. Gori, G.B. (1982) Regulation of cancer-causing substances:
Utopia or reality. Chem. Eng. News 6:25-32.

16. Horn, L., J. Kaldor, and J. McCann (1983) A comparison of
alternative measures of mutagenic potency in the Salmonella
(Ames) test. Mutation Res. 109:131-141.

17. Interagency Regulatory Liaison Group, Work Group on Risk
Assessment (1979) Scientific bases for identification of
potential carcinogens and estimation of risks. J. Natl.
Cancer Inst. 63:242.

18. Krewski, D., and C. Brown (1981) Carcinogenic risk assessment:
A guide to the literature. Biometrics 37:353-366.

19. Linsell, C.A., and F.G. Peers (1977) Field studies on liver
cell cancer, In Origins of Human Cancer, H.H. Hiatt et al.,
Eds., Cold Spring Harbor, New York.

20. Mantel, N., and M. Schneiderman (1975) Estimating "safe"
levels. A hazardous undertaking. Cancer Res. 35:1379-1386.

21. Maugh, T.H. (1978) Chemical carcinogens: How dangerous are low
doses? Science 202:37-41.

22. McCann, J., L. Horn, G. Litton, J. Kaldor, R. Magaw,
L. Bernstein, and M. Pike (1983) Short-term tests for
carcinogens and mutagens: A data base designed for
comparative quantitative analysis, In Structure Activity as a
Predictive Tool in Toxicology. Fundamentals, Methods, and
Applications., L. Goldberg, Ed., Hemisphere Press, D.C.,
pp. 229-240.

23. Munro, I.C., and D.R. Krewski (1981) Risk assessment and
regulatory decisionmaking. Food Cosmet. Toxicol. 19:549-560.

24. National Academy of Sciences, Food Protection Committee, Food
and Nutrition Board (1970) Evaluating the safety of food
chemicals. National Academy Press, Washington, D.C.

25. National Academy of Sciences, Commission on Natural Resources
(1975) Pest control: An assessment of present and alternative
technologies. Vol. 1, Contemporary pest control practices
and prospects. The Report of the Executive Committee,
Washington, D.C.

26. National Academy of Sciences, Committee for a Study on
Saccharin and Food Safety Policy (1979) Food safety policy:
Scientific and social considerations. National Academy
Press, Washington, D.C.

27. National Academy of Sciences, Committee on the Biological
Effects of Ionizing Radiation (1980) The effects on

populations of exposure to low levels of ionizing radiation. National Academy Press, Washington, D.C.

28. National Academy of Sciences, Committee on Risk and Decision Making (1982) Risk and decision making: Perspectives and research. National Academy Press, Washington, D.C.

29. National Academy of Sciences, Committee on the Institutional Means for Assessment of Risks to Public Health (1983) Risk assessment in the federal government: Managing the process. National Academy Press, Washington, D.C.

30. Occupational Safety and Health Administration (19980) Final rule: Identification, classification, and regulation of potential occupational carcinogens. Fed. Reg. 45:5001.

31. Office of Science and Technology Policy, Regulatory Work Group on Science and Technology, Executive Office of the President (October 1, 1982) Potential human carcinogens: Methods for identification and characterization. Part 1: Current Views: Discussion Draft.

32. Office of Technology Assessment (1981) Assessment of the technologies for determining cancer risks from the environment. Washington, D.C.

33. Peto, R., M. Pike, L. Bernstein, L.S. Gold, and B. Ames (1984) The TD_{50}: A proposed general convention for the numerical description of the carcinogenic potency of chemicals in chronic-exposure animal experiments. Enivron. Health Perspectives (submitted).

34. Ricci, P.F., and L.S. Molton (1981) Risk and benefit in environmental law. Science 214:1096-1100.

35. Saffiotti, U. (1980) The problem of extrapolating from observed carcinogenic effects to estimates of risk for exposed populations. J. Tox. & Environ. Health 6:1309-1326

36. Sontag, J.M., M.P. Page, and U. Saffiotti (1976) Guidelines for carcinogenic bioassay in small rodents. National Cancer Institute, National Institutes of Health, Bethesda, Maryland.

37. Squire, R.A. (1981) Ranking animal carcinogens: A proposed regulatory approach. Science 214:877-880.

38. Tomatis, L., et al. (1978) Evaluation of the carcinogenicity of chemicals: A review of the monograph program of the International Agency for Research on Cancer, Cancer Res. 38:877-885.

39. Weisburger, J.H., and G.M. Williams (1980) Carcinogen testing: Current problems and new approaches. Science 214:401-407.

40. Weisburger, J.H., and G.M. Williams (1981) The decision-point approach for systematic carcinogen testing. Food Cosmet. Toxicol. 19:561-566.

41. Whittemore, A.S. (1982) Estimating human risks from animal data. Presented at EPRI Workshop on Risk Assessment for Toxic Substances, Carmel, California. December 1-2 (unpublished).

ALTERNATE METHODS FOR INTEGRATED EVALUATION

OF TOXICITY AND RISK ASSESSMENT

Laila A. Moustafa

World Health Organization, International Programme on
Chemical Safety, Interregional Research Unit
Research Triangle Park, NC 27709, USA

SUMMARY

Animal bioassays by themselves can yield ambiguous results,
especially in relation to human risk assessment. Three major
alternates are proposed for integration in toxicology evaluations.
The first is mathematical modeling and expression, viz.,
statistical and quantitative structure/activity relationship
(QSAR). Mathematical modeling can run indefinitely after animal
experiments have terminated, and can serve as a powerful adjunct
to animal experiments. Quantum mechanical analysis (QMA) can help
researchers to study, on the molecular level, the interactions of
chemicals and the body and to understand and predict side effects.
QSAR assumes that a functional dependence exists between the
observed biological response and certain physiological properties
of molecules. With QMA implementation in QSAR, one can obtain
reactivity characteristics in order to relate molecular structure
to the observed biological activity. Stereology (morphometry) can
be used to attach quantitative values to complex biological
structures identified in light and electron micrographs. By
integrating structural/functional data one expects to pinpoint the
specific cellular responses to one or several chemicals, and even
to predict genetic events from cytoplasmic changes. The second
alternate is a battery of in vitro toxicity tests. For example,
plasticity of synapse formation with cultured nerve cells should
yield rich dividends when coupled with monoclonal antibody and
recombinant DNA technology. Monoclonal antibody technology, when
applied to studying nerve cell biochemistry, should be extremely
useful in identiying the structure of various surface components
on nerve cells. Recombinant DNA technology is likely to yield
important information with regard to specific gene expression. An

application of <u>in vitro</u> systems to teratogenicity testing has been
actively pursued. The third alternate is the use of plants.
Plant systems show promise for immediate use in laboratory short-
term bioassays for toxicity evaluation of specific chemicals or
chemical mixtures. For testing under field conditions, few test
organisms offer the advantages provided by plants.

INTRODUCTION

     The risks to health that abound in the environment are often
beyond the capacity of the individual to control, and few would
adhere to the belief that human beings can achieve absolute
safety. No one has the exact figures, but approximately 4.5
million chemicals have been isolated from natural products or
synthesized (33), and probably about 2 million mixtures,
formulations, and blends of chemicals are in commercial use.
About 63,000 chemicals are in everyday use, some 48,000 of them in
commercially significant amounts. Of these, about 6,000 have thus
far been tested for toxicity (40,41,49). The number of chemicals
in use increases rapidly. The American Chemical Society's
Chemical Abstract Service (CAS) identifies some 6,000 chemicals
each day, about 1,000 of which are new (41). Most of them,
however, are synthesized for specific research purposes. The
toxicological implications and statistical permutations of this
number of chemicals are almost beyond comprehension and the task
of proper safety evaluation and monitoring is very formidable.
Even assuming that the appropriate technologies were available to
monitor these chemicals properly, the resources required to
complete this job satisfactorily seem overwhelming. In FY 1980,
approximately $262 million was spent in the United States on
toxicological studies of various sorts (27). With the escalation
of inflation and greater demands for more extensive tests in more
than one species and multiple observational endpoints, there is a
tremendous need to use available facilities as efficiently as
possible and to reduce duplication of tests.

     At our present stage of scientific knowledge, there is a
tendency to rely on guidelines for studies on laboratory animals
designed to predict toxic potential of chemicals in humans.
Animal bioassays by themselves can yield ambiguous results,
especially in relation to human risk assessment. In addition, it
is very difficult to take into account the synergistic or
antagonistic interactions of the tested chemical with other
chemicals either in the environment or present simultaneously in
the organism. Similarly, there are problems with variables such
as age, sex, nutrition, pregnancy, and genetic or temporary
predisposition (20,48,66). The ability of animals to reflect
probable human reactions to chemicals is limited, (42,47,61,64).
For instance, subjective reactions to chemicals, such as

gastrointestinal discomfort, nausea, headache, irritability and depression cannot be adequately reported in animal experiments. So far no animal has been found that resembles humans in most, or all, relevant aspects, which makes the choice of the experimental model all the more dependent on the problem that the scientist wants to solve.

Efforts are under way within the International Programme on Chemical Safety (IPCS) of the UNEP, ILO and WHO to integrate alternate methods for toxicological evaluation with the existing ones as well as to harmonize methodologies and validate tests which involve the statistical treatment of data to determine precision, accuracy, sensitivity and reproducibility of the procedure from laboratory to laboratory in order that results generated by one Member State institution may be accepted by another. Validation provides a common denominator for agreement on just what an analytical result really means. Knowledge of the chemical being tested and increased credibility of tests may provide a protocol with which to minimize and optimize the necessary testing and, it is to be hoped, to create a scientific network of communication.

Alternate methods for integrated toxicological evaluation (AMITE) of new and existing chemicals warrant international concern and action. An essential feature of an AMITE project is that a broad range of scientific disciplines have an opportunity to interdigitate their knowledge and technology in such a way that the results can be evaluated at specific key points in the test series and decisions can be made regarding the potential toxicity of a given chemical. A classification of chemical toxicity may serve as a basis for a rational, sequential system of detecting chemicals that might present a health risk. At the same time, scientists will have the opportunity of achieving a better understanding of the mechanism of action of a test chemical.

An evaluation is made at the end of each test (or stage) to determine whether the data available are significant enough to reach a definitive conclusion or whether to recommend further tests. In addition, we will deal here with some, but not all, AMITE that should be considered in such programs.

MATHEMATICAL ADJUNCTS TO BIOASSAY

1. Quantitative Structure/Activity Relationship (QSAR)

The spectacular advancement in computer technology has greatly stimulated the development of sophisticated mathematical models of statistical and quantitative structure/activity relationship (QSAR). It has been the goal of scientists in

diverse areas to correlate structural features of compounds with
desired effects. This has been most successfully pursued in the
drug and pesticide industries. Of equal importance has been the
quest for the predictive tools to correlate structure with
toxicity and environmental effects including stability,
persistence, transport, bioaccumulation and biomagnification.
Although QSAR models have at present limited applicability for
explaining or predicting biological response—on the basis of
either physicochemical properties or of some index calculated
directly from structure—they are capable of further refinement as
a greater understanding of the complex interactions which occur
between molecules is reached.

QSAR is an actuarial or statistical method in which only
objective data are used with no intrusion of models or mechanistic
hypotheses. The equation that is obtained not only accounts for
the relative potencies of the compounds, but from it are deduced
predictions of the potencies of untested compounds—if the
question is valid, the predictions are ineluctable (50).

The prediction of toxic response is a great challenge to
mathematical modeling, and is probably its most difficult
objective. This difficulty arises because acute toxicity can be
the result of many mechanisms, simultaneously involving several
target organs and enzyme systems, whereas successful modeling
requires that the mechanisms by which the response occurs should
involve one rate-determining step common to all compounds.
Despite this difficulty, models for predicting acute toxicity have
been proposed, although at present they can be used only to set
priorities for further testing or to eliminate from further
evaluation certain compounds predicted to be highly toxic.

In areas other than that of acute toxicity, mathematical
models have proven to be of considerable value in rationalizing
and predicting biological data for congeneric series of compounds.
The attempt to correlate biological activity with chemical
structure in quantitative terms assumes that a functional
dependence exists between the observed biological response and
certain physiological properties of molecules (50,51).

Methods Used in QSAR

Most structure activity data (50) have been analyzed by:

A. The empirical methods of Hansch (28). This method
relates the observed biological activity to extrathermodynamic
(ETD) parameters that are assumed to represent the electronic,
steric and hydrophobic properties of the compounds responsible for
the biological effect.

The purpose of the Hansch relationship is to find a quantitative functional dependence of the biological activity on independently obtained parameters. From this one can predict biological activities of compounds that have not been tested.

B. The mathematical method of Free and Wilson (23). This method assumes that the biological activity of a molecule can be represented as the sum of the activity contributions of a definite substructure and the corresponding substituents. This implies that the contributions of the parent fragment (a hypothetical structure that has no substituents) and of the substituents are constant and independent of substituents at other positions on the fragment. Cammarata and Yau (8) and Fujita and Ban (24) proposed some modifications to the Free-Wilson method. The Free-Wilson method and its modifications are all based on the linear additivity assumption.

It has been suggested (38) that both the Hansch and the Free-Wilson methods be used in QSAR studies, but in general the methods are used exclusively.

C. Pattern recognition analysis. A collection of methods used to detect relationships (i.e., patterns) within a large number of observations which simultaneously depend on many variables (9,15,17,35,39,62,65,67). These methods usually do not assume any explicit form of relationship between the observables and the independent variables, and thus are not subject to a statistical test of their quality. They are very helpful in classification and reduction of large amounts of information, e.g., biological activity and chemical structure, to patterns that can be then treated by standard regression methods.

D. Cluster analysis. The relationships between the observations that are associated with several properties can be qualified in this analysis (16,29). For example, many substituents can be analyzed to determine which of the subgroups of substituents represent similar properties. Cluster analysis can be helpful in minimizing the number of compounds that need to be synthesized in order to obtain the maximum variation in the effects of substituents on biological activity.

E. Discriminant analysis distinguishes between groups that confer different properties on the predictor variables (14,37,39,54). It could be used, for example, to identify among a group of compounds those structural parameters that contribute to activity or lack of activity or to agonism and antagonism. This method is helpful in classification of compounds and in the construction of a set of compounds to which one of the quantitative methods (e.g., Hansch or Free-Wilson) can be applied.

More work should be carried out to establish just how
appropriate are the presently employed parameters--both
physicochemical and structural--for the development of reliable
QSAR (50). This requires a better understanding of the steric and
conformational aspects of interaction between compounds and
biological macromolecules.

2. QUANTUM CHEMICAL MECHANICS

With the discovery of quantum mechanics (chemistry),
physicists realized that molecular behavior could ultimately be
understood in terms of the boundaries of each atom in the molecule
and the energy shifts of each atom's electrons. Today,
accomplishments in this field are assuming greater and greater
importance for biology as well as for physics and chemistry. They
have proven particularly useful in recent work in pharmacology,
helping researchers to study, on the molecular level, the
interactions of chemicals and the body (26).

No-one claims that such complex and multimolecular phenomena
as blood flow and the diffusion of compounds across cellular
membranes can be described readily in quantum mechanical terms.
Still, quantum mechanics does apply to the chemical interactions
upon which every living thing ultimately depends. It can be used
in the way a biologist employs other sophisticated tools, such as
infrared spectroscopy, or radioimmunological assays, to gain
insight into the first steps of life: metabolism; catabolism; the
breaking down of chemicals within a living system; and the
molecular behavior controlling biosynthesis, which is the
production of chemical compounds by an organism (26).

Until recently, the major stumbling block in applying quantum
mechanical analysis to molecular aspects of life processes has
been the magnitude of the calculations involved. Now, methods for
calculating the properties of complex molecules have been
simplified, and new generations of computers have proved capable
of handling the calculations.

Parameters Derived from Quantum Chemical Calculations

The quantum chemical computation methods are now able to
predict good relative values of physicochemical properties that
can be determined by experiment either with great difficulty or
only by inference. These include multipole movements, molecular
polarizabilities, ionization potentials, electron affinities,
charge distribution, scattering potentials, spectroscopic
transitions, geometric and energies of transition status, and the
relative populations of various conformations of molecules. Some
of these properties are directly related to molecular reactivity

(e.g., charge distributions, molecular polarizabilities, scattering potentials), and they can be implemented in QSAR studies (50).

Quantum Chemistry and Biological Response

Because of quantum mechanics, it is possible for the first time to discuss chemical affinity for certain receptors in terms of components of molecular reactivity, which are more relevant than structural components. With quantum mechanics, side effects become easier to understand and to predict. By simulating the dynamics of the interaction, the effects on charge redistributions, and the structural rearrangements that follow compound-receptor interactions, such calculations can provide a glimpse into some of the steps of triggering mechanisms that give rise to biological response. To do more, however, additional information about the chemical structures of receptors is needed (72,73).

Investigations of the enzyme-substrate interactions are of special interest in the application of quantum chemistry to the study of biological mechanisms because the structures and the properties of enzymes have been studied extensively and in detail by many experimental techniques (72,73). Since the intermediate stage of the mechanisms proposed for the function of many enzymes is often well delineated, theoretical studies can contribute directly to the elucidation of these fundamental processes by combining the information obtained experimentally from a variety of sources and by analyzing it in a unified formalism. This can be achieved by modeling the structural components and by simulating mechanisms of interaction between the enzyme and substrates or inhibitors. These calculations can be expected to reveal the nature of the intermolecular forces involved in these interactions and to provide a useful basis both for the comparison of proposed mechanisms and for the description of the roles played by the functional groups.

Many problems remain in applying quantum mechanics to the study of biological reactions. At present some of these problems may seem real and forbidding, but the rapid developments of the past decade, particularly in computer software, make it likely that such problems will soon be solved.

Pharmacokinetic Modeling

Pharmacokinetic modeling is a method which attempts to calculate the distribution of chemicals in the body over various time periods. With the techniques of interspecies scaling, it is possible to model the disposition of drugs/chemicals in humans quite well, given some basic animal data for use as a benchmark in the scaling calculations. Many parameters exhibit relationships which are, to some extent, species-independent. Metabolic rate, for example, is in a general sense a function of body weight, and facts of this sort are of great help in the scaling process. Chemical structure must always be considered in relationship to species' metabolic parameters. The guinea pig, for example, in contrast to other rodents or humans, has only limited amounts of the necessary enzymes to carry out N-hydroxylation, and yields almost exclusively detoxified metabolites. Therefore, the arylamines so far tested are not carcinogenic in this species (1).

Information on structure and metabolism also provides a guide to the selection among limited in vivo bioassays and, as more information accrues, may eventually contribute to selection of specific short-term tests.

Mechanisms of Reactions Between Chemicals and Nucleic Acids

There appears to be a reasonable correlation between the distortion of the Watson-Crick double helix for DNA, caused by chemical reaction, and carcinogenic potency. Two types of distortion of DNA that cause miscoding are envisaged: (i) introduction of substituent groups at sites on the pairing faces of the bases thereby breaking a number of H-bonds interconnecting the bases; or (ii) introduction of bulky substituents at sites on the back side of the purine bases tending to interfere with the stacking of the bases inside the double helix by grossly distorting the sugars and phosphate groups at the periphery, and in extreme cases causing "unstacking" (30).

Thirteen years ago Miller, whose idea arose from the discovery of positively charged intermediates in N-(2-fluorenyl)-acetamide and amino-azo dye metabolism (vide infra, 44), attempted to account for the interactions of chemical carcinogens by the electrophilicity of the ultimate, reactive forms. The argument, based on the work of Price (56,57), that S_N2 reactions or reactions of a type intermediate between S_N1 and S_N2 appear to predominate over pure S_N1 or free radical reactions with the reactive forms of chemical carcinogens would make available a range of electrophilic reactivities, which would be expected to span a range of probable (nucleophilic sites in genetic material) receptors in vivo. As the electrophilic properties of chemical carcinogens or their reactive forms were held to be important to their carcinogenic mode of action, it followed that nucleophilic mutagens such as hydroxylamine would not be expected to be carcinogenic, and this supposition proved to be correct (Poirier, cited in Hathway and Kolar, 1980 [30]).

3. STEREOLOGY

Stereology is a body of mathematical methods relating three-dimensional parameters defining the structure of two-dimensional measurements obtained on sections of the structure (69,70). Quantitative stereology is based on geometric probability theory. It is postulated that (1) the structural components of a biological system--organs, tissues, cells, organelles, and molecules--are organized mathematically as coherent sets; and (2) the structures within each set represent specific biochemical compositions and arrangements. These assumptions are used to design a mathematical model for a specific organ (i.e., testis, liver, pancreas, etc.) (4,6,52,53).

The narrative description of the nature of stereological procedures reveals one important point: all stereological measurements are in principle obtained as relative measurements, more precisely as a ratio of at least two joint measurements, one relating to the components, the other to the structure as a whole. The latter is called the "reference system". The stereological principles establish the precise relationship between such ratios measured on sections, and the corresponding ratios in the spatial structure (69,70). Automatic methods of image analysis were introduced into stereology as soon as they became available. In connection with the development of such instruments important progress has been made in the field of mathematical image analysis and pattern recognition.

Stereology and Its Uses in Cell Biology

Previous work in this area, labeled by such terms as morphometry, quantitative microscopy, or metallography, stereometric metallography, and micrometric or modal analysis, has been applied in a relatively restricted sense (12,68). In its broadest context, stereology includes not only the quantitative study and characterization of any spatial structure, but also its qualitative interpretation, since exact representations are not yet possible in all cases. For toxicological evaluations, the effects of experimental treatments on cell structure and the time course of morphological events can now be precisely charted on the basis of cell or organelle volume, membrane area, organelle number or size. Subjectivity is thereby greatly reduced and firm kinetic data can result. Mechanistic hypotheses may then follow. Stereological methods represent, therefore, a major advance in microanatomy (5,32).

Quantitative microanatomical estimations, furthermore, provide an improved framework on which to found cytochemical observations. Immunocytochemical reactions are greatly enhanced in interest if, for example, a statement in words about the sites

of antibody binding can be upgraded to a figure stating the number
of antigenic sites that occur per unit area of a particular
species of membrane.

There are two sides to stereology, and accordingly two
complementary groups of stereologists. The first group, made up
of mathematicians of varying denominations, deals with the
theoretical foundations of stereology, of the relations between
structure and sections; the second, made up of biologists and
material scientists, seeks to extract from these foundations
methods that help solve some practical problems of microscopic
morphometry.

Stereological methods have proved to be most effective in
gaining access to structural information at the cellular level.
The strategy underlying these (stereological) methods consists
first of choosing a small but representative sample of the object
that contains the cells (a set of sections) and then collecting
counts of points, intersections, and profiles from the sections
with a test grid. Equations based on geometric probability theory
can then be used to transform these counts into three-dimensional
information that characterizes the structure of the original solid
object. The solutions to these equations provide estimates for
the volumes, surface areas, lengths, and frequencies of structures
found within an average unit of reference volume (12,68,69,70,75).

Most stereological projects commence at the light microscope
(LM) level. It is often necessary to assess changes in cell
population composition, cell numbers and cell size before
proceeding to quantitative subcellular observations. The more
firmly the LM base line is established, the more valuable the
electron microscope (EM) observations are likely to be (75).

Stereological experiments at the EM level are ideal for
objectively assessing changes which may be hard to describe in
words or illustrate with one or two micrographs. Frequently, they
reveal quite unsuspectedly precise changes or differences, and
often confirm just how precisely determined cell composition is.
In association with classical biochemical techniques and/or with
autoradiography or immunocytochemistry they are a formidable
combination indeed.

Stereology: Applications to Toxicology

The trend in toxicological problem solving is clearly moving
in the direction of collecting more and more detailed information,
which in turn is being drawn upon to evaluate the effectiveness
and safety of chemicals.

Chemically induced changes in cells represent very complex events. Complex information in this form can be accommodated in a variety of ways, but one that has gained increasing attention in recent years has been that of system analysis. The basic idea of this approach is that a mathematical model for a system is constructed, then an analysis of the model is performed, and finally, the results are applied to the original system. Ideally, there should be extensive interactions between the construction, the analysis, and the interpretation of the model (5,6).

Stereology is of interest to a systems analysis approach because it can be used to attach quantitative values to complex biological structures identified in light and electron micrographs. When combined, the structural and functional data can be used to assemble relatively simple, yet surprisingly powerful, information networks. In effect, networks may be thought of as a means of establishing an experimental position from which one can compose questions with improved confidence. For example, integrated structure-function networks are expected to satisfy the dimensional requirements needed to map the molecular topographies of organelles, to pinpoint the specific cellular responses to one or several xenobiotics, and even to predict genetic events from cytoplasmic changes.

Biochemical data can be fitted into stereological data, thus increasing experimental accuracy. It was demonstrated that changes in the quantities of sulfated proteoglycans detected chemically in colon carcinoma can be associated with distinct morphologic changes in the size, location, and distribution of the intercellular matrix granules. Thus, by integrating the biochemical and stereological data it has been possible to define the nature of the proteoglycan changes in colon carcinoma more precisely than would have been possible by using either one of these methods alone (32).

A rapidly developing area of research is the study of intracellular distributions of membrane bound marker enzymes. Although the sensitivity of the methods used for locating the positions of marker enzymes on membranes is often excellent, the information has not yet been used to construct areal distribution maps. Cytochemical approaches provide similar qualitative information, and are successful in locating a single marker enzyme at several different morphological locations. However, this information has not been used topographically.

Interhepatocytic heterogenetics are likewise being studied using a cell separation system. Stereologists, however, have approached the problem of interhepatocytic heterogenetics by collecting data directly from the lobular zones: i.e., periportal, midzonal, and centrolobular. The major problem here seems to be

knowing (a) where these zones actually are in the lobule under
both control and experimental conditions and (b) how to weigh the
data of each zone in order that they may be representative of the
entire liver lobule (3,7,31,36).

BIOTECHNOLOGY

We are now in a period of especially rapid progress in
applied biology. Important advances have already been made
employing recombinant DNA, hybridomas and protein engineering in a
wide range of living cells (60). For example, plasticity of
synapse formation with cultured nerve cells should yield rich
dividends when coupled with two additional areas of development in
biotechnology: those of monoclonal antibodies, and recombinant
DNA. Monoclonal antibody technology, when applied to the study of
nerve cell biochemistry, should be extremely useful in
investigating the structure of various surface components on nerve
cells. Recombinant DNA technology is likely to yield important
information with regard to specific gene expression.

1.   Monoclonal Hybridomas

Monoclonal antibodies have provided a means to recognize
specific molecules and to inhibit their biologic activity.
Hybridoma techniques provide a means of producing antibodies which
was not considered possible even a short time ago. The uses of
such antibodies are myriad and include the detection and
neutralization of chemical and biological toxins. With
appropriate immunization, an animal will mount an antibody
response to an antigen or hapten and these antibody-producing
cells can be immortalized by the hybridoma technique (13). It is,
therefore, possible in principle to use monoclonal antibodies to
detect toxins in the serum or tissue of an animal or person.
There are in fact already monoclonal antibodies to digitalis. It
is hoped such antibodies will be useful in treating digitoxicity.
There is a human monoclonal antibody derived from a human-mouse
fusion to tetanus that neutralizes the toxin in mice. Antibodies
such as these could be used therapeutically to neutralize toxins
in vivo. Even if the antibody were not directed to the active
site of the toxin or to the site that is bound to a cellular
receptor, such an antibody might be used to eliminate the toxin
from the circulation. If the immune complex of antibody coupled
to toxin proved not to be harmful to the reticuloendothelial
system, such complexes could be formed and removed in vivo by
macrophages. If the complex did injure macrophages, the
extracorporeal removal of the toxin by antibody would be necessary
(Diamond, 1983, personal communication). In addition, it might be
possible to generate monoclonal antibodies not only to the toxin,
but also to the cellular receptor for the toxin. Such antibodies

might be used to prevent a toxin from injuring its target cell until it was excreted or metabolized into a nontoxic form.

In vitro immunization of human lymphocytes may make possible the production of human antibodies and avoid the possible complications of using murine antibodies in people. Diamond expressed that the greatest difficulty will be in generating an antibody response to poor immunogens, but with chemical alteration of the toxin, or coupling the toxin to an immunogenic carrier, or with manipulation of the animals' suppressor cell response, she indicated that it should be possible to generate monoclonal antibodies to most chemical and biological toxins (Diamond, 1983, personal communication).

2.   In Vitro Systems for Toxicological Evaluation of Chemicals

Apart from subcellular fractions (e.g., mitochondria, lysosomes, nuclei, etc.) there are four different types of in vitro systems of interest that are characterized by a different level of complexity: whole embryo, organ, tissue, and cell cultures. Recently these systems have become widely used for toxicological evaluations (34,63).

Whole embryo cultures. It is possible to investigate adverse developmental effects of chemicals and metabolites on preimplantation embryonic stages of development. In vitro treated preimplanted embryos can be transferred post-treatment into the uteri of foster mothers to detect post-implantation and/or post-natal induced developmental and/or functional alterations (44,45). Chemicals which can alter the processes of cellular and morphogenetic differentiation might have a potential to produce developmental toxicity. With post-implantation embryo culture techniques it is possible to follow morphogenetic cell movements, organizational patterning and cell differentiation. Thus, increased understanding of the underlying processes will aid in extrapolating laboratory studies to humans, estimating risks and analyzing human health hazards associated with exposure to teratogens.

A number of investigators have begun establishing in vitro systems that may serve as teratogenicity screens (34). Unfortunately, the methods for validating such systems are as yet undefined. The wide variety of in vitro systems used to study normal and abnormal development has a correspondingly large number of endpoints that can be used to evaluate adverse developmental effects. These range from such endpoints as blastulations of treated preimplantation embryos, implantation in foster uteri, somite number and neural tube closure in cultured post-implanted whole embryos, to characteristics of growth and attachment in cell culture of the latter and live birth of normal neonates from the

former. Consequently, to define particular endpoints as
"acceptable" is a difficult task, given the current state of our
knowledge (34).

Kimmel and colleagues postulated that the correct basis for
dose selection in vitro and its relationship to in vivo exposure
is dependent on the specific system being considered (34). Once
data on dose-response relationships are collected, defining a
"most appropriate" basis for dosing might be possible, although
determination of such a basis might still be difficult. For
example, how does one relate dosing on the basis of milligrams/
kilogram in the intact animal with micrograms/milliliter in a cell
culture system? Any dosing design that would "predict the
findings of generally accepted safety evaluations of
teratogenicity" would be relevant. This issue may be better
resolved when comparisons can be made of dose-response data for
the same compound using different methods.

Organ cultures. With organ cultures it is possible to follow
in vitro development of an organ fragment. The ability of in
vitro differentiation has been proven with several isolated
embryonal organs including vertebra, tibia, liver, pancreas,
intestines, skin, thyroid, and gonads. Moreover, it is known that
organ explants can retain at least some of the functions they
perform in vivo, such as secretions of specific products (e.g.,
the liver secretes glycogen, the pancreas secretes glucagon, and
the gonads secrete sexual hormones). De Ritis and colleagues (10)
have shown that differentiation and maturation of small intestinal
mucosa from a rat fetus may take place in vitro in a way
comparable to what happens in vivo. This system was tested by
detecting the effect of wheat proteins and peptide fractions (11)
and the results were compatible with those of in vivo studies. It
is hoped that further investigation with the organ culture
system(s) will provide better understanding of the mechanism(s)
involved.

Tissue cultures. Tissue cultures represent a transition
system between organ cultures, in which an organized structure and
a complete growth control are better maintained, and cell
cultures, where this control is lost. These systems may give
information on regulation of intercellular contacts rather than on
tissue organization.

Cell cultures. During recent years cell cultures have been
increasingly used for toxicological investigations (63,74).
Culture techniques have been substantially improved (21) and
applied to almost any type of cells. Cells can be cultured in
monolayer or suspension. A number of cell types are capable of
surviving and dividing in vitro. Available cell culture systems
include: (a) primary cell cultures, obtained by culturing for more

than 24 h dispersed cells from tissues or organs taken directly from organisms; (b) cell lines, subcultures derived from a primary culture; they may be diploid, established or clonal; (c) clonal cell lines, derived from the mitosis of a single cell; and (d) a cell strain, obtained from a primary culture, diploid or established cell line by selecting a small number of cells that have a common biological characteristic which is useful as a marker (e.g., a specific enzyme or high sensitivity to a specific virus). The biological marker must persist during subsequent culturing (22). Embryonic and tumor cells are more easily cultured than cells from normal adult tissues because they have a higher growth capability and adapt more readily to variations of external factors. Recently, a human embryonic palatal mesenchyme (HEPM) cell line was established (77). These cells can grow in either a serum-free hormone-supplemented medium or a serum-containing medium. The growth of these cells is quite rapid in culture and inhibited in a dose-dependent manner by most teratogens thus far tested, such as dexamethasone. These cells are highly sensitive to a variety of DNA synthetic and mitotic inhibitors. Pratt and colleagues (55) proposed the use of these cells for screening assays for teratogens. Although cell cultures, and establishment of cell lines in particular, imply some loss of differentiation, there are numerous examples of cell types displaying highly specialized in vitro biological activity that is characteristic of their original tissues and/or organs. For example, brain cells and neuronal established cells (e.g., neuroblastoma C 1300) display in vitro electrophysiological activities (25,59) and the presence of characteristic enzymes (e.g., acetylcholinesterase and tyrosine-hydroxylase).

One of the most widely used parameters for assessing the in vitro toxicity of chemical substances is observing effects on cell viability. Baur and coworkers (2) suggested stimulating cellular respiration with succinate as a sensitive test for damaged cell membranes. Recently, a rapid screening test on hamster CHO-K1 cells was described, which measured labile energy metabolites, such as AMP, ATP and NADH in cells by isotachophoresis (76). Changes were noted at 1/1000 of the lethal dose. Cell growth and multiplication are other important parameters in evaluating the effects of toxicants in culture. Growth and multiplication in tissue cultures can be quantitatively evaluated through changes in cell number, total protein and DNA content. Analysis of changes in cell number, coupled with cell mass or cell size determinations, will aid in determining toxic compounds which affect cell division and/or multiplication. Plating efficiency of a cell population may also provide useful information. Cytological studies accompanied by stereological analysis should determine toxic substances damaged at cellular and subcellular levels (46).

The need for representative metabolism in an in vitro test system is generally viewed as a requirement. It is recognized that in a battery or tiered approach not every test may need to meet this requirement, since the time and resources required might be prohibitive. Nevertheless, the overall evaluation of the screen must account for aspects of metabolism if this evaluation is to have any utility in prioritization for further testing.

Kimmel and colleagues (34) proposed that all test compounds should be tested both with and without a metabolic activating system and in those test systems where such activity was not endogenous, an exogenous source should be added. A suggested modification of this approach was to include representative metabolic capabilities only in the case where no effect was seen at high concentrations or at the limit of solubility. It should be noted that while structure-activity relationships have not proven successful in determining teratogenic activity, consideration of the structure of a compound can be helpful in determining the type of metabolic activating system that may be required. For example, while an oxidative environment may very well be the most appropriate metabolic system for certain compounds, other compounds may require reductive reactions. It is also possible that activation is mediated through a multistep reaction that requires steps either before or after the oxidative activity. In both cases, the teratogenic potential would be missed by addition of only the standard S-9 fraction. It was also pointed out that cellular subfractions may overrepresent the production of reactive intermediates while underrepresenting reactions that tend to detoxify the reactive intermediates. In vitro systems that maintain their endogenous capacity for metabolic activation and detoxification are important for addressing some of these concerns. For example, some systems are amenable to the use of serum from animals that have been exposed in vivo to particular agents, allowing endogenous metabolism to be affected prior to exposure of the in vitro subject (34).

3.   Plants in Biotechnology

Tissue culture techniques for growing plant cells have been very successfully developed. The techniques of culture in vitro were extended to many species and, aided by advances in the knowledge of plant hormones that were made in part through use of tissue culture, regeneration of plants from cultured tissues was achieved in the late 1950's. The first application of these developments was to the clonal multiplication of plants, which proved more efficient than conventional methods of asexual plant propagation.

In the 1960's, research in plant cell and tissue culture produced a number of achievements that individually represented

significant technical advances and refinements (60) which could be
utilized for toxicological evaluation.

TOXICITY AND SAFETY EVALUATION WITH PLANTS

Although plants have historically been important in the
development of genetic principles and the illustration of the
hazards of ionizing radiation, they have not been adequately
utilized when evaluating the toxicity of environmental chemicals.
For testing under field conditions (in situ monitoring), few test
organisms offer the advantages provided by plants. Because of
plants' ability to metabolically activate several classes of
promutagens and in view of their role in the human food chain,
Sandhu (1983, personal communication) indicated that plant systems
warrant consideration in toxicity and safety evaluation.

Plant systems show promise for immediate use in short-term
bioassays used in the laboratory for toxicity evaluation of
specific chemicals or chemical mixtures. These bioassays can be
organized into cytogenetic and specific locus systems.

1.   Cytogenetic Test Systems

These systems are more commonly employed in the laboratory to
evaluate the clastogenic potential of environmental toxicants.
Somatic tissues, usually root tips, emeristematic cells, or
germinal tissues such as pollen mother cells (PMC) may be used to
determine chemical toxic effects. Root tips are relatively
inexpensive to obtain in large numbers, and they handle easily and
display a wide range of sensitivity to low chemical concentration.
The species most commonly used for this assay are broad beans
(vicia faba), onions (Allium cepa), barley (Hordeum vulgara), corn
(zea mays), and Tradescantia paludosa.

The use of PMC in cytogenetics offers the advantage of
evaluating potential heritability of genetic lesions. PMC as a
test system has been developed so far in these plants species:
Tradescantia paludosa (2n=12), Vicia faba (2n=12), and Hordeum
vulgara (2n=14).

2.   Specific Locus Test Systems

Point mutations, as in the case of chromosomal aberrations,
could be detected either in somatic cells or in germ cells. The
most commonly used assays for detecting environmental mutagens in
plant somatic cells are the Tradescantia stamen hair assay, the
Arabidopsis multilocus assay, the Hordeum chlorophyll deficiency
assay, and the Glycine leaf spot assay. The most often used germ
cell gene mutation assays in plants utilize waxy locus in cereals

(especially maize and barley) and adh locus in maize. The use of
pollen grains for mutagenesis offers unique advantages: 1) The
unit of measurement is a pollen grain, and millions are available,
thus the resolving power of the test system is equivalent to a
microbial test system, but it retains the advantages of eukaryotic
organization; 2) the endpoints are specific gene products; and 3)
the functional unit is a haploid germinal cell indicating
potential of heritable damage.

Seed germination and anchorage of a plant in the soil or
other medium entails a series of complex processes. Any
toxicant(s) that may interfere with these vital developmental
processes will affect the survival of the plant. These criteria
have been used for determining the toxicological potentials of
pesticides, heavy metals, toxic chemicals and allopathic
substances (19,58).

3.   In situ Monitoring of Ecosystem Pollutants

Plants such as tobacco, petunia species, and pinto beans have
been used to monitor the levels of photochemical oxidants,
especially ozone hydrogen, fluoride and sulfates. In most of
these studies, specific toxic effects were designated to a
particular plant species. For example, Bel-W3 strain of tobacco
(Nicotiana tobaccum), which is sensitive to ozone, responds to
increasing levels of ozone by producing more flecks or lesions.
The lesion size was positively related to ozone level (18).

A number of plant species such as lichens and mosses among
the lower eukaryotes and corn, barley and citrus among the higher
eukaryotes, have been employed for detecting the levels of heavy
metals such as zinc, lead, cadmium, nickel, copper and magnesium.
Hyacinth tops can be used for monitoring river streams when
cadmium levels are 0.10 mg/l or greater.

In summary, plant bioassays offer the advantages of
simplicity, low cost, short study time, reliability and
sensitivity. A strong concordance of response has been observed
between animal bioassays and mammalian cells in culture on the one
hand and plant bioassays on the other. Plant bioassays provide
unique opportunities for monitoring acute as well as chronic toxic
effects under in situ conditions.

CONCLUSION

The immense scale of the Alternate Methods for Integrated
Toxicological Evaluation is well recognized. It is hoped that the
approaches discussed here will not be taken as the only AMITE and
will not be considered as proposals for eliminating in vivo test

methods. With AMITE, evaluation can be made at the end of each
test (or stage) to determine whether the data available are
significant enough to reach a definite conclusion or whether to
recommend further tests. We believe that AMITE could lead to a
more constructive and efficient utilization of our knowledge and
resources.

ACKNOWLEDGEMENTS

I should like to express my sincere gratitude to the
following colleagues: Dr. Robert P. Bolender, University of
Washington, Seattle, Washington, USA; Dr. Betty Diamond, Albert
Einstein College of Medicine, New York, USA; Dr. Jack Peter Green,
The Mount Sinai School of Medicine of the City University of New
York, New York, USA; Dr. Robert Pratt, National Institute of
Environmental Health Sciences, USA; Dr. Shahbeg Sandhu,
Environmental Protection Agency, USA; and Dr. Vittorio Silano,
Istituto Superiore di Sanita, Italy. I wish to thank them for
their friendly cooperation and their kind assistance in providing
scientific information, which contributed greatly to this
manuscript.

REFERENCES

1. Asher, I.M., and C. Zervos, Eds. (1977) Structural Correlates
   of Carcinogenesis and Mutagenesis. A Guide to Testing
   Priorities?, Proceedings of 2nd FDA Office of Science Summer
   Symposium, 31 August - 2 September, 1977.
2. Baur, H., S. Kasperek, and E. Pfaff (1975) Criteria of
   viability of isolated liver cells. Z. Physiol. Chem.
   356:827-838.
3. Blouin, A., R.P. Bolender, and E.R. Weibel (1977) Distribution
   of organelles and membranes between hepatocytes and
   nonhepatocytes in the rat liver parenchyma. A stereological
   study. J. Cell Biol. 72(2):441-445.
4. Bolender, R.P. (1974) Stereological analysis of the guinea pig
   pancreas. I. Analytical model and quantitative description
   of nonstimulated pancreatic exocrine cells. J. Cell
   Biol. 61:269-289.
5. Bolender, R.P. (1981) Stereology: Applications to
   pharmacology. Ann. Rev. Pharmacol. Toxicol. 21:549-573.
6. Bolender, R.P. (1982) Stereology and its uses in cell biology.
   Ann. N.Y. Acad. Sci 383:1-16.
7. Bolender, R.P., D. Baumgartner, G. Losa, D. Muellener and
   E.R. Weibel (1978) Integrated stereological and biochemical
   studies on hepatocytic membranes. I. Methods and membrane
   recoveries. J. Cell Biol. 77:565-583.

8. Cammarata, A., and S.T. Yau (1970) Predictability of
    correlations between in vitro tetracycline potencies and
    substituent indices. J. Med. Chem. 13:93-97.
9. Chu, K.C., R.J. Feldman, M.B. Shapiro, G.F. Hazard, and
    R.I. Geran (1975) Pattern recognition and structure-activity
    relationship studies. Computer-assisted prediction of
    antitumor activity in structurally diverse drugs in an
    experimental mouse brain tumor system. J. Med. Chem.
    18(6):539-545.
10. De Ritis, G., Z.M. Falchuk, and J.S. Trier (1975)
    Differentiation and maturation of cultured fetal rat jejunum.
    Develop. Biol. 45:304-313.
11. De Ritis, G., P. Occorsio, S. Auricchio, F. Gramenzi,
    G. Morisi, and V. Silano (1979) Toxicity of wheat flour
    proteins and protein-derived peptides for in vitro developing
    intestine from rat fetus. Pediat. Res. 13:1255-1261.
12. Dettoff, R.T., and F.N. Rhines, Eds. (1968) Quantitative
    Microscopy, McGraw-Hill Book Company, New York.
13. Diamond, B., and M.D. Scharff (1982) Monoclonal antibodies.
    J. Am. Med. Assoc. 249:3165-3169.
14. Dove, S., R. Franke, O.L. Mndshojan, W.A. Schkuljev, and
    L.W. Chashakjan (1979) Discriminant-analytical investigation
    on the structural dependence of hyperglycemic and
    hypoglycemic activity in a series of substituted o-
    toluenesulfonylthioureas and o-toluene-sulfonylureas. J.
    Med. Chem. 22(1):90-95.
15. Dunn, W.J., III, and S. Wold (1978) A structure-
    carcinogenicity study of 4-nitroquinoline 1-oxides using the
    SIMCA method of pattern recognition. J. Med. Chem.
    21:1001-1007.
16. Dunn, W.J., III, M.J. Greenberg, and S.S. Callejas (1976) Use
    of cluster analysis in the development of structure-activity
    relations for antitumor triazenes. J. Med. Chem.
    19:1299-1301.
17. Dunn, W.J., III, S. Wold, and Y.C. Martin (1978) Structure-
    activity study of β-adrenergic agents using the SIMCA method
    of pattern recognition. J. Med. Chem. 21:922-930.
18. EHP (1978) Higher plants as monitors of environmental
    mutagens--workshop, January 1978. Environm. Health
    Persp. 27:1-206.
19. EHP (1981) Pollen systems to detect biological activity of
    environmental pollutants. Proceedings of conference.
    Environm. Health Persp. 37:1-200.
20. Elias, P.S. (1978) General Guidelines for the Toxicological
    Evaluation of Chemical Substances, Commission of the European
    Communities, Doc. No. V/F/1/78/26, Luxembourg, April 1978.
21. Feder, J., and W.R. Tolbert (1983) The large-scale cultivation
    of mammalian cells. Scientific American 248(1):36-43.
22. Fedoroff, S. (1977) Primary cultures, cell lines and cell
    strains: Terminology and characteristics, In Cell, Tissue and

Organ Cultures in Neurobiology, S. Fedoroff and L. Hertz, Eds., Academic Press, New York, pp. 265.

23. Free, S.M., Jr., and J.W. Wilson (1964) A mathematical contribution to structure-activity studies. J. Med. Chem. 7:395-399.

24. Fujita, T., and T. Ban (1971) Structure-activity study of phenethylamines as substrates of biosynthetic enzymes of sympathetic transmitters. J. Med. Chem. 14:148-152.

25. Giller, E.L., X.O. Breakefield, C.N. Christian, E.A. Neale, and P.G. Nelson (1975) Expression of neuronal characteristics in culture: Some pros and cons of primary cultures and continuous cell lines, In Golgi Centennial Symposium, M. Santini, Ed., Raven Press, New York, p. 603.

26. Green, J.P., and H. Weinstein (1981) Quantum mechanics can account for the affinities of drugs and receptors. The Sciences (September 1981) 21:27-29.

27. Griesemer, R. (1981) Whole animal methods for toxicity testing, In Trends in Bioassay Methodology, In Vivo, In Vitro and Mathematical Approaches Symposium, Organized by National Institutes of Health and National Toxicology Program, 18-20 February 1981. NIH Publication No. 82-2382, Washington, D.C.

28. Hansch, C. (1971) Quantitative structure-activity relationships in drug design. In Drug Design (1), E.J. Ariens, Ed., Volume 11, part 1 of Medicinal Chemistry: A Series of Monographs, George DeStevens, Gen. Ed., Academic Press, New York, pp. 271-337.

29. Hansch, C., A. Leo, S.H. Unger, K.H. Kim, D. Nikaitani, and E. Lien (1973) "Aromatic" substituent constants for structure-activity correlations. J. Med. Chem. 16:1207-1216.

30. Hathway, D.E., and G.F. Kolar (1980) Mechanisms of reaction between ultimate chemical carcinogens and nucleic acid. Chem. Soc. Rev. 9(2):241-264.

31. Hirota, N., and G.M. Williams (1979) The sensitivity and heterogeneity of histochemical markers for altered foci involved in liver carcinogenesis. Amer. J. Pathology 95(2):317-324.

32. Iozzo, R.V., R.P. Bolender, and T.N. Wight (1982) Proteoglycan changes in the intercellular matrix of human colon carcinoma. An integrated biochemical and stereological analysis. Lab. Invest. 47(2):124-138.

33. IRPTC (1982) International Register of Potentially Toxic Chemicals, Bulletin, Vol. 5, No. 2, October 1982.

34. Kimmel, G.L., K. Smith, D.M. Kochhar, and R.M. Pratt (1982) Overview of in vitro teratogenicity testing: Aspects of validation and application to screening. Teratog. Carcinog. Mutagen. 2:221-229.

35. Kowalski, B.R., and C.F. Bender (1974) The application of pattern recognition to screening prospective anticancer drugs. Adenocarcinoma 755 biological activity test. J. Amer. Chem. Soc. 96:916-918.

36. Losa, G., E.R. Weibel, and R.P. Bolender (1978) Integrated stereological and biochemical studies on hepatocytic membranes. III. Relative surface of endoplasmic reticulum membranes in microsomal fractions estimated on freeze fracture preparations. J. Cell Biol. 78:289-308.

37. Martin, Y.C. (1974) Proceedings: Extrathermodynamic approach to drug design (supp.), In Cancer Chemother. Rep. part 2, 4(4):35-36.

38. Martin, Y.C. (1978) Quantitative drug design, a critical introduction, In Medicinal Research Series, Volume 8, G.L. Grunewald, Ed., Marcel Dekker, Inc., New York.

39. Martin, Y.C., J.B. Holland, C.H. Jarboe, and N. Plotnikov (1974) Discriminant analysis of the relationship between physical properties and the inhibition of monoamine oxidase by aminotetralins and aminoindans. J. Med. Chem. 17:409-413.

40. Maugh, T.H. (1978) Chemicals: How many are there? Science 199:152.

41. Maugh, T.H. (1983) Chemicals: How many are there? Science 220:293.

42. Meyler, L., and A. Herscheimer, Eds. (1972) Side Effects of Drugs, Volume 7: A Survey of Unwanted Effects of Drugs, Reported in 1968-1971 Excerpta Medica Amsterdam.

43. Miller, J.A. (1970) Carcinogenesis by chemicals: An overview. The G.H.A. Clowes Memorial Lecture. Cancer Res. 30:559-576.

44. Moustafa, L.A. (1976) New observations on rabbit blastocysts after in vitro exposure to thalidomide--some correlated SEM and TEM studies, In Scanning Electron Microscopy 1976, part VI, G. Johari, and R.P. Becker, Eds.., IIT Research Institute, Chicago, pp. 385-391.

45. Moustafa, L.A., and J. Hahn (1978) Untersuchungen über die Brauchbarkeit der Kultivierung von befruchteten Eizellen in Mini-Pailletten. Zuchthygiene 13:61-67.

46. Nardone, R.M. (1977) Toxicity testing in vitro, In Growth, Nutrition and Metabolism of Cells in Culture, Vol. 3., G.H. Rothblat, and V.J. Cristofalo, Eds., Academic Press, New York, p. 471.

47. NAS (1975) Principles for Evaluating Chemicals in the Environment, National Academy of Sciences, Washington, D.C.

48. Nixon, G.A., C.A. Tyson, and W.C. Wertz (1975) Interspecies comparisons of skin irritancy. Toxicol. Appl. Pharmacol. 31(3):481-490.

49. NTP (1982) National Toxicology Program: Fiscal Year 1983 Annual Plan, US Department of Health and Human Services, Public Health Service.

50. Osman, R., H. Weinstein, and J.P. Green (1979) Parameters and methods quantitative structure-activity relationships, In Computer-Assisted Drug Design, E.C. Olson, and R.E. Christofferson, Eds., ACS Symposium Series No. 112, pp. 21-77.

51. Osman, R., H. Weinstein, and S. Topiol (1981) Models for
    active sites of metalloenzymes. II. Interactions with a
    model substrate. Ann. N.Y. Acad. Sci. 367:356-369.
52. Pieri, C., I.Zs. Nagy, C. Mazzufferi, and C. Giuli (1975a) The
    aging of rat liver as revealed by electron microscopy.
    Morphometry - I. Basic parameters. Exp. Gerontol.
    10(5):291-304.
53. Pieri, C., I.Zs. Nagy, G. Mazzufferi, and C. Giuli (1975b) The
    aging of rat liver as revealed by electron microscopy. II.
    Parameters of regenerated old liver. Exp. Gerontol.
    10(6):341-349.
54. Prakash, G., and E.M. Hodnett (1978) Discriminant analysis and
    structure-activity relationships. 1. Naphthoquinones. J.
    Med. Chem. 21:369-374.
55. Pratt, R.M., R.I. Grove, and W.D. Willis (1982) Prescreening
    for environmental teratogens using cultured mesenchymal cells
    from the human embryonic palate. Teratog. Carcinog. Mutagen.
    2:313-318.
56. Price, C.C. (1958) Fundamental mechanisms of alkylation. Ann.
    N.Y. Acad. Sci. 68:663-668.
57. Price, C.C., G.M. Gaucher, P. Koneru, R. Shibakawa, J.R. Sowa,
    and M. Yamaguchi (1969) Mechanism of action of alkylating
    agents. Ann. N.Y. Acad. Sci. 163:593-600.
58. Sandhu, S. (1983) Monitoring for mutagenicity with plants, In
    Methods for the Detection of Environmental Mutagens,
    Carcinogens, and Teratogens in Developing Countries,
    A. Massoud, R. Tica, and M. Waters, Eds., Plenum Press, New
    York (in press).
59. Sato, G., Ed. (1973) Tissue Culture of the Nervous System -
    Current Topics in Neurobiology, Volume 1, Plenum Press, New
    York.
60. Science (1983) Biotechnology issue. Science
    219(4585):611-747.
61. Sharrat, M. (1977) Objective Evaluation from Animal Data of
    the Risks to Human Health from Chemical Agents, Commission of
    the European Communities, Doc. No. 2144/77e, Luxembourg, June
    1977.
62. Soltzberg, L.J., and C.L. Wilkins (1977) Molecular transforms:
    A potential tool for structure-activity studies. J. Amer.
    Chem. Soc. 99:439-443.
63. Stammati, A.P., V. Silano, and F. Zucco (1981) Toxicology
    investigations with cell culture systems. Toxicology
    20:91-153.
64. Stewart, R.D., J.E. Peterson, P.E. Newton, C.L. Lake,
    M.J. Hosko, A.J. Lebrun, and G.M. Lawton (1974) Experimental
    human exposure to propylene glycol dinitrate. Toxicol. Appl.
    Pharmacol. 30(3):377-395.
65. Stuper, A.J., and P.C. Jurs (1975) Classification of
    psychotropic drugs as sedatives or tranquilizers using

pattern recognition techniques. <u>J</u>. <u>Amer</u>. <u>Chem</u>. <u>Soc</u>. 97(1):182–187.

66. Szabo, K.T., M.E. DiFebbo, Y.J. Kang, A.K. Palmer, and R.L. Brent (1975) Comparative embryotoxicity and teratogenicity of various tranquilizing agents in mice, rats, rabbits, and rhesus monkeys. Fourteenth Annual Meeting of the Society of Toxicology, Williamsburg, Virginia, 9–13 March 1975. Abstract in: <u>Toxicol</u>. <u>Appl</u>. <u>Pharmacol</u>. 33:124.

67. Ting, K.H., R.C.T. Lee, G.W.A. Milne, H. Shapiro, and A.M. Gaurino (1973) Applications of artificial intelligence: Relationships between mass spectra and pharmacological activity of drugs. <u>Science</u> 180:417–420.

68. Underwood, E.E. (1970) <u>Quantitative Stereology</u>, Addison–Wesley Publishing Company, Reading, Massachusetts.

69. Weibel, E.R. (1979) <u>Stereological Methods</u>. Volume 1: Practical Methods for Biological Morphometry, Academic Press, London.

70. Weibel, E.R. (1980) <u>Stereological Methods</u>. Volume 2: Theoretical Foundations, Academic Press, London.

71. Weinstein, H., R. Osman, and J.P. Green (1979) The molecular basis of structure–activity relationships: Quantum chemical recognition mechanisms in drug–receptor interactions, In <u>Computer–Assisted Drug Design</u>, ACS Symposium Series 112, E.O. Olson, and R.E. Christofferson, Eds., American Chemical Society, Washington, D.C., pp. 161–187.

72. Weinstein, H., S. Topiol, and R. Osman (1981a) On the relation between charge redistribution and intermolecular forces in models for molecular interactions in biology, In <u>Intermolecular Forces</u>, B. Pullman, Ed., D. Reidel Publishing Company, pp. 383–396.

73. Weinstein, H., R. Osman, S. Topiol, and J.P. Green (1981b) Quantum chemical studies on molecular determinants for drug action. <u>Ann</u>. <u>N.Y</u>. <u>Acad</u>. <u>Sci</u>. 367:434–451.

74. Weisburger, J.H., and G.M. Williams (1981) Carcinogen testing: Current problems and new approaches. <u>Science</u> 214:401–407.

75. Williams, M.A. (1977) Quantitative methods in biology, In <u>Practical Methods in Electron Microscopy</u>, Vol. 6, A.M. Glauert, Ed., North Holland Publishing Company, Amsterdam.

76. Wininger, M.T., J.M. Lavoie, and W.D. Ross (1982) <u>In</u> <u>vitro</u> sub–lethal toxicity indicated by changes in cellular energy metabolite levels. 33rd Annual Meeting of the Tissue Culture Association, 6–10 June 1982, San Diego.

77. Yoneda, T., and R.M. Pratt (1981) Mesenchymal cells from the human embryonic palate are highly responsive to EGF. <u>Science</u> 213:563–565.

SELECTED POSTER ABSTRACTS FROM THE INTERNATIONAL WORKSHOP

ON ENVIRONMENTAL MUTAGENESIS, CARCINOGENESIS

AND TERATOGENESIS

Shanghai

May 25 – June 1, 1983

LYMPHOCYTE CHROMOSOME ABERRATIONS IN PERSONS
EXPOSED TO LOW LEVELS OF RADIATION

S.Y. Cao, Z. Y. Zhou, C. F. Yu, C. L. Wang, and Y. X. Liu

North China Institute of Radiation Protection, Taiyuan

The dose-effect response of chromosome aberrations induced by low levels of chronic radiation in lymphocytes of 456 persons was studied.

1. The larger the level of background radiation, the higher the frequency of spontaneous chromosome-type aberrations. In the farmer group (53) and worker group (40), the cells with chromosome-type aberrations were 0.076% and 0.33% respectively. There were certain differences in background exposure between the village (17 uR/hr) and the town (40–60 uR/hr).

2. Three groups of exposed workers showed statistically significant higher yields of chromosome-type aberrations than the control group. The yield was 0.91% (in 71 uranium miners), 0.61% (in 97 reactor plant workers), 0.83% (in 167 radiologists) and 0.31% (in 86 control workers) respectively.

3. Correlation and regression analyses demonstrated that the incidence of acentric fragments and Cu cells in 80 nuclear industry workers was positively correlated with individual accumulated doses. This could be described by a linear dose response model, $Y = a + bX$.

811

CYTOGENETIC FOLLOW-UP STUDIES IN PERSONS ACCIDENTALLY
EXPOSED TO ^{60}Co γ-RAYS,
10 YEARS AFTER EXPOSURE

C.Z. Jin, J. Yang, X.L. Liu, and Y.N. Mu

Institute of Radiation Medicine, Beijing

We have periodically analyzed the chromosome aberrations in
peripheral blood lymphocytes in 8 men exposed to γ-radiation
during an accident in a cobalt-60γ-ray therapeutic facility which
occurred in Wuhan in Dec. 1972. Physical doses were estimated to
be within a range of 5-245 rads. In previous studies, we took the
unstable aberrations as the indicators for cytogenetic changes.
However, the frequencies of abnormalities dropped progressively as
the time went on. Recently, by using microscopic and karyotypic
analysis, we measured the chromosome aberrations from 8 persons 10
years after the accident. The FPG technique was also used to
determine the cell division cycle and to measure the rate of
sister chromatid exchanges SCE. The following results were
obtained:

1. 20 aberrant cells were scored in 720 metaphases and a
total of 16 cells contained stable types. Most of the persistent
aberrations were translocations and deletions. The highest
frequencies of aberrations were observed in 3 cases who received
the highest radiation doses.

2. The distribution of metaphase in first, second and third
divisions in cultures were determined. As compared with the
control, no delay in cell cycle of lymphocytes of the exposed
cases was observed.

3. The analysis of SCE in blood lymphocytes of the exposed
persons was performed. It failed to show any increase in SCE rate
in comparison with the control.

STUDIES ON MICRONUCLEUS TEST BY HUMAN SKIN PUNCTURE

K.X. Xue, Y.Y. Cai, B.Y. Ding, X.J. Sun, P. Zhou,
G.J. Ma, and S. Wang

Cancer Institute of Jiangsu Province, Nangjing

Our work on scoring the frequency of micronucleus (MNF)
directly in 1-2 drops of human blood obtained from skin puncture
was reported in a previous paper. The results of further study
are given as follows:

1. MNF in 117 health donors (male 58, female 59) were in the range of 0-1%, with the mean of 0.14%, and about 20.5% of the samples with ≥ 0.5%.

2. The MNF in 100 workers dealing with radioactive materials was 0-8%, with a mean of 1.3%; the rate of detected positive was 69%, and 28% of them were ≥ 1%. The MNF in the exposed group differed significantly from the normal control.

3. Surveillance of workers exposed to technical PBNA (phenyl B-Naphthylamine), an antioxidant D incorporated into rubber, and a known animal carcinogen) was done and the result was positive.

COMPARISON OF THE RESPONSES OF THE MICRONUCLEUS
TEST IN DIFFERENT STRAINS OF MICE

N.J. Huang

National Institute for the Control of Pharmaceutical
and Biological Products, Beijing

The purpose of this work was to compare susceptibility to the micronucleus induction and the care needed to maintain six strains of mice, to determine which one is the most suitable model for laboratory tests.

One dose of 60 mg/kg cyclophosphamide (CP) was given by i.p. injection to every mouse in groups of inbred male NIH, C3H, 615, C57BL, LACA and hybrid male KM (Kun-Ming) strains. Saline served as placebo in the control. A total of 120 mice were studied.

The frequencies of micronuclei in polychromatic erythrocyte (MNPCE) in 6 control groups were all below 4%. However, the frequencies of MNPCE in the CP treated groups were all above 18%, i.e. 36.80, 28.04, 24.95, 26.30, 23.39, 18.45 in NIH, KM, C3H, 615, C57BL and LACA respectively. The difference in frequencies of MNPCE between NIH and LACA was statistically significant. In addition, the micronucleus in normochromatic erythrocyte (MNNCE) and the total MN frequency (MNPCE+MNNCE) were also the highest in NIH strain (9.18%, 26.98%) and the lowest in LACA strain (1.12%, 15.84%).

Though the frequencies were not low, the range of variation of MNPCE in KM strain mice was wide, 6-51%. This is the reason why KM strain was not considered suitable for micronucleus test.

The results indicate that there are remarkable differences in the induction of MNPCE and MNNCE among 6 strains tested. As the number of micronuclei that can be induced is the highest and the

animal husbandry is convenient, the NIH strain is ranked at the
top priority for further studies.

STUDY ON THE MICRONUCLEUS IN NUCLEATED ERYTHROCYTES IN
VERTEBRATES AND ITS APPLICATION TO MONITORING
ENVIRONMENTAL POLLUTION

C.J. Zhang, Y.N. Cai, L.J. Chen, and B.Z. Liu

Department of Biology, Hua Chung Normal University, Wuhan

204 well developed individuals of vertebrates belonging to 14
families and/or species (Pisces, Amphibia, Reptilia and Aves) were
obtained from apparently non-polluted habitats. The background
mean frequency of micronuclei (MNF) was 0.04%, but it increased to
0.94% and 1.33%, respectively, in Elaphe rufodorsata (Reptilia)
and Columbia livia domestica (Aves) after treatment with
antioxidant D (N-phenyl-2-naphthylamine). The results suggest
that MNF scoring can serve as a simple indicator for environmental
monitoring. It has the advantage of convenient sample collecting
in organisms exposed to a variety of atmospheric and aquatic
contaminations.

CHROMOSOME ABERRATIONS IN HUMAN LYMPHOCYTES INDUCED BY TRITIUM

X.Z. Zhang, L.C. Dong, F. Qiao, and X.Z. Zhao

Institute of Atomic Energy, Academic Sinica, Beijing

Human lymphocytes at various phases of the cell cycle were exposed to tritium-containing compounds in vitro. A dose-response relationship was observed.

Chromatid-type aberrations were induced by ^{3}H-TdR or tritiated water (HTO) in the S and G_2 phases of dividing lymphocytes. The relationship between the dose (D) and the yield of chromatid breaks per cell (Y) gave the best fit to the linear model. $Y3_{H-TdR} = (2.62 \pm 0.07) \cdot 10^{-1}$ D and $Y_{HTO} = (2.49 \pm 0.03) \cdot 10^{-2}$ D. The results show that the yield of the chromatid breaks induced by ^{3}H-TdR incorporated into DNA is higher than that of HTO distributed in the cell by 10 fold.

Chromosomal-type aberrations were induced by HTO or ^{60}Co-γ-ray irradiation in the G_0 phase of lymphocytes. The data of dicentrics plus centric rings gave the best fit to the linear-quadratic model for both types of radiation: $Y_{HTO} = (5.79 \pm 1.92) \cdot 10^{-4}$ D + $(4.61 \pm 1.10) \cdot 10^{-6}$ D^2, $Y60_{Co-\gamma-ray} = (0.52 \pm 0.41) \cdot 10^{-4}$ D + $(3.81 \pm 0.24) \cdot 10^{-6} + D^2$. The RBE value of dicentrics plus centric rings for H-β-ray irradiation versus that of ^{60}Co-γ-rays is not a constant. The RBE values decrease as the dose increases in the range of the experimental dose; it varies from 3.7 to 1.4.

THE EFFECT OF GOSSYPOL ON HUMAN LYMPHOCYTES IN VITRO:
CELL KINETICS, TRADITIONAL CHROMOSOME BREAKAGE,
MICRONUCLEI, AND SISTER CHROMATID EXCHANGE

Y.C. Tsui[1], M.R. Creasy[2], and M.A. Hulten[2]

[1]National Research Institute for Family Planning,
Beijing

[2]West Midlands Regional Cytogenetics Laboratory,
East Birmingham Hospital, Birmingham, U.K.

The male antifertility agent gossypol at test concentrations
of 4, 20, 30 and 40 µg/ml appears to have very little adverse
effect on lymphocyte chromosomes in vitro. In conventionally
stained preparations there was no difference in the frequency of
chromatid and chromosome aberrations between gossypol treated
cultures (1.6%) and controls (1.2%). The incidence of micronuclei
was nearly the same in the gossypol treated (0.27%) and untreated
(0.28%) cultures. There was a very slight dose-dependent increase
in micronuclei and SCEs in gossypol cultures. Even at the highest
concentration (40 µg/ml) used, which was presumed to be 10 times
the serum level in users, the SCE rate was still within the normal
range (1-21 SCE/metaphase).

Gossypol does, however, affect lymphocyte kinetics, either
reducing the response to phytohemagglutinin, or retarding it, or
both. This can be demonstrated from the reduction in mitotic
index, the decrease in the proportion of stimulated interphase
nuclei and the dose-dependent decrease in second and third
metaphases after BudR treatment in 66 hour cultures, as well as
from the progressive reduction in 2nd and 3rd metaphases in
cultures between 66 and 96 hours.

INDUCTION OF SISTER CHROMATID EXCHANGES WITH "COMPOUND
PRESCRIPTION PROGESTOGEN NO. 1" IN THE
MATERNAL AND EMBRYONIC TISSUES OF MICE

H.Z. Zheng, Y.N. Ding, X.R. Jiang, and Z.S. Zhang

Shanghai Institute of Planned Parenthood Research, Shanghai

Analyses of sister chromatid exchanges (SCE) were conducted
in the maternal and embryonic tissues substituted in vivo with 5-
iodo-deoxyuridine (IudR) by implantation of a IudR tablet (40 mg/
tablet) in pregnant mice at mid-gestation. Following maternal

exposure to 4.2 mg/kg, 2.1 mg/kg, 0.525 mg/kg "Progestogen No. 1" and 20 mg/kg cyclophosphamide, SCE analyses were conducted in maternal bone marrow and embryonic liver cells.

The results of this study showed that none of the doses of "Progestogen No. 1" increased the SCE frequency compared with the normal control. Following maternal exposure to 4.2 mg/kg, 2.1 mg/kg, and 0.525 mg/kg, the mean of SCEs of maternal bone marrow and embryonic liver cells were 2.88 ± 1.48, 2.67 ± 1.19; 2.67 ± 1.687, 2.45 ± 1.20, 2.85 ± 1.67, 2.73 ± 2.56, ranging from 0-7, 1-6; 0-10 1-7 and 0-9 0-8, respectively. Those of the control were 2.69 ± 1.49 and 2.44 ± 1.39, ranging from 0-9 and 0-8 SCEs. The differences between the treated and the control were not statistically significant (P > 0.05).

In the positive control group administered with cyclophosphamide 20 mg/kg b.w., the mean SCE values were 37.55 ± 10.77 in maternal bone marrow and 20.64 ± 8.04 in embryonic liver cells, ranging from 11-66 and 12-13 SCEs per cell, respectively, and were higher than in the untreated groups. The differences between them were all statistically significant (P > 0.001).

The results showed that "Progestogen No. 1" may not be transformed of pregnant mice. Thus it is appropriate to consider that "Progestogen No. 1" used as a long acting oral contraceptive is a comparatively safe drug.

INDUCTION OF SISTER CHROMATID EXCHANGES WITH GOSSYPOL ACETATE IN MOUSE SPERMATOGONIAL CELLS IN VIVO

R.L. Wang[1], M.M. Wang[2], Q. Lu[3], H.Z. Zhing[1], Y.L. Yao[2], X.R. Jiang[1], and Z.S. Zhang[1]

[1] Shanghai Institute of Planned Parenthood Research
[2] Shanghai Institute of Pharmaceutical Industry
[3] Institute of Genetics, Fudan University, Shanghai

This paper presents the results of testing gossypol acetate for the possible induction of sister chromatid exchanges (SCEs) in mouse spermatogonial cells in vivo following a single intraperitoneal injection of BudR adsorbed on charcoal. Our purpose was to elucidate the mutagenicity of gossypol. Animals were randomly divided into five groups: the lower, medium, and higher dosage groups (5, 20, and 50 times as high as gossypol clinical dosage, respectively), the negative control group and the positive control group (Mitomycin C 0.5 mg/kg b.w.). The results are:

1.   Comparing the SCE frequencies of the lower dosage group
(1.42 ± 0.006/cell) and the medium dosage group (1.43 ± 0.06/cell)
with the negative control group (1.43 ± 0.07/cell), the
differences were not statistically significant (P > 0.05).

2.   The difference between the higher dosage group (1.84 ±
0.07/cell) and the negative control group was highly significant
(P < 0.001).

Since positive results were obtained from a dosage 50 times
as high as the clinical dosage, it is reasonable to assume that
gossypol used as fertility regulator at a much lower dosage should
be safer.   However, it still should not be neglected to take into
account some protective measures to avoid the probable
mutagenicity to workers who may contact with high levels of
gossypol over a long period.   The course of spermatogenesis in man
is 64 days.   To avoid DNA damage in spermatogonia which may affect
the offspring, we suggest that gossypol users take measures to
prevent procreation for at least two months after the treatment.

THE INFLUENCE OF ORAL CONTRACEPTIVE ON SILVER-STAINED ACROCENTRIC ASSOCIATION

X.Z. Li, and X.T. Zhou

Institute of Genetics, Academic Sinica, Beijing

The silver-stained acrocentric chromosome association (Ag-AA)
is the remnant of the nucleolus.   Its frequency reflects the
transcriptional activity of cellular rRNA genes.

The Ag-stained NOR (nucleolar organizer region) and Ag-AA in
lymphocytes taken from 14 normal women (the control group) and 9
healthy women (the experimental group) who had received a
methylnorethindrone compound oral contraceptive for over 6 months
were studied.   Metaphases per individual were analyzed by double-
blind scoring of the number of Ag-NOR's and Ag-AA's.   The mean Ag-
AA frequencies of the control and experimental groups are 0.81 ±
0.22 and 1.07 ± 0.29, respectively (t = 2.30, DF = 21, P > 0.05).
There is no significant difference in Ag-NOR frequencies between
the two groups (t = 0.40, DF = 21, P > 0.05).

In the control group the percentage of cells with 4 Ag-AAs is
0.87 ± 0.25 ($p ± \sigma_p$%*), while in the experimental group it is 3.41
± 0.62 ($p ± \sigma_p$%).   t-test indicates that the percentage of the
cells with 4 Ag-AAs in the experimental group is significantly
higher than that of the control's (t = 3.79, p < 0.001).

It is concluded that the Ag-AA was significantly increased in lymphocytes under the influence of the methylnorethindrone compound oral contraceptive. The increased percentage of the cells with higher Ag-AA number would be more significant than the elevated mean Ag-AA frequency in this study.

CYTOTOXICITY AND INDUCTION OF SISTER CHROMATID EXCHANGES (SCE) IN VITRO BY SEVERAL ANTITUMOR DRUGS

B. Xu, X.W. Wang, J.X. Han, and C.C. Huang[*]

Shanghai Institute of Materia Medica, Academia Sinica

This paper summarizes our results on tests of cytotoxicity and the induction of SCE in V79 Chinese hamster cell line treated with the following antitumor drugs: camptothecin (CPT), hydroxycamptothecin (HCPT), harringtonine (HAR), homoharringtonine (HHAR), lycobetaine (LBT) and oxalysine (OLS).

1. For cytotoxicity tests, V79 cells were seeded in a Petri dish and treated with a drug for one hour. Viable cell counts were made at different times after treatment. Plating efficiency (PE) was determined one week after. All the drugs exhibited a dose-dependent toxicity as measured by viable cell counts or PE. However, the degree of inhibition varied greatly among the drugs tested. For instance, the doses which resulted in 50% inhibition of PE for HAR was 2 µg/ml of medium, while for OLS was 10 mg/ml. Continuous treatments had much more pronounced effect than one hour pulse.

2. V79 cells treated with CPT or HCPT showed a significant increase of SCE frequencies in a dose-dependent manner. The metabolic activation system of S9 mix had no influence on SCE in such experiments. Treatment of LBT with or without S9 mix showed a relatively low level increase of SCE. HAR, HHAR and OLS induced no increase of SCE in V79 cells with or without S9 mix.

[*] Visiting scientist from Roswell Park Memorial Institute, Buffalo, New York, USA.

THE INDUCTION OF SISTER CHROMATID EXCHANGES IN HUMAN GASTRIC CARCINOMA CELLS AND LYMPHOCYTES EXPOSED TO ANTI-TUMOR DRUGS

Z.R. Ma, Y.L. Liu, C.B. Li, and S.Y. Zhao

Institute of Genetics, Fudan University, Shanghai

It is generally believed that the occurrence of SCE is related to DNA primary damage. Using sister chromatid exchanges as an index it may be possible to detect the mutagenicity of anti-tumor drugs and to use this as a method for screening new drugs. In this paper we report the effect of Camptothecin, Harringtonine, Flurouracil and Mytomycin C on SCEs in human gastric carcinoma cells (SGC-7901) and lymphocytes. It was observed that all of them could increase SCE frequency in both kinds of cells. However more SCEs were induced by these drugs in SGC-7901 cells than in the normal lymphocytes and the differences were statistically significant. At a lower concentration (10^{-7} μg/ml), Camptothecin and Flurouracil could remarkably increase SCEs in the tumor cells but not in the lymphocytes. This result is in good agreement with the fact that they are specifically used for treating stomach cancer. Therefore we consider that the method we described above may be useful for screening new anti-stomach cancer drugs.

INFLUENCE OF 2450 MHz MICROWAVE RADIATION ON SISTER
CHROMATID EXCHANGE (SCE) OF HUMAN LYMPHOCYTES
IN SHORT TERM CULTURE

Z. Er, Y.Y. Lu, C.J. Wu, C. Li, and M.L. Hu

Department of Cytogenetics, Peking Institute for Cancer Research, Beijing

The peripheral blood of 27 normal adult individuals of both sexes was collected for study. Every sample was 5 ml which was further divided into several parts for culture. The cultured cells were exposed once or twice to microwave radiation (MWR) at power densities of 1 mW/cm^2, 10 mW/cm^2, 20 mW/cm^2, and 30 mW/cm^2.

The single irradiation group was exposed immediately after the culture was assembled; a separate group of cultures were irradiated twice. Every exposure lasted 20 min, and the culture flask was kept at a certain distance from the microwave source in the dark at room temperature. The control group was handled the same way but without microwave irradiation (MWI).

At least 20 metaphases were scored for each case under oil immersion microscopy. A total of 1,960 mitotic figures were counted. The SCE of cells exposed to MWI represents as follows: 1 mW/cm^2 - 7.77 ± 0.4/cell; 10 mW/cm^2 - 6.77 ± 0.39/cell; 20 mW/cm^2 - 8.05 ± 0.85/cell and 30 mW/cm^2 - 6.88 ± 0.37/cell. The results demonstrate that the frequency of SCE of the exposed group is significantly increased over that of the control (5.83 ± 0.25; P <

0.05). In addition, we failed to find any regularity of SCEs at power densities from 1 mW/cm^2 to 30 mW/cm^2 and could not display any difference between single and double exposures to MWI. All these observations need to be further studied.

SISTER CHROMATID EXCHANGES IN LYMPHOCYTES OF
BUFO BUFO GARGARIZANS AS AN
ENVIRONMENTAL MUTAGENICITY ASSAY

Q. Lu, C.X. Wen, S.E. Lin, S.H. Jiang, and W. Xiang

Institute of Genetics, Fudan University, Shanghai

Our experimental results demonstrated that the SCE frequencies of two known positive controls, cyclophosphamide and methyl methane sulphonate, were 13.67 ± 0.86 and 11.87 ± 0.74, respectively, in female toads. The background SCE frequency of Lushan's sample was the lowest, i.e. 7.7 ± 0.23 in female, and 7.6 ± 0.30 in male. Generally, the females were more sensitive than the males. The SCE frequency of Huongshan's sample and that of Fudan University campus' sample were higher than that of Lushan's, but the differences were not statistically significant. Nevertheless, the SCE frequencies of the samples from the waste water treatment plant (11.78 ± 0.27 in female) and the oxidation pond (13.91 ± 0.33 in female) of a chemical factory were markedly higher.

We propose that SCE in lymphocytes of Bufo bufo gargarizans, an animal widely distributed all over China, could be used as a sensitive and reliable in vivo screening system in genetoxicological studies.

A MODIFIED METHOD OF UDS ESTIMATION IN VITRO SUITABLE FOR
SCREENING THE DNA-DAMAGING EFFECTS OF CHEMICALS

Y.N. Yu[1], C. Ding[1], Q.G. Li[2], and X.G. Chen[1]

[1]Zhejiang Medical University, Hangzhou
[2]Zhejiang College of Traditional Chinese Medicine, Hangzhou

Unscheduled DNA synthesis (UDS) induced in cultured human FL cells by exposure to chemicals is measured as hydroxyurea (HU)-resistant-incorporation of tritium labeled thymidine (^{3}H-TdR) in the acid insoluble fraction of the ^{14}C-TdR-prelabeled cells, synchronized by the combination of arginine starvation and HU pretreatment. As the cells had previously incorporated ^{14}C-TdR into the parent DNA, the final counts recorded as the ratios of $^3H/^{14}C$ radioactivities are the measures of specific activities of ^3H, and therefore represent the levels of UDS. The relative $^3H/^{14}C$ ratios as compared with the solvent controls are used for evaluating the results. In validation studies, it was demonstrated that in the absence of a liver metabolizing system, 2 direct acting alkylating agents, methyl methanesulfonate and N-methyl-N'-nitro-N-nitrosoguanidine, and a cross-linking agent, mitomycin C, elicited UDS, but the 3 tested procarcinogens, i.e., benzo(a)pyrene, aflatoxin B_1 and cyclophosphamide, did not. In the latter cases, however, if rat liver microsomal metabolic system was included in the incubate, a significant increase of HU-resistant incorporation of ^3H-TdR also occurred. Three chemicals of unknown carcinogenicity were also found to be able to induce UDS in this assay system, i.e., bis-(0,0-diethylphosphinothioyl)-disulfide, 4-chlorophenoxy acetic acid (sodium salt) and caramelized malt sugar. With the exception of 4CPA, they were also active in the Ames test performed in our laboratory.

EFFECTS OF SEVERAL KINDS OF RADIOSENSITIZERS ON γ-RAY-INDUCED DNA SINGLE STRAND BREAKS AND THEIR REJOINING IN MAMMALIAN CELLS

Y.P. Zhang, S.X. Xia, and H.Y. Xu

Institute of Radiation Medicine, Beijing

By means of improved membrane filtration methods the effects of six kinds of radiosensitizers on γ-ray-induced DNA single strand breaks and their rejoining in Chinese hamster ovary cells were compared. It was found that: (1) 5-bromodeoxyuridine (an analog of a DNA precursor) increased the number of single strand breaks but did not inhibit their rejoining; (2) KI and N-ethylmaleimide (a powerful SH-blocking agent) inhibited break rejoining but did not enhance the number of breaks; (3) quinacrine (an inhibitor of DNA repair) and Misonidazole (a radiosensitizer for hypoxic cells) both increased the breaks and inhibited their rejoining; (4) actinomycin D (a DNA-breaking agent) had neither effect; (5) only under hypoxic conditions could Misonidazole sensitize the DNA macromolecule of CHO cells to radiation. The mechanisms of action of these radiosensitizers were discussed in connection with DNA damage and repair.

This method might also be used in the detection of carcinogenic substances or other agents in the environment noxious to DNA.

EFFECTS OF INTERFERON ON MITOMYCIN-INDUCED SISTER
CHROMATID EXCHANGES AND ON ULTRAVIOLET-LIGHT-INDUCED
UNSCHEDULED DNA SYNTHESIS

X.P. Wang, Z.S. Jiang, Y.P. Hu, H.J. Chen, and L.S. Yang

Second Military Medical College, Shanghai

We studied the effects of interferon (IFN) on sister
chromatid exchanges (SCE) induced by mitomycin (MMC) and/or on
unscheduled DNA synthesis (UDS) induced by ultraviolet light (UV)
in peripheral lymphocytes taken from 4 normal individuals and 3
patients with xeroderma pigmentosum (XP) in whom the existence of
congenital DNA excision repair defect had been verified.

The lymphocytes from 2 normal individuals and 3 patients with
XP were treated with IFN 8 hours before MMC was added. SCE
frequencies in IFN groups were lower than those in the controls.
Statistical computation showed that all differences were highly
significant ($P < 0.01$).

Measurement of UDS was undertaken by the method of Burk et
al. (1971). The lymphocytes from 2 normal individuals and 2
patients with XP were treated with IFN 8 hours before MMC was
added. Ultraviolet-induced ^{3}H-TdR incorporation in the IFN group
was similar to that of the control in all cases. No significant
differences exist between the different groups in each individual
($P > 0.05$).

It is suggested that 1) IFN may be considered as an
antimutagen; 2) the anti-mutation effect of IFN probably is not
related to the DNA excision repair, but to the inhibited DNA
recombination repair and SOS repair; 3) the anti-mutation effect
of IFN may be of multiple significance as shown by our results and
the report of Shvetsova, T. P., et al. (Genetica, USSR 17(7):1290,
1981).

ISOLATION AND IDENTIFICATION OF DIRECT-ACTING MUTAGENS
IN DIESEL EXHAUST

X.B. Xu[1], Z.L. Jin[1], E.T. Wei[2], J.P. Nachtman[2], S.M. Rappaport[2],
and A.L. Burlingame[3]

[1]Institute of Environmental Chemistry, Academia Sinica, Beijing
[2]School of Public Health, University of California, Berkeley, USA
[3]School of Pharmacy, University of California, San Francisco, USA

The mutagenicity obtained in the absence of activation by mammalian oxidative enzymes in the Ames Salmonella assay of particulates from air or from engine exhaust indicated that these particulates contain chemicals of unrecognized toxic potentials other than unsubstituted polynuclear aromatic hydrocarbons (PAH). Bioassay (strain TA 98, without S-9) directed fractionation and characterization studies of a combined diesel exhaust sample of 225 g organic extracts had been carried out over the last two years in our laboratories for the purpose of identifying those so called "direct-acting" mutagens.

Attention had been concentrated on those most active fractions derived from sequential fractionation by both low and high resolution liquid chromatography, and the successive increase of their mutagenicities after each step showed the concentration of active components in those fractions.

Of the 50 NO_2-PAH tentatively identified by direct-probe high resolution mass spectrometry (HRMS), mutagenic and carcinogenic 5-nitroacenaphthene, 2-nitrofluorene, 9-nitroanthracene and 1-nitropyrene were positively confirmed by high resolution gas chromatography/low resolution mass spectrometry (HRGC/LRMS) and HRGC/HRMS. The presence of compounds with other interesting compositions has also been suggested. As more nitro-aromatics are being proved to be animal carcinogens, their presence in atmospheric and the possible health hazard to mankind should be of concern.

APHIDICOLIN INDUCES ENDOREDUPLICATION IN CHINESE HAMSTER CELLS

Y.Q. Huang[1], C.C. Chang[2], and James E. Trosko[2]
[1]Department of Biology, Nanjing Teacher's College, Nanjing
[2]Department of Pediatrics and Human Development, College of Human Medicine, Michigan State University, East Lansing, Michigan, USA

Aphidicolin, an inhibitor of DNA polymerase α, was shown to induce a very high frequency of endoreduplication in Chinese hamster V79 cells. The aphidicolin-induced endoreduplication was inhibited when cells were incubated at 41° C. Since it is known that DNA polymerase β is more thermally labile than DNA polymerase α, the data are consistent with the hypothesis that DNA polymerase β might be responsible for endoreduplication.

SEARCH FOR CARCINOGENIC AGENTS IN THE SHANGHAI RUBBER INDUSTRY

H.W. Wang[1], X.J. You[2], Y.H. Qu[3], W.F. Wang[4], D. Wang[1], Y.M. Long[5], J.A. Ni[3], and R.W. Dzeng[1]

[1] Shanghai Institute of Cell Biology, Academia Sinica
[2] Shanghai Institute of Occupational Hygiene
[3] Shanghai Institute of Cancer Research
[4] Zhong-shan Hospital, Shanghai First Medical College
[5] Shanghai First Rubber Tyre Factory

Preliminary studies on crude cancer incidences (1961-1970) among workers of 89 factories in Shanghai revealed excessive cancer risks to workers in certain rubber tyre factories. Three 2-year in situ animal exposures staged in the First Workshop of Rubber Tyre Factory A showed that compounding and banbury mills for mastication and mixing were the origins of carcinogenic pollutants. These results prompted reconstruction of the workshop and suspension of the use of antioxidant D in Shanghai.

All 8 chronic experiments carried out in different institutes by feeding, injection and inhalation of technical and pure phenyl-2-naphthylamine (PBNA), especially to the lungs, of rats and mice, indicated carcinogenicity.

Repeated subcutaneous injections of both technical and chemically pure PBNA dissolved in DMSO resulted in 45% of male ICR mice developing malignant tumors within 10 months. Results of the determination of atmospheric concentration of PBNA at different sites in two rubber tyre factories suggested a relationship between PBNA in air and lung cancers of the workers.

The α-isomer of PBNA, N-phenyl-1-naphthaylamine (PANA) is the main constituent of antioxidant A, also a widely used rubber additive, the carcinogenicity of which had never been suspected before. Its biological effect was compared with PBNA in chronic injection experiments with ICR and TA1 mouse strains. Both tests showed a similar carcinogenic potency. PANA had a tendency to induce haemangiosarcomas. Previous unilateral nephrectomy enhanced induction of renal haemangiosarcomas by both chemicals.

The approximately similar carcinogenic potency of PANA (technical and pure) and PBNA suggests routes of metabolic activation other than dephenylation of PANA and PBNA in mice.

INDUCTION OF LUNG CANCER IN WISTAR RATS BY
TIN ORE DUST CONTAINING ARSENICALS

Z.M. Pan

Institute of Labor Protection, Yun-Nan Tin Corporation, Kuenming

As the crude death rate of lung cancer of tin miners is up to $338.13/10^5$ person-years, etiological research in this field is much imperative. A long-term carcinogenicity assay was conducted using intra-tracheal injection of mining dusts into Wistar rats. Five groups of rats, each of 20–70 of 200–300 gm body weight, were treated with the following dust samples:

1) dust from tin oxide ore, 1.98% As

2) dust from another tin oxide ore, 11.39% As

3) combination of dust from tin oxide and Pb-As ore, 16.05% As

4) mixture of dusts from tin oxide and tin sulfide ores, 30.22% As

5) normal saline serving as negative control

The administration of dust preparations was carried out bi-weekly in doses of 6–10 mg each time, the total doses being 90–150 mg. On necrospy and pathological examination, precancerous (P) and malignant (M) changes were noticed as follows: 1) P: 5%, M: 15%; 2) 35.6%, 17.7%; 3) 18.8%, 18.8%; 4) 45.2%, 22.6%; 5) 0.0%, 0.0%.

The results suggest lung cancer in rats was induced by dust of tin ores containing arsenicals. Though not significant statistically, the incidences of total malignancies in different groups seem to be dependent on the percentage of As contained in the dust.

CARCINOGENICITY OF EXTRACT OF SOOT FROM XUAN WEI COUNTY
AFTER ADMINISTERING SUBCUTANEOUSLY TO MICE

C.K. Liang, N.Y. Guan, F. Ma, Y. Zhang, E.M. Wang, and X.R. Yin

Institute of Health, Chinese Academy of Medical Sciences, Beijing

According to the data of a retrospective epidemiological survey, the annual standardized lung cancer death rate in Xuan Wei County, Yuen-Nan Province was 25.95 per 100,000 during 1973–75. Sex ratio of the death rates is 0.98 to 1. The highest rate, up to 151.78 per 100,000, was recorded in one commune. As it has been customary to burn locally produced smoke coal in shallow pits indoors without chimneys, concentrations of coal smoke are usually high. The preliminary studies on the carcinogenicity of soot extract from Xuan Wei are presented.

Indoor soot collected from local farmers' families was extracted with cyclohexane and then concentrated. The benzo(a)pyrene concentration in the extract was found to be 2 mg/ g. Male Kunming mice of 18-26 g were divided into the following groups with different treatments: solvent control; 1000 mg of extract per mouse (total dose); 500 mg of extract; 2 mg of benzo(a)pyrene (BaP). The number of animals in each group was 38, 56, 57 and 38 respectively. Solvent and extract were administered subcutaneously to mice once a week for 10 weeks.

At the end of the 10 months' experiment, the incidence of total lung tumor (including lung cancer, fibrosarcoma and adenoma) in 4 groups mentioned above was 15.8% (6/38), 73.2% (41/56), 82.5% (47/51) and 34.2% (13/38), respectively. The incidence of lung cancer was 2.6% (1/38), 64.3% (36/56), 77.2% (44/57), and 34.2% (13/38), respectively. It is thus evident that incidences of total lung tumor and lung cancer in animals exposed to soot extract were significantly higher than that of the control ($P <$ 0.001) and BaP groups ($P < 0.001$).

No tumors were found at the injection sites in the control group. Two dermal squamous cell carcinomas and one carcinosarcoma were observed in the 1000 mg of extract group. Five dermal squamous cell carcinomas, one carcinosarcoma and two subcutaneous fibrosarcomas were seen in the 500 mg of extract group. However, the incidence of subcutaneous fibrosarcoma in the BaP group was as high as 84.2% (32/38) which was significantly higher than that of the two soot-extract-treated groups ($P < 0.001$).

The average latencies of lung tumor in the control, 500 mg of extract, 1000 mg of extract and BaP group were at 301, 201, 144, and 238 days, respectively.

The results indicate that the soot extract contained potent carcinogens responsible for the induction of lung cancer in treated mice. The high risk of lung cancer in Xuan Wei County might be associated with certain carcinogens other than BaP in the indoor air derived from the burning of local coal.

STUDY ON THE CARCINOGENICITY OF CHLORDIMEFORM
III. LONG-TERM FEEDING ASSAY IN MICE

F. Lee, S.Z. Xue, and Y.L. Wang

Department of Occupational Health, Shanghai First Medical College, Shanghai

The carcinogenicity of purified Chlordimeform (CDM) was tested in a two-year feeding assay in male Swiss mice. Five groups, 50 each of weaning mice were used in the test. Besides one group of non-treated negative control and a positive control group (300 ppm p-Chloro-o-Toluidine), three dose levels of 20, 100 and 300 ppm of CDM were added into the feeds prepared every two weeks. On necropsy and pathological examination, no neoplastic growth was found in the natural control group, 31 tumor growths with 13 malignancies were found in the positive control group. The number of animals bearing neoplasms in the tested mice were as follows: 8 (0 malignancy); 22 (5) and 36 (16) respectively in the groups receiving the increasing level doses. The regression equation calculated is:

$$Y_{(\text{incidence of tumor})} = 0.0976 + 0.0022\ X_{(\text{ppm of CDM})}$$

Major findings were hemangioma and hemangiosarcoma mostly located in the abdominal or peri-bladder adipose tissue varying from pinpoint to the size of an almond. Metastases in liver were seen in some of the specimens of the highest dose group. Dose-dependent pulmonary adenocarcinoma was also noticed (0,3,6), as were liver carcinoma, renal adenocarcinoma, intestinal carcinoma etc. The result described above is comparable to that reported by Dr. Hess (Director, Toxicol. Lab., Georgy, Ciba). There is no doubt of the carcinogenicity of CDM in mice as verified by different reports; however, several results from assays in rats are suspicious or inconclusive. Long-term assays on an animal other than these two species, such as hamster, and knowledgeable extrapolation from animal to man are imperative for providing a scientific basis for safety evaluation. Most short-term tests seem ineffective owing to their negative responses, as in the case of DDT.

INDUCTION OF TUMOR IN THE PROGENY OF DINITROSOPIPERAZINE-
TREATED PREGNANT RATS

B.X. Ou, Y.F. Lu, and G.S. Zheng

Department of Tumor Etiology, Cancer Institute, Zhongshan Medical College, Guangzhou

The hypothesis of this study is that if initiation occurs during the embryonic life, tumors might develop early in those animals exposed to promoters after birth. Four groups of rats were used in this experiment:

Group PN: A subthreshold dose of dinitrosopiperazine (DNP) was given (H) to the rats on the 18th day of pregnancy. $NiSO_4$ was given (O) daily to the rats of the next generation for 6 months starting at the age of 1 month. Five tumors developed in 21 rats. They were: 2 squamous cell carcinomas, 2 undifferentiated carcinomas and 1 carcinoma in situ in the nasal cavity (NC).

Group P: The maternal rats were treated as those in Group PN, while the rats of the second generation did not receive any $NiSO_4$. Three tumors were found in 11 rats. They were: squamous carcinoma of nasopharynx, neurofibrosarcoma of peritoneal cavity and granulosa-theca cell tumors of ovary.

Group N: No DNP was injected into the maternal rats during pregnancy, but $NiSO_4$ was given to the rats of the next generation. No tumor was found.

Group O: Rats of both generations were not treated with either DNP nor $NiSO_4$. No tumor was observed.

In summary, (1) DNP is a carcinogen which could be trans-placental and induce tumors in the next generation, (2) Fetal tissues of rats may be more sensitive to DNP than those of adults. The subthreshold dose of DNP for fetal rats should be smaller than that for the adult rats; (3) DNP and $NiSO_4$ possess synergetic action in carcinogenesis of the nasal cavity and the nasopharynx of rats.

PROMOTING EFFECT AND RELATED ACTIONS OF TPA ON C3H/10T1/2 CELLS

S.J. Cheng

Cancer Institute, Chinese Academy of Medical Sciences, Beijing

Treatment of 3-methylcholanthrene (MC, 0.1 µg/ml) did not transform C3H/10T1/2 cells, but transformed foci were induced if the cells were first initiated with MC (0.1 µg/ml) and then followed by treatment with phorbal myristate acetate (TPA, 0.1 µg/ml).

In another parallel experiment, the alterations of sister chromatid exchanges (SCE) and DNA synthesis in the same cells were checked at intervals. SCEs were significantly increased in the cells 8 and 13 days after MC treatment. However, no further increase of SCEs was noted by TPA treatment in the MC-initiated cells checked on the same days. A transient inhibition followed by a significant stimulation of DNA synthesis was demonstrated after TPA treatment in C3H/10T1/2 cells, whether the cells were initiated with MC or not.

The results indicate that TPA enhances transformation, but it does not increase SCE, in the MC-initiated C3H/10T1/2 cells. It seems that the initiation of MC may have some relation to the induction of SCE, but the promotion of TPA may have nothing to do with the SCE induction. The alterations of DNA synthesis in the cells may be the result of a series of biochemical reactions induced by TPA.

MALIGNANT TRANSFORMATION OF MOUSE BONE MARROW CELLS IN VITRO
INDUCED BY ^{60}Co γ-RAY IRRADIATION

F.M. Gao, X.L. Li, M.J. Qian, W.H. Wang, F.Q. Qi, and H.Y. Jiang

Department of Biological Effects of Radiation, Laboratory of Industrial Hygiene, Ministry of Public Health, Beijing

Male LACA mouse bone marrow cells in long term culture were irradiated in vitro by ^{60}Co γ-ray at 300, 200 and 100 rads, respectively. Morphological and malignant transformation took place only in the cells receiving 300 rad irradiation 42 and 66 days following exposure. Some biological characteristics of the malignantly transformed cells were as follows:

1) maximum starvation density was $7.4 \pm 2.5 \times 10^4/cm^2$;
2) anti-ouabain inhibition test was negative (no mutation);
3) they were agglutinated by concanavalin A;
4) cytochalasin B induced the formation of multinuclear cells with 22 nuclei per cell at the maximum;
5) chromosome aberrations in number and structure appeared;
6) fibrosarcomas were induced in local subcutaneous tissues following inoculation of the transformed cells.

From the above results, we believe that the malignantly transformed cells were derived from interstitial cells of bone marrow and the latter were not as sensitive to the carcinogenic effects of radiation as mouse embryo cells.

SOME BIOLOGICAL CHARACTERISTICS OF MNNG-TRANSFORMED
CELLS IN VITRO

Y.X. Du, J.K. Chen, Z.L. Wu, J.W. Feng, H.W. Chen, and G.Q. Chen

Department of Hygiene, Guangzhou Medical College, Guangzhou

The early passage diploid Syrian hamster embryo (SHE) cells
were treated with N-methyl-N'-nitro-N-nitrosoguanidine (MNNG).
The treated cells proliferated rapidly; the doubling times
shortened; colonies appeared in solid agar medium and transformed
foci formed in tissue culture--all of these phenomena suggest that
malignant transformation of SHE cell has occurred.

Faster cell division rate and multipolar mitosis were
demonstrated by time lapse cinemicrography and scanning electron
microscopy. Multipolar mitosis has two forms: triangle of direct
division and moniliform of indirect division. The transformed
cells were more abundant in microvilli, the number of which
increased in accordance with the degree of malignancy.

In comparison with the controls, the transformed cells
expressed a greater tritiate thymidine incorporation, greater DNA
contents and more chromosomes, but no difference in nuclear area.

The determination of amino acid changes in media due to the
growth of transformed cells showed that the decrease in arginine
and increase in ornithine are significant.

The results of allogenic animal inoculation suggest that the
transformed cells can be characterized into several different
stages in the process of transformation.

STUDY OF CARCINOGENIC MECHANISM FOR POLYCYCLIC AROMATIC
HYDROCARBONS: EXTENDED BAY REGION THEORY AND
ITS QUANTITATIVE MODEL

L.S. Yan

Department of Environmental Ecology, Chinese Academy of
Environmental Sciences, Beiyuan, Beijing

The author suggests the concept of an extended bay region and
thinks that the essential agent of carcinogenesis for polycyclic
aromatic hydrocarbons (PAH) should be, as a rule, the highest
delocalization energy (β unit) of carbonium ion at the aromatic
angular ring (A region), which is obtained by the perturbational
molecular orbit (PMO) method. The carcinogenic activity exhibited

by a PAH is determined by the competition between the carcinogenesis and detoxification, in which it participates. The detoxificative efficacy of each kind of the hinder carcinogenic factors has been evaluated, including the biologic factor B and three structural factors of PAH's molecule: K, A and L. After making necessary approximations, K = 0.228, A = 0.5, L = 1.22 and B = 0.7 are obtained. This paper also suggests a concept of a carcinogenic constant. For the PAH with the same N, which is the number of aromatic rings, C is a constant. The relationship between C and N is called the "Pyramid Regulation". The quantitative equation of the extended bay region theory is:

$$\log R = C[E^3_{veloc}/(0.7 + 0.228n_k + 0.5n_a + 1.22n_l)]$$

50 PAHs which had been tested by long-term laboratory animal assays were calculated. Of these, a quantitative estimation of 92% of the PAHs tallies with the carcinogenic activities reported from animal assay.

THE COMBINED ACTION OF POLYCYCLIC AROMATIC HYDROCARBONS (PAH) ON THE ACTIVITY OF ARYL HYDROCARBON HYDROXYLASE (AHH): IN VITRO ASSAY

Y.K. Cheng[1], Z.Y. Wang[2], X.L. Xiao[2], and L.S. Yan[3]

[1]Institute of Industrial Health, Anshan Iron and Steel Co., Anshan
[2]Institute of Labor Protection Science, Ministry of Labor and Personnel, Beijing
[3]Chinese Research Academy of Environmental Sciences, Beijing

PAHs are widespread contaminants in our environment and may be a major cause of cancer in human beings. It is accepted that PAHs are not active per se, but require some kind of metabolic activation by monooxygenase enzymes for the manifestation of their carcinogenic activity. A wide variety of PAHs are present simultaneously in the environment, so the potential for interaction with two or more PAHs in a biological system is great. We have recently studied the combined action of two or three PAHs on AHH activity of rat methylcholanthrene-induced liver microsomes. The experiment is designed according to the orthogonal L_4 (2^3) and L_8 (2^7) tables. Determination of AHH activity is based on the use of radioactivity-labelled substrate—[^{3}H]BaP, and the radioactivity of the lower phase of the reaction medium is quantified by scintillation counting. Ten PAHs are divided into 6 groups for studying the combined action of two or three factors. The medium containing [^{3}H]BaP only serves as the control group, and its scintillation counting value is 100 when compared with the test groups. The results show that all the test

groups exhibit a varying degree of inhibitory action on AHH
activity. The following regression equations are obtained from
the experiment:

1. An+Phe group: $y=85.5-1.15x_1-5.05x_2+0.08x_1x_2$ (F=3.79 a=0.10)
2. BaA+Chry group: $y=90.6-4.14x_1-1.36x_2+1.71x_1x_2$ (F=11.69
   a=0.05)
3. BeP+Per group: $y=96.6+1.63x_1-4.55x_2+4.4x_1x_2$ (F=1.56 a>0.25)
4. BghiP+Cor group: $y=90.0-2.29x_1+1.79x_2-0.44x_1x_2$ (F=3.15
   a=0.25)
5. DMBA+MCA group: $y=61.8-8.63x_1-6.93x_2-2.88x_1x_2$ (F=9.52 a=0.05)
6. DMBA+MCA+BeP group:
   $y=43.8-2.83x_1-3.41x_2-3.38x_3-2.88x_1x_2-0.91x_1x_3-0.41x_2x_3$
   $-0.91x_1x_2x_3$ (F=9.52 a=0.05)

EPIDEMIOLOGICAL SURVEY ON THE TERATOGENIC EFFECT OF MATDA [METHYLENE-BIS-(2-AMINO-1,3,4-THIADIAZOLE)]

X.Q. Gu[1], C. Qin[2], Q.M. Lu[1], and Z.Y. Hong[2]

[1]Department of Occupational Health, School of Public Health,
Shanghai First Medical College, Shanghai
[2]Health-Antiepidemic Stations and Maternal-Child Health Stations
of Yuen-Cheng District, Jiang-Su Province

MATDA is a very effective bacteriocide against Xanthomonus
oryzae ("bacterial leaf blight") of rice for both prevention and
cure, and no substitutes have been found. Although it has been
proven to be a potent teratogen in rodents with a threshold dose
of 0.1 mg/kg to 1.0 mg/kg given in mid-gestation, the pesticide
has been applied to a vast number of rice fields since the late
period of the "cultural revolution" (1966-1976). The field survey
designed to check the effects of this pesticide was of great
importance and interest.

End of pregnancy and birth defects among the 2.3 million
residents living in 4 counties during 1979-1981 were carefully
investigated. 15-18 tons of MATDA had been applied annually on
the field in the past 7-8 years preceding this survey, but none on
the rest (control). The maximal residue amount of MATDA in
dehusked rice was measured as 0.04 mg/kg. The level of residue in
the polished rice consumed by local residents was estimated as
approximately 0.0120 mg/kg. The amount actually intaken is 0.128
μg/kg body weight each day.

To avoid the loss of subjects from the survey, files and
records of the birth control program were used as a guide. All of
the birth defects were reported by rural paramedical personnel,

and then, diagnosed and confirmed by physicians in commune and/or county hospitals. Data were collected and checked stepwise from brigade to commune, county and district to conform to uniform criteria.

The incidence of birth defects in the observed group was 9.66% (713/73,790), 14.41% (520/36,098) in the control, and 11.22% (1233/109,888) overall. Although there is a noticeable difference between the figures for the two groups, they are both towards the lower end of the range of statistics in the general birth defects registry in China (and, incidentally, indicate a far lower incidence of birth defects than in the U.S.). There was no marked difference in the incidence of spontaneous abortion or in the sites of the birth defects recorded.

The results indicate that MATDA, under the conditions described, does not increase the frequency of birth defects or spontaneous abortion. Underlying mechanisms may include species differences, an adequate NAA level in the human body serving as an effective antidote, and the low amounts of residue ingested.

MUTAGENICITY STUDY OF V_2O_5

R.S. Shi, C.L. Zhan, and S.Q. Li

Toxicology Research Laboratory, School of Public Health, Sichuan Medical College, Chengdu

The mutagenicity of V_2O_5 was assayed by a battery of 6 tests. Following are the results:

1. **E. coli reverse mutation test.** Among the 5 tester strains used, WP2, WP2uvrA, CM891, ND160 and MR2-102, the number of revertants increased as the dosage increased between 10–100 μg/plate with or without S9 in WP2, WP2uvrA, and CM891. Hence, V_2O_5 can induce base pair substitution mutations.

2. **Ames test.** V_2O_5 at 50–200 μg/plate, revertant counts in the 5 standard tester strains were similar to that of the negative control with or without S9. Inhibition occurred at doses above 200 μg/plate.

3. **Micronucleus test.** "615" inbred mice susceptible to cancer were selected for exposure to V_2O_5. Micronucleated polychromatophil rate increased as the dosage increased between 0.25–4.0 mg/kg and was significantly higher than that of the negative control. Chromosome breakages were suggested.

4. Mouse dominant lethal test. Results were negative (dosage range 0.25–4.0 mg/kg).

5. B. Subtilis rec-assay. Bacterial inhibition took place in M45 (rec⁻), and the inhibition zone difference between H17 and M45 extended to 5.5 mm. This suggested primary DNA damage and repair induction by V_2O_5.

6. Sister chromatid exchange. the result of human lymphocyte SCE test was negative at doses between 2.6×10^{-6} – 2.6 $\times 10^{-4}$ M V_2O_5, and cell mitosis was inhibited when concentration was increased to 2.6×10^{-3} M.

Three out of six tests were positive, and induction of gene mutation, chromosome aberration and primary DNA damage or repair by V_2O_5 were suggested. Therefore, V_2O_5 is a potential mutagen.

MUTAGENICITY AND TERATOGENICITY OF THE FUNGICIDE ETHYL DISULFIDE

X.S. Huang[1], X.R. Chen[2], W.A. Xu[1], Y.N. Yu[2], C. Ding[2], Z.C. Jin[2], and M. Wang[2]

[1]Pesticide Toxicology Unit
[2]Department of Pathophysiology, Zhejiang Medical University, Hangzhou

Ethyl disulfide is an organophosphorus fungicide with the chemical name bis-(O,O-diethylphosphinothioyl)-disulfide. A battery of tests was carried out in an attempt to evaluate its mutagenicity and teratogenicity:

1. Ames test revealed that ethyl disulfide was mutagenic to strain TA 100 and that S-9 seemed to show an inactivating effect.

2. DNA repair synthesis test using the human amnion cell strain FL and double label ³H-TdR revealed DNA damage by ethyl disulfide.

3. Dominant lethal test with mice revealed that ethyl disulfide at doses of 62.5 or 250 mg/kg caused dominant lethality in mice during the second and the fifth week, but the mutagenicity was weak.

4. The No. 2 sample of ethyl disulfide caused chromosome aberrations in the bone marrow cells and micronuclei in fetal liver polychromatic erythrocytes of the mouse.

5. Teratology study in rat showed neither teratogenicity nor embryo-toxicity.

THE FREQUENCY OF SISTER CHROMATID EXCHANGE IN PERIPHERAL LYMPHOCYTES OF WORKERS EXPOSED TO TNT

Q.Z. Liu, L.F. Zheng[1], F. Liu[1], S.Z. Fu[1], and Y.L. Yang[2]

Medical Genetics Laboratory, Harbin Medical University, Harbin
[1]Tsitsihar Medical School, Tsitsihar
[2]Hua-An Mechanical Factory, Tsitsihar

Using the technique of 5-bromodeoxyuridine incorporation and differential staining, the SCE frequencies of 20 male TNT workers and 20 healthy males (control group) were scored and analyzed. The subjects of study were either the workers who were exposed to TNT for more than six years with complaints of clinical symptoms, or those who had been diagnosed as suffering from TNT poisoning. The history of cigarette smoking and medication received as well as irradiation of these subjects and controls was noted and taken into account.

The mean and standard deviation of SCE frequency in TNT workers is 13.59 ± 1.44 per cell and that of the control group is 8.12 ± 0.58 per cell. A highly significant difference with $p < 0.001$ was observed statistically. Results suggest that the increment of SCE frequency in the workers may be related to the exposure to TNT.

STUDIES ON THE MALIGNANT TRANSFORMATION OF MOUSE BONE MARROW CELLS IN VITRO INDUCED BY ^{60}CO γ-RAY IRRADIATION

F.M. Gao, X.L. Li, M.J. Qian, W.H. Wang, F.G. Qi, and H.Y. Jiang

Department of Biological Effects of Radiation, Laboratory of Industrial Hygiene, Ministry of Public Health, Beijing

Long-term cultures of the male LACA mouse bone marrow cells in Fischer liquid medium were irradiated in vitro with 300, 200 and 100 rads of ^{60}Co γ-rays, respectively. During the course of culture, morphological transformation began to emerge 42 days after the irradiation in cells treated with 300 rads: formation of multiple cell conglobation and multiple layers with cell crisscross on the feeder layer. Malignant transformation took place in the third subculture passage (66 days after exposure), as

evidenced by colony formation in semi-agar medium. The cells of
the control group gradually degenerated after the 91st day, then
died within 120 days.

Biological characteristics of malignant transformation were
observed:

1. Plating efficiency was 40.0% ($2 \times 10^5/5 \times 10^5$ cells per dish);
    the malignant transformation rate was 0.10 - 1.45%.
2. Maximum saturation density was $7.4 \pm 2.5 \times 10^4/cm^2$.
3. Anti-ouabain inhibition test: there was no cell proliferation
    or colony formation in the growth medium containing 1 mmole
    of ouabain for three times. It implies that no mutation at
    this locus took place in the transformed cells.
4. Transformed cells of 1×10^6 cells/ml or higher densities were
    agglutinated by concanavalin A (concentration in 1.0 mg/ml)
    within 30 minutes at room temperature. It implies that the
    glucoproteins in the cell membrane had changed.
5. Cytochalasin B in the medium at a final concentration of 1.7
    μg/ml caused the transformed cells to become multinucleated.
6. Aneuploidy, mainly hyper-triploid (61-70) cells, and
    aberrations including rings, acentrics, dicentrics,
    micromeres, disintegrations etc. were found.
7. Fibrosarcomas were induced 2-3 weeks after subcutaneous
    inoculation of $1-2 \times 10^6$ cells in immunosuppressed mice, with
    an incidence from 55.6 to 100.0%.

DOMINANT LETHAL MUTATIONS INDUCED BY TRITIATED WATER
AND ^{60}CO GAMMA RAYS IN FEMALE MICE

X.Y. Zhou, J.C. Dong, X.S. Geng, S.Y. Zhou, and C.F. Zhang

Laboratory of Industrial Hygiene, Ministry of Public Health,
Beijing

Female mice of the LACA strain at the age of 11 weeks (about
25-30 g.b.w.) were divided into five groups and injected
intraperitoneally with 4.7, 17, 37, 75, 110 ci tritium/g.b.w. of
tritiated water, respectively. The doses absorbed by the ovaries
in 10 days after injection were 3.9, 14.1, 30.7, 62.2, 91.2 rads,
respectively. The cumulative doses exposed continuously to gamma
rays for 10 days were obtained, in comparison with an exponential
decrease trend in the ovary tritium concentration. The cumulative
doses received in 10 days were 52.6, 74.2, 112.2, 161.0, 210.0,
299.0 rads, respectively.

Dominant Lethal Testing: every two of the female mice from
each group and one male mouse were housed together in plastic

cages for a 5-day-breeding starting from the 10th day after irradiation. The females were sacrificed, the number of corpora lutea and uterine contents were examined at the 18th day after mating and the number of dominant lethal mutations were calculated. The calculation was based on the corpora luteum and implants (classified as viable embryos, early embryonic deaths and late embryonic deaths).

Our preliminary results showed that when the doses absorbed by the ovaries after injection of tritiated water in 10 days were 4-90 rads and the cumulative doses delivered by gamma rays in 10 days were 52-299 rads, a linear relationship exists between the number of dominant lethal mutations and the cumulative doses. The cumulative doses for 2.0 dominant lethal mutations were 78 rads for tritium beta rays and 224 rads for ^{60}Co gamma rays. The RBE values calculated were 2.87; tritium beta rays were more effective than ^{60}Co gamma rays in the induction of dominant lethal mutations.

Dominant lethal mutations result from both chromatid- and chromosome-type aberrations in the germ cells while heritable translocations result primarily from the latter type. Thus, further research on the relationship between dominant lethal mutations and heritable translocations is needed.

INHIBITORY EFFECTS OF PHENOLIC OR HYDROXY COMPOUNDS
ON THE METABOLISM OF BENZO(a)PYRENE

F.Z. Hou[1], Z.P. Bao[1], Z.Y. Wang[2], and P.T. Liu[1]

[1]Institute of Environmental Science, Beijing Normal University
[2]Institute of Labor Protection, Ministry of Labor and Personnel,
Beijing

The natural products cucurbitacin B from the fruit of Cucumis
melo L. cucurbitaceae, Paeonol from the root of Pycnostelma
paniculatum (Bge.) K. Shun, the synthetic compound p–ethoxyphenol,
and the intermediate of vitamin E, 2,3,5-trimethylhydroquinone
were used as inhibitors of the metabolites of benzo(a)pyrene
(BaP). Butylated hydroxyanisole (BHA), the known inhibitor of
neoplastic induction by BaP and dimethylbenzanthracene, served as
a reference compound. The test compounds were added individually
to BaP and each sample was incubated with the liver microsomes
isolated from rats induced with 3–methylcholanthrene according to
Kinoshita's procedure. The metabolites of BaP were analyzed by
high performance liquid chromatography.

Each inhibitor at the same concentration could inhibit the
formation of diols and phenolic metabolites of BaP to a certain
extent. The preliminary results showed that the four compounds
tested possess similar inhibitory effects as BHA on BaP
metabolites. They could inhibit the formation of 7,8–diol–BaP,
9,10–diol-BaP, and 9-OH-BaP; the last two metabolites were more
severely inhibited. It is interesting to perform further research
to determine whether they might possess chemoprevention potential
against chemical carcinogens.

A PRELIMINARY OBSERVATION ON THE MUTAGENIC EFFECT OF
DIBROMOCHLOROPROPANE (DBCP) IN DROSOPHILA
MELANOGASTER SEX-LINKED RECESSIVE LETHAL TEST

E.C. Ye, and Y.H. Ai

Department of Hygiene, Shanghai Railway Medical College, Shanghai

LD_{50} of DBCP in adult male Drosophila fed for 3 days was
tested according to routine method and estimated to be 300 ppm.
One to two day old male fruitflies of Oregon K stock, which had
been pretreated by feeding for 72 hr with a 2% sucrose solution
containing dissolved DBCP at concentrations 50,500 ppm, were mated
to 3-4 day old Basc females. In the meantime, those individuals
raised on a blank 2% sucrose solution (DBCP-free) were counted as
a negative control; those on 2% sucrose solutions containing
methyl methanesulfonate (MMS) of known concentrations served as
positive controls. The mating scheme and the scoring criteria in
F_2 were all in accordance to the standard Muller-5 test. The
results are summarized as follows:

Induction of SLRL in Drosophila exposed to DBCP

	Brood I (0-2)[*]		Brood II (2-5)		Brood III (5-8)	
	Leth/Chr.	%	Leth/Chr.	%	Leth/Chr.	%
DBCP: 50 ppm	1/99	1.01	1/99	1.11	1/90	1.11
200 ppm	7/99	7.07	10/99	10.10	2/80	2.50
2% Sucrose	0/498	0	0/488	0	0/485	0
MMS: 1.0 mM	6/93	6.5	4/100	4.0	3/27	
2.0 mM	16/86	18.6	9/92	9.8	2/12	

[*]days

The pronounced differences between the percentage of lethality from DBCP-treated series and the controls were observed and the peak of mutagenicity appeared in treated spermatids. Preliminary results suggest that DBCP is a mutagen of moderate potency.

Acknowledgements: We are indebted to Prof. E. Vogel and Dr. W.G.H. Blijleven for their kindness in furnishing our laboratory with Oregon K and Basc stocks.

REGRESSION ANALYSIS OF ^{60}CO IRRADIATION AND
PINYANGMYCIN INDUCED MICRONUCLEI AND CHROMOSOMAL
ABERRATIONS IN THE ROOT TIP CELLS OF VICIA FABA

Y.Y. Wang, A. Zhao, R.X. Li, and D.Y. Tang

Beijing Municipal Institute of Environmental Protection, Beijing

The induction of micronuclei and chromosome aberrations in the root tip cells of Vicia faba by ^{60}Co irradiation and Pinyangmycin (PYM, Bleomycin A_5) was studied. The relationships between the dose and the effects with the two agents were directly proportional. The linear regression equations for ^{60}Co irradiation were: $Y_1 = 5.23 + 0.0199X_1$ and $Y_2 = 1.92 + 0.0086X_2$. The equations for PYM were: $Y_3 = 32.0 + 1903.5X_3$ and $Y_4 = 6.38 + 551.5X_4$. The correlation between the micronuclei and chromosomal aberrations was also good. The linear regression equation for ^{60}Co was: $Y_5 = 2.036 + 2.219X_5$; that for PYM was: $Y_6 = 10.38 + 3.48X_6$. The results of t test indicate that the correlations in parameters in the six above-mentioned equations were statistically significant ($P < 0.01$). As compared to metaphase analysis, the result of micronucleus induction by ^{60}Co irradiation was 100 rads, but that of chromosomal aberrations was 500 rads, the ratio of both regression coefficients being 2.1:1. The initial dose of PYM to induce micronuclei was 0.003 g, but the frequency of micronuclei observed was higher than that of chromosomal aberrations. Hence, the induction of micronuclei by irradiation or the chemical action appears more sensitive and the test is easy to perform and inexpensive. Thus, the micronucleus test in this plant cell system may replace the more tedious chromosome aberration analysis for the detection and assessment of genotoxicity of environmental agents.

LUNG CANCER AND EXPOSURE TO CHLORO-METHYL-ETHERS
--AN OCCUPATIONAL EPIDEMIOLOGICAL SURVEY

S.Z. Hsueh[1], G.F. Tong[2], J.Z. Zhou[3], C. Qie[1], and J. Dang[4]

[1]Department of Occupational Health, Shanghai First Medical College, Shanghai
[2]Department of Occupational Disease, Huashan Hospital, SFMC., Shanghai
[3]Institute of Occupational Disease Prevention, Bureau of Chemical Industry of Shanghai
[4]Institute of Oncology, Shanghai

There were four factories engaged in manufacturing and processing chloro-methyl-ethers (CME, including BCME and technical CMME) in Shanghai, prior to 1975, when the carcinogenicity of CME in humans was recognized and generally accepted. A cohort of 318 workers (approximately 2/3 male and 1/3 female), with 3980.5 person-years, exposed to CME at least one year during the period of 1958-1981 was established. Among them, 21 deaths, 16 cancer deaths and 12 deaths of lung cancer were recorded. Based on the sex, age and calendar-year-specific mortality derived from the death and cancer registry in the urban population of Shanghai, relative risks (rate ratios) were calculated. These include SMR (Standardized Mortality Ratio), PMR (Proportional Mortality Ratio), and PCMR (Proportional Cancer Mortality Ratio): All Cancer SMR 485 (16/3.3); Lung Cancer SMR 2,296 (12/0.52); All Cancer PMR 219 (16/7.29); Lung Cancer PMR 855 (12%/1.41%); Lung Cancer PCMR 390 (75%/19%). Figures in parentheses are the observed and expected numbers.

All the results are highly significant statistically. Because all the cancer deaths occurred in male workers, the SMR and PMR computerized from this fraction of workers are bigger. Exposed males bear a 25-fold greater risk of lung cancer death than the general population. Average exposure preceding illness is 10.5 (2-18) years. On histopathological examination 70% were undifferentiated cell type carcinomas. Survival time after diagnosis is 10 months on average.

The results confirm that CME is a potent human carcinogen and suggest that CME must be handled with great care. Serious preventive measures have to be taken. Fortunately, the escape of these noxious vapors has been more stringently controlled since 1976, and though still not satisfactory, the extent and degree of exposure have been greatly reduced.

STUDIES ON THE MUTAGENICITY AND TERATOGENICITY OF METHYL-PARATHION

Y.D. Yu, Y.C. Jia, C.F. Hong, Y.H. Yang, Y.H. Luo, and J.Y. Le

Zhejiang Institute of Experimental Medicine and Hygiene, HongZhou

Organophosphorous pesticides are rapidly replacing organochlorine insecticides, owing to their rapid decomposition and disappearance from the environment. Methyl-parathion has recently been widely used. It is imperative to test the mutagenic and teratogenic potentials of the compound for risk assessment.

Physical examinations of 22 workers exposed to methyl-Parathion in two pesticide plants were carried out. Significant increase was found in the SCE frequency of the cultured peripheral lymphocytes taken from the exposed workers, as compared with local controls (7.14 versus 4.7 per cell). However, there was no statistical difference in the frequencies of the cells with chromosomal aberrations between the exposed and control groups. An increment in the numbers of the chromosomal aberrations in the bone marrow cells of mice receiving 10 mg/kg of methyl-parathion in vivo was also noticed.

The reverse mutation bioassay with Salmonella typhimurium strains TA 100 and TA 98 was carried out according to the protocol introduced by Ames et al. Methyl-parathion gives a positive response on TA 100 either in the presence (300 µg/pl) or absence (400 µg/plate) of S9 mix, but not on the TA 98.

Results were positive in the assay of unscheduled DNA synthesis in human peripheral lymphocytes in culture incubated with methyl-parathion. The results of DNA synthesis inhibition test on mouse testes, micronucleus scoring in mouse, and mouse dominant lethal mutation test were also positive.

Teratogenicity tests by mid-gestation administration in rats and mice were conducted. One fetus with extroversion of intestine was found in the 3.56 mg/kg dose level. A similar result was noticed in a repeat test. Increased incidence of cleft palate in the fetal mice was recorded in the group receiving a higher dose (6.32 mg/kg).

STUDY ON THE CARCINOGENICITY OF γ-BENZENE-HEXACHLORIDE IN RATS

J.K. Xu, F.D. Xu, S.L. Xu, Z.L. Liu, S.Q. Wang, and Y. Dai

Department of Nutrition and Hygiene, Institute of Hygiene, Chinese Academy of Medical Science, Beijing

Four groups of Wistar rats, each consisting of 20 of both sexes were tested for carcinogenic potential at dose levels of 0 (negative control), 5, 25, and 125 ppm of pure γ-BHC (99.9%

purity, determined by gas liquid chromatography) respectively by feeding for 27 months. The background level of γ-BHC was 0.005 ppm and 0.012 ppm for α-BHC in the semi-synthetic feeds. In addition to routine necropsy and histopathology procedures, the BHC contents in the liver, kidney, brain, testes and adipose tissue were analyzed.

Major findings were: The number of overall tumor-bearing animals was 15 out of 160, with the distribution as follows:

Group fed 0 ppm BHC   6/40
Group fed 5 ppm BHC   1/40
Group fed 25 ppm BHC   4/40
Group fed 125 ppm BHC   4/40

and the histopathological diagnoses were:

Cerebral choriomeningo-endothelioma   8
Bronchial papilloma   1
Bronchial adenoma   1
Bronchial squamous cell carcinoma   1
Adrenoid tumor   1
Osteosarcoma   1
Myosarcoma   1

Livers were intact; no tumor or precancerous change was found there.

On the basis of these findings, no dose response correlation was indicated, and the highest tumor incidence occurred in the group of negative control. It is, therefore, unlikely that the neoplasms observed in the test animals is associated with the γ-BHC added to the feeds. However, pathological lesions in the kidney characterized by the appearance of eosinophilic transparent granules in the epithelial cells of convulated tubules and vacuole degeneration of hepatocytes were noticed and seemed to be dose-dependent in the male rats which received BHC.